A New Handbook of

# Political Science

# A New Handbook of
# Political Science

edited by
**Robert E. Goodin**
and
**Hans-Dieter Klingemann**

**Oxford University Press**

*Oxford University Press, Great Clarendon Street, Oxford* OX2 6DP
*Oxford New York*
*Athens Auckland Bangkok Bogota Bombay*
*Buenos Aires Calcutta Cape Town Dar es Salaam*
*Delhi Florence Hong Kong Istanbul Karachi*
*Kuala Lumpur Madras Madrid Melbourne*
*Mexico City Nairobi Paris Singapore*
*Taipei Tokyo Toronto Warsaw*
*and associated companies in*
*Berlin Ibadan*

*Oxford is a registered trade mark of Oxford University Press*

*Published in the United States by*
*Oxford University Press Inc., New York*



*First published 1996*



*British Library Cataloguing in Publication Data*
*Data available*

*Library of Congress Cataloging in Publication Data*
*A new handbook of political science / edited by Robert E. Goodin and Hans-Dieter Klingemann.*
*Includes bibliographical references and index.*
*1. Political science—Handbooks, manuals, etc. I. Goodin, Robert E. II. Klingemann, Hans-Dieter.*
*JA66.N49 1996 320—dc20 96-21871*
*ISBN 0-19-828015-7*

10 9 8 7 6 5 4

*Printed in Great Britain*
*on acid-free paper by*
*Bookcraft (Bath) Ltd.*
*Midsomer Norton, Avon*

# Contents

# Foreword

THE International Political Science Association is proud to have contributed to the making of the *New Handbook of Political Science.* Participants in the IPSA XVI World Congress in Berlin in 1994 had an opportunity to hear initial presentations of many of the chapters in a series of sessions, organized by the Editors, on 'The State of the Discipline'.

This is an appropriate time at which to take stock of political science. The discipline faces new challenges in understanding and assessing recent dramatic, sometimes tragic, world events and sweeping political changes. The *New Handbook* covers familiar, staple topics in political science, such as political institutions, political behavior, public policy and political theory, but the political context, in North as well as South, East as well as West, is no longer so familiar. New questions are being asked about such fundamentals as the nation state and sovereignty, and there is growing interest in the politics of religion, ethnicity and varieties of pluralism.

Intellectual developments, too, during the last twenty years have brought a wide range of new theoretical frameworks and methodologies into political science. Some scholars, for example, now use highly technical and sophisticated mathematical models, others have moved under the philosophical umbrella called post-modernism, many are proponents of the 'new institutionalism', and feminism is a major presence.

Political science is also changing in other ways. As part of the current wave of democratization, the discipline is being established or strengthened in many countries, and now exists more widely than ever before. The numbers of women and younger scholars from around the world participating in the IPSA World Congress, and the fact that the Congress was addressed by the Association's first woman President, reflected the democratization that has been taking place within political science itself.

The *New Handbook* provides an unusually comprehensive and systematic discussion of all the major areas of the discipline. In this era of specialization, practitioners and their students will turn to contributions beyond their own particular expertise to find out about developments elsewhere in political science. But interested non-specialists and citizens will find this a very accessible, and extremely well organized, volume from which to learn about the state of the art in contemporary political science, and about its history and relation to other social sciences.

The contributors, all leading authorities in their fields, reflect the growing strength of political science today outside the USA, where the first professional association was formed. They include both elder statesmen and a new generation of men and women scholars, writing from a variety of perspectives. The *New Handbook of Political Science* offers a sure, reliable and expert guide through the broad expanses and the thickets of the discipline and its sub-divisions. There could be no better volume to take political science into the new century.

Carole Pateman
President IPSA 1991–1994

# Preface

THE *New Handbook*, by its very title, pays explicit homage to the truly Herculean efforts of our predecessors, Fred Greenstein and Nelson Polsby, in compiling the original *Handbook of Political Science* (1975). Though that eight-volume work is now two decades old, it remains a landmark in the discipline and an essential reference. We have set our task as the examination of what has happened in the discipline in the twenty years since publication of the Greenstein–Polsby original. Inevitably, some contributors have needed to go slightly beyond those bounds to tell a coherent story (the story of contemporary political theory, for example, clearly starts with the publication of Rawls's *Theory of Justice,* four years before the Greenstein–Polsby *Handbook*). Basically, however, the first three contributors to each section have been held to that remit, with the fourth ("Old and New") being invited explicitly to reflect upon how these newer developments articulate with older traditions within each subdiscipline.

The *New Handbook* is conspicuously more international than the old, with just under half of our 42 contributors having non-North American affiliations. That is due in some small part to its origins in a meeting of the International Political Science Association (see our Acknowledgments, immediately following). But it is due in much larger part to genuine internationalization of the discipline over the past two decades. American political science undoubtedly remains *primus inter pares*—but it now has many equals, most of whom actually see themselves as collaborators in some shared enterprise. These and various other new voices make political science a richer discourse today than twenty years ago, albeit a discourse which is clearly continuous with that earlier one.

The *New Handbook* is also conspicuously organized around subdisciplines in a way that the old was not. Some such subdisciplinary affliations are, and virtually always have been, the principal points of allegiance of most members of our discipline. The particular subdisciplines around which we have organized the *New Handbook* represent what seem to us to constitute the dominant configuration of the discipline at present. Subdisciplines are far from being hermetically sealed, however. Work across and between subdisciplinary divides is increasingly common in its frequency and compelling in its quality.

The original *Handbook of Political Science* was loosely inspired by the model of Lindzey and Aronson's (1954/1985) *Handbook of Social Psychology* (Greenstein and Polsby 1975: vol. i, p. vi). While social psychology remains central to much political science, it is a mark of the broadening scope of the contemporary discipline that the *New Handbook* was loosely inspired, in like fashion, by the *New Palgrave: A Dictionary of Economics* (Eatwell *et al.* 1987). Again, our modest single volume cannot compare with the four-volume sweep of that latter work, nor does it face quite the same challenge of catching up on a century's worth of developments since publication of the original. But like the *New Palgrave*, the *New Handbook of Political Science* strives to encourage cutting-edge practitioners to stand back from the fray and reflect upon where, collectively, we have been and where, collectively, we are going in their corner of the discipline. And like the *New Palgrave*, the *New Handbook of Political Science* takes that disciplinary remit broadly to embrace cognate work in economics and sociology, psychology and statistics, anthropology and area studies.

In addition to these masterly surveys of cognate disciplines, we should also acknowledge our debt—and our profession's—to various other interim assessments of the state of political science itself. Although the surveys of *Political Science: The State of the Discipline* (Finifter 1983; 1993) are not nearly as comprehensive in their aspirations as the handbook, old or new, several of their chapters have become classics that now stand alongside corresponding chapters in the original handbook as authoritative statements upon which any subsequent work must build. Another four-volume collection, *Political Science: Looking to the Future* (Crotty 1991), also contains many truly excellent chapters which repay careful study. Subfields of political science have also been well-served, a particular landmark being *Public Administration: The State of the Discipline* (Lynn and Wildavsky 1990)—a joint venture between the American Society for Public Administration and APSA. Looking beyond the Anglophone orbit, there are also ambitious and excellent handbooks of political science published in French (Leca and Grawitz 1985), German (von Beyme 1986) and Italian (Graziano 1987). The *New Handbook* aims not to supplant any of those previous efforts but, rather, to extend and supplement them.

Greenstein and Polsby felt compelled to remark upon the inevitable incompleteness of their original eight-volume handbook. So too must we emphasize, all the more strongly, the inevitable incompleteness of our one-volume successor. Authors of the lead chapter in each section have been asked to provide an overview of recent developments, as best they are able—but within the tightly limited number of pages they have been allowed, inevitably there is much that they have had to leave out. We have attempted to supplement each of those overview chapters with other shorter ones from particular perspectives—but with only two per subdiscipline, there are again many perspectives that are inevitably left out. While we cannot hope to have provided a comprehensive coverage of all recent developments of consequence, we nonetheless hope to have touched upon most of the main currents in the disciplines. It is a lively and thriving enterprise, of which we are proud to be part.

## References

Crotty, W., ed. 1991. *Political Science: Looking to the Future.* 4 vols. Evanston, Ill.: Northwestern University Press.

Eatwell, J.; Milgate, M.; and Newman, P., eds. 1987. *The New Palgrave: A Dictionary of Economics.* 4 vols. New York: Stockton Press.

Finifter, A. W., ed. 1983. *Political Science: The State of the Discipline.* Washington, D.C.: American Political Science Association.

—— ed. 1993. *Political Science: The State of the Discipline II.* Washington, D.C.: American Political Science Association.

Graziano, L., ed. 1987. *La Scienca Politica Italiana.* Milano: Felterinellei.

Greenstein, F. I., and Polsby, N. W., eds. 1975. *Handbook of Political Science.* 8 vols. Reading, Mass.: Addison-Wesley.

Leca, J., and Grawitz, M., eds. 1985. *Traité de Science Politique.* 4 vols. Paris: Presses Universitaires de France.

Lindzey, G., and Aronson, E. 1985. *Handbook of Social Psychology.* 2 vols. 3rd edn. Reading, Mass.: Addison-Wesley; originally published 1954.

Lynn, N. B., and Wildavsky, A., eds. 1990. *Public Administration: The State of the Discipline.* Chatham, N.J.: Chatham House, for the American Political Science Association and the American Society for Public Administration.

von Beyme, K., ed. 1986. *Politikwissenschaft in der Bundesrepublik Deutschland.* PVS Sonderheft 17. Opladen: Westdeutscher Verlag.

# Acknowledgments

THE *New Handbook of Political Science* has its origins in a set of panels we organized on the "state of the discipline" for the XVIth World Congress of the International Political Science Association, which met in Berlin in August 1994. Some contributors were unable to come at the last minute; others who came have fallen by the wayside, for one reason or another, in the course of transforming the Congress papers into a coherent book. But most of the contributors to this volume had the invaluable opportunity to discuss their draft chapters with one another, and with others from adjacent subdisciplines, in Berlin. It is a much more unified and cohesive book than it would otherwise have been.

We want to take this opportunity to thank the many people who made those meetings possible—most particularly our secretaries and assistants, Norma Chin, Frances Redrup, and Judith Sellars in Canberra, and Gudrun Mouna and Hubertus Buchstein in Berlin. We should also take this opportunity to pay tribute to the larger IPSA organization—including most especially the then President, Carole Pateman; the then Secretary-General, Francesco Kjellberg, and his able assistant, Lise Fog; and the Local Organizer of the Berlin Congress, Gerhard Göhler.

We should also thank the many people who have offered valuable advice on the substance of the *New Handbook*, ranging from suggestions on the selection of topics and contributors to detailed comments on the substance of particular chapters. First and foremost, once again, are the members of the IPSA Program Committee—most particularly Carole Pateman, Jean Leca, Ted Lowi, and Luigi Graziano. Among the many others who have been invaluable sources of excellent advice, the contributions which particularly stand out are those of John Dryzek, John Uhr, Barry Weingast, and members of Research Unit III at the Wissenschaftszentrum Berlin.

We ought to pay more than the usual passing tribute to our publishers. Tim Barton and Dominic Byatt have been constant sources of advice and assistance, encouragement and admonitions. Putting together a high-profile reference book such as this is a process which is bound to try any editor's nerves: theirs have held remarkably firm, throughout. We are grateful to them for their constant support.

Transoceanic collaborations can rarely be easy. With the advent of the electronic age, they are easier than they used to be. Still, there are strict limits to what can be done in "virtual reality," and co-editors must literally come together from time to time. Our respective home institutions have generously facilitated this, on several occasions and in various connections. Goodin would like to pay particular tribute to the Wissenschaftszentrum Berlin for bringing him to Berlin early in the planning stages of this project, and for providing him with office space and secretarial assistance in the run-up to the Berlin Congress. Klingemann would similarly like to thank the Research School of Social Sciences at the Australian National University for a Visiting Fellowship to facilitate final editing of the complete manuscript. Most of all, we would like to thank our partners—Diane Gibson and Ute Klingemann—for suffering such an intrusive presence as this project with such good humor for so long.

R.E.G.
H.-D.K.

*Canberra, October 1995*

Part I

# The Discipline

Chapter 1

# Political Science: The Discipline

Robert E. Goodin
Hans-Dieter Klingemann

RETROSPECTIVES are, by their nature, inherently selective. Many fascinating observations are contained within the wide-ranging surveys which constitute the *New Handbook of Political Science.* Many more emerge from reading across all of its chapters, collectively. But, inevitably, the coverage is incomplete and, equally inevitably, somewhat idiosyncratic. All authors are forced to leave out much of merit, often simply because it does not fit their chosen narrative structure. The *New Handbook*'s contributors tell a large part of the story of what has been happening in political science in the last two decades, but none would pretend to have told the whole story.

It is the task of this introductory chapter to set those chapters in a larger disciplinary context and to pull out some of their more interesting common threads. Just as the coverage of each of the following chapters is inevitably selective, that of this overview of the overviews is, inevitably, all the more so. Of the several themes and subthemes which emerge, looking across these chapters as a whole, we shall focus upon one in particular.

The *New Handbook* provides striking evidence of the professional maturation of political science as a discipline. This development has two sides to it. On the one side, there is increasing differentiation, with more and more sophisticated work being done within subdisciplines (and, indeed, within sub-specialities within subdisciplines). On the other side, there is increasing integration across all the separate subdisciplines.

Of the two, increasing differentiation and specialization is the more familiar story, integration the more surprising one. But clearly it is the case that there is, nowadays, an increasing openness to and curiosity about what is happening in adjacent subdisciplines. An increasingly shared overarching intellectual agenda across most all of the subdisciplines makes it possible for theoretical innovations to travel across subdisciplinary boundaries. An increasingly shared methodological tool-kit makes such interchange easy. All of this is facilitated, in turn, by an increasing band of synthesizers of the discipline, often intellectually firmly rooted in one particular subdiscipline but capable of speaking to many subdisciplines in terms which they find powerfully engaging. Among the many things which strike us, reading across the chapters of the *New Handbook* as a whole, these are the ones that strike us most forcefully and which we will elaborate upon in this chapter.

## I Political science as a discipline

A central claim of this chapter is that political science, as a discipline, has become increasingly mature and professionalized.[1] As an important preliminary to that discussion, we must address, necessarily briefly, a few threshold questions. What is it for political science to constitute a discipline? What is politics? In what sense can the study of politics aspire to the status of a science?

### A The nature of a discipline

Inured as we are to speaking of the subdivisions of academic learning as "disciplines," it pays to reflect upon the broader implications of that phrase. According to the *Concise Oxford English Dictionary* a discipline is variously defined as: "a branch of instruction; mental and moral training, adversity as effecting this; military training, drill . . . ; order maintained among schoolboys, soldiers, prisoners, etc.; system of rules for conduct; control exercised over members of church; chastisement; (Ecclesiastical) mortification by penance."

The last dictionary definition would seem to have only marginal application to academic disciplines, but most of the others have clear counterparts. An academic "discipline" may enjoy minimal scope to "punish," at least in the most literal senses (Foucault 1977). Still, the community of

[1] Once "professionalized" might have equated, readily and narrowly, to "Americanized." But as alluded to in our Preface and as is evident from *New Handbook* contributors' affiliations, the profession itself is becoming more internationalized, both in its personnel and in its professional concerns.

scholars which collectively constitutes a discipline does exercise a strict supervisory function—both over those working within it and, most especially, over those aspiring to do so. The "order maintained" is not quite the same as that over soldiers or schoolboys, nor is the training strictly akin to military drill. Nonetheless, there is a strong sense (shifting over time) of what is and what is not "good" work within the discipline, and there is a certain amount of almost rote learning involved in "mastering" a discipline.

All the standard terms used to describe academic disciplines hark back to much the same imagery. Many, for example, prefer to think of political analysis as more of an "art" or "craft" than a "science," strictly speaking (Wildavsky 1979). But on that analogy the craft of politics can then only be mastered in the same manner in which all craft knowledge is acquired, by apprenticing oneself (in academic craftwork, "studying under") a recognized "master." Others like to speak of politics, as well as the academic study thereof, as a "vocation" (Weber 1919/1946) or a "calling".[2] But, tellingly, it is a vocation rather than an avocation, a job rather than a hobby; and as in the core religious meaning so too in the academic one, the "calling" in question is to service of some higher power (be it an academic community or the Lord). Most of us, finally, talk of academic disciplines as "professions." In Dwight Waldo's (1975: 123) delightful phrase "sciences know, professions profess." What scientists profess, however, are articles of the collective faith.

Any way we look at them, then, disciplines are construed at least in large part as stern taskmasters. But the same received disciplinary traditions and practices which so powerfully mould and constrain us are at one and the same time powerfully enabling. The framework provided by the structure of a discipline's traditions both focuses research and facilitates collaboration, unintentional as well as intentional. A shared disciplinary framework makes it possible for mere journeymen to stand, productively, on the shoulders of giants. It also makes it possible for giants to build, productively, on the contributions of legions of more ordinarily gifted practitioners.[3]

Discipline, academic or otherwise, is thus a classic instance of a useful self-binding mechanism. Subjecting oneself to the discipline of a discipline—or in the case of Dogan's (below: chap. 3) hybrid scholars, of several—is conducive to more and indisputably better work, both individually

[2] Both Berger's *Invitation to Sociology* (1963) and Medawar's *Advice to a Young Scientist* (1989: esp. chap. 2) verge on this. Much the finest work in this genre remains F. M. Cornford's justly celebrated *Microcosmographia Academia* (1908).

[3] For powerful evidence of the way that certain discoveries are "on the cards" at some point in time, consider the cases of "multiple discoveries" discussed in Merton (1973).

and collectively. That is as true for the "chiefs" as the "indians" of the discipline, as true for the "Young Turks" as the "greybeards."

Branches of academic learning are "professions" as well as disciplines. "Professional" connotes, first of all, a relatively high-status occupational grade; and the organization of national and international "professional associations" doubtless has to do, in no small part, with securing the status and indeed salaries of academics thus organized. But the term "professional" also, and more importantly, indicates a certain attitude toward one's work. A profession is a self-organizing community, oriented toward certain well-defined tasks or functions. A professional community is characterized by, and to a large extent defined in terms of, certain self-imposed standards and norms. Incoming members of the profession are socialized into those standards and norms, ongoing members are evaluated in terms of them. These professional standards and norms not only form the basis for evaluation of professionals by one another; they are "internalized," with professionals themselves taking a "critical reflective attitude" toward their own performances in light of them.[4]

The specific standards and norms vary from profession to profession, of course. But across all professions there is a sense of "minimal professional competence," captured by the ritual of "qualifying examinations" for intending political scientists in North American post-graduate training programs. And across all professions there is a notion of particular "role responsibilities" attaching to membership in a profession. The professional ethics of academics do not touch on issues of life and death in quite the same way as those of doctors or lawyers, perhaps. But virtually all academic professions have increasingly formal codes of ethics, touching largely on matters to do with integrity in the conduct and promulgation of research; and all professionals are expected to adhere to them faithfully (APSA 1991).

One of our themes in this chapter is the increasing "professionalism" within political science as a whole. By this we mean, firstly, that there is increasing agreement to a "common core" which can be taken to define "minimal professional competence" within the profession. Secondly, there is an increasing tendency to judge work, one's own even more than others', in terms of increasingly high standards of professional excellence.

While the minimal standards are largely shared ones, the higher aspirations are many and varied. But as in medicine so too in political science, each sub-speciality within the larger profession has its own internal stan-

[4] In much the same way Hart (1961) depicts the norms of legal systems, more generally, being internalized. On the nature of professions and members orientation toward them, see Hughes (1958) and Parsons (1968).

dards of excellence, by which each member of that fraction of the profession is properly judged. And in political science just as in medicine, there is some broad sense across the profession as a whole as to how all the subspecialities sit together to form a coherent larger whole.

## B What is politics?

The foregoing observations, by and large, pertain to academic disciplines quite generally. Disciplines are differentiated one from another in many ways, principally among them by their substantive concerns and by the methodologies that they have made their own. Although there are, as we shall argue, a number of useful "tricks" in political science's tool-kit which are shared by most members of most of its subdisciplines, Alker (below: chap. 35) is undeniably correct in saying that political science does not have—much less define itself in terms of—a single big methodological device all its own, the way that many disciplines do. Rather, political science as a discipline is defined by its substantive concerns, by its fixation on "politics" in all its myriad forms.

"Politics" might best be characterized as the *constrained use of social power.* Following on from that, the study of politics—whether by academics or practical politicians—might be characterized, in turn, as the study of the *nature and source of those constraints* and the *techniques for the use of social power* within those constraints.[5]

When defining politics in terms of power, we follow many before us.[6] "Power" is, by now, well known to be a fraught conceptual field.[7] Respectful though we are of its complexities, we decline to let ourselves get bogged down in them. Dahl's (1957) old neo-Weberian definition still serves well enough. In those terms, *X* has power over *Y* insofar as: (i) *X* is able, in one way or another, to get *Y* to do something (ii) that is more to *X*'s liking, and (iii) which *Y* would not otherwise have done.

Where our analysis departs from tradition is in defining politics in terms of the *constrained* use of power. To our way of thinking, unconstrained power is force, pure and simple. It is not a political power play at all, except perhaps in some degenerate, limiting-case sense. Pure force, literally

[5] This in turn gives rise to the dual foci of the discipline, identified by Almond (below: chap. 2), on "the properties of political institutions and the criteria we use to evaluate them."

[6] Notable among them: Weber (1922/1978); Lasswell (1950; Lasswell and Kaplan 1950), Dahl (1963) and Duverger (1964/1966). We, like them, focus specifically on "social" power, the power of people over other people.

[7] To classic texts such as Russell (1938), Jouvenel (1945/1948) and Dahl (1957; 1961*b*; 1963) have recently been added Lukes (1974), Barry (1989: esp. chaps. 8–11) and Morriss (1987).

speaking, is more the province of physics (or its social analogues: military science and the martial arts) than of politics.[8] It is the constraints under which political actors operate, and strategic maneuvering that they occasion and that occurs within them, that seems to us to constitute the essence of politics.[9] It is the analysis of those constraints—where they come from, how they operate, how political agents might operate within them—that seems to us to lie at the heart of the study of politics.[10]

We talk, broadly, about the *use* of social power (rather than, more narrowly, about its "exercise") as a gesture toward the multitude of ways in which political agents might maneuver within such constraints. We mean the term to cover intentional acts as well as unintended consequences of purposeful action (Merton 1936). We mean it to cover covert manipulatory politics as well as overt power plays (Schattschneider 1960; Goodin 1980; Riker 1986). We mean it to cover passive as well as active workings of power, internalized norms as well as external threats (Bachrach and Baratz 1963; Lukes 1974). The infamous "law of anticipated reactions," non-decisions and the hegemonic shaping of people's preferences (Laclau and Mouffe 1985) must all be accommodated in any decently expansive sense of the political.

One further comment on concepts. In defining politics (and the study of it) as we do, we explicitly depart from the purely *distributional* tradition of Lasswell's (1950) classic formulation of "politics" as "who gets what, when and how."[11] Perhaps it is true that all political acts ultimately have distributional consequences; and perhaps it is even true that therein lies most of our interest in the phenomenon. But in terms of the meaning of the act to the actor, many political acts are at least in the first instance distinctly non-distributional. And even in the final analysis, much of the social significance—objective as well as subjective—of certain political interactions might never be reducible to crass questions of dividing up the social pie. Distributive, regulative, redistributive (Lowi 1964) and identity (Sandel 1982) politics may all have their own distinctive styles.

[8] Thus, an absolute dictator in quest of complete, unconstrained power would rightly be said to be engaged in an (inevitably futile) attempt to transcend politics.

[9] Consider the following analogy drawn from a cognate discipline. Philosophers talk in terms of "strong" considerations, "compelling" arguments, and such like (Nozick 1981: 4–6). But consider an argument such that if we did believe it we would die: that is about as compelling as an argument can get; but winning a point by means of such an argument seems the antithesis of real philosophical disputation, the essence of which is give-and-take. By the same token, the very essence of politics is strategic maneuvering (Riker 1986); and irresistible forces, insofar as they leave no scope for such maneuvering, are the antithesis of politics (however successful they are at getting others to do what you want).

[10] In saying this, we are following (loosely) Crick 1962.

[11] Or Easton's (1965) of politics as the authoritative allocation of values—at least insofar as that is construed, first and foremost, as a matter of the allocation of "valued things" in a society.

Distributional struggles are characterized, in welfare economists' terms, as squabbles over where we sit on the Paretian frontier; but getting to the Paretian frontier is itself a tricky problem, involving a lot of politicking of quite a different sort which is often distinctively non-distributional, at least in the first instance. Important though it undeniably is that our understanding of politics should be attuned to distributive struggles, then, it is equally important that it not be committed in advance to analyzing all else exclusively in terms of them.

## C The several sciences of politics

Much ink has been spilt over the question of whether, or in what sense, the study of politics is or is not truly a science. The answer largely depends upon how much one tries to load into the term "science." We prefer a minimalist definition of science as being just "systematic enquiry, building toward an ever more highly-differentiated set of ordered propositions about the empirical world."[12] In those deliberately spartan terms, there is little reason to think that the study of politics cannot aspire to be scientific.

Many, of course, mean much more than that by the term. A logical positivist might cast the aspirations of science in terms of finding some set of "covering laws" so strong that even a single counter-example would suffice to falsify them. Clearly, that sets the aspirations of science much too high ever to be attained in the study of politics. The truths of political science, systematic though they may be, are and seem inevitably destined to remain essentially probabilistic in form. The "always" and "never" of the logical positivist's covering laws find no purchase in the political world, where things are only ever "more or less likely" to happen.

The reason is not merely that our explanatory model is incomplete, not merely that there are other factors in play which we have not yet managed to factor in. That will inevitably be true, too, of course. But the deeper source of such errors in the positivist model of political science lies in a misconstrual of the nature of its subject. A covering law model may (or may not: that is another issue) work well enough for billiard balls subject to the sorts of forces presupposed by models of Newtonian mechanics: there all actions can be said to be caused, and the causes can be exhaustively traced to forces acting externally upon the "actors." But human beings, while they are undeniably subject to certain causal forces as well, are also in

[12] After the fashion of the older *Wissenschaft* traditions of the German universities from which 19th-century Americans imported political science into their own country (Waldo 1975: 25–30)—and to which contemporary "policy scientists" are now harking back (Rivlin 1971).

part intentional actors, capable of cognition and of acting on the basis of it. "Belief," "purpose," "intention," "meaning" are all potentially crucial elements in explaining the actions of humans, in a way that they are not in explaining the "actions" of a billiard ball. The subjects of study in politics, as in all the social sciences, have an ontological status importantly different from that of billiard balls; and that, in turn, makes the logical positivist's covering law model deeply inappropriate for them, in a way it would not be for billiard balls themselves.[13]

To say that scientific understanding in politics must crucially include a component relating to the meaning of the act to the actor is not, however, necessarily to deny political science appropriate access to all the accoutrements of science. Mathematical modelling and statistical testing remain as useful as ever.[14] The interpretation of the results is all that has to change. What we are picking up with those tools are seen, now, not as inexorable workings of external forces on passive actors, but rather as common or conventional responses of similar people in similar plights. Conventions can change, and circumstances all the more, so the truths thereby uncovered are less "universal" perhaps than those of Newtonian physics. But since we can, in like fashion, aspire to model (more or less completely) changes in conventions and circumstances themselves, we might eventually aspire to closure even in this more amorphous branch of science.

## II The maturation of the profession

What the *New Handbook*'s chapters taken together most strongly suggest is the growing maturity of political science as a discipline. Whether or not "progress" has been made, after the optimistic fashion of Gabriel Almond (below: chap. 2), is perhaps another matter. But maturity, understood in ordinary developmental terms of a growing capacity to see things from the other's point of view, does indeed seem to have been substantially achieved across most quarters of the discipline.

It was not always so. The "behavioral revolution," in its heyday, was from many perspectives a thoroughly Jacobin affair; and it would hardly be

[13] For good surveys of these issues, see Hollis (1977), Taylor (1985) and, with specific reference to politics, Moon (1975) and Almond and Genco (1977). Post-positivist sensitivity to such hermeneutic concerns is much in evidence across many chapters of the *New Handbook*, as discussed in Sect. IIIC below.

[14] Indeed, some of the most highly mathematical developments in recent political science have come in elaboration of the "rational actor" model; and the basic forces driving those models are the rational choice of individuals themselves rather than any causal forces acting externally upon them.

pressing the analogy too far to say that the reaction was decidedly Thermidorian to boot. Early behavioral revolutionaries, for their part, were devoted to dismissing the formalisms of politics—institutions, organizational charts, constitutional myths and legal fictions—as pure sham. Those whom the behavioral revolution left behind, just as those who would later try to leave it behind in turn, heaped Olympian scorn upon the scientific pretensions of the new discipline, calling down the wisdom of sages and of the ages.[15]

A generation later, the scenario replayed itself with "rational choice" revolutionaries imposing formal order and mathematical rigor upon loose logic borrowed by behavioralists from psychology. Once again, the struggle assumed a Manichean Good-versus-Evil form. No halfway houses were to be tolerated. In the name of theoretical integrity and parsimony, rational choice modellers strove (at least initially) to reduce all politics to the interplay of narrow material self-interest—squeezing out, in the process, people's values and principles and personal attachments as well as a people's history and institutions.[16] In the rational choice just as in the behavioral revolution, many famous victories were scored (Popkin *et al.* 1976). But while much was gained, much was also lost.

In contrast to both those revolutionary moments, we now seem to be solidly in a period of rapprochement. The single most significant contribution toward that rapprochement, running across a great many of the following chapters, is the rise of the "new institutionalism." Political scientists no longer think in the either/or terms of agency or structure, interests or institutions as the driving forces: now, virtually all serious students of the discipline would say it is a matter of a judicious blend of both (Rothstein below: chap. 5; Weingast below: chap. 6; Majone below: chap. 26; Alt and Alesina below: chap. 28; Offe below: chap. 29). Political scientists no longer think in the either/or terms of behavioral propensities or organization charts: again, virtually all serious students would now say it is a matter of analyzing behavior within the parameters set by institutional facts and

[15] Sober statements of the behavioralist agenda can be culled from Dahl (1961*a*) and Ranney (1962). Judicious statements of the institutionalist reaction are found in Ridley (1975) and Johnson (1989), with the more philosophical wing of the anti-scientific reaction being perhaps best represented by Oakeshott (1951/1956) and Stretton (1969). For "post-behavioralist" statements, see particularly Wolin (1960), McCoy and Playford (1968) and Easton (1969); the philosophy-of-science face of this tendency is well represented in the Greenstein–Polsby *Handbook* by a singularly judicious chapter from Moon (1975).

[16] Classic early manifestos include Mitchell (1969) and Riker and Ordeshook (1973). The criticisms here canvassed in the text come from friendly critics (Goodin 1976; Sen 1977; North 1990), and refined rational choice models now regularly go some way (albeit perhaps not far enough—cf. Offe below: chap. 29) toward taking most of them on board (Kiewiet 1983; Mansbridge 1990; Monroe 1991).

opportunity structures (Pappi below: chap. 9; Dunleavy below: chap. 10). Political scientists no longer think in the either/or terms of rationality or habituation: virtually all serious rational choice modellers now appreciate the constraints under which real people take political actions, and incorporate within their own models many of the sorts of cognitive shortcuts that political psychologists have long been studying (Pappi below: chap. 9; Grofman below: chap. 30). Political scientists no longer think in the either/or terms of realism or idealism, interests or ideas as driving forces in history: virtually all serious students of the subject carve out a substantial role for both (Goldmann below: chap. 16; Sanders below: chap. 17; Keohane below: chap. 19; Nelson below: chap. 24; Majone below: chap. 26). Political scientists no longer think in either/or terms of science or story-telling, wide-ranging cross-national comparisons or carefully crafted case studies unique unto themselves: virtually all serious students of the subject now see the merit in attending to local detail and appreciate the possibilities of systematic, statistically compelling study even in small-N situations (Whitehead below: chap. 14; Ragin *et al.* below: chap. 33). Political scientists no longer think in either/or terms of history or science, mono-causality or hopeless complexity: even hard-bitten econometricians have now been forced to admit the virtues of estimation procedures which are sensitive to "path" effects (Jackson below: chap. 32), and simplistic early models of politico-economic interactions have now been greatly enriched (Hofferbert and Cingranelli below: chap. 25; Alt and Alesina below: chap. 28).

The point is not just that rapprochement has been achieved on all these fronts. What is more important is the way in which that has been achieved and the spirit pervading the discipline in its newly configured form. Although each scholar and faction would place the emphasis differently on the elements being combined, the point remains that the concessions have been made gladly rather than grudgingly. They have been made, not out of a "live and let live" pluralism, still less out of postmodern nihilism. Rather, concessions have been made and compromises struck in full knowledge of what is at stake, what alternatives are on offer and what combinations make sense.[17] The upshot is undoubtedly eclectic, but it is an ordered eclecticism rather than pure pastiche.

Political scientists of the present generation come, individually and col-

[17] Consider , for example, Fiorina's (1995) *modus operandi*: "I teach my students that rational choice models are most useful where stakes are high and numbers low, in recognition that it is not rational to go to the trouble to maximize if the consequences are trivial and/or your own actions make no difference . . . Thus, in work on mass behavior I utilize minimalist notions of rationality (Fiorina 1981: 83), whereas in work on elites I assume a higher order of rationality (Fiorina 1989: chaps. 5, 11)."

lectively, equipped with a richer tool-kit than their predecessors. Few of those trained at any of the major institutions from the 1970s forward will be unduly intimidated (or unduly impressed, either) by theories or techniques from behavioral psychology, empirical sociology or mathematical economics. Naturally, each will have his or her own predilections among them. But nowadays most will be perfectly conversant across all those methodological traditions, willing and able to borrow and steal, refute and repel, as the occasion requires.[18]

There are many ways of telling and retelling these disciplinary histories, with correspondingly many lessons for how to avoid the worst and achieve the best in the future. One way to tell the tale would be in terms of the rise and decline of the "guru." Unproductive periods in the prehistory of modern political science, just as in mid-century political philosophy, were characterized by gurus and their camp followers, the former engaging with one another minimally, the latter hardly at all.[19] These dialogues of the deaf are transformed into productive, collaborative engagements only once factional feuds have been displaced by some sense of a common enterprise and of some shared disciplinary concerns.[20]

Another lesson to draw from that tale concerns the bases upon which a sufficiently overlapping consensus is most likely for founding such a common enterprise. As in liberal politics itself (Rawls 1993), so too in the liberal arts more generally: a *modus vivendi* sufficient for productive collaboration is likely to emerge within an academic discipline only at lower levels of analysis and abstraction. It is sheer folly to seek to bully or cajole a diverse and dispersed community of scholars into an inevitably false and fragile consensus on foundational issues—whether cast in terms of the one true philosophy of science (logical positivism or its many alternatives) or in terms of the one true theory of society (structural-functionalism, systems theory, rational choice or whatever).

[18] Outstanding examples of such dexterity include Elster's (1983) *Explaining Technical Change* (1983) and Putnam's *Making Democracy Work* (1993). Elster and Putnam are uncommonly gifted practitioners of the art, emblematic, if not exactly representative, of political science at the turn of the century.

[19] Compare Dogan's (below: chap. 3) discussion of the "mutual disregard" among turn-of-the-century sociologists such as Durkheim, Weber, Töennies and Simmel and Waldo's (1975: 47–50) account of the ongoing wars between Chicago and Harvard in the 1930s with Parekh's (below: chap. 21) account of mid-century political philosophy.

[20] Indeed, judging from Warren Miller's (chap. 10 below) account, the early history of past breakthroughs—in his case, the behavioral revolution—was similarly characterized by cross-disciplinary conversations of just this sort. One might say the same of the "public choice" movement, growing out of collaborations among public-finance economists (Buchanan, Olson), lawyers (Tullock), political scientists (Riker, Ostrom) and sociologists (Coleman)—to tell the subdiscipline's history just in terms of the early presidents of its peak body, the Public Choice Society. Testimony to the frustrating strength of subdisciplinary enclaves is found in Almond (1990) and Easton and Schelling (1991).

Endless disputation over foundations is as unnecessary as it is unproductive, however. The simple sharing of "nuts and bolts"—the building-blocks of science—goes a long way toward consolidating a shared sense of the discipline (Elster 1989). Tricks and tools and theories which were initially developed in one connection can, as often as not, be transposed into other settings—*mutatis mutandis.* Much mutation, adaptation and reinterpretation is, indeed, often required to render borrowed tools appropriate to their new uses. But the borrowing, cross-fertilization and hybridization, and the conceptual stretching which it imposes on both sides of the borrowing and lending relationship is what scientific progress today seems principally to be about (Dogan below: chap. 3).

Whether it is a "science," strictly speaking, which has been achieved is an open question—and one best left as open, pending the ultimate resolution of interminable disputes among philosophers of science themselves over the "true" nature of science. But by the standards of the spartan definition of science we proposed in Section IC above—"systematic enquiry, building toward an ever more highly-differentiated set of ordered propositions about the empirical world"—our discipline has indeed become more scientific. Certainly it is now more highly differentiated, both in its own internal structure and in its propositions about the world.

It is yet another open question, however, whether the growth of science thus understood is a help or a hindrance to genuine scientific understanding. It is an open question whether we know more, or less, now that we have carved the world up into increasingly smaller pieces. More is not necessarily better. Metaphysicians cast their aspirations in terms of "carving reality at its joints." Scientists, in their theory-building, are always in danger not only of carving at the wrong places but also, simply, of taking too many cuts. Niche theorizing and boutique marketing could well prove a serious hindrance to genuine understanding in political science, as in so many of the other social and natural sciences.

It is the job of the integrators of the profession's subdisciplines to overcome these effects, to pull all the disparate bits of knowledge back together. On the evidence of Section IV, below, it seems that they do so admirably.

## III Professional touchstones

The increasing professionalization of the profession is manifest in many ways. Perhaps the most important are the extent to which practitioners, whatever their particular specialities, share at least a minimal grounding in broadly the same methodological techniques and in broadly the same core literature. These have been acquired in myriad ways—in postgraduate training, at Michigan or Essex summer schools, or on the job, teaching and researching. The depth and details of these common cores vary slightly, depending on country and subfield.[21] But virtually all political scientists nowadays can make tolerably good sense of regression equations, and virtually everyone is at least loosely familiar with broadly the same corpus of classics in the field.

## A Classic texts

Political science, like virtually all the other natural and social sciences, is increasingly becoming an article-based discipline. But while some classic journal articles never grow into a book, and some whole debates have been conducted on the pages of journals alone, the most lasting contributions still come predominantly in book form.[22] Almond and Verba's *Civic Culture* (1963), Campbell, Converse, Miller and Stokes' *American Voter* (1960), Dahl's *Who Governs?* (1961*b*), Dahrendorf's *Class and Class Conflict in Industrial Society* (1959), Deutsch's *Nerves of Government* (1963), Downs's *Economic Theory of Democracy* (1957), Easton's *Systems Analysis of Political Life* (1965), Huntington's *Political Order in Changing Societies* (1968), Key's *Responsible Electorate* (1966), Lane's *Political Ideology* (1962), Lindblom's *Intelligence of Democracy* (1965), Lipset's *Political Man* (1960), Moore's *Social Origins of Dictatorship and Democracy* (1966), Neustadt's *Presidential Power* (1960), Olson's *Logic of Collective Action* (1965): all these are the common currency of the profession, the *lingua franca* of our shared discipline and the touchstones for further contributions to it.[23]

One of the defining manifestations of the new professionalism within political science is the phenomenon of the "instant classic."[24] These are books which, almost immediately upon publication, come to be incorporated into the canon—books which everyone is talking about and

[21] That common methodological core might best be seen as being bracketed between Galtung (1967) and King, Keohane and Verba (1994).

[22] Marshall's *In Praise of Sociology* (1990) similarly virtually defines that discipline in terms of ten "classic" texts in post-war empirical (in his case, British) sociology.

[23] Just as had, a generation previously, books (to name a few) like: Duverger's *Political Parties* (1951/1954); Key's *Politics, Parties and Pressure Groups* (1942) and *Southern Politics* (1950); Schumpeter's *Capitalism, Socialism and Democracy* (1943); and Simon's *Administrative Behavior* (1951).

[24] Initially remarked upon by Brian Barry (1974), in connection with a book just outside this period—Hirschman's *Exit, Voice and Loyalty* (1970).

presumed to know, at least in passing. Whether these instant classics will have the same staying power as those older pillars of the profession is, perhaps, an open question. One of the striking findings to emerge from an analysis of the references contained within the rest of the chapters of the *New Handbook* is how short a shelf-life most work in political science actually enjoys. More than three-quarters of the works cited in the *New Handbook* have, as Appendix 1A shows, been published in the twenty years since the 1975 publication of the old *Handbook*; and over 30 per cent have been published in the last five years alone.[25] Cynics may say that is a reflection of pure faddishness. Others may say, more sympathetically, that it is an inevitable reflection of the way in which the next building-block inevitably fits on the last in any cumulative discipline. Whatever the source of the phenomenon, it is transparently true that several books which were much discussed, at some particular period, have now been substantially superseded in professional discourse.[26]

Still, for conveying a quick impression of substantive developments within the discipline over the past quarter-century, we could hardly do better than simply list "great books" produced over that period which have initiated a professional feeding frenzy of just that sort. The list is long, inevitably incomplete and disputable at the margins. Like the larger profession, it is also strongly Anglophone and largely U.S.-oriented. But by almost any account, these contemporary classics would probably have to include:

- Graham Allison's *Essence of Decision* (1971);
- Robert Axelrod's *Evolution of Co-operation* (1984);
- Samuel Barnes, Max Kaase *et al.*'s *Political Action* (1979);
- Morris Fiorina's *Retrospective Voting in American National Elections* (1981);
- Ronald Inglehart's *Silent Revolution* (1977);
- James March and Johan Olsen's *Rediscovering Institutions* (1989);
- Elinor Ostrom's *Governing the Commons* (1990);

[25] The former finding might be explained by the fact that authors of the first three chapters in each section of the *New Handbook* have been explicitly instructed to focus on developments since the 1975 publication of the Greenstein–Polsby *Handbook*. (Only the authors of the last "Old and New" chapter in each section have been encouraged to range temporally further afield.) But the latter fact cannot be explained away in that fashion, and it is so continuous with the former that it seems unlikely that the former can be wholly explained away in that fashion either.

[26] Perhaps the two most conspicuous examples, within the twenty years here under review, are Lindblom's *Politics and Markets* (1977) and Tufte's *Political Control of the Economy* (1978)—both of which were much discussed toward the beginning of the period but which now figure surprisingly peripherally in the *New Handbook* chapters touching upon the literatures they have spawned.

- Theda Skocpol's *States and Social Revolution* (1979);
- Sidney Verba and Norman Nie's *Participation in America* (1972);

Among the much-discussed books of the past two or three years which seem set to join this list are King, Keohane and Verba's *Designing Social Inquiry* (1994) and Robert Putnam's *Making Democracy Work* (1993).

## B Recurring themes

At the outset, we defined politics as the constrained use of social power. As we noted there, any novelty that that definition might claim lies in its emphasis upon *constraint* as a key to politics. But that novelty is not ours alone. Politics as (and the politics of) constraints has, in one way or another, been a recurring theme of political science over the past quarter-century.[27]

Running across virtually all the following chapters is, as has already been remarked upon, a renewed recognition of the importance of institutional factors in political life. With the rise of this "new institutionalism" comes a renewed appreciation of history and happenstance, rules and regimes as constraining forces in political life. Of course, it has long been a commonplace in some corners of the discipline that "history matters": for those who cut their professional teeth on Lipset and Rokkan's (1967) notions of "frozen cleavages" or Moore's (1966) developmental models of communism, fascism or parliamentary democracy or Burnham's (1970) theories of "critical realignments," there is little novelty in the thought that the coalition structure at crucial moments in the past might have shaped political life for years to come. But these new institutionalist themes are now central to the discipline as a whole, across its several subfields. Sterling examples include two contemporary classics in policy history: Skocpol's *Protecting Soldiers and Mothers: The Political Origins of Social Policy in the United States* (1992) and Orren's *Belated Feudalism: Labor, the Law and Liberal Development in the United States* (1991).

The legacy of history, then, is one of the constraints to which new institutionalism points us. Another is the nested and embedded nature of social rules and regimes, practices and possibilities. In this Russian-doll model of social life, ordinary machinations occur relatively near the surface. But, to take just the most straightforward legalistic example, the rules we invoke in enacting ordinary legislation are embedded higher-order principles, rules

[27] In another sense, too, there has been increasing interest in the decreasing capacity of the state apparatus. See Rose and Peters (1978), Nordlinger (1981) and Flora (1986).

of a constitutional sort. And, as many have recently discovered, even constitution-writers do not enjoy a completely free hand: even those "highest" laws are embedded in some other even higher-order principles, rules and procedures, albeit of an extra-legal sort. The same is true of all the other practices and procedures and rules and regimes that collectively frame social life. None is free-standing: all are embedded in, defined in terms of, and work on and in relation to, a plethora of other similar practices and procedures and rules and regimes. None is ultimate: each is nested within an ever-ascending hierarchy of yet-more-fundamental, yet-more-authoritative rules and regimes and practices and procedures (North 1990; Tsebelis 1990; Easton 1990; Weingast below: chap. 5; Alt and Alesina below: chap. 28; cf. Braybrooke and Lindblom 1963).

Standing behind all those nested rules and regimes, practices and procedures are, of course, socio-economic constraints of a more standard sort. Perhaps the more deeply nested aspects of social organization are as robust as they are only because they are sociologically familiar and materially productive: therein, in the final analysis, may well lie the ultimate source of their strength as constraints on the use of social power. Most of the time, however, those most deeply nested aspects of the social order exercise their influence unobtrusively, passing unnoticed and unquestioned. The ultimate sources of their strength as constraints are therefore virtually never on display (Granovetter 1985).

At other times, the use of social power is shaped and constrained by socio-economic forces that work right at the surface of social life. This may seem to be a tired old theme, reworked endlessly from the days of Marx (1852/1972*b*; 1871/1972*a*) and Beard (1913) forward. However, these themes have been powerfully formalized and elaborated in contemporary classics such as Lindblom's *Politics and Markets* (1977) and Tufte's *Political Control of the Economy* (1978). And surprisingly much remains to be said on these themes, judging from recent works such as Przeworski and Sprague's *Paper Stones* (1986), on the socio-economic logic limiting the prospects of electoral socialism, and Rogowski's *Commerce and Coalitions* (1989), grounding the structure of domestic coalitions in the terms of international trade.

The use of social power is also constrained in yet another way which has recently come to the fore across several subfields of political science. The constraints here in view are of a cognitive sort, constraints on the exercise of pure (and, more especially, practical) reason. Political sociologists and psychologists have long been sensitive to the irrational and arational aspects of political life: the workings of socialization and ideology within mass belief systems (Jennings and Niemi 1981; Converse 1964). But even rational choice modellers are now coming to appreciate the analytic possibilities that open up when relaxing heroic assumptions of complete information and perfect rationality (Simon 1954; 1985; Bell, Raiffa and Tversky 1988; Popkin 1991; Pappi below: chap. 9, Grofman below: chap. 30). Of course, what political economists see as informational shortcuts others construe as psychological imprints, and for all sorts of purposes that difference clearly still matters. But it is the convergence that has been achieved rather than the differences that remain which, from the present perspective, seems the more remarkable. Political scientists of virtually every ilk are once again according a central role to people's beliefs and what lies behind them.

What people believe to be true and important, what they believe to be good and valuable, not only guides but also constrains their social actions (Offe below: chap. 29). Those beliefs, in turn, are framed around past teachings and past experiences. Shaping those teachings and experiences can shape people's beliefs and values and thereby their political choices (Neustadt and May 1986; Edelman 1988). The manipulation of these constraints, like the manipulation of people within those constraints, is a deeply political act meriting—and, increasingly, receiving—as much analytic attention as any other. Among notable recent contributions are Allison's (1971) work on "conceptual maps," March's (1972) on "model bias," Axelrod's (1976) schema theory, Jervis's (1976) on the role of perceptions in international relations, and much work in political communications (Nimmo and Sanders 1981; Swanson and Nimmo 1990; Graber 1993).

Another recurring theme in the new political science which this *New Handbook* maps is the increasing appreciation that ideas have consequences. The point pops up, time and again, in discussions of public policy. Getting new perspectives on old problems, seeing new ways of doing things, seeing new things to do: all these, as applied to public problems, are quintessentially political activities (Olsen 1972; Nelson below: chap. 24; Majone below: chap. 26). But the same is true cross-nationally: the spread of the idea of democratization, as well as of particular ideas for how to democratize particular sorts of regimes, was undeniably central to some of the most dramatic recent political developments worldwide (Whitehead below: chap. 14). Within international relations, too, an idealism of almost Hegelian proportions is once again rightly on the rise (Goldmann below: chap. 16; Sanders below: chap. 17; Keohane below: chap. 18). Within the framework of "politics as the constrained use of social power," those maneuvers amount to moving or removing constraints; that makes them less obviously confrontational than other exercises of power, but exercises of power nonetheless.

Finally, there has been a virtual meltdown of the fact vs. value distinction, that old bugbear of the behavioral revolution in its most insistently positivist phase. There are meta-theoretical reasons aplenty for resisting the distinction; and insofar as the distinction can be defensibly drawn at all, there are ethical reasons for insisting upon the primacy of values, for insisting upon a "political science with a point" (Goodin 1980; 1982). But what eventually proved compelling was the simple recognition that political agents themselves are ethical actors, too (Taylor 1967; 1985). They internalize values and act upon them; and occasionally they find themselves persuaded (occasionally, perhaps, even by academic political philosophers) to internalize other, better values.

If we want to understand such people's behavior, we must incorporate values into our analysis—both the values that they actually have, and the ones that they might come to have. Thus, James Scott's *Moral Economy of the Peasant* (1976) explains peasant rebellions in Southeast Asia to perplexed policy-makers in simple terms of people's reactions against policies they perceive to be unjust, according to conventional local understandings of what justice requires; and Barrington Moore's *Injustice* (1978) aspires to generalize the proposition. The spread of the democratic ideal across southern Europe, then Latin America, then eastern Europe might similarly be seen as political action inspired by a vision of what was good, combined with a vision of what was possible (Dalton below: chap. 13; Whitehead below: chap. 14). Trying to disentangle facts and values in the mental processes and political dynamics underlying these developments would be pure folly.

Similarly, political scientists find themselves increasingly wanting to employ complex research designs systematically relating structures, processes and outcomes. To do so, they need a theoretical framework which can straddle and integrate all these levels of analysis. Therein arguably lies the great power of rational choice analysis and new institutionalism; and that, in turn, may go some way toward explaining the predominance of those intellectual agendas across contemporary political science as a whole (see Section IV below). At the same time, however, those complex research designs also aspire to the *normative* evaluation of structures, processes and outcomes, and in so doing they integrate normative political philosophy into their designs in ways that would have been an anathema to previous generations. Therein arguably lies the explanation for the prevalence of Rawls's works on justice (1971; 1993), among the most commonly referenced books, and for the presence of normative theorists like Barry, Dahl and Rawls among the most the most frequently referenced and important integrators of the discipline (see Appendices 1C, 1D and 1E).

## C New voices

We have learned, from feminists, deconstructionists and postmodernists more generally, to be attentive to "silences"—to what is left out and what goes unsaid. When surveying a whole discipline, trying to think what is not there but should be is always a daunting task.

Certainly it is true enough that whole subfields wax and wane. Of late, there has been rather less public law and rather less public administration finding its way into mainstream political science than once was the case (Wildavsky 1964; 1979; Wilson 1973)— although there is evidence that that is now changing, once again (Drewry below: chap. 6; Peters below: chap. 7; Peters and Wright below: chap. 27). Certain once-prominent subfields are thinly represented in the *New Handbook*—as perhaps they are, too, in the recent history of the profession which contributors have been asked to track. Commentators on public policy in general nowadays find far fewer occasions than once they would have done to reflect upon urban politics (Banfield and Wilson 1963; Banfield 1970; Katznelson 1981); commentators on international relations say less nowadays than they would have done, only a few years ago, about strategic studies (Schelling 1960; Freedman 1981); writers on institutions say rather little nowadays about the once-thriving field of representation (Eulau and Wahlke 1978; Fenno 1978); and writers on behavior say rather less than once they would have done about political influence (Banfield 1961), political communication and political participation more generally (cf. Pappi below: chap. 9; Dalton below: chap. 13; Grofman below: chap. 30; and McGraw below: chap. 34). Finally, there has always been too little attention paid, within the Anglocentric political science mainstream, to Marxian theories and foreign-language sources—although, again, there is evidence that that, too, is now changing (Whitehead below: chap. 14; Apter below: chap. 15; von Beyme below: chap. 22; Offe below: chap. 29).

Among the most notable new voices clearly represented in political science today, compared to a quarter-century ago, are those of postmodernists and of feminists themselves. Not only is there now a large literature on the distinctive roles played by women in politics (Nelson and Chowdhury 1994); there is now a distinctively feminine voice to be heard, particularly in political theory (Pateman 1988; Shanley and Pateman 1991; Young below: chap. 20), international relations (Tickner below: chap. 18) and public policy (Nelson below: chap. 24).

Postmodernism more generally has made rather more modest inroads, in part because its central precepts are cast on such a high theoretical plane (White 1991). Political theorists, however, have certainly shown an interest

(Young below: chap. 20; von Beyme below: chap. 22). Moreover, such theories prove to be a rich source of inspiration and insights for those studying the so-called "new political movements" (Dunleavy below: chap. 10; Dalton below: chap. 13; Young below: chap. 20) and the fracturing of the old international order (Tickner below: chap. 18). Wherever once there were clearly defined structures, and now there are none (or many disconnected ones), the post-structural theoretical arsenal may well offer insights into how that happened and why.

Whether fully postmodern, contemporary political science is decidedly substantially post-positivist, in that it certainly has taken lessons of the hermeneutic critique substantially on board. Subjective aspects of political life, the internal mental life of political actors, meanings and beliefs and intentions and values—all these are now central to political analysis across the board (Edelman 1964; 1988; Scott 1976; Riker 1986; Popkin 1991; Kaase, Newton and Scarbrough 1995). These developments are much in evidence throughout the *New Handbook*.[28]

Political methodology, more generally, seems to be entering something of a postmodern phase. Perhaps few methodologists would embrace that self-description quite so enthusiastically as Alker (below: chap. 35). Many now do, however, emphasize the need for contextualized and path-dependent explanations (Jackson below: chap. 32; Ragin *et al.* below: chap. 33). That, in turn, represents something of a retreat away from generality and toward particularity, away from universality and toward situatedness, in the explanatory accounts we offer for political phenomena. In that sense, these recent developments in political methodology might be seen as a "postmodern turn."

Indeed, treating the history of the discipline as a whole as our "text," postmodern techniques might help us see many possible narratives in our collective past—and correspondingly many possible paths open for future development (Dryzek, Farr and Leonard 1995). Those fixated on a "big science" vision of linear progress might be disappointed by that prospect of development along disparate trajectories.[29] But on Dogan's (below: chap. 3) account of progress within the discipline, this proliferation of "new breeds" among political scientists is to be greatly welcomed for the fruitful possibilities for hybridization that it creates.

[28] Weingast below: chap. 5; Pappi below: chap. 9; Dunleavy below: chap. 10; Whitehead below: chap.14; Tickner below: chap. 18; von Beyme below: chap. 21; Hofferbert and Cingranelli below: chap. 25; Majone below: chap. 26; Offe below: chap. 29; Grofman below: chap. 30; Alker below: chap. 35.

[29] As is clearly revealed by looking at the disparate paths of development of political science within different national communities. Compare the U.S. story, as classically related by Somit and Tanenhaus (1967), with the stories contained in, for example: Easton, Gunnell and Graziano 1991; Wagner, Wittrock and Whitley 1991; Dierkes and Biervert 1992; Rokkan 1979; and Chester 1986.

## IV The shape of the profession: a bibliometric analysis

Perhaps the best way to substantiate these broader claims about the nature of the discipline, as revealed in the *New Handbook*, is through a closer bibliometric analysis of the references contained herein. The conventional style of bibliometric analysis counts how frequently works or, more commonly, works by particular authors are cited. Albeit inevitably flawed in various respects, these are nonetheless useful measures for all sorts of purposes: for gauging the reputation and standing of individuals and departments within the profession, for assessing the intensity of use of any particular piece or type of work or of works by any particular individual, and so on.[30]

What is of more interest in the present context, however, is the penetration of works by members of one subdiscipline into other subdisciplines, and the resulting integration across the discipline provided by those authors and works. For those purposes, we have preferred to concentrate, not on counts of the number of times authors or pieces of work are cited in the body of the text, but rather on the number of times authors or pieces of work are cited in the reference lists of the other chapters of the *New Handbook*. (To avoid biasing the results, these counts systematically exclude references in our own reference list at the end of this chapter.[31]) This approach, while not without its distortions, seems to be the most appropriate to our task.[32]

Various things emerge relatively clearly from the resulting bibliometric counts. The first is that the vast majority of political scientists are specialists contributing primarily to their own subdisciplines. A great majority of the authors and works mentioned at all are found in the reference lists of only one subdisciplinary section of the *New Handbook*. Indeed (as Appendix 1B

[30] For assessments of U.S.-based individuals and U.S. departments along these lines, see Klingemann (1986). More recent data are available from Klingemann upon request.

[31] We do so to avoid "cooking the books" in favor of the generalizations we hope to establish through our own pattern of referencing. We have also, as is conventional, excluded all self-references from our counts (which imposes a greater hardship than usual on *New Handbook* contributors, since they are in effect excluded from a quarter of the chapters in which their own names would most naturally appear, whoever was writing them). We have counted all co-authors equal and in full (i.e., as if each of them were the author of a single-authored work); although less conventional, that seemed more appropriate given our focus upon assessing authors as potential integrators rather than upon apportioning credit to reputations.

[32] Most notably, counting the number of times an author appears in reference lists, rather than counting citations in the text of the chapters, introduces a bias against Berlin's (1953) "hedgehogs" (who know one big thing—or have written one big book) and in favor of his "foxes" (who know many little things—or have written many books or articles to which people refer).

shows) almost two-thirds of authors are mentioned only once, in a single chapter's reference list.[33]

At the other extreme, a handful of scholars reappear frequently in *New Handbook* chapters' reference lists. Some thirty-five authors (listed in Appendix 1C) are mentioned more than ten times in various chapters' reference lists. No particular importance should be attached to anyone's precise standing in the resulting league of honor: we are dealing with a small sample, here, in looking across only thirty-four chapters' reference lists. While exact rankings within this list may therefore be imprecise, and the membership of the list itself somewhat unreliable at the margins, in broad outline this list seems to have both surface plausibility and broad reliability as an indicator of whose work is of broad interest across several subfields of the discipline.

Inspection of the names on that list—and of the most frequently referenced books (Appendix 1D), more especially—reveals with remarkable clarity the intellectual agendas currently dominating the political science community as a whole. We see quite strikingly the residues of the "two revolutions," first the behavioral revolution and then the rational choice one, on the contemporary profession. Looking at the list of most widely referenced books, the old classics of the behavioral revolution—Campbell, Converse, Miller and Stokes's *American Voter*; Almond and Verba's *Civic Culture*; Lipset and Rokkan's *Party Systems and Voter Alignments*—are still there, albeit in the lower tiers. But sweeping all three top tiers are the classics of the subsequent rational choice revolution: Downs's *Economic Theory of Democracy* and Olson's *Logic of Collective Action*, joined recently by Ostrom's *Governing the Commons*. The rational choice putsch has been remarkably successful, not so much in pushing out the old behavioral orthodoxy, as in carving out a predominant role for itself alongside it.[34] That the residue of the older revolution is still so strongly in evidence is in itself an impressive fact about the discipline. Cynics say that scientific revolutions are simply the product of fad and fashion. If so, we would expect one fad to disappear completely when another takes over. However, that clearly has not happened. Whether knowledge is strictly cumulative is, per-

[33] A dispiriting interpretation of this result, together with that in Appendix 1A, is that most scholars make minor contributions which will be soon forgotten. Remember, though, that the *New Handbook* is a highly selective survey of major contributions over the past two decades; having made a contribution meriting mention there is in itself a major accomplishment, and in those terms it is an encouraging sign that there are so many scholars working at the many cutting edges of our discipline.

[34] Barry's *Sociologists, Economists and Democracy* (1970/1978), written at the cusp of this change, subjects both to a merciless logical critique; he remarks, in the preface to the 1978 edition, upon the remarkable waning of the "sociological" (behavioral) paradigm over the intervening eight years.

haps, another issue. But at least the older insights have not been lost as new ones are added, in successive revolutions within political science.

Inspecting those same tables, we also see the growing evidence of the next revolution on its way: the "new institutionalist" movement. That movement is partly in league with the rational choice movement—an alliance represented, among most frequently referenced books, by Ostrom's *Governing the Commons* and North's *Institutions, Institutional Change and Economic Performance.* In other authors' hands, new institutionalism takes on a decidedly sociological and anti-rational choice cast. This strand is represented, among most frequently referenced books by March and Olsen's *Rediscovering Institutions* and Skocpol's *States and Social Revolutions.* Admitting as it does of either interpretation—and of both at once—new institutionalism thus has great power to provide an integrative framework for the sorts of complex research designs discussed above.

The next step in our bibliometric profile of the profession is to search among those frequently-referenced members of the discipline for "integrators" of the discipline as a whole. We define as an "integrator" anyone who appears at least once in the reference lists in more than half (that is, five or more) of the eight subdisciplinary parts of the *New Handbook.* Of the 1630 authors represented in the *New Handbook*'s references, only seventy-two (4.4 percent) appear in five or more chapters. Of those, only twenty-one constitute "integrators" of the discipline as a whole—in the sense that their influence spreads across more than half the subdisciplinary parts of the *New Handbook.* These twenty-one "integrators" are listed in Appendix 1E.[35]

Using much the same techniques, we see how well each subdiscipline itself is integrated into the larger discipline. Here we focus on each subdiscipline's first three tiers of most frequently referenced authors (these are listed in Appendix 1F). To see how well the subdiscipline is integrated into the larger discipline, we then ask (in Appendix 1G) two questions. To what extent are that subdiscipline's most frequently referenced authors also the most frequently referenced in the discipline as a whole (defined as being among its first ten tiers of most frequently referenced)? And to what extent are that subdiscipline's most frequently referenced authors among the integrators of the discipline as a whole?

Two subdisciplines (Comparative Politics and Political Economy) are, on both measures, particularly well-integrated into the profession as a whole. There are other subdisciplines (Public Policy and Administration and Political Theory) whose own most referenced authors serve as integra-

[35] Having only twenty-one integrators among the hundreds of scholars currently active may make political science seem like a relatively unintegrated enterprise. Conversely, having an entire discipline fix its collective focus on so few individuals and their work might make for more integration.

tors for the discipline as a whole, while there are others (notably, Political Institutions) which largely lack integrators but whose own most frequently referenced authors are among the most frequently referenced within the discipline as a whole. There is one other subdiscipline (Political Methodology) whose own most frequently referenced authors figure neither among the larger discipline's integrators nor among its most-frequently-referenced. This latter subdiscipline seems to stand substantially outside and to develop relatively independently of the larger discipline.[36]

A good composite view of the shape of the discipline emerges from combining all these criteria: who are the "integrators" of the profession, who are "most frequently referenced in the discipline as a whole," and who are "most frequently referenced within their own subdisciplines." As Appendix 1H shows, there are some ten key scholars—"powerhouses" of the discipline, we call them—who score highly on all three criteria. These ten individuals (listed as "group 1" in Table A1.H) are among the "most frequently referenced" authors, both within the discipline as a whole and within their respective subdisciplines; and, at the same time, they are integrators of the discipline as a whole. Another twenty-eight scholars (groups 2–5 in Table A1.H) play one or the other of those discipline-wide roles, with yet another thirty-nine playing similarly key roles within subdisciplines alone.

The general pattern is clear enough: there are highly differentiated subdisciplinary communities making great advances. But there is also a small band of scholars at the peak of the profession who genuinely do straddle many (in a few cases, most) of those subdisciplinary communities and integrate them into one coherent disciplinary whole.

## V Conclusion

The picture that emerges from this analysis, and from the remaining thirty-four chapters of the *New Handbook* upon which it is based, is a happy one of a fractious discipline of bright and enterprising scholars constantly looking over the fences that used to separate subdisciplines. The old aspiration of a Unified Science might still remain a chimera (Neurath, Carnap and Morris 1955). But at the turn of the century, ours looks to be at least a potentially unifiable science. The intellectual energy, curiosity and openness that has been required to carry us even that far is, in itself, surely something to celebrate.

[36] What we cannot analyze using these data are relations between the subdisciplines of political science and other disciplines. On these connections, see Dogan (below: chap. 3).

# Appendix 1A

## The Continuing Impact of Works in Political Science

How much continuing impact is any given work in political science likely to have? To answer this question, we have categorized publications in all the reference lists of all the chapters in the *New Handbook* according to their year of original publication. The results are shown in Table A1.A.

**Table A1.A** Publication years of references

| *Year* | *Number* | *Percent* | *Cumulative Percent* |
|---|---|---|---|
| pre-1900 | 22 | 0.6 | 0.6 |
| 1900–20 | 11 | 0.3 | 1.0 |
| 1921–40 | 59 | 1.7 | 2.7 |
| 1941–50 | 45 | 1.3 | 4.0 |
| 1951–60 | 155 | 4.6 | 8.6 |
| 1961–65 | 147 | 4.3 | 12.9 |
| 1966–70 | 165 | 4.8 | 17.7 |
| 1971–75 | 214 | 6.3 | 24.0 |
| 1976–80 | 320 | 9.4 | 33.4 |
| 1981–85 | 441 | 13.0 | 46.4 |
| 1986–90 | 792 | 23.3 | 69.7 |
| 1991–95 | 1032 | 30.3 | 100.0 |
| Total | 3403 | | |

The upshot of this analysis is that more than half of the works mentioned in the *New Handbook* have been published in the last decade, and two-thirds having been published in the last two decades. Less than one-tenth were published before the publication of the *American Voter* (Campbell, Converse, Miller and Stokes 1960).

# Appendix 1B

## Frequency of Appearances of Authors Reference Lists

All in all, there are some 1630 authors mentioned in the reference lists of chapters 2 through 35 of the *New Handbook.* As is standard practice in such analyses, we exclude self-references. We also omit our own reference list (chapter 1) from this analysis, to avoid biasing the results of the analysis in favor of the propositions we hope to prove through our own pattern of referencing. In cases of multiple authorship, each author was counted as if the author of single-authored work.

The total number of references, thus defined, is 3341. The mean number of times an author is referenced in these lists is 2.1. However, the variance is large (5.8) and the distribution is skewed (3.9). Thus, it makes more sense to take the median as a descriptor of the distribution: the median is 1. And this is what a look at Table A1.B clearly shows. The vast majority of the authors (1063 = 65.2 percent) are referenced only once.

In our analysis we are interested in differentiation and integration of the discipline, which we have divided in this *New Handbook* into eight subdisciplines. Put in that context, this finding points to differentiation: almost two-thirds of the authors, appearing as they do in only one chapter's reference list, must perforce have been cited in only one of the subdisciplinary parts of the *New Handbook.* Other authors are cited more frequently, some much more frequently. These are further analyzed in the Appendices that follow.

**Table A1.B** Frequency of references

| *Number of references* | *Number of authors* | *Percent* |
|---|---|---|
| 1 | 1063 | 65.2 |
| 2 | 266 | 16.3 |
| 3 | 93 | 5.7 |
| 4 | 55 | 3.4 |
| 5 | 52 | 3.2 |
| 6 | 23 | 1.4 |
| 7 | 14 | 0.9 |
| 8 | 12 | 0.7 |
| 9 | 10 | 0.6 |
| 10 | 7 | 0.4 |
| 11 | 10 | 0.6 |
| 12 | 4 | 0.2 |
| 13 | 5 | 0.3 |
| 14 | 2 | 0.1 |
| 15 | 3 | 0.2 |
| 16 | 3 | 0.2 |
| 17 | 2 | 0.1 |
| 18 | 3 | 0.2 |
| 19 | 2 | 0.1 |
| 25 | 1 | 0.1 |
| Total | 1630 | 100 |

# Appendix 1C

## Most Frequently Referenced Authors across the Discipline

Authors who are referenced frequently have the potential for integrating the subfields of the discipline. We define the "most frequently referenced authors" as those who occupy the first ten tiers in the number appearance in reference lists in chapters 2 through 35 of the *New Handbook.* That cut-off point gives us the thirty-five authors (2.1 percent of all authors referenced) listed in Table A1.C.

While frequently referenced authors might integrate the discipline, it is also possible that frequently referenced authors may be referenced mostly in their own specific subfield of the discipline. In that case, the frequency of references to that author would not be counted as evidence of integration but rather of differentiation. In order to investigate that dimension, we have to look at the most frequently referenced authors subdiscipline by subdiscipline (see Appendix 1F).

**Table A1.C** Most frequently referenced authors, discipline-wide

| *Tier* | *Authors* | *Number of times appearing in reference lists* |
|---|---|---|
| 1 | Verba, S. | 25 |
| 2 | Lipset, S.M. | 19 |
| | Shepsle, K. | |
| 3 | Almond, G. | 18 |
| | Dahl, R. | |
| | Riker, W. | |
| 4 | Lijphart, A. | 17 |
| | Skocpol, T. | |
| 5 | Keohane, R. | 16 |
| | McCubbins, M. | |
| | Weingast, B. | |
| 6 | March, J. | 15 |
| | North, D. | |
| | Ostrom, E. | |
| 7 | Elster, J. | 14 |
| | Inglehart, R. | |
| 8 | Barry, B. | 13 |
| | Downs, A. | |
| | Olson, M. | |
| | Przeworski, A. | |
| | Simon, H. | |
| 9 | Converse, P. | 12 |
| | Fiorina, M. | |
| | Ferejohn, J. | |
| | Schmitter, P. | |
| 10 | Buchanan, J. | 11 |
| | Easton, D. | |
| | Lasswell, H. | |
| | Moe, T. | |
| | Olsen, J. | |
| | Ordeshook, P. | |
| | Rawls, J. | |
| | Rokkan, S. | |
| | Sartori, G. | |
| | Wildavsky, A. | |

# Appendix 1D

## Most Frequently Referenced Books

Table A1.D shows the books appearing most frequently in reference lists in chapters in the *New Handbook.*

**Table A1.D** Most frequently referenced books

| *Tier* | *No. of refs.* | *Author* | *Title* | *Date of pub.* |
|---|---|---|---|---|
| 1 | 11 | Anthony Downs | *An Economic Theory of Democracy* | 1957 |
| 2 | 9 | Mancur Olson | *The Logic of Collective Action* | 1965 |
| 3 | 8 | Elinor Ostrom | *Governing the Commons* | 1990 |
| 4 | 7 | Douglass North | *Institutions, Institutional Change and Economic Performance* | 1990 |
| 5 | 6 | Gabriel A. Almond and Sidney Verba | *The Civic Culture* | 1963 |
| | | Angus Campbell, Philip E. Converse, Warren E. Miller and Donald Stokes | *The American Voter* | 1960 |
| | | James G. March and Johan P. Olsen | *Rediscovering Institutions* | 1989 |
| | | John Rawls | *A Theory of Justice* | 1971 |
| 6 | 5 | Brian Barry | *Sociologists, Economists and Democracy* | 1970/1978 |
| | | Morris P. Fiorina | *Retrospective Voting in American National Elections* | 1981 |
| | | Seymour Martin Lipset and Stein Rokkan, eds. | *Party Systems and Voter Alignments* | 1967 |
| | | John Rawls | *Political Liberalism* | 1993 |
| | | William Riker and Peter C. Ordeshook | *An Introduction to Positive Political Theory* | 1973 |
| | | Theda Skocpol | *States and Social Revolutions* | 1979 |

# Appendix 1E

## The Integrators

Who are the integrators? In order to answer this question we have looked at all authors who have had at least five references. Given our division of the world of political science into eight substantive subfields, these authors might in principle be present in more than half of the subfields distinguished (one reference in five parts of the *New Handbook*). Our starting-point, then, is with the seventy-two authors who appear in reference lists of at least five chapters; they make up 4.4 percent of the 1630 authors.

We define an "integrator" as an author who is to be found at least once in the reference sections of more than half of the eight subfields (that is in five or more of them). Out of the total of 1630 authors twenty-one (or 1.3 percent) qualify as integrators. Their names are listed in Table A1.E.

**Table A1.E** Integrators of the profession

| *Number of parts in which they appear* | *Author* | *Appears in parts* | | | | | | | |
|---|---|---|---|---|---|---|---|---|---|
| | | *2* | *3* | *4* | *5* | *6* | *7* | *8* | *9* |
| 8 | Ostrom, E. | x | x | x | x | x | x | x | x |
| 6 | Barry, B. | x | x | x | x | x | | x | |
| | Downs, A. | | x | x | | x | x | x | x |
| | March, J. | x | x | x | x | | x | x | |
| | Olson, M. | x | x | x | | x | x | x | |
| 5 | Almond, G. | x | x | x | | | | x | x |
| | Dahl, R. | x | x | x | | x | | | x |
| | Erikson, R. | | x | x | | | x | x | x |
| | Laver, M. | x | x | x | | x | | x | |
| | Lijphart, A. | x | x | x | | x | | | x |
| | Lipset, S. M. | x | x | x | | | x | | x |
| | Olsen, J. | x | | x | x | | x | x | |
| | Ordeshook, P. | x | | | | x | x | x | x |
| | Polsby, N. | x | | | x | x | x | | x |
| | Riker, W. | x | | | | x | x | x | x |
| | Scharpf, F. | x | | x | | x | x | x | |
| | Shepsle, K. | x | | x | | x | x | x | |
| | Simon, H. | x | x | | | x | x | x | |
| | Skocpol, T. | x | | x | | x | x | | x |
| | Verba, S. | | x | x | x | x | | | x |
| | Weingast, B. | x | | x | x | | x | x | |

# Appendix 1F

## Most Frequently Referenced Authors, by Subdiscipline

The rank-order of most frequently referenced authors contained in Appendix 1C might reflect either strong prominence in one of the subfields (differentiation) or across subfields (integration) or both. In order to sort out that issue, we look next at the top three ranks for the subfields.

Who, then, dominates the subdisciplines? We define the group of authors who are prominent in any particular subfield as those among the top three tiers of most frequently referenced within the part of the *New Handbook* devoted to that subfield. By this criterion, we find that fifty-nine authors (3.6 percent of the total authors referenced) are prominent in one—or, in the case of three authors, more (McCubbins, Sects. II and VIII; Stokes, Sects. III and VII; Verba, Sects. I and IV)—subfields. They are listed in Table A1.F below.

**Table A1.F** Most frequently referenced authors, by subdiscipline

| *Tiers* | *Authors* | *Number of times referenced* |
|---|---|---|
| Part I **The Discipline** | | |
| 1 | Dahl, R. | 10 |
| 2 | Lipset, S. | 9 |
| | Verba, S. | |
| 3 | Lasswell, H. | |
| Part II **Political Institutions** | | |
| 1 | North, D. | 9 |
| 2 | Elster, J. | 8 |
| | McCubbins, M. | |
| 3 | Ferejohn, J. | 7 |
| Part III **Political Behavior** | | |
| 1 | Converse, P. | 9 |
| | Sprague, J. | |
| 2 | Campbell, A. | 6 |
| | Sniderman, P. | |
| | Stokes, D. | |

Table A1.F (*cont.*)

| *Tiers* | *Authors* | *Number of times referenced* |
|---|---|---|
| Part III **Political Behavior** *(cont.)* | | |
| 3 | Heath, A. | 5 |
| | Miller, W. | |
| Part IV **Comparative Politics** | | |
| 1 | Almond, G. | 11 |
| 2 | Verba, S. | 8 |
| 3 | Inglehart, R. | 7 |
| | Lijphart, A. | |
| Part V **International Relations** | | |
| 1 | Keohane, R. | 10 |
| 2 | Waltz, K. | 8 |
| 3 | Holsti, K. | 5 |
| | Krasner, S. | |
| Part VI **Political Theory** | | |
| 1 | Goodin, R. | 6 |
| | Habermas, J. | |
| | Kymlicka, W. | |
| 2 | Barry, B. | 5 |
| | Cohen, J. | |
| | Gutmann, A. | |
| | Rawls, J. | |
| | Taylor, C. | |
| 3 | Dowding, K. | 4 |
| | Galston, W. | |
| | Hardin, R. | |
| | Miller, D. | |
| | Pateman, C. | |
| | Walzer, M. | |
| Part VII **Public Policy and Administration** | | |
| 1 | Lindblom, C. | 5 |
| | Wildavsky, A. | |
| 2 | Merriam, C. | 4 |
| | Skocpol, T. | |
| | Wilson, J. Q. | |
| 3 | Derthick, M. | 3 |
| | deLeon, P. | |
| | Esping-Andersen, G. | |
| | Flora, P. | |
| | Klingemann, H.-D. | |
| | Lowi, T. | |
| | Olson, M. | |
| | Sharkansky, I. | |
| | Stokes, D. | |

| *Tiers* | *Authors* | *Number of times referenced* |
|---|---|---|
| Part VIII **Political Economy** | | |
| 1 | Weingast, B. | 10 |
| 2 | McCubbins, M. | 8 |
| 3 | Shepsle, K. | 7 |
| Part IX **Political Methodology** | | |
| 1 | Achen, C. | 6 |
| | King, G. | |
| 2 | Beck, N. | 5 |
| | Brady, H. | |
| | Campbell, D. | |
| | Palfrey, T.R. | |
| 3 | Kinder, D. | 4 |
| | Lodge, M. | |

# Appendix 1G

## The Integration of Subdisciplines into the Discipline

Among the fifty-nine most frequently referenced authors within subdisciplines (listed in Appendix 1F), twenty (or 34 percent) of them are among those most frequently referenced in the discipline as a whole (see Table A1.C). Nearly two-thirds of those prominent in subdisciplines are prominent predominantly in those subfields, which constitutes a measure of differentiation within the discipline as a whole. This point is underscored by the fact that only ten (or 17 percent) of the fifty-nine most frequently referenced authors within subdisciplines are also among the twenty-one integrators of the discipline as a whole listed in Appendix 1E. A more detailed picture is given in Table A1.G.

Looking at this result from the perspective of differentiation and integration, it seems that Political Behavior, International Relations, Political Theory, Public Policy and Administration, and Political Methodology in particular are subfields with quite a lot of independent development. Relatively few of those subfields' most frequently referenced authors figure among the most frequently referenced authors of the discipline as a whole (Appendix 1C), and relatively few of them are among the integrators of the discipline as a whole (Appendix 1E).

**Table A1.G** Integration of subdisciplines into the discipline

| Part | *Column 1 number of most frequently referenced authors within the subdiscipline* | *Column 2 number of col. 1 who are also most frequently referenced within the discipline as a whole* | *Column 3 number of col. 1 who are also disciplinary integrators* |
|---|---|---|---|
| I Discipline | 4 | 4 | 3 (Dahl, Lipset, Verba) |
| II Political Institutions | 4 | 4 | 0 |
| III Political Behaviour | 7 | 1 | 0 |
| IV Comparative Politics | 4 | 4 | 3 (Almond, Lijphart, Verba) |
| V International Relations | 4 | 1 | 0 |
| VI Political Theory | 14 | 2 | 1 (Barry) |
| VII Public Policy and Administration | 14 | 3 | 2 (Skocpol, Olson) |
| VIII Political Economy | 3 | 3 | 2 (Weingast, Shepsle) |
| IX Political Methodology | 8 | 0 | 0 |
| Subtotal | 62 | 22 | 11 |
| Less repeated names | –3 | –2 | –1 |
| Total | 59 | 20 | 10 |
| % | 100 | 33 | 17 |

# Appendix 1H

## Summary of Leading Figures in the Discipline

Table A1.H is a summary of results from Appendices C through G and combines three types of information.

- Column 1 addresses the question, "Is the author among the most frequently referenced in the discipline as a whole?" Scholars listed in Appendix 1C get an *x* in that column for being "prominent in the discipline."
- Column 2 addresses the question, "Is the author among the most frequently referenced in one or more of the eight subdisciplinary parts of the *New Handbook*?" Scholars listed in Appendix 1F get an *x* in that column for being "prominent in their subdisciplines."
- Column 3 addresses the question, "Is the author an integrator within the discipline as a whole?" Scholars listed in Appendix 1E get an *x* in that column for being "integrators."

**Table A1.H** Leading figures in political science

**Group 1**

The "powerhouses" are those authors who are integrators and who are also among the most frequently referenced both across the discipline as a whole and within one or more of its subdisciplines. According to these criteria we find ten (0.6 percent) powerhouses. These are:

| | *Prominent in:* | | |
|---|---|---|---|
| | *Discipline* | *Subdiscipline* | *Integrator* |
| Almond, G. | x | x | x |
| Barry, B. | x | x | x |
| Dahl, R. | x | x | x |
| Lijphart, A. | x | x | x |
| Lipset, S. M. | x | x | x |
| Olson, M. | x | x | x |
| Shepsle, K. | x | x | x |
| Skocpol, T. | x | x | x |
| Weingast, B. | x | x | x |
| Verba, S. | x | x | x |

**Group 2**

The next group are the "highly visible integrators." These are defined as integrators who are among the most frequently referenced across the discipline as a whole but not in any particular subfield. There are seven (0.4 percent) highly visible integrators. These are:

| | *Prominent in discipline* | *Integrator* |
|---|---|---|
| Downs, A. | x | x |
| March, J. | x | x |
| Olsen, J. | x | x |
| Ordeshook, P. | x | x |
| Ostrom, E. | x | x |
| Riker, W. | x | x |
| Simon, H. | x | x |

**Group 3**

There are four (0.2 percent) integrators with a lower degree of visibility, meaning that they qualify as integrators but are not among the most frequently referenced either across the discipline as a whole or within any particular subfield. These are:

| | *Integrator* |
|---|---|
| Erikson, R. | x |
| Laver, M. | x |
| Polsby, N. | x |
| Scharpf, F. | x |

**Group 4**

The first three groups exhaust the "integrators." Next we come to a group of "generally prominent subfield representatives," defined as those who are among the most frequently referenced both across the discipline as a whole and within their own subfields. We have ten (0.6 percent) such scholars:

| | *Prominent in:* | |
|---|---|---|
| | *Discipline* | *Subdiscipline* |
| Converse, P. | x | x |
| Elster, J. | x | x |
| Ferejohn, J. | x | x |
| Inglehart, R. | x | x |
| Keohane, R. | x | x |
| Lasswell, H. | x | x |
| McCubbins, M. | x | x |
| North, D. | x | x |
| Rawls, J. | x | x |
| Wildavsky, A. | x | x |

**Group 5:**
Another seven (0.4 percent) authors are just "generally prominent." That is, they are among the most frequently referenced in the discipline overall but neither among the most frequently referenced within any particular subfield nor among the integrators. These are:

| | *Prominent in discipline* |
|---|---|
| Buchanan, J. | x |
| Easton, D. | x |
| Fiorina, M. | x |
| Moe, T. | x |
| Rokkan, S. | x |
| Sartori, G. | x |
| Schmitter, P. | x |

**Group 6**
Groups 1–5 exhaust those who are integrators of the discipline or who are most frequently referenced across the discipline as a whole. Finally, we have a group of authors who are among the most frequently referenced within their own subfields but do not qualify according to the two other criteria. They may be called the "special subfield representatives." There are thirty-nine (2.4 percent) authors of this type. They are:

| | *Prominent in subdiscipline* |
|---|---|
| Achen, C. | x |
| Beck, N. | x |
| Brady, H. | x |
| Campbell, A. | x |
| Campbell, D. | x |
| Cohen, J. | x |
| deLeon, P. | x |
| Derthick, M. | x |
| Dowding, K. | x |
| Esping-Andersen, G. | x |
| Flora, P. | x |
| Galston, W. | x |
| Goodin, R. | x |
| Gutmann, A. | x |
| Habermas, J. | x |
| Hardin, R. | x |
| Heath, A. | x |
| Holsti, O. | x |
| Kinder, D. | x |
| King, G. | x |
| Klingemann, H.-D. | x |
| Krasner, S. | x |
| Kymlicka, W. | x |

| | *Prominent in subdiscipline (cont.)* |
|---|---|
| Lindblom, C. | x |
| Lodge, M. | x |
| Lowi, T. | x |
| Merriam, C. | x |
| Miller, D. | x |
| Miller, W. | x |
| Palfrey, T. | x |
| Pateman, C. | x |
| Rawls, J. | x |
| Sharkansky, I. | x |
| Skocpol, T. | x |
| Sniderman, P. | x |
| Stokes, D. | x |
| Sprague, J. | x |
| Taylor, C. | x |
| Walzer, M. | x |
| Waltz, K. | x |
| Wilson, J. Q. | x |

**Group 7**

The seventy-six (4.7 percent) scholars in groups 1–6 above exhaust the list of those who, by our criteria, count as integrators of the discipline as a whole or the most frequently referenced authors either across the discipline as a whole or within any particular subfield of it. There are another 1523 authors referenced in the *New Handbook* whose contributions to the discipline are sufficiently substantial as to merit notice in what is, in and of itself, a very selective list.

## Acknowledgments

We are grateful to Frank Castles, Mattei Dogan, John Dryzek, Dieter Fuchs, Richard I. Hofferbert, Giandomenico Majone and to a seminar of Research Unit III of the Wissenschaftszentrum Berlin, more generally, for comments on earlier drafts of this chapter. We are especially grateful to Nicolas Schleyer for assistance with the bibliometric research discussed in Sect. IV and the Appendices.

## References

Allison, G. T. 1971. *Essence of Decision*. Boston: Little, Brown.

Almond, G. A. 1990. *A Discipline Divided*. Newbury Park, Calif.: Sage.

—— and Genco, S. J. 1977. Clocks, clouds and the study of politics. *World Politics*, 29: 489–522.

—— and Verba, S. 1963. *The Civic Culture*. Princeton, N.J.: Princeton University Press.

APSA (American Political Science Association). 1991. *A Guide to Professional Ethics in Political Science*. 2nd edn. Washington, D.C.: APSA.

Axelrod, R., 1984. *The Evolution of Cooperation*. New York: Basic Books.

—— ed. 1976. *The Structure of Decision*. Princeton, N.J.: Princeton University Press.

Bachrach, P. and Baratz, M. S. 1963. Decisions and non-decisions: an analytic framework. *American Political Science Review*, 57: 632–42.

Banfield, E. C. 1961. *Political Influence*. Glencoe, Ill.: Free Press.

—— 1970. *The Unheavenly City*. Boston: Little, Brown.

—— and Wilson, J. Q. 1963. *City Politics*. New York: Vintage.

Barnes, S.; Kaase, M.; *et al.* 1979. *Political Action: Mass Participation in Five Western Democracies*. Beverly Hills, Calif.: Sage.

Barry, B. 1974. Review article: exit, voice and loyalty. *British Journal of Political Science*, 4: 79–107. Reprinted in Barry 1989: 186–221.

—— 1978. *Sociologists, Economists and Democracy*. 2nd edn. Chicago: University of Chicago Press; originally published 1970.

—— 1989. *Democracy, Power and Justice*. Oxford: Clarendon Press.

Beard, C. A. 1913. *An Economic Interpretation of the Constitution of the United States*. New York: Macmillan.

Bell, D. E.; Raiffa, H.; and Tversky, A., eds. 1988. *Decision Making*. Cambridge: Cambridge University Press.

Berger, P. L. 1963. *Invitation to Sociology*. New York: Doubleday.

Berlin, I. 1953. *The Hedgehog and the Fox*. London: Weidenfeld and Nicolson.

Braybrooke, D., and Lindblom, C. E. 1963. *A Strategy of Decision*. New York: Free Press.

Burnham, W. D. 1970. *Critical Elections and the Mainsprings of American Politics*. New York: Norton.

Campbell, A.; Converse, P. E.; Miller, W.; and Stokes, D. 1960. *The American Voter*. New York: Wiley.

Chester, N. 1986. *Economics, Politics and Social Studies in Oxford, 1900–85*. London: Macmillan.

Converse, P. E. 1964. The nature of belief systems in mass publics. Pp. 206–61 in *Ideology and Discontent*, ed. D. E. Apter. New York: Free Press.

Cornford, F. M. 1908. *Microcosmographia Academia*. Cambridge: Bowes and Bowes.

Crick, B. 1962. *In Defence of Politics*. London: Weidenfeld and Nicholson.

Dahl, R. A. 1957. The concept of power. *Behavioral Science*, 2: 201–15.

—— 1961*a*. The behavioral approach in political science: epitaph for a monument to a successful protest. *American Political Science Review*, 55: 763–72.
—— 1961*b*. *Who Governs*? New Haven, Conn.: Yale University Press.
—— 1963. *Modern Political Analysis*. 3rd edn. Englewood Cliffs, N.J.: Prentice-Hall.
Dahrendorf, R. 1959. *Class and Class Conflict in Industrial Society*. Stanford, Calif.: Stanford University Press.
Deutsch, K. 1963. *The Nerves of Government*. Glencoe, Ill.: Free Press.
Dierkes, M., and Biervert, B., eds. 1992. *European Social Science in Transition*. Frankfurt am Main/Boulder, Colo.: Campus Verlag/Westview.
Downs, A. 1957. *An Economic Theory of Democracy*. New York: Harper.
Dryzek, J.; Farr, J.; and Leonard, S., eds. 1995. *Political Science in History*. Cambridge: Cambridge University Press.
Duverger, M. 1954. *Political Parties*, trans. B. and R. North. London: Methuen; originally published 1951.
—— 1966. *The Idea of Politics: The Uses of Power in Society*, trans. R. North and R. Murphy. London: Methuen; originally published 1964.
Easton, D. 1965. *A Systems Analysis of Political Life*. New York: Wiley.
—— 1969. The new revolution in political science. *American Political Science Review*, 63: 1051–61.
—— 1990. *The Analysis of Political Structure*. New York: Routledge.
—— Gunnell, J. G.; and Graziano, L., eds. 1991. *The Development of Political Science: A Comparative Survey*. London: Routledge.
—— and Schelling, C., eds. 1991. *Divided Knowledge*. Newbury Park, Calif.: Sage.
Edelman, M. 1964. *The Symbolic Uses of Politics*. Urbana: University of Illinois Press.
—— 1988. *Constructing the Political Spectacle*. Chicago: University of Chicago Press.
Elster, J. 1983. *Explaining Technical Change*. Cambridge: Cambridge University Press.
—— 1989. *Nuts and Bolts for the Social Scientist*. Cambridge: Cambridge University Press.
Eulau, H., and Wahlke, J. C., *et al.* 1978. *The Politics of Representation*. Beverly Hills, Calif.: Sage.
Fenno, R. 1978. *Home Style*. Boston: Little, Brown.
Ferejohn, J. A. 1974. *Pork Barrel Politics*. Stanford, Calif.: Stanford University Press.
Fiorina, M. P. 1981. *Retrospective Voting in American National Elections*. New Haven, Conn.: Yale University Press.
—— 1989. *Congress: Keystone of the Washington Establishment*. 2nd edn. New Haven, Conn.: Yale University Press.
—— 1995. Rational choice, empirical contributions and the scientific enterprise. *Critical Review*, 9: 85–94.
Flora, P., ed. 1986. *Growth to Limits*. 4 vols. Berlin: de Gruyter.
Foucault, M. 1977. *Discipline and Punish*, trans. A. Sheridan. Harmondsworth, Mddx.: Allen Lane.
Freedman, L. 1981. *The Evolution of Nuclear Strategy*. London: Macmillan.
Galtung, J. 1967. *Theory and Methods of Social Research*. Oslo: Universitetsforlaget.
Goodin, R. E. 1976. Possessive individualism again. *Political Studies*, 24: 488–501.
—— 1980. *Manipulatory Politics*. New Haven, Conn.: Yale University Press.
—— 1982. *Political Theory and Public Policy*. Chicago: University of Chicago Press.
Graber, Doris. 1993. Political communication: scope, progress, promise. Pp. 305–34 in *Political Science: The State of the Discipline, II*, ed. A. Finifter. Washington, D.C.: American Political Science Association.
Granovetter, M. 1985. Economic action and social structure: the problem of embeddedness. *American Journal of Sociology*, 91: 481–510.

Greenstein, F. I., and Polsby, N. W., eds. 1975. *Handbook of Political Science.* 8 vols. Reading, Mass.: Addison-Wesley.

Hart, H. L. A. 1961. *The Concept of Law.* Oxford: Clarendon Press.

Hirschman, A. O. 1970. *Exit, Voice and Loyalty.* Cambridge, Mass.: Harvard University Press.

Hollis, M. 1977. *Models of Man.* Cambridge: Cambridge University Press.

Hughes, E. C. 1958. *Men and their Work.* Glencoe, Ill.: Free Press.

Huntington, S. P. 1968. *Political Order in Changing Societies.* New Haven, Conn: Yale University Press.

Inglehart, R. 1977. *The Silent Revolution.* Princeton, N.J.: Princeton University Press.

Jennings, M. K., and Niemi, R. G. 1981. *Generations and Politics.* Princeton, N.J.: Princeton University Press.

Jervis, R. 1976. *Perception and Misperception in International Politics.* Princeton, N.J.: Princeton University Press.

Johnson, N. 1989. *The Limits of Political Science.* Oxford: Clarendon Press.

Jouvenel, B. de. 1948. *On Power,* trans. J. F. Huntington. London: Hutchinson; originally published 1945.

Kaase, M.; Newton, K.; and Scarbrough, E., eds. 1995. *Beliefs in Government.* 5 vols. Oxford: Oxford University Press.

Katznelson, I. 1981. *City Trenches.* Chicago: University of Chicago Press.

Key, V. O. Jr. 1942. *Politics, Parties and Pressure Groups.* New York: Crowell.

—— 1950. *Southern Politics in State and Nation.* New York: Knopf.

—— 1966. *The Responsible Electorate: Rationality in Presidential Voting, 1936–1960.* Cambridge, Mass.: Harvard University Press.

Kiewiet, D. R. 1983. *Micropolitics and Macroeconomics.* Chicago: University of Chicago Press.

King, G.; Keohane, R. D.; and Verba, S. 1994. *Designing Social Inquiry.* Princeton, N.J.: Princeton University Press.

Klingemann, H.-D. 1986. Ranking the graduate departments in the 1980s: toward objective qualitative indicators. *PS,* 29: 651–61.

Laclau, E., and Mouffe, C. 1985. *Hegemony and Socialist Strategy.* London: Verso.

Lane, R. E. 1962. *Political Ideology.* Glencoe, Ill.: Free Press.

Lasswell, H. D. 1950. *Politics: Who Gets What, When, How?* New York: P. Smith.

—— and Kaplan, A. 1950. *Power and Society: A Framework for Political Inquiry.* New Haven, Conn.: Yale University Press.

Lindblom, C. E. 1965. *The Intelligence of Democracy.* New York: Free Press.

—— 1977. *Politics and Markets.* New York: Basic.

Lipset, S. M. 1960. *Political Man.* New York: Doubleday.

—— and Rokkan, S., eds. 1967. *Party Systems and Voter Alignments.* New York: Free Press.

Lowi, T. J. 1964. American business, public policy, case-studies and political theory. *World Politics,* 16: 676–715.

Lukes, S. 1974. *Power: A Radical View.* London: Macmillan.

Mansbridge, J. J., ed. 1990. *Beyond Self-Interest.* Chicago: University of Chicago Press.

March, J. G. 1972. Model bias in social action. *Review of Educational Research,* 42: 413–29.

—— and Olsen, J. P. 1989. *Rediscovering Institutions.* New York: Free Press.

Marshall, G. 1990. *In Praise of Sociology.* London: Unwin Hyman.

Marx, K. 1972*a*. The civil war in France. Pp. 526–77 in *The Marx-Engels Reader,* ed. R. C. Tucker. New York: Norton; originally published 1871.

—— 1972*b*. The eighteenth brumaire of Louis Napoleon. Pp. 436–525 in *The Marx-Engels Reader,* ed. R. C. Tucker. New York: Norton; originally published 1871.

McCoy, C. A., and Playford, J., eds. 1968. *Apolitical Politics: A Critique of Behavioralism.* New York: Thomas Y. Crowell.
Medawar, P. B. 1979. *Advice to a Young Scientist.* New York: Harper & Row.
Merton, R. K. 1936. The unintended consequences of purposive social action. *American Sociological Review* 1: 894–904.
—— 1973. *The Sociology of Science,* ed. N. W. Storer. Chicago: University of Chicago Press.
Mitchell, W. C. 1969. The shape of political theory to come: from political sociology to political economy. Pp. 101–36 in *Politics and the Social Sciences,* ed. S. M. Lipset. New York: Oxford University Press.
Monroe, K. R., ed. 1991. *The Economic Approach to Politics.* New York: HarperCollins.
Moon, J. D. 1975. The logic of political inquiry: a synthesis of opposed perspectives. In Greenstein and Polsby 1975: vol. i, pp. 131–228.
Moore, B., Jr. 1966. *The Social Origins of Dictatorship and Democracy.* Boston: Beacon.
—— 1978. *Injustice: The Social Bases of Obedience and Revolt.* London: Macmillan.
Morriss, P. 1987. *Power.* Manchester: Manchester University Press.
Nelson, B. J., and Chowdhury, N., eds. 1994. *Women and Politics Worldwide.* New Haven, Conn.: Yale University Press.
Neurath, O.; Carnap, R.; and Morris, C. L., eds. 1955. *International Encyclopedia of Unified Science: Foundations of the Unity of Science.* 2 vols. Chicago: University of Chicago Press.
Neustadt, R. E. 1960. *Presidential Power.* New York: Wiley.
—— and May, E. R. 1986. *Thinking in Time.* New York: Free Press.
Nimmo, D. D., and Sanders, K. R., eds. 1981. *Handbook of Political Communication.* Beverly Hills, Calif.: Sage.
Nordlinger, E. A. 1981. *On the Autonomy of the Democratic State.* Cambridge, Mass.: Harvard University Press.
North, D. C. 1990. *Institutions, Institutional Change and Economic Performance.* Cambridge: Cambridge University Press.
Nozick, R. 1981. *Philosophical Explanations.* Cambridge, Mass: Harvard University Press.
Oakeshott, M. 1956. Political education. Pp. 1–21 in *Philosophy, Politics and Society.* 1st series. Oxford: Blackwell; originally delivered as an LSE Inaugural Lecture in 1951.
Olsen, J. P. 1972. Public policy-making and theories of organizational choice. *Scandinavian Political Studies,* 7: 45–62.
Olson, M. Jr. 1965. *The Logic of Collective Action.* Cambridge, Mass.: Harvard University Press.
Orren, K. 1991. *Belated Feudalism.* Cambridge: Cambridge University Press.
Ostrom, E. 1990. *Governing the Commons.* Cambridge: Cambridge University Press.
Parsons, T. 1968. Professions. Vol. xii, pp. 536–47 in *International Encyclopedia of the Social Sciences,* ed. D. L. Sills. London: Macmillan.
Pateman, C. 1988. *The Sexual Contract.* Oxford: Polity.
Popkin, S. L. 1991. *The Reasoning Voter.* Chicago: University of Chicago Press.
—— *et al.* 1976. What have you done for me lately? *American Political Science Review,* 70: 779–805.
Przeworski, A., and Sprague, J. 1986. *Paper Stones: A History of Electoral Socialism.* Chicago: University of Chicago Press.
Putnam, R. D. 1993. *Making Democracy Work.* Princeton, N.J.: Princeton University Press.
Ranney, A., ed. 1962. *Essays on the Behavioral Study of Politics.* Urbana: University of Illinois Press.
Rawls, J. 1971. *A Theory of Justice.* Cambridge, Mass.: Harvard University Press.
—— 1993. *Political Liberalism.* New York: Columbia University Press.
Ridley, F. F. 1975. Political institutions: the script not the play. *Political Studies,* 23: 365–80.

Riker, W. H. 1986. *The Art of Political Manipulation.* New Haven, Conn.: Yale University Press.
—— and Ordeshook, P. C. 1973. *An Introduction to Positive Political Theory.* Englewood Cliffs, N.J.: Prentice-Hall.
Rivlin, A. M. 1971. *Systematic Thinking for Social Action.* Washington, D.C.: Brookings Institution.
Rogowski, R. 1989. *Commerce and Coalitions.* Princeton, N.J.: Princeton University Press.
Rokkan, S., ed. 1979. *A Quarter Century of International Social Science.* New Delhi: Concept, for the International Social Science Council.
Rose, R., and Peters, B. G. 1978. *Can Government Go Bankrupt?* New York: Basic.
Russell, B. 1938. *Power.* London: Allen and Unwin.
Sandel, M. 1982. *Liberalism and the Limits of Justice.* Cambridge: Cambridge University Press.
Schattschneider, E. E. 1960. *The Semi-Sovereign People.* New York: Holt, Rinehart and Winston.
Schelling, T. C. 1960. *The Strategy of Conflict.* Cambridge, Mass.: Harvard University Press.
Schumpeter, J. A. 1943. *Capitalism, Socialism, and Democracy.* London: Allen and Unwin.
Scott, J. C. 1976. *The Moral Economy of the Peasant.* New Haven, Conn.: Yale University Press.
Sen, A. 1977. Rational fools: a critique of the behavioral foundations of economic theory. *Philosophy and Public Affairs,* 6: 317–44.
Shanley, M. L., and Pateman, C., eds. 1991. *Feminist Interpretations and Political Theory.* Oxford: Polity.
Simon, H. A. 1951. *Administrative Behavior.* New York: Macmillan.
—— 1954. A behavioral theory of rational choice. *Quarterly Journal of Economics,* 69: 99–118.
—— 1985. Human nature is politics: the dialogue of psychology and political science. *American Political Science Review,* 79: 293–304.
Skocpol, T. 1979. *States and Social Revolutions.* Cambridge: Cambridge University Press.
—— 1992. *Protecting Soldiers and Mothers.* Cambridge, Mass.: Harvard University Press.
Somit, A., and Tanenhaus, J. 1967. *The Development of American Political Science.* Boston: Allyn & Bacon.
Stretton, H. 1969. *The Political Sciences.* London: Routledge and Kegan Paul.
Swanson, D. L., and Nimmo, D., eds. 1990. *New Directions in Political Communication.* Newbury Park, Calif.: Sage.
Taylor, C. 1967. Neutrality in political science. Pp. 25–57 in *Philosophy, Politics and Society,* ed. P. Laslett and W. G. Runciman. 3rd series. Oxford: Blackwell. Reprinted in Taylor 1985: vol. ii, pp. 58–90
—— 1985. *Philosophical Papers.* 2 vols. Cambridge: Cambridge University Press.
Tsebelis, G. 1990. *Nested Games.* Berkeley: University of California Press.
Tufte, E. R. 1978. *The Political Control of the Economy.* Princeton, N.J.: Princeton University Press.
Verba, S., and Nie, N. H. 1972. *Participation in America.* New York: Harper and Row.
Wagner, P.; Wittrock, B.; and Whitley, R., eds. 1991. *Discourses on Society: The Shaping of the Social Science Disciplines.* Dordrecht: Kluwer.
Waldo, D. 1975. Political science: tradition, discipline, profession science, enterprise. In Greenstein and Polsby 1975: vol. i, pp. 1–130.
Weber, Max. 1946. Politics as a vocation. Pp. 77–128 in *From Max Weber: Essays in Sociology,* ed. H. Gerth and C. W. Mills. New York: Oxford University Press; originally published 1919.

—— 1978. *Economy and Society*, ed. G. Roth and C. Wittich, trans. E. Fischoff *et al.* Berkeley: University of California Press; originally published 1922.

White, S. K. 1991. *Political Theory and Postmodernism.* Cambridge: Cambridge University Press.

Wildavsky, A. 1964. *Politics of the Budgetary Process.* Boston: Little, Brown.

—— 1979. *Speaking the Truth to Power: The Art and Craft of Policy Analysis.* Boston: Little, Brown.

Wilson, J. Q. 1973. *Political Organizations.* New York: Basic Books.

Wolin, S. 1960. *Politics and Vision.* Boston: Little, Brown.

Chapter 2

# Political Science: The History of the Discipline

Gabriel A. Almond

## I Introduction

IF we were to model the history of political science in the form of a curve of scientific progress in the study of politics over the ages, it would properly begin in Greek political science, make some modest gains in the Roman centuries, not make much progress in the Middle Ages, rise a bit in the Renaissance and the Enlightenment, make some substantial gains in the 19th century, and then take off in solid growth in the 20th century as political science acquires genuine professional characteristics. What would be measured by this curve is the growth and qualitative improvement in knowledge concerned with the two fundamental questions of political science: the properties of political institutions, and the criteria we use in evaluating them.

We would record three rising blips in the 20th-century growth curve. There was the Chicago blip in the interwar decades (1920–1940), introducing organized empirical research programs, emphasizing psychological and sociological interpretations of politics, and demonstrating the value of quantification. A second much larger blip in the decades after World War II would measure the spread of "behavioral" political science throughout the world, improvements in the more traditional subdisciplines, and professionalization (in the sense of the establishment of multi-membered, meritocratically recruited, relatively non-heirarchic, departments; the establishment of associations and specialist societies, refereed journals; and so on). A third blip would register the entry of deductive and mathematical methods, and economic models in the "rational choice/methodological individualist" approach.

We might call this view of disciplinary history, the "progressive-eclectic" view. It would be shared by those who accept as the criterion of political science scholarship the search for objectivity based on rules of evidence and inference. This criterion would be applicable not only to studies we call "behavioral" but also to political philosophy (both historical and normative), empirical case studies (both historical and contemporary), systematic comparative studies, statistical studies involving survey and aggregate quantitative data, as well as research involving formal mathematical modelling and experiments (both real and simulated). In this sense it is an eclectic and non-hierarchical, rather than an integral, standard.

It is "progressive" in the sense that it imputes the notion of improvement to the history of political studies, in the quantity of knowledge, and its quality in terms of both insight and rigor. With respect to insight, most colleagues would agree that Michael Walzer (1983), has a better grasp of the concept of justice than does Plato, and with respect to rigor (and insight as well) Robert Dahl (1989) gives us a better theory of democracy than did Aristotle.[1]

There are four opposing views of the history of political science. Two of them would challenge its scientific character. There is an "anti-science" position as well as a "post-science" position. Two more of them—Marxists and the "rational choice" theorists—would challenge its eclecticism in favor of a purist, hierarchical monism. The Straussians express the "anti-science" view, that the introduction of scientific methodology is a harmful illusion, that it trivializes and clouds understanding and that the basic truths about politics are to be uncovered through direct colloquy with the classics and old texts. The "post-empirical," "post-behavioral" approach to disciplinary history takes a deconstructive view; there is no privileged history of the discipline. There is a pluralism of disciplinary identities, each with its own view of disciplinary history.

The Marxist, neo-Marxist and "critical theory" approaches challenge our eclecticism, arguing that political science or rather social science (since there can be no separable political science) consists of the unfalsifiable truths discovered and stated in the works of Marx and elaborated by his associates and followers. This view rejects the notion of a political science separable from a science of society. The science of society reveals itself in the course of its own dialectical development. Rational choice theory rejects our eclecticism in favor of a hierarchic model of political science as moving toward a parsimonious set of formal, mathematical theories applicable to the whole of social reality, including politics.

[1] See, on a more modest scale, Riker 1982.

This chapter also assumes that political science has both scientific and humanistic components, both governed by the same imperatives of scholarly inquiry—the rules of evidence and inference. Contributions to knowledge may come from great insight or great virtuosity. We also assume that, within the ontology of the families of sciences, it is on the "cloud" side of Karl Popper's (1972) "clouds and clocks" continuum. That is to say, the regularities it discovers are probabilistic rather than lawlike, and many of them may have relatively short half-lives.

## II Themes of a progressive-eclectic history

The essential object of political science, which it shares with all of scholarship, is the creation of knowledge, defined as inferences or generalizations about politics drawn from evidence. As King, Keohane and Verba (1994: 7) put it in their recent book, "Scientific research is designed to make . . . inferences on the basis of empirical information about the world." This criterion is evident even in such explicitly "anti-scientific" work as that of the Straussians. That is, they consider evidence, analyze it and draw inferences from it. It is impossible to conceive of a scholarly enterprise that does not rely on this evidence-inference methodological core. It would include Marxist and neo-Marxist studies, even though these studies are based on assumptions about social processes that are unfalsifiable and hence not fully subject to the rules of evidence or logical inference. It would include Clifford Geertz's (1973) "thick description" style of political science, exemplified by Womack's (1968) study of the Mexican peasant leader Zapata, at the simple display-of-evidence extreme; and it would include the work of Downs (1957) Riker (1962), and Olson (1965) at the inferential-deductive extreme. In *Zapata* we seem to have only evidence without inference and in the *Economic Theory of Democracy*, inference without evidence. But Hirschman (1970) tells us that the biography of the peasant leader is teeming with explanatory and policy implications; and the axioms and theorems of Downs generate a whole family of propositions testable by evidence. Both are falsifiable, through contrary evidence or logical flaws.

# III A historical overview

## A The Greeks and Romans

Though heroic efforts have been made to include writings of the ancient Near East in the political science chronicle, they are more properly viewed as precursors. Love for the Bible cannot convert the advice given to Moses by his father-in-law as to how he might more efficiently adjudicate the conflicts among the children of Israel, or the Deuteronomic doctrine of kingship, into serious political science.[2] But when we reach the Greece of Herodotus (c. 484–425 BCE) we are in a world in which analysis of political ideas and ideals, and speculation about the properties of different kinds of polities and the nature of statesmanship and citizenship, have become part of conventional wisdom. Informed Greeks of the 5th century BCE—living in the many independent Greek city states, in which the same language is spoken and the same or similar gods are worshipped, sharing common historical and mythological memories, engaged in inter-city trade and diplomacy, forming alliances and carrying on warfare—provided an interested audience for information and speculation about varieties of governmental and political arrangements, economic, defense and foreign policies.

The history of political science properly begins with Plato (428–348 BCE) whose *Republic*, *Statesman* and *Laws* are the first classics of political science.[3] In these three studies, Plato sets out propositions about justice, political virtue, the varieties of polity and their transformation which have survived as political theories well into the 19th century and even until the present day. His theories of political stability and of performance optimization, modified and elaborated in the work of Aristotle and Polybius, anticipate contemporary speculation about democratic transition and consolidation. In his first political typology, in the *Republic*, Plato presents his ideal regime based on knowledge and possession of the truth, and hence exemplifying the rule of virtue, and he then presents four other developmentally related regimes in descending order of virtue—Timocracy, Oligarchy, Democracy, and Tyranny. Timocracy is a corruption of the ideal state in which honor and military glory supplant knowledge and virtue; oligarchy is a corruption of Timocracy, replacing honor with wealth as the principle of recruitment; democracy arises out of the corruption of oligarchy, and in turn is corrupted into tyranny.

In *The Statesman*, written much later than *The Republic*, and in *The Laws*, written in his old age (after the sobering experiences of the

[2] Cf. Wildavsky 1984; 1989.

[3] See further: Sabine and Thorson 1973: chaps. 4, 5; Strauss and Cropsey 1987: 33 ff.

Peloponnesian War and the failure of his mission to Syracuse), Plato distinguishes between the ideal republic and the realistically possible varieties of polity. To classify real regimes, he introduces the famous three-by-two table, marrying quantity and quality: the rule of the one, the few, the many; each in its pure and impure versions. That generated the six-fold classification of regimes—monarchy, tyranny, aristocracy, oligarchy, democracy, ochlocracy—which Aristotle perfected and elaborated in his *Politics*, and which has served as a basic taxonomy through the ages and into the 19th century.

In *The Laws* Plato presented the first version of the "Mixed Constitution" as the realistically best and most stable regime, designed to halt the cycle of development and degeneration implicit in the six-fold scheme. The Mixed Constitution, as formulated by Plato, attains stability by combining principles which might otherwise be in conflict—the monarchic principle of wisdom and virtue, with the democratic principle of freedom. This scheme was adopted and improved upon by Aristotle. It is the first explanatory theory in the history of political science, in which institutions, attitudes, and ideas, are related to process and performance. It is the ancestor of separation of powers theory.

Aristotle (384–322 BCE) spent 20 years as a member of Plato's Academy. Then after a period of tutoring of Alexander of Macedon, Aristotle returned to Athens and formed his own Lyceum, a teaching institution cum library-museum and research institute. The method of the Lyceum was inductive, empirical and historical, in contrast to the predominantly idealist and deductive approach stressed in Plato's Academy. The Lyceum is said to have collected 158 constitutions of Greek city-states, only one of which—that of Athens—has survived. The lectures which make up Aristotle's *Politics* were apparently drawn from the analyses and the interpretations of these data.

While Plato's metaphysics led him to depreciate the real world and the human capacity to perceive and understand it, and to posit a world of ideal forms of which reality was a pale approximation, Aristotle, in contrast, was more of a hands-on empiricist viewing political reality as a physician might view illness and health. Sir Ernest Barker points out,

> It is perhaps not fanciful to detect a special medical bias in a number of passages of the *Politics*. This is not merely a matter of the accumulation of "case records," or of the use of the writings of the school of Hippocrates such as the treatise of "Airs, waters, and places." It is a matter of recurring comparison between the art of the statesman and the art of the good physician; it is a matter of the deep study of the pathology of constitutions, and of their liability to the fever of sedition, which we find in Book V of the *Politics*;

it is a matter of the preoccupation with therapeutics which we also find in the same book—a preoccupation singularly evident in the passage (at the end of chapter XI) which suggests a regimen and cure for the fever of tyranny (Barker, introduction to Aristotle 1958 edn.: xxx).

While in his theory of the polity Aristotle begins from Plato's six-fold classification of states, from a realistic point of view he argues that there are really four important types: oligarchy and democracy, the two types into which most of the Greek city states might be classified; "polity" or constitutional or "mixed" government, which is a combination of oligarchy and democracy, and which (because it reconciles virtue with stability) is the best attainable form of government; and tyranny, which is the worst. To back up his argument he points out that, while the social structures of cities vary according to the economies, occupations, professions and statuses contained in them, these differences are reducible into different distributions of rich and poor citizens. Where the rich dominate, we have oligarchy; where the poor dominate, we have democracy. Where the middle class dominates, we may have "mixed" or constitutional government, tending to be stable since extreme interests are out-weighed by moderate ones. Political structures and patterns of recruitment are classified according to the arrangements of the deliberative, magistrative and judicial organs, and according to the access of different classes to them.

A modern political scientist—a Dahl, Rokkan, Lipset, Huntington, Verba, or Putnam—would be on quite familiar ground with Aristotle's analysis, in *The Politics* and *Ethics*, of the relation of status, occupation, profession and class to varieties of political institutions, on the one hand, and of the relation between political socialization and recruitment to political structure and process, on the other. The metaphysics and ontology would be shared. But had these chapters, or something like them, been submitted by contemporary graduate students in search of dissertation topics, one can visualize marginal comments of a Dahl or Verba: "What cases are you generalizing about?"; "What about using a scale here?"; "How would you test the strength of this association?"; and the like. Aristotle presents a whole set of propositions and hypotheses—on what makes for political stability and what makes for breakdown, on developmental sequences, on educational patterns and political performance—that cry out for research designs and careful quantitative analysis. The Aristotelian method consists essentially of a clinical sorting out of specimens, with hypotheses about causes and sequences, but without systematic tests of relationships.

The Greek political theory of Plato and Aristotle was a combination of universalistic and parochial ideas. The world about which they generalized was the world of the Greek city-states. They were generalizing about Greeks and not about humankind. Citizens were differentiated from slaves, alien residents and foreigner barbarians. With Alexander's conquests, and the intermingling of Greek and oriental cultures, two notions generated by the Stoical philosophical school gained in authority. These were the idea of a universal humanity and of an order in the world based on natural law. These ideas were first advanced by the Stoic philosopher, Chrysippus, in the last third of the 3rd century BCE. Their clearest formulation was in the work of Panaetius (185–109 BCE) and Polybius (203–120 BCE), two Stoic philosophers of the second century who in turn transmitted these ideas to the Roman intellectual élite of the late republic. While Panaetius developed the philosophical and ethical aspects of late Stoicism, Polybius adapted Platonic and Aristotelian ideas to the history of Rome and to the interpretation of Roman institutions.

Polybius attributes the remarkable growth and power of Rome to its political institutions. He makes more explicit the developmental ideas of Plato and Aristotle, offering simple social psychological explanations for the decay of the pure forms of monarchy, aristocracy and democracy and for their degeneration into the impure forms of tyranny, oligarchy and ochlocracy. According to Polybius, the Roman state builders had, through a process of trial and error, rediscovered the virtues of the mixed constitution—the combination of the monarchic, aristocratic and democratic principles implemented in the Consulate, the Senate and the Assembly. It was these institutions which made possible the conquest of the world in the course of half a century, and which according to Polybius guaranteed a future of stable and just world rule under Roman law.[4]

Three-quarters of a century later, the Roman lawyer Cicero (106–43 BCE), applied "mixed constitution" theory to Roman history at a time when the institutions of the Roman Republic were already in deep decay. This part of his work was an appeal for a return to the structure and culture of the earlier Roman Republic prior to the populist and civil war decades of the Gracchi, Marius and Sulla. More significant and lasting was his development of the Stoic doctrine of natural law. This was the belief that there is a universal natural law resulting from the divine order of the cosmos and the rational and social nature of humanity. It was his formulation of this natural law idea which was taken up in the Roman law, and passed from it into Catholic church doctrine and ultimately into its Enlightenment and modern manifestations.[5]

Thus we find formulated in Greek thought by the end of the third century BCE, and in Roman thought in the following centuries, the two great

[4] See further Sabine and Thorson 1973: chaps. 4–9.
[5] See further Sabine and Thorson 1973: chaps. 9, 10.

themes of political theory, themes that carry through the history of political science into the present day. These are: "What are the institutional forms of polity?" and "What are the standards we use to evaluate them?" The answer to the first was the Platonic and Aristotelian six-fold classification of pure and impure organizational forms, and the mixed constitution as the solution to the problem of degeneracy and cyclicalism. The answer to the question of evaluation—legitimacy, justice—was the doctrine of natural law. These ideas were transmitted to Rome by the late Stoics (particularly Panaetius and Polybius) and from the work of Romans (such as Cicero and Seneca) into Catholic political theory.

## B Mixed constitutions and natural law theory in history

Mixed constitution theory and the theory of law receive their fullest medieval codification in the work of Thomas Aquinas (1225–1274), who relates the mixed constitution to justice and stability through its conformity to divine and natural law. His exemplars of the mixed constitution are the divinely ordained political order of the Israel of Moses, Joshua and the judges, balanced by elders and tribal leaders, and the Roman Republic in its prime, with its mix of Assembly, Senate and Consulate. He follows the arguments of Aristotle on the weaknesses and susceptibility to tyranny of the pure forms of monarchic, aristocratic and democratic rule. Combining the pure forms is the antidote to human weakness and corruption.[6]

In the late middle ages and in the Renaissance, mixed government and natural law provide the theoretical coinage according to which governments were valued. Just as Israel of the pre-monarchic period and Rome of the Republican age were viewed by Thomas Aquinas and those whom he influenced as approximating the ideal of mixed government in the past, for the Italian political theorists of the late middle ages and Renaissance the exemplar was Venice, with its monarchic Doge, its aristocratic Senate and democratic Great Council. The stability, wealth and power of Venice were taken as proof of the superiority of the mixed system.

The variety of principalities and republics in northern Italy in these centuries, the overarching and rival claims of Church and Empire, the warfare, conquest, revolution, diplomatic negotiation and institutional innovation in which they were constantly engaged, stimulated several generations of political theorists who reflected and wrote on this political experience.[7]

[6] See further Blythe 1992: chap. 3.
[7] See further Blythe 1992; Pocock 1975; Skinner 1978.

Central to their discussions were the ideas of the mixed constitution as expressed in Aristotle and Thomas Aquinas. With the translation of his *History of Rome* in the 16th century, Polybius became influential particularly in Florence and on the work of Machiavelli (1469–1527). In the Florentine crises of the late 15th and early 16th centuries, Machiavelli engaged in a polemic with the historian Guicciardini in which the principal authorities cited were Aristotle, Polybius and Thomas Aquinas, and the issues turned on which countries were the best exemplars of the mixed constitution. Guicciardini favored an Aristotelian, Venetian-Spartan aristocratic bias; Machiavelli favored a somewhat greater role for the popular element, relying more on Polybius for support.[8]

The breakthrough of Renaissance political theory lay in Machiavelli's treatment of the legitimacy of regimes and political leaders. Prior to *The Prince* and the *Discourses,* writers treated political regimes dichotomously as pure and corrupt, normative or non-normative, in the original Platonic and Aristotelian senses.[9] Machiavelli, viewing politics as practised in Italy in the 15th and 16th centuries, legitimized non-normative politics as unavoidable, as survival-related, as part of reality. A Prince who failed to employ problematic means when necessary to survival would be unable to do good when that was possible. Machiavelli touched the nerve of political science with this "value-free" orientation, and his name became a synonym for moral indifference and political cynicism. The issues raised by this venture into realism are still fluttering the dovecotes of political philosophy.

The theory of sovereignty, so important a theme in the Middle Ages, Renaissance, and the Enlightenment, receives its first full formulation in the work of Jean Bodin (1529–1596). His doctrine of absolutism as a solution to the problem of instability and disorder is formulated in polemic with the theory of the mixed constitution. Employing a realistic, historical method he makes the argument that the classic cases of mixed government, Rome and Venice, were actually concentrated and centralized regimes: indeed, every important and long-lasting regime concentrated the legislative and executive powers under a central authority. His appreciation of the influence of environmental and social structural conditions on the characteristics of states anticipates Montesquieu in its anthropological sensitivity.[10]

While there was substantial progress in the development of political science in the Enlightenment, such writers as Hobbes, Locke, Montesquieu, Hume, Madison and Hamilton were pursuing the same themes that con-

[8] See further Blythe 1992: 292 ff.

[9] See further Skinner 1978: 131 ff.

[10] See further Sabine and Thorson 1973: chap. 21.

cerned Plato, Aristotle, Polybius, Cicero, Aquinas, Machiavelli and Bodin—the forms and varieties of rule, and the standards by which one judged them. In considering the progress made by the Enlightenment philosophers we look for improvements in the gathering and evaluation of evidence and in the structure of inference.

The first scholarly project completed by Thomas Hobbes (1588–1679) was a translation of Thucydides' *Peloponnesian Wars*, a history of a disorderly and tragic epoch, just as England of the 17th century was disturbed by civil war, regicide, dictatorship and exile. Hobbes's view of the state of nature, of the reasons for humankind's consent to be governed, the nature of political obligation, and the legitimacy of different forms of government, was influenced by reflections on the fall of Athens and the violence and moral confusion of 17th-century England. In his later books *De Cive* and especially *Leviathan*, Hobbes concluded that sovereign authority in a society is required if the deliverance of its members out of a disorderly and violent state of nature is to be secured. In exchange for obligation and obedience, the subject gets safety and security. The best form of government—logically derived from these premises, because it is rational and unambiguous—is monarchic absolutism, limited by the ruler's obligation to provide for the security and welfare of the members of the society. Hobbes's achievement was his logical derivation of conclusions about the best form of government from what he viewed as material conditions and human needs. He advanced the argument by restricting assumptions to what he viewed, and what he thought history confirmed, as "material" evidence of the human condition. He drew uncompromising logical inferences from these assumptions.[11]

John Locke's conclusions about the origins and legitimacy of government, in his *Second Treatise of Government*, are derived from a different set of contractual assumptions than those of Hobbes. People consent to government to assure their welfare and liberty. The Lockean state of nature is not so abysmal as that of Hobbes. There are inconveniences and costs, and the consent to government is a conditional one, measured by the extent to which government performs these limited functions. In moving from the state of nature people cede to the community their right of enforcing the law of reason so as better to preserve life, liberty and property. There are the beginnings of "separation of powers" theory in John Locke. The power granted to the community is divisible into three components—the legislative, the executive, and the federative, the last a relatively unspecified power pertaining to foreign relations. In Locke as well as in Hobbes, the progress

[11] See further Sabine and Thorson 1973: chap. 24; Strauss and Cropsey 1987: 396–420.

in political science scholarship lies in the logical derivation of the nature and forms of government, and of the bases of authority, liberty and obligation, from sociological and psychological assumptions. Their strength lies in their logical rationalism, rather than in the gathering of evidence.

Though it is an exaggeration to describe Montesquieu's evidence as rigorously gathered and accumulated, surely he takes this step beyond Hobbes and Locke. While he recognizes laws of nature, and derives the formation of government from these laws, he emphasizes above all the variety of human political experience and the pluralism of causation. Montesquieu goes to "Persia," and back in time to Rome, so to speak, to Venice, to many other European countries, and especially to England, to compare their institutions with those of France. He is a comparativist and causal pluralist. To explain varieties of polity and public policy he considers climate, religion, customs, economy, history and the like. He founds the best form of government in his notion of separation of powers, and a kind of Newtonian balance among these powers, which he views as most likely to preserve liberty, and promote welfare. And, in Book XI of his *Spirit of Laws*, he finds his best exemplification of separation of powers in post-Petition of Right England.

Montesquieu's classification of governments includes republics, monarchies and despotisms, with the republican category being further divisible into aristocracies and democracies. He finds exemplified in the government of England the ideal of mixed government combining democratic, aristocratic and monarchic institutions in a dialectic-harmonic balance. His political theory is an explanatory, system-functional, conditions–process–policy theory.

It had great influence on the framers of the American Constitution. And it may have been in Hamilton's mind when he wrote in *Federalist 9*, "The science of politics . . . has received great improvement. The efficacy of the various principles is well understood, which were either not known at all, or imperfectly known to the ancients" and, in *Federalist 31*, "Though it cannot be pretended that the principles of moral and political knowledge have, in general, the same degree of certainty with those of the mathematics, yet they have much better claims in this respect than . . . we should be disposed to allow them" (Hamilton 1937 edn.: 48, 189). What led Madison and Hamilton to view themselves as such good political scientists was through having tested the theories of Montesquieu, Locke and other European philosophers against the experience of the thirteen colonies and of the United States under the Articles of Confederation. They had the confidence of engineers in applying laws of politics, derived from empirical and laboratory-like examinations of individual cases. Separating executive, legislative and judicial power (which they had learned from Montesquieu) and mixing powers through checks and balances (which they had learned from practical experience with the thirteen colonies) enabled them to treat politics in equation-like form: "Separation + checks and balances = liberty."

## C The 19th century

In the 17th and 18th centuries, the philosophers of the Enlightenment forecast the improvement in the material, political and moral condition of humanity as a consequence of the growth of knowledge. In the 19th and 20th centuries, scholars and intellectuals elaborated this theme of progress and improvement, predicting different trajectories, and causal sequences. In the first part of the 19th century there were the great historicists (or historical determinists)—Hegel (1770–1831), Comte (1798–1857) and Marx (1818–1883)—who, in the Enlightenment tradition, saw history as unilinear development in the direction of freedom and rational rule. In Hegel, reason and freedom are exemplified in the Prussian bureaucratic monarchy. In Comte, the constraints of theology and metaphysics are broken by science as it enables humanity to exercise rational control over nature and social institutions. In Marx, capitalism supplants feudalism, and is supplanted in turn, first by proletarian socialism and then by the truly free, egalitarian society.

Hegel departs from Enlightenment notions by his dialectical view of history as the clash of opposites and the emergence of syntheses. The Prussian bureaucratic monarchy as rationalized and modernized in the post-Napoleonic decades was viewed by Hegel as the exemplification of an ultimate synthesis.[12] In Marx, the Hegelian dialectic became the principle of class struggle leading to the ultimate transformation of human society. According to Marx, the nature of the historical process was such that the only social science that is possible is one that is discovered in, and employed in, political action. In Marxism, this science of society became a fully validated, economy-ideology-polity driven scheme. Armed with this powerful theory an informed vanguard would usher in a world of order, justice and plenty.[13]

Auguste Comte, the originator with Saint-Simon (1760–1825) of philosophical positivism, inaugurated the new science of "sociology" in his

[12] See further Sabine and Thorson 1973: chap. 17; Strauss and Cropsey 1987: 732 ff.
[13] See further Sabine and Thorson 1973: chap. 34; Strauss and Cropsey 1987: 802 ff.

six-volume *Cours de Philosophie Positive* (Koenig 1968). He made the argument that all the sciences went through two stages—first, the theological; second, the metaphysical—before becoming, in the third stage, scientific or positive. Thus, argued Comte, astronomy first passed through these three stages, then physics, then chemistry, then physiology. Finally, social physics (the social sciences inclusive of psychology) was in process of maturing as a science. Comte saw this new scientific sociology as furnishing a blueprint for the reform of society.

There was a wave of empiricism in reaction to these sweeping, abstract, monistic theories. This reaction produced a large number of descriptive, formal-legal studies of political institutions and several monumental, pedestrian, descriptive political ethnographies, such as Theodore Woolsey's *Political Science; Or the State Theoretically and Practically Considered* (1878); Wilhelm Roscher's *Politik: Geschichtliche Naturlehre der Monarchie, Aristokratie, und Demokratie* (1892); and Woodrow Wilson's *The State: Elements of Historical and Practical Politics* (1889/1918). These were essentially ponderous classificatory exercises, employing some variation of the Platonic-Aristotelian system of classification.

Similar to the historicists, but more empirical in approach and more pluralistic in explanation, were a group of writers in the second half of the 19th century who might be characterized as "evolutionists," and who influenced modern sociology in a variety of ways. These included Herbert Spencer (1820–1903), Sir Henry Sumner Maine (1822–1888) and Ferdinand Töennies (1855–1936). Spencer (1874/1965), an early post-Darwinian social evolutionist, avoids a simple unilinearism. He is concerned with accounting for cultural and political variation, as well as generic improvement. He explains political decentralization and centralization by physical features of the environment, such as mountainous versus open prairie terrain. He also makes the argument, backed up by historical example, that democratization is the consequence of socio-economic changes resulting in urban concentration, and the proliferation of interests due to the growth of manufactures and the spread of commerce.

There was a common dualistic pattern among the later 19th-century writers on the historical process. Maine (1861/1963) distinguishes ancient from modern law in terms of the shift from status relationships of a diffuse character to specific contractual ones. Tönnies (1887/1957) introduces the distinction between *Gemeinschaft und Gesellschaft* (*Community and Society*). At the turn of the century Weber (1864–1920) and Durkheim (1858–1917) contrast modern rationality with traditionality (Weber 1922/1978: vol. i, pp. 24 ff.), organic with mechanical solidarity

(Durkheim 1893/1960). This theme of "development," of "modernization," continues into the 20th century and to the present day, with efforts at defining, operationalizing, measuring, and interpreting socio-economic-political "modernization" discussed below.

It was common throughout the 19th century to speak of the study of politics and society as sciences, for knowledge about politics to be described as consisting of lawful propositions about political institutions and events based on evidence and inference. Collini, Winch and Burrow document this in great depth and detail in their book, *That Noble Science of Politics* (1983). As in earlier times, the historians and publicists of the 19th century looked for "lessons" from history, but with increasing sophistication. Recalling his "method" in writing *Democracy in America*, Tocqueville (1805–59) observed, "Although I very rarely spoke of France in my book I did not write one page of it without having her, so to speak, before my eyes"; and in appreciation of the comparative method more generally, he said, "Without comparisons to make , the mind does not know how to proceed" (Tocqueville 1985: 59, 191).

Collini, Winch, and Burrow point out that in the 19th century propositions about the nature and explanation of political phenomena increasingly came to be based on historical inductions rather than from assumptions about human nature. In part this was attributable to the simple growth of knowledge about contemporary and historical societies. Imperialism and colonialism brought vast and complex cultures such as India, as well as small-scale and primitive societies such as the American Indian and the African cultures, into the intellectual purview of European scholars and intellectuals. Exotic parts of the world became accessible and invited more cautious and controlled efforts at inferring cause and effect than was the case with Machiavelli and Montesquieu. At Oxford and Cambridge, at the very end of the 19th century, under the leadership of E. A. Freeman (1874), Frederick Pollock (1890) and John Seeley (1896) comparative history came to be viewed somewhat sanguinely as the basis for a genuinely scientific study of politics. It was introduced into the History Tripos at Cambridge in 1897 in the form of two papers—one on Comparative or Inductive Political Science, and a second on Analytical and Deductive Politics (Collini *et al.*: 341 ff.). As early as 1843, John Stuart Mill (1806–73) had recognized in his *System of Logic* (1843/1961) that the comparative method in the human sciences was in some sense equivalent to the experimental method in the natural sciences. A century and a half ago Mill had in effect anticipated the "most similar systems strategy" of Przeworski and Teune. (1970).

For John Stuart Mill, Tocqueville, Ostrogorski, Wilson and Michels, democracy as an alternative to other regimes is a major preoccupation. Each in his own way continues the debate about "mixed government." Mill wants the educated, the informed, the civicly responsible to play a pre-eminent role in democracy to avoid the corrupt and mass potentialities latent in it. Tocqueville found in the American legal profession an aristocratic admixture to moderate the "levelling" propensities of democracy. Ostrogorski (1964: vol. ii, Conclusion) and Michels (1949) both see fatal flaws in democracy, and inevitable oligarchy, resulting from the bureaucratization of mass political parties.

These 19th-century trends readily fall under our organizing concept of the advancing rigor and logical coherence of the study of political phenomena defined as the properties and legitimacy of rule.

Linking European political theory with American political science of the first decades of the 20th century was the concept of "pluralism," a variation on the "mixed government" theme. The concept of state sovereignty, associated with the ideology of absolutist monarchy, was challenged in the late 19th and early 20th centuries by "pluralists" of both the right and the left. Otto Gierke (1868) in Germany and Leon Duguit (1917) in France question the complete authority of the central state. Conservative political theorists such as Figgis (1896) asserted the autonomy of churches and communities; left-wing theorists such as Harold Laski (1919) made such claims for professional groups and trade unions.

With the seminal figures of Marx and Freud, and the great sociological theorists of the turn of the 19th century—Pareto, Durkheim, Weber—and with the polemic about sovereignty and pluralism, we are already in the immediate intellectual background of 20th-century political science.

## D The professionalization of political science in the 20th century

In the latter half of the 19th century and the first decades of the 20th, the rapid growth and concentration of industry and the proliferation of large cities in the United States, populated in considerable part by immigrants from the countryside or from foreign countries, created a situation prone to corruption on a major scale. It took political entrepreneurs with resources to organize and discipline the largely ignorant electorates that swarmed into such urban centers as New York, Boston, Philadelphia, Chicago, St. Louis, Kansas City and the like. The "boss" and the "machine" and intermittent reform movements were the most visible American political phenomena of the late 19th and early 20th centuries. Reform movements inspired by an ideology of efficiency and integrity, and supported by urban business and professional élites, drew on the talents of journalists of the quality media and academic communities. The corruption of politics by business corporations seeking contracts, franchises and protection from governmental regulation became the subject of a journalistic "muck-raking" literature, which brought to public view a political infrastructure and process—"pressure groups" and the "lobby," deeply penetrative and corrupting of local, state and national political processes.

In the interwar years American political scientists were challenged by this political infrastructure, and by the muckraking literature which exposed it, and began to produce serious monographic studies of pressure groups and lobbying activities. Peter Odegard (1928) wrote on the American Anti-Saloon League, Pendleton Herring (1929) on pressure groups and the Congress, Elmer Schattschneider (1935) on politics and the tariff, Louise Rutherford (1937) on the American Bar Association, Oliver Garceau (1941) on the American Medical Association, and there were many others. They put their stamp on the political science of the interwar years. The realism and empiricism of these early students of what some called "invisible" or "informal" government drew on the ideas of an earlier generation of American political theorists including Frank Goodnow (1900) and Woodrow Wilson (1887).

### 1 The Chicago School

Thus in the first decades of the 20th century the notion of a "scientific" study of politics had put on substantial flesh. Europeans such as Comte, Mill, Tocqueville, Marx, Spencer, Weber, Durkheim, Pareto, Michels, Mosca, Ostrogorski, Bryce and others had pioneered, or were pioneering, the development of a political sociology, anthropology and psychology, in which they moved the study of politics into a self-consciously explanatory mode. Empirical studies of governmental and political processes had made some headway in American universities. But in major part the study of politics in American universities in these decades was still essentially legal, philosophical and historical in its methodology. The significance of the University of Chicago school of political science (*c.* 1920–40) lay in its demonstration through concrete, empirical studies that a genuine enhancement of political knowledge was possible through an interdisciplinary research strategy, the introduction of quantitative methodologies and through organized research support. Other writers spoke a language similar to Merriam's (1931*b*) in "The Present State of the Study of Politics" (for example, Catlin 1964). But the school which Merriam founded in the 1920s, and staffed in part with his own students, made a quantum leap in empirical investigative rigor, inferential power in the study of things political, and in institutional innovation.

What led him to become the great political science entrepreneur of his generation was the dynamic setting of the city of Chicago, booming with wealth and aspiring toward culture in the early 20th-century decades, and the interplay of his academic life and his political career. His hopes for high political office had been dashed by his defeat in the Chicago mayoralty campaign in 1919. It was no longer possible for him to aspire to become the "Woodrow Wilson of the Middle West" (Karl 1974: chap. 4). At the same time he was unable to settle for a quiet academic career. His years in municipal politics, and his wartime experience with foreign affairs and propaganda, sensitized him to "new aspects" in the study of politics. Not long after returning to the University of Chicago from his "public information" post in Italy, he issued his *New Aspects* (1931*b*) declaration and began his build-up of the Chicago department and the various research programs which identified it as a distinctive "school." He was an institutional innovator: first creating the Social Science Research Committee at the University of Chicago to dispense financial support for promising research initiatives among the Chicago social science faculty; and then pioneering the formation of the Social Science Research Council to provide similar opportunities on the national scale.

The first major research program to be initiated at Chicago was built around Harold Gosnell, who received his doctorate under Merriam in 1921 and was appointed to an assistant professorship in 1923. He and Merriam collaborated in a study of the attitudes toward voting of a selection of some six thousand Chicagoans in the mayoral election of 1923 (Merriam and Gosnell 1924). The selection was made prior to the introduction of "probability sampling" and was carried out through "quota control" which sought to match the demographic characteristics of the Chicago population by quotas of the principal demographic groups. Quota control, discredited in the Truman–Dewey election of 1948, was then the "state of the art" approach to the sampling of large populations. The interviewers were University of Chicago graduate students trained by Merriam and Gosnell. Gosnell followed up this study with the first experiment ever to be undertaken in political science. This was a survey of the effects on voting of a non-partisan mail canvass in Chicago that was intended to get out the vote in the national and local elections of 1924 and 1925. The experimental technique Gosnell (1927) devised was quite rigorous: there were carefully matched experimental and control groups, different stimuli were employed, and the results were analyzed according to the most sophisticated statistical techniques then available. Follow-up research was done by

Gosnell in Britain, France, Germany, Belgium and Switzerland. Nothing like this had been done by political scientists before.

Harold Lasswell (1902–78), a young prodigy from small-town Illinois, brilliantly implemented Merriam's interest in political psychology. His accomplishments when he was in his 20s and 30s were extraordinary. Between 1927 and 1939 he produced six books, each one innovative, exploring new dimensions and aspects of politics. The first, *Propaganda Technique in the World War* (1927), introduced the study of political communication (to be followed in 1935 by a book-length annotated bibliography called *Propaganda and Promotional Activities*), identifying the new literature of communications, propaganda and public relations. The second book, *Psychopathology and Politics* (1930) explored the "depth psychology of politics" through the analysis of the case histories of politicians, some of them mentally disturbed. The third book, *World Politics and Personal Insecurity* (1935), speculated about the psychological bases and aspects of individual political behavior, different kinds of political regimes, and political processes. The fourth book, the celebrated *Politics: Who Gets What, When, and How* (1936), was a succinct exposition of Lasswell's general political theory, emphasizing the interaction of élites, competing for such values as "income, deference and safety." In 1939 he published *World Revolutionary Propaganda: A Chicago Study*, in which he and Blumenstock examined the impact of the world depression on political movements among the Chicago unemployed, exemplifying the interaction of macro and micro factors in politics at the local, national and international levels. Lasswell also published some twenty articles during these years in such periodicals as *The American Journal of Psychiatry*, *The Journal of Abnormal Psychology*, *Scientific Monographs*, *The American Journal of Sociology*, *The Psychoanalytical Review* and the like. He was the first investigator of the interaction of physiological and mental-emotional processes to use laboratory methods. He published several articles during these years reporting the results of his experiments in relating attitudes, emotional states, verbal content and physiological conditions as they were reported or reflected in interview records, pulse rates, blood pressure, skin tension and the like.

While Gosnell and Lasswell were the full-time makers of the Chicago revolution in the study of political science, the senior scholars in the department—including Merriam himself, and his colleagues Quincy Wright in international relations and L. D. White in public administration—were also involved in major ways in the making of the reputation of the Chicago School. Merriam (1931*a*) sponsored and edited a series of books on civic education in the US and Europe, a forerunner of contemporary studies of political socialization and culture. During these same years

Quincy Wright (1942) carried on his major study of the causes of war, which involved the testing of sociological and psychological hypotheses by quantitative methods. Leonard White took on Lord Bryce's (1888) problem of why in America the "best men do not go into politics." His book *The Prestige Value of Public Employment*, based on survey research, appeared in 1929.

## 2 World War II and the post-war behavioral revolution

The Chicago School continued its productivity up to the late 1930s, when the University administration under Hutchins attacked the value of empirical research in the social sciences. Several of the leading professors in the Department of Philosophy, including George Herbert Mead and others of its leading "pragmatists" resigned and went to other universities. In political science, Lasswell and Gosnell resigned, and Merriam's retirement brought the productivity of the Chicago Department of Political Science almost to a halt. However, the Chicago School had reached a mass which assured its future in the country at large. Herman Pritchett continued his innovative work in public law at the University of Chicago; Lasswell continued his work at Yale, inspiring Dahl, Lindblom and Lane in their transformation of the Yale department. V. O. Key, Jr., at Harvard, produced several generations of students with empirical and quantitative research interests in political parties, elections and public opinion. David Truman and Avery Leiserson brought the study of interest groups to theoretical fruition. William T. R. Fox, Klaus Knorr and Bernard Brodie and the present author and their students brought University of Chicago international relations and comparative politics to Yale, Princeton, Columbia, Stanford, MIT and the Rand Corporation.

World War II turned out to be a laboratory and an important training experience for many of the scholars who would seed the "behavioral revolution." The problems of how to insure the high rate of agricultural and industrial production on the part of a reduced labor force, how to recruit and train soldiers, sailors and airmen, and later how to discharge and return them to civilian life, how to sell war bonds, how to control consumption and inflation, how to monitor internal morale and the attitudes of allies and enemies, created demand for social science personnel in all the branches of the military and civilian services. The war effort created pools of social science expertise which, on the conclusion of the war, were fed back into the growing academic institutions of the post-war decades.

Working for the Department of Justice, Lasswell developed systematic quantitative content analysis for the monitoring of the foreign language press, and the study of allied and enemy propaganda in the United States.

He also participated with social scientists such as Hans Speier, Goodwin Watson, Nathan Leites and Edward Shils in the work of an analysis division in the Foreign Broadcast Intelligence Service of the Federal Communications Commission, which among other things analyzed the content of Nazi communications for information on internal political and morale conditions in Germany and occupied Europe. Survey research techniques, other kinds of interviewing methods, statistical techniques, especially sampling theory, were brought to bear on the war-related problems of the various military services, the Departments of Agriculture, Treasury and Justice and such agencies as the Office of Price Administration and the Office of War Information. Anthropology, then in its psychiatric-psychoanalytic phase, was similarly drawn into the war effort. The causes of Fascism and Nazism, the reasons for the French political breakdown, the cultural vulnerabilities of Russia, Britain and the United States, were sought in family structure, childhood socialization and cultural patterns. The Office of War Information and the War Department drew on the anthropological and psychological expertise of Ruth Benedict, Margaret Mead, Cora Dubois, Clyde Kluckhohn, Ernest Hilgard, Geoffrey Gorer and others. Social psychologists and sociologists specializing in survey research and experimental social psychology—including Rensis Likert, Angus Campbell, Paul Lazarsfeld, Herbert Hyman, Samuel Stouffer and Carl Hovland—were employed by the Army, Navy and Air Force in dealing with their personnel problems, by the Department of Agriculture in its effort to increase food production, by the Treasury in its effort to market bonds, and by the various intelligence services, including the OSS. A younger generation of political scientists working in these various agencies during the war years experienced something like post-doctoral internships under the supervision of leading scholars in the social science disciplines.

The rapidly growing academic enterprise in the postwar and Cold War world drew on these war-time interdisciplinary experiences. The curriculum of political science and departmental faculties expanded rapidly in response to this broadened conception of the discipline and the spread of higher education. The study of international relations, stimulated by the important American role in the postwar and Cold War world, was fostered in mostly new research institutes at Yale, Princeton, Columbia, MIT, Harvard, spreading into the middle western and western universities in the 1950s and 1960s. New subspecialities such as security studies, international political economy, public opinion and political culture studies joined with the older subspecialities of international law, organization and diplomatic history in the staffing of these research institutes and political science departments. The new and developing nations of Asia, Africa, the Middle

East and Latin America, now seen as threatened by an aggressive Soviet Union, required area specialists and specialists in economic and political development processes and problems. Departments of political science expanded rapidly to accommodate these new area specialties and international relations programs.

The survey research specialists of World War II found themselves to be in great demand. Business wanted to know how best to market and merchandize its products; and politicians wanted to know the susceptibilities and intentions of their constituents. From small beginnings in the 1930s and 1940s, the field of survey and market research exploded in the post-war decades (Converse 1987). It had both commercial and academic components. The main academic institutions involved in this development were: the University of Michigan, with its Institute of Social Research and its Survey Research Center founded by the psychologists Rensis Likert, Angus Campbell and Dorwin Cartwright; the Bureau of Applied Social Research at Columbia, founded by sociologists Paul Lazarsfeld and Robert Merton; and the National Opinion Research Center at the University of Chicago, headed in its early years by sociologist Clyde Hart. These three organizations in the postwar decades produced a literature and a professoriate that contributed substantially to the "behavioral revolution."

Among these three university centers, the University of Michigan turned out to be the most important in the recruitment and training of political scientists. Its Institute of Social Research established a Summer Training Institute in the use of survey methods open to young political and other social scientists as early as 1947. Over the years this program has trained hundreds of American and foreign political scientists in survey and electoral research techniques. In 1961 it established an Interuniversity Consortium for Political and Social Research (ICPSR), supported by subscribing universities and maintaining in machine-readable form a rapidly growing archive of survey and other quantitative data. This archive has served as the database for a large number of doctoral dissertations, articles in learned journals and important books illuminating various aspects of the democratic process. It has administered its own summer training program in quantitative methods.

In 1977 the University of Michigan, Survey Research Center Election Studies became the American National Election Studies supported by a major grant from the National Science Foundation, with an independent national Board of Overseers drawn from American universities. This organization—based at the Center of Political Studies of the Institute of Social Research of the University of Michigan, directed by Warren Miller, and with its Board of Overseers chaired by Heinz Eulau of Stanford

University—has regularly conducted national election studies, with input from the larger national political and social science community, and whose findings are available to the scholarly community as a whole (Miller 1994; below: chap. 11).

If we can speak of the University of Chicago school of political science as the agency which sparked the scientific revolution in the study of politics in the inter-war decades, surely the University of Michigan Institute of Social Research deserves a major credit for the spread of this scientific culture in the post-World War II decades into most of the major academic centers in the United States and abroad. Several hundreds of young scholars have been trained in survey and statistical methods in its Summer Training Institutes; scores of articles and dozens of books have been produced by scholars using its archival materials; the Michigan election studies have served as models for sophisticated election research in all the rest of the world.

The spread and improvement in empirical political theory involved more than election research technique and theory. Such fields as international relations and comparative politics grew as rapidly as did the field of American politics, and their newer growth involved quantification and interdisciplinary approaches. The major university centers of graduate training during the post-war decades—Yale, the University of California at Berkeley, Harvard, the Universities of Michigan, Wisconsin, Minnesota, Stanford, Princeton, MIT and others—turned out hundreds of political science PhDs to staff the proliferating and growing political science departments in American and in many foreign colleges and universities. Most of these centers of graduate training provided instruction in quantitative methods in the decades after World War II (Somit and Tannenhaus 1967; Crick 1959; Eulau 1976).

Under the leadership of Pendleton Herring, the Social Science Research Council in the 1940s to 1960s facilitated and enriched these developments through its graduate and post-doctoral fellowship and research support programs. Two of its political science research committees—the Committee on Political Behavior, and its spin-off Committee on Comparative Politics, were particularly active in spreading these ideas and practices. The Committee on Political Behavior provided direction and support in American election and legislative studies. The Committee on Comparative Politics led in the development and sophistication of area and comparative studies.[14] While most of the participants in these programs were American political and social scientists, around one-fifth of the

[14] For details, see especially its 1972 report.

participants in the conferences of the Committee on Comparative Politics during the years 1954–1972 were foreign scholars. Some of these—Stein Rokkan, Hans Daalder, Samuel Finer, Richard Rose, Giovanni Sartori, among others—were in turn leaders in movements in Europe and in their particular countries to expand and improve the quality of the work in political and social science.

The discipline of political science was becoming a modern "profession" over these years. Departments of Political Science, Government and Politics had first come into existence at the turn of the 19th century, when they began to be formed by an alliance of historians, lawyers and philosophers. By the first decades of the 20th century, there were free-standing departments in many American universities. The American Political Science Association was formed in 1903 with a little more than 200 members. It reached around 3,000 members at the end of World War II, exceeded 10,000 in the mid-1960s, and now includes more than 13,000 individual members. Most of these members are instructors in institutions of higher education, organized in a large number of subspecialties. Most political science teachers and researchers have obtained degrees as Doctors of Philosophy in political science in one of the major centers of graduate training. Qualifications for the degree normally involve passing a set of field and methodological examinations, and the completion of a major research project. Scholarly reputations are based on the publication of books and articles screened for publication by "peer review." Advancement in scholarly rank normally requires evaluation by external reviewers specialized in the field of the candidate. There are dozens of political science journals, specialized by field and governed by the processes of peer review.

The half-century of political science training and research since the end of World War II has created a major academic profession, with many subspecialties, and has made many substantive contributions to our knowledge and understanding of politics in all its manifestations. Area-studies research on Western and Eastern Europe, East, Southeast and South Asia, the Middle East, Africa and Latin America, carried on by literally thousands of trained scholars organized in "area study" centers in scores of universities and colleges, with their own professional organizations and journals—has produced libraries of informative and often sophisticated monographs.

A quick and selective review of substantive research programs may help us appreciate this growth of political knowledge. We have already described the spread and sophistication of election research. Its forecasting record may be compared with that of meteorology and of seismology. We have made major progress in our understanding of political culture as it affects political institutions and their performance, as well as the cultures of important élite and other social groups. Examples from survey research include the work of Gabriel Almond, Sidney Verba, Alex Inkeles, Ronald Inglehart, Samuel Barnes and Robert Putnam.[15] More descriptive-analytical studies of political culture are exemplified in the work of Lucian Pye (1962; 1985; 1988; Pye and Verba 1965). Our understanding of political participation has been brought to a high level through a series of studies carried on over the last decades by Verba and his associates.[16]

In the early decades of the postwar period Talcott Parsons and others developed "system" frameworks for the comparison of different types of societies and institutions, building on the work of such European sociological theorists as Weber and Durkheim.[17] Drawing on these and other sources David Easton pioneered the introduction of the "system" concept into political science (Easton 1953; 1965; 1990; Almond and Coleman 1960; Almond and Powell 1966).

Through aggregate statistical methods, we now have vastly improved understanding of the processes of modernization and democratization[18] and of governmental performance.[19] Significant progress has been made in our understanding of interest groups and of "corporatist" phenomena,[20] and in our appreciation of the key importance of political parties in the democratic process.[21]

Theories of representation and of legislative behavior and process have been explored and codified in studies by Eulau, Wahlke, Pitkin and Prewitt.[22] Herbert Simon, James March and others, beginning from studies of governmental organizations, have created a new interdisciplinary field of organization theory generally applicable to all large-scale organizations including business corporations.[23] Public policy research, pioneered jointly in Europe and the United States, has taken off in recent decades and promises the development of a new political economy.[24]

[15] Almond and Verba 1963. Verba 1987. Inkeles and others 1950; 1959; 1974. Inglehart 1977; 1990. Barnes and Kaase *et al.* 1979. Putnam 1973; 1993.

[16] Verba and Ahmed 1973. Verba and Nie 1972. Verba, Nie, and Kim 1978. Schlozman and Verba 1979. Schlozman, Verba and Brady 1995.

[17] Parsons 1951. Parsons and Shils 1951. Parsons and Smelser 1956.

[18] Lerner 1958; Deutsch 1961; Lipset 1959; 1960; 1994; Diamond and Plattner 1993.

[19] Hibbs 1978; Cameron 1978; Alt and Chrystal 1983.

[20] Goldthorpe 1978; Schmitter and Lehmbruch 1979; Berger 1981.

[21] Lipset and Rokkan 1967; Sartori 1976; Lijphart 1968; 1984; Powell 1982.

[22] Wahlke and Eulau 1962; 1978. Eulau and Prewitt 1973. Eulau 1993. Pitkin 1967.

[23] Simon 1950; 1953; 1957. March and Simon 1958. March 1965; 1988.

[24] Wildavsky 1986. Flora and Heidenheimer 1981. Heidenheimer, Heclo and Adams 1990. Castles 1989.

The theory of democracy has been significantly advanced by the work of Robert Dahl, Arend Lijphart and Giovanni Sartori.[25] That of democratization has been developed by Juan Linz, Larry Diamond, Phillipe Schmitter, Guillermo O'Donnell, Samuel Huntington and others.[26] The life-long dedication of Robert Dahl to the study of democracy is an example of how normative and empirical political theory may mutually enrich each other (Dahl 1989).

While we have stressed the growth and spread of empirical, explanatory and quantitative political science in this chapter, there has been "progress" in the older branches of the discipline as well. The propositions and speculations of the political historians, political philosophers and legal scholars have been increasingly based on improvements in scholarly methodology—rigorous accumulation of information, and refinements in the logic of analysis and inference. Comparative political history has made important contributions to the theory of the state, political institutions and public policy (Moore 1966; Skocpol 1979; 1984). Refinements in case study methodology have been made by Harry Eckstein and Alexander George, and these have increased the rigor of historical studies in comparative politics and foreign policy.[27] The methodology of comparison has been refined and improved through the work of Almond and his collaborators, Adam Przeworski and James Teune, Arend Syphart, Neil Smelser, Mattei Dogan, David Collier, and Gary King, Robert Keohane and Sidney Verba.[28]

With the work of Rawls, Nozick, Barry, Walzer, Fishkin and others, normative political philosophy has made substantial progress, and not entirely without influence from empirical studies.[29] William Galston (1993), in the recent edition of *Political Science: The State of The Discipline II,* points out that political philosophy and theory are moving in the direction of increasing reliance on empirical evidence, much of it drawn from the research of political science and the other social science disciplines. Galston urges political theorists to take on the task of codifying the findings of empirical research as they may bear on political philosophy, as

[25] Dahl 1956; 1961; 1966; 1970; 1971; 1973; 1982; 1985. Lijphart 1968; 1984; 1994. Sartori 1987.

[26] Linz and Stepan 1978. Diamond and Plattner 1993. Schmitter, O'Donnell and Whitehead 1986. Huntington 1991.

[27] On the methodology, see Eckstein 1975 and George and McKeown 1982. For applications, see: George and Smoke 1974; George 1980; George *et al.* 1983; George and Simons 1994.

[28] Almond and Coleman 1960; Almond, Flanagan and Mundt 1973. Przeworski and Teune 1970. Syphart 1971. Smelser 1976. Dogan and Pelassy 1990. Collier 1993. King, Keohane and Verba 1994.

[29] Rawls 1971. Nozick 1974. Barry 1970. Walzer 1983. Fishkin 1992.

Robert Dahl (1956), Dennis Thompson (1970) and James Q. Wilson (1993) have done.

Martin Shapiro's (1993) evaluation of the contemporary study of the courts and public law similarly urges a closer integration of legal studies with institutional and processual political science. Political science without legal analysis is seriously lacking in explanatory power; and legal analysis without the political institutional and processual context is formalistic and sterile. The work of Shapiro and that of a growing band of students of the courts and public law demonstrates the validity of this proposition (see Drewry below: chap. 6).

Thus, our account of the history of political science is inclusive of progress made by the earlier traditional subdisciplines, measured by the same criteria. As the scientific revolution of the last century has impinged on the study of politics the response of the discipline of political science has been multivocal and ambivalent. Some parts of the discipline responded earlier to these challenges; and some parts saw the face of science as lacking in all compassion and empathy, and as a threat to humane scholarship. One ought not overlook the fear of obsolescence generated by the introduction of statistics, mathematics and diagrammatic virtuosity. But the newer generations cultivating the study of political history, philosophy and law have overcome these anxieties, discovered the vulnerabilities and shortcomings of the behavioral approach, developed their own arsenal of mystifications, and have proven to be quite as competent in the employment of smoke and mirrors as their behavioral brethren.

### 3 Political science in Europe

While political science had its origins and first growth in the Mediterranean world of antiquity, in medieval Catholic, Renaissance, Reformation, Enlightenment and 19th-century Europe,[30] this was a matter of individual scholarship—whether in institutional settings, as the Greek academies, or the European universities of the Middle Ages and later. Many early political philosophers and theorists operated as part-time scholars within the framework of the Church, its bureaucracy and orders, supported by kingly and aristocratic patrons, or were themselves aristocrats or persons of wealth. In the 19th century with the growth of European universities, scholarship on the state, administration, politics and public policy was increasingly based in universities. Until recently the typical unit of European universities consisted of a single professorial chair held by a single scholar surrounded by lesser docents and assistants. In the postwar

[30] And indeed in Indian antiquity (Rangavajan 1987) and in medieval Islam (Rabi 1967).

decades some of these university chairs have been broadened into departments with a number of professorial billets assigned to different teaching and research specialties.

A recent issue of the *European Journal of Political Research* (Vallès and Newton 1991) is devoted to the post-war history of West European political science. An introductory essay by the editors argues that the progress of political science in Europe has been associated with democratization, for obvious reasons, and with the emergence of the welfare state, because an activist, open, penetrative state requires large amounts of information about political processes and political performance. While recognizing that the impact of American political science on European has been very substantial, they point to the fact that there already was a "behavioral" election study tradition in Europe prior to World War II (Siegfried 1930), with Duverger (1951/1976) in France and Tingsten (1937/1963) in Sweden. The great 19th- and early 20th-century figures in the social sciences who inspired the creative developments in America were European, as we have suggested. Richard Rose (1990) points out that, while the major development of modern political science took place in the United States after World War II, the founders of American political science—the Woodrow Wilsons, the Frank Goodnows, the Charles Merriams—took their degrees or spent post-graduate years at European universities, principally the German ones. Learning, culture and professional skill were concentrated in the Old World, and it thinned out as one went west. In the period prior to World War I, American scholars still viewed themselves as provincials. In the interwar years, and in such an innovating center as the University of Chicago, Merriam still urged his most promising students to spend a post-graduate year in Europe and provided the financial support to do so.

The conquests of Nazism and Fascism and the devastation of World War II disrupted university life in continental Europe for almost a decade. Much of German social science scholarship was effectively transplanted to the United States, where it contributed to the American war effort and enriched American sociological, psychological and political science teaching and research. There was an entire "exiled" Graduate Faculty in the New School for Social Research in New York; and hardly a major university was without one or more exiled professors in its social science faculties. Scholars such as Paul Lazarsfeld, Kurt Lewin, Wolfgang Kohler, Hans Speier, Karl Deutsch, Hans Morgenthau, Leo Lowenthal, Leo Strauss, Franz Neumann, Henry Ehrmann, Otto Kirchheimer, Herbert Marcuse made important contributions to the behavioral revolution in the United States, as well as to the various trends which attacked it. Hence, the political science which was planted in Europe after World War II was in part the product of a political science root-stock that had originally come from Europe.

In the first decades after World War II, as the physical plant of Europe was renewed and its institutions put back in place and staffed, what was new in the social sciences was mostly American in origin. The break from legalism and the historical approach in the study of governmental institutions, political parties and elections, interest groups, public opinion and communications had been accomplished in American universities and research centers. Along with the Marshall Plan for the shattered European economy, American scholars backed up by American philanthropic foundations were missionaries for the renewal of European scholarship and for the assimilation of the American empirical and quantitative approaches. Young European scholars supported by Rockefeller and other foundation fellowships visited and attended American universities by the dozens. America-based research programs—the SSRC Committee on Comparative Politics, the University of Michigan election studies, the Inglehart studies of political values—sought out European collaborators, trained them and often funded them.

This one-sided dependency only lasted for a short period of time. Social science scholarship and traditions were too deeply rooted in European national cultures to have been thoroughly destroyed in the Nazi period. By the 1960s, old universities had been reconstituted and many new ones established. European voices were increasingly contributing to the significant research output in the social sciences. The Committee on Political Sociology of the International Sociological Association, though joining American with European efforts, was predominantly European in participation. Its impact in Europe was much like that of the American Committee on Comparative Politics before it. Comparative European studies, such as the Smaller European Democracies project led by Dahl, Lorwin, Daalder and Rokkan, helped contribute to the development of a European political science professionalism. The Survey Research Center of the University of Michigan began its active role in the development of sophisticated election research in Europe with a study in England in the early 1960s, followed by other European countries. Each such national election study left a cadre of trained professionals to carry on future work in election research.

A European Consortium for Political Research (ECPR) was founded in 1970 with funds from the Ford Foundation (Rose 1990), with an agenda similar to that of the political science committees of the American Social Science Research Council. It provided funds for the establishment of a summer school training program in social science methodology (located at the University of Essex), workshops held in different national centers concerned with particular research themes, actual joint research projects. Among the activities which it has fostered are a Data Archive and a professional journal, *The European Journal for Political Research.* Membership in the ECPR is by department and institution. By 1989 the ECPR had 140 member departments. By 1985 the Directory of European Political Scientists listed just under 2,500 members. The strength of political science in individual European countries is suggested by the number of national departments affiliated with the ECPR. Of the 140 members as of 1989, 40 were in the United Kingdom, 21 in Germany, 13 in the Netherlands, 11 in Italy, and 5 in France (Rose 1990: 593). The influence of American political science on European and international political science is reflected to some extent by the number of foreign members of the American Political Science Association, and hence subscribers to the *American Political Science Review*: the United Kingdom, Germany, and Japan each have well over a hundred members; Israel, South Korea, and the Netherlands each have around fifty members; Norway, Sweden, and Taiwan have around thirty members; France has 27 (APSA 1994: 327 ff.).

By the 1990s, organized in the International Political Science Association, in various national and subnational organizations, as well as in many different functional specializations, the profession of political science along with a common conception of scholarship was well established globally.

## IV Opposing perspectives on disciplinary history

Those who would disagree with this progressive-eclectic account of the history of political science may be sorted out in four groups. There are those who reject the notion of a progressive political science—from an anti-science perspective (the Straussians); or from a post-science, deconstructive perspective. Then there are those who reject the eclecticism of our position. Among those are the Marxists and neo-Marxists, who argue that the basic laws of human society have been discovered by Marx and his associates and that these laws show that historic, economic, social and political processes, as well as the human action effecting these processes, are one inseparable unity; hence the Marxists would reject both the progressiveness and eclecticism of our approach. The second group rejecting the methodological eclecticism of our approach are the maximalists among the "rational choice" political scientists, whose view of disciplinary history is one which culminates in a parsimonious, reductive, formal-mathematical stage.

## A Anti-science

The Straussian version of the history of political science harks back to the German intellectual polemics of the late 19th and early 20th centuries. As a young German PhD in the immediate post-World War I years, Leo Strauss shared in the general admiration of Max Weber for "his intransigent devotion to intellectual honesty . . . his passionate devotion to the idea of science . . ." (Strauss 1989: 27). On his way north from Freiburg where he had heard the lectures of Heidegger in 1922, Strauss describes himself as having experienced a Damascan disillusionment with Weber and a conversion to Heideggerian existentialism. Strauss's mode of coping with the pessimism of the Heidegger view of the nature of "being" was through an affirmative political philosophy, seeking the just society and polity through the recovery of the great exemplars of the canon of political philosophy, through dialogue and deliberation, and through the education of a civic élite.

According to Strauss, Weber was the problematic intellectual figure who legitimated modern positivistic social science, its separation of fact and value, its "ethical neutrality," its effort to become "value free." Strauss attributes to Max Weber the belief that all value conflicts are unsolvable. "The belief that value judgments are not subject, in the last analysis to rational control, encourages the inclination to make irresponsible assertions regarding right and wrong or good and bad. One evades serious discussion of serious issues by the simple device of passing them off as value problems." This search for objectivity produces an

> emancipation from moral judgments . . . a moral obtuseness . . . The habit of looking at social or human phenomena without making value judgments has a corroding influence on any preferences. The more serious we are as social scientists, the more completely we develop within ourselves a state of indifference to any goal, or of aimlessness and drifting, a state which can be called nihilism.

A bit later he qualifies this statement, "Social science positivism fosters not so much nihilism, as conformism and philistinism" (Strauss 1959: 21 ff.).

This attack on Weber has been extended by Strauss and his followers to the contemporary social sciences, and in particular to the "behavioral" trends in political science which Weber is said to have inspired. In contrast to this "positivistic," Weberian social science, Strauss presents a model of a "humanistic social science" in which scholarship is intimately and passionately engaged in dialogue with the great political philosophers over the meaning of the central ideas and ideals of politics—justice, freedom, obligation and the like. The history of political science time-line, which the Straussians offer in the place of the one presented here, characterizes contemporary "behavioral" political science as the product of a heresy which assumed palpable form in the 19th century and was fully formulated in the work of Max Weber at the turn of the century.[31]

Its characterization of Weber as the arch-positivist and separator of fact and value, and of "behavioral" political science as pursuing this erroneous course of "ethical neutrality," is mistaken both with respect to Max Weber and with respect to most of the contemporary practitioners of so-called behavioral political science. Weber's views of the relation between "fact and value" are much more complex, and involve a deeper concern for value issues, than the caricature contained in the writings of Strauss and his students. We draw attention to two contexts in which Weber deals with these questions: in his lecture "Politics as a Vocation" (1949), and in his essay on "Objectivity in Social Science" (1958). In the lecture on "Politics as a Vocation" he refers to two kinds of ethically oriented political action—the ethics of absolute ends, and the ethics of responsibility (*Gesinnungsethik und Verantwortungsethik*). Science would have little to contribute to the ethics of absolute ends, other than examining the adequacy of the relation of means to ends. Since the chosen end is sacred or absolute, there can be no opportunity-cost analysis of the consequences of pursuing this end for other ends. But if one takes a rationally responsible view of the effect of means on ends, scientific analysis makes possible an "opportunity-cost" analysis of political action, that is, how a given choice of policy or action may, on the one hand, transform the end one is seeking, and on the other preclude the choice of other options. "We can in this way," says Weber (1949: 152), "estimate the chances of attaining a certain end by certain available means . . . we can criticise the setting of the end itself as practically meaningful . . . or as meaningless in view of existing conditions." Elaborating his argument about the ways in which means may effect ends in "unintended ways," Weber (1958: 152) says,

> we can answer the question: what will the attainment of the desired end "cost" in terms of the predictable loss of other values. Since in the vast majority of cases, every goal that is striven for does "cost" . . . something in

[31] For the full flavor of the Straussian challenge, see the essays in Storing 1962 and the debate they aroused in the *American Political Science Review* (Schaar and Wolin 1963; Storing *et al.* 1963).

> this sense, the weighing of the goal in terms of unintended consequences cannot be omitted from the deliberation of persons who act with a sense of responsibility. . . . [Science can make one] realize that all action and naturally . . . inaction, imply in their consequences the espousal of certain values, and . . . what is so frequently overlooked, the rejection of certain others.

But in addition to this twofold means-end analysis, Weber (1958: 152) points out that science can enable us to clarify our goals, and comprehend their meaning. "We do this through making explicit and developing in a logically consistent manner the 'ideas' which . . . underlie the concrete end. It is self evident that one of the most important tasks of every science of cultural life is to arrive at a rational understanding of these 'ideas' for which men . . . struggle."

"But," Weber (1958: 152) goes on, "the scientific treatment of value judgments may not only understand and empathically analyze the desired ends and the ideals which underlie them; it also can judge them critically . . ." according to their internal consistency. "The elevation of these ultimate standards . . . to the level of explicitness is the utmost that the scientific treatment of value judgments can do without entering into the realm of speculation . . . An empirical science cannot tell anyone what he should do but rather what he can do—and under certain circumstances—what he wishes to do."

The reality of the Weberian fact-value formulation is as far from the Straussian caricature, as is its depiction of the state of contemporary empirical political science. We therefore reject the view of the history of the discipline implied in the Straussian perspective. On the other hand, we would include much of the substantive work done by these political theorists—and that of Strauss himself, in the work which we include in the progressive-eclectic account which we give here, to the extent that it has increased the body of logically drawn inferences about politics, from reliable accumulations of evidence.

## B Post-science, post-behavioralism

Among contemporary political scientists, there is a prevailing, perhaps predominant view of the history of the discipline, that we are now, in a "post-positivist, post-scientific, post-behavioral" stage. Saxonhouse (1993: 9) speaks of the

> demise of positivism and the demands for verification as the only philosophic stance for the human sciences, with the rejuvenation of normative

> discourse in a society concerned with the dangers of an unleashed science . . . [P]olitical scientists in general and political theorists in particular are no longer willing to adopt uncritically the distinction of fact and value that controlled the social sciences for several generations . . .

A small subdiscipline in political science specializing in the "history of political science" pursues this theme. David Ricci, in a 1984 book called *The Tragedy of Political Science*, argues that the naïve belief in political "science" that had emerged in American political science in the 1920s to the 1960s had been thoroughly discredited in the disorders of the 1960s and 1970s. He concludes that political science as empirical science without the systematic inclusion of moral and ethical values and alternatives, and a commitment to political action, is doomed to disillusion. Political science has to choose sides or become a "precious" and irrelevant field of study. Even more sharply, Raymond Seidelman (1985) rejects political science professionalism, saying that modern political science must bridge this separation of knowledge and action "if [these professional] delusions are to be transformed into new democratic realities."

There has been a substantial exchange of ideas about the "identity" and history of political science in the decade bounded by the two editions of Ada Finifter's, *Political Science: State of the Discipline* (1983; 1993). In the first, John Gunnell (1983, p.12 ff.) presents a picture of the history of political science marked by a "scientistic" revolution in the half-century, from the 1920s until the 1970s, followed by a "post-empiricist" period continuing into the present. In the second edition, Arlene Saxonhouse (1993) makes the comments about the "demise of behavioralism," quoted above. In the interval between these two volumes there has been a further exchange of views in the *American Political Science Review* among a number of historians of political science. In an article appearing in the December 1988 issue, "History and Discipline in Political Science," John Dryzek and Stephen Leonard (1988: 1256)

> conclude that there is no neutral stance for evaluating, accepting, or rejecting disciplinary identities. Rather, standards can only emerge in the conflicts and debates within and between traditions of inquiry. It is in this conflict and debate that the relationship between disciplinary history and identity crystallizes . . . [P]lurality is going to be the essence of, rather than an obstacle to, the progress of political science.

The view expressed here is that there will be as many disciplinary histories as there are "disciplinary identities" and that there is no "neutral" way of choosing among them.

A flurry of responses to this pluralist approach to the history of political science appeared under the general title "Can Political Science History be Neutral?" (Dryzek *et al.* 1990). Contributions came from James Farr, John Gunnell and Raymond Seidelman, with a reply by Dryzek and Leonard. All three of the respondents support the "pluralist" view of disciplinary history expressed by Dryzek and Leonard, with some qualifications. In two recent collections of articles and papers treating the history of political science, James Farr and his associates (Farr and Seidelman 1993; Dryzek, Farr and Leonard 1995) codify this pluralist perspective.

We have to conclude from these exchanges that, at least among this group of contemporary writers on the history of political science, there is a "deconstructionist, postmodernist" consensus arguing that there is no privileged canon of political science. While each one of the major competing schools of political science history—the so-called "behavioral" or political "science" perspective, the anti- and post-science perspectives, and the Marxist and rational choice ones—makes claim to being the valid approach to disciplinary history, this consensus argues that no one of them has a valid claim. Our account of the growth of political knowledge defined as the capacity to draw sound logical inferences from an increasing body of reliable evidence, which these "historians" of political science refer to as "neo-positivism," would only be one of several accounts no one of which would have any special claim to validity.

Our treatment in this chapter advances and demonstrates in its historical account that there is indeed a "privileged" version of our disciplinary history and that this is a progressive one, measured by the increase of knowledge based on evidence and inference. It would include the work of the opposing schools, insofar as it meets these standards. It would exclude those claims and propositions not founded on evidence, or not falsifiable through evidence and logical analysis. Objective, rigorous scholarship is indeed the privileged thread in our disciplinary history.

## C Integralism and maximalism: anti-pluralism

### 1 Theory and praxis

There are several schools which would challenge the approach to the history of political science as the progress of "objective" scholarship, on the grounds that objectivity is both impossible to achieve and, if sought, leads to "scientism" and the embrace of the status quo. From this point of view even the search for professional objectivity is to be eschewed. One has to choose political sides and self-consciously employ scholarship in the service of good political goals. For the various neo-Marxist schools this meant hooking scholarship up to socialism.

In the history of Marxist scholarship there was a point at which one branch of this tradition rejected this dialectical view of scholarship. Karl Mannheim in *Ideology and Utopia*, concluded that objectivity in political science was possible. "The question, whether a science of politics is possible and whether it can be taught, must if we summarize all that we have said thus far, be answered in the affirmative." He attributes to Max Weber the demonstration that objective social science scholarship is possible (Mannheim 1949: 146). But while objectivity becomes possible for Mannheim, this capacity is only likely to be developed "by a relatively classless stratum which is not too firmly situated in the social order . . . This unanchored, relatively classless stratum is, to use Alfred Weber's terminology, the 'socially unattached intelligentsia' " (1949: 171). For contemporary political science scholarship, "professionalism" has taken the place of Mannheim's "unattached intelligentsia" as the guarantor of the obligation of the search for objectivity—professionalism in the sense of affiliation to professional associations, peer accreditation and reviewing of recruitment and scholarship and the like. At the time that Weber and Mannheim were presenting these ideas, professional associations in the social sciences and particularly in political science and sociology were in their infancy. And it is of interest that it is precisely this notion of the search for objectivity through professionalism that continues to be the target of both contemporary neo-Marxist and of other "left" critics.

This polemic against "ethical neutrality" and the "search for objectivity" has been carried on from a number of perspectives. The Frankfurt School out of which "critical theory" emerged—inspired by the Marxist theorist Lukács and led by Max Horkheimer, Theodor Adorno, Herbert Marcuse and currently led by Jürgen Habermas—view the conduct of political inquiry as an aspect

> of a total situation caught up in the process of social change. . . . Positivists fail to comprehend that the process of knowing cannot be severed from the historical struggle between humans and the world. Theory and theoretical labor are intertwined in social life processes. The theorist cannot remain detached, passively contemplating, reflecting and describing "society" or "nature" (Held 1980: 162 ff.).

A recent formulation by Habermas (1992: 439 ff.) reaffirms this unity of theory and "praxis" perspective. The influence of this point of view was reflected in the deep penetration of views such as these into Latin American, African and other area studies, under the name of "dependency theory," during the 1970s and 1980s (Packenham 1992).

How may we treat Marxist and neo-Marxist scholarship in this progressive-eclectic account of the history of political science? These literatures are very substantial indeed, running into the many hundreds of volumes and learned articles in very large numbers. Exemplary of the very important place some of this work must have in the history of political science are the important empirically based studies of class and politics which were largely the product of Marxist and neo-Marxist scholarship. Nevertheless, while Marxism directed attention to the explanatory power of economic development and social structure, it also diverted scholarly attention away from other important explanatory variables such as political institutions, religion, ethnicity, the international setting, individual leadership, contingency and chance. Its conception of economic development was oversimplified and primitive. As the modern economy produced an increasingly diversified labor force, and internationalized, the capacity of Marxist scholarship to perceive and properly weight economic, social, and political variables attenuated. Thus, while these various Marxist schools greatly increased the quantity and kind of evidence available to historical and social science scholarship, their inferential logic was seriously faulty and not properly open to falsification. Eric Hobsbawm (1962; 1987; 1994) and other Marxist historians (Hill 1982; Hilton 1990; Thompson 1963) make a great contribution to the historical scholarship on the 19th and earlier centuries, but have difficulties in their efforts to interpret and explain the 20th (Judt 1995).

**2 Scientific maximalism: the rational choice approach**

The rational choice approach—variously called "formal theory," "positive theory," "public choice theory" or "collective choice theory"—is predominantly a lateral entry into political science from economics. Economic metaphors had been used by political scientists such as Pendleton Herring, V. O. Key, Jr., and Elmer Schattschneider (Almond 1991: 32 ff.). But it was the economists—Kenneth Arrow, Anthony Downs, Duncan Black, James Buchanan and Gordon Tullock, and Mancur Olson—who first applied economic models and methods in the analysis of such political themes as elections, voting in committees and legislative bodies, interest group theory and the like.[32] In the 1993 edition of *Political Science: The State of the Discipline* the chapter dealing with "formal rational choice theory" describes this approach as promising "a cumulative science of politics." Its co-authors claim that "rational choice theory has fundamentally changed how the discipline ought to proceed in studying politics and training students" (Lalman *et al.* 1993).

[32] Arrow 1951. Downs 1957. Black 1958. Buchanan and Tullock 1962. Olson 1965.

This approach holds out the prospect of a unified, cumulative political science theory—part of a unified, formal social science theory—based on common axioms or assumptions derived essentially from economics. These assumptions are that human beings are rational, primarily short-term, material self-interest maximizers. Its advocates argue that from such premises it is possible to derive hypotheses regarding any sphere of human activity—from decisions about what to buy and how much to pay for it, and whom to vote for, to decisions about whom to marry, how many children to have, how political parties should negotiate and form coalitions, how nations should negotiate and form alliances and the like. The theory is parsimonious, logically consistent, mathematical and prefers experimental methods to observational and inductive ones for the testing of hypotheses.

This is the maximal, aspirational version of the approach—encountered in the contribution to the *State of the Discipline II* volume cited above (Lalman *et al.* 1993), in Peter Ordeshook's "The Emerging Discipline of Political Economy" (1990), in William Riker's "Political Science and Rational Choice" (1990), in Mancur Olson's "Toward a Unified View of Economics and the Other Social Sciences" (1990), as well as in other writers in this genre. This approach argues a discontinuity in the history of political science, in which everything that went before had to be viewed as pre-scientific. Its vision of the future of the discipline is of a cumulating body of formal theory, internally logical and consistent, capable of explaining political reality with a relatively small number of axioms and propositions.

Some very eminent writers in this movement do not share in these maximal expectations. On such a question as the content of utility, some economists reject the model of Economic Man as the rational, material self-interest maximizer. Milton Friedman (1953) long ago took the position that it made no difference whether this assumption was correct or incorrect, just as long as it produced valid predictions. Just as long as it proved relevant at all, it could serve a heuristic function in testing the usefulness of different versions of utility. It is of interest that one of the pioneers of rational choice political theory, Anthony Downs, has long since moved his away from Political Man modelled on Economic Man; and he is now engaged in a major work on social values and democracy which assumes the importance of political institutions in shaping political choices, and the importance of the political socialization of élites and citizens in the utilization and improvement of political institutions (Downs 1991). Having lost contact with institutions through the reductionist strategy followed by this movement, now most of its practitioners are in search of institutions (Weingast below: chap. 5; Alt and Alesina below: chap. 28).

Robert Bates (1990), a pioneer in the application of rational choice theory in the study of developing countries, now favors an eclectic approach to political analysis. "Anyone working in other cultures knows that people's beliefs and values matter, so too do the distinctive characteristics of their institutions. . . ." He wants to combine the political economy approach with the study of cultures, social structures, and institutions. "A major attraction of the theories of choice and human interaction, which lie at the core of contemporary political economy, is that they offer the tools for causally linking values and structures to their social consequences."

This less heroic version of rational choice theory is quite continuous with so-called "behavioral" political science. And it is so viewed in this version of the history of political science. Its formal deductive approach to generating hypotheses has distinct uses, but it is not inherently superior to the process of deriving hypotheses from deep empirical knowledge, as some of its devotees claim. Green and Shapiro (1994: 10) argue that

> formalism is no panacea for the ills of social science. Indeed, formal exposition does not even guarantee clear thinking. Formally rigorous theories can be inexact and ambiguous if their empirical referents are not well specified. Formalization, moreover, cannot be an end in itself; however analytically tight and parsimonious a theory might be, its scientific value depends on how well it explains the relevant data.

In a major critique of the empirical literature produced by the rational choice approach, Green and Shapiro (1994: 10) conclude:

> exceedingly little has been learned. Part of the difficulty stems from the sheer paucity of empirical applications: proponents of rational choice seem to be most interested in theory elaboration, leaving for later, or others, the messy business of empirical testing. On our reading, empirical failure is also importantly rooted in the aspiration of rational choice theorists to come up with universal theories of politics. As a consequence of this aspiration, we contend, the bulk of rational-choice-inspired empirical work is marred by methodological defects.

To escape from this sterility Green and Shapiro advise rational choice theorists to

> resist the theory-saving impulses that result in method driven research. More fruitful than asking "How might a rational choice theory explain X?" would be the problem driven question: "What explains X?" This will naturally lead to inquiries about the relative importance of a host of possible explanatory variables. No doubt strategic calculation will be one, but there will typically be many others, ranging from traditions of behavior, norms, and cultures to differences in peoples' capacities and the contingencies of

> historical circumstance. The urge to run from this complexity rather than build explanatory models that take it into account should be resisted, even if this means scaling down the range of application. Our recommendation is not for empirical work instead of theory; it is for theorists to get closer to the data so as to theorize in empirically pertinent ways.

Responding to the Green and Shapiro critique, Ferejohn and Satz (1995: 83) tell us, "The aspiration to unity and the quest for universalistic explanations have spurred progress in every science. By ruling out universalism on philosophical grounds, Green and Shapiro surrender the explanatory aspirations of social science. Such a surrender is both premature and self-defeating." On the other hand Morris Fiorina (1995: 87), a member of the more moderate, eclectic camp of the rational choice school, in answer to the Green–Shapiro critique minimizes the extent of universalism and reductionism in the rational choice community. He acknowledges, "Certainly, one can cite rational choice scholars who write ambitiously—if not grandiosely—about constructing unified theories of political behavior." But these, according to Fiorina are a small minority. And in making extravagant claims, rational choicers are no different in their over-selling, from the functionalists, systems theorists and other innovators in the social sciences and other branches of scholarship. Thus two of the most important contributors to the rational choice approach adopt very different positions on the question of scientific maximalism—one defends it as an aspiration without which scientific progress would be compromised, the other offers a half-apology for its hubris, the other half of the apology being withheld since "everybody does it."

The polemic regarding the larger aspirations of the rational choice approach leads us to subsume its accomplishments under our progressive-eclectic view of disciplinary progress, rejecting its maximal claims and view of political science and recognizing its positive contribution of a formal deductive approach to the arsenal of methodologies, hard and soft, which are available to us in our efforts to interpret and explain the world of politics. The movement to penetrate political science laterally, so to speak, without in many cases acquiring knowledge of the substantive fields that are proposed to be transformed, has led inevitably to a method-dominated strategy and an illustrative record of accomplishment, rather than a problem-focused strategy in which formal, deductive methods find their appropriate place.

## V Conclusion

The recent historians of political science, cited above, ask us to adopt a pluralist view of political science. The *methodenstreit*—the methodological war—of the 1970s and 1980s, they tell us, has ended in a stalemate. The idea of a continuous discipline oriented around a shared sense of identity has been rejected. There are as many histories of political science, they say, as many distinct senses of identity, as there are distinct approaches in the discipline. And the relations among these distinct approaches are isolative. There is no shared scholarly ground. We are now, and presumably into the indefinite future, according to these writers, in a post-behavioral, post-positivist age, a discipline divided, condemned to sit at separate tables.

What we propose in this chapter on the history of political science is a view based on a search of the literature from the ancients until the present day, demonstrating a unity of substance and method, and cumulative in the sense of an increasing knowledge base, and improvements in inferential rigor. There is a pluralism in method and approach, but it is eclectic and synergistic, rather than isolative. It acknowledges the substantive contributions of Marxist scholarship as exemplified in its history of social classes, the contribution of Straussians to the history of political ideas, the contribution of rational choice political science to analytical rigor, and the like. This pluralism is not "isolative," it is eclectic and interactive, governed ultimately by its uncompromising commitment to rules of evidence and inference.

## Acknowledgments

I want to acknowledge most helpful criticism from Robert E. Goodin (and his anonymous referees), Heinz Eulau, Alex Inkeles, S. M. Lipset, Robert Packenham, Neil Smelser and Kaare Strom.

## References

Almond, G. A. 1991. Rational choice theory and the social sciences. Pp. 32–52 in *The Economic Approach to Politics*, ed. K. R. Monroe. New York: HarperCollins.

—— and Coleman, J., eds. 1960. *The Politics of the Developing Areas*. Princeton, N.J.: Princeton University Press.

—— and Powell, G. B. 1966. *Comparative Politics: A Developmental Approach*. Boston: Little, Brown.

Almond, G. A. and Verba, S. 1963. *The Civic Culture.* Princeton, N.J.: Princeton Ulniversity Press.
—— Flanagan, S.; and Mundt, R. 1973. *Crisis, Choice, and Change.* Boston, Mass.: Little, Brown.
Alt, J., and Chrystal, A. 1983. *Political Economics.* Berkeley: University of California Press.
—— and Shepsle, K., eds. 1990. *Perspectives on Positive Political Economy.* Cambridge: Cambridge University Press.
(APSA) American Political Science Association. 1994. *Directory of Members; 1994–96.* Washington, D. C.: APSA.
Aristotle. 1934. *The Ethics of Aristotle.* London: Everymans' Library, Dent.
—— 1958. *The Politics of Aristotle,* trans. E. Barker. Oxford: Oxford University Press.
Arrow, K. J. 1951. *Social Choice and Individual Values.* New Haven, Conn.: Yale University Press.
Barnes, S.; Kaase, M.; *et al.* 1979. *Political Action.* Beverly Hills, Calif.: Sage.
Barry, B. 1970. *Sociologists, Economists and Democracy.* London: Collier-Macmillan.
Bates, R. H. 1990. Macropolitical economy in the field of development. In Alt and Shepsle 1990: 31–54.
Berger, S., ed. 1981. *Organizing Interests in Western Europe.* Cambridge: Cambridge University Press.
Black, D. 1958. *The Theory of Committees and Elections.* Cambridge: Cambridge University Press.
Blythe, J. M. 1992. *Ideal Government and the Mixed Constitution.* Princeton, N.J.: Princeton University Press.
Bryce, J. 1888. *The American Commonwealth.* 3 vols. London: Macmillan.
Buchanan, J., and Tullock, G. 1962. *The Calculus of Consent.* Ann Arbor: University of Michigan Press.
Cameron, D. R. 1978. Social democracy, corporatism, labor quiescence, and the representation of economic interest in advanced capitalist society. In Goldthorpe 1978: 143–78.
Castles, F. G., ed. 1989. *The Comparative History of Public Policy.* Oxford: Polity.
Catlin, G. E. G. 1964. *The Science and Method of Politics.* Hamden, Conn.: Archon Books; originally published 1927.
Collier, D. 1993. The comparative method. In Finifter 1993: 105–20.
Collini, S.; Winch, D.; and Burrow, J. 1983. *That Noble Science of Politics: A Study in Nineteenth-Century Intellectual History.* Cambridge: Cambridge University Press.
Committee on Comparative Politics, Social Science Research Council (SSRC). 1972. *A Report on the Activities of the Committee.* New York: SSRC.
Converse, J. M. 1987. *Survey Research in the United States: Roots and Emergence, 1890–1960.* Berkeley: University of California Press.
Crick, B. 1959. *The American Science of Politics.* Berkeley: University of California Press.
Dahl, R. A. 1956. *A Preface to Democratic Theory.* Chicago: University of Chicago Press.
—— 1961. *Who Governs?* New Haven, Conn.: Yale University Press.
—— ed. 1966. *Political Oppositions in Western Democracies.* New Haven, Conn.: Yale University Press.
—— 1970. *After the Revolution.* New Haven, Conn.: Yale University Press.
—— 1971. *Polyarchy.* New Haven, Conn.: Yale University Press.
—— ed. 1973. *Regimes and Oppositions.* New Haven, Conn.: Yale University Press.
—— 1982. *Dilemmas of Pluralist Democracy.* New Haven, Conn.: Yale University Press.
—— 1985. *A Preface to Economic Democracy.* Berkeley: University of California Press.
—— 1989. *Democracy and its Critics.* New Haven, Conn.: Yale University Press.

Deutsch, K. 1961. Social mobilization and political development. *American Political Science Review*, 55: 494–514.
Diamond, L., and Plattner, M., eds. 1993. *The Global Resurgence of Democracy.* Baltimore, Md.: Johns Hopkins University Press.
Dogan, M., and Pelassy, D. 1990. *How to Compare Nations.* Chatham, N.J.: Chatham House.
Downs, A. 1957. *An Economic Theory of Democracy.* New York: Harper.
—— 1991. Social values and democracy. Pp. 143–70 in *The Economic Approach to Democracy*, ed. K. R. Monroe. New York: HarperCollins.
Dryzek, J.; Farr, J.; and Leonard, S., eds. 1995. *Political Science in History.* Cambridge: Cambridge University Press.
—— and Leonard, S. 1988. History and discipline in political science. *American Political Science Review*, 82: 1245–60.
—— —— Farr, J.; Seidelman, R.; and Gunnell, J. 1990. Can political science history be neutral? *American Political Science Review*, 84: 587–607.
Duguit, L. 1917. *The Law and the State.* Cambridge, Mass.: Harvard University Press.
Durkheim, E. 1893/1960. *The Division of Labor in Society*, trans. G. Simpson. Glencoe, Ill.: Free Press.
Duverger, M. 1976. *Political Parties*, trans. B. and R. North. 3rd edn. London: Methuen; originally published 1951.
Easton, D. 1953. *The Political System.* New York: Knopf.
—— 1965. *A Systems Analysis of Political Life.* New York: Wiley.
—— 1990. *The Analysis of Political Structure.* New York: Routledge.
Eckstein, H. 1975. Case study and theory in political science. Vol. vii, pp. 79–137 in *Handbook of Political Science*, ed. F. I. Greenstein and N. S. Polsby. Reading, Mass.: Addison-Wesley.
Eulau, H. 1976. Understanding political life in America: the contribution of political science. *Social Science Quarterly*, 57: 112–53.
—— 1993. The Congress as research arena: an uneasy partnership between history and political science. *Legislative Studies Quarterly*, 18: 569–92.
—— and Prewitt, K. 1973. *Labyrinths of Democracy.* Indianapolis, Ind: Bobbs-Merrill.
Farr, J., and Seidelman, R., eds. 1993. *Discipline and History: Political Science in the U. S.* Ann Arbor: University of Michigan Press.
Ferejohn, J., and Satz, D. 1995. Unification, universalism, and rational choice theory. *Critical Review*, 9: 71–84.
Figgis, J. N. 1896. *The Divine Right of Kings.* Cambridge: Cambridge University Press.
Finifter, A., ed. 1983. *Political Science: The State of The Discipline.* Washington, D.C.: APSA.
—— ed. 1993. *Political Science: The State of The Discipline II.* Washington, D.C.: APSA.
Fiorina, M. 1995. Rational choice, empirical contributions, and the scientific enterprise. *Critical Review*, 9: 85–94.
Fishkin, J. 1992. *The Dialogue of Justice.* New Haven, Conn.: Yale University Press.
Flora, P., and Heidenheimer, A., eds. 1981. *The Development of Welfare States in Europe and America.* New Brunswick, N.J.: Transaction.
Freeman, E. A. 1874. *Comparative Politics.* New York: Macmillan.
Friedman, M. 1953. *Essays in Positive Economics.* Chicago: University of Chicago Press.
Galston, W. 1993. Political theory in the 1980s: perplexity amidst diversity. In Finifter 1993: 27–55.
Garceau, O. 1941. *The Political Life of the American Medical Association.* Cambridge, Mass.: Harvard University Press.

GEERTZ C. 1973. Thick description: toward an interpretive theory of culture. Pp. 3–30 in Geertz, *The Interpretation of Cultures.* New York: Basic.

GEORGE, A. L. 1980. *Presidential Decision-making in Foreign Policy: The Effective Use of Information and Advice.* Boulder, Colo: Westview Press.

—— and MCKEOWN, T. 1982. Case studies and theories of organizational decision-making. *Advances in Information Processes in Organization,* 2: 21–58.

—— *et al.* 1983. *Managing U.S.–Soviet Rivalry: Problems of Crisis Prevention.* Boulder, Colo: Westview Press,

—— and SIMONS, W., eds. 1994. *The Limits to Coercive Diplomacy.* Boulder, Colo.: Westview Press.

—— and SMOKE, R. 1974. *Deterrence in American Foreign Policy.* New York: Columbia University Press.

GIERKE, O. 1868. *Die Deutsche Genossenschaftsrecht.* Berlin: Weidman.

GOLDTHORPE, J. H., ed. 1978. *Order and Conflict in Contemporary Capitalism.* Oxford: Clarendon Press.

GOODNOW, F. 1900. *Politics and Administration.* New York: Macmillan.

GOSNELL, H. F. 1927. *Getting Out The Vote.* Chicago: University of Chicago Press.

GREEN, D. P., and SHAPIRO, I. 1994. *Pathologies of Rational Choice Theory.* New Haven, Conn.: Yale University Press.

GUNNELL, J. G. 1983. Political theory: the evolution of a sub-field. In Finifter 1983: 3–45.

HABERMAS, J. 1992. Further reflections on the public sphere/ . . . /Concluding remarks. Pp. 421–46 and 462–79 respectively in *Habermas and the Public Sphere,* ed. C. Calhoun. Cambridge, Mass.: MIT Press.

HAMILTON, A.; MADISON, J.; and JAY, J. 1937. *The Federalist.* Washington, D.C.: National Home Library; originally published 1787–8.

HEIDENHEIMER, A.; HECLO, H.; and ADAMS, C. 1990. *Comparing Public Policy.* 3rd edn. New York: St. Martins.

HELD, D. 1980. *Introduction to Critical Theory.* Berkeley: University of California Press.

HERRING, P. 1929. *Group Representation Before Congress.* Washington, D.C.: Brookings Institution.

HIBBS, D. 1978. On the political economy of long run trends in strike activity. *British Journal of Political Science,* 8: 153–75.

HILL, C. 1982. *The Century of Revolution.* New York: Norton.

HILTON, R. 1990. *Class Conflict and the Crisis of Feudalism.* London: Verso.

HIRSCHMAN, A. 1970. The search for paradigms as a hindrance to understanding. *World Politics,* 22: 329–43.

HOBSBAWM, E. J. 1962. *The Age of Revolution.* New York: New American Library.

—— 1987. *The Age of Empire.* New York: Pantheon Books.

—— 1994. *The Age of Extremes: A History of The World, 1914–1991.* New York: Pantheon.

HUNTINGTON, S. 1991. *The Third Wave: Democratization in the 20th Century.* Norman: University of Oklahoma Press.

INGLEHART, R. 1977. *The Silent Revolution.* Princeton, N.J.: Princeton University Press.

—— 1990. *Culture Shift in Advanced Industrial Society.* Princeton, N.J.: Princeton University Press.

INKELES, A. 1950. *Public Opinion in Soviet Russia.* Cambridge, Mass.: Harvard University Press.

—— 1959. *The Soviet Citizen.* Cambridge, Mass.: Harvard University Press.

—— and SMITH, D. 1974. *Becoming Modern: Individual Change in Seven Countries.* Cambridge, Mass.: Harvard University Press.

JUDT, T. 1995. Downhill all the way. *New York Review of Books,* 42 (#9: May 25): 20–5.

KARL, B. 1974. *Charles E. Merriam and the Study of Politics*. Chicago: University of Chicago Press.
KING, G.; KEOHANE, R. O.; and VERBA, S. 1994. *Designing Social Inquiry*. Princeton, N.J.: Princeton University Press.
KÖNIG, R. 1968. Auguste Comte. Vol. iii, pp. 201–6 in *International Encyclopedia of the Social Sciences*, ed. D. L. Sills. New York: Macmillan/Free Press.
LALMAN, D.; OPPENHEIMER, J.; and SWISTAK, P. 1993. Formal rational choice theory: a cumulative science of politics. In Finifter 1993: 77–105.
LASKI, H. 1919. *Authority in the Modern State*. New Haven, Conn.: Yale University Press.
LASSWELL, H. D. 1927. *Propaganda Technique in the World War*. New York: Knopf.
—— 1930. *Psychopathology and Politics*. Chicago: University of Chicago Press.
—— 1935. *World Politics and Personal Insecurity*. New York: McGraw-Hill.
—— 1936. *Politics: Who Gets What, When, and How*. New York: McGraw-Hill.
—— and BLUMENSTOCK, D. 1939. *World Revolutionary Propaganda*. New York: Knopf.
—— CASEY, R. D.; and SMITH, B. L. 1935. *Propaganda and Promotional Activities*. Minneapolis: University of Minnesota Press.
LERNER, D. 1958. *The Passing of Traditional Society*. Glencoe, Ill.: Free Press.
LIJPHART, A. 1968. *The Politics of Accomodation*. Berkeley: University of California Press.
—— 1971. Comparative politics and the comparative method. *American Political Science Review*, 65: 3.
—— 1984. *Democracies*. New Haven, Conn.: Yale University Press.
—— 1994. *Electoral Systems and Party Systems*. Oxford: Oxford University Press.
LINZ, J., and STEPAN, A. 1978. *The Breakdown of Democratic Regimes*. Baltimore, Md.: Johns Hopkins University Press.
LIPSET, S. M. 1959. Some social requisites of democracy. *American Political Science Review*, 53: 69–105.
—— 1960. *Political Man*. New York: Doubleday.
—— 1994. The social requisites of democracy revisited. *American Sociological Review*, 59: 1–22.
—— and ROKKAN, S., eds. 1967. *Party Systems and Voter Alignments*. New York: Free Press.
MAINE, H. 1963. *Ancient Law*. Boston: Beacon; originally published 1861.
MANNHEIM, K. 1949. *Ideology and Utopia*, trans. L. Wirth and E. Shils. New York: Harcourt Brace.
MARCH, J. G., ed. 1965. *Handbook of Organizations*. Chicago: Rand McNally.
—— 1988. *Decisions and Organizations*. New York: Blackwell.
—— and SIMON, H. A. 1958. *Organizations*. New York: Wiley.
MERRIAM, C. E. 1931*a*. *The Making of Citizens*. Chicago: University of Chicago Press.
—— 1931*b*. *New Aspects of Politics*. 2nd edn. Chicago: University of Chicago Press.
—— and GOSNELL, H. F. 1924. *Non-Voting: Causes and Methods of Control*. Chicago: University of Chicago Press.
MICHELS, R. 1949. *Political Parties*, trans. E. and C. Paul. Glencoe, Ill.: Free Press; originally published 1915.
MILL, J. S. 1843–1961. *A System of Logic, Ratiocinative and Inductive*. London: Longman.
MILLER, W. 1994. An organizational history of the intellectual origins of the American National Election Studies. *European Journal of Political Research*, 25: 247–65
MOORE, B. 1966. *Social Origins of Dictatorship and Democracy*. Boston: Beacon.
NOZICK, R. 1974. *Anarchy, the State, and Utopia*. New York: Basic.
ODEGARD, P. 1928. *Pressure Politics: The Story of the Anti-Saloon League*. New York: Columbia University Press.
OLSON, M. 1965. *The Logic of Collective Action*. Cambridge, Mass.: Harvard University Press.

Olson, M. 1990. Toward a unified view of economics and the other social sciences. In Alt and Shepsle 1990: 212–32.

Ordeshook, P. C. 1990. The emerging discipline of political economy. In Alt and Shepsle 1990: 9–30.

Ostrogorski, M. 1964. *Democracy and the Organization of Political Parties*, ed. S. M. Lipset. New York: Doubleday.

Packenham, R. A. 1992. *The Dependency Movement*. Cambridge, Mass.: Harvard University Press.

Parsons, T. 1951. *The Social System*. Cambridge, Mass.: Harvard University Press.

—— and Shils, E. 1951. *Toward a General Theory of Action*. Cambridge, Mass.: Harvard University Press.

—— and Smelser, N. 1956. *Economy and Society*. London: Routledge and Kegan Paul.

Pitkin, H. 1967. *The Concept of Representation*. Berkeley: University of California Press.

Pocock, J. G. A. 1975. *The Machiavellian Moment*. Cambridge: Cambridge University Press.

Pollock, F. 1890. *The History of the Science of Politics*. London: Macmillan.

Popper, K. 1972. *Objective Knowledge: An Evolutionary Approach*. Oxford: Clarendon Press.

Powell, G. B. 1982. *Contemporary Democracies*. Cambridge, Mass.: Harvard University Press.

Przeworski, A., and Teune, H. 1970. *The Logic of Comparative Social Inquiry*. New York: Wiley.

Putnam, R. 1973. *The Beliefs of Politicians*. New Haven, Conn.: Yale University Press.

—— 1993. *Making Democracy Work*. Princeton, N.J.: Princeton University Press.

Pye, L. 1962. *Politics, Personality, and Nation-Building*. New Haven, Conn.: Yale University Press.

—— 1985. *Asian Power and Politics*. Cambridge, Mass.: Harvard University Press.

—— 1988. *The Mandarin and the Cadre*. Ann Arbor: University of Michigan Press.

—— and Verba, S., eds. 1965. *Political Culture and Political Development*. Princeton, N.J.: Princeton University Press.

Rabi, M. M. 1967. *The Political Theory of Ibn Khaldun*. Leiden: E. J. Brill.

Rangavajan, L. N. 1987. *Kautilya: The Arthashastra: The Science of Statecraft*. New Delhi: Penguin Books.

Rawls, J. 1971. *A Theory of Justice*. Cambridge, Mass.: Harvard University Press.

Ricci, D. 1984. *The Tragedy of Political Science*. New Haven, Conn.: Yale University Press.

Riker, W. H. 1962. *The Theory of Coalitions*. New Haven, Conn.: Yale University Press.

—— 1982. The two-party system and Duverger's law: essay on the history of political science. *American Political Science Review*, 76: 753–66.

—— 1990. Political choice and rational choice. In Alt and Shepsle 1990: 163–81.

Roscher, W. 1892. *Politik: Geschichtliche Naturlehre der Monarchie, Aristokratie, und Demokratie*. Stuttgart: J. G. Colta.

Rose, R. 1990. Institutionalizing professional political science in Europe: a dynamic model. *European Journal of Political Research*, 18: 581–603.

Rutherford, L. C. 1937. *The Influence of the American Bar Association on Public Opinion and Legislation*. Philadelphia: Foundation Press.

Sabine, G., and Thorson, T. 1973. *A History of Political Theory*. 4th edn. New York: Holt, Rinehart and Winston.

Sartori, G. 1976. *Parties and Party Systems*. Cambridge: Cambridge University Press.

—— 1987. *Theory of Democracy Revisited*. 2 vols. Chatham, N.J.: Chatham House.

Saxonhouse, A. 1993. Texts and canons: the status of the "great books" in political science. In Finifter 1993: 3–26.

Schaar, J. H., and Wolin, S. S. 1963. Essays on the scientific study of politics: a critique. *American Political Science Review*, 57: 125–50.

Schattschneider, E. E. 1935. *Politics, Pressures, and the Tariff.* Englewood Cliffs, N.J.: Prentice-Hall.

Schlozman, K. L., and Verba, S. 1979. *Injury to Insult: Unemployment, Class, and Political Response.* Cambridge, Mass: Harvard University Press.

—— —— and Brady, H. 1995. Participation is not a paradox: the view from American activists. *British Journal of Political Science*, 25: 1–36.

Schmitter, P., and Lehmbruch, G., eds. 1979. *Trends Toward Corporate Intermediation.* Beverley Hills, Calif.: Sage.

—— O'Donnell, G.; and Whitehead, L. 1986. *Transitions From Authoritarian Rule.* Baltimore, Md.: Johns Hopkins University Press.

Seeley, J. R. 1896. *An Introduction to Political Science.* London: Macmillan.

Seidelman, R. 1985. *Disenchanted Realists: Political Science and the American Crisis, 1884–1984.* Albany: State University of New York Press.

Siegfried, A. 1930. *Tableau des Partis en France.* Paris: Grasset.

Shapiro, Martin. 1993. Public law and judicial politics. In Finifter 1993: 365–81.

Simon, H. A. 1950. *Public Administration.* New York: Knopf.

—— 1953. *Administrative Behavior.* New York: Macmillan.

—— 1957. *Models of Man.* New York: Wiley.

Skinner, Q. 1978. *The Foundations of Modern Political Thought.* Cambridge: Cambridge University Press.

Skocpol, T. 1979. *States and Social Revolutions.* Cambridge: Cambridge University Press.

—— ed. 1984. *Vision and Method in Historical Sociology.* Cambridge: Cambridge University Press.

Smelser, N. 1976. *Comparative Methods in the Social Sciences.* Englewood Cliffs, N.J.: Prentice-Hall.

Spencer, H. 1965. *The Study of Sociology.* Ann Arbor: University of Michigan Press; originally published 1874.

Somit, A., and Tannenhaus, J. 1967. *The Development of American Political Science.* Boston: Allyn and Bacon.

Storing, H. J., ed. 1962. *Essays on the Scientific Study of Politics.* New York: Holt, Rinehart and Winston.

—— Berns, W.; Strauss, L.; Weinstein, L.; and Horowitz, R. 1963. Replies to Schaar and Wolin. *American Political Science Review*, 57: 151–60.

Strauss, L. 1959. *What is Political Philosophy?* Chicago: University of Chicago Press.

—— 1989. *The Rebirth of Classical Political Rationalism.* Chicago: University of Chicago Press.

—— and Cropsey, J. 1987. *A History of Political Philosophy.* Chicago: University of Chicago Press.

Thompson, D. 1970. *The Democratic Citizen.* Princeton, N.J.: Princeton University Press.

Thompson, E. P. 1963. *The Making of the English Working Class.* New York: Vintage.

Tingsten, H. 1963. *Political Behavior.* Totowa, N.J.: Bedminton Press; originally published 1937.

Tocqueville, A. de. 1985. *Selected Letters on Politics and Society*, ed. Roger Boesche. Berkeley: University of California Press.

Tönnies, F. 1957. *Community and Society*, ed. and trans. C. Loomis. East Lansing: Michigan State University Press.

Vallès, J. M., and Newton, K., eds. 1991. Special issue: political science in Western Europe, 1960–90. *European Journal of Political Research*, 20: 225–466.

Verba, S. 1987. *Elites and the Idea of Equality*. Cambridge, Mass.: Harvard University Press.
—— and Ahmed, B. 1973. *Caste, Race and Politics*. Beverly Hills, Calif.: Sage.
Verba, S., and Nie, N. 1972. *Participation in America*. New York: Harper and Row.
—— —— and Kim, J. 1978. *Participation and Political Equality*. Cambridge: Cambridge University Press.
Wahlke, J., and Eulau, H. 1962. *The Legislative System*. New York: Wiley.
—— —— 1978. *The Politics of Representation*. Beverly Hills, Calif.: Sage.
Walzer, M. 1983. *Spheres of Justice*. New York: Basic.
Weber, M. 1949. *The Methodology of The Social Sciences*, trans. E. Shils and H. Finch. Glencoe, Ill.: Free Press.
—— 1958. *From Max Weber*, trans. and ed. H. Gerth and C. W. Mills. New York: Oxford University Press.
—— 1978. *Economy and Society*, ed. G. Roth and C. Wittich, trans. E. Fischoff *et al.* Berkeley: University of California Press.
White, L. D. 1929. *The Prestige Value of Public Employment*. Chicago: University of Chicago Press.
Wildavsky, A. 1984. *The Nursing Father: Moses as a Political Leader*. Birmingham: University of Alabama Press.
—— 1986. *A History of Taxation and Public Expenditure in the Western World*. New York: Simon and Schuster.
—— 1989. What is permissible so that this people may survive? Joseph the administrator. *PS: Political Science and Politics*, 12: 779–88.
Wilson, J. Q. 1993. *The Moral Sense*. New York: Free Press.
Wilson, W. 1887. The study of administration. *Political Science Quarterly*, 2: 197–222.
—— 1918. *The State: Elements of Historical and Practical Politics*. Boston: Heath; originally published 1889.
Womack, J. 1968. *Zapata and the Mexican Revolution*. New York: Knopf.
Woolsey, T. 1878. *Political Science: The State Theoretically and Practically Considered*. New York: Scribner.
Wright, Q. 1942. *A Study of War*. Chicago: University of Chicago Press.

Chapter 3

# Political Science and the Other Social Sciences

Mattei Dogan

THE discipline of political science is "ill-defined, amorphous and heterogeneous." With this diagnosis, editors Fred I. Greenstein and Nelson W. Polsby open their preface to the first *Handbook of Political Science* (1975: 1). Twenty years later, the main features of political sciences are: specialization, fragmentation and hybridization. Its frontiers are open and moving and need not be defined. The process of specialization has generated an increasing fragmentation in subfields, which are not "amorphous" but rather well-organized and creative. The "heterogeneity" has been greatly nourished by exchanges with neighbouring disciplines through the building of bridges between specialized fields of the various social sciences. This process of cross-fertilization is achieved by hybridization.

The relations between political science and the other social sciences are in reality relations between sectors of different disciplines, not between whole disciplines. It is *not* an "interdisciplinary" endeavor. Since there is no progress without specialization, the creative interchanges occur between specialized subfields, most of the time at the margins of the formal disciplines. The current advancement of the social sciences can be explained in large part by the hybridization of segments of sciences. It would be impossible to conceive of a history of political science and of its current trends without reference to the other social sciences.

# I Specialization, fragmentation, hybridization

A distinction has to be drawn between specialization within a formal discipline and specialization at the intersection of monodisciplinary subfields. The latter, hybridization, can occur only after the former has become fully developed. In the history of science a twofold process can be seen: on the one hand, a fragmentation of formal disciplines and, on the other, a recombination of the specialities resulting from fragmentation. The new hybrid field may become independent, like political economy; or it may continue to claim a dual allegiance, like political geography. In the latter case, we cannot be sure whether to place a work in the category of geography or of political science.

The criterion could be the predominance of one or the other component or the formal affiliation of the author. Political anthropology is a branch of anthropology but also a subfield of political science. Where does historical sociology end and social history begin? We may feel even more unsure when faced with a case of threefold recombination. As the relative proportions are not always obvious, it remains somewhat arbitrary where the essential affiliation may be said to lie, since the degree of kinship between specialities varies greatly.

## A Interdisciplinary research or recombination of fragments of sciences?

Some scholars praise "interdisciplinarity." Such a recommendation often comes from the most creative scientists because they are the first to see the problems caused by gaps between disciplines. But this recommendation is not realistic. Nowadays, it is no longer possible for anyone to have a thorough knowledge of more than one discipline. It is utopian to want to master two or more whole disciplines. Given that it implies the ability to be familiar with and combine entire disciplines, the idea of interdisciplinary research is illusory.

Because it is so difficult for a single scholar to be truly multidisciplinary, some methodologists are led to advocate teamwork. This is what is proposed by Pierre de Bie in the monumental work published by UNESCO (1970). Teamwork is productive in the big science laboratories, but where the social sciences are concerned it is difficult to achieve in practice. The only examples of successful teamwork concern data production or collection, and very seldom interpretation or synthesis (archaeology being the exception, here).

The multidisciplinary approach is illusory because it advocates the slicing up of reality. Some researchers proceed piecemeal, with philological, anthropological, historical, ethnological, psychological and sociological approaches. This alternation of approaches, which almost never allows disciplines to meet, results at best in a useful parallelism—but not in a synthesis. In fact, research enlisting several disciplines involves a combination of segments of disciplines, of specialities and not whole disciplines. The fruitful point of contact is established between sectors, and not along disciplinary boundaries. Considering the current trends in the social sciences, the word "interdisciplinarity" appears inadequate. It carries a hint of superficiality and dilettantism, and consequently should be avoided and replaced by hybridization of fragments of sciences.

## B Specialization and fragmentation

In Cartesian thought, analysis means breaking things into parts. All sciences, from astronomy to zoology, have made progress from the 16th century on by internal differentiation and cross-stimulation among emergent specialities. Each speciality developed a patrimony of knowledge as its understanding of the world advanced. With the growth of these patrimonies, specialization became less a choice and more a necessity. Increasingly focused specialization has led to the creation of subdisciplines, many of which have gone on to become autonomous.

There are in the literature dozens of lamentations and jeremiads about the fragmentation of political science. I cite here just two recent complaints. "Today there is no longer a single, dominant point of view . . . the discipline is fragmented in its methodological conception . . . students are no longer certain what politics is all about" (Easton and Schelling 1991: 49). In the Nordic countries, "political science showed tendencies to disintegrate into subfields, but these were still subfields of political science. However, the disintegration has continued and has lately taken on different forms which renounce the identity of political science" (Anckar 1987: 72).

In reality, fragmentation results from specialization. The division of the discipline into subfields tends to be institutionalized, as can be seen in the organization of large departments of political science in many American and European universities.

A good indication of the fragmentation of the discipline is the increasing number of specialized journals. In the last twelve years one hundred specialized journals in English relevant to political science have been launched. Most of these journals cross the borders of two or three disciplines, and many of them are located in Europe. Some other new hybrid journals have appeared in French and in German. European unification has had an impact on the development of cross-national journals focusing on special fields.

Increasing specialization may have consequences for the role of national professional associations and of the general journals.

> As political scientists have become more specialized, some members [of the American Political Science Association, APSA] have concluded that their interests are better served by other organizations. A comparative government area specialist, for instance, may find that he/she has more in common with economists, sociologists and anthropologists working in the same area than with other political scientists. This may also decrease the value of the *American Political Science Review* . . . Specialization has devalued the reasons for joining APSA (Lynn 1983: 114–15).

The same phenomenon can be observed in Europe. The national professional associations are losing ground in favor of cross-national organizations that represent topical specializations across disciplines.

## C Specialization into hybridization

It is necessary to stress both parts of the process: fragmentation into special fields and specialization by hybridization. It is the interaction of these two processes, and not each one in isolation, that has led to the remarkable advance of the natural as well as the social sciences. The continuous restructuring of political science, like that of the other social sciences, has been the result of these two contending processes. However, both fragmentation and its correlate, hybridization, have developed much more recently in political science than elsewhere. In the distant past, hybrid fields were the result of gaps between full disciplines. Today the gaps appear between specialized subfields among neighbouring subdisciplines. As a result, the fragmentation of disciplines into specialized subfields in the last few decades has led to the development of hybrid specialities. The hybrid specialities do not necessarily stand midway between two sovereign disciplines. They may be enclaves of a section of political science into a sector of another discipline. They combine two delimited domains, not entire disciplines. These domains do not need to be adjacent.

Hybridization appears in the list of research committees sponsored by the International Political Science Association. Among the forty recognized groups in 1995 a majority are related to specialities of other disciplines and are therefore hybrid: Political Sociology, Political Philosophy, Political Geography, Psycho-politics, Religion and Politics, Political and Social Elites, Armed Forces and Politics, Political Alienation, Politics and Ethnicity, Political Education, International Political Economy, International Economic Order, Comparative Judicial Studies, Biology and Politics, Business and Politics, Science and Politics, Socio-political Pluralism, Health Policy, Sex Roles and Politics, Global Environmental Change, Conceptual and Terminological Analysis, etc. Each of these groups is in contact with specialists belonging formally to other disciplines.

Sociometric studies show that many specialists are more in touch with colleagues who belong officially to other disciplines than with colleagues in their own discipline. The "invisible college" described by Robert Merton, Diana Crane and other sociologists of science is an eminently interdisciplinary institution because it ensures communication not only from one university to another and across all national borders, but also and above all between specialists attached administratively to different disciplines. The networks of cross-disciplinary influence are such that they are obliterating the old classification of the social sciences.[1]

## II Borrowing from neighboring disciplines

The process of hybridization consists first of all in borrowing and lending concepts, theories and methods. A review of the lending process would

[1] Indeed, we might construct a "Genealogical Tree of Political Science," cross-nationally. "The content of Swedish political science research before 1945 was dominated by three main currents: each of these currents was oriented toward another academic discipline: constitutional law, history, philosophy" (Ruin 1982: 299). In India, "while in the past political science has been heavily irrigated by thought streams originating in disciplines like philosophy, law and history . . . no political science teacher in India today can afford to be out of touch with the latest advances in disciplines like sociology, social anthropology, economics, management and public administration" (Narain and Mathur 1982: 197). In the Netherlands "about half of the present full and associate professors of political science studied initially in a different field than political science, usually sociology or law" (Hoogerwerf 1982: 227). In Scandinavia, "the bulk of the efforts at theorizing remained heavily sociological in style and in orientation. Explicitly sociological frameworks for political analysis were developed by Erik Allardt in Finland, Ulf Himmelstrand in Sweden, Wilhelm Aubert, Johan Galtung, Stein Rokkan, Ulf Torgersen, Francesco Kjellberg and Øyvind Østerud in Norway. This work parallels other endeavors on the border between sociology and politics" (Kuhnle 1982: 259). In earlier times in the United States, political science "had no distinctive methodology. It had no clearly-defined subject matter that could not be encompassed within one or more of its sister disciplines. Its various parts could have survived simply as political history, political sociology, political geography, political philosophy, and political psychology—subfields in the other disciplines. Other parts could have remained constitutional law, public law and international law. Indeed, they have done so. Each of the other social science disciplines claims a piece of political science" (Andrews 1988: 2).

take us too far. I have to forgo to such a review here. In any case, political science has always borrowed much more than it has lent.

## A The diffusion of concepts across disciplines

For a century and a half, from Sir George Cornewall Lewis' 1832 *The Use and Abuse of Some Political Terms* to Giovanni Sartori's edited collection on *Social Science Concepts* in 1984, numerous scholars have denounced the conceptual confusion and the polysemy of terms in various disciplines and particularly in political science. One of the reasons for this polysemy is indicated by Sartori (1984: 17): "We cannot form a sentence unless we already know the meanings of the words it contains . . . It is not that words *acquire* their meaning via the sentences in which they are placed, rather, the meaning of a word is *specified* by the sentence in which it is placed."

Another important reason for this semantic problem comes from the peregrination of concepts from one discipline to another. Borrowed concepts need some adaptation to the context of the new discipline, because a concept is not only a term, it is also a notion or an idea. A recent study of more than 400 concepts used in the social sciences has found few neologisms (de Grolier 1990: 271), and this can be explained by the fact that more concepts are borrowed than created. Some concepts are reanimated after a long oblivion. Max Weber resurrected the concept of charisma after centuries of neglect. David Apter made use of the concept of consociational organization, which was originally applied to Presbyterian institutions in Scotland. He used it to analyze political conflict in Uganda. Arendt Lijphart and many others have developed it further with respect to small European democracies, Canada and South Africa.

We can neglect the etymology of concepts in order to stress how borrowing fertilizes imagination. The word "role" comes from the theatre, but Max Weber gave it a sociological meaning. From sociology this concept spread everywhere. The word "revolution" was proposed by Copernicus, but it was first applied to politics by Louis XIV. Historians adopted it, sociologists articulated it, before offering it to political science.

The patrimony of political science is full of borrowed concepts, which are hybrid in the sense that they were concocted in other disciplines and replanted skilfully in the garden of political science. This discipline has nevertheless generated for its own use a long series of important concepts, the oldest being "power," formulated by Aristotle, and the youngest, "implosion," suggested by the fall of the Soviet Union.

Using the *International Encyclopaedia of Social Sciences* (Sills 1968) and the analytical indexes of some important books, I have compiled an inventory of more than two hundred concepts "imported" into political science. In the process of adoption and adaptation many of these concepts have changed their semantic meaning. Political science has borrowed the following important concepts (excluding "lay" terms):

- *From sociology*: accommodation, aggregate, assimilation, élite circulation, clique, cohesion, collective behavior, hierarchy, ideal-type, individualism, legitimacy, mass media, mass society, militarism, nationalism, pattern variables, Protestant ethic, secular, segregation, social class, social control, social integration, social structure, socialization, status inconsistency, working class, *Gemeinschaft–Gesellschaft.*
- *From psychology*: affect, alienation, ambivalence, aspiration, attitude, behavior, consciousness, dependency, empathy, personality, social movement, stereotype, *Gestalt.*
- *From economics*: allocation of resources, cartel, corporatism, diminishing returns, industrial revolution, industrialization, liberalism, mercantilism, gross national product, scarcity, undeveloped areas.
- *From philosophy and the ancient Greeks*: anarchism, aristocracy, consensus, democracy, faction, freedom, general will, idealism, monarchy, oligarchy, phratry, pluralism, tyranny, value, *Weltanschauung.*
- *From anthropology*: acculturation, affinity, caste, nepotism, patriarchy, plural society, *rites de passage.*
- *From theology*: anomie (disregard of divine law), charisma.
- *From journalists and politicians*: imperialism, internationalism, isolationism, Left and Right, lobbying, neutralism, nihilism, patronage, plebiscite, propaganda, socialism, syndicalism.

Many concepts have multiple origins. Authoritarianism has two roots, one psychological and one ideological. It is often inadvertently interchangeable with despotism, autocracy, absolutism, dictatorship, etc. Authority has been analyzed from different disciplinary perspectives by Malinowski, Weber, Parsons, Lasswell, Kaplan, B. de Jouvenel, and C. J. Friedrich, among others. The concept of culture (civic, political, national) has many variants: cultural convergence, cultural configuration, cultural evolution, cultural integration, cultural lag, cultural parallelism, cultural pluralism, cultural relativity, cultural system, post-materialist culture. In the last two decades political scientists have been very productive in this subfield.

Max Weber and Karl Marx, both hybrid scholars, were the most prolific generators of concepts. Only Aristotle is comparable to them. Almond and Parsons are also the fathers of an impressive number of concepts. Concepts are often germinal grains of theories: structure generates structuralism, system becomes systemism, capital engenders capitalism, and so on.

## B Theories across disciplinary borders

Paradigm is a word often used or abused in political science, as much as in sociology, instead of the words theory or grand theory. Thomas Kuhn, who concocted this word, has explicitly acknowledged that in the social sciences its use is not justified. He explains in his preface to *The Structure of Scientific Revolutions* (Kuhn 1957: viii) that it was during a stay at Palo Alto Center for Advanced Studies, in the company of social scientists, including political scientists, that he was led to formulate the concept of paradigm with the very aim of making clear the essential difference between natural sciences and the social sciences. The reason given by Kuhn was the absence of a theoretical consensus in any discipline of social sciences. Today, if someone "wants to legitimate his theory or model as a revolutionary achievement, there are always some who do not rally round the flag" (Weingart 1986: 270).

Are there in the social sciences instances of paradigmatic upheavals comparable to those created by Copernicus, Newton, Darwin or Einstein? Can the theories of Keynes, Chomsky or Parsons be described as paradigmatic? In the social sciences, does progress occur through paradigmatic revolutions or through cumulative processes? Are there really paradigms in the social sciences?

Within a formal discipline, several major theories may cohabit, but there is a paradigm only when one testable theory alone dominates all other theories and is accepted by the entire scientific community. When Pasteur discovered the microbe, the theory of spontaneous generation collapsed: contagion became the new paradigm. In the social sciences, however, we see at best a confrontation between several non-testable theories. Most of the time there is not even a confrontation but careful mutual avoidance, superb disregard on all sides; this is relatively easy owing to the size of scientific communities, and its division into schools. This is true for all countries, big or small.

This mutual disregard is an old practice in the social sciences. At the turn of the century, the great scholars did not communicate, or very little. In the writings of Weber there is no reference to his contemporary

Durkheim. Yet Weber was acquainted with Durkheim's journal, *L'Année sociologique.* For his part, Durkheim, who could read German, makes only one, fleeting reference to Weber. Yet they worked on a number of the same subjects, such as religion. Durkheim does no more than mention in passing Simmel and Töennies. Harshly criticized by Pareto, Durkheim never alluded to Pareto's work. Pareto's judgment of Durkheim's book on suicide was unfavorable. "Unfortunately," he wrote, "its arguments lack rigour" (quoted in Valade 1990: 207).

Weber seems to have been unaware of Pareto's theory on the circulation of élites and Pareto in his turn says nothing about the Weberian theory of political leadership. Weber and Croce met only once, and that briefly. There was no exchange between Weber and Freud. Ernst Bloch and George Lukács met regularly with Weber in Heidelberg, but their work shows no sign of his influence. Nor was there any communication between Weber and Spengler. Of Weber's contemporaries the only one who referred to him was Karl Jaspers, but he was a philosopher (cf. Mommsen and Osterhammel 1987). As was noted by Raymond Aron, each of the three great scholars followed a "solitary path."

Many examples could be cited of scholars co-existing without influencing one another, such as Angus Campbell and Paul Lazarsfeld, who nevertheless devoted a large part of their lives to studying the same political behavior. The same remark can be made with reference to other topical fields. It is not a bad thing to pit theories one against the other. But there must be debate. There are no paradigms in the social sciences because each discipline is fragmented.

For there to be a paradigm, one other condition must be met: theories must refer to essential aspects of social reality. However, the more ambitious a theory is, the less it can be directly tested by the data available. In the social sciences there are no "fundamental discoveries," as there sometimes are in the natural sciences. Instead, unverifiable theories are constructed, partly because social reality itself changes. Also, and more importantly, the mistakes made by the giants of the natural sciences are most of the time methodological errors; in the social sciences they are basic mistakes.

Consider Malthusianism for instance. Is it a theory or a paradigm? Malthusianism is one of the major theories in the history of the social sciences. Malthus influenced many scientists, primarily Charles Darwin, who acknowledged it to be one of his main sources of inspiration. A host of sociologists, political scientists, demographers and economists took their cue from him, either to agree or to disagree with him. But when demographic conditions changed in the West, his projections were invalidated, and he was condemned as a false prophet. However, if we consider today the gap between economic development and population growth in Africa, Asia or Latin America, he could be hailed as a great visionary. We need only agree to an asynchronous comparison between the England of his time and the Third World to admit the asynchronous validity of his theory. Should we go further and talk of a Malthusian paradigm?

Is there at least cumulative progress in political science? Clearly, there is such progress since the discipline has its heritage of concepts, methods, theories and praxis. It can soon be recognized whether someone is a professional or an amateur. There is cumulative progress even in the theoretical field. If a theory becomes outdated, or is invalidated, something nevertheless remains of it, which is incorporated into new theories, for a great deal is learned by making mistakes. We do not repeat a mistake that has been denounced. In recent times, progress in political science has been ensured through a long series of sectoral empirical discoveries. For example, the correlation established by D. Lerner (1958: 63) between degrees of urbanization, literacy and communication is a proven fact that remains valid. In these specialized sectors—whether hybrid or monodisciplinary—there is no need for ambitious theories, it is enough for them to be what Merton (1973) has called "middle-range theories."

Let us take a concrete example of a cumulative process. One of the great findings in political science is the influence of electoral techniques on party systems. A bibliography, even a very selective one, on this theme could easily comprise two or three hundred titles in English, not to mention the many varied observations derived from the direct experience of politicians in numerous countries. From Condorcet, Bachofen, John Stuart Mill, Hare and Hondt to Hermens, Downs, Duverger, Sartori, Lijphart, the theory is based on the contributions and successive improvements made by a very large number of specialists. The consequences of proportional representation had already been described by Bachoven in 1850.

It is now recognized that "no paradigm seeks any more to order, and even less to unify, the field of the social sciences" (*Annales* 1989: 1322). The word paradigm should be excluded from the vocabulary of social sciences unless it is placed between quotation marks.

Having thus cleared up the apparent theoretical contradiction between hybridization of specialities and the disciplinary paradigm, let us now look to some hybrid theories. Examples of theoretical cross-fertilization abound. Interest group theory's most-cited work, David B. Truman's *The Governmental Process*, draws heavily on sociological theories of groups. Mancur Olson's attack on traditional interest group theory, *The Logic of Collective Action*, was based on economics. Meanwhile, sociologists and economists have borrowed from interest group theories developed by political scientists.

The theories of sister disciplines have often confronted one another on the ground of political science, with results beneficial to all concerned. "Rational choice analysis" is a case in point. This approach has proved to be quite impervious to empirical criticism: an argument that a given politician was irrational, for instance, is not usually taken to present a threat to the theory. Instead, modifications to, or attacks on, rational choice have tended to come either from within or from theoreticians of other disciplines. The strongest criticisms have been the construction of theoretical alternatives. A theory is discredited only by replacing it, usually with the aid of theories from outside the discipline. Psychology has provided the foundation for several of these attacks. Herbert Simon's theory draws not only on economics but also on psychology and on the study of public administration within political science.

Theorists of political systems have often used extensive analogies with biological systems. Biology first developed the concept of "system" as a way to organize life and of organic systems as phenomena not reducible to their constituent chemistry. Some structural functionalists have argued that social systems are like biological systems in that they are self-regulating and homeostatic. These theorists also noted that certain functions have to be performed in any biological system and used the analogy to ask what functions were vital to social systems. "Functionalism was well established in biology in the 1920s and had been used independently in Freudian analysis of the personality and in the study of primitive societies. Thence it spread throughout the social sicences, and with it spread logical scepticism about the exact status of the word function" (Mackenzie 1967: 91). Systems theory, whether that of David Easton in comparative politics, or of Morton Kaplan, Richard Rosecrance, and Kenneth Waltz in international relations, drew primarily upon such sources for some sectors of sociology.

Dependency theory, which seduced so many Latin America specialists, originates in the work of a group of economists, sociologists and demographers, in co-operation with statisticians from the United Nations. Among them are: Fernando H. Cardoso and Enzo Faletto (authors of *Dependencia y Desarrollo en Latin America),* André Gunder Frank, Theotonio Dos Santos, Ruy Mauro Marini.

Theories decay. How old theories are superseded by new ones is a good question. But there is another, raised by Daniel Bell, the phenomenon of theories going wrong or turning into a blind alley: "Why does what was once regarded as an advance become a cul-de-sac?" (D. Bell, in Deutsch *et al.* 1986: 220). One could read today with great interest dozens of political philosophers and grand theorists of the past and cite them with pleasure. But only a handful of theories formulated before World War II are still alive. Theories survive more easily in linguistics and economics. Castles built on sands by political scientists are ruined at the first rain. In 1912, Gustave Le Bon wrote in *La Psychologie Politique* that the rules formulated by Machiavelli in *The Prince* were not valid any more because the society observed by the great Florentine no longer existed.

But we are not going to make a pilgrimage to the cemetery of political theories. It is sufficient to note that in this necropolis there are fewer tombs in the alley of hybrid theories than in the alley of monodisciplinary theories.

Specialized domains need theoretical orientations, but the discipline of political science as a whole cannot have a universal and monopolistic theory. Methods have a much longer life expectancy and some are even perpetual acquisitions across the boundaries of formal disciplines.

## C Borrowing methods

Distinctions should be made between scientific reasoning—in the tradition of J. S. Mill, Durkheim, Claude Bernard or Hubert Blalock—strategy of investigation, method of research and technological ability. All four are cross-disciplinary. I shall concentrate on the borrowing of methods by political scientists, who rarely import directly from logic, mathematics or statistics. Usually they find an intermediary in certain sectors of psychology, economics or sociology, all of which have played a crucial role in the methodological enrichment of political science. Tabular demonstration, graphic presentation, summation, measures of variability, ratios, rates, sampling distribution, statistical inference, ecological fallacy, binominal distribution, multiple regression, linear correlation, contingency, factor analysis and so on—none of these methods has been imagined by political scientists. All have been imported, and some, after improvement, have been exported in refined forms.

The borrowing of methods has not diminished since Oliver Benson admitted, in 1963, that "most mathematical literature relevant to political science is by outsiders, by those who could not identify themselves as primarily students of political phenomena" (Benson 1963: 30). Borrowing methods is easy. Once the difficult process of invention and initial elaboration is completed, a method can be used by anyone, with or without imagination.

A substantial number of political scientists are familiar with scaling methods elaborated by psychologists, path-analysis imported from biology via economics, the multivariate reasoning and measuring of the sociologist Paul Lazarsfeld, the linear structural relation forged by the statistician Jöreskog. With the rich methodology of the *American Soldier* edited by Samuel Stouffer many representatives of other disciplines have collaborated.

Up to a certain point, the introduction of mathematics into political science has been valuable not only for its own contributions, but as an entrée for additional borrowing. Adoption of these mathematical methods and models has paid several dividends: the rigor necessary for modelling, for example, has also been invaluable in developing compelling and logical arguments, even for work which forgoes mathematical presentation.

Because there is no need to obtain a license in order to adopt a method or a research technique, the importation has sometimes been indiscriminate. What is needed is good sense in applying the method to a new field. Too many political scientists are still confusing scientific reasoning, research strategy and technological tools. Today the main source of disputes among political scientists is not, as many people believe, ideology, but methodology, most of it exogenous to political science. Debates between ideologues are possible, even if often sterile; but between methodological schools, they are inconclusive.

The borrowing of statistical methods and techniques is not always beneficial. Many political scientists who use quantitative methods extend the borders of political knowledge. However, others are motivated mainly by an interest in technique, rather than substance. They routinely build unverifiable models, over-quantify, and over-model. They often choose to discuss minor issues, spending much talent and energy to improve a correlation coefficient, to split a hair into four by factor analysis. They are productive scholars—any input into the computer will result in an output, mechanically. Few of their papers see the light of day in major mainstream journals, because most are characterized by a painful contrast between highly sophisticated analytical techniques and poor imagination in research design, or data too weak to support the powerful techniques utilized (Dogan 1994).

Interdisciplinary free trade methodology needs to be guided by scientific strategy and not by mechanical facilities, particularly in some great universities where many graduate students in political science complain that they are "oppressed" by a heavy program of imported statistical techniques, to the detriment of scientific reasoning.

## III Hybrid domains

If each of the twelve principal social sciences were crossed with all the others, we would in theory obtain a grid with 144 squares. Some squares would remain empty, but more than three-quarters of them would be filled by hybridized specialities enjoying some autonomy (Dogan and Pahre 1990). These hybrid specialities branch out in turn, and give rise, at the second generation, to an even larger number of hybrids. A full inventory of all the existing combinations cannot be obtained by crossing the disciplines two by two at the level of the second generation, since some hybrid fields among the most dynamic ones are of multiple origin. In addition, hybrid fields such as prehistory which are partly rooted in the natural sciences would not appear in the 144-square grid, confined as it is to recombinations of segments of the social sciences. The configuration of hybrid fields is changing constantly. Political psychology, political sociology and political economy have long been recognized, whereas political anthropology is not yet autonomous.

### A Political psychology

Between psychology and political science there is a hybrid domain flying its own flag: political psychology. This is a third generation hybrid, because psychology itself was born as a hybrid discipline, rooted partly in the natural sciences and partly in the social sciences. Political psychology has two sisters: an older one, social psychology, formally recognized in all major universities of the world; and a younger one, cognitive science, today the best endowed of the young sciences on both sides of the Atlantic. Political psychology rarely meets cognitive science, but is in permanent contact with social psychology.

In a recent survey D. O. Sears and C. L. Funk (1991: 346) write that political psychology, being "an interdisciplinary endeavor, runs the danger of falling between the cracks in academic institutions" because of pressures for "disciplinary orthodoxy induced by bureaucratic inertia." But the inventory they make by showing how political psychology penetrated into political science departments does not justify this fear. The journal *Political Psychology* is a good window onto this hybrid field.

In its territory we find the provinces of political socialization, role theory, alienation, psycho-biography, personality analysis, political attitudes and beliefs, small groups, typological analysis of political leaders, national character, mass participation, generations, political dissatisfaction, and a rich methodological area (attitude measurement, sociometric measurement, content analysis, clinical method, quasi-experimental approach, and particularly, survey research).

Very few hybrid domains celebrate a founding father. But American political psychology has one: Harold Lasswell. His progeny include Fred I. Greenstein, Robert Lane, Herbert Hyman, Erik Erikson, Sidney Verba and James C. Davies, among many others.

In Western Europe the hybrid field of political psychology is institutionalized in few universities, but the literature related to the field is rich and of great variety as is illustrated in France for instance by the work of Philippe Braud, and in Germany by the contributions of Erwin K. Scheuch to the methodology of sample surveys and the problems of comparability in politics and social psychology. Scheuch has the merit of having discovered the "individualistic fallacy" (Scheuch 1966; 1969). Among the books belonging to the field of political psychology, *Political Action*, edited by Samuel Barnes and Max Kaase, should be singled out. Their typology of protesters, activists, reformists, conformists and inactives is pertinent for many countries.

## B Political geography

Geography—a master discipline in the past—today has no core. It is divided into many subfields: biogeography, social geography, urban, historical, economic, political geography. There are multiple encounters between political science and geography: geopolitics, electoral geography, urban politics, territorial bases of federalism, spatial organization of society, core–periphery, city–hinterland, environmental problems, urban–rural differences, territorial aspects of social mobilization, etc. Demography is an intervening dimension in political geography.

From the "Geographical Pivot of History" by H. J. Mackinder in 1904 to Stein Rokkan's "conceptual map of Europe" (see the special issue dedicated to his conceptions by *Revue Internationale de Politique Comparée* in 1994), many essays have been published in the field of political geography, and not only in Europe. F. J. Turner's *The Significance of the Frontier in American History* is dealing with geography as much as with history.

In Kasperson and Minghi's collection, *The Structure of Political Geography* (1969), many chapters are of interest even for political scientists who are not oriented toward geography (Ratzel's laws of the spatial growth of states, geopolitical regions, transaction-flow analysis, heartland and rimland, the impact of negro migration, and so on). The concept of center–periphery obviously has a geographical dimension (Rokkan, Urwin *et al.* 1987)

Political science and geography also meet in the domain of electoral geography, particularly for the analysis of aggregate data in countries characterized by a great territorial diversity, and for which information is available at the level of small administrative units. The privileged countries from this point of view are, or were until recently: France, Italy, Spain, Portugal, Belgium, Norway, Finland, Austria, Canada. André Siegfried (1913) did research on the North-West of France, V. O. Key on *Southern Politics* (1949), Rudolf Herberle (1963) on Schleswig-Holstein during the Weimar Republic, Erik Allardt (1964) on Finland, Mattei Dogan (1968) on Italy, Stein Rokkan and H. Valen (1964) on regional contrasts in Norwegian politics, Juan Linz and Amando de Miguel (1966) on the "Eight Spains," R. E. De Smet and R. Evalenko (1956) and Frognier *et al.* (1974) on Belgium. This geographical approach has, however, been challenged in an analysis by deciles, where the territory disappears in favor of a sociological reordering of the territorial units and of the variables (Dogan and Derivry 1988). This hybrid field has a series of specialized journals which are interdisciplinary bridges: *Economic Geography*, *Urban Geography*, *International Journal of Urban and Regional Research* and, particularly, *Political Geography*.

Political scientists are still adopting the nation-state as a unit of analysis at a time when there are in the world more giant cities with over four million inhabitants than independent states which reach this level. The world is increasingly dominated by giant cities (Dogan and Kasarda 1988). Geographers and urbanists are in this domain at the front rank, proposing theoretical frameworks, concepts and methods of measurement. Urban studies are expanding; they may soon become an independent discipline. Today in almost all countries, advanced and developing, the number of specialists in "urbanology" is higher than the number of political scientists. "Urban politics" is a growing field.

## C Political sociology

Political science and sociology have a condominium: political sociology. This is an old hybrid, recognized as early as the 1950s, as was testified by Neil Smelser:

> In the newer branches of political science that have grouped loosely under the heading of behavioral approach, the methods of research are, except for

> relative emphasis, almost indistinguishable from the methods of sociology ... political scientists have employed a vast array of methods of data gathering, statistical manipulation and comparative methods that are also commonly used in sociology (Smelser 1967: 27).

The overlap is obvious.

Giovanni Sartori makes a distinction between political sociology and the sociology of politics. For him, the latter is a branch of sociology, like the sociology of religion. A dividing line could be traced by considering the emphasis on the dependent or on the independent variables. "The independent variables—causes, determinants or factors—of the sociologist are, basically, *social* structures, while the independent variables of the political scientist are, basically *political* structures" (Sartori 1969: 67). He concludes that "political sociology is an interdisciplinary hybrid attempting to combine social and political explanatory variables, i.e. the inputs suggested by the sociologist with inputs suggested by the political scientist" (Sartori 1969: 69).

Many of the best-known scholars in political science are leading sociologists. Quite a number of scholars have or have had a dual appointment in political science and in sociology, among them R. Aron, S. M. Lipset, R. Bendix, J. Linz, G. Sartori, M. Kaase, J. D. Stephens, Mildred A. Schwartz, Ch. Ragin, and M. Dogan. Today political economy tends to weaken the privileged relations between sociology and political science.

## D How political science conquered the territories of economics

Some economists advocate an "imperialistic expansion of economics into the traditional domains of sociology, political science, anthropology, law and social biology" (Hirschleifer 1985: 53). Several of these imperialists are famous scholars, including a few Nobel laureates. A kind of manifesto has been published in the *American Economic Review*, which is worth quoting:

> It is ultimately impossible to carve off a distinct territory for economics, bordering upon but separated from other social disciplines. Economics interpenetrates them all, and is reciprocally penetrated by them. There is only one social science. What gives economics its imperialist invasive power is that our analytical categories—scarcity, cost, preferences, opportunities, etc.—are truly universal in applicability ... Thus economics really does constitute the universal grammar of social science. But there is a flip side to this. While scientific work in anthropology and sociology and political science and the like will become increasingly indistinguishable from economics, economists will reciprocally have to become aware of how constraining

has been their function. Ultimately, good economics will also have to be good anthropology and sociology and political science and psychology (Hirschleifer 1985: 53).

This view is anachronistic, and contrasts with the perception of economics as a shrinking discipline: "Economics as a formal discipline is suffering because its main achievements—conceptualization, theory, modelling and mathematization—have been accompanied by an excessive isolation from the other social sciences" (Beaud 1991: 157).

In reality, the recent history of the social sciences shows that enormous areas of scientific knowledge have been abandoned by the science of economics. These areas have been taken over by neighboring disciplines. At one particular moment, economics reached a fork in the path: it could have chosen intellectual expansion, the penetration of other disciplines, at the cost of diversification, and at the risk of dispersal (a risk taken by political science); instead it chose to remain unflinchingly pure, true to itself, thereby forfeiting vast territories. Yet many economists consider that the choice of purity, methodological rigor and hermetic terminology was the right choice.

It is thus clear that self-sufficiency, to use a word familiar to economists, leads sooner or later to a shrinking of borders. But this does not imply general impoverishment, since the lands abandoned by the economists were soon cultivated by others. Those abandoned lands now have their own flags: management, political economy, development science, the comparative study of Third World countries, economic and social history. The position of economics in the constellation of the social sciences might have been more enviable today had it not withdrawn into itself.

This situation is particularly surprising, in that few classical scholars—from Marx and Weber to Schumpeter, Polanyi, Parsons and Smelser (Martinelli and Smelser, 1990), not forgetting Pareto—have failed to assign a central place in their theories to the relationship between economy, society and politics. A whole army of famous American economists have given priority to the study of political phenomena, even if they have kept one foot in economics. Among them are Kenneth Arrow, Anthony Downs, Kenneth Boulding, Charles Lindblom, James Buchanan, Gordon Tullock, Albert Hirschman, John Harsanyi, Herbert Simon, Duncan Black, Jerome Rothenberg, Thomas Schelling, Richard Musgrave, Mancur Olson and others.

Some eclectic economists denounce the reductionism advocated by others, particularly with reference to research on development: development is reduced to economic development; this is reduced to growth; which in turn is reduced to investment, in other words to accumulation. It has taken several decades to dethrone per capita GDP as a composite indicator of development. Gunnar Myrdal railed against economists who were in favor of unidisciplinary models.

In many countries large numbers of economists have locked themselves up in an ivory tower, and as a result whole areas have escaped their scrutiny. Their contribution to the problem of the development of the Third World, for instance, is rather modest when compared with the work of political scientists and sociologists. This is particularly true in the United States, Latin America and India.

If a discipline has a tendency to turn in upon itself, if it does not open up enough, if its specialities do not hybridize, the neighboring territories do not remain barren. Many economists have had a somewhat condescending attitude towards political science. This has resulted in the development, alongside and in competition with economics, of a new corporate body, with an extremely active and large membership in the United States, the UK and Scandinavia: political economy was protected by only one of its parents and renamed through the revival of an old name from the French nomenclature of the sciences. Political economy is currently one of the main provinces of American political science—with a large output and renowned journals. It is one of the most popular sectors among Ph.D. students in political science. Political science is the greatest beneficiary of the monodisciplinary self-confinement of economics.

Thirty years ago F. A. Hayek wrote that "nobody can be a great economist who is only an economist—and I am even tempted to add that the economist who is only an economist is likely to become a nuisance if not a positive danger" (Hayek 1956: 463). It may now be too late for economics to reconquer the territories conquered by political science, sociology, economic history and particularly by political economy. Some economists are still hoping: "It is necessary to reduce the use of the clause *ceteris paribus*, to adopt an interdisciplinary approach, that is to say to open economics to multidimensionality" (Bartoli 1991: 490). Abandonment of reasoning by assumptions and by theorems would not be enough, because the reality has changed: "economic issues become politicized and political systems become increasingly preoccupied with economic affairs" (Frieden and Lake 1991: 5).

## E From political anthropology to hybrid area studies

In a few years towards the end of the 1950s and the beginning of the 1960s, about fifty colonies achieved national independence. At that time some three thousand American social scientists, among them many political scientists, were sent—with the financial help of American foundations—to Asia, Africa and Latin America in order to study the newly independent nation-states. They covered the planet with hundreds of books and articles. They have become area specialists. They have replaced the European scholars who returned home after the withdrawal of Britain, France, Belgium, the Netherlands and Portugal from their colonies.

This spontaneous generation of area specialists was born hybrid. The topics of their research blurred the disciplinary boundaries. They and their successors were confined to non-Western underdeveloped countries, to stateless societies, to what Joel S. Migdal calls "weak states and strong societies"—which is to say, to the privileged territory of an old discipline, anthropology, which had flourished in Western Europe around the turn of the century. The European anthropologists had discovered these "primitive" societies long before the American area specialists had done so.

There is a basic difference between the two. The European anthropologists were monodisciplinary scholars with a clear identity, vocabulary and theoretical framework. They were exporters of knowledge to the entire spectrum of social sciences. Some of them had imperialistic ambitions, proclaiming that anthropology was the master science. All other disciplines, including political science and sociology, were considered by these academic imperialists as provinces of anthropology.

But when the European empires which covered half of the planet started to disintegrate, these anthropologists lost their research fields. Anthropology shrank. The abandoned territories were delivered to specialists in area studies. In contrast to their predecessors, the new invaders did not come under a disciplinary flag. Few of them were trained in anthropology, and most of them were neither theoreticians nor methodologists. The most famous exceptions are David Apter, Leonard Binder, James Coleman, Lucian Pye, Fred Riggs, Dankwart Rustow, Richard Sklar and Myron Wiener.

David Easton was then eager to establish a new subfield: political anthropology. He published an essay under this title in 1959. Retrospectively it can be said that this was a sickly child, born at a moment when the new hegemonic power needed non-disciplinary specialists of these new countries—not experts in anthropology, a discipline which began to be colonized by other disciplines. It is significant that at the same moment Margaret Mead (1961: 475) was frightened of seeing her discipline "swallowed" and "isolated from the community of scientists and scholars." Good old anthropology fell from imperialism to being an "unsuitable scientific repository" (Mead 1961: 476).

Political anthropology does not flourish today, because it is too anthropological and insufficiently political, at a time when the poor countries are developing, except in Africa, and are experiencing an increasing internal diversification in facing the global economic world. The seminal essay by Lucian Pye in 1958 "The Non-Western Political Process" needs a serious updating by reducing the scale of dichotomies. The field of political anthropology seems to be the only hybrid field in decline.

Meanwhile, a French demographer-economist-sociologist Alfred Sauvy (1952; 1956) suggested calling these underpriviledged new countries "the Third World," by analogy with the Third Estate before the French Revolution. This label has survived even though the "second world" imploded in 1989. It is probable that sooner or later this label will be abandoned, because it includes an enormous variety of countries: old civilizations like China and artificial states in Africa; rich countries like Saudi Arabia and extremely poor ones. Which discipline will propose the new labels?

Area studies in the Third World give priority to topics which seem important to understanding a particular country. "They do not respect disciplinary boundaries" (Lambert 1991: 190). In the area studies, humanities are well represented. "Area specialists who are in the social sciences are likely to have a great deal more contact and shared intellectual activity with human sciences than do most of their non-area-oriented disciplinary colleagues"; it is at the conjunction of anthropology, history, literature and political science that "much of the genuinely interdisciplinary work in area studies occurs" (Lambert 1991: 192).

Describing the struggle between the conventional disciplines and area studies, which has affected the self-identity of scholars, Lucian W. Pye (1975: 3) writes: "The emergence of area specialization has changed perspectives and raised questions which go to the foundations of the social sciences." These foundations have been altered much more by the hybrid fields at the interstices of disciplines than by the transversal hybrid area studies.

## F Political development across natural and social sciences

The geographical distribution of various types of political regime is a striking phenomenon. But it has been absent from the literature during the last few decades, as a reaction against the exaggerations of the sociologist Ellsworth Huntington, who was severely and rightly criticized by the sociologist Pitirim Sorokin in 1928. This criticism dissuaded an entire generation of American sociologists and political scientists from taking into consideration environmental and climatic factors.

But many prominent economists did not remain silent. In 1955, W. Arthur Lewis noted in his *Theory of Economic Growth*: "It is important to identify the reasons why tropical countries have lagged during the last two hundred years in the process of modern economic growth" (Lewis 1955: 53). John Kenneth Galbraith wrote in 1951: "If one marks off a belt a couple of thousand miles in width encircling the earth at the equator one finds within it no developed countries . . . Everywhere the standard of living is low and the span of human life is short" (Galbraith 1951: 39–41). Charles Kindleberger (1965: 78) wrote some fifteen years later: "The fact remains that no tropical country in modern times has achieved a high state of economic development." Kenneth Boulding (1970: 409) goes one step further: "The principal failure of economics, certainly in the last generation, has been in the field of economic development [which] has been very much a temperate zone product."

These economists are cited by Andrew Kamarck, director of the Economic Development Institute of the World Bank, in his *The Tropics and Economic Development* (1976). There is no reference at all to politics in that book, but it nevertheless manages to challenge our perception of politics in tropical areas. Trypanosomiasis, carried by the tsetse fly, prevented much of Africa from progressing beyond subsistence level: "For centuries, by killing transport animals, it abetted the isolation of tropical Africa from the rest of the world and the isolation of the various African peoples from one another" (Kamarck 1976: 38). Twenty years ago an area of Africa greater than that of the United States was thereby denied to cattle (Kamarck 1976: 39). Agricultural production in humid tropics is limited by the condition of the soil, which has become laterite (Kamarck 1976: 25). Surveys in the 1960s by the World Health Organization and by the World Food Organization estimated that parasitic worms infected over one billion people throughout the tropics and sub-tropics. Hookworm disease, characterized by anemia, weakness and fever, infected 500 million in these areas (Kamarck 1976: 75).

These ecological factors are confirmed by a considerable amount of research in tropical areas during the last two decades by geologists, geographers, biologists, zoologists, botanists, agronomists, epidemiologists, parasitologists, climatologists, experts of the World Bank and several agencies of the United Nations, and also by hybrid political scientists well-versed in tropical agriculture, the exploitation of minerals, and sanitary conditions of these countries. The situation has improved during the last generation, according to dozens of reports prepared by international organizations.

Translating these economic and social conditions into political terms, it is worth asking questions such as these:

- Why are almost all pluralist industrial democracies in temperate zones?
- Why has India—which according to some theories "should not be democratic," and which is a relatively poor tropical country—nevertheless had, for a long period, a democratic regime?
- Is there any relationship between the fact that most of the 30 million square kilometers of continental Africa (excluding the Mediterranean rim) are in the tropics, and the facts that this continent is the poorest and is without a single truly pluralist democracy capable of surviving more than a few years?
- To what extent should ecological factors be included in the parameters of economic, social and political development?

Such questions can be asked not only by the old "school of development," but also by its successor, the new "school of transition." One team (G. O'Donnell, P. C. Schmitter and L. Whitehead) gave to their book the prudent title: *Transitions from Authoritarian Rule*, which does not indicate the final outcome. Another team (led by L. Diamond, J. J. Linz and S. M. Lipset) took a risk by suggesting, in the title of their book *Democracy in Developing Countries*, that democratic institutions are indeed taking root in these countries, which were previously considered by one of these co-editors as not responding to the "requisites of democracy."

Neither of these two teams make an explicit and functional distinction between the genuine pluralist democracy, Dahl's polyarchy, and the limited, partial, façade or embryonic kind of democracy. Processes of democratization, stages of modernization, liberalization, electoral games, respect for human rights are only steps toward the "western model." Today the word "democracy" without an adjective can be misleading. As anyone would admit there is a large variety of democratic regimes. Democracy comes by degrees, as is shown by the data collected by Raymond Gastil in his series *Freedom in the World*. Only by a clear distinction between types of democracy would it be possible to frame a tentative reply to the first question asked above: why until now have truly advanced democracies tended to flourish in temperate zones?

India as a democratic country is a clinical case, a scientific "anomaly" in the sense given to this word by Thomas Kuhn. Comparativists interested in this case should proceed as biologists do when they have the good fortune to discover an abnormality, they could follow the advice of Claude Bernard in *Introduction à la médecine expérimentale* (1865), which is still a pertinent book. They could start with one of the best indicators that we have in comparative politics: small agricultural ownership. The Indian peasant is poor, but he is a proprietor![2]

Concerning tropical Africa and other similar areas, natural sciences and demography should be brought into the picture when asking, as Samuel Huntington does: how many countries will become democratic? Dependency theory may be of some help for Latin America and Eastern Europe, but it is much less so for tropical Africa. The literature on the ecological parameters of the tropics can be contrasted with the literature about the transfer of flora and fauna from one temperate zone to another. For instance, Alfred Crosby's 1986 work on *Ecological Imperialism: The Biological Expansion of Europe 900–1900* casts new light on the building of American power. If the eminent comparativist Charles Darwin were still alive, he would criticize monodisciplinarity, in particular W. W. Rostow whose theory of "stages of growth" does not admit any physical or environmental constraints on growth.

## G Comparative politics as a hybrid domain

The process of hybridization appears not only in exchanges of concepts, theories and methods among disciplines and between subfields. It is also evident in exchanges of information, substance, indicators, statistical data and in the daily praxis of empirical research. This trade is excedentary for some disciplines and deficitary for others. Social geography borrows information from physical geography, which borrows in turn from geology, rather than the reverse. Political science has contracted an enormous foreign debt, because politics cannot be explained exclusively by politics. Political phenomena are never produced *in vitro*, artificially in the laboratory. They are always related to a variety of factors behind politics. Dozens of non-political variables are used to explain politics. This is one of the main reasons why political science is interwoven with the other social sciences.

The storage of information produced by other social sciences is particularly important in the domain of comparative politics, to such a degree that it could be said that a comparison across nations necessarily encompasses several disciplines. In the history of comparative politics there was a privileged time of co-operation and convergence during the 1960s. During the fifteen years between 1958 and 1972, three dozen important books and

[2] On land ownership, see the data collected by Tatu Vanhanen.

articles were published, which shared three characteristics: comparison by quantification, by hybridization and by cumulative knowledge. "Such a combination had never previously been achieved in the history of political science" (Dogan 1994: 39). This privileged moment also marks a break with European classical comparisons in the sociological style of Tocqueville, J. S. Mill, Marx, Spencer, Weber and Pareto.

At that particular moment sociology was no longer at the center of the constellation of social sciences. For the first time in the history of social sciences, it was political science. In the new constellation a number of stars are visible—it is unnecessary to name them. What should be emphasized is the process of cumulative knowledge, in which several dozen specialized scholars and experts have participated.

The alarm concerning the parochial state of comparative politics—after the subjugation of all social sciences during the period of totalitarianism in Europe (Scheuch 1991) and before their renaissance in the United States—was raised by Roy Macridis in 1955. Around the same time (1954) the Statistical Bureau of the United Nations started to publish "social statistics," none of which was political. They concerned demographic variables, income, standard of living, social mobility, sanitary conditions, nutrition, housing, education, work, criminality.

In 1957, *Reports on the World Social Situation* began to be published by the Department of Economic and Social Affairs of the United Nations. Chapters in these publications on "The Interrelations of Social and Economic Development and the Problem of Balance" (in the 1961 volume) or on "Social-economic Patterns" (in the 1963 volume) are contributions that can be read today with great interest, even if the political data so important for comparative politics are absent from these analyses.

Two years after that series began came Lipset's *Political Man* (1959), the most cited book by political scientists for two decades. In fact, however, this book borrows from all social sciences and very little from political science. One year later Karl Deutsch produced his "manifesto" (Deutsch 1960), followed by a seminal article a few months later (Deutsch 1961). Both articles deal with indicators which are not directly political. The following year an important article by Phillip Cutright (1963) was published which appears in retrospective to have been prophetic: it is the only article of that time to give priority to political variables. In the same year Arthur Banks and Robert Textor published their *A Cross-Polity Survey* (1963) in which the majority of the fifty-seven variables proposed and analyzed are not political. Shortly thereafter, the first *World Handbook of Political and Social Indicators* (Russett *et al.* 1965) discussed seventy-five variables, of which only twelve are strictly political and eight others economic-cum-political.

A year later G. Almond and G. Bingham Powell published a fundamental book, *Comparative Politics* (1966), in which several social sciences, particularly social anthropology, are seen in the background. From that moment on the field of international comparisons becomes bifurcated. One road continues with quantitative research, in which contributors constantly use non-political factors in their analysis of "correlates of democracy." An important recent input comes again from the Development Program of the United Nations, the *Human Development Report* (1990 *et seq.*). In this publication GNP per capita is dethroned and is replaced by a new indicator: Purchasing Power Parity (PPP).

The other road gave priority to sectoral comparisons, for instance the eight volumes on political development, published by Princeton University Press, where politics is for most of the time a dependent variable explained by non-political factors. There are several good reviews of the "political development" school (Almond 1990; Wiarda 1989). Today this school seems to have reached its limits, to be out of steam, to have exhausted the theme. It is a good example of crowded fields subjected, after a period of productivity, to diminishing marginal returns: "the higher the density of scholars in a given field, the less likely innovation is per capita" (Dogan and Pahre 1990: 36). This "paradox of density" designates creative marginality as the opposite of density of scholars.

In the recent period, the field of comparative politics has expanded in all directions, penetrating into the territories of other disciplines: transition to democracy, values and beliefs, crisis of confidence, public corruption, ungovernability, limits to growth and so on. (These new directions figure throughout many other chapters of the *New Handbook*.) Is the field of comparative politics becoming imperialistic?

As we can see, comparative politics does not consist only in cross-national analysis. It is also necessarily a cross-disciplinary endeavor, because in comparative research we are crossing units (nations) and variables (numerical or nominal). The variables are usually more numerous than the units. The relations between variables are often more important for theoretical explanations than the discovery of analogies and differences between nations.

In comparative politics there is no single major book that attempts to explain politics strictly by political variables—except in constitutional matters. But of course the dose of hybridization varies according to the subject and to the ability of the author to leave in the shadow what should be implicitly admitted. For instance, in their comparisons of political systems, scholars like Klaus von Beyme or Giovanni Sartori might not need to discuss at length social structure or cultural diversity. By contrast, Arend

Lijphart (in his comparison of consociational democracies) and Ronald Inglehart (in his analysis of beliefs and values) do have to stress the importance of social, religious, linguistic and historical variables. In these cases, Lijphart and Inglehart cross the disciplinary boundaries more than von Beyme and Sartori do.

Comparative politics across disciplines means first of all crossing history. The relation between comparative history and comparative politics merits a lengthy discussion. Here it is sufficient to admit that the two subfields do not co-operate along their common frontiers, but only at certain gates, and usually on the territory of other hybrid fields: historical sociology, social history, economic history, cultural history, asynchronic comparisons. Some of the most important books in comparative politics also belong to this "hyphened history," from Dumont's *Homo Hierarchicus*, Bairoch's *De Jericho à Mexico, villes et économie dans l'histoire*, Wittfogel's *Oriental Despotism*, Wallerstein's *Modern World System*, to Lipset's *The First New Nation* or Bendix's *Kings or People*. Ironically, these contributors to comparative politics and hyphened history are neither political scientists, nor historians; they are, administratively, sociologists.

## IV Conclusion

Different disciplines may proceed to examine the same phenomenon from different foci. This implies a division of territories between disciplines. On the contrary, hybridization implies an overlapping of segments of disciplines, a recombination of knowledge in new specialized fields. Innovation in the various sectors of political science depends largely on exchanges with other fields belonging to other disciplines. At the highest levels of the pyramid of political science, most researchers belong to a hybrid subdiscipline: political sociology, political economy, political psychology, political philosophy, political geography, public administration, area studies and so on. Alternatively, they may belong to a hybrid field or subfield: mass behavior (related to social psychology), élite recruitment (related to sociology and history), urban politics (related to social geography), welfare states (related to social economy and social history), values (related to philosophy, ethics, and social psychology), governmental capabilities (related to law and economics), poverty in tropical countries (related to agronomy, climatology and economic geography), development (related to all social sciences and to several natural sciences).

There is probably as much communication with outsiders as between internal subfields. For instance, a political psychologist who studies protest movements and alienation interacts only a little with the colleague who uses game theory to study the same topic. He may find intellectual common ground with the social historian who studies the phenomenon in previous times, or with the sociologist who studies the impact of unemployment or immigration on violence and delegitimation in some European countries. There is no communication between two political scientists analysing the crisis of the social security system, one by abstract modelling, the other in vernacular language. The first is in contact with modellers in economics, and the second cites scholars from other disciplines.

All major issues are crossing the formal borders of disciplines: breakdown of democracy, anarchy, war and peace, generational change, the freedom–equality nexus, individualism in advanced societies, fundamentalism in traditional societies, ruling class, public opinion. Most specialists are not located in the so-called core of the discipline. They are in the outer rings, in contact with specialists of other disciplines. They borrow and lend at the frontiers. They are hybrid scholars. The number of "general" political scientists is rapidly decreasing. Everyone tends to specialize in one or several domains. When two political scientists meet for the first time, the spontaneous question they ask each other is: "What is your field?" This is true also for other disciplines. At professional congresses, scholars meet according to specialities. Congresses that bring together crowds of people who have little in common consume a lot of energy which could be better invested in the organization of meetings by fields bringing together specialists from various disciplines.

Suppose it were possible to select, from all political scientists in the various countries, the five or six hundred scholars who are doing the most creative research, those who advance knowledge, the most renowned of them. Suppose further that we except from this upper-stratum of eminence the scholars who specialize in the study of constitutional matters and the governmental process of their own country, some of whom are famous in their own field. After making this double delimitation, we would discover that among this body of scholars, the majority are not "pure" political scientists. They are specialists of a research domain which is not exclusively political. Those who shut themselves within the traditional frontiers of political science are narrowing their perspective and reducing their opportunities to innovate—except in constitutional matters and the organization of the state apparatus.

At the other extreme are the enthusiastic imitators. In some domains borrowing is too much simple imitation and not enough imaginative adaptation. If it were possible to rank the various subfields and schools on a scale of eclecticism it would appear that sophisticated statistical analysis and economic heuristic assumptions are the two most imitative schools. I have commented already about the over-quantifiers. I refer to Neil J. Smelser, a specialist of economic sociology, for a solomonic judgment about economic modelling: "Anthony Downs's model of political behavior imitates economic theory by postulating a version of political rationality and building a theory of political process on this and other simplifying assumptions" (Smelser 1967: 26).

Political science lives in symbiosis with the other social sciences, and will continue to be a creative science only if it remains extrovert. In fact, this science has no choice, because it is genetically programmed to generate grandchildren who will talk different tongues and who will sit, as Almond says, "at distant tables." These tables are distant because they are placed at the interstices of disciplines in the enormous hinterland of political science.

## References

Those interested in pursuing these issues are referred particularly to these further readings: Dogan and Pahre's *Creative Marginality* (1990) and Mackenzie's *Politics and the Social Sciences* (1967); collections by Lipset on *Politics and the Social Sciences* (1969) and by Deutsch, Markovits and Platt on *Advances in the Social Sciences* (1986); the OECD report *Interdisciplinarity, Problems of Teaching and Research in Universities* (1972); and the "Social Sciences" section of UNESCO's *Main Trends of Research in the Social and Human Sciences* (1970).

ALLARDT, E. 1964. Patterns of class conflict and working class consciousness in Finnish Politics. Pp. 97–131 in *Cleavages, Ideologies and Party Systems*, ed. E. Allardt and Y. Littunen. Helsinki: Westermarck Society.

ALMOND, G. A. 1990. *A Discipline Divided: Schools and Sects in Political Science*. Newbury Park, Calif.: Sage.

—— and POWELL, G. B. 1966. *Comparative Politics: A Developmental Approach*. Boston: Little, Brown.

ANCKAR, D. 1987. Political science in the Nordic countries. *International Political Science Review*, 8: 73–84.

ANDREWS, W. G. 1982. *International Handbook of Political Science*. Westport, Conn.: Greenwood Press.

—— 1988. The impact of the political context on political science. Paper presented to the World Congress of the International Political Science Association, Paris.

ANDRAIN, C. A. 1980. *Politics and Economic Policy in Western Democracies*. Belmont, Calif.: Wadsworth.

*ANNALES*. 1989. Special issue: Histoire et Sciences Sociales. *Annales—Economie Sociétés Civilisations*, 44 (#6).

Apter, D. E. 1968. *Conceptual Approaches to the Study of Modernization.* Englewood Cliffs, N.J.: Prentice-Hall.

Bairoch, P. 1988. *De Jericho à Mexico, villes et économie dans l'histoire* (*Cities and Economic Development*), trans. C. Braider. Chicago: University of Chicago Press.

Banks, A. S., and Textor, R. B. 1963. *A Cross-Polity Survey.* Cambridge, Mass.: MIT Press.

Barnes, S. H.; Kaase, M.; *et al.* 1979. *Political Action, Mass Participation in Five Western Democracies.* Beverly Hill, Calif.: Sage Publications.

Bartoli, H. 1991. *L'Economie Unidimensionnelle.* Paris: Economica.

Beaud, M. 1991. Economie, théorie, histoire: essai de clarification. *Revue Economique*, 2: 155–72.

Bendix, R., ed., 1968. *State and Society: A Reader in Comparative Political Sociology.* Boston: Little, Brown.

—— 1978. *Kings or People.* Berkeley: University of California Press.

Benson, O. 1963. The mathematical approach to political science. Pp. 30–57 in *Mathematics and the Social Sciences*, ed. J. Charlesworth. New York: American Academy of Political and Social Science.

Bernard, C. 1865. *Introduction à la Médecine Expérimentale.* Various editions.

Beyme, K. von. 1979. *Die Parlamentarischen Regierungssysteme in Europe.* Munich: Piper.

—— 1985. *Political Parties in Western Democracies.* Aldershot: Gower.

Boulding, K. 1970. Is economics culture-bound? *American Economic Review (Papers and Proceedings)*, 60 (May): 406–11.

Cardoso, F. H., and Enzo Faletto, E. 1979. *Dependencia y desarrollo en America Latina* (*Dependency and Development in Latin America*), trans. M. M. Urquidi. Berkeley: University of California Press.

Clastres, P. 1974. *La Société contre l'Etat, Recherches d'anthropologie politique.* Paris: Editions Minuit.

CNRS. 1990. *Carrefour des Sciences: L'Interdisciplinarité.* Paris: CNRS.

Cutright, P. 1963. National political development: measurement and analysis. *American Sociological Review*, 28: 253–64.

Dahl, R. A., and Lindblom, C. E. 1953. *Politics, Economics and Welfare.* New York: Harper and Row.

Deutsch, K. W. 1960. Toward an inventory of basic trends and patterns in comparative and international politics. *American Political Science Review*, 54: 34–56.

—— 1961. Social mobilization and political development. *American Political Science Review*, 55: 493–514.

—— Markovits, A. S.; and Platt, J., eds. 1986. *Advances in the Social Sciences.* New York: University Press of America.

Diamant, A. 1990. If everybody innovates, will we all sit at separate tables? Paper presented to the World Congress of the International Sociological Association, Madrid.

Diamond, L.; Linz, J.; and Lipset, S. M., eds. 1988. *Democracy in Developing Countries.* 3 vols. Boulder, Colo.: Lynne Rienner.

Dogan, M. 1968. Un fenomeno di atassia politica. Pp. 465–88 in *Partiti Politici e Strutture Sociali in Italia*, ed. M. Dogan and O. M. Petracca. Milano: Edizioni Comunità.

—— 1994. Limits to quantification in comparative politics. Pp. 35–71 in *Comparing Nations*, ed. M. Dogan and A. Kazancigil. Oxford: Blackwell.

—— and Derivry, D. 1988. France in ten slices: an analysis of aggregate data. *Electoral Studies*, 7: 251–67.

—— and Kasarda, J. D., eds. 1988. *The Metropolis Era.* 2 vols. Newbury Park, Calif: Sage.

—— and Pahre, R. 1989*a*. Fragmentation and recombination of the social sciences. *Studies in Comparative International Development*, 24 (#2): 2–18.

—— —— 1989*b*. Hybrid fields in the social sciences. *International Social Science Journal*, 121: 457–70.

—— —— 1990. *Creative Marginality: Innovation at the Intersections of Social Sciences*. Boulder, Colo.: Westview Press.

—— —— 1993. *Las Nuevas Ciencias Sociales*. Mexico City: Grijalbo.

Dowse, R. E., and Hughes, J. A. 1975. *Political Sociology*. New York: Wiley.

Dumont, L. 1966. *Homo Hierarchicus: Le Système des Castes et ses Implications*. Paris: Gallimard.

Easton, D. 1959. Political anthropology. Pp. 211–62 in *Biennial Review of Anthropology*, ed. B. J. Siegel. Stanford, Calif.: Stanford University Press.

—— and Schelling, C. S., eds. 1991. *Divided Knowledge Across Disciplines, Across Cultures*. Newbury Park, Calif.: Sage.

—— Gunnell, J. G.; and Graziano, L., eds. 1991. *The Development of Political Science*. London: Routledge.

Finifter, A. W., ed. 1983. *Political Science: The State of the Discipline*. Washington, D.C.: American Political Science Association.

Frieden, J. A., and Lake, D. A. 1991. *International Political Economy*. New York: St. Martin.

Frognier, A. P., *et al*. 1974. *Vote, Clivages Socio-Politiques*. Louvain: Vander.

Galbraith, J. K. 1951. Conditions for economic change in underdeveloped countries. *Journal of Farm Economics*, 33 (November): 255–69.

Gastil, R. 1980–1989. *Freedom in the World*. New York: Freedom House.

Greenstein, F. I., and Lerner, M. 1971. *A Source Book for the Study of Personality and Politics*. Chicago: Markham.

—— and Polsby, N. W., eds. 1975. *Handbook of Political Science*. Reading, Mass.: Addison-Wesley.

Grolier, E. de. 1990. Des théories aux concepts et des faits aux mots. *Revue Internationale des Sciences Sociales*, 124: 269–79.

Hayek, F. A. 1956. The dilemma of specialization. Pp. 462–73 in *The State of the Social Sciences*, ed. L. White. Chicago: University of Chicago Press.

Heberle, R. 1963. *Landbevölkerung und Nationalsozialismus: Eine Soziologische Untersuchung der Politischen Willensbildung in Schleswig-Holstein*. Stuttgart: Deutsche Verlagsanstalt.

Hirschleifer, J. 1985. The expanding domain of economics. *American Economic Review (Papers and Proceedings)*, 75 (#6): 53–68.

Hoogerwerf, A. 1982. The Netherlands. In Andrews 1982: 227–45.

Huntington, E. 1924. *Civilization and Climate*. New Haven, Conn.: Yale University Press.

Huntington, S. 1984. Will more countries become democratic? *Political Science Quarterly*, 99: 193–218.

Inglehart, R. 1990. *Culture Shift in Advanced Industrial Society*. Princeton, N.J.: Princeton University Press.

Intriligator, M. D. 1991. Some reflections about the interactions between the behavioral sciences. *Structural Changes and Economic Dynamics*, 1: 1–9.

Johnston, R. J. 1994. Geography journals for political scientists. *Political Studies*, 42: 310–17.

Kamarck, A. M. 1976. *The Tropics and Economic Development*. Baltimore, Md.: Johns Hopkins University Press for the World Bank.

Kasperson, R. E., and Minghi, J. V., eds. 1969. *The Structure of Political Geography*. Chicago: Aldine.

Kavanagh, D. 1991. Why political science needs history. *Political Studies*, 39: 479–95.

Key, V. O. 1950. *Southern Politics*. New York: Knopf.

Kindleberger, C. P. 1965. *Economic Development.* 2nd edn. New York: McGraw-Hill.
Knutson, J. N., ed. 1973. *Handbook of Political Psychology.* San Francisco, Calif.: Jossey-Bass.
König, R. 1973. *Handbook der Empirischen Sozialforschung.* 3rd edn. Stuttgart: Enke.
Kuhn. T. S. 1957. *The Structure of Scientific Revolutions.* 1962 edn. Chicago: University of Chicago Press.
Kuhnle, S. 1982. Norway. In Andrews 1982: 256–74.
Lambert, R. D. 1991. Blurring the disciplinary boundaries: area studies in the United States. In Easton and Schelling 1991: 171–94.
Laponce, J. 1989. Political science: an import-export analysis of journals and footnotes. *Political Studies,* 28: 401–19.
Le Bon, G. 1912. *La Psychologie Politique.* Paris: Flammarion.
Leca, J. La science politique en France. *Revue Française de science politique,* 32/4: 653–18.
Lepsius, R. M., ed. 1981. *Soziologie in Deutschland und Osterreich.* Opladen: Westdeutscher Verlag.
Lerner, D. 1958. *The Passing of Traditional Society.* New York: Free Press.
Lewis, G. C. 1832. *Remarks on the Use and Abuse of Some Political Terms.* London: B. Fellowes.
Lewis, W. A. 1955. *The Theory of Economic Growth.* Homewood, Ill.: Irwin.
Linz, J. J., and Miguel, A. de. 1966. Within-nation differences and comparisons: the eight Spains. Pp. 267–320 in *Comparing Nations: The Use of Quantitative Data in Cross-National Research,* R. L. Merritt and S. Rokkan. New Haven, Conn.: Yale University Press.
Lipset, S. M. 1959. *Political Man.* New York: Doubleday.
—— ed. 1969. *Politics and the Social Sciences.* New York: Oxford University Press.
—— 1979. *The First New Nation: The United States in Historical and Comparative Perspective.* New York: Norton.
—— and Hofstadter, R. 1968. *Sociology and History: Methods.* New York: Basic.
Loewenberg, P. 1983. *Decoding the Past: The Psychohistorical Approach.* New York: Knopf.
Lynn, N. B. 1983. Self-portrait: profile of political scientists. In Finifter 1983: 114–15.
Mackenzie, W. J. M. 1967. *Politics and the Social Sciences.* Harmondsworth: Penguin.
Mackinder, H. J. 1904 The geographical pivot of history. *Geographical Journal,* 23: 421–37; reprinted in *The Scope and Methods of Geography, and The Geographical Pivot of History,* ed. E. W. Gilbert. London: Royal Geographical Society, 1951.
Macridis, R. C. 1955. *The Study of Comparative Government.* New York: Random House.
Martinelli, A., and Smelser, N. J. 1990. Economic sociology. *Current Sociology,* 38 (#2): 1–49.
Marvick, D., ed. 1977. *Harold Lasswell on Political Sociology.* Chicago: University of Chicago Press.
Mead, M. 1961. Anthropology among the sciences. *American Anthropologist,* 63: 475–82.
Merton, R. M. 1973. *The Sociology of Science.* Chicago: University of Chicago Press.
Migdal, J. S. 1983. Studying the politics of development and change: the state of the art. In Finifter 1983: 309–38.
Mitchell, W. C. 1969. The shape of political theory to come: from political sociology to political economy. In Lipset 1969: 101–36.
Mommsen, W. J., and Osterhammel, J. 1987. *Max Weber and His Contemporaries.* London: Allen and Unwin.
Narain, I., and Mathur, P. C. 1982. India. In Andrews 1982: 194–206.
O'Donnell, G.; Schmitter, P.; and Whitehead, L., eds. 1986. *Transitions from Authoritarian Rule.* 4 vols. Baltimore, Md.: Johns Hopkins University Press.

Olson, M. 1965. *The Logic of Collective Action*. Cambridge, Mass.: Harvard University Press.
Organisation for Economic Cooperation and Development (OECD). 1972. *Interdisciplinarity, Problems of Teaching and Research in Universities*. Paris: OECD.
Pye, L. W. 1958. The non-western political process. *Journal of Politics*, 20: 468–86.
—— 1975. The confrontation between discipline and area studies. Pp. 3–22 in *Political Sciences and Area Studies: Rivals and Partners*, ed. L. W. Pye. Bloomington: Indiana University Press.
Radnitzky, G., and Bernholz, P., eds. 1986. *Economic Imperialism: The Economic Approach Outside the Traditional Areas of Economics*. New York: Paragon House.
Randall, V. 1991. Feminism and political analysis. *Political Studies*, 39: 513–32.
Rokkan, S., *et al.* 1995. Special issue dedicated to Rokkan's geo-economic-political model. *Revue Internationale de Politique Comparée*, 2(#1): 5–170.
—— ed., 1979. *A Quarter Century of International Social Science*. New Delhi: Concept Co.
—— Urwin, D.; *et al.* 1987. *Center-Periphery Structures in Europe*. Frankfurt: Campus Verlag.
—— and Valen, H. 1967. Regional contrasts in Norwegian politics. Pp. 162–238 in *Cleavages, Ideologies and Party Systems*, ed. E. Allardt and Y. Littunen. Transactions of the Westermarck Society, vol. x. Helsinki: Academic Rockstore.
Rostow, W.W. 1963. *The Stages of Economic Growth*. Cambridge: Cambridge University Press.
Ruin, O. 1982. Sweden. In Andrews 1982: 299–319.
Russett, B. M., *et al.* 1965. *World Handbook of Political and Social Indicators*. New Haven, Conn.: Yale University Press.
Sartori, G. 1969. From the sociology of politics to political sociology. In Lipset 1969: 65–100.
—— ed. 1984. *Social Science Concepts: A Systematic Analysis*. London: Sage.
Sauvy, A. 1952. Trois mondes, une planète. *L'Observateur*, 14 August 1952.
—— ed. 1956. Preface. *Le Tiers Monde, sous-développement et développement*. Paris: Institut National d'Etudes Demographiques.
Scheuch, E. K. 1966. Cross-national comparison using aggregate data. Pp. 131–67 in *Comparing Nations*, ed. R. Merritt and S. Rokkan. New Haven, Conn.: Yale University Press.
—— 1969. Social context and individual behavior. Pp. 133–55 in *Quantitative Ecological Analysis in the Social Sciences*, ed. M. Dogan and S. Rokkan. Cambridge, Mass.: MIT Press.
—— 1988. Quantitative analysis of historical material as the basis for a new co-operation between history and sociology. *Historical Social Research*, 46: 25–30.
—— 1991. German sociology. Vol. iv, pp. 762–72 in *Encyclopedia of Sociology*, ed. E. F. and M. L. Borgatta. New York: Macmillan.
Sears, D. O., and Funk, C. L. 1991. Graduate education in political psychology. *Political Psychology*, 2: 345–62.
Siegfried, A. 1913. *Tableau Politique de la France de l'Ouest sous la III République*. Paris: Colin.
Sills, D. L., ed. 1968. *International Encyclopaedia of Social Sciences*. New York: Macmillan.
Simon, H. A. 1982. *Models of Bounded Rationality*. Cambridge, Mass.: MIT Press.
Smelser, N. J. 1967. Sociology and the other social sciences. Pp. 3–44 in *The Uses of Sociology*, ed. P. Lazarsfeld, W. H. Sewell, H. L. Wilensky. New York: Basic.
Smet (De), R. E., and Evalenko, R. 1956. *Les Elections Belges, Explication de la répartition géographique des suffrages*. Bruxelles: Institut de Sociologie Salvay.
Sorokin, P. A. 1928. *Contemporary Sociological Theories*. New York: Harper and Row.

STOUFFER, S. A. 1950. *American Soldier.* Princeton, N.J: Princeton University Press.

TÖNNIES, F. 1887. *Gemeinschaft und Gesellschaft.* Berlin: Curtius.

TRENT, J. E. 1979. Political science beyond political boundaries. In Rokkan 1979: 181–99.

TRUMAN, D. B. 1955. *The Governmental Process.* New York: Knopf.

TURNER, F. J. 1959. *The Significance of the Frontier in American History.* Gloucester, Mass.: P. Smith; originally published 1893.

UNITED NATIONS. Department of International Economic and Social Affairs. 1957 *et seq. Reports on the World Social Situation.* New York: United Nations.

—— United Nations Development Program (UNDP). 1990 *et seq. Human Development Report* New York: Oxford University Press for the UNDP.

—— UNESCO., 1970. Part 1: Social Sciences. *Main Trends of Research in the Social and Human Sciences.* Paris: Mouton.

VALADE, B. 1990. *Pareto: La naissance d'une autre sociologie.* Paris: P.U.F.

VANHANEN, T. 1984. *The Emergence of Democracy: A Comparative Study of 119 States, 1850–1979.* Helsinki: Finnish Society of Sciences and Letters.

WALLERSTEIN, I. 1974. *Modern World System.* New York: Academic Press.

WEINGART, P. 1986. T. S. Kuhn: revolutionnary or agent provocateur. In Deutsch *et al.* 1986: 265–85.

WIARDA, H. J. 1989. Rethinking political development: a look backward over thirty years and a look ahead. *Studies in Comparative International Development,* 24 (#4): 65–82.

WIESE, L. VON. 1926. *Soziologie, Geschichte und Hauptprobleme.* Berlin: De Gruyter.

WIRTH MARVICK, E., ed. 1977. *Psychopolitical Analysis: Selected Writing of Nathan Leites.* London: Sage.

WITTFOGEL, K. A. 1957. *Oriental Despotism.* New Haven, Conn.: Yale University Press.

Part II

# Political Institutions

Chapter 4

# Political Institutions: An Overview

Bo Rothstein

## I Introduction

AMONG political scientists the basic understandings of the origins of any kind of formalized political power come in two different variants. The first, we can call it the "good" or "democratic" or "community-based" type, embodies the following story. A group of people share some common characteristics. They live in the same area, for example, or they work at the same place, or they are dependent on the same type of natural resources. In their daily lives, they soon discover that they have not only individual interests but also a number of common interests. As a geographical community, they realize a common need for laws regulating conflicts about property and other types of individual rights, and for the effective enforcement of such laws. Or they discover a need for an organization to pursue their common interests of better wages and working conditions, or a need to regulate use of natural resources to avoid "the tragedy of the commons" (Hardin 1982; Ostrom 1990). So, they get together as equals and form an organization to solve their collective interests, which is to say, they form a government. Or, per the other two examples, they establish a union or an economic co-operative of some kind—which from this perspective are to be seen as just different sorts of governments.

In all three cases, the members of the community soon discover that, in order to pursue their common interests, they need four basic types of political institutions. One type of institution is needed for making collectively binding decisions about how to regulate the common interests (a rule-making institution). A second type of institution is needed for implementing these decisions (rule-applying institutions). A third sort of institution is needed for taking care of individual disputes about how to interpret the general rules laid down by the first institution in particular cases (rule-adjudicating institutions). Lastly, a fourth type of institution is needed to take care of and punish rule-breakers, whether outsiders or insiders (rule-enforcing institutions).

Thus, in this "good" story, what the members of the community do to pursue their collective interest is to create four basic types of political institutions. Each institution contains a number of sub-institutions, with rules specifying the process of decision-making. These rules dictate how to appoint members of the decision-making assembly, how voting procedures should be organized, who should be appointed as judges, civil servants, policy-makers, army officers.

The other story—we can call it the "bad" or "dictatorial" or "adversarial" story—is that the specific geographical area (or the workplace, or the natural resources) is historically controlled by supreme force by an individual ruler or a united ruling group. They—the clan, the political élite, the ruling class, the managers, the feudal lord or whoever—want to extract resources from the ruled group as efficiently as possible (North and Thomas 1973; Levi 1988; Gambetta 1993). In this story, ruler(s), in order to enforce his (their) will with maximum efficiency, also need four basic types of institutions. One is for creating legitimacy from his subordinates; one for implementing the ruler'(s) will; one to take care of disputes among subjects and subordinates; and one to take action against those who contest the power of the ruler(s). In short, what ruler(s) need to do is to create the same four basic types of institutions as discussed above.

We need not go into the question of which story is more correct than the other. What type of story different political scientists want to tell is determined by basic ideological worldviews and differences in the understanding of human nature (Mansbridge 1990). In any case, this is an empirical question; and, considering the known historical and geographical variation, the answer surely has to be that both are valid. There are also several cases where a "good" story has turned into a "bad" one, and vice versa.

The crucial point is that, in both stories, the political institutions are basically the same. Both in the bad and the good story about the origins of government, we get the same four basic types of political institutions. Both the democratic and the undemocratic union needs an institution for making and implementing policy, and for deciding what to do with and how to handle rule-breakers. That goes for the feudal mansion as much as for the farming economic co-operative.

The point is that whichever story political scientists want to tell, it will be a story about institutions. A central puzzle in political science is that what we see in the real world is an enormous variation, over time and place, in the specifics of these institutions. We do not only get democratic and undemocratic political institutions. In both cases there exists a great variation in the specific form of institutions. This is readily seen from the fact that, although Italian fascism and German Nazism shared many ideological features, they were remarkably different in their institutional set-up (Payne 1990). In respect of present western industrialized democracies, I hardly need remind the reader of the differences that exist. Despite the fact that they share a common basic structure, economically and politically, there is indeed great variation in political institutions. If we take ten of the most basic aspects of the political system and make very a crude distinction between existing opposite institutional forms (ignoring all the possible variation between these two forms), we get $2^{10} = 1{,}024$ possible ways of configuring the political institutions of a modern western capitalist democracy (see Table 4.1 below, cf. Schmitter and Karl 1991; Lijphart 1984). Since even on a generous count there are only about thirty cases of such democracies, the possible institutional variation as shown in this example is much greater than the number of empirical cases.

**Table 4.1** Institutional variation among western capitalist democracies

| | |
|---|---|
| *Party-system* | Two-party vs. multi-party system |
| *Electoral system* | Proportional vs. majoritarian |
| *Legislative assembly* | Unicameral vs. bicameral |
| *Government structure* | Unitarian vs. federalist |
| *Central authority* | Parliamentarism vs. presidentialism |
| *Court system* | Judicial review vs. judicial preview |
| *Local government* | Weak vs. strong autonomy |
| *Civil service* | Spoils recruitment vs. merit recruitment |
| *Armed forces* | Professional vs. conscription |
| *State–economy relation* | Liberal vs. corporatist |

Explicating why institutional differences such as these occur and what difference they make is, more or less, how modern political scientists make a living. More specifically, political scientists ask three different but interrelated questions about political institutions. One is normative: Which institutions are best suited for creating "good" government and societal relations. The other two are empirical: What explains the enormous variation in institutional arrangements? And what difference those differences make for political behavior, political power and the outcome of the political process?

One of the problems with political science as a discipline is the way subfields are organized. The standard list (political behavior, comparative politics, policy and administration, theory, and so on) has one great disadvantage, namely, that what counts as comparative politics is decided by geography, not by the focus of the subject. In the United States, "British politics"—be it behavior, policy or adminstration—counts as "comparative politics" (and conversely), although the design of such studies are not really comparative in their methodological orientation. Organizing our thinking in terms of Political Institutions as a subfield, rather than institutions of the particular geographical area, is theoretically a more logical way of partitioning the discipline. Studies of domestic politics and institutions of specific countries may in this way be set along side one another, so that more common theoretical questions about the role of differences in institutional arrangements can be asked. By inviting us to theorize Political Institutions as a distinct area of study, this heading helps us to render a more coherent and systematic account of the functions of these institutions as such.

This overview will be divided into five parts. Section II contains short remarks about the role of institutions in normative political theory, classical as well as modern. The third section discusses possible reasons behind the increasing interest in institutions in positive political science, contrasting two different views—the rational-economic versus the structural-historical. Section IV takes up a number problematic issues in the contemporary analysis of political institutions. Lastly, Section V speculates about the role of institutional analysis in the future of political science.

## II Institutions in political theory

Classical political theory does not just consist of discussions of political ideals and individual duties. For Plato and Aristotle—as well as for Machiavelli, Locke, Rousseau, Hobbes and others in that tradition—one of the major problems was to ascertain which political institutions produce the best type of society and individual. Plato's *Republic* is largely a comparison of different forms of government—timocracy, oligarchy, democracy and tyranny. In the *Laws*, the 'Stranger from Athens' (Plato) compares the constitutions of Sparta and Crete, arguing that the state is needed not only for defense against foreign enemies but also to secure internal peace and to promote civic virtue: a first-rate law-maker protects society from bad government (mistakes, bad judgment, corruption on the part of the rules), which can only be done through certain institutional arrangements. Another example, also from the *Dialogue*, is when the "Stranger from Athens" uses a typology based on Sparta's constitution to discuss what type of laws the men from Knossos should enact in their new colony on the island of Crete (called Magnesia) to produce the best form of government.

Aristotle's *Politics* continues the discussion about political institutions and the "good" society. In Book IV he launches a comprehensive program for the study of political institutions, asking: What is the ideal form of government? What kind of constitution is best under various non-ideal circumstances? What characterizes the constitutions which are not good (not ideal) under various non-ideal circumstances? And what type of constitution is suitable in most cases (presumably, the non-ideal cases)? Aristotle also points to the need for different institutions in government, the deliberative, the executive and the judicial authority. Aristotle's empirically based discussion of the possible variation in the specifics of these institutions shows that the old Hellas served somewhat like a laboratory for the study of the effects of variation in political institutions (Malnes and Midgaard 1993; Sabine 1961).

And so on down through the ages. Indeed, when political science started to emerge as a modern academic discipline in the late 19th and early 20th centuries, the classical tradition had a great impact (Almond above: chap. 2). Both in Europe and the U.S., students were to a great extent concerned with what may be called "constitutional architecture." A central question was what kinds of constitutions are to be preferred. This often resulted in detailed configurative studies of the specific origins and operations of different national constitutions and other political institutions. This was, at least in Northern Europe, a consequence of the close connection between political science and constitutional law, especially the German tradition of *Allgemeine Staatslehre* (Andrén 1928).

Institutions have come to enjoy a similarly central place in contemporary political theory. Rawls's 1971 *Theory of Justice* initiated a wave of renewed interest in this area. Rawls's theory broke with the utilitarianism that dominated political philosophy, demonstrating the significance of a markedly normative discourse in which concepts such as justice, equality, and rights again took center-stage Addressing some of the criticisms made of his original theory, in a 1985 article Rawls made the (partly self-critical) comment that the basic principles on which he had built his theory should be regarded not as metaphysical postulates but rather as being based "solely upon basic intuitive ideas that are embedded in the *political institutions* of a constitutional democratic regime and the public traditions of their interpretations" (Rawls 1985: 226, emphasis added). In his latest book, Rawls (1993) distinguishes between two different logics of action—"the rational" and "the reasonable"—the essential difference being that "the reasonable is public in a way that the rational is not" (Rawls 1993: 51). Instead of an anonymous, strategic, game-like situation, Rawls's "original position" is intended, now, to be a discursive political institution, in which representatives of different currents of opinion meet in order to try—through discussion, deliberation and negotiation—to find the common principles of social and political order that should prevail in society (Rawls 1993: 135 ff.; cf. Soltan 1987; Barry 1995). The Rawlsian goal is a society in which an overarching consensus prevails on the principles of justice to which the political institutions should conform—a consensus which is in no way undermined by the fact that different groups and individuals hold fundamentally divergent views as regards cultural, religious and ideological questions (Rawls 1993: 131–72).

According to Rawls, then, it is just political institutions that can generate a just society, not a just society that generates just institutions. As should be obvious, this accords with the view of institutions held in classical political theory: institutions are not only "the rules of the game." They also affect what values are established in a society, that is, what we regard as justice, collective identity, belonging, trust, and solidarity (March and Olsen 1989: 126; cf. Dworkin 1977: 160 ff.). Jon Elster (1987: 231), for instance, has argued that "one task of politics is surely to shape social conditions and institutions so that people behave honestly, because they believe that the basic structure of their society is just." If social norms (about justice, for example) vary with the character of political institutions, then we can at least to some extent decide which norms shall prevail in the society in which we live, because we can, at least sometimes, choose how to design our political institutions.

In *The Tragedy of Political Science*, David Ricci points out that classical political science was dominated by normatively charged concepts (justice, nation, rights, patriotism, society, virtue, tyranny and so on). The tragedy of political science today, according to Ricci, lies in the fact that such concepts have largerly disappeared—replaced, in the the brave new world of mass data and policy analysis, by bloodless technical concepts like "attitude," "cognition," "socialization" and "system." Ricci argues that this shift in interest—from critical, normatively charged questions about the foundations of politics and democracy to the empirically manageable and the politically useful—helps to explain the crisis in which, he claims, modern political science finds itself (Ricci 1984: 296; cf. Held 1987: 273). The renewed focus on political institutions, in both positive and normative theory, may serve to bridge the gap indicated by Ricci. As is clear from Rawls's theory, political institutions are both empirical and normative entities, which cannot be understood fully unless their normative foundations are laid bare.

## III Forgetting and rediscovering political institutions

If the construction as well as the operation of political institutions were central in the classical traditions in political theory, they largely evaporated during the post-war era of behavioral and group-theoretic political science of various sorts. In theories such as structural-functionalism, system analysis, group theory (whether pluralist or élitist)—and, later, economic approaches such as neo-marxism—formal political institutions played little or no role. The reductionist tendency to reduce the explanation of political processes to social, economic or cultural variables meant that the institutional and organizational forms of political life made little or no difference (March and Olsen 1984; 1989). In structural-functionalism, the systemic needs of the social system (for political stability or social stratification, for example) tended to produce political institutions that were more-or-less automatically functional in relation to these needs. Every society tended to give rise to a set of institutions needed to solve "the perennial, basic problems of any society" (Eisenstadt 1968: 410; cf. Parsons 1964). In group theory in its various forms (including Marxism), institutions were largely seen as arenas where political battles between groups with pre-defined interests took place; but the specific construction of the arenas, as such, was not considered as an important variable for determining the outcomes of such battles (March and Olsen 1984; 1989; Steinmo and Thelen 1992). A typical formulation of this reductionist understanding of institutions can be found in Nelson Polsby's contribution to the 1975 *Handbook of Political Science*: in his chapter about legislatures, the major question was "how a peculiar organizational form, the legislature, *embeds* itself in a variety of environmental settings" (Polsby 1975: 257, emphasis added).

In traditional Marxist political analysis, political institutions *ex hypothesi* could have no independent causal function, because they were seen as mirroring the basic economic structure of society (Poulantzas 1968;

Wright 1978; cf. Offe and Preuss 1991). In other, more individualistic instrumental views of politics, rational actors would simply construct the type of institutions that would serve their pre-defined goals. Occasionally, institutions were seen as intermediate variables where political actors, if successful, could invest power which later could be used to enhance that power (Korpi 1983). In the early forms of economic approaches to political analysis (rational- or public-choice and game theory), too, analyses were carried out in an institution-free world, where autonomous agents pursued their pre-defined self-interests (Levi 1991: 132; Ostrom 1991; Moe 1990).

The general neglect of the analysis of political institutions during the behavioral period can be understood as a reaction against the lack of ambition towards positive theory in earlier political studies. The concentration either on formal, legal aspects of institutions, or on their individual historical trajectories, or on their "internal logic" seemed to block the development of any form of explanatory theory (cf. Wheeler 1975; King 1975). But it also seems reasonable to argue that in face of many major historical events (such as the Bolshevik revolution, the fall of the Weimar Republic, the rise of Nazism and the Communist victory in China) formal political institutions seemed to have played little or no role. Instead, scholars went after variables such as the "authoritarian personality," or the different distribution of economic power between social classes, or the persuasive power of certain political ideologies.

In many respects, this picture is an exaggeration. Many scholars continued to analyze political institutions during the era dominated by behavioralism and group theory. Scholars studying comparative historical paths to development, state-building and democracy, especially, continued to pay attention to political institutions (Duverger 1954; Bendix 1977; Eisenstadt 1965; Crozier 1964; Lipset and Rokkan 1967). A classic formulation from this period is Schattschneider's (1960): "Organization is the moblization of bias." In 1975, Ridley, for example, argued against behavioralism with the metaphor that the way a building is constructed is, to a great extent, determined by who will inhabit that building and what they might be expected to do in that building. Functionalism was criticized for its ambition to explain institutional variation by reference to general social needs, leading to a tendency to "explain the particular by the more general" and to neglect the historical fact that "in any given situation adequate institutional arrangements may fail to crystallize" (Eisenstadt 1968: 410, 414).

The increased focus on the importance of political institutions during the 1980s had many sources. The first thing to notice is that this enhanced interest in institutions took place at about the same time in parts of the discipline that are theoretically and methodologically very different. Secondly, this simultaneous reorientation happened independently in separate subfields (compare, e.g., Hall 1986 and Shepsle 1986). Neo-marxists were among the first to pay attention to the specific organizational forms of the state (Therborn 1978; Wright 1978). Organizational theorists discovered the importance of organizational culture (Meyer and Rowan 1977), scholars in comparative-historical analysis argued for the need to "bring the state back in" (Evans, Rueschemeyer and Skocpol 1985; Hall 1986), students of international relations put emphasis on "regimes" (Krasner 1983) and economically oriented scholars discovered the importance of "the rules of the game" (North 1990; Levi 1988).

Theoretically, this renewed interest had several sources. One is perhaps intra-disciplinary—the failure of grand theory such as behavioralism, structural-functionalism, Marxism and so on to come up with workable hypotheses. Either they were false or, more often, too general to take into account the variation between different political entities that was readily observed. But also in this case, it is reasonable to argue that events outside the discipline were important for the rediscovery of institutions. One such event may have been the collapse of the what can be called the meta-hypothesis of convergence in the social sciences. It seems to have been the case that, during the 1950s and 1960s, it was expected that most Third World countries would follow a similar path towards modernization, and that western capitalist democracies would converge over time (Stepan and Skach 1993). Early comparative studies of public policy seemed to confirm that (*a*) there were important signs of convergence and (*b*) politics played little or no role. Thus, despite important institutional variation between, for example, the social democratic countries of Scandinavia and the liberal United States, political behavior and political outcome seemed to be heading towards the same paths (Hofferbert and Cingranelli below: chap. 25).

As is well known, this view changed dramatically with the second and third generations of comparative public policy studies (not to mention studies of modernization). Instead of convergence along similar paths, persistent and even increasing differences between countries were discovered. At first, these differences seemed to be explicable by reference to differences in class structure and/or political mobilization (Esping-Andersen 1990). But first it was discovered that, even if such behavioral variables could account for parts of the variation, important differences were left unexplained (see Hofferbert and Cingranelli below: chap. 25). Secondly, it could be shown that differences in political mobilization and the organizational strength of social classes could, to a large extent, be explained by differences in political institutions. Formal political institutions determined political mobilization, thus they were not to be seen only as intermediate variables where already powerful agents invested power to enhance their future political strength (Hattam 1993; Rothstein 1990); instead, they were "social forces in their own right" (Grafstein 1992: 1). Thirdly, it could be shown that, while trying to explain important differences in public policy or interest group mobilization between countries with reference to class and group theory or behavioral categories gave paradoxical results that did not square with the theories, a focus on how formal political institutions historically structured the political process could do the job. To give just a few recent examples:

- Comparing Britain, Sweden and the US, Steinmo showed that constitutions influenced the distribution of tax burdens more than did the organizational strength of different social classes (Steinmo 1993).
- In a comparison of health policy in France, Sweden and Switzerland, the political institutionalization of possible veto points were shown to explain the way in which interests groups influenced policy more than the initial strength of these interest groups (Immergut 1993).
- The political position of the courts *vis-à-vis* the legislatures around the turn of the century has been shown to be an important factor behind the very different choices of organization and strategy taken by the British and American union movements (Hattam 1993).
- The administrative capabilities of the government have been shown to be crucial for explaining different paths of social and labor-market policy (Weir and Skocpol 1985; Weir 1992; King 1995).
- Comparing eighteen countries, it has been shown that the institutional structure of the governments' unemployment policy explains, to a significant degree, the variation in the strength of the national union movements, which in turn explains a large part of the variation in social insurance and labor-market policies (Rothstein 1992).

In this approach—labelled "historical institutionalism"—studies were usually directed at the way in which power over political process was structured by the specific character of the state, where historical traits played a central role (Birnbaum 1988). But political institutions, it was argued, not only distribute power and influence strategies. Contrary to what was assumed (in group theory and Marxism, for example) they also influenced how various groups came to define their political interests (Steinmo and Thelen 1992; Dowding and King 1995; for a critical view see Pontusson 1995).

In the economic approaches, such as rational choice and game theory, institutions were "rediscovered" by both endogenous and exogenous factors. Initially, this type of theory predicted that social systems consisting of only utility-maximizing rational individuals engaged in strategic interaction would stabilize in an equilibrium.[1] The problem, however, was that the theory left room for an abundant number of such equilibria and that it could not predict which equilibrium would be the one for any particular system (Bianco and Bates 1990: 137; Bicchieri 1993). Both complete co-operation as well as complete defection could be the result. The theory thus showed that "almost any outcome can occur" (Ferejohn 1991: 284). The job of predicting which type of outcome would be most likely was then given to existing institutions setting constraints on actors' choice of strategy, that is, the rules of the game (Shepsle 1986).

One of the exogenous factors behind the increased interest in institutions was that, according to rational choice theory, a legislative assembly such as the U.S. Congress would be a difficult forum in which to establish stable majorities (Arrow 1951). If every congressman voted according to his or her first-order preferences, this would lead to "cycling" where the majority for one bill would quickly be overturned by a new majority for another bill. The theory, then, predicted collective irrationality and political chaos; but empirical findings showed rather stable majorities and long-term rationality. Again, the existence of stable procedural rules was given as the explanation (Riker 1980; Shepsle 1989; see further Weingast below: ch. 5).

Another such disjuncture arose between what theory predicted and what empirical observation reveals, namely, how different collective action problems are resolved. Rational choice theorists were able to show that, given their assumptions about rational utility-maximizing behavior, it is incomprehensible that individuals ever choose to collaborate to solve common problems. Briefly put, the theory says there is no reason for rational, self-interested individuals to act co-operatively in the pursuit of common objectives,[2] for the goods produced by such co-operation are "public" in nature. That is, such goods benefit those who pull their weight and those who do not, alike (Olson 1965; Hardin 1982). Since people do in fact solve collective action problems and organize action groups (interest organizations, political parties, states and even supranational organs), the question posed by rational choice and game theory was indeed both fundamental and counterintuitive to the way political scientists used to think about

[1] Defined as a situation in which social forces balance because no actor or coalition of actors can improve their position by defecting from the specified pattern of action (Green and Shapiro 1994: 25).

[2] For example defending the country (Levi 1990), preserving scarce natural resources (Ostrom 1990), paying taxes (Laurin 1986) or establishing a union (Olson 1965; Bendor and Mookherjee 1987).

political organizations (cf. Bicchieri 1993).[3] The solution to the problem was attributed to the pre-existence of political institutions or leaders that could provide different forms of "selective incentives" to those and only those agents who contributed to the collective purpose (Olson 1965; North 1990; cf. Levi 1988).[4]

In many disciplines closely related to political science—such as economic history (North 1990), economics (Williamson 1985), the sociology of organizations (Powell and DiMaggio 1991), social anthropology (Douglas 1988) and industrial relations (Streeck 1992; Thelen 1991)—institutional analysis has also been "rediscovered" recently and come to play a central role in important studies. Many of them point to the fact that, to understand important variation in the lines of development between different social systems, it is not social or economic structural variables but political institutions that are the most important explanatory factors (see for example North 1990). One could therefore say that politics, understood as institutional engineering, has regained much of the centrality that it lost during the earlier era of behavioral and group theory. The renewed focus on political institutions has to some degree turned upside down behavioral, group-theoretical or Marxist modes of explanation stressing social and economic forces as explanations for political processes. On the other hand, it is far from certain that political science as a profession will come up with the most interesting and precise answers as to what role political institutions really play, how all the differences in institutional structure can be explained, and what causes change in political institutions.

## IV Issues in contemporary institutional analysis

If the institutions of political life have regained much of their central role in political analysis, this does not mean that the basic disagreements in the discipline have diminished. On the contrary, a number of important theoretical and conceptual problems have come up or been reformulated to fit the institutionalist agenda. In general the traditional controversies between, for example, holistic versus atomistic, structural versus individual, and formal versus historical forms of analysis prevail in the contempor-

[3] Or as it has been put by Fritz Scharpf: "maximin strategies are not used in most real-world encounters, and profitable games are, in fact, often played" (1990: 476).

[4] Later in this chapter, I will try to show why this solution is not a solution compatible with the premises of rational choice theory.

ary debates about the analysis of political institutions (Koelbe 1995; Hall and Taylor 1996). Below, there follow a few notes about some of the most debated issues.

## A What are political institutions?

There seems to be a general agreement that, at their core, political institutions are "the rules of the game." The question, however, is what should be included in the concept of rules. One such classic division is between "formal" and "informal" rules. Most people most of the time follow predefined rules of behavior, and most of these rules are not formalized as laws or other written regulations. Instead they are "routines," "customs," "compliance procedures," "habits," "decision styles" or even "social norms" and "culture" (March and Olsen 1989; Scharpf 1989; Hall 1986). On the other hand, *political* institutions in a narrower sense can be defined as "formal arrangements for aggregating individuals and regulating their behavior through the use of explicit rules and decision processes enforced by an actor or set of actors formally recognized as possessing such power" (Levi 1990: 405). Obviously, "culture," "norms," and so on are neither explicit nor formalized.

The obvious problem here is where to draw the line. Are we to understand political institutions as any kind of repetitive behavior that influences political processes or outcomes? Or should we reserve the term "political institutions" for formal rules that have been decided upon in a political process? Including "habits," "culture" and the like has the advantage of incorporating most of the things that guide individual behavior. The drawback is that "institution," as a concept, then risks becoming too diluted. If so, it risks the same fate as that of other popular concepts in the social sciences (such as "planning" and "rationality"): if it means everything, then it means nothing.[5]

If culture is nothing but informal institutional rules, as Douglass North (1990) claims, then there is no possibility left for distinguishing between the importance of formal political institutions and other social facts. There would, for example, be no point in analyzing the role of different constitutions in political outcomes, because such a broad conceptualization of political institutions would conflate formal constitutions and the overall culture of society. If every type of repetitive behavior can be explained by institutional rules broadly defined, than there is no chance of singling out

[5] As Etzioni (1988: 27) has put it: "Once a concept is defined so that it encompasses *all* the incidents that are members of a given category . . . it ceases to enhance one's ability to explain."

what role political institutions more narrowly defined play in social processes, because the answer is given already in the definition.[6]

On the other hand, if you limit the definition of institutional rules to the formal ones, you risk missing a lot of non-formalized but nonetheless "taken-for-granted" rules that exist in any political organization and which determine political behavior. The advantage, however, is that you may be able to give an answer to what "changes in formal political institutions" mean—that is, if "politics," in the narrow sense of the specific design of political institutions, "matters." Comparing political entities (states, regions, cities) with similar historical and cultural traits, but with differences in formal political institutions (such as constitutions), may give important results (Weaver and Rockman 1993; Stepan and Skach 1993; Elster and Slagstad 1988). This is of course a classical question for any analysis of the importance of *political* institutions.

One way out of this dilemma is to acknowledge a third type of rule, what in public administration has been called "standard operating procedures" (Hall 1986).[7] By this, scholars have tried to identify the rules actually agreed upon and followed by the agents involved. The advantage of a definition of this sort is obvious. While "culture" and "social norms" are excluded from the definition, rules that are *political* (in the sense that they have been established by either an explicit or a tacit agreement) are included, whether or not they have been written down and decided upon in a formal procedure. Any theory of political institutions must find a way to conceptualize these, alongside and in distinction to other institutions and general social facts (Levi 1990; Moe 1990; Steinmo and Thelen 1992).

## B What do political institutions do? Strategy, preferences and social capital

Most approaches would agree that institutions influence actors' strategies, that is, the way they try to reach their goals. This is obvious from the fact that institutions determine: (*a*) who are the legitimate actors; (*b*) the number of actors; (*c*) the ordering of action; and, to a large extent, (*d*) what information actors will have about each other's intentions (Steinmo and Thelen 1992: 7; Weingast below: chap. 5). This is a very important part of institutional analysis, because it has been shown that even small and seem-

[6] Another problem is that reductionism is brought in through the back door only relabeled as institutions (Levi 1990: 404).

[7] Other labels of this type of rule are "work rules" (Ostrom 1990), or "rules-in-use" as opposed to "rules-in-form" (Sproule-Jones 1993).

ingly unimportant changes in institutional rules affecting strategy (for example, who "moves" first) greatly influence the outcome of political processes (Ostrom 1995).

The problem is what institutions do with preferences. At the end of a continuum, two very different views exist. One is the view, common in the economic approaches, which holds that preferences can only be held by individuals (Riker's (1990: 171) theoretical individualism) and that they are exogenous to institutions. The actors come to the institutionalized "game" with a fixed set of preferences which, moreover, they are able rank in a rational manner.[8] Institutions determine the exchanges that then occur among actors, but the institutions as such do not influence preferences. As utility maximizers, actors rank their preferences and engage in a strategic "logic of exchange" with other agents within the constraints set by prevailing institutional rules. If the institutions change, actors usually change their strategy, but not their preferences. Note that in this "logic of exchange" approach, the calculative nature of action is universal, as the agents preferences are always to maximize expected individual utility. The problem is how to design institutions so that an effective *aggregation* of individuals' preferences to collective choice can be made (March and Olsen 1989: 119 ff.).

At the other end is the more cultural or sociological approach, which holds that institutions dictate a "logic of appropriateness," that is, they tell the actors what they ought to prefer in that specific situation in which they are placed (Douglas 1988; March and Olsen 1989; DiMaggio and Powell 1991). Without denying that individual action is purposive, this approach assumes that individuals cannot be considered to have the computational or cognitive ability necessary to be fully rational in their interaction with other agents (Simon 1955). Instead, they tend to follow "scripts" or "templates" given to them by the institutions in which they are acting. In a sense, in this cultural understanding of institutions, institutions not only determine actors' preferences but also to some extent create them. Institutions create or socially construct the actors' identities, belongings, definitions of reality and shared meanings. In a given institutional setting, the agent usually does not calculate what action would enhance his or her utility the most. Instead, by reference to the institutional setting, she asks "Who am I?" (a judge, a stockbroker, a nurse, a prisoner, a scientist) and what is the appropriate action for such an individual in this situation (to be impartial, or wealth-maximizing, or caring, or escape justice, or search for the truth). Note that in this "logic of appropriateness" approach, action is not

[8] Meaning basically just transitively: if they prefer A over B, and B over C, then they also prefer A over C.

universal but rather situational, as the individual's preferences vary in different institutional settings. The problem is how to construct institutions that *integrate* the individual with society (March and Olsen 1989: 124 ff.; cf. Jeppersen 1991; Scott 1991).

The advantage with the economic approach is that it provides us with clearly defined and universal micro-foundations of how individuals will act in different institutional settings (viz., they will maximize their expected utility). The problem is that, because it has no theory of where preferences come from (of what is "expected utility" for different actors), it must generally deduce preferences from behavior: the dependent variable is thus used to explain the independent, which in turn is re-used to explain the same dependent variable. This is not only a serious shortcoming from the viewpoint of sound scientific methodology (cf. Shapiro and Wendt 1992), it also creates problems because, at least in empirical research, it will be impossible to single out what type of behavior reflects genuine or true preferences and what type is the result of strategic action (cf. Weingast below: chap. 5). A rational utility-maximizing agent will not reveal true information about his or her preferences to someone whom the agents knows will use this information to contruct institutions that will work against his or her interests. When using this approach in designing real-world institutions to solve collective action problems, this problem of asymetries in information severely cripples the practical use of the approach, because with the wrong type of information about agents' genuine preferences the designer will create the wrong type of institutions, thereby creating pathological dilemmas of collective action (Miller 1992; Hurwicz 1977).

The problem with the "cultural" approach is that: if institutions determine preferences, then how can one explain why agents acting in the same type of institutions sometimes hold different preferences? While the economic approach may present an undersocialized view of how agents' establish preferences, the cultural view may be termed over-socialized (Granovetter 1985). If institutions determine preferences, the cultural approach has a long way to go in specifying what type of institutions give rise to what sort of preferences for what type of actor(s). In other words, this approach is in great need of clear and sound micro-foundations for its basic propositions.

Of particular interest here is Elinor Ostrom's analysis of common pool resources to which "everyone" has access. In such situations (for example, fresh water, grazing lands, fishing waters and so on), the following applies: if "everyone" sees only to his own self-interest and exploits the resource to the maximum, it will soon be exhausted. Some form of regulation preventing the over-exploitation of the resource must therefore be applied.

According to rational choice theory, however, such regulation will never arise, because the users cannot trust each other to restrict their individual consumption to what the common resource can bear (Bates 1988; Miller 1992). The two classical solutions have been either to grant the sovereign (for which read, "the state") control over the common resource and the prerogative to regulate its consumption in hierarchical fashion; or to divide the resource into shares over which individual control can be exercised, in a "market" solution (which in Ostrom's cases cannot be done, on account of "technical" characteristics of the resources in view).

What Ostrom has found is that people often solve problems of common resource management in an altogether different way from that which game theory predicts. Analyzing how the water-dependent farmers of southern California managed to regulate the limited water supplies, she writes:

> In each basin, a voluntary association was established to provide a forum for face-to-face discussions about joint problems and potential strategies . . . The provision of a forum for discussion transformed the structure of the situation from one in which decisions were made independently without knowing what others were doing to a situation in which individuals discussed their options with one another (Ostrom 1990: 138).

Neither centralized compulsion nor market strategies but, rather, a political institution solved the collective action problem. As Ostrom shows, it is the structure of the decision-making institution that plays the decisive role in changing individuals' view of wherein their self-interest lies (cf. Dowding and King 1995: 2). To use Rawls's terminology as presented above, the institution in which they have to act changes them from narrowly self-interested into reasonable actors. Their short-term self-interest furnishes them with a motivation to act unco-operatively. When such persons are placed in a situation in which they must argue on behalf of their actions and take responsibility for them, however, social norms become decisive. Co-operative solutions of various types then result. This does not mean the persons in question change from egoists into pure altruists; rather, they redefine their self-interest so that it accords with the collective interest in not draining the common resource. The discursive and public character of the institution "launders" the individuals' preferences (Goodin 1986; Offe and Preuss 1991; Miller 1993).

This is an important empirical result because it contradicts one of the basic theorems in the economic approach to politics, namely, that "the representative . . . acts on the basis of the same over-all value scale when he participates in market activity and in political activity" (Buchanan and Tullock 1962: 20). On the other hand, it is in accordance with a general result from experimental studies, which is that if the institutional structure is such that a possibility to communicate exists, this greatly increases agents' co-operative and solidaristic behavior (Frohlick and Oppenheimer 1992; Dawes *et al.* 1977).

In Ostrom's analysis, the specific solutions to the problem of collective action vary across the cases, some being more successful than others. The successful solutions involve neither pure hierarchy nor pure market, but rather a mixture of voluntary organization and public direction, in which the role of the authorities (which is to say, the degree of compulsion) is often very limited. Co-operation thus arises, contrary to the predictions of rational choice and game theory, largely on a voluntary basis. And it is political institutions, again, that do the job.

Another recent study on this topic is Robert Putnam's *Making Democracy Work: Civic Traditions in Modern Italy* (1993). Putnam investigated why public institutions, such as the democratic system, function so differently in Italy's twenty different regions. With minor exceptions this is a North–South question: that is, democracy (and the economy) works a great deal better in the north than in the south. The book is based on a great range of data collected over no less than twenty years.

What, then, is it that enables democracy to function in some areas of Italy but not others? Putnam's study yields the surprising result that it is the density and weight of the local organizational network that is decisive for establishing and securing efficient political institutions. The more people have been organized in such bodies as choirs, bird-watching clubs, sports associations and so on, the better democracy works. This is an answer in line with Tocqueville's classical analysis of the young American republic—that a functioning democracy requires a developed civic spirit. Citizens must, when deciding on common affairs (that is, when engaging in politics), be prevailed upon to see not just to their own short-term interest, but also to that of the whole—to the common good (Offe and Preuss 1991: 169). This capacity, according to Putnam, is something people can develop by taking part in voluntary associations. Putnam's feat involves demonstrating that this factor is more significant than traditional socio-economic variables for explaining democracy's manner of operation. His analysis goes a step beyond this, however. He claims that this civic spirit, as expressed in a dense organizational network in the civil society, actually explains why certain regions have enjoyed higher economic growth than others. The differential development of civic spirit in the various regions better accounts for their present economic standing than does their original economic position. It is not economic growth that produces civic spirit, but rather civic spirit that produces economic growth (and functioning democratic institutions). Putnam's study thus stands in a sharp contrast to Mancur Olson's *The Rise and Decline of Nations* (1982), where it was argued that countries with declining economies were characterized by having overly strong interests groups and associations seeking to use their position to get subsidies and thereby hindering economic competition.

According to Putnam, participation in organizational life creates *social capital*, which enables interaction between citizens to be built on *trust*. That is to say, people choose to co-operate with their neighbors because they trust that the latter will co-operate, too. In the various networks of associational life, a binding element arises in the form of norms facilitating co-operation. Expressed in economic terms, social capital reduces transaction costs in the economy, costs associated with ensuring that contracts are kept (Coleman 1990). Concretely, this is a matter of whether agreements can be confirmed with a handshake, or whether scores of lawyers and stacks of insurance policies are needed instead. It seems, then, that we have the solution here to the problem to which game theory has called our attention. For by taking part in a multitude of social networks and associations, individuals can build up social capital, which solves the problem of collective action. This is the good news from Putnam's pioneering research. The bad news is that this state of affairs cannot easily be brought into being. After all, the situation that has become institutionalized in southern Italy—a badly functioning democracy and economy, as a result of the shortage of social capital—appears to be a stable equilibrium as well. It is, in other words, a long way from Palermo to Milan.

Another important result in this area is delivered by Margaret Levi who, comparing voluntary military service in wartime in five different countries,[9] has shown that not only micro- but also macro-institutions may change preferences.[10] The way the government's institutions operate has a great impact on the percentage "young males" who are willing to risk their life for their country. Not only must they be convinced that the war in question is a "just war," they must also be convinced that (*a*) the institutions for implementing policies are fair and (*b*) that other citizens take a fair share of the burden (for example, the government must have institutions that can inform agents correctly what others are doing). It is, in other words, very heroic but also meaningless to be the only one who stands up and defends a country. In this classic collective action problem, the character of macro-institutions serves to change individual preferences about whether to

[9] Indeed a very "irrational" type of behavior.

[10] A typical micro-institution is a rule set by agents who are so few that they can meet in a face-to-face situation and regulate their own common interests. A typical macro-institution is a general law decided by a government, and that is supposed to steer the behavior of agents who did not participate directly in the decision of the rule.

volunteer or not (Levi 1991; cf. Margolis 1984). This type of analysis can be used equally well for other government programs, such as social insurance programs and the collection of taxes to finance them (Laurin 1986; Rothstein 1994).

## C Institutional stability

A central idea in all accounts of institutions is that they are enduring entities: they cannot be changed at once at the will of the agents. This is, in fact, a central tenet of all schools of institutional analysis. If institutions changed as the structure of power or other social forces surrounding them changed, then there would simply be no need for a separate analysis of institutions (Krasner 1984). Obviously, some institutions, such as constitutions, get an almost sacred status. But why do institutions endure, even if they influence power and strategy in ways contrary to the interests of the actors involved? As could be expected, different approaches give different answers to this question.

The economic approach analyzes institutions as points of equilibria which endure, in the first instance, because no one has an interest in changing them. They are, in that sense, self-reinforcing. This may take several forms. One is the idea of "sunken costs." So much has been invested by the agents in learning how to operate within a known institutional setting that, although there may theoretically be more efficient solutions, the costs of learning will prohibit change. Another idea is that, while a change might allow actors a short-term gain, uncertainty about long-term consequences prevents them from trying to change the institutions at hand (Shepsle 1986). A third argument about stability points to the cost of engaging in the processes of change. Even if every rational actor may be able to calculate that a certain change would be in his or her interest, engaging in the presumably costly process of changing the institutions in itself poses a standard collective action problem.

A more cultural or historical approach argues that institutions are in most cases not chosen. Instead, they are more like coral reefs with layer upon layer, and with no one agent or group of agents having decided how to arrange the structure of the whole. In different periods, different agents with different interests have established institutions, taking as their starting-point the already existing system of institutions. Individuals do not choose their institutions. Instead, institutions choose their individuals. At any one time only a few variables can be loosened from the historical situation and altered (Crouch 1986).

A third idea, derived from historical institutionalism, argues that power is what causes institutional stability. One thing powerful agents will try do is to change "the rules of the game" in ways which they believe will serve their interests in retaining and extending that power. The power an agent derives from more basic assets (such as the control of capital or the control of an organization of producers) is increased by the power coming from the institutions which he has designed (property laws favoring holders of capital; industrial laws favoring unionization) (Rothstein 1992). This "doubling" of power will of course make it extra costly, or extra risky, for less powerful agents to challenge the established institutional order.

## D Institutions and change

Admittedly, this is the weakest and most difficult point in political institutionalist analysis. As regards the conclusion, the analysis has not moved beyond Marx's statement that "men make history . . . but not under circumstances of their own making" (Przeworski 1985). But who are these "men," and what are the "circumstances" under which they may be able to create or change political institutions? Institutions may, according to Goodin (1996), change for three different reasons. One is by sheer accident or unforeseen circumstances: the interaction of different institutions may result in totally unforeseen new types of institutions. A second type is evolutionary change: the institutions that best suit certain stages of social development simply survive, through the operation of some kind of selective mechanism. Third, institutions may change by way of intentional design by strategic agents.

The general question of "agency versus structure" as the cause of social change pervades institutional as much as other social analysis (Mouzelis 1988; Koelbe 1995). Economic approaches to institutions, despite their claims to provide the micro-foundation for institutional analysis, usually resort to stringent functionalist explanations of why institutions arise and change. The "need" for this or that "explains" a certain institutional order (Hall and Taylor 1996).[11] Legal institutions, for example, arose because of merchants' need for information about other merchants' credibility when doing business (Milgrom, North and Weingast 1990). Changes in relative prices resulting from population dynamics or technological change, for example, induce institutional change (North 1990). The focus in this approach on institutions as mainly constraining individual choices makes

[11] Arguments against the use of functionalist explanations in the social sciences have been provided by Elster (1985).

it difficult to account for institutional change within the theoretical constraints of methodological individualism. As Bates (1988) has argued, while this approach might explain the demand for institutions, it does not explain why they are supplied.

Historical or cultural approaches fare no much better. While they have often given interesting explanations of the causes and consequences of institutional change in specific historical cases, so far there is no sign of a general theory coming out of this.[12] It has been shown that, at certain moments, specific political actors have been able to create institutions that will greatly enhance their future political power (cf. Rothstein 1995). To some extent, political agents can structure the political future: that is, men do make history; political agents are not "structural dopes," simply serving the "needs" of the greater social system (cf. Giddens 1979). The problem, however, is that history is also abundant with example of mistakes, where agents miscalculate what type of institution they should establish to further their future interests (Lewin 1988, chap. 3; Immergut 1993; Rothstein 1992).

The structure–agency problem in the analysis of institutional change, however, motivates a more narrow conceptualization of political institutions as formally decided rules. The reason is that these are rules where the impact of distinct agency, and thereby their possible effects on broader political and social outcomes, can most clearly be isolated. This problem of identifying the sources behind institutional change may be a symptom of a greater problem in institutional analysis, namely, its theoretical emptiness. To say that "institutions matter" does not tell us anything about *which institutions* are more important than others and *for what issues.* The value of the institutional approach may only emerge when it is combined with a more substantial theory from which we can draw hypothesis about why some agents, resources and institutions are more important than others (Rothstein 1992). At any one time in post-Stone Age history, the number of political institutions has been immense and therefore the analysis of political institutions is badly in need of a theory telling us what kind of institutions are important for what issues.

Understanding the implications of institutional change is probably one of the most challanging problems for political science. The question is whether we know enough about the outcomes of different political institutions that political science can become something like an architectonic discipline. The question has come up recently when so many of the former

[12] It should be added that this may not necessarily illustrate the weakness of the approach as such; but instead, as with "holes," there simply may be no such thing as a general theory for the creation or change of institutions.

East European, African and Latin American countries have abandoned authoritarianism and embarked upon routes to democratization. A reasonable question posed to political scientists is whether we know what type of institutional arrangements are most likely to secure for them democratic rule (Przeworski 1991). Should they opt for presidentialism or parliamentarism, unicameralism or bicameralism, multi-party or two-party electoral systems, and so on. As could be expected, two very different opinions exists. One answer is that we do know a great deal about the effects of different institutions, enough to be able to engage in this type of social and political engineering. For example, presidential constitutions have been shown to be a far less suitable for the consolidation of democracy than are parliamentary systems (Stepan and Skach 1993).[13] Parliamentary systems, combined with other institutions for political accommodation between opposing groups, may insure democratic rule and bring social stability to highly divided societies (Lijphart 1984; Lewin 1992). Various corporatist arrangements may insure peaceful industrial relations and at the same time secure economic growth (Goldthorpe 1984; Katzenstein 1985).

The opposing view is the well-known warning from Hayek about engaging in social engineering at all (Hayek 1949; cf. Lundström 1994). Our knowledge about the different effects of deliberate institutional change is simply too limited and the risk for mistakes leading to unexpected, contradictory and even perverse results too great. According to Hayek, institutions should not be designed but should, rather, evolve. It may simply be beyond our present knowledge to calculate, or even estimate, the differences between local versus global effects, between long-term versus short-term effects, between partial versus net effects of institutional change (Elster 1991). Another major complication is that institutional arrangements which work well under some social and economic conditions may prove disastrous under others (Przeworski 1991: 35 ff.). Much of the so-called "established wisdom" about the effects of political institutions has indeed been shown to be very fragile. One example is the type of corporatist institution mentioned above, which far from being the solution has more often become a problem for economic growth and peaceful industrial relations (Lewin 1992). Another is that single-district winner-take-all electoral systems produce stable two-party systems.

According to Elster (1991), this lack of knowledge should not lead us to refrain from making deliberate institutional changes. One reason is that at certain historical moments, such as transitions to democracy, political institutions simply must be designed (cf. Przeworski 1991). Elster has

[13] But for a partly opposite argument, see Sartori (1994).

suggested a third option for engaging in institutional engineering, namely, that instead of basing such changes on rational instrumental or consequentialist assumptions (of which we know too little), we enter a more deontological or normative discourse in our reasoning about institutions. This means that institutions should be constructed according to certain moral standards—such as shared conceptions of justice, for example, or norms about equality and fair treatment—as proposed in normative theory (Rawls 1971; Dworkin 1985; Barry 1995).

From a game-theoretical perspective, it has been argued that what secures democratic rule in new democracies is the new political institutions offering the losing forces behind the earlier authoritarian regime a *fair chance* of furthering their political interests or even (through fair elections) winning back political power in the future (Przeworski 1991). Thus, if the forces behind the change to democratic rule design the new democratic political institutions (the electoral system, for example) so as to deny the old anti-democratic groups any chance of ever coming back to political power—which is to say, if they act as the utility-maximizing rational agents which the theory presumes them to be—they may end up with a political take-over by the very authoritarian forces they tried to "design" out of power. Thus, principles of fairness, not instrumental rationality, should guide us in the construction of political institutions.

## V Institutional analysis and the future of political science

It should by now be clear that, in the analysis of political institutions, one of the major divisions in political science stands out very clearly. That is the division between the "hard" type of analysis aiming at universal laws (as in formal rational choice and behavioral theories) and the "soft" historically oriented analysis of political events and lines of cultural development (Shapiro and Wendt 1992; cf. Green and Shapiro 1994).[14] The final argument of this chapter is, however, that in the analysis of political institutions these two camps may engage in a fruitful exchange.

This argument starts from the above-mentioned problem created by the rational choice and game-theoretical approaches in political science: why

[14] The terms "hard" and "soft" should not be confused with the type of data used in these approaches. Instead, it only refers to the type of methodology that is used. In fact, the "hard" approaches often uses very soft (if any!) data, as e.g., mass data about attitudes, while "soft" approaches often use rock-hard data such as, for example, archival records.

do self-interested rational actors organize at all to produce collective goods, such as effective political institutions? If everyone acted in the self-interested way presumed by rational choice theory, institutions such as these would never come into being as a result of voluntary contributions (Scharpf 1990). Or as Pierre Birnbaum puts it: "The economic theory of individualism, . . . renders every theory of mobilization by atomization alone null and void" (Birnbaum 1988: 18; cf. Bicchieri 1993).

Simply put, in a situation in which co-operation for a collective good requires the contribution of many rational and self-interested agents, it will always pay for the individual to defect, rather than to collaborate (Tsebelis 1990: 74 ff.; Scharpf 1990: 476). Bendor and Mookherjee similarly conclude, on the basis of rational choice theory itself, that while the rational-actor model might explain why patterns of collective action persist it does "not explain how they arise. . . . [T]he emergence of co-operation is a hard problem—one that may require other methods of analysis" (Bendor and Mookherjee 1987: 146). To explain why co-operation sometimes takes place and sometimes does not, they opt for various explanations rather alien to this type of "hard" theorizing: "idiosyncratic events, for example, an unlucky early defeat that creates widespread suspicion of shirking" (Bendor and Mookherjee 1987: 146), or causes of another type arising from the specific historical circumstances at hand (cf. Scharpf 1990: 484). Yet even where organized co-operation for one or another (irrational or arational) reason does appear, according to the rational-actor model it should be highly unstable, on account of the constant temptation facing members of the organization to stop paying for its upkeep, while continuing to draw benefits from its operation (Hechter 1987: 10 ff.; Bendor and Mookherjee 1987).

The rational choice position is not strengthened by the claim that co-operative action can be explained by the previous existence of cultural, social, or political bonds. Such bonds, it is thought, would increase each actor's confidence that others will co-operate as well and, therefore, that the situation will stabilize in an equilibrium. The problem, as Michael Hechter puts it, is that "institutions such as these represent the very Pareto-efficient equilibria that game theorists presumably are setting out to explain" (Hechter 1992: 47; cf. Bicchieri 1993: 128). As mentioned above, the rational choice approach can explain why an equilibrium is stable, but not why different equilibria arise, or indeed why any specific equilibria come into being at all (Shepsle 1989; Scharpf 1990: 474). As George Tsebelis states: "rational choice theory cannot describe dynamics; it cannot account for the paths that actors will follow in order to arrive at the prescribed equilibria" (Tsebelis 1990: 28).

Thus, establishing a political institution to overcome a collective action problem itself presents a collective action problem. The formation of such an institutions is, according to Robert Bates, "subject to the very incentive problem it is supposed to solve" (Bates 1988: 395; cf. Scharpf 1990: 477 ff.). For among rational self-interested actors, such institutions (and the selective incentives necessary for their creation) would never arise. In other words, "logically, the game starts only after the actors have been constituted, and their order of preferences has been formed as a result of processes that cannot themselves be considered as part of the game" (Berger and Offe 1982: 525; cf. Grafstein 1992: 77 ff.).

Rational choice theorists have sometimes stressed the role of iteration (repeated play) in explaining why collective action occurs so often in the real world, notwithstanding the motivation of rational self-interested agents not to collaborate (Axelrod 1984). However, iteration can only play a very limited role in explaining co-operation. While iteration can indeed lead to stable co-operation between the parties, it can result in stable non-co-operation between them as well; and rational choice and game theory is at a loss to explain why the outcome is sometimes the one and sometimes the other. The role of iteration in solving the collective action problem is also very sensitive to asymmetries in information between the actors, especially when more than two players take part in the game (Bianco and Bates 1990; Bendor and Mookherjee 1987; Molander 1994). In a two-person game, it is possible to acquire information about whether or not one's opponent will choose to co-operate; this is very difficult, however, when one faces many opponents. The unfortunate result is that the "participants outsmart themselves into a suboptimal outcome" (Grafstein 1992: 71; cf. Scharpf 1990: 477 ff.).

Instead of iteration, political leaders have been shown to be important in solving collective action problems. Leaders must enjoy a reputation for trustworthiness among would-be members. They must have both the incentive and the capacity to reward those who contribute their fair share and to punish those who do not (Bianco and Bates 1990). The problem is that rational choice or game theory is not particularly helpful for identifying such leaders or specifying what it is that makes certain individuals in certain situations become leaders. Historical analysis seems to show that political leaders are more likely to create inefficient than efficient social institutions (North 1981).

There is quite some irony in this state of affairs. The "hard" rational choice and game-theoretical revolution in political science was, together with the similarly "hard" behavioral revolution, very much a reaction against the old "soft" historically oriented political science (Shepsle 1989).

In their search for general laws and parsimonious theories, the champions of the "hard" approach dismissed much traditional historically oriented political science as "soft story-telling" (cf. Miller 1992).

As previously discussed, pre-behavioral political science focused strongly on analyzing the establishment and operation of institutions, such as legislatures, constitutions and bureaucracies—not to mention political leadership (cf. March and Olsen 1984). The problem, however, was that the traditional approach had no way of telling, except by intuition, which institutions were really important, or what their effects were on political behavior.

Rational choice theory's great contribution has been to furnish us with such an idea: namely, that the important institutions are the ones which are capable of solving the problem of collective action, which is to say, the ones that make co-operation seem possible and rational for the agents involved. The problem at the moment, however, is that there is no way to understand why such institutions arise other than undertaking "soft" historical case-studies of their origins. If the explanation of why efficient institutions sometimes occur or fail to occur is path-dependent, we must study the specific single steps in history when the different paths were taken (Scharpf 1990; cf. Rothstein 1992). To understand why society at all is possible (that is, why mutually beneficial games are in fact sometimes played), the economic approaches in political science must take socially, historically and culturally established norms about co-operation, trust, honor, obligation and duty into account (Ostrom 1995; Bates 1988; cf. Gambetta 1988).

If political systems are usually tightly structured by institutions, changes can only occur at certain times. It is only during such formative moments that political actors are able to change the institutional parameters or the nature of the "game." These *formative moments* of political history are distinguished by the fact that existing political institutions are so incapacitated as to be unable to handle a new situation (Krasner 1984). In such situations, political actors not only play the game: they can also change the rules of the game. In other words political actors are, at such times, able to shape the political institutions of the future, and sometimes they are even able to establish rules favoring themselves (Rothstein 1992).

One could say that political science has thereby come full circle—back to the need for detailed analyses of the cultural origins and historical procedures of change in political institutions (Ostrom 1991: 242). Even so, we have learned at lot during this tour. We now understand better why some political institutions are more important than others and also, more specifically, what it is in the operational logic of such institutions that helps rational agents solve the problem of collective action. The simultaneous focus on institutions from the various approaches in the discipline may, in fact, change it from a situation were scholars work at "separate tables" into a much more unified enterprise (Ostrom 1995). As with information technology, if the product is going to be useful, the hardware and the software must be compatible and those who design the hardware must learn how to co-operate with those who produce the software and vice versa. Critical breakthroughs are most likely to happen when a combination of these different approaches can be accomplished.[15]

A second irony of the current state of affairs is that, if agents seek to establish institutions capable of overcoming the problem of collective action (that is, if they ever hope to create efficient social institutions), then they must take care not to behave according only to the individualistic and utilitarian premises of rational choice theory. If, as Jon Elster (1989) has argued, rational choice theory is not only to be understood as a reasonable empirical description of behavior[16] but also as a normative theory telling us how we *ought* to behave, then political science is indeed in big trouble. The reason is that if we were to take the utilitarian micro-foundation of rational choice and game theory as our normative basis for (recommended) social action, socially efficient institutions would in all likelihood never be established. As Bicchieri has argued, "common knowledge of the theory of the games makes the theory inconsistent and hence self-defeating" (Bicchieri 1993: 128).[17] Socially efficient institutions for overcoming sub-optimal equilibria simply cannot be established by persons who act in the way players in rational choice or game theory are supposed to act. Situations would be abundant where "everyone knew" that they would all be better off if they co-operated to establish efficient institutions, but that this was impossible because "everyone" knew that others would not co-operate because of a lack of trust. The analysis of the importance of institutions not only calls for a combination of formal modelling and the study of detailed historical cases of construction (or destruction) of political institutions (cf. Ostrom 1995), it also shows a need for the integration of positive and normative political theory. The importance of the classical as well as modern discourse for the relations between political institutions and civic virtue is shown by a question recently posed by Jane Mansbridge (1990: 20): "Can we design institutions to encourage motivations that we

[15] Cf. Ferejohn 1991; Bates 1988; Steinmo and Thelen 1992; Rothstein 1992; Hall and Taylor 1996.

[16] As such, it is simply wrong. For example, there is more compliance than selective incentives explain: "More people refrain from jaywalking or littering, pay their taxes, or join the armed forces than the theory predicts" (Levi 1990: 408; cf. Mansbridge 1990: 19 ff.).

[17] But as Jon Elster has stated, "Much of the social choice and public choice literature, with its assumptions of universally opportunistic behavior, simply seems out of touch with the real world, in which there is a great deal of honesty and sense of duty" (Elster 1991: 120).

believe on normative grounds are either good in themselves or will lead to good and just outcomes?"

## References

For further reading see particularly: Mary Douglas, *How Institutions Think* (1988); Gary Miller, *Managerial Dilemmas: The Political Economy of Hierarchy* (1992); James March and Johan Olsen, *Democratic Governance* (1995); and collections by Steinmo, Thelen and Longstreth on *Structuring Politics: Historical Institutionalism in a Comparative Perspective* (1992), Weaver and Rockman on *Do Institutions Matter? Government Capabilities in the United States and Abroad* (1993) and Goodin on *The Theory of Institutional Design* (1996).


ALMOND, G. A. 1960. Introduction. Pp. 7–33 in *The Politics of Developing Areas*, ed. G. A. Almond and J. S. Coleman. Princeton, N.J.: Princeton University Press.

ANDRÉN, G. 1928. *Huvudströmningar i tysk statsvetenskap.* Lund: Fahlbeckska stiftelsen.

ARROW, K. J. 1951. *Social Choice and Individual Values.* New York: Wiley.

AXELROD, R. A. 1984. *The Evolution of Cooperation.* New York: Basic Books.

BARRY, B. 1995. *Justice as Impartiality.* Oxford: Clarendon Press.

BATES, R. H. 1988. Contra contractarianism: some reflections on the new institutionalism. *Politics and Society*, 18: 387–401

BENDIX, R., ed. 1977. *Nation-Building and Citizenship.* Berkeley: University of California Press.

BENDOR, J., and MOOKHERJEE, D. 1987. Institutional structure and the logic of ongoing collective action. *American Political Science Review*, 81: 129–54.

BERGER, J., and OFFE, C. 1982. Functionalism vs. rational choice? *Theory and Society*, 11: 521–26.

BIANCO, W. T., and BATES, R. H. 1990. Cooperation by design: leadership, structure and collective dilemmas. *American Political Science Review*, 84: 133–47.

BICCHIERI, C. 1993. *Rationality and Coordination.* Cambridge: Cambridge University Press.

BIRNBAUM, P. 1988. *States and Collective Action: The European Experience.* Cambridge: Cambridge University Press.

BUCHANAN, J. M., and TULLOCK, G. 1962. *The Calculus of Consent.* Ann Arbor: University of Michigan Press.

COLEMAN, J. S. 1990. *Foundations of Social Theory.* Cambridge, Mass.: Harvard University Press.

CROUCH, C. 1986. Sharing public space and organized interests in Western Europe. Pp. 162–84 in *States in History*, ed. J. A. Hall. Oxford: Blackwell.

CROZIER, M. 1964. *The Bureaucratic Phenomenon.* Chicago: University of Chicago Press.

DAWES, R. M.; ORBELL, J. M.; and SIMMONS, R. T. 1977. Behavior, communication, and assumptions about other people's behavior in a commons dilemma situtation. *Journal of Personality and Social Psychology*, 35: 1–11.

DIMAGGIO, P. J., and POWELL, W. W. Introduction. Pp. 1–40 in Powell and DiMaggio.

DOWDING, K., and KING, D., eds. 1995. *Preferences, Institutions and Rational Choice.* Oxford: Clarendon Press.

DOUGLAS, M. 1988. *How Institutions Think.* London. Routledge and Kegan Paul.

—— 1954. *Political Parties: Their Organization and Activity in the Modern State*, trans. B. and R. North. London: Methuen.

Dworkin, R. 1977. *Taking Rights Seriously.* London: Duckworth.
—— 1985. *A Matter of Principle.* Cambridge, Mass.: Harvard University Press.
Eisenstadt, S. N. 1965. *Essays on Comparative Institutions.* New York: Wiley.
—— 1968. Social institutions. Vol. xiv, pp. 409–29 in *International Encyclopedia of the Social Sciences*, ed. D. I. Sills. New York: Macmillan and Free Press.
Elster, J. 1985. *Making Sense of Marx.* Cambridge: Cambridge University Press.
—— 1987. The possibility of rational politics. *Archives Europennées de Sociologie,* 28: 67–103.
—— 1989. *The Cement of Society.* Cambridge: Cambridge University Press.
—— 1991. The possibility of rational politics. Pp. 115–42 in *Political Theory Today*, ed. D. Held. Oxford: Polity Press.
—— and Slagstad, R., eds. 1988. *Constitutionalism and Democracy.* Cambridge: Cambridge University Press.
Esping-Andersen, G. 1990. *The Three Worlds of Welfare Capitalism.* Oxford: Polity Press.
Etzioni, A. 1988. *The Moral Dimension: Towards a New Economics.* New York: Free Press.
Evans, P. B; Rueschemeyer, D.; and Skocpol, T., eds. 1985. *Bringing the State Back In.* Cambridge: Cambridge University Press.
Ferejohn, J. 1991. Rationality and interpretation: parliamentary elections in early Stuart England. Pp. 279–305 in *The Economic Approach to Politics: A Critical Reassessment of the Theory of Rational Choice*, ed. K. R. Monroe. New York: HarperCollins.
Frochlich, N., and Oppenheimer, J. A. 1992. *Choosing Justice.* Berkeley: University of California Press.
Gambetta, D., ed. 1988. *Trust.* Oxford: Blackwell.
—— 1993. *The Sicilian Mafia: The Business of Protection.* Cambridge, Mass.: Harvard University Press.
Giddens, A. 1979. *Central Problems in Social Theory: Action, Structure and Contradiction in Social Analysis.* London: Macmillan.
Goldthorpe, J. H., ed. 1984. *Order and Conflict in Capitalist Societies.* Oxford: Clarendon Press.
Goodin, R. E. 1986. Laundering preferences. Pp. 75–102 in *Foundations of Social Choice Theory*, ed. J. Elster and A. Hylland. Cambridge: Cambridge University Press.
—— 1996. Institutions and their design. Pp. 1–53 in *The Theory of Institutional Design*, ed. R. E. Goodin. Cambridge: Cambridge University Press.
Grafstein, R. 1992. *Institutional Realism: Social and Political Constraints on Rational Actors.* New Haven, Conn.: Yale University Press.
Granovetter, M. 1985. Economic action and social structure: the problem of embeddedness. *American Journal of Sociology*, 91: 481–510.
Green, D. P., and Shapiro, I. 1994. *Pathologies of Rational Choice.* New Haven, Conn.: Yale University Press.
Hall, P. A. 1986. *Governing the Economy: The Politics of State Intervention in Britain and France.* New York: Oxford University Press.
—— and Taylor, R. C. R. 1996. Political science and the four new institutionalisms. *Political Studies*, forthcoming.
Hardin, R. 1982. *Collective Action.* Baltimore, Md.: Johns Hopkins University Press.
Hattam, V. C. 1993. *Labor Visions and State Power: The Origins of Business Unionism in the United States.* Princeton, N.J.: Princeton University Press.
Hechter, M. 1987. *Principles of Group Solidarity.* Berkeley, Calif.: University of California Press.

—— 1992. The insufficiency of game theory for the resolution of real-world collective action problems. *Rationality and Society*, 4: 33–40.

Held, D. 1987. *Models of Democracy*. Oxford: Polity Press.

Hurwicz, L. 1977. The design of resource allocation mechanisms. Pp. 3–37 in *Studies in Resource Allocation Processes*, ed. K. J. Arrow and L. Hurwicz. Cambridge: Cambridge University Press.

Immergut, E. M. 1993. *Health Politics: Interests and Institutions in Western Europe*. New York: Cambridge University Press.

Jepperson, R. L. 1991. Institutions, institutional effects and institutionalism. In Powell and DiMaggio 1991: 143–63.

Katzenstein, P. J. 1985. *Small States in World Markets*. Ithaca, N.Y.: Cornell University Press.

King, A. 1975. *Executives*. Vol. v, pp. 173–256 in *Handbook of Political Science*, ed. F. I. Greenstein and N. W. Polsby. Reading, Mass.: Addison-Wesley.

King, D. 1995. *Actively Seeking Work: The Politics of Unemployment Policy and Welfare Policy in the United States and Great Britain*. Chicago: Chicago University Press.

—— and Rothstein, B. 1993. Institutional choices and labor market policy: a British-Swedish comparison. *Comparative Political Studies*, 26: 147–77

Koelbe, T. A. 1995. The new institutionalism in political science and sociology. *Comparative Politics*, 37: 231–43.

Korpi, W. 1983. *The Democratic Class Struggle*. London: Routledge and Kegan Paul.

Krasner, S. D., ed. 1983. *International Regimes*. Ithaca, N.Y.: Cornell University Press.

—— 1984. Approaches to the state: alternative conceptions and historical dynamics. *Comparative Politics*, 16: 223–46

Laurin, U. 1986. *På heder och samvete. Skattefuskets utbredning och orsaker*. Stockholm: Norstedts.

Levi, M. 1988. *Of Rule and Revenue*. Berkeley: University of California Press.

—— 1990. A logic of institutional change. Pp. 402–19 in *The Limits of Rationality*, ed. K. S. Cook and M. Levi. Chicago: University of Chicago Press.

—— 1991. Are there limits to rationality? *Archives Europennées de Sociologie*, 32: 130–41.

Lewin, L. 1988. *Ideology and Strategy: A Century of Swedish Politics*. Cambridge: Cambridge University Press.

—— 1991. *Self-Interest and Public Interest in Western Politics*. Oxford: Oxford University Press.

—— 1992. The rise and decline of corporatism: the case of Sweden. *European Journal of Political Research*, 26: 59–79.

Lijphart, A. 1984. *Democracies: Patterns of Majoritarian and Consensus Government in Twenty-One Countries*. New Haven, Conn.: Yale University Press.

—— and Crepaz, M. M. L. 1991. Corporatism and consensus democracy in eighteen countries: conceptual and empirical linkages. *British Journal of Political Science*, 21: 235–56.

Lipset, S. M., and Rokkan, S., eds. 1967. *Party Systems and Voter Alignments*. New York: Free Press.

Lundström, M. 1994. *Politikens moraliska rum. En studie i F. A. Hayeks politiska filosofi*. Uppsala: Almqvist and Wiksell International.

Malnes, R., and Midgaard, K. 1993. *De politiska idéernas historia*. Lund: Studentlitteratur.

Mansbridge, J. J. 1990. The rise and fall of self-interest in the explanation of political life. Pp. 2–24 in *Beyond Self Interest*, ed. J. J. Mansbridge. Chicago: University of Chicago Press.

March, J. G., and Olsen, J. P. 1984. The new institutionalism: organizational factors in political life. *Amcerican Political Science Review*, 78: 734–49.

March, J. G., and Olsen, J. P. 1989. *Rediscovering Institutions: The Organizational Basis of Politics*. New York: Free Press.
—— —— 1995. *Democratic Governance*. New York: Free Press.
Margolis, H. 1984. *Selfishness, Altruism and Rationality*. Chicago: University of Chicago Press.
Meyer, J. W., and Rowan, W. R. 1983. *Organizational Environments: Rules and Rationality*. Beverly Hills, Calif.: Sage.
Miller, D. 1993. Deliberative democracy and social choice. Pp. 74–92 in *Prospects for Democracy: North, South East, West*, ed. D. Held. Oxford: Polity Press.
Miller, G. J. 1992. *Managerial Dilemmas: The Political Economy of Hierarchy*. New York: Cambridge University Press.
Milgrom, P. R.; North, D. C.; and Weingast, B. R. 1990. The role of institutions in the revival of trade: the law merchant, private judges and the Champagne fairs. *Economics and Politics*, 2: 1–23.
Moe, T. 1990. Political institutions: the neglected side of the story. *Journal of Law, Economics and Organization*, 6: 213–53.
Molander, P. 1994. *Akvedukten vid Zaghouan*. Stockholm: Atlantis.
Mouzelis, N. 1988. Marxism and post-Marxism. *New Left Review*, 167: 107–23.
North, D. C. 1981. *Structure and Change in Economic History*. New York: Norton.
—— 1990. *Institutions, Institutional Change and Economic Performance*. Cambridge: Cambridge University Press.
—— and Thomas, R. P. 1973. *The Rise of the Western World: A New Economic History*. Cambridge: Cambridge University Press.
Offe, C., and Preuss, U. K. 1991. Democratic institutions and moral resources. Pp. 143–72 in *Political Theory Today*, ed. D. Held. Oxford: Polity Press.
Olson, M. C. 1965. *The Logic of Collective Action*. Cambridge, Mass.: Harvard University Press.
—— 1982. *The Rise and Decline of Nations: Economic Growth, Stagflation and Social Rigidities*. New Haven, Conn.: Yale University Press.
Ostrom, E. 1990. *Governing the Commons*. Cambridge: Cambridge University Press.
—— 1991. Rational choice theory and institutional analysis: toward complementarity. *American Political Science Review*, 85: 237–43.
—— 1995. New horizons in institutional analysis. *American Political Science Review*, 89: 174–8.
Parsons, T. 1964. *The Social System*. London: Routledge and Kegan Paul.
Payne, S. G. 1990. *Fascism: Comparison and Definition*. Madison: University of Wisconsin Press.
Polsby, N. W. 1975. Legislatures. Vol. v, pp. 257–320 in *Handbook of Political Science*, ed. F. I. Greenstein and N. W. Polsby. Reading, Mass.: Addison-Wesley.
Pontusson, J. 1995. From comparative public policy to political economy. *Comparative Political Studies*, 28: 117–47.
Poulantzas, N. 1968. *Political Power and Social Classes*. London: New Left Books.
Powell, W. W., and Di Maggio, P. J., eds. 1991. *The New Institutionalism in Organizational Analysis*. Chicago: University of Chicago Press.
Przeworski, A. 1991. *Democracy and the market: political and economic reforms in Eastern Europe and Latin America*. Cambridge: Cambridge University Press.
—— 1985. Marxism and rational choice. *Politics and Society*, 14: 379–410.
Putnam, R. B. 1993. *Making Democracy Work: Civic Traditions in Modern Italy*. Princeton, N.J.: Princeton University Press.
Rawls, J. 1971. *A Theory of Justice*. Cambridge, Mass.: Harvard University Press.

—— 1985. Justice as fairness: political not metaphysical. *Philosophy and Public Affairs*, 14: 223–51.

—— 1993. *Political Liberalism*. New York: Columbia University Press.

Ridley, F. F. 1975. *The Study of Government*. London: Allen and Unwin.

Ricci, D. 1984. *The Tragedy of Political Science*. New Haven, Conn.: Yale University Press.

Riker, W. H. 1980. Implications from the disequilibrium of majority rule for the study of institutions. *American Political Science Review*, 74: 432–47.

—— 1990. Political science and rational choice. Pp. 163–81 in *Perspectives on Positive Political Economy*, ed. J. E. Alt and K. A. Shepsle. Cambridge: Cambridge University Press

Rothstein, B. 1990. Marxism, institutional analysis and working class strength: the Swedish case. *Politics and Society*, 18: 317–45.

—— 1992. Labor market institutions and working-class strength. In Steinmo, Thelen and Longstreth 1992: 33–56.

—— 1994. *Vad bör staten göra. Den generella välfärdspolitikens moraliska och politiska logik.* Stockholm: SNS Förlag.

—— 1996. *The Social Democratic State: Bureaucracy and Social Reforms in Swedish Labor Market and School Policy*. Pittsburgh: University of Pittsburgh Press.

Sabine, G. H. 1961. *A History of Political Theory*. London: George G. Harrap.

Sartori, G. 1994. *Comparative Constitutional Engineering*. London: Macmillan.

Scharpf, F. W. 1989. Decision rules, decision styles and policy choices. *Journal of Theoretical Politics*, 2: 149–76.

—— 1990. Games real actors could play. *Rationality and Society*, 2: 471–94.

Schattschneider, E. E. 1960. *The Semi-Sovereign People*. New York: Holt Renhardt.

Schmitter, P. C., and Karl, T. 1991. What democracy is . . . and is not. *Journal of Democracy*, 2: 136–72.

Scott, W. R. 1991. Unpacking institutional arguments. In Powell and DiMaggio 1991: 164–93.

Shapiro, I., and Wendt, A. 1992. The difference that realism makes. *Politics and Society*, 20: 197–223.

Shepsle, K. A. 1986. Institutional equilibrium and equilibrium institutions. Pp. 51–81 in *Political Science: The Science of Politics*, ed. H. A. Weisberg. New York: Agathon.

—— 1989. Studying institutions: some lessons from a rational choice approach. *Journal of Theoretical Politics*, 1: 131–47.

Simon, H. A. 1955. *Administrative Behavior*. New York: Macmillan.

Soltan, K. 1987. *The Causal Theory of Justice*. Berkeley: University of California Press.

Sproule-Jones, M. 1993. *Governments at Work: Canadian Parliamentary Federalism and Its Public Policy Effects*. Toronto: University of Toronto Press.

Stepan, A., and Skach, C. 1993. Constitutional frameworks and democratic consolidation: parliamentarism versus presidentialism. *World Politics*, 46: 1–22.

Steinmo, S. 1993. *Taxation and Democracy: Swedish, British and American Approaches to Financing the Modern State*. New Haven, Conn.: Yale University Press.

—— and Thelen, K. 1992. Historical institutionalism in comparative politics. In Steinmo, Thelen and Longstreth 1992: 1–32.

—— —— and Longstreth, F, eds. 1992. *Structuring Politics: Historical Institutionalism in a Comparative Perspective*. New York: Cambridge University Press.

Streeck, W. 1992. *Social Institutions and Economic Performance*. Beverly Hills, Calif.: Sage.

Thelen, K. 1991. *Union of Parts: Labor Politics in Postwar Germany*. Ithaca, N.Y.: Cornell University Press.

Therborn, G. 1978. *What Does the Ruling Class Do When It Rules?* London: New Left Books.

Tsebelis, G. 1990. *Nested Games: Rational Choice in a Comparative Perspective.* Berkeley: University of California Press.

Weaver, R. K., and Rockman, B. A., eds. 1993. *Do Institutions Matter? Government Capabilities in the United States and Abroad.* Washington D.C.: Brookings.

Weir, M. 1992. *Politics and Jobs: The Boundaries of Employment Policy in the United States.* Princeton, N.J.: Princeton University Press.

—— and Skocpol, T. 1985. State structures and the possibilities for 'Keynesian' responses to the Great Depression in Sweden, Britain and the United States. In Evans, Rueschemeyer and Skocpol 1985: 107–68.

Wheeler, H. 1975. Constitutionalism. Vol. v, pp. 1–92 in *Handbook of Political Science*, ed. F. I. Greenstein and N. W. Polsby. Reading, Mass.: Addison-Wesley.

Williamson, O. 1985. *The Economic Institutions of Capitalism.* New York: Free Press.

Wright, E. O. 1978. *Class, Crisis and the State.* London: New Left Books.

Chapter 5

# Political Institutions: Rational Choice Perspectives

Barry R. Weingast

## I Introduction

RATIONAL choice theory provides a distinctive set of approaches to the study of institutions, institutional choice, and the long-term durability of institutions. Rooted in the economic theory of the firm (Williamson 1985; Milgrom and Roberts 1991), economic history (North 1990), and positive political theory (Enelow and Hinich 1984; McKelvey 1976; Riker 1982), this approach provides a systematic treatment of institutions. Although it has much in common with other approaches to institutions, rational choice theory has its distinctive features, most importantly, providing the micro-foundations of institutional analysis.[1] Applications range across all political and social problems, from the effects of the major political institutions of the developed West (legislatures, courts, elections, and bureaucracies) to more recent studies of developing countries (for example, corruption, production and exchange, and revolution).

The rational choice approach to institutions can be divided into two separate levels of analysis (Shepsle 1986). In the first, analysts study their effects, taking institutions as fixed and exogenous. In the second, analysts study why institutions take particular forms, thus allowing institutions to be endogenous. The former analysis is clearly antecedent to the latter and is far more well developed. The latter provides a deeper approach to institutions. In combination, these approaches not only provide a method for analyzing the effects of institutions and social and political interaction, but they provide a means for understanding the long-term evolution and survival of particular institutional forms.

[1] Hall and Taylor (1996) review the varied approaches to institutional analysis.

Four features distinguish rational choice approaches to institutions. First, they provide an explicit and systematic methodology for studying the effects of institutions. The latter are modeled as constraints on action (North 1990), typically by how they affect the sequence of interaction among the actors, the choices available to particular actors, or the structure of available information.

Second, the methodology is explicitly comparative, affording predictions of two different sorts: (1) models often compare two related but distinct institutional constraints, predicting differences in behavior and outcomes; and (2) because this approach relies on equilibrium analysis, it often yields comparative statics results about how behavior and outcomes will change as the underlying conditions change.[2] In combination, these predictions not only allow empirical tests but provide the basis for a new and systematic approach to comparative politics. The two types of predictions afford comparisons of the behavior and outcomes under related institutions within a given country (for example, two regulatory agencies or two ministries) and of the effects of similar institutions across countries.

Third, the study of endogenous institutions yields a distinctive theory about their stability, form, and survival. In contrast to approaches that take institutions as given, this approach allows scholars to study how actors attempt to affect the institutions themselves as conditions change.

Fourth, the approach provides the micro-foundations for macro-political phenomena such as revolutions and critical elections (see, e.g., Poole and Rosenthal 1995). Until recently, these phenomena remained largely the domain of macro-sociologists and historical institutionalists (for example, Skocpol 1979). Although applications of rational choice theory are relatively new to these questions, its approach provides links with micro-behavior, potentially affording a new methodology for comparison across cases. Explicit models of discontinuous political change provide an exciting new set of applications of rational choice theory.

The purpose of this short survey is to reveal the logic of rational choice analyses of institutions, to suggest the range of applications, and to guide the reader to additional work. It proceeds in three stages: first, providing examples of the effects of institutions; second, discussing the newer and burgeoning literature on endogenous institutions; finally, identifying a series of frontier issues.

[2] The two sorts of predictions differ in that the first holds constant for conditions, studying the effect of changing institutions; the second holds the institutions constant, studying the effect of changing conditions.

## II The effects of institutions

The rational choice approach to institutions begins with a set of individuals, each with well-defined preferences. Strategic interaction of individuals within a well-defined context is the hallmark of the approach. Institutions are modeled via their effects on the set of actions available to each individual, on the sequence of actions, and on the structure of information available to each decision-maker (the last topic is discussed in Section IV). Rational choice analysis has been applied to virtually every major democratic institution, including constitutions, legislatures, executives, the bureaucracy, courts, and elections.[3] A host of applications show how institutions affect policy choice, including macro-economic policy-making, welfare, budgets, regulation and technology.[4]

### A The setter model

The setter model illustrates the type of logic employed in rational choice analyses.[5] This model begins with the standard spatial model of voting among alternatives from a one-dimensional continuum (see Figure 5.1).[6] The configuration represents the range of possible policy choices, which might be the degree of environmental protection or the proportion of GNP devoted to social welfare. Each individual is represented by a preference function which attains a maximum at the individual's "ideal point," the policy which the individual prefers to all other alternatives. An individual prefers policies that are closer to her ideal point than to those further away. All individuals are assumed to act strategically, that is, to maximize their goals given constraints.

[3] On constitutions, see: Buchanan and Tullock (1962), Brennan and Buchanan (1984), Elster (1989), Hardin (1989), Ordeshook (1993), Weingast (1995*b*). On legislatures, see: Austen-Smith and Banks (1988), Cox and McCubbins (1993), Fiorina (1977), Kiewiet and McCubbins (1991), Krehbiel (1991), Laver and Schofield (1990), Laver and Shepsle (1995). On presidents see: Eskridge (1991), Kiewiet and McCubbins (1991), Matthews (1989), and Moe (1985*a*). On bureaucracies see: Ferejohn and Shipan (1989), Fiorina (1977, 1981), McCubbins, Noll, and Weingast (1987), McCubbins and Schwartz (1984), Moe (1987*b*, 1989), Noll (1987, 1989), Rothenberg (1994), and Weingast and Moran (1983). On courts, Rodriguez (1994) provides a recent review. On elections see: Bawn (1993), Cox and Rosenbluth (1993), and Kousser (1974); a host of specialty topics are also studied, e.g., campaign finance (Austen-Smith 1987; Baron 1991; Snyder 1991) and the effects of the initiative and referendum processes (Gerber 1996).

[4] On macro-economic policy-making, see Persson and Tabellini (1990); on welfare Ferejohn (1989); on budgets Cogan (1994) and McCubbins (1989); on technology Cohen and Noll (1991). For recent surveys of work on regulation, see Noll (1989) and Romer and Rosenthal (1985); see also references in note 3 under "bureaucracy."

[5] Rosenthal (1990) surveys the setter model's range of applications.

[6] For further details of the spatial framework, see Enelow and Hinich (1984).

Central to the spatial model with a single issue is the "median voter," that individual who divides the distribution of voters into two equal parts. Policy *M* in Figure 5.1 represents the median voter's ideal policy. The status quo, or the policy currently in effect, is labeled *Q*. The spatial framework affords predictions about voting outcomes under a series of alternative institutional specifications.

**Fig. 5.1** The setter model

Suppose that any alternative may be proposed and that individuals wishing to offer proposals are recognized randomly. Each proposal is pitted in a majority vote against the status quo, the winner becoming the new status quo. The process continues until no more proposals are offered. What policy will result? In the presence of a single issue, the median voter theorem applies, holding that the only stable alternative is the median voter's ideal point (Enelow and Hinich 1984). To see this, consider another alternative, *X*. Stability requires that there exist no alternatives that can command a majority against *X*. Yet, *M* commands a majority against *X*. If *X* is to the right of *M*, then every voter to *M*'s *left* prefers *M* over *X*; so too does the median; and by definition, the median plus all voters to one side constitute a majority. The same logic applies, *mutatis mutandis*, if *X* is to the left of *M*.

To see the influence of institutions, notice that there is no unique manner of forming the "agenda," the process determining the set of alternatives that arise for a vote. In the setter model, a distinguished individual (or organization or committee) called the "setter" holds exclusive or monopoly power over the agenda. The setter chooses a proposal, and then the set of voters vote for either the proposal or the status quo, *Q*.

Returning to Figure 5.1, suppose that the monopoly agenda setter's ideal is policy *S*. Define the "winset," $W(x)$, of alternative *x* as the set of policy alternatives that command a majority against *x*. With status quo, *Q*, the set of feasible policies for the setter is given by $W(Q)$, the interval $(Q,Q')$.[7] Given $W(Q)$, the setter will propose the policy from this set that she most prefers, $Q'$. When faced with a vote between *Q* and $Q'$, a majority will vote

[7] To see why $W(Q) = (Q,Q')$, notice that any policy in this set is preferred by a majority to *Q* (everyone to the right of the median prefers this policy to *Q*, as does the median). Notice also that no policy outside this set can beat the status quo (the median and all voters to the right prefer *Q* to any policy left of *Q*; and the median and all voters to the left prefer $Q'$ to any policy to the right of $Q'$).

for *Q'*.[8] Unless *S* equals *M*, the monopoly power of the agenda-setter results in an outcome different from the median voter's ideal. The divergence in outcomes follows from the setter's institutionalized power over the agenda. Although a majority prefers *M* to *Q'*, the setter's agenda control allows her to prevent *M* from being considered.

The power of rational choice analysis is revealed by comparative statics results showing how the equilibrium choice changes with various parameters of the model. The setter model depends on three parameters: the location of *Q*, *M*, and *S*. Holding constant for *M* and *Q*, setters located at different policies will make different proposals.[9] A more interesting comparative statics result reveals the relationship between *Q* and *M* (Romer and Rosenthal 1978). Return to the configuration in Figure 5.1. Notice that as the status quo moves to the left, the size of the feasible set (policies preferred to the status quo by the median) increases. Thus, in Figure 5.2, if the new status quo is $R < Q$, then the set of feasible policies expands from $(Q,Q')$ to $(R,R')$. Given the setter's ideal of *S*, she will now propose *R'* instead of *Q'*. This yields a paradox: the worse the status quo for the median, the worse the outcome from a monopoly setter whose ideal is extreme.

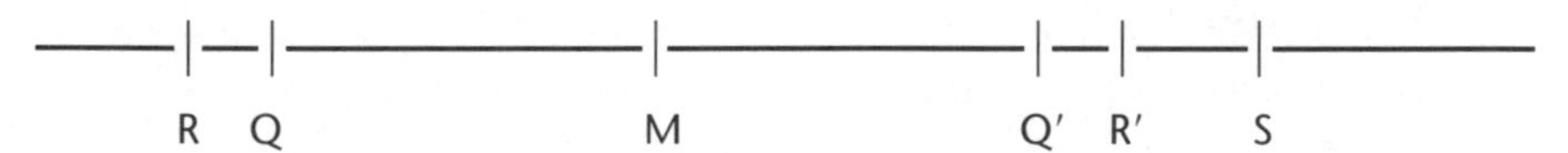


**Fig. 5.2** Comparative statics in the setter model

Scholars have applied this model to a range of institutional and policy contexts. Romer and Rosenthal (1978), for example, studied school board financing in the state of Oregon in which the school board had exclusive rights to make proposals for school bonds. Given that the board preferred high levels of expenditure (that is, it had an extreme ideal point such as *S* in Figure 5.1), the board sought to propose the maximum level of expenditures that would command the support of the median voter. In Oregon, institutions also determined the "reversion point," the policy that would go into effect if a bond issue failed.[10] If a bond issue failed, school funding reverted to the level of funds spent in 1911, woefully inadequate for current

[8] There are technicalities, involving the boundary of this set, beyond the scope of this survey. Suffice it to say that these determine that the boundary is included within the winset even though, in this illustration, the median voter is technically indifferent between *Q* and *Q'*.

[9] The outcome of the game is easy to calculate under different specifications of the setter's ideal point, *S*. Thus, if *S* is located anywhere to the left of *Q*, the outcome will be *Q*; anywhere within $W(Q)$, the outcome will be *S*; and anywhere to the right of *Q'*, the outcome will be *Q'*.

[10] Notice that the reversion point here differs from the status quo.

needs. Moreover, over time, the value of the reversion policy deteriorated. The setter model shows how the institutional structure of Oregon's school financing gave power to the school board. As the reversion level, $Q$, deteriorated, it moved to the left. This implied that the maximum expenditure policy commanding a majority, $Q'$, increased.

A second application of the setter model focuses on Congress. Denzau and Mackay (1983) and Shepsle and Weingast (1987) sought to show how the rules governing the control over the content and sequence of the agenda affected congressional policy choice. These models assumed that the congressional committee with jurisdiction over the policy issue was the setter. This approach predicted policy results consistent with those described in the behavioral literature on Congress in the 1960s and 1970s: namely, that the committee system biased policy in favor of the interests represented on that committee. The approach also predicts that, as the composition of interests on the committee changed, so too would policy.

A final application concerns the influence of Congress on the bureaucracy. Assuming that Congress, via its oversight committees, has a direct impact on regulatory agency policy, then changes in the interests on the relevant committees ought to be reflected in changes in regulatory agency choice. Weingast and Moran (1983), studying the Federal Trade Commission, showed that the set of cases opened by the commission were remarkably sensitive to congressional preferences.[11]

The setter model, focusing on a single agent interacting with a set of passive voters, is particularly simple. It nonetheless reveals the power of the rational choice approach to institutions: a precise analysis of the mechanisms underlying policy choice, and predictions about how outcomes change as various parameters change.

## B Separation of powers

The setter model paved the way for deeper and more realistic models of political institutions where control over the agenda is more dispersed, such as the separation of powers in the United States. Not only must the President bargain with Congress over legislation in this setting, but policy

[11] A substantial empirical literature shows similar results (as well as the influence of other political actors such as the President). See, e.g., Grier (1991), Moe (1985*b*), and O'Halloran (1994). An important debate concerns whether Congress alone *dominates* the bureaucracy (e.g, Moe 1987*a*). More recent work emphasizes the influence of multiple political actors, e.g., Ferejohn and Shipan (1989), McCubbins, Noll, and Weingast (1989), Moe (1989), Rothenberg (1994), and Snyder and Weingast (1994).

implementation is subject to interpretation by both the bureaucracy and the courts.

Recent models of the interaction among Congress, the President and the courts illustrate the power of this approach.[12] The dominant approach in both political science and legal scholarship views courts as the final mover in the legislative sequence: once legislation is passed, courts interpret it, often markedly altering its implementation. In this view, judicial power is paramount.

Rational choice models emphasize the interaction between courts and elected officials. These models begin with the observation that the legislative and judicial processes are on-going: not only may courts re-interpret legislation, but elected branches can react to judicial decisions with legislation (Eskridge 1991). These models capture an intimate relationship between the courts and the elected branches. First, they show how potential judicial rulings alter elected officials' choices over legislation. Second, they show how the prospect of legislation overturning court decisions has a direct and constraining influence on those decisions.

To see the effect of on-going interaction on courts, consider the simple model of policy choice depicted in Figure 5.3. The model represents Congress and the President as unitary actors with ideal points located at *C* and *P*. The court's ideal policy is located at *J*. Legislation, *L*, is assumed to arise from a bargaining process between *C* and *P*. For simplicity, the court is assumed to have complete powers of interpretation, that is, the ability to move legislation to any policy.

**Fig. 5.3** The separation of powers system

If the court is viewed as moving last, its choice is unrestricted and draws no reaction, so it will reinterpret legislation at *J*. Courts acting in this capacity are omnipotent and politically unconstrained by the elected branches. Viewing courts as participating in an on-going legislative process yields a different view. Suppose that legislation is located at *L*, that courts choose their interpretation, and that Congress and the President have the

[12] Studies of the separation of powers include: Eskridge (1991), Ferejohn and Shipan (1989), Miller and Hammond (1989), Kiewiet and McCubbins (1991), McCarty and Poole (1995), McCubbins, Noll, and Weingast (1989), Moe (1987*b*), Rothenberg (1994), Snyder and Weingast (1994), Gely and Spiller (1992).

opportunity to overturn the court's interpretation with legislation. Given these assumptions and the configuration in Figure 5.3, the court will not amend the legislation to policy *J*, for *J* will be overturned by legislation (both *C* and *P* prefer a range of policies to *J*). The best the court can do is to interpret the legislation at policy *P*. Clearly the President will not participate in legislation to overturn this interpretation.

This view also yields comparative statics predictions: the further the distance between Congress and the President, the wider the latitude afforded to the courts. This implies that courts have considerable influence during periods of sustained differences between Congress and the President; for example, during periods of divided government. When Congress and the President are ideologically united, however, courts have far less latitude; for example, during Reconstruction or the New Deal.

Although highly stylized, this model illustrates the implications of recent rational choice approaches to the separation of powers. In contrast to the traditional approach to studying American politics—viewing Congress, the President, the courts, and the bureaucracy, in isolation—recent rational choice models emphasize the interaction among the different branches of American government. Political scientists have long understood that the branches interact (see, e.g., Polsby 1986). Nonetheless, the study of American politics remains dominated by studies of one branch in isolation from the other. The models discussed above show how decisions made by actors in one branch systematically depend on the sequence of interaction; and the preferences, actions, and potential actions of actors in the other branches.[13] The potential result is a genuine theory of interaction of the major institutions of American national politics, a mature theory of the separation of powers.

## C Conclusion

The hallmark of the rational choice approach to institutions is its ability to analyze how institutions influence outcomes. Often apparently minor, micro-level details have dramatic effects on outcomes. Thus, the rules governing the agenda process within each house of Congress have a critical effect on legislative policy choice. Similarly, the sequence of interaction among strategic agents has a dramatic effect on outcomes. Scholars have exploited these techniques in a variety of settings. The promise of the ra-

[13] Recent studies add a third factor influencing the interaction among the branches, the distribution of information (see, e.g., Epstein and O'Halloran 1994, and Matthews 1989). The topic of information is discussed in Section III.

tional choice approach is that it provides the technology for keeping track of many different parameters simultaneously, thus affording testable propositions about how the results of political interaction will vary with underlying circumstances. For example, because congressional circumstances vary, there is no one pattern of congressional influence over the courts—and vice versa. The advantage of theory is that it shows how, as circumstances change, this mutual interaction varies.

## III Endogenous institutions

The most promising and far-reaching aspect of the study of institutions concerns questions about why institutions take one form instead of another. The finding that institutions have powerful effects on outcomes forces us to ask what makes institutions resistant to change. Most studies of institutions ignore this question by assuming that institutions are fixed and hence cannot be altered by individuals. For demonstrating the effects of particular institutions, this assumption is useful. But this assumption precludes addressing deeper questions: if institutions can be changed, why are they altered in some circumstances but not others?

The rational choice approach to institutional stability provides one of the few systematic approaches to these questions.[14] It begins with the concept of self-enforcement. Because institutions limit the flexibility of decision-makers, it must be in the interests of actors to abide by the limits imposed by institutions. A model of institutional stability must meet two conditions: first, the model must allow institutions to be altered by particular actors, and second, it must show why these actors have no incentive to do so. When these conditions hold, we say that institutions are "self-enforcing."

### A Endogenous institutions: political stability in antebellum America

Examining the mechanisms underlying the political stability during antebellum America illustrates the approach to self-enforcing institutions.[15] Students of American politics before the Civil War largely rely on political

[14] Works in this tradition include: North (1990), Ordeshook (1993), Weingast (1995*b*).

[15] No short discussion of this complex and long-disputed topic can be convincing, let alone do it justice. My purpose here is only to illustrate how the concepts are used. This discussion draws on my larger work on the American Civil War era, Weingast (1995*a*).

culture as one of their principal paradigms. Hartz (1955), for example, argues that most Americans believed in private property rights, markets, and limited government, especially limits on a remote national government. Even individuals who did not hold property held these ideals, in part because they believed that they or their children could do so. As a second example, students of the "new political history" rely on ethnocultural approaches: these scholars emphasize that most Americans cared more about local issues than about national ones (e.g., Silbey 1985).

The focus on the relationship of American political culture to actual limits on government appears to rest on an implicit premise: if nearly all voters held a particular belief about their relationship to government, then that belief caused this relationship to hold in practice. The antebellum United States was therefore characterized by federalism with a limited national government because most citizens preferred strong limits on the national government.

Unfortunately, the literature gives little attention to the mechanisms underlying its implicit premise, obscuring the link between citizen values and governmental behavior. The views and attitudes of Americans were not self-implementing. Although most Americans held similar general beliefs about government and property, they differed on scores of specific policy issues, sufficiently so that they fought for control of the national government. Differences occurred on a host of economic issues such as the tariffs; moreover differences about slavery could have profound impacts on the Southern economy. The general consensus about limited national government could not protect Americans from potential intrusions on specific issues.

The problem concerns how federalism was sustained during these years, specifically, what made federalism's constraints self-enforcing on national political officials? I argue that institutions were required to implement and maintain federalism, especially a national government limited in scope. Further, throughout most of the antebellum era, a particular set of institutions made federalism self-enforcing.

To see this, consider the episodic crises during these years. Deep problems with the national government's policies arose early in the 19th century. During the War of 1812, mercantile interests of the northeast were sufficiently harmed by wartime economic policies that they called a conference of secession at Hartford in late 1814. Although secession did not occur—the war ended shortly after the conference—the conference reflected many Northerners' deep questioning of the legitimacy of policy outcomes chosen under the Constitution.[16]

[16] This convention was not an isolated event; individuals in nearly every state considered secession at some point prior to the Civil War.

A related event occurred a few years later. The proposal to admit Missouri without a corresponding free state led to a crisis in 1818. Northerners, fearing Southern dominance of the national government, sought to amend the Missouri statehood bill with a provision to emancipate Missouri's slaves. The amendment passed in the House of Representatives, where population gave Northerners a majority, but failed in the Senate, where Southerners had equal representation and hence a veto.

The Missouri Compromise resolved this crisis. The compromise held that the "balance rule," affording equal representation in the Senate to both North and South, would govern territorial expansion and statehood. The balance rule not only provided each region with a veto over policies inimical to their interests, but was central to most of the major national political events of the antebellum years (Roback, forthcoming; Weingast 1995): it underpinned the formation of the second party system in the late 1820s and early 1830s; and it was central to the crisis years during the 1850s, underlying two of the Democratic Party's débâcles, the Kansas–Nebraska Act in 1854 and the attempt to admit Kansas as a slave state in 1858.

Importantly, the balance rule also underpinned American federalism. Although most Americans feared an intrusive national government, they did so for different reasons. The various crises during the antebellum years demonstrated that regional interests could use the national government for their own ends, to the detriment of others. Americans' strong feelings about limited national government did not protect them against these potential intrusions on federalism by that government because the general consensus did not extend to many specific issues, such as slavery. Instead, Americans' attitudes and preferences worked in combination with the balance rule to make federalism with a limited national government self-enforcing.

The balance rule effectively implemented an agreement among Americans of different political persuasions and policy preferences: if you'll give up the potential to control the national government, I will also do so. The system of regional vetoes, afforded by the balance rule, was thus central to maintaining limits on the national government. Only policies preferred by a large majority in the nation could pass the national government. Moreover, the balance rule was self-enforcing. As long as it remained in effect, each region held the power to prevent attempts to alter it.

To see its power, consider the balance rule's remarkable implications for slavery. By granting Southerners a veto over national policy in the Senate, the balance rule allowed Southerners to prevent anti-slavery initiatives. The balance rule was thus central to the self-enforcing nature of the states' rights philosophy underpinning federalism during the second party system. This version of federalism granted states the right to handle issues concerning slavery, ensuring that it could be maintained throughout the South without fear of encroachment by the national government. The episodic success of anti-slavery initiatives in the House of Representatives, where Northerners held a majority, reveals that, in the absence of a Southern veto, such measures may well have become national policy.[17] The dual system of vetoes was central to maintaining a limited national government.

In sum, institutions such as the balance rule were necessary to implement and maintain American federalism during the antebellum era. A limited national government was not self-enforcing based on citizen preferences alone, and, at several points during the antebellum era, regional interests threatened to dominate the national government for regional purposes. When anti-slavery initiatives passed the Northern dominated House of Representatives, Southerners vetoed these in the Senate. Thus, the system of dual vetoes implied that policies promoting regional interests could not succeed. The balance rule provided the principal institutional underpinning of federalism, not only helping to maintain a states' rights philosophy and a limited national government, but making them self-enforcing.

## B Institutions provide durability

Another component of the literature provides a related answer to why institutions exist and take on specific forms. Several scholars argue that institutions are designed to endow particular policy outcomes with durability. An important application focuses on policies delegated to bureaucracies. One problem to be solved in any legislation concerns the ability of tomorrow's elected officials to influence bureaucratic decisions. Politicians creating a new bureaucracy thus face a dilemma, for they want the bureaucracy to be responsive to *their* interests, not those of future politicians. Elected officials respond to the dilemma by manipulating bureaucratic structure and process, shaping the incentives of bureaucrats and blunting the influence of future politicians (McCubbins, Noll and Weingast 1989; Moe 1989).

[17] The controversy over and the demise of sectional balance in the 1850s led in part to the Civil War (see Weingast 1995*a*). Not only did the system's demise leave the South unprotected, but the fast-paced growth of the North led to demands that the "majority rule," implying that the South risked losing all influence over the national government. This factor alone hardly explains Southern secession, let alone a devastating Civil War. But it did serve as an important political factor underlying the sectional crisis.

Recent work on environmental protection and occupational safety regulation illustrates this theme (for example, McCubbins, Noll and Weingast 1987; Moe 1989). These studies begin with a question about, how we understand why bureaucratic structure and administrative procedures are so complicated and cumbersome, often defeating the expressed purpose of the agency subject to them.

Moe's (1989) study of the EPA begins with the agency's strikingly rigid procedures that limit its ability to pursue its mandate. How are these to be explained? Moe's answer is that, when the environmentalists held sufficient power in 1970 to produce the new and radical legislation, they did not expect to remain in power for the next three decades. Moe argues that environmentalists believed that the opponents of regulation would dominate regulatory decisions, as had occurred over the previous decade.

Given their expectations, the environmentalists sought to provide durability to their policies by enacting a series of structures and procedures that would force even opponents of environmental regulation to pursue clean air. The trade-off was that, during periods when environmentalists held power, these procedures would compromise environmental goals; but during periods when the opponents held power, it would facilitate those goals. After the fact, we know that the environmentalists' rise to power in 1970 represented a permanent and dramatic shift in power away from the interests being regulated. Paradoxically, the unexpected shift implies that the constraints imposed on regulation hindered the environmentalists' pursuit of their own goals.

## C Explaining institutional change: North and Thomas on the demise of serfdom in Western Europe

North and Thomas' (1973) study of the "Rise of the Western World" provides one of the most creative applications of rational choice analysis. The demise of feudalism in Western Europe was an important step in the development of Western Europe, helping to propel it to world leadership. How did this occur? For a variety of reasons during this period—notably, the lack of secure property rights due to the inability of the state to police extensive areas—the economy developed in very small units centered around the manor. The extent of the manor's territory was determined by the area that could reasonably be defended through withdrawal into the manor. One of the central determinants of this process was the ability to feed the set of local inhabitants inside the castle while protecting them from assault and siege.

Underpinning the manor economy was a system of production exchange based on tradition rather than on explicit markets. High transport costs combined with the difficulty in defending property rights made markets ancillary to the lives of most individuals living in the interior of Europe. Largely self-sufficient, the manor society produced nearly all it consumed.

Wages and rents were not paid with money. Instead, peasants contributed labor dues to the manor lord.[18] North and Thomas show that, although no explicit markets existed, the labor time due the lord adjusted, if slowly, to the forces of supply and demand. As the population began to rise relative to the supply of land, from the 11th through the 14th centuries, the peasants' time due to the lord increased.

The manor society suffered a large shock with the Black Death in the mid-14th century, instantly transforming the scarce economic factor from land to labor. In a market economy, wages would quickly rise relative to land rents to reflect the reversal in scarcity. The medieval economy's absence of markets implied that adjustments had to occur through other means. Labor scarcity induced competition among lords who had to attract and retain labor, forcing a dramatic change in institutions in Western Europe. The absence of an easy ability to increase wage rates forced lords to grant labor more attractive rights. Scarce labor implied that, following the Black Death, a runaway serf could typically find a willing lord, putting a *de facto* (and, later, *de jure*) end to feudalism.

The importance of North and Thomas's model is twofold. First, institutions are the endogenous variable, adjusting as exogenous circumstances change. Feudalism could be sustained within a large range of population growth, as long as it did not change too quickly. The sudden and unexpected effects of the Black Death ended the system, as competitive pressures forced lords to adjust labor rights in response to the newly scarce labor. Second, North and Thomas's approach provides a method for modeling the comparative performance of different states over time.[19]

## D Conclusions

This section discussed several related approaches to the question why institutions take specific forms. For institutions to have durable effects on

[18] One set of fascinating debates generated by the North and Thomas model concerns the monopoly and discriminatory aspects of this system, advantaging the lords of the manor.

[19] A host of such studies have recently appeared, also exploiting comparative statics. For example, Greif's (1995) creative analysis of merchant institutions suggests why Western traders had an advantage over traders from the Islamic world in the late middle ages.

behavior and outcomes, they too must be durable. Because all institutions are subject to potential change, stable institutions must be self-enforcing in the sense that those with the power to change institutions have no incentive to do so.

The case of the balance rule during antebellum America illustrates this principle. Political interests in both the North and the South were worried about a national government devoted to regional ends. Yet the widespread desire for a limited national government was not self-implementing. The balance rule, a system of dual vetoes over national policy-making, made limited national government self-enforcing. By ensuring that policies inimical to either region could be vetoed, it changed the incentives of national politicians to pursue such initiatives.

North and Thomas's argument about the rise of the West illustrates another aspect of this argument. During the middle ages, feudalism remained stable as long as land remained the scarce resource. Although lords could offer more attractive rights to serfs, it was not in their interests to do so. Following the Black Death, labor became the scare resource. Lords facing competition for labor for the first time attempted to attract labor, offering more attractive bargains and, in the process, destroying feudalism in Western Europe.

## IV The frontiers

This section discusses some of the exciting innovations in rational choice approaches to institutions. These works promise a large pay-off to future research. This brief discussion is intended to indicate the range of new applications rather than provide a discussion of their logic.

### 1 The central role of information on politics and political organization

A growing literature studies the role of information in politics. Among the most influential work is that of Gilligan and Krehbiel (1990) and Krehbiel (1991).[20] Although much of this development has taken place within the context of American political institutions, the range of potential applications encompasses not only legislatures throughout the developed world, but also institutions in developing countries.

Although these models are technically difficult, part of the intuition

[20] Banks (1991) and Calvert (1986) provide excellent surveys of this literature, through the mid-1980s. See also Austen-Smith and Riker (1987), Banks (1989), Epstein and O'Halloran (1994), and Matthews (1989).

motivating the analysis can be easily understood. Gilligan and Krehbiel begin with the problem of endemic uncertainty about the relationship between policies and outcomes, implying an uncertain divergence between a policy's intended and actual effects. At a cost, members of the legislature can learn about that relationship, reducing the divergence between a policy's intended effects and its actual effects.

Obtaining this knowledge creates a tension for legislators. Efficiency dictates that only a few legislators undertake the investment to become experts and learn about the relationship between policies and outcomes. Yet that specialized knowledge potentially allows them to bias outcomes in their favor. The tension is that some legislators have to be given sufficient incentive to make costly investments in expertise and information, while not extracting too much of the value of information for themselves. Gilligan and Krehbiel show that, under certain conditions, legislators can design institutions that convey the information without too much bias toward the experts.

A host of other studies show that problems of information occur throughout politics. Austen-Smith and Riker (1987), for example, use similar models to study legislative rhetoric and persuasion. Models of campaign finance show that money may play a remarkably different role in elections than buying influence, for example, indicating whether candidates are vulnerable.[21]

## 2 Extending the range of applications

Rational choice theorists developed their models through applications to American politics. This narrow focus helped generate a range of theoretical techniques whose implications could readily be compared. Unfortunately, the focus on American politics also gave the approach a parochial flavor, making it difficult to distinguish general results from those reflecting specialized assumptions required by American institutions.

Extending the approach beyond American politics therefore represents a promising new development. I briefly mention four applications. First, studies comparing different electoral systems illustrate the new and growing range of applications.[22] Studies of electoral systems demonstrate that differences in electoral rules have systematic effects on the number of parties, the existence of a majority party within the legislature, and on policy outcomes. Recent studies also show how those designing a new electoral system often attempt to bias electoral institutions in their favor. Second, a

[21] See the citations in note 3, above.

[22] See, e.g., Brady and Jongryn (1990), Bawn (1993), Cox and Rosenbluth (1993), Kousser (1974), and Rogowski (1989).

number of studies extend models of legislatures to various parliamentary systems in Western Europe and Japan.[23] Third, a range of applications compare bureaucratic structure and performance across nations.[24] Finally, a number of recent studies investigate problems in international relations: for example, the effects of international institutions (e.g., Keohane and Martin 1994); and an exploding literature on the effects of domestic institutions on foreign policy-making (e.g., Bueno de Mesquita and Lalman 1992; O'Halloran 1994).

The above works focus on institutions in developed countries. An equally promising development, though one far more in its infancy, represents application of the approach to problems outside of the developed world, focusing on political problems in developing states.[25]

### 3 Extending the range of phenomena studied

Many political scientists approach phenomena from a non-instrumentalist perspective. In the past, the interaction among scholars using different perspectives has tended to emphasize their seeming irreconcilability, as if Kuhn's "competing paradigms" provides the unique program for interaction among different approaches in the social sciences.

In recent years, an alternative program has emerged, emphasizing the complementarities among different approaches. This new program acknowledges differences not as competing paradigms but as complementary approaches to complex phenomena. This suggests a more fruitful interaction among scholars of differing approaches, where not only the tools and techniques of the other become relevant, but so too do the phenomena under study.

Recently, some rational choice scholars have begun exploring links with phenomena studied and insights generated by scholars using alternative approaches. Ferejohn (1992), for example, studies the parallel nature of interpretivist and instrumentalist perspectives for understanding elections in 17th-century England. Goldstein and Keohane (1993) explore different approaches, including rational choice ones, to understanding the role of ideas. Bates and Weingast (1995), Fearon (1994), and Laitin (1988) also explore a range of phenomena normally viewed as outside the purview of the rational choice approach: riots, ethnic violence, and the politics of language usage.

[23] See, e.g., Cain, Ferejohn, and Fiorina (1981), Cowhey and McCubbins (1996), Cox (1986), Laver and Shepsle (1995), Laver and Schofield (1990), Lohmann (1994), Ramseyer and Rosenbluth (1995), Strom (1990), and Tsebelis (1990).

[24] See, e.g., some of the studies in Cowhey and McCubbins (1996).

[25] See, for example: Bates (1989), Bates and Weingast (1995), Ames (1987), Fearon (1994), Greif (1995), Hardin (1995), Levi (1988), North (1990), Przeworski (1991), Root (1994), and Rosenthal (1992).

### 4 Relaxing behavioral foundations: cognitive limits on human action

Critics of the rational choice approach have long argued that its assumptions about individuals are highly unrealistic, compromising its ability to study a range of phenomena. One of the most foundational aspects of frontier research concerns rational choice theory's behavioral assumptions.[26] Until recently, no attempt to use a more realistic set of assumptions about individuals has held the promise of similar power to the conventional set of assumptions.

A host of scholars focus on the implications of cognitive science for rational choice theory. Denzau and North (1994), Frank (1989), Lupia and McCubbins (1995), Noll and Krier (1990), and North (1990) explore the implications of a range of assumptions of cognitive limits on individuals. In contrast to much of the earlier work on "bounded rationality," the hallmark of this work is to attempt to use less arbitrary assumptions and to explore their implications within the larger corpus of rational choice theory. These theories emphasize, for example, not only the importance of learning (long an aspect of rational choice theory), but also different assumptions about how individuals learn.

## V Conclusions

This brief survey of rational choice approaches to institutions emphasizes its distinctive features. Rational choice theory provides a systematic method for studying the effects of institutions. Applications not only include all the major elements of modern democracy but a range of topics in comparative politics, such as ethnic warfare, riots, democratic stability, and revolutions. New studies of the central importance of information hold considerable promise for extending this approach.

The approach to institutions yields new insights into a host of problems. For the problem of bureaucracy, it suggests that bureaucrats may have less influence over policy than previously believed. Observing behavior alone suggests that, because bureaucrats make so many important decisions without direct input from politicians, the latter have little direct influence. Emphasizing the importance of sequence, rational choice approaches suggest that, when politicians are able to punish bureaucrats for deviating from desired policies, bureaucrats may implement those policies without the need for direct input from politicians. This doesn't prove that politi-

[26] This contrasts to approaches that use a different type of rationality altogether, e.g., earlier notions of "satisficing" or Kahneman and Tversky's (1979) "prospect theory."

cians dominate bureaucrats; but it does suggest that we cannot infer the lack of influence from the lack of direct political input. This general principle—that decision-makers anticipate the consequences of their actions and therefore take into account the concerns of actors who may intervene later—has been shown to work in a variety of contexts, for example, in the relationship between the courts and the political branches and between parties and their leaders.

Rational choice approaches also provide the basis for a conversation across the tradition boundaries within political science. As the setter model illustrates, this basic principle has applications for problems in bureaucracy, the interaction of political branches and the courts, and within legislatures. As the emerging literature on the American system of separation of powers indicates, this has the potential to transform the study of American politics across traditional field boundaries. A similar conversation has begun between Americanists and comparativists, holding the potential for new theory to politics, instead of one for American and one for each major country or region. Finally, the historical examples show that the approach is not confined to modern, advanced industrialized countries. Applications include a range of macro-political phenomena, such as revolution, civil war, and the demise of serfdom.[27]

Most studies of institutions, including those relying on approaches other than rational choice, assume institutions are fixed and study their effects. This begs the issue of why institutions endure. Recent rational choice models provide an approach for studying institutional survival and durability. If institutions directly influence outcomes, then something must prevent individuals from altering these institutions to achieve different ends. The concept of self-enforcing institutions affords a technique for studying how institutions survive.

## Acknowledgments

The author gratefully acknowledges Michael Bailey, Bruce Bueno de Mesquita, Randy Calvert, Kelly Chang, Rui deFigueiredo, and Andrew Rutten for helpful comments; Douglas Grob for research assistance; and Ilse Dignam for editorial assistance.

[27] No discussion about the rational choice approach is complete without reference to Green and Shapiro's (1994) recent critique. Although space constraints do not allow a discussion of this work, I recommend that it be read in tandem with the responses in the special issue of the *Critical Review* (1995).

## References

Alt, J. E., and Chrystal, A. 1983. *Political Economics.* Berkeley: University of California Press.

Ames, B. 1987. *Political Survival.* Berkeley: University of California Press.

Austen-Smith, D. 1987. Interest groups, campaign contributions, and probabilistic voting. *Public Choice,* 54: 123–39.

—— and Banks, J. 1988. Elections, coalitions, and legislative outcomes. *American Political Science Review,* 82: 405–22.

—— and Riker, W. H. 1987. Asymmetric information and the coherence of legislation. *American Political Science Review,* 81: 897–918.

Banks, J. S. 1989. Agency budgets, cost information, and auditing. *American Journal of Political Science,* 33: 670–99.

—— 1991. *Signalling Games in Political Science.* New York: Harwood.

Baron, D. 1989. *Design of Regulatory Mechanisms and Institutions.* Vol. ii, pp. 1347–447 in *Handbook of Industrial Organization,* ed. R. Schmalensee and R. D. Willig. Amsterdam: North-Holland.

—— 1991. Service-induced campaign contributions and the electoral equilibrium. *Quarterly Journal of Economics,* 104: 45–72.

—— 1993. Government formation and endogenous parties. *American Political Science Review,* 87: 34–47.

—— and Ferejohn, J. 1989. Bargaining in legislatures. *American Political Science Review,* 83: 1181–206.

Bates, R. H. 1989. *Beyond the Miracle of the Market.* New York: Cambridge University Press.

—— and Weingast, B. R. 1995. A new comparative politics. Working Paper, Hoover Institution, Stanford University.

Bawn, K. 1993. The logic of institutional preferences: German electoral law as a social choice outcome. *American Journal of Political Science,* 37: 965–89.

Brady, D., and Jongryn, M. 1990. Strategy and choice in the 1988 national Assembly election of Korea. Unpublished working paper, Hoover Institution, Stanford University.

Brennan, G., and Buchanan, J. M. 1984. *The Power to Tax.* New York: Cambridge University Press.

Buchanan, J. M., and Tullock, G. 1962. *The Calculus of Consent.* Ann Arbor: University of Michigan Press.

Bueno de Mesquita, B., and Lalman, D. 1992. *War and Reason.* New Haven, Conn.: Yale University Press.

Cain, B.; Ferejohn, J.; and Fiorina, M. 1981. *The Personal Vote.* Cambridge, Mass.: Harvard University Press.

Calvert, R. L. 1986. *Models of Imperfect Information in Politics.* New York: Harwood.

—— 1995. The rational choice theory of social institutions: co-operation, co-ordination, and communication. In *Modern Political Economy,* ed. J. Banks and E. Hanushek. New York: Cambridge University Press.

Cogan, J. 1994. The congressional budget process and the federal budget deficit. Pp. 1333–45 in *Encyclopedia of the American Legislative System,* ed. J. H. Silbey. New York: Scribner's.

Cohen, L., and Noll, R. G. 1991. *The Technology Porkbarrel.* Washington, D.C.: Brookings Institution.

Cowhey, P. F., and McCubbins, M. D., eds. 1996. *Political Structure and Public Policy in Japan and the United States.* New York: Cambridge University Press.

Cox, G. 1986. *The Efficient Secret.* New York: Cambridge University Press.
—— and McCubbins, M. D. 1993. *Legislative Leviathan.* Berkeley: University of California Press.
—— and Rosenbluth, F. 1993. The electoral fortunes of legislative factions in Japan. *American Political Science Review,* 87: 577–89.
*Critical Review.* 1995. "Special Issue: Rational Choice Theory and Politics." (Winter–Spring) Vol. 9, nos. 1–2.
Denzau, A., and Mackay, R. 1983. Gate keeping and monopoly power of committees. *American Journal of Political Science,* 27: 740–62.
—— and North, D. C. 1994. Shared mental models: ideologies and institutions. *Kyklos,* 47: 3–31.
Elster, J. 1989. *The Cement of Society.* New York: Cambridge University Press.
Enelow, J. M., and Hinich, M. J. 1984. *The Spatial Theory of Voting.* New York: Cambridge University Press.
Epstein, D., and O'Halloran, S. 1994. Administrative procedures, information, and agency discretion. *American Journal of Political Science,* 38: 697–722.
Eskridge, W. N., Jr. 1991. Reneging on history? Playing the Court/Congress/President civil rights game. *California Law Review,* 79: 613–84.
—— and Ferejohn, J. 1992. The Article I, Section 7 game. *Georgetown Law Journal.* 80: 523–64.
Fearon, J. 1994. Ethnic warfare as a commitment problem. Unpublished manuscript, University of Chicago.
Ferejohn, J. 1989. Changes in welfare policy in the 1980s. Pp. 123–42 in *Politics and Economics in the Eighties,* ed. A. Alesina and G. Carliner. Chicago: University of Chicago Press.
—— 1992. Rationality and interpretation: parliamentary elections in early Stuart England. Pp. 279–305 in *The Economic Approach to Politics,* ed. K. Monroe. New York: HarperCollins.
—— and Shipan, C. 1989. Congressional influence on telecommunications. Pp. 393–410 in *Congress Reconsidered,* ed. L. Dodd and B. Oppenheimer. 4th edn. Washington: CQ Press.
Fiorina, M. P. 1989. *Congress: Keystone of the Washington Establishment.* 2nd edn. New Haven, Conn.: Yale University Press; originally published 1977.
—— 1981. Congressional control of the bureaucracy. Pp. 332–48 in *Congress Reconsidered,* ed. L. Dodd and B. Oppenheimer. 2nd edn. Washington: CQ Press.
Frank, R. 1989. If homo-economicus could choose his own utility function, would he want one with a conscience? *American Economic Review,* 79: 594–6.
Gely, R., and Spiller, P. T. 1992. The political economy of Supreme Court constitutional decisions: the case of Roosevelt's Court packing plan. *International Review of Law and Economics,* 12: 45–67.
Gerber, E. 1996 Legislative response to threats of popular initiatives. *American Journal of Political Science,* 40: forthcoming.
Gilligan, T. W., and Krehbiel, K. 1990. Organization of informative committees by a rational legislature. *American Journal of Political Science,* 34: 531–64.
Goldstein, J., and Keohane, R. O. 1993. *Ideas and Foreign Policy.* Ithaca, N.Y.: Cornell University Press.
Green, D., and Shapiro, I. 1994. *Pathologies of Rational Choice.* New Haven: Yale University Press.
Greif, A. 1995. The institutional foundations of Genoa's economic growth: self-enforcing political relations, organizational innovations, and economic growth during the commercial revolution. Mimeo., Department of Economics, Stanford University.

GREIF, A.; MILGROM, P.; and WEINGAST, B. R. 1994. Commitment, coordination, and enforcement: the case of the merchant guilds. *Journal of Political Economy*, 102: 745–76.

GRIER, K. B. 1991. Congressional influence on U.S. monetary policy: an empirical test. *Journal of Monetary Economics*, 28: 201–20.

HALL, P., and TAYLOR, R. 1996. Political science and the four new institutionalisms. *Political Studies*, forthcoming.

HARDIN, R. 1989. Why a constitution? Pp. 100–20 in *The Federalist Papers and the New Institutionalism*, ed. B. Grofman and D. Wittman. New York: Agathon Press.

HARDIN, R. 1995. *One for All: The Logic of Group Conflict*. Princeton, N.J.: Princeton University Press.

HARTZ, L. 1955. *The Liberal Tradition in American*. New York: Harcourt, Brace.

KAHNEMAN, D., and TVERSKY, A. 1979. Prospect theory: an analysis of decision under risk. *Econometrica*, 47: 163–91.

KEOHANE, R. O., and MARTIN, L. 1994. Delegation to International Organizations. Paper given at the conference, "Where is the New Institutionalism Now?" University of Maryland, College Park, Oct. 15, 1994.

KIEWIET, D. R., and MCCUBBINS, M. D. 1991. *The Logic of Delegation*. Chicago: Chicago University Press.

KOUSSER, J. M. 1974. *The Shaping of Southern Politics*. New Haven, Conn.: Yale University Press.

KREHBIEL, K. 1991. *Information and Legislative Organization*. Ann Arbor: Michigan University Press.

LAITIN, D. 1988. Language games. *Comparative Politics*, 20: 389–402.

LAVER, M., and Schofield, N. 1990. *Multiparty Government*. New York: Oxford University Press.

—— and SHEPSLE, K. A. 1995. *Making and Breaking Governments*. Cambridge: Cambridge University Press.

LEVI, M. 1988. *Of Rule and Revenue*. Berkeley: University of California Press.

LOHMANN, S. 1994. Federalism and central bank autonomy: the politics of German monetary policy, 1960–1989. Mimeo., Department of Political Science, UCLA.

LUPIA, A., and MCCUBBINS, M. D. 1995. *Knowledge, Power, and Democracy*. Mimeo., University of California, San Diego.

MATTHEWS, S. 1989. Veto threats: rhetoric in bargaining. *Quarterly Journal of Economics*, 104: 347–69.

MCCARTY, N. M., and POOLE, K. T. 1995. Veto power and legislation: an empirical analysis of executive and legislative bargaining from 1961 to 1986. *Journal of Law, Economics, and Organization*, 11: 282–312.

MCCUBBINS, M. D. 1989. Party governance and U.S. budget deficits: divided government and fiscal stalemate. Pp. 83–111 in *Politics and Economics in the Eighties*, ed. A. Alesina and G. Carliner. Chicago: University of Chicago Press.

—— and SCHWARTZ, T. 1984. Oversight overlooked: police patrols vs. fire alarms. *American Journal of Political Science*, 28: 165–79.

—— NOLL, R. G.; and WEINGAST, B. R. 1987. Administrative procedures as instruments of political control. *Journal of Law, Economics, and Organization*, 3: 243–77.

—— —— —— 1989. Structure and process, politics and policy: administrative arrangements and the political control of agencies. *Virginia Law Review*, 75: 431–82.

MCKELVEY, R. 1976. Intransitivities in multidimensional voting models and some implications for agenda control. *Journal of Economic Theory*, 12: 472–82.

MILGROM, P., and ROBERTS, J. 1991. *Economics, Organizations, and Management*. Englewood Cliffs, N.J.: Prentice-Hall.

MILLER, G. J., and HAMMOND, T. H. 1989. Stability and efficiency in a separation-of-powers constitutional system. In *The Federalist Papers and the New Institutionalism*, ed. B. Grofman and D. Wittman. New York: Agathon Press.

MOE, T. M. 1985*a*. The politicized presidency. Pp. 235–71 in *The New Direction in American Politics*, ed. J. E. Chubb and P. E. Peterson. Washington, D.C.: Brookings Institution.

—— 1985*b*. Control and feedback in economic regulation: the case of the NLRB. *American Political Science Review*, 79: 1094–116.

—— 1987*a*. An assessment of the positive theory of congressional dominance. *Legislative Studies Quarterly*, 12: 475–520.

—— 1987*b*. Interests, institutions, and positive theory: the politics of the NLRB. *Studies in American Political Development*, 2: 236–99.

—— 1989. The political structure of agencies. Pp. 267–329 in *Can the Government Govern?*, ed. J. E. Chubb and P. E. Peterson. Washington, D.C.: Brookings Institution.

MONTINOLA, G.; QIAN, Y.; and WEINGAST, B. R. 1995. Federalism, Chinese style: the political basis for economic reform in China. *World Politics*, 48: 50–81.

NOLL, R. G. 1987. The political foundations of regulatory policy. Pp. 462–92 in *Congress: Structure and Policy*, ed. M. D. McCubbins and T. Sullivan. New York: Cambridge University Press.

—— 1989. *Economic perspectives on the politics of regulation.* Vol ii, pp. 1253–87 in *Handbook of Industrial Organization*, ed. R. Schmalensee and R. D. Willig. Amsterdam: North-Holland.

—— and KRIER, J. 1990. Some implications of cognitive psychology for risk regulation. *Journal of Legal Studies*, 29: 747–79.

NORTH, D. C. 1990. *Institutions, Institutional Change, and Economic Performance.* New York: Cambridge University Press.

—— and THOMAS, R. 1973. *The Rise of the Western World.* New York: Norton.

—— and WEINGAST, B. R. 1989. Constitutions and commitment: the evolution of institutions governing public choice in 17th century England. *Journal of Economic History*, 49: 803–32.

O'HALLORAN, S. 1994. *Politics, Process, and American Trade Policy.* Ann Arbor: University of Michigan Press.

ORDESHOOK, P. C. 1993. Constitutional stability. *Constitutional Political Economy*, 3: 137–75

OWEN, G., and GROFMAN, B. 1988. Optimal partisan gerrymandering. *Political Geography Quarterly*, 7: 5–22.

PERSSON, T., and TABELLINI, G. 1990. *Macroeconomic Policy, Credibility, and Politics.* London: Harwood.

POLSBY, N. W. 1986. *Congress and the Presidency.* 4th edn. Englewood Cliffs, N.J.: Prentice-Hall.

POOLE, K., and ROSENTHAL, H. 1995. *Congress: A Quantitative History of Roll Call Voting: 1789–1989.* New York: Oxford University Press.

PRZEWORSKI, A. 1991. *Democracy and the Market.* New York: Cambridge University Press.

RAMSEYER, J. M., and ROSENBLUTH, F. M. 1995. *The Politics of Oligarchy: Institutional Choice in Imperial Japan.* New York: Cambridge University Press.

RIKER, W. H. 1980. Implications from the disequilibrium of majority rule for the study of institutions. *American Political Science Review*, 74: 432–46.

—— 1982. *Liberalism Against Populism.* San Francisco: W.H. Freeman.

ROBACK, J. forthcoming. *An Imaginary Negro in an Impossible Place: The Territories and Secession.* Princeton, N.J.: Princeton University Press.

Rodriguez, D. 1994. The positive political dimensions of regulatory reform. *Washington University Law Quarterly*, 72: 1–150.

Rogowski, R. 1989. *Commerce and Coalitions.* Princeton, N.J.: Princeton University Press.

Romer, T., and Rosenthal, H. 1978. Political resource allocation, controlled agendas, and the status quo. *Public Choice*, 33: 27–43.

—— —— 1985. Modern political economy and the study of regulation. Pp. 73–116 in *Public Regulation: New Perspectives on Institutions and Politics*, ed. E. E. Bailey. Cambridge, Mass.: MIT Press.

Root, H. L. 1994. *The Fountain of Privilege: Political Foundations of Market in Old Regime France and England.* Berkeley: University of California Press.

Rosenthal, H. 1990. The setter model. In *Advances in the Spatial Theory of Voting*, ed. J. M. Enelow and M. J. Hinich. New York: Cambridge University Press.

Rosenthal, J.-L. 1992. *The Fruits of Revolution: Property Rights, Litigation, and French Agriculture, 1700–1860.* New York: Cambridge University Press.

Rothenberg, L. S. 1994. *Regulation, Organizations, and Politics.* Ann Arbor: University of Michigan Press.

Shepsle, K. A. 1986. Institutional equilibrium and equilibrium institutions. Pp. 51–81 in *Political Science: The Science of Politics*, ed. H. Weisberg. New York: Agathon.

—— 1991. Discretion, institutions, and the problem of government commitment. Pp. 245–63 in *Social Theory for a Changing Society*, ed. P. Bourdieu and J. Coleman. Boulder, Colo.: Westview.

—— and Weingast, B. R. 1987. The institutional foundations of committee power. *American Political Science Review*, 81: 85–194.

Silbey, J. H. 1985. *The Partisan Imperative: The Dynamics of American Politics Before the Civil War.* New York: Oxford University Press.

Skocpol, T. 1979. *States and Social Revolutions.* New York: Cambridge University Press.

Snyder, J. 1991. On buying legislatures. *Economics and Politics*, 3: 93–109.

Snyder, S. K., and Weingast, B. R. 1994. The American system of shared powers: the president, Congress, and the NLRB. Working Paper, Hoover Institution, Stanford University.

Strom, K. 1990. *Minority Government and Majority Rule.* New York: Cambridge University Press.

Tsebelis, G. 1990. *Nested Games.* Berkeley: University of California Press.

Weingast, B. R. 1995*a*. *Institutions and Political Commitment: A New Political Economy of the American Civil War Era.* Mimeo., Stanford University.

—— 1995*b*. The economic role of political institutions: federalism, markets, and economic development. *Journal of Law, Economics, and Organization*, 11: 1–31.

—— and Moran, M. 1983. Bureaucratic discretion or congressional control? *Journal of Political Economy*, 91: 765–800.

Williamson, O. E. 1985. *The Economic Institutions of Capitalism.* New York: Free Press.

—— 1994. The institutions and governance of economic development and reform. Working Paper, University of California, Berkeley.

Chapter 6

# Political Institutions: Legal Perspectives

Gavin Drewry

IT is a truism, though sadly one that both political scientists and lawyers sometimes seem to forget, that law and politics—in both their theoretical and their practical aspects—are very closely interconnected. In a lecture published in 1882 entitled "The History of English Law as a Branch of Politics," the English jurist Sir Frederick Pollock wrote that "law is to political institutions as the bones to the body" (Pollock 1882: 200–1).

Pollock's metaphor (taken perhaps rather misleadingly out of its context: see below) seems at least as apt today as it was then, for a wide variety of interrelated reasons. Consequently, it is not surprising to discover that the literature of political science is peppered with legal concepts and terminology, and that some of it addresses more or less directly and explicitly the relationships between law and politics and between legal and political actors and institutions. Constitutional issues, in particular, are frequently addressed from a hybrid legal and political science perspective, the balance depending on whether the writer is primarily a lawyer or a political scientist. Public law is woven tightly into the fabric of public administration, albeit more tightly in some countries than in others. The rise of "new institutionalism" within political science promises to strengthen all these ties (Smith 1988).

To date, however, the quantum of legal peppering found in political science has not been nearly as abundant as it might be. This in part because there has also been a countervailing tendency among some political scientists to reject legal approaches to their discipline as unpalatably formalistic and old-fashioned. The famous "behavioral revolution" in American political science was substantially a reaction to this; and behaviorists, having thrown away the formalistic bath-water, are not easily persuaded that there might be good arguments for retrieving and reviving the discarded legal baby. For their part, many academic lawyers—even ones specializing in public law—have tended to regard their subject as pre-eminently practice-relevant and to have perceived excursions into the unfamiliar territory of political science as a waste of time.

In Britain, this antipathy towards legal and constitutional aspects of political science has been compounded by the absence of a codified constitution and by the underdevelopment of public law. Membership of the European Union, and the growing impact of the European Court of Justice on the British courts and on British politics—coinciding with continuing debates about constitutional reform, including the possibility of incorporating the European Convention on Human Rights into U.K. law—have highlighted the relevance of legal approaches to political science. Still, there remains a very big lacuna in the literature and a continuing absence of cross-disciplinary academic research.

## I The nature of the affinity between law and politics

Why and in what respects are politics and law so closely interrelated? Some of the links are obvious. For one thing, constitutions and rules of public law, and the courts that interpret and apply them, set the formal ground-rules of political practice and provide important mechanisms for governmental accountability and constraint. Any quantitative study of, for instance, voting behavior or legislative behavior must be informed by a working knowledge of electoral or legislative rules if the research is to be done properly and the research findings made sense of.

International relations are underpinned by international law. Courts can be arenas for pressure group activity, via constitutional and legal challenges to government, public interest litigation and test-case strategies (Harlow and Rawlings 1992). The appointment of judges to such courts is a matter of recurrent interest to politicians and political scientists; the academic and media attention given to U.S. Senate hearings on the ratification of presidential nominees to the Supreme Court are a particularly high profile instance of this (for a recent discussion, see Vieira and Cross 1990).

Law-making is a clear manifestation of state power. Laws are the medium through which policy is translated into action. Laws are an important resource for public policy-making (Rose 1986). Legislatures are not only, by definition, law-making bodies but also arenas in which politicians operate—and in many countries a high proportion of legislators themselves have legal backgrounds. The convergence of legal and political and official careers has been a well-recognized phenomenon, certainly since the time of de Tocqueville, who in 1835 observed (without irony) that "as the lawyers constitute the only enlightened class which the people does not mistrust they are naturally called upon to occupy most of the public stations" (Tocqueville 1946: 206–7).

Like another Frenchman, Moliére's M. Jourdain, who was astonished to find that he had been talking prose all his life without realizing it, even the most anti-legalistic political scientist may, on thinking about it, be surprised to find just how deeply legal ideas have penetrated the fabric and the day-to-day vocabulary of political theory and discourse. The social contracts discussed by Hobbes, Locke and Rousseau have legal resonance; so do ubiquitous concepts such as legitimacy, rights and justice. Many of the "great masters" of political thought—Machiavelli, Bentham, Marx, Hegel and so on—feature as prominently on the reading lists of university courses in theoretical jurisprudence as on ones for courses on the history of political thought. Moreover, they are often joined nowadays by more modern names like Rawls, Nozick and Foucault. A quick glance at many current jurisprudence textbooks (see, for instance, Freeman 1994) will serve to verify this. It is by no means unknown for non-lawyer political scientists to venture an important contribution to legal theory (e.g. Shklar 1964).

## II The neglect of legal aspects of politics in the UK

Having made some play with Pollock's "bones of the body" metaphor, I must acknowledge the apparent irony of its having been coined by the native of a country with no codified constitution—and in the same decade that A. V. Dicey was cautioning right-thinking Englishmen about the threat to the rule of law that would arise from any attempt to transplant French-style administrative law into the well-tilled soil of the English common law. Dicey had principally in mind that aspect of the rule of law which meant, in his words, "equality before the law, or the equal subjection of all classes to the ordinary law of the land administered by the ordinary law courts" (Dicey 1959: 193).

Pollock was not writing for a modern audience of political scientists but for political and constitutional historians, encouraging them to develop an understanding of the legal ramifications of their subject. At the turn of the century (and indeed for a long time after), most lawyers who read Dicey accepted his views as axiomatic. There is no reason to suppose that Pollock would have been any exception.

But if we take liberties with the context and apply Pollock's metaphor to modern British political science, we find the latter to be a singularly invertebrate discipline. Textbooks on British government and politics seldom refer, even in passing, to courts or judges or to public law. British political science journals contain few articles on law-related subjects. There is plenty of politics-related material in British legal journals but little evidence that political scientists make much use of it (Drewry 1991). The sharpness of the contrasts with other countries—particularly the United States—can hardly be exaggerated. The legal systems of Britain and the U.S. are both rooted in the same common-law inheritance, but the constitutional and legal differences between the two countries are immense, and this is reflected in their respective political science literatures. The absence in any textbook of U.S. government and politics of a substantial discussion of judicial review and the role of the Supreme Court would be a serious omission, as will be discussed below.

Yet there was a pre-modern era of British, and American, political studies which did draw heavily and explicitly upon legal scholarship. W. J. M. Mackenzie has observed that, "in the generation before 1914 it would have been inconceivable that one should discuss political systems without also discussing legal systems" (Mackenzie 1967: 278). Bernard Crick suggests that the foundations of political science in the United States, laid in the mid-19th century, "in part arose to fill th[e] gap in learning left by the decay of Jurisprudence in American law schools" (Crick 1959: 13). Likewise, Fred Ridley has suggested that, in America, jurisprudence was "the godfather of political science" (Ridley 1975: 179).

So far as English political scholarship is concerned, Mackenzie cites the important intellectual foundations laid by writers such as Maine, Pollock, Maitland and Vinogradoff as exemplars of this kind of legally based approach. But the tradition died in the interwar years, and UK political science ceased to draw much of its inspiration from the discipline of law. We have already noted that this neglect was later compounded by the anti-formalist sentiments implicit in the "behavioral revolution" reaching Britain in the 1970s.

The ever-growing entanglement of British legal and governmental institutions with those of other European countries—all of which have codified constitutions, and most of which have more markedly juridified political cultures and better developed systems of administrative law—is beginning to erode the contrast between Britain and its European neighbors. Constitutional and legal issues are thereby being brought more prominently onto the agenda not only of British political science but also of British politics itself.[1] The process has been slow and patchy, however.

## III Politics and the judicial role

As we noted earlier, law is a product of the legislative process. Judges too have a role in that process and more generally (particularly when sitting in a Supreme Court) as political actors. Even in the UK, where judicial creativity is discouraged by a doctrine of parliamentary sovereignty and by the absence of a codified constitution and a Supreme Court, the role of judges both in curbing abuses of state power and as interstitial law-makers has come increasingly to be recognized, even by the judges themselves (Drewry 1992*a*).

There is a very considerable international and cross-national comparative literature on the political role and behavior of judges.[2] Some of this literature takes a positive line about judicial activism in political contexts, as protection against abuses of executive power and against bureaucratic oppression and maladministration.

But other writers stress the dysfunctional consequences of substituting judicial decisions for political ones, or allowing judges to second-guess politically accountable decision-makers in the context of litigation. Thus British public lawyer John Griffith has written in deeply skeptical vein about the wisdom of encouraging non-accountable, activist judges to constrain politicians (for example, via a U.K. Bill of Rights).[3] And these are themes which are echoed even by jurisprudential theorists of rights themselves (Waldron 1993).

From a very different but also skeptical perspective, the American academic Donald Horowitz (1977*a*; 1977*b*; 1982) has written about the role of the judiciary in social policy and has also highlighted the dangers of over-reliance on the courts as guardians of the public interest. He notes that

[1] Note particularly the work of Charter 88 and discussions of a written constitution for the United Kingdom growing out of it: Allan 1993; Barnett, Ellis and Hirst 1993; Institute for Public Policy Research 1993; and Jowell and Oliver 1994.

[2] See for instance: Abraham 1993; Bell 1983; Friendly 1963; Jaffe 1969; Katzmann 1988; Schmidhauser 1987; Volcansek 1992.

[3] Griffith is also the author of a much-discussed scholarly polemic about ideological predilections of the British judiciary (Griffith 1979; 1991; for a critique, see Lee 1989) and has written, from a similar left-of-center standpoint, a historical outline of 20th-century judicial politics in Britain (Griffith 1993).

judges may have assumed new roles in administrative law litigation, but they continue to act very much within the framework of an old process—a process that evolved, not to devise new programs or to oversee administration, but to decide controversies—and the constraints of that process operate to limit the range of what can reasonably be expected from courts (Horowitz 1977*b*: 151). Thus, he continues:

> Courts are public decision makers, yet they are wholly dependent on private initiative to invoke their powers: they do not self-start. Parties affected by administrative action choose to seek or not to seek judicial redress on the basis of considerations that may bear no relation to the public importance of the issues at stake, to the recurring character of the administrative action in question, or to the competence of courts to judge the action or change it.

Variants of Horowitz's arguments are echoed by many other writers, including the American Lon Fuller (1978) and the English administrative lawyer Peter Cane (1986). Cane argues that:

> Because judicial proceedings are essentially bipolar, they are designed to resolve disputes in terms of the interests of only two parties or groups represented by those parties. And, because judicial proceedings are adversarial, disputes are to be decided only on the basis of material which the parties choose to put before the courts. If the problem is one which is felt to require, for its proper resolution, the consideration of interests of parties not before the court and not in formal dispute with one another, of persons who will be affected consequentially or incidentally by any resolution of the dispute between the parties, then a court is not the ideal body to resolve that dispute (Cane 1986: 149).

The problem is exacerbated by the incapacity of the courts to assimilate highly technical evidence and by their lack of a facility for monitoring consequences.

## IV Public law and public administration

Whether or not one subscribes to Weberian orthodoxy about the inherently rational-legal basis of bureaucratic organization, law in its many and diverse manifestations is self-evidently a core element in any developed public administration system.[4] Law in various shapes and forms is the

[4] Consider, for instance: legal constraints and entitlements; a presumed respect by public servants for legality and due process and the rule of law; provision of adjudicative and investigative procedures for dispute resolution; and the consistency of practice and the following of precedent.

medium through which much public policy is delivered. Public bodies and public officials are subject to the jurisdictions of courts and tribunals, auditors and ombudsmen. Law is an important and costly public service, just like public health or welfare, and it has to be managed and administered by public departments (in most countries, ministries of justice) set up for the purpose: this can (as has recently been the case in the U.K.) give rise to tensions about perceived executive insensitivity to the importance of judicial independence (Browne-Wilkinson 1988; Drewry 1992*b*). The administration of justice is part of the substantive agenda of academic public administration.

Public law may have its limitations as a means of constraint and redress against public administrators, but it has always been an essential element of the study and the practise of public administration. In his essay on the linkage between public law and public administration in the US, Phillip Cooper shows that the pioneers of the academic study of public administration—scholars like Frank Goodnow, Woodrow Wilson, Ernst Freund and John Dickinson—saw the discipline of public administration as being firmly rooted in public law. Indeed, Wilson saw himself as a professor of public law and nursed ambitions "to found a school of public law with public administration as a major unit within that institution" (Cooper 1990: 256). As Andrew Dunsire (writing from a British perspective) points out, Wilson's famous essay on "The Study of Administration," first published in 1887, sets up a working definition of public administration as "detailed and systematic execution of public law" (Dunsire 1973: 88; Wilson 1887).

Following Cooper, this writer has written an essay on the contribution of public law to the development of the discipline of public administration in the U.K. (Drewry 1995) and lamented the unrealized potential for closer collaboration between lawyers and public administrators (Drewry 1986). These arguments need not be repeated here.

In this context, the contrast between Britain and continental Europe is particularly evident. In most European countries, much political discourse is couched in a legalistic vocabulary, and public bureaucracies are regulated by legal codes interpreted and applied by administrative courts. As the authors of a respected 1950s textbook on British central government observed:

> in all the countries of Western Europe except Britain it has been the tradition for centuries that the most important posts in central administration should be filled by men [*sic*] trained in the Law Faculties of the Universities . . . [and, some exceptions notwithstanding] a fair contrast can be drawn between the position of lawyers in British bureaucracy and in that of

> Western Europe. In the former they are advisers to the administration, in the latter they are the administration itself (Mackenzie and Grove 1957: 96).

In general, the continental administrator is (as C. H. Sisson (1959: 39) wrote at about the same time as Mackenzie and Grove) "a lawyer, specialising in that branch of law—namely administrative law—which is mostly concerned with the functions of government." The British generalist civil servant, untrained in law or indeed in any other professional discipline, manifestly is not.

Enduring Diceyan prejudices may account at least in part for the tardy development of administrative law in Britain. Dicey himself died in 1922, and it was in the 1920s that the writings of jurists like Sir Cecil Carr (Carr 1921), W. A. Robson (Robson 1928) and F. J. Port (Port 1929) began to lay the intellectual foundations of the process by which administrative law gradually won acceptance as a distinctive and professionally respectable area of legal theory and practice. This process accelerated after the Second World War, and the 1950s saw a major upsurge of interest in various aspects of the subject.

The early British writers on administrative law were preoccupied with judicial control of government. However, as Fred Ridley (1984: 4) has observed, "the idea of 'political' rather than 'legal' protection of citizens against administration is deeply embedded in British political traditions and has imprinted itself on British ways of thought." And the substance of public debate, particularly in the 1940s and 1950s, about the state of administrative law had relatively little to do with the judicial process as such.

Much of the discussion about redress of citizens' grievances that took place in this period had to do with non-judicial aspects of administrative law: in particular with administrative tribunals and inquiries and with the possibility of setting up an ombudsman to investigate complaints of maladministration by central government. The latter debate culminated in the passing of the Parliamentary Commissioner Act 1967 which established an ombudsman to investigate complaints of maladministration in central government (see Stacey 1971; Gregory and Hutchesson 1975). The Parliamentary Commissioner for Administration operates within the constitutional framework of ministerial responsibility to Parliament and was always perceived as an adjunct to and not a substitute for the traditional role of the Member of Parliament in taking up constituents' grievances. Many legal academics writing about administrative law during this period tended to regard such developments as being more the business of political science than of law.

Though they still lean heavily in the direction of judicial review, most current British textbooks on administrative law (of which there are a great many) do include substantial coverage of non-judicial remedies, and some of them take pains to cover the political angles. Public lawyers have done useful work at the boundary between legal and political theory (for instance Loughlin 1992); indeed one such writer, Paul Craig (1990), writing from an Anglo-American comparative standpoint and undertaking an encyclopedic review of a large literature, suggests that no one should tackle the subject of public law at all until they have sorted out the exact nature of the political theory upon which their analysis is founded. A book by Birkinshaw (1985) and an edited volume by Richardson and Genn (1994) are two good examples of writing by contemporary British public lawyers which approaches substantive issues of administrative law with a good awareness of the wider political and theoretical context.

Judicial review of administrative action has been a minor growth industry in Britain in recent years, with various procedural reforms facilitating access to the courts and administrative law cases being tried mainly by a semi-specialist group of "Crown Office List" judges in the High Court. There has been some research into the use and impact of judicial review, mainly by legal academics, such as Maurice Sunkin, under the auspices of the Public Law Project.[5] But the subject has yet to catch the imagination of many political scientists.

In many parts of the world, old-style public administration is being displaced by a new market-orientated style of public management—involving smaller public bureaucracies and the privatization and contracting out of public services. Such developments have substantial legal and constitutional implications which have not yet been fully worked out, and which are only just beginning to be discussed in the (mainly legal) academic literature (e.g. Baldwin and McCrudden 1987; Harden 1992).

## V The United States: law and the constitution

In *Democracy in America*, part I of which was published in 1835, Alexis de Tocqueville observed that "there is hardly a political question in the United States which does not sooner or later turn into a judicial one." And he opined that "the power vested in the American courts of justice of pronouncing a statute to be unconstitutional forms one of the most

[5] Sunkin 1987; 1991*a*, *b*; Le Sueur 1991; Le Sueur and Sunkin 1992; Sunkin *et al.* 1993.

important barriers which has ever been devised against the tyranny of political assemblies" (Tocqueville 1946: 83). Herein lies the main explanation for the difference between the U.K. and the U.S., noted above. As the authors of an important, though now rather dated, Anglo-American comparative study of public law have observed: "Constitutional issues permeate American law and life to an extent that foreign observers find incredible. Americans have become a people of constitutionalists, who substitute litigation for legislation and see constitutional issues lurking in every case" (Schwartz and Wade 1972: 6). A propensity for extensive and expensive litigiousness, in both public and private law, is a well-marked if not universally admired feature of American life and culture. And the US Supreme Court, which has a central role in government that compels close attention by political scientists, has no equivalent in Britain.[6]

The academic literature on the U.S. Supreme Court is substantial and diverse: there is even an "Oxford Companion" (Hall 1992). The court has been written about from a very wide range of perspectives and direction;[7] and there is a vast journal literature. The American literature on the judicial role and behavior is substantially orientated towards the Supreme Court, as is the literature on constitutional interpretation (Tribe 1988; Wellington 1990; Sunstein 1990; 1993). Apart from works specifically devoted to the Court, the constitutional and democratic implications of judicial review crop up throughout the wider political science literature, including modern classics such as Dahl's 1956 *Preface to Democratic Theory* and Lowi's *The End of Liberalism*—which calls *inter alia* for the Supreme Court to restore the rule of law by striking down powers delegated to an administrative agency that are not accompanied by clearly defined standards of implementation (Lowi 1979: 300).[8]

As with the Supreme Court, so much has been written about the U.S. Constitution—by both political scientists and lawyers—that it is hard to know where to begin. There are many important political-historical studies (Bailyn 1967; Wills 1978); and many more modern interpretations (Corwin 1978; Tribe 1988; Sunstein 1990; 1993). Aspects of constitutionalism have generated a substantial output of writing from scholars of United States government—from Vile's essay on the separation of powers (Vile

[6] Though there has been a handful of major studies of the judicial role of the House of Lords (Blom-Cooper and Drewry 1972; Stevens 1979; Paterson 1982). Paterson's study is based on unique data from interviews with serving law lords—his success underlining the difficulty that British researchers into judicial behavior (far more than their American counterparts) have experienced in gaining access to the judges.

[7] See for instance: Hodder-Williams 1980; McKeever 1993; O'Brien 1993; Schmidhauser and Berg 1972; Wolfe 1986; Woodward and Armstrong 1979.

[8] For recent developments in US administrative law, see: Stewart 1975; Breyer and Stewart 1979; Mashaw 1985; Sunstein 1990; Lessig and Sunstein 1994.

1967) to more recent work by Jon Elster and others (Elster and Slagstad 1988). A recent collection of essays examines comparatively the constitutional implications of political change in Eastern Europe and elsewhere (Greenberg *et al.* 1993), building on earlier comparative work on the political role of courts more generally (Shapiro 1981).

## VI Conclusions

There is a natural affinity between law and politics, which takes many forms. Almost every aspect of political activity and political change—at sub-national, national, international or global levels—has a legal or a constitutional aspect to it. This chapter, which has concentrated on two contrasting national examples, has sought to illustrate the point but does not claim to have done more than scratch the surface of such a vast subject.

Political scientists ignore the legal dimensions of their discipline at their peril. But having said that, it is hardly realistic to expect political scientists to transform themselves into lawyers, or vice versa. The way forward lies in including adequate coverage of legal aspects in social science curricula; in more cross-disciplinary research collaboration; and, more modestly, in lawyers and political scientists reading one another's literatures. This writer proffers his views with due diffidence, from his own national perspective in a country where the gulf between law and political science is particularly wide. Still, it is hoped that this chapter has provided some modest encouragement and guidance in that direction.

## References

The interplay between law and politics has so many aspects that it is difficult to identify an appropriate point of entry for the non-lawyer. Perhaps the best place to start is by browsing in *The Oxford Companion to Law* (1980), edited by D. M. Walker; the anglocentric bias of this can be offset by referring to *The Oxford Companion to the US Supreme Court* (Hall 1992). Essays referred to in the text by P. J. Cooper (Cooper 1990) and this writer (Drewry 1995) outline the links between public administration and law in the United States and the UK respectively. The political theory underlying public law is usefully discussed by Martin Loughlin's *Public Law and Political Theory* (1992).

Abraham, H. J. 1993. *The Judicial Process.* 6th edn. New York: Oxford University Press.
Allan, T. R. S. 1993. *Liberty and Justice: The Legal Foundations of British Constitutionalism.* Oxford: Clarendon Press.

Bailyn, B. 1967. *The Ideological Origins of the American Revolution.* Cambridge, Mass.: Harvard University Press.

Baldwin, R., and McCrudden, C., eds. 1987. *Regulation and Public Law.* London: Weidenfeld and Nicolson.

Barnett, A.; Ellis, C.; and Hirst, P., eds. 1993. *Debating the Constitution.* Oxford: Polity.

Bell, J. 1983. *Policy Arguments in Judicial Decisions.* Oxford: Clarendon Press.

Birkinshaw, P. 1985. *Grievances, Remedies and the State.* London: Sweet and Maxwell.

Blom-Cooper, L. and Drewry, G. 1972. *Final Appeal: A Study of the House of Lords in its Judicial Capacity.* Oxford: Clarendon Press.

Breyer, S. G., and Stewart, R. B. 1979. *Administrative Law and Regulatory Policy.* Boston: Little, Brown.

Browne-Wilkinson, N. 1988. The independence of the judiciary in the 1980s. *Public Law,* 1988: 44–57.

Cane, P. 1986. *An Introduction to Administrative Law.* Oxford: Clarendon Press.

Carr, C. T. 1921. *Delegated Legislation.* London: Oxford University Press.

Cooper, P. J. 1990. Public law and public administration: the state of the union. Pp. 256–84 in *Public Administration: The State of the Discipline,* ed. N. B. Lynn and A. Wildavsky. Chatham, N.J.: Chatham House.

Corwin, E. S. 1978. *The Constitution and What it Means Today,* rev. H. W. Chase and C. R. Ducat. 14th edn. Princeton, N.J.: Princeton University Press.

Craig, P. P. 1990. *Public Law and Democracy in the United Kingdom and the United States of America.* Oxford: Clarendon Press.

Crick, B. 1959. *The American Science of Politics.* London: Routledge.

Dahl, R. A. 1956. *A Preface to Democratic Theory.* Chicago: University of Chicago Press.

Dicey, A.V. 1959. *Introduction to the Study of the Law of the Constitution.* 10th edn. London: Macmillan; originally published 1885.

Drewry, G. 1986. Lawyers and public administrators: prospects for an alliance? *Public Administration,* 64: 173–88.

—— 1991. British law journals as a resource for political studies. *Political Studies,* 39: 560–7.

—— 1992*a.* Judicial politics in Britain: patrolling the boundaries. *West European Politics,* 15: 9–28.

—— 1992*b.* Justice and public administration: some constitutional tensions. Vol. xlv, pt. 2, pp. 187–212 in *Current Legal Problems,* ed. R. W. Rideout and B. A. Hepple. Oxford: Oxford University Press.

—— 1995. Public Law (Special Issue on British public administration: the state of the discipline). *Public Administration,* 73: 41–57.

Dunsire, A. 1973. *Administration.* London: Martin Robertson.

Elster, J., and Slagstad R., eds. 1988. *Constitutionalism and Democracy.* Cambridge: Cambridge University Press.

Freeman, M. D. A. 1994 . *Lloyd's Introduction to Jurisprudence.* 6th edn. London: Sweet and Maxwell.

Friendly, H. J. 1963. The gap in lawmaking—judges who can't and legislators who won't. *Columbia Law Review,* 63: 787–807.

Fuller, L. 1978. The forms and limits of adjudication. *Harvard Law Review,* 92: 353–409.

Greenberg, D. *et al.,* eds. 1993. *Constitutionalism and Democracy: Transitions in the Contemporary World.* New York: Oxford University Press.

Gregory, R., and Hutchesson, P. 1975. *The Parliamentary Ombudsman.* London: Allen and Unwin.

Griffith, J. A. G. 1979. The political constitution. *Modern Law Review,* 42: 1–21.

—— 1991. *The Politics of the Judiciary.* 4th edn. London: Fontana.

—— 1993. *Judicial Politics since 1920.* Oxford: Blackwell.
HALL, K. L., ed. 1992. *The Oxford Companion to the Supreme Court of the United States.* New York: Oxford University Press.
HARDEN, I. 1992. *The Contracting State.* Buckingham: Open University Press.
HARLOW, C., and RAWLINGS, R. 1992. *Pressure Through Law.* London: Routledge.
HODDER-WILLIAMS, R. 1980. *The Politics of the US Supreme Court.* London: Allen and Unwin.
HOROWITZ, D. L. 1977*a*. *The Courts and Social Policy.* Washington D.C.: Brookings Institution.
—— 1977*b*. The courts as guardians of the public interest. *Public Administration Review*, 37: 148–54.
—— 1982. The judiciary: umpire or empire. *Law and Human Behavior*, 6: 129–43.
INSTITUTE FOR PUBLIC POLICY RESEARCH. 1993. *A Written Constitution for the United Kingdom.* London: Mansell.
JAFFE, L. 1969. *English and American Judges as Lawmakers.* Oxford: Clarendon Press.
JOWELL, J., and OLIVER, D. 1994. *The Changing Constitution.* 3rd edn. Oxford: Clarendon Press.
KATZMANN, R. A., ed. 1988. *Judges and Legislators: Towards Institutional Comity.* Washington D.C.: Brookings Institution.
LE SUEUR, A. P. 1991. Is judicial review undemocratic? *Parliamentary Affairs*, 44: 283–97.
—— and SUNKIN, M. 1992. Applications for judicial review: the requirement of leave. *Public Law*: 102–29.
LEE, S. 1989. *Judging Judges.* London: Faber and Faber.
LESSIG, L., and SUNSTEIN, C. R. 1994. The president and the administration. *Columbia Law Review*, 94, 1–124.
LOUGHLIN, M. 1992. *Public Law and Political Theory.* Oxford: Clarendon Press.
LOWI, T. J. 1979. *The End of Liberalism.* 2nd edn. New York: Norton.
MCKEEVER, R. 1993. *Raw Judicial Power.* Manchester: Manchester University Press.
MACKENZIE, W. J. M. 1967. *Politics and Social Science.* Harmondsworth: Penguin.
—— and GROVE, J. W. 1957. *Central Administration in Britain.* London: Longmans.
MASHAW, J. L. 1985. *Due Process in the Administrative State.* New Haven, Conn.: Yale University Press.
O'BRIEN, D. M. 1993. *Storm Center.* New York: Norton.
PATERSON, A. 1982. *The Law Lords.* London: Macmillan.
POLLOCK, F. 1882. *Essays in Jurisprudence and Ethics.* London: Macmillan.
PORT, F. J. 1929. *Administrative Law.* London: Oxford University Press.
RICHARDSON, G., and GENN, H. 1994. *Administrative Law and Government Action.* Oxford: Clarendon Press.
RIDLEY, F. F. 1975. *The Study of Government.* London: Allen and Unwin.
—— 1984. The citizen against authority: British approaches to the redress of grievances. *Parliamentary Affairs*, 37: 1–32.
ROBSON, W. A. 1928. *Justice and Administrative Law.* London: Macmillan.
ROSE, R. 1986. Law as a resource of public policy. *Parliamentary Affairs*, 39: 297–314.
SCHMIDHAUSER, J. R., and BERG, L. L. 1972. *The Supreme Court and Congress: Conflict and Interaction, 1945–1968.* New York: Free Press.
—— ed. 1987. *Comparative Judicial Systems.* London: Butterworth.
SCHWARTZ, B., and WADE, H. W. R. 1972. *Legal Control of Government.* Oxford: Clarendon Press.
SHAPIRO, M. M. 1981. *Courts: A Comparative and Political Analysis.* Chicago: University of Chicago Press.

Shklar, J. 1964. *Legalism.* Cambridge, Mass.: Harvard University Press.
Sisson, C. H. 1959. *The Spirit of British Administration.* London: Faber and Faber.
Smith, R. M. 1988. Political jurisprudence, the "new institutionalism" and the future of public law. *American Political Science Review,* 82: 89–108.
Stacey, F. 1971. *The British Ombudsman.* Oxford: Clarendon Press.
Stewart, R. B. 1975. The reformation of American administrative law. *Harvard Law Review,* 88: 1669–813.
Stevens, R. 1979. *Law and Politics: The House of Lords as a Judicial Body, 1800–1976.* London: Weidenfeld and Nicolson.
Sunkin, M. 1987. What is happening to applications for judicial review? *Modern Law Review,* 50: 432–67.
—— 1991*a.* The judicial review case-load, 1987–1989. *Public Law,* 1991: 490–9.
—— 1991*b.* Can government control judicial review? *Current Legal Problems,* 44: 161–83.
—— *et al.* 1993. *Judicial Review in Perspective.* London: Public Law Project.
Sunstein, C. R. 1990. *After the Rights Revolution.* Cambridge, Mass.: Harvard University Press.
—— 1993. *The Partial Constitution.* Cambridge, Mass.: Harvard University Press.
Tocqueville, A. de. 1946. *Democracy in America.* Abridged World Classics edn. London: Oxford University Press.
Tribe, L. H. 1988. *American Constitutional Law.* 2nd edn. Mineola, N.Y.: Foundation Press.
Vieira, N., and Cross, L. E. 1990. The appointment clause: Judge Bork and the role of ideology in judicial confirmations. *Journal of Legal History,* 11: 311–52.
Vile, M. J. C. 1967. *Constitutionalism and the Separation of Powers.* Oxford: Clarendon Press.
Volcansek, M. L., ed. 1992. *Judicial Politics and Policy Making in Western Europe.* London: Frank Cass.
Waldron, J. 1993. A right-based critique of constitutional rights. *Oxford Journal of Legal Studies,* 13: 18–51.
Walker, D. M., ed. 1980. *The Oxford Companion to Law.* Oxford: Oxford University Press.
Wellington, H. H. 1990. *Interpreting the Constitution.* New Haven, Conn.: Yale University Press.
Wills, G. 1978. *Inventing America.* New York: Doubleday.
Wilson, W. 1887. The study of administration. *Political Science Quarterly,* 2: 197–222.
Wolfe, C. 1986. *The Rise of Modern Judicial Review.* New York: Basic Books.
Woodward, B., and Armstrong, S. 1979. *The Brethren: Inside the Supreme Court.* New York: Simon and Schuster.

Chapter 7

# Political Institutions, Old and New

B. Guy Peters

## I Introduction: so what's new?

THE new institutionalism has become one of the growth areas of political science. It is now difficult to pick up a journal or attend a conference without coming across one or more papers written from the perspective of the new institutionalism. The frequent use of the term of "new institutionalism" implies first that there was an old institutionalism, and second that the contemporary version is different from the older version. Both of those assumptions are correct, and the differences between the old and new versions of institutionalism are crucial for understanding the development of contemporary political theory.

The "old institutionalism" characterized political science until at least the early 1950s, and to some extent never really died out among many students of politics. Scholars (Eckstein 1963; Macridis 1955) advocating the newer, more scientific approaches to politics generally associated with the "behavioral revolution" maligned the old institutionalism and pointed out a number of deficiencies in that body of research. While understandable at the time, those attacks may have undervalued the work of major scholars such as E. A. Freeman, Taylor Cole, Gwendolyn Carter and even Carl Friedrich. Their scholarship made a definite contribution to a literature that enabled the researchers who came after to better understand the dynamics of politics and policy-making. Indeed, the need to engage in more micro-level analysis was so evident in part because of how well the formal institutions had been described.[1]

[1] Kuhn (1970) makes the point that paradigms change in ways to compensate for the strengths as well as weaknesses of the earlier paradigms.

The old institutionalism did make some definite contributions to comprehension of governance. One contribution came from the attention given to the details of structures which is, to some degree, returning to academic fashion—particularly in historical institutionalism (see Sect. II.C below). The fundamental point of that descriptive research was that seemingly insignificant details could have a pervasive impact on the actual behavior of the institution and individuals within it. This perspective could be contrasted with some of the vague characterizations of government as the "black box" in systems analyses of politics so much in fashion in comparative politics during the height of the behavioral revolution.[2] Further, some scholars working in the institutionalist tradition may have been analogous to Molière's famous character and were "speaking theory" without knowing it. Scholars such as James Bryce (1921) and even Woodrow Wilson (1906) began to make generalizations about politics while vigorously rejecting any overt claims of theorization.

The new institutionalism differs from its intellectual precursor in several ways, all reflecting its development after the behavioral revolution in political science. First, that movement was characterized by an explicit concern with theory development and by the use of quantitative analysis. Although the new institutionalism focuses on structures and organizations rather than on individual behavior, the concern with theory and analytic methods is shared with behavioral approaches to politics. Whereas the older version of institutionalism was content to describe institutions, the newer version seeks to explain them as a "dependent variable" and, more importantly, to explain other phenomena with institutions as the "independent variables" shaping policy and administrative behavior.

Further, contemporary institutional analysis looks at actual behavior rather than only at the formal, structural aspects of institutions. Does having the format of a parliamentary or presidential regime really matter for how actors within the regime behave, or are these only formalistic differences—are all governments really "divided"? (Pierce 1991; Stepan and Skach 1993; Fiorina 1992; Jones 1995). Finally, the newer approaches to institutional analysis focus on outcomes in the form of public policies or other decisions. While the old institutionalism might follow a bill through Congress (Bailey and Samuel 1952) and utilize the legislative process as a window on the dynamics of institutions, it was not particularly concerned with what government did. The new institutionalism, however, reflects the

[2] In comparative politics the period of advocacy for scientific politics was characterized by systems approaches intended to be sufficiently general to be useful for all types of government (Easton 1965).

public policy movement in political science and its concerns with what benefits and burdens governments actually produce for their citizens.[3]

One of the virtues of the newer versions of institutionalism is that they enable the discipline to talk about institutions in more genuinely comparative ways. Each country tends to look at its own politics as special and at politics elsewhere as being just another comparative case. American politics, for example, has been obsessed with the impact of divided government on our policy decisions (Fiorina 1992; Jones 1995). British politics has become more concerned with the role of the Prime Minister in the wake of Mrs Thatcher (Foley 1993). Swedes have become aware of the changing nature of their own political system which had served as a model for other systems (Lane 1995).

The above list could be extended, but none of these national concerns can really advance comparative analysis, no matter how interesting and important they may be on their own. That is what the "new institutionalism" is able to do, if utilized properly and sensitively. It can enable researchers to ask whether institutional dynamics are more similar across regimes than they might appear from the particularistic descriptions provided by nationally (or even regionally) based scholars. Adopting one of the particular institutionalist stances discussed below does not guarantee success at comparison, but it does provide the framework for such analysis.

## II Varieties of new institutional theory

The new institutionalism is often discussed as if it were a single thing, but in reality there are a number of different strands of thinking contained within it. Indeed, some versions of the approach campaign actively against other versions, although all believe that they are concerned with the same fundamental approach to politics. While to the external observer it is clear that all these different versions of institutionalism do share a common concern with the structures of the public sector, at times there appears to be little else that unifies them. Therefore, one role for a chapter of this sort is to attempt to identify the various strands of a literature and look at its internal contradictions and reinforcement. We should also point to what

[3] First, we should note that this concern with outcomes does have some earlier precursors in the work of Lasswell (1936) and some of his colleagues. Second, the policy movement has at times been described as "post-behavioral" given its less explicit concern with individual behavior and its more explicit value concerns.

differentiates these versions of the "new" institutionalism from the older versions of institutionalism.

## A Normative institutionalism

The phrase "new institutionalism" was coined by James March and Johan Olsen in their seminal (1984) article. While they stressed the organizational factors in political life, they also stressed the importance of norms and values in defining how those organizations should and would function. Thus, one important approach to institutions defines those entities in terms of the "logic of appropriateness" that guides the actions of their members. In this view the most important element defining an institution is the collection of values by which decisions and behaviors of members are shaped, not any formal structures, rules or procedures.

This normative view of institutions is closely related to several approaches in organization theory. For example, the organizational culture perspective on organizational theory (Ott 1989) argues many of the same points concerning the importance of values in understanding organizational behavior. These two approaches differ in part because the institutional approach assumes (or at least implies) uniformity in values, while organizational culture permits a number of different cultures within the organization (Siehl and Martin 1984). For institutionalism the uniformity of values is definitional, while for organizational culture they are in essence a variable within the structures.

March and Olsen provided a general perspective on institutions and they also provided a more differentiated view of the internal dynamics of institutions. For example, they distinguished between aggregative and integrative institutions (March and Olsen 1989: 118–42). The former institutions are characterized by internal bargaining and exchange, with decisions arising from that political process. These institutions correspond to the predictions about behavior that might come from rational choice analysis. Integrative decision-making styles, on the other hand, "presume an order based on history, obligation and reason" (March and Olsen 1989: 118). Thus, the value base that they argue is so central to institutional analysis may not be an appropriate description of market organizations, but then in their view the economic nexus would not be appropriate for understanding non-market organizations.

## B Rational choice approaches

One primary target of March and Olsen's advocacy of their version of the new institutionalism was the growing dominance of rational choice models of politics (March and Olsen 1984: 736–7). They argued that the micro-level focus of rational choice theories and their emphasis on utilitarian calculations of individuals devalued many normative and collective aspects of governing. March and Olsen further argued that the economic orientation of these models tended to assume an individualistic dominance of decisions that could better be explained by collective, institutional factors. That superiority was certainly normative and also was argued to be empirical.

Despite the rejection of rational choice approaches by the apparent founders of contemporary institutional analysis in political science, there is a burgeoning literature on rational choice approaches to institutions. This orientation to institutions to some extent comes from economics itself and the resurgence of institutional analysis.[4] Douglass North's award of the Nobel Prize in Economics in 1993 was one indication of that resurgence, as was James Buchanan's Nobel Prize in 1986 for his work in public choice. This discipline has also begun to accept the importance of structural constraints on the behavior of individuals and the necessity of rational actors to maximize within boundaries set by exogenous structures.

Within political science the adherents of rational choice have developed their own version of institutional analysis. In many ways institutionalism can be seen as solving one of the major problems in the economic analysis of politics. This problem concerns the difficulties of achieving an equilibrium in a world composed of rational individualists. The rules imposed through institutions constrain individual maximizing behavior and enable stable and predictable decision-making.

Even this is not a unitary approach to institutions but within itself contains several alternative views. For example, Elinor Ostrom (1990; 1991) has developed an approach to institutions based upon rules and the enforcement of rules that "permit, prescribe and proscribe" actions by the members of the institution. Margaret Levi (1988; 1996) has employed a similar approach in her analysis of the importance of rules in explaining the choices made by government in extracting resources (through taxation or conscription) from its citizens. In many ways this view is not dissimilar from that of March and Olsen given that it argues for the crucial role of collective values in shaping behavior.

[4] Institutional economics has a long and honorable history manifested in the work of such scholars and practitioners as Rexford Guy Tugwell and John Commons (1934), and John Kenneth Galbraith. See Rutherford 1994.

As we will point out below, the central element in this approach to institutions—rules—presents some conceptual problems. The question which emerges is, "When is a rule a rule?" The obvious answer is that a rule is something that is obeyed, but then what about the formal rules of an institution that are not obeyed, or are obeyed by some members and not by others? Does this mean that an institution really does not exist? If we are forced to fall back on informal organizational rules, is this approach really any different than well-established approaches to organizations (Scott 1995)?

Kenneth Shepsle (1989) and Barry Weingast (above: chap. 5) offer another perspective on the role of rational choice in institutional analysis. They argue that institutions should be understood as means of aggregating the preferences of individuals who are each attempting to pursue their own self-interests. As such, institutions become a self-selected mode of constraint, at least for the first generation of members of the organization. Further, they can be seen as ways of avoiding some of the pitfalls of more individualistic models of decision-making, such as Arrow's Impossibility Theorem.

## C **Historical institutionalism**

Yet another strand of thinking about institutions can be labelled "historical institutionalism." This group of scholars emphasize the role of institutional choices made early in the development of policy areas, or even of political systems. The argument is that these initial choices (structural as well as normative) will have a pervasive effect on subsequent policy choices. This approach takes the basic tenets of institutionalism and points to the importance of the structural choices made at the inception of a policy. It appears in these arguments that even if subsequent structural changes are made, the initial choices have an enduring impact.

The discussion among political theorists in the United States about "bringing the state back in" (Almond 1988) can be seen as one of the precursors of the historical institutionalists.[5] As well as arguing for the importance of the public sector itself—as opposed to primarily societal, "input" influences such as interest groups and political parties—this strand of theory argued for a differentiated conception of the state. That is, the state is not discussed as a single entity but rather as an aggregation of organizations and institutions, each with its own interests. Policies could then be

[5] Many Europeans find the discussion of bringing the state back in to be extremely peculiar, given that in the *Staatslehre* style of political science the state never left, and indeed remains the central focus for analysis.

explained by the ideas and interests of those institutional actors within a differentiated public sector (see Hall 1986) rather than merely a response to external pressures.

Several significant research studies have demonstrated the importance of original institutional choices. For example, Douglas Ashford (1986) developed these points in reference to the European welfare state. Also, the several substantive studies contained in the Thelen, Longstreth and Steinmo (1992) volume argue that institutions once developed shape goals, define means, and provide evaluative criteria for policies. These authors argue effectively for the importance of institutions across a range of substantive policies and a range of political systems. Desmond King's (1995) comparative study of labor-market policy in the United States and the United Kingdom similarly argues that the initial choices about how institutionally to pay unemployment benefits have had a profound effect on the success of labor-market policies in those two countries.

Historical institutionalism goes a long way to address another of the critiques of political science offered by March and Olsen in their original article—the ahistoricism that had become characteristic of much of the discipline. While valuable in attempting to reunite political science with some of its roots, in theoretical terms the approach may encounter some problems. In particular, knowing how a particular policy has developed over time it may be difficult to imagine any other sequence of development. Thus, refutation of the institutionalist approach may be difficult. In overly simplified terms, the argument appears to be that there was a set institution, there was a policy outcome, and the two must be linked.

## D Social institutionalism

Institutional analysis can be extended to cover the relationships between state and society as well as the institutions of the public sector itself. This encounters the risk of conceptual stretching and reductionism (see below) but may also be a useful extension of the argument. Some of the most important characterizations of the relationship between state and society, e.g. corporatism and networks, have many relevant features that could be classified as structural or institutional. For example, Schmitter's initial (1974) discussion of corporatism makes it clear that this is a system governed by rules. Likewise, the discussion of networks by scholars such as Marsh and Rhodes (1992) points to the role of both shared values (normative institutionalism) and exchange (rational choice institutionalism) in defining the relationships among groups.

The language that has been developed to describe interest group behavior in contemporary societies, and their relationship to government, emphasizes their institutional character. For example, concepts such as "networks" (Rhodes 1988; Laumann and Knoke 1987) and "communities" (Coleman and Skogstad 1990) and "implementation structures" (Hjern and Porter 1980) all emphasize a patterned interaction of organizations among themselves. As well as being a useful metaphor, structure is a useful analytic perspective for understanding the public sector. These concepts point to the stable patterns of interaction that exist between private sector groups and between those groups and the public sector.[6] The availability of powerful means of analysis to describe these linkages further strengthens the utility of this approach to institutional analysis.

## E Structural institutionalism

Finally, what I am labeling "structural institutionalism" to some extent hearkens back to the old institutionalism, but presents it in a much refined and more "scientific" manner. This approach uses many categories that would have been familiar to the older institutional scholars such as Bryce or Friedrich. Interest in the differences between presidential and parliamentary regimes, and between federal and unitary systems characterize this body of work. This work also reflects, however, the development of political and social theory that has been amassed in the intervening years.

The leading work in this school is Kent Weaver and Bert Rockman, *Do Institutions Matter?* (1993). This work asks the fundamental question about institutions contained in its title and does so with special reference to the difference between presidential and parliamentary governments. This is by no means the only recent research to focus on the differences between presidential and parliamentary government (Shugart and Carey 1992; Lijphart 1984) but it does provide a much more complete framework for understanding the impact of institutions on policy.[7]

Another, more general version of the structural approach to institutions discusses those entities as a collection of "veto points" (Immergut 1992). These points are points in the decision chain at which an actor can prevent

[6] The relationships that exist among the groups themselves are relevant to political institutionalism only to the extent that they structure the interactions of the groups and public sector organizations.

[7] Also, unlike some work addressing this question (e. g., Riggs 1994 and Linz 1990) the Weaver and Rockman (1993) book is not polemical but rather very analytical. They do not assume that presidential institutions are inherently inappropriate but look more at their probable effects.

action. More complex institutional structures, e.g. the divided government of the United States, have more points at which action can be delayed or prevented than do parliamentary governments. This model is not dissimilar to Pressman and Wildavsky's (1979) description of implementation going through a series of "clearance points." In both cases institutions are conceptualized (rather negatively) as a set of interconnected nodes at which action can be blocked.

## III Theoretical problems

No approach to politics is without its problematic elements, and institutional analysis is no different. There are a number of theoretical questions that emanate from the orientation of this body of literature utilizing collective and structural modes of explanation, and critics have been quick to point out those problems (Pedersen 1991; Jordan 1990). Thus, the problems that plague institutional analysis tend to be the reverse of those that often afflict micro-level analysis. Whereas micro-level analysis faces the problems of methodological individualism, institutionalism encounters the problem of the ecological fallacy.

### A Individuals and institutions

Perhaps the most fundamental theoretical problem faced by institutionalism might be deemed the paradox of constraint (Grafstein 1992). On the one hand institutions gain much of their explanatory power from the argument that they impose constraints on the behavior of their members, and that individuals cannot function effectively in unrestrained, market-like situations (Granovetter 1985). This constraint is important whether researchers employ a normative, historical, or rational choice orientation toward institutions. On the other hand, if institutions are the products of human choices then there are few real constraints on behavior. If this is true then the decision by each individual to accept the restraint on behavior is a more important predictive factor than the rules themselves.

The problem of constraint is to some extent a part of a much broader issue of how we link individual behavior and organizational behavior. How do institutions shape behavior, or is it really behaviors that shape institutions? The fundamental point of institutional analysis is that there are entities which, even if not physically identifiable, do have a reality that reduces the diversity of policy choices that might otherwise be made.[8] How are those collective values transmitted, learned and reinforced in societies that appear increasingly atomistic (Pedersen 1989)? How is deviation punished, and when is deviation sufficient to argue that deinstitutionalization has taken place? Individuals play different roles in different institutions, and how do we account for difficulties that institutions may have in enforcing their own rules in that complex aggregation of rules and values (Douglas 1986)?

## B Definitions and existence

A second fundamental theoretical issue is how we know when an institution exists, or indeed what one is. Although the intention was to develop institutional analysis as a means of getting around the problems of micro-level and rationalistic analysis of political phenomena, it may be difficult to identify the point at which the principal element—the institution—comes into existence and when it ceases to exist. All the various approaches to institutional analysis offer some form of definition, but all contain sufficient vagueness to make identification of an institution problematic. DiMaggio and Powell (1991: 1), for example, argue that, "Institutionalism purportedly represents a distinctive approach to the study of social, economic and political phenomena; yet it is often easier to gain agreement about what it is *not* than what it is." This absence of clear definitions does not appear to be the most fruitful way in which to embark on a scientific enterprise.

Pedersen (1991: 131–2) argues that institutional theories are incapable of providing clear definitions of an institution. He argues that the three major theorists he discusses tend to treat institutions as independent and dependent variables simultaneously. While this judgment may not be applicable to the full range of institutional theorists—especially the rational choice approaches—it does raise some important questions about the general capability of this approach to separate endogenous and exogenous factors and to separate institutions *per se* from surrounding social economic and political structures.

[8] For a complete discussion of the methodological problems implied by the absence of clearly identifiable entities, see Easton (1990) and Cerny (1990).

## C Tautologies?

At the other extreme, institutionalism could become a tautology. If the rules that shape behavior are expanded to include implicit rules and vague understandings, in order to cover instances in which observed behaviors do not correspond to the formal rules of an institution, then the theory may not be falsifiable. If we observe behaviors that do not conform to the strictures of the formal rules then there must be other rules that were not identifiable. Furthermore, what is a rule? A rule is something that is obeyed, so that if it is not obeyed then by definition it ceases to be a rule. This is perhaps an unkind characterization of some styles of institutionalist reasoning, but is not much of an exaggeration.

The tautology problem threatens most of the approaches to institutionalism mentioned above. For example, normative institutionalism tends to assume that any observed behaviors are a function of collective values but there is no external means of verifying that claim, plausible though it may be. The definition of an institution appears sufficiently vague that the empirical problems of verification could be defined out of existence.

The rational choice approach to institutions also encounters problems with becoming a tautology. Here the problem of defining and identifying rules becomes crucial. If institutions are defined by rules, and rules are those demands on individuals that are obeyed, then there appears to be little place for non-compliance in the closed system and the claim that institutions determine behavior cannot be refuted. As noted above, if it is not obeyed an attempt to mold behavior loses its status as a rule and hence by definition the institution is effective.

Finally, historical institutionalism is perhaps the most subject to the lack of falsifiability. Given that each historical development is unique and there are no counter-factuals, it is difficult to say that the particular institutional choices made early in the process were not determinate. The chances for testing the assumptions of historical institutionalism could be enhanced by careful research design, choosing different institutional structures confronting similar policy demands.[9]

## D Design and intentionality

Institutions are often discussed as they exist, but less attention is given to the capacity of individuals, groups, or other institutions to design

[9] Immergut's (1992) analysis of health policy fits this characterization to some degree.

institutions in a manner which will produce the desired outcomes. In some instances, e.g. the use of electoral rules to shape electoral outcomes, the capacity to predict the outcomes of a design decision is quite high (Taagepera and Shugart 1989). For other structural variables, e.g. the choice of presidential or parliamentary regime types, or federal versus unitary regimes, that predictive capacity is much less. Often, institutional consequences arise when attempting to "solve" other problems (Shepsle 1989: 141–2). Some scholars (e.g. Elster 1989) are less concerned with predictability *per se* than with a robust capacity to structure institutions that are likely to produce socially desired outcomes.

On both empirical and normative grounds the capacity of institutional designs to produce particular outcomes is somewhat questionable (Komesar 1994). The conscious manipulation of institutions often produces either no discernible outcomes, or even those which are the opposite of those intended. Further, there is often a pronounced trade-off between internal and external values that may be achieved through institutional design (Goodin 1996). An institution that functions well internally may not produce the outcomes desired by the broader society, and one that serves the public may not function efficiently. For example, contemporary efforts to empower members of public organizations may reduce the likelihood of their making decisions that meet the needs of their clients. In short, our knowledge base for institutional design is often much weaker than people changing government in settings such as the former Soviet Union would desire.

## E **Reductionism**

Finally, institutionalism runs the risk—common to many approaches to politics—of conceptual stretching (Sartori 1970) and reductionism. Adherents of the institutional approach tend to explain most phenomena through their institutional characteristics. Just as micro-level analysts attempt(ed) to explain all politics through social or psychological variables, the institutionalists tend to assume that political phenomena, and especially policy choices, could be explained by the institutions responsible for shaping and implementing them. Institutional analysis gains some greater credence through the observation that the same individuals behave differently in different institutional settings. This means that individuals do accept the different roles sent to them by the different institutions, and that therefore institutions do appear important in understanding behavior.[10]

[10] Role theory appears to be one way in which to integrate the macro level of structures and the micro-level of individual behavior.

The danger is that by attempting to explain, or perhaps even claiming to explain everything, institutional analysis may explain nothing (see Rothstein above: chap. 4). For behavioralists and the economic analysis of political life individual choices are virtually the only phenomena of relevance, and therefore these approaches have difficulty explaining collective decisions or individual choices that appear irrational. Institutionalism also has a certain domain to which its claims should be restricted. The most appropriate range of explanation appears to be for public policy and for the "bureaucratic politics" that occurs among government organizations. Even within the domains of policy and administration, however, emerging perspectives such as policy entrepreneurship (Kingdon 1995) and policy framing (Schon and Rein 1994) may return more emphasis to the places of individuals and individual values.

## IV Conclusion

The argument about institutionalism is in some ways an exceedingly parochial discussion. For political science in much of the world the behavioral revolution was merely an overlay on a continuing concentration on the central importance of government institutions and the state in the lives of citizens. For many analysts of political phenomena, however, politics and policy had become more a function of mass behavior; government was little more than the reflection of the activities of the mass public and political organizations. Therefore, the restatement of the central role of institutions in shaping policy has been important in redefining political theory.

Institutional analysis has proven very effective in explaining a particular range of phenomena, especially those found in the "black box" of government, largely ignored by the theories developed during the 1960s and 1970s. Despite those successes, it has a number of thorny problems as an approach to politics and governance. Most of these problems have to do with the ability to offer unambiguous definitions of the principal phenomena under investigation and the difficulties in relating individual and collective behavior. There is also the difficulty in isolating the impact of institutions as an independent variable as opposed to their reflection of social forces or indeed their reflection of the policies they implement.

The problems discussed above are significant, but should not lead to the rejection of this mode of analysis, but rather to its improvement. The development of this approach has been a useful antidote to the rampant individualism that has characterized both behavioral and rational choice approaches to political science. There is a good deal to do to enhance this approach to politics, but there is also a good deal of intellectual progress already to its credit.

## References


ALMOND, G. 1988. The return to the state. *American Political Science Review*, 82: 853–74.

ASHFORD, D. E. 1986. *The Emergence of Welfare States.* Oxford: Blackwell.

BAILEY, S. K., and SAMUEL, H. D. 1952. *Congress at Work.* New York: Holt.

BRYCE, J. 1921. *Modern Democracies.* New York: Macmillan.

CERNY, P. G. 1990. *The Changing Architecture of Politics.* London: Sage.

COLEMAN, W., and SKOGSTAD, G. 1990. *Policy Communities and Policy-Making in Canada.* Toronto: Copp Clark Pitman.

COMMONS, J. R. 1934. *Institutional Economics.* New York: Macmillan.

DIMAGGIO, P. J., and POWELL, W. W. 1991. Introduction. Pp. 1–38 in *The New Institutionalism in Organizational Analysis*, ed. W. W. Powell and P. J. DiMaggio. Chicago: University of Chicago Press.

DOUGLAS, M. 1986. *How Institutions Think.* New York: Syracuse University Press.

EASTON, D. 1965. *A Framework for Political Analysis.* Englewood Cliffs, N.J.: Prentice-Hall.

—— 1990. *The Analysis of Political Structure.* London: Routledge.

ECKSTEIN, H. 1963. A perspective on comparative politics, past and present. Pp. 3–32 in *Comparative Politics*, ed. H. Eckstein and D. E. Apter. New York: Free Press.

ELSTER, J. 1989. *The Cement of Society.* New York: Cambridge University Press.

FIORINA, M. 1992. *Divided Government.* New York: Macmillan.

FOLEY, M. 1993. *The Rise of the British Presidency.* Manchester: Manchester University Press.

GOODIN, R. E. 1996. Institutions and their design. Pp. 1–53 in *The Theory of Institutional Design*, ed. R. E. Goodin. New York: Cambridge University Press.

GRAFSTEIN, R. 1992. *Institutional Realism: Social and Political Constraints on Rational Actors.* New Haven, Conn.: Yale University Press.

GRANOVETTER, M. 1985. Economic action and social structure: the problem of embeddedness. *American Journal of Sociology*, 91: 481–510.

HALL, P. A. 1986. *Governing the Economy: The Politics of State Intervention in Britain and France.* New York: Oxford University Press.

HJERN, B., and PORTER, D. O. 1980. Implementation structures: a new unit of organizational analysis. *Organization Studies*, 2: 211–34.

IMMERGUT, E. 1992. *Health Politics: Interests and Institutions in Western Europe.* Cambridge: Cambridge University Press.

JONES, C. O. 1994. *The Presidency in a Separated System.* Washington, D.C.: Brookings Institution.

—— 1995. *Separate but Equal Branches.* Chatham, N.J.: Chatham House.

JORDAN, A. G. 1990. Policy community realism versus 'new' institutionalism ambiguity. *Political Studies*, 38: 470–84.

KING, D. S. 1995. *Actively Seeking Work: The Politics of Unemployment and Welfare Policy in the United States and Great Britain.* Chicago: University of Chicago Press.

KINGDON, J. 1995. *Agendas, Alternatives and Public Policies.* 2nd. edn. Boston: Little, Brown

Komesar, N. K. 1994. *Imperfect Alternatives: Choosing Institutions in Law, Economics and Public Policy.* Chicago: University of Chicago Press.
Kuhn, T. S. 1970. *The Structure of Scientific Revolutions.* 2nd. edn. Chicago: University of Chicago Press.
Lane, J.-E. 1995. The decline of the Swedish model. *Governance,* 8: 579–90.
Lasswell, H. D. 1936. *Politics: Who Gets What, When, How.* New York: McGraw-Hill.
Laumann, E. O., and Knoke, P. 1987. *The Organizational State.* Madison: University of Wisconsin Press.
Levi, M. 1988. *Of Rule and Revenue.* Berkeley: University of California Press.
—— 1996. *Contingencies of Consent.* Forthcoming.
Lijphart, A. 1984. *Democracies: Patterns of Majoritarian and Consensus Government in Twenty-One Countries.* New Haven, Conn.: Yale University Press.
Linz, J. 1990. The perils of presidentialism. *Journal of Democracy,* 1: 51–69.
Macridis, R. C. 1955. *The Study of Comparative Government.* New York: Random House.
March, J. G., and Olsen, J. P. 1984. The new institutionalism: organizational factors in political life. *American Political Science Review,* 78: 734–49.
—— —— 1989. *Rediscovering Institutions.* New York: Free Press.
Marsh, D., and Rhodes, R. A. W. 1992. Policy communities and issue networks: beyond typology. Pp. 249–68 in *Policy Networks in British Government,* ed. D. Marsh and R. A. W. Rhodes. Oxford: Clarendon Press.
North, D. C. 1990. *Institutions, Institutional Change and Economic Performance.* Cambridge: Cambridge University Press.
Ostrom, E. 1990. *Governing the Commons: The Evolution of Institutions for Collective Action.* New York: Cambridge University Press.
—— 1991. Rational choice theory and institutional analysis: toward complementarity. *American Political Science Review,* 85: 237–43.
Ott, J. S. 1989. *The Organizational Culture Perspective.* Pacific Grove, Calif.: Brooks/Cole.
Pedersen, O. K. 1989. Fra individ til aktor i struktur: tilbivelse af en juridisk rolle, *Statsvetenskaplig Tidskrift,* 92: 173–92.
—— 1991. Nine questions to a neo-institutional theory in political science. *Scandinavian Political Studies,* 14: 125–48.
Pierce, R. 1991. The executive divided against itself: cohabitation in France, 1986–1988. *Governance,* 4: 270–94
Pressman, J. L., and Wildavsky, A. 1979. *Implementation.* Berkeley: University of California Press.
Rhodes, R. A. W. 1988. *Beyond Westminster and Whitehall.* London: Unwin Hyman.
Riggs, F. W. 1994. Presidentialism in comparative context. Pp. 72–152 in *Comparing Nations,* ed. M. Dogan and A. Kazancigil. Oxford: Blackwell.
Rutherford, M. 1994. *Institutions in Economics: The Old and the New Institutionalism.* Cambridge: Cambridge University Press.
Sartori, G. 1970. Concept misinformation in comparative politics. *American Political Science Review,* 64: 1033–53.
Schmitter, P. C. 1974. Still the century of corporatism? *Review of Politics,* 36: 85–131.
Schon, D., and Rein, M. 1995. *Frame Reflection: Resolving Intractable Policy Issues.* New York: Basic.
Scott, W. R. 1995. *Institutions and Organizations.* Thousand Oaks, Calif.: Sage.
Shepsle, K. 1989. Studying institutions: lessons from the rational choice approach. *Journal of Theoretical Politics,* 1: 131–47.
Shugart, M. S., and Carey, J. M. 1992. *Presidents and Assemblies: Constitutional Design and Electoral Dynamics.* Cambridge: Cambridge University Press.

Siehl, C., and Martin, J. 1984. The role of symbolic management: how can managers effectively transmit organizational culture. Pp. 227–39 in *Leaders and Managers*, ed. J. G. Hunt *et al.* New York: Pergamon.

Stepan, A., and Skach, C. 1993. Constitutional frameworks and democratic consolidation: parliamentarianism versus presidentialism. *World Politics*, 46: 1–22.

Taagepera, R., and Shugart, M. S. 1989. *Seats and Votes: The Effects and Determinants of Electoral Systems.* New Haven, Conn.: Yale University Press .

Thelen, K.; Longstreth, T.; and Steinmo, S., eds. 1992. *Structuring Politics: Historical Institutionalism in Comparative Perspective.* Cambridge: Cambridge University Press.

Weaver, R. K., and Rockman, B. A., eds. 1993. *Do Institutions Matter?* Washington, D.C.: Brookings Institution.

Wilson, W. 1906. *The State: Elements of Historical and Practical Politics.* Boston: D.C. Heath.

Part III

# Political Behavior

Chapter **8**

# Political Behavior: An Overview

Edward G. Carmines
Robert Huckfeldt

THE fiftieth anniversary of the modern era in political behavior research was celebrated (quite silently) in 1994. We mark 1944 as the birth of the modern era because in that year Paul Lazarsfeld and his colleagues from the Bureau of Applied Social Research at Columbia University published the first academically inspired study of an election that focused primarily on individual voters. Their study, which reported on fieldwork carried out in the 1940 presidential campaign in Elmira, New York, was quite primitive in some respects and quite advanced in others. Moreover, it established an enduring intellectual paradigm in political behavior research—an intellectual paradigm that we will consider more extensively below. But the fundamental significance of their study for the modern era was that it focused on individual voters, and in so doing it helped transform the study of citizenship and democratic politics.

The Columbia sociologists were not the only scholars during this general period who turned their focus to the individual citizen. Two other streams of intellectual research, both of which understood democratic politics in the context of individual voters, locate their origins in the same postwar period. The American National Election Study series and the work of Campbell, Converse, Miller and Stokes (1960) trace their origins to fieldwork conducted in the 1948 election. And economic theories of democracy have a genesis which dates, perhaps most notably, to the work of Anthony Downs (1957). Taken together, these efforts have established three schools of research: the political sociology tradition that flowed from the work of the Bureau of Applied Research at Columbia, the political psychology tradition with origins at the University of Michigan's Center for Survey Research, and the political economy tradition which seriously began to apply concepts of rationality and self-interest to the study of citizen behavior.

The danger in identifying these separate traditions is that we ignore their commonalities to focus on their differences, thereby missing the forest for the trees. The thesis of this chapter is that each of the three traditions addressed a distinct challenge to democratic theory. Each tradition has made substantial progress in formulating a response to these distinctive challenges, but in moving toward a response, they have also tended to converge on a unified view of the citizen in democratic politics. We might indeed say that the convergence has produced a new empirical model of the democratic citizen, and hence an updated vision of democratic politics. None of this is meant to suggest that important differences do not continue to exist—much of this chapter will be devoted to an examination of their implications. We do intend to point out the revolution that the three traditions commonly fostered in the study of electoral politics, the common intellectual problems that they addressed, and the convergence that has marked their development through time.

Where do the commonalities originate? Most important, all these studies share a common focus on individual voters—a focus that is enduring in political behavior research. Moreover, each in a different way called into question the capacity of individual citizens to function in democratic politics, thereby presenting a challenge to democratic theory. Indeed, it is no exaggeration to say that this challenge has supplied much of the intellectual agenda for political behavior research during the past 50 years. Political sociologists have been forced to confront social determinism as a challenge to the model of the independent citizen. Political psychologists have confronted the political knowledge vacuum that characterizes typical citizens, and the implications that arise for theories of democracy that depend on astute and politically sophisticated citizen decision-makers. Political economists have posed perhaps the ultimate riddle of citizenship: why would a rational individual invest in the duties of citizenship when the effect of each individual investment is likely to be trivial and unimportant?

At the same time that each tradition has addressed a distinctive challenge to the capacity of individual citizens and hence to the potential of democratic politics, all three have produced a high degree of convergence in their responses. Each tradition has arrived at a reconciliation, of sorts, between the capacities of flesh-and-blood citizens and the theoretical requirements imposed by democratic politics. But in arriving at the reconciliation, a new empirical model of citizenship has evolved which, in many respects, is common across the three traditions. And the point of convergence among the three traditions is the purposeful, instrumentally motivated citizen.

The purposeful, instrumentally motivated citizen is, of course, the citizen embodied in political economy traditions. Thus, in a crucial sense, it was this model of citizenship and this tradition of scholarship that posed the fundamental challenge to political behavior research. In other words, the political economy tradition encouraged both the political sociology and political psychology traditions to reconsider—and in the process, rediscover—their own roots in the analysis of purposeful citizens and political behavior.

We begin this chapter by considering briefly the challenge to traditional views of democratic citizenship posed by the political economy tradition. We then show that, somewhat ironically, in spite of the dead end to which the model of democratic citizenship ultimately led, it fostered a revival of interest in the study of purposeful, instrumentally motivated citizenship in political behavior research more generally. Our purpose is not to review the entire field of political behavior research but instead to identify emergent streams in this research that exemplify this renewed view of democratic citizenship.

## I The economists' challenge

Perhaps the most fundamental challenge to traditional visions of democratic citizenship was posed by Anthony Downs (1957) in his articulation of an "economic theory of democracy." If citizens act rationally on the basis of individually defined self-interest, Downs showed, they might very well abstain from voting in elections. Rational abstention is a quite sensible response when the benefits that any voter obtains due to participation are compared to the costs of participation, after these benefits are discounted by the microscopically small likelihood that any single voter will play a pivotal role in an election outcome. Indeed, based on such a calculus, it becomes quite clear that the only reasonable course of action is to refrain from voting.

The same calculation can be invoked in the analysis of other citizenship activities as well. Most particularly, the costs of becoming informed about politics are often much higher than the costs of voting, and hence the same underlying analysis comes into play. Citizens might act more strategically to realize their interests if they based their actions on better, more complete information. But given the minimal likelihood that any single act will be pivotal, what is the use? Why should anyone bother to become informed?

Downs's analysis of rational abstention and the costs of becoming informed foreshadowed a more general scholarly assault on a wide range of collective action problems more broadly conceived, most particularly Olson's (1965) analysis of political organization and the challenge of securing a membership. Perhaps the ultimate tribute to both analyses is that they have stimulated so much intellectual activity, both within and beyond the political economy tradition. A significant portion of this intellectual effort has been aimed at a reconciliation between the expectations that are generated by the political economy models and the empirical reality as we know it. That is, given the strong individually based incentives for free-riding and shirking responsibility, why does *anyone* participate in elections, join organizations, or stay informed about politics? Perhaps ironically, the task of reconciling theoretical expectations with empirical reality is more readily accomplished with respect to political organization. The importance of coercion, selective incentives, and institutional arrangements (Ostrom 1990) for the long-term maintenance and survival of political co-operation is now widely accepted, albeit with some dissenters (Knoke 1990; Chong 1991).

The more difficult challenge has arisen with respect to the most widespread and basic citizenship duties—the decision to participate in elections and stay informed about politics. At the same time that Downs makes a persuasive case for rational abstention—and more than a few political scientists have been converted to personal non-participation as a result!—we witness the riddle of relatively widespread participation. Moreover, and particularly in the American case, levels of information and participation are highest among the best educated—those who should be best able to understand and comprehend that the exercise of citizenship duties is not in their own individual best interests. In short, the simple and persuasive logic of rational abstention runs opposite to some straightforward empirical evidence with respect to the functioning of democratic politics.

A great deal of ink has been spilled in the effort to resolve this riddle. Beginning in the 1960s, political economists began an effort to reconcile the facts of widespread participation with the logic of rational abstention—an effort that was perhaps made more urgent by a chorus of critics who viewed the idea of rational citizenship as a crackpot idea. We do not possess the inclination, and this is not the place, to engage in an extensive review of these efforts (see Aldrich 1993; Jackman 1993), except to say that they have mostly failed. Finally and fundamentally, widespread information-seeking and voting in elections cannot be understood from the vantage-point of short-term rational calculation because, as Fiorina (1990) argues, participation and the acquisition of information are activities that are valued intrinsically. People perform citizenship duties because they like to, or because they feel guilty if they don't, or because it is embarrassing to look stupid during lunchtime conversation. The important point is that the performance of citizenship duties takes on value as an end in itself.

Does this mean that voters are not rational? Or more importantly, does it mean that we cannot profitably view the decision to become informed and vote within the context of a rational calculus? These are fairly absurd questions, and their absurdity might be highlighted by analogy with an example of pure (rational) consumer behavior. One of the authors drives a Volvo while the other drives General Motors products. The GM buyer believes that his Volvo co-author has perverse preferences that lead to the expenditure of a great deal of money that might be used more productively elsewhere—at least more productively with respect to the GM buyer's preferences. But the co-author likes Volvos. He might even be willing to make an argument that Volvos are worth the additional expense, but finally he just likes Volvos. Now if the exchange rate changed and the price of Volvos went even higher, the co-author might get smart and buy a Chevrolet. And hence, the Volvo co-author is not irrational. He just likes Volvos.

By demonstrating that rational abstention is the best solution for citizens who base their decision to vote on narrowly defined self-interest, Downs showed us that widespread participation is best understood as a manifestation of an activity that is valued intrinsically. Citizens who read the *New York Times* and vote are not irrational—they just like to vote, or maybe they feel bad when they don't vote. Their preferences are not determinate, however. If the election campaign is particularly nasty, or if the candidates are both unattractive, or if the outcome is a foregone conclusion, or if there is a blizzard on election day, they may stay home. In short, their preference for voting is not absolute, it is simply one of the factors they take into account when deciding whether to vote.

Unfortunately, the discussion of rational citizenship took an unproductive turn that led down a blind alley. Arguments regarding whether citizens are rational tend to be very unproductive affairs. One measure of the success realized by the political economy tradition is the extent to which it has encouraged other traditions of political behavior to rediscover their own roots in the analysis of purposeful citizens and political behavior. Much of the development of these traditions must be seen against the backdrop of the rediscovery of purposeful citizenship by political scientists.

Indeed, much of the progress in political behavior research during the past forty years, beginning with Simon's (1957) work, has come in response to the political economist's challenge—the challenge to rediscover the role of purposeful citizens in the workings of democratic politics. We are not suggesting that game theory has or should become the vocabulary of choice among political scientists, or that other traditions in political behavior research have been submerged in a rational choice paradigm. Quite to the contrary, we argue that the other traditions in political behavior research have relocated the role of purposeful citizens in their own unfolding intellectual histories.

## II Political sociology responses

A central insight of the Columbia school was that social attributes are important not because they translate directly and deterministically to a set of interests and concomitant preferences, but rather because they locate individuals in social structure and hence affect exposure to political information (McPhee *et al.* 1963). While this is a sophisticated argument with powerful consequences, the early Columbia work, and in particular *The People's Choice*, provided its critics with a ready target. Consider the following formulation of the authors' argument: "a person thinks, politically, as he is, socially. Social characteristics determine political preference" (1944: 27).

Framed in this way, and perhaps taken out of context, it may be difficult to imagine a more straightforward statement of social determinism. Where did such a statement of the problem leave political science? Was politics simply a residue of social life? Was political science best seen as a subfield within sociology? In their classic response, Key and Munger (1959) seek to establish the unique, independent, and idiosyncratic character of politics. Based on political histories within Indiana counties, they argue that the development of political loyalties takes peculiar and unexpected turns, and hence there is no simple one-to-one correspondence between political preference and social characteristics. Political preferences are connected to social characteristics in one way at one place, but in quite another way at another place. Hence, politics is not a simple residue of social life, and political preference is not simply and socially determined.

The Key and Munger response is often cited as a refutation of the social determinism thought to be inherent in the Columbia studies, but more importantly it provides a political scientist's reconstruction of the political sociological tradition. The Key and Munger argument is a direct assault on the idea that political preference is the direct consequence of political interest defined in terms of individual characteristics. If we want to understand Indiana voters, the inference becomes, we must understand them where they reside, in the circumstances that surround them, and such an undertaking cannot be accomplished if we study them simply as individuals who are divorced from a place and time.

Key's contribution to the political sociological tradition is perhaps best understood with respect to his analysis of white voters in the context of *Southern Politics* (1949). Key provided a compelling account of racial politics in his demonstration that white racial hostility in the south varied as a function of black population concentrations within counties. The central thesis of *Southern Politics* was that the historical unfolding of southern politics depended on the relative dominance of white and black citizens in locally defined populations. (For updates on this argument see: Matthews and Prothro 1963; Wright 1976; Giles and Evans 1985; Huckfeldt and Kohfeld 1989, and Alt 1994). But in showing that racial hostility among whites was a direct function of local black population concentrations, how does Key escape his own indictment? What keeps politics in this analysis from becoming a socially determined side-show of social life more broadly defined?

Key sees racial conflict as the product of politics and purpose, where the interests of the participants in the conflict are locally defined. In his own words (1949: 5) the central problem in the high black concentration counties was "the maintenance of control by a white minority." In Key's analysis, political self-interest is used to explain racial hostility, even though racial hostility must still be understood with respect to a particular set of socially structured circumstances. Thus, the device that saves Key is individual purpose as it unfolds within a particular time and place. Key takes it as a given that, overall, whites in the postwar south had as a primary goal the maintenance of white political dominance. The actions taken to realize this goal depended on situational specifics, however. The maintenance of white dominance was assumed in the hill country of Mississippi because there were very few blacks, but it was problematic in the delta where the concentration of black citizens was much higher. In summary, we see the merger of individual purpose and environmental circumstance quite early in the modern emergence of the political sociological tradition—a theme that is carried forward in more recent studies of the political consequences that arise due to the contexts and networks surrounding individual citizens.

## A Social contexts

Understanding the politics of individual citizens within the political and social settings where they are located fits well within the political sociological traditions established by Key and Lazarsfeld and others. Indeed, a significant literature within modern voting studies addresses the contexts and networks of individual citizens—a literature inaugurated by the effects of Warren Miller (1956) and Robert Putnam (1966). While there is significant overlap in the study of contexts and networks, the fact remains that they present significantly different conceptual devices for understanding the impact of social structure on politics. More recent effects have provided a combined focus on the interplay between networks and contexts, but such a strategy requires that they be understood as separate constructs.

Social contexts are politically consequential because they influence the probabilities of social interaction within and across group boundaries, thereby affecting the social flow of politically relevant information. Hence, social contexts are perhaps best defined in terms of social composition (Eulau 1986), and in this sense, a context can be seen as the social composition of an environment. The environment might be defined on varying bases—a club, a neighborhood, a church, a country, a province—and it has a multitude of defining properties. Some environments are old, big, geographically based, isolated, and attractive, while others are not. Similarly, the social context of an environment can be defined in terms of its various compositional properties—the proportion of inhabitants who are Social Democrats, liberals, Catholic, politically active, and well educated. Moreover, some efforts have been made to assess the consequences of heterogeneity within contexts, conceived as dispersion around a central compositional tendency.

Defined in this way, contexts lie beyond the reach of individual control—they are not manufactured by way of individual preference. This is not to say, however, that people do not locate themselves within particular contexts while they avoid others. Indeed, such locational decisions may be strategically motivated on the basis of social, political and economic criteria. The important point is, however, that once a person is located within a context, heroic efforts will be required to change the context. There are, of course, heroic exemplars: the party worker who delivers his or her precinct, the Jehovah's Witness canvasser who tries to convert his or her neighborhood, and so on. But most of us probably take the contexts as given. Hence, absent the problem of self-selection, contexts are treated as exogenous to the actor.

When can the issue of self-selection be ignored? Most particularly, self-selection becomes a less plausible hypothesis when it involves large units of aggregation, when the motive underlying locational or membership choice is politically irrelevant, when the demonstration of contextual effects is linked empirically to a dynamic process, when the contextual effect is linked to a process of social or political conflict, and in general when a more sophisticated mechanism of contextual influence is specified. We consider each of these in turn.

First, consider the size of the aggregation unit. In a fascinating study, Erikson, Wright, and McIver (1993) have demonstrated the importance of state-level political cultures for the structure of political coalitions and the behavior of individual voters. There are a great many other reasons for rejecting self-selection arguments in their work, but at the most basic level, would we really expect that people locate themselves in states (or provinces or counties) for politically relevant reasons? As the units of aggregation become larger, the plausibility of self-selection as an account of contextual effects grows smaller.

Second, an increasing number of studies have turned toward the demonstration of contextual effects in time. Huckfeldt and Sprague (1990; 1995) show that political campaigns enhance contextual effects—the effects are larger at the end of the campaign—and that the magnitude of the enhancement varies in systematic ways across different groups defined in terms of individual characteristics. Grofman *et al.* (1993) and Carmines *et al.* (1995) show that the previously documented "Key effect" can be demonstrated in time. Both efforts show that the white exodus from the Democratic Party has been accelerated by events of the 1960s in counties where black populations are most concentrated. Dynamic formulations such as these are quite difficult to recast in terms of a self-selection argument: do people relocate themselves between observations in a way that produces the spurious appearance of a time-structured effect?

Third, the "Key effect" is insulated from the self-selection argument on another basis as well. When a contextual effect is rooted in social or political hostility, self-selection once again lacks plausibility. If southern whites who live in black counties were racially hostile toward blacks, how can we lay the blame on self-selection? Do racially hostile whites locate in black counties in order to be near the objects of their hostility? In such an instance, the political behavior is quite clearly predicated, in perhaps complex ways, on the surrounding social context (Huckfeldt 1986; Huckfeldt and Kohfeld 1989).

Finally, more sophisticated specifications of contextual influence tend in general to undermine the self-selection argument. One of the most important contributions in this regard was made by Erbring and Young (1979), who argued that contextual effects should be seen as the endogenous feedbacks of behavioral interdependence. In this same regard, Przeworski (1974), Sprague and Westefield (1979), Huckfeldt (1983; 1984; 1986), and Przeworski and Sprague (1986) produced arguments that generally moved the contextual effects literature toward behavioral and informational interdependence as opposed to arguments rooted in shared social characteristics. By focusing on socially communicated information and interdependent behavior as the medium of contextual influence, the contextual effects literature has moved away from social reference group explanations for contextual influence. Consider Langton and Rapoport's (1975) explanation of support for Allende in 1964 among Santiago workers: workers who lived among other workers came to identify with these workers and hence to support Allende. In spite of its attractiveness, such an explanation is vulnerable to the self-selection counter explanation: only workers who identify as workers choose to live among workers.

As these more recent efforts suggest, there is more to the self-selection dispute than simple issues of temporal precedence. More fundamentally, these are debates regarding the role of individual purpose, the exogeneity of individual preferences, and the form of independent structural effects on individual behavior. More complete specifications of contextual influence allow a reconciliation between individual purpose and the independent exogenous effect of social structure. But such an explanation typically involves a marriage between social contexts, social networks, and interdependent citizens who act purposefully on the basis of political goals.

Finally, self-selection is not a repudiation of environmental influence. If people decide against joining the local Sierra Club chapter because it is full of knee-jerk environmentalists, it becomes difficult to suggest that political choice is independent of the social context. People choose to be part of some environments, and they avoid others, but neither choice is readily interpreted apart from the intersection of the individual and the environment.

## B Social networks

While contexts are structurally defined and external to the individual, social networks are created as a consequence of individuals and their choices. Even though individuals may typically have little control over the political and social composition of their surroundings, they do exercise discretion over the networks of contacts they establish within various settings. In this way the social context serves as an environmental constraint on the individual—particularly on the individual's patterns of social interaction—and citizens make their choices among alternatives that are environmentally imposed. In some instances individual choice is swamped by these constraints. At a typical Peoria cocktail party, even committed socialists have little opportunity to talk politics with other socialists. Yet, in other instances, individual choice is imposed quite readily.

Consider the empirical results from Finifter's (1974) study of politics and the workplace, where she examined the social interaction patterns of Detroit automobile workers at the factories where they were employed. The partisan composition of these factories was overwhelmingly Democratic, and the Democratic majority was relatively indiscriminate in their patterns of social interaction and association. In contrast, members of the Republican miniority were more likely to turn inward and form friendship groups which served as "protective environments for political deviants." (For parallel findings see Gans (1967) and Berger (1960).) In short, members of the minority imposed their own political preferences on the social context in the creation of networks of social and political relations. In other words, social networks are created as the product of an intersection between the externally imposed social context and the citizen's own exogenous preference. And thus, at this particular level, social structure is neither deterministic nor a simple residue of individual preference (see Boudon 1986). Moreover, the exercise of individual choice in the construction of a network is contextually conditioned—minorities are more likely to be selective in their associational patterns.

Huckfeldt and Sprague engage in a series of analyses aimed at understanding the construction of social and political networks as a process that occurs through time, as individuals (1) realize repeated encounters within particular contexts and (2) make decisions regarding whether to convert these encounters into sources of political information (Huckfeldt and Sprague 1987; 1988; 1995; also Huckfeldt 1983; 1986). Such a logic, which is inspired by the work of Coleman (1964) and McPhee *et al.* (1963), takes account of both exogenous preference and independent, structurally imposed constraints on social interaction. People do not roll over and accept whatever political information the environment happens to provide, but neither are they able to wholly escape the consequences of an environmentally supplied stream of information.

In keeping with Downs's (1957) early analysis, people reduce the costs of information by seeking out information sources that are biased in keeping with their own political preferences. The extent to which individuals pursue such a strategy is likely to vary across individuals, depending on their sophistication, attentiveness, majority or minority standing, and so on. But one cannot avoid the conclusion that more politically sophisticated citizens should be more concerned to obtain biased information as opposed to information that is in some sense deemed to be politically objective—a conclusion that is in keeping with the analysis of Calvert (1985). In short, and perhaps ironically, information which carries a clearly understood political bias is more likely to be useful to the citizen.

Political preference is not determinate in the choice of information, however, because it operates within the context of an incoming information stream that is environmentally supplied. That is, the menu of choices lies beyond the control of the individual, and hence individual demand for information must be understood with respect to environmental supply. The key elements are twofold. First, people select information sources on the basis of their own preferences. Second, the individual's set of informational alternatives reflects the bias of an environmental source. People are not unthinking products of their surroundings. Rather, their opinions become the complex products of environmentally biased information and their own pre-existing orientations.

The crucial element in recovering the independent citizen is recognizing that both individual choice and the environmental supply of information operate probabilistically. The environment operates probabilistically because social contexts are seldom politically homogeneous. Even families disagree about politics, and hence people obtain information and viewpoints that reflect a less than determinate bias. If you live in a Democratic neighborhood, you are more likely to see Democratic bumper stickers, but on occasion you will see one supporting a Republican candidate.

Why is choice probabilistic? First, individuals form social relationships based on a variety of preferences, not all of which can be easily realized in any single relationship. Moreover, many relationships are less than wholly specialized—we talk politics, sports, and weather with overlapping sets of associates. Everything else being equal, Republican auto-workers would choose to eat lunch with another Republican, but everything else is not equal. If they are also fans of the Chicago Bears, they might prefer to avoid fans of the Detroit Lions, and it may be difficult to find a lunch partner who satisfies both preferences, particularly in a Detroit automobile factory. In other words, realizing political preferences in the construction of a social network may be quite costly, particularly if multiple preferences are being realized in a single choice. Inevitably, people reduce their costs by compromising their preferences, and hence the control of information flow is incomplete.

Second, even if the Republican auto-worker is quite careful to avoid political conversation with a Democrat, a great deal of information is unsolicited—the person who works next to our Republican may wear a Democratic lapel pin. And hence, people are exposed to information inadvertently and unwillingly, and much of this exposure carries an environmental bias—people who work in banks see more Republican bumper stickers than people who work in automobile factories.

Finally, the choice of information sources is often incomplete because the underlying preference, upon which choice might be based, is uninformed. People often lack preferences, and thus they do not exercise informational control because they have no preference to use in arriving at selection criteria. Indeed, we are at the edge of a "catch-22" conundrum. I may wish to select information sources that are coincidental with my own political preferences, but how can I know what to prefer absent the necessary information? While information is often more useful if it carries an identifiable bias, there are times when people lack the necessary basis for either understanding or evaluating the bias that is carried.

## C Models of influence

How are the political consequences of social and political structure realized? What are the mechanisms that tie together citizens with their surroundings? What are the alternative models of individual influence?

Methodological individualism lies at the core of the political sociological tradition we are describing—individual citizens and their political choices constitute the primary objects of study (Przeworski 1985). The individual actor is not, however, seen in isolation, but rather in the context of surrounding constraints and opportunities that operate on patterns of social interaction, the acquisition of political information, and the formulation of political choice. At the same time, however, the individualistic impulse requires that the pathways of influence between the individual and the environment be clearly specified. It is not enough, in the case of Tingsten's Swedish workers (1963), to show that those who live among other workers are more likely to vote for the Socialists. The vehicle of influence must be specified as well. Did these workers vote for the socialists because they talked to their Socialist neighbors or because they talked to the local precinct worker? Why were the neighbors (or party workers) influential?

Hence, an important part of the sociological tradition revolves around the development, articulation, and evaluation of alternative models of influence—models of influence that place the individual at the center of the action. Not surprisingly, many of these models are derived from the political psychology and political economy traditions. And in this sense, the political sociological tradition has become agnostic with respect to epistemological issues regarding individual impulse—there is no official micro-theory that is inseparable from the political sociology tradition. This is not to say that any particular practitioner of the craft lacks an intellectual commitment to a particular individual-level model. Indeed, those who engage in multi-level analyses connecting citizens and environments include game theorists and cognitive psychologists, rational choice theorists and learning theorists. The tradition does not, however, necessarily dictate a micro-level motor that drives political and social processes. In other words, the political economist need not denounce rationality and rational actors to invoke a multi-level explanation because particular micro-sociological models might have roots that lie either in political psychology or in political economics.

The distinctive feature of micro-theory in the political sociological tradition is its purpose. The explanation of individual behavior is not an end in itself, but rather a vehicle to attain a quite different goal —the explanation of interdependence among citizens and a more complete understanding of politics in its collective representations (Eulau 1986). For political sociologists, the slippery slope of methodological individualism leads to methodological reductionism, and therein lies the antithesis of the political sociological tradition. A commitment to methodological individualism does not negate the sociological impulse, but an analysis that begins and ends either with the individual psyche or with an individual's rational calculation ceases to be sociological.

Boudon (1986) argues for an individualistic tradition in sociology and a micro-sociological model that he traces to the work of Weber (1966). The tradition is individualistic in its emphasis on individual purpose and individual motives, but it is sociological in its focus on the individually external factors that constrain individual behavior. Hence, Boudon presents a micro-sociological model that places a (more or less) rational actor within a set of macro-sociological constraints.

Boudon's individualistic tradition might be illustrated quite well with developments in the models that seek to explain the nature of influence in social relationships. Particularly in the context of path-breaking family socialization studies (Jennings and Niemi, 1974; 1981), it was perhaps natural for political scientists to adopt a model of social influence that was built upon intimacy and the interaction that occurs within cohesive social groups. The implicit or explicit assumption is that the influential communication of political information takes place between people who hold one another in high esteem and close personal regard, and these are the factors that make them trusted sources of political information and guidance. According to this view, intimacy and trust become the defining ingredients of political influence among citizens, and hence influence is unrelated to motive and purpose.

While this model was enhanced by the political socialization studies, its history extends to an earlier period. Indeed, the Elmira and Erie County studies articulated social influence in terms which Burt (1987) calls the social cohesion model. The social cohesion model portrays social influence in politics as occurring among people who hold each other in high personal regard and hence share a normative climate of opinion. Social influence becomes the by-product of close, intimate ties—citizens are more likely to influence one another to the extent that they hold each other in high personal regard, both as friends and as knowledgeable sources of political information (Katz 1957).

For many purposes, such a model may be entirely appropriate—there is no reason to expect that intimacy and cohesive social groups are unrelated to important forms of political influence. At the same time, the intimacy model may disregard important consequences that derive from the social communication of political information across the boundaries of social groups. Moreover, even the family socialization studies have experienced difficulty in demonstrating the importance of intimacy to the flow of influential political communication (Jennings and Niemi 1974; 1981; Tedin 1974). Burt reanalyzes evidence from the medical innovation study (Coleman *et al.* 1966) and calls into question the role of social communication through cliques and cohesive groups in encouraging doctors to adopt a new drug. Huckfeldt and Sprague (1991) demonstrate that people engage in political discussion with others who are less than intimate associates, and intimacy does not necessarily translate into higher levels of influence.

Findings such as these push for a reconstruction of the micro-theory of political influence arising through social communication. Burt (1987) advocates a structural equivalence model, conceived in terms of network ties. Two individuals are structurally equivalent if they are similarly located in social structure—if, in network terms, they are connected to the same individuals in the same ways. In terms of politics, such a model might suggest that an individual's capacity for exercising influence is rooted in coincidental interests. One citizen is more likely to be influenced by another to the extent that the citizen being influenced recognizes the existence of commonalties and shared interests, and hence the existence of a congenial bias (Downs 1957; Huckfeldt and Sprague 1991).

But the relevance of communication occurring across the boundaries of cohesive social groups extends beyond the explanation of individual behavior. If intimacy is a pre-condition for influence, then methodological reductionism is not seriously called into question as a method of political analysis. It only becomes important to relax the tightly constructed focus on individuals to incorporate a slightly expanded vision of the cohesive social cells that surround voters—their families and their most intimate associates. In contrast, if information is conveyed through less intimate social contacts, important implications arise for the nature and consequence of social communication in politics.

First, if social communication beyond cohesive social groups is important, then the focus of political analysis can be profitably expanded to include the multiple and intersecting environments where social communication occurs—workplaces, neighborhoods, and so on. Second, if important political communication occurs beyond the boundaries of cohesive groups, it becomes important to invoke Granovetter's analysis of weak ties (1973). According to Granovetter, information that is communicated through weak ties travels more widely because it is less likely to feed back to the point of origin. Information obtained from one best friend is likely to be shared with another best friend, who is, in turn, quite likely to report it back to the source. Information obtained from a casual acquaintance might be shared with a second acquaintance, but it is unlikely to find its way back to the source because the source is less likely to be connected to the second acquaintance. As a consequence, information communicated through weak ties tends to spread and disseminate. In terms of politics, information communicated through weak ties is likely to create a public opinion that is more than a simple aggregation of individual opinions, but instead is a product of complex patterns of social communication.

In short, social communication that occurs through casual, less intimate social interaction is capable of creating a public opinion that is truly public—a public opinion that extends beyond the confines of individual calculation and cohesive social groups (Stimson 1991; MacKuen *et al.* 1989). This is not to suggest that individuals play no role in these public opinion dynamics. Some citizens play more crucial roles than others in the dissemination of opinion, and an important task is the identification of these individuals. The important point for present purposes is that micro-theoretical constructs such as these have implications that extend beyond the explanation of individual behavior to include the consequences of behavioral interdependence for the larger political community.

## III Political psychology responses

Having seen that the model of the purposeful, instrumentally motivated citizen emerges quite clearly in a wide variety of sociologically oriented studies, we now show that the same citizen has been rediscovered by political psychologists in research as diverse as media studies, racial politics, and heuristically oriented political decision-making.

## A Media effects

An important focus of the early Columbia studies was directed toward media effects on individual citizens and their choices. But after nearly twenty years of research, Klapper (1960) summarized the effort as the "minimal effects model." Supporters of this general thesis have argued that the political impact of the media is likely to be attenuated for a variety of reasons:

1) Citizens are not sufficiently attentive to be affected.
2) Conflicting and intermittent messages tend to cancel out one other.
3) Individual level processes—selective attention and retention based on pre-existing political predispositions—tend to distort media messages and hence nullify their effects.
4) Any message that does leak through to the individual has been processed through patterns of social interaction and communication.

And thus, especially to the extent that patterns of social communication are contained within cohesive social groups that reflect an individual's own interests and characteristics, another element of the external political environment might be safely ignored.

Much of the early work on the media and media effects focused primarily on the political consequence of bias: How are people's political preferences altered by a biased media? Two interdependent processes—one at the level of individuals and the other at the level of the media—made the pursuit of bias effects a problematic undertaking. First, particularly in the United States, the media has self-consciously moved toward a politically detached and objective style of reporting. In practice this has meant that the partisan press often has been replaced by an adversarial press (Iyengar and Kinder 1987). Rather than continuing the 19th-century role of partisan champion for particular candidates and parties, the modern press has created the role of the professional, politically detached, critical observer. Such a role frequently places the news media in an adversarial position with respect to both sides in partisan struggles. Indeed, we might argue that it creates a new bias—the bias of objectivity.

Second, and returning to our earlier discussion of the greater utility of biased information for sophisticated individuals, this trend has produced information that is frequently more costly and hence less useful for many citizens. If information without a well-understood bias is difficult to integrate and incorporate within political decision-making, then the modern media have become increasingly less relevant as they have become increasingly detached and adversarial. The modern newspaper, for example, finds itself in a position where it has no political following, but only an audience of political junkies who are willing and able to pay the high price of sifting, sorting, and evaluating the politically detached information that it presents. In other words, because the media has come to occupy a permanently adversarial position with respect to *all parties* in the political arena, the citizen cannot use the media as a cost-cutting source of information. And thus it is not surprising that the utility of media reports as sources of political information has tended to decline for many citizens, at the same time that the impact of the media on citizen viewpoints has been difficult to establish. In short, bias-free information is often less useful and hence less influential.

These problems are perhaps illustrated best in the work of Beck (1991). As he shows, Democratic partisans accuse the press of a Republican bias, and Republican partisans accuse the press of a Democratic bias. Such effects are particularly noteworthy because they operate in an opposite direction from the effects that arise on social communication. At the same time that citizens (often wrongly) project their own partisan preferences onto their associates, they also (often wrongly) project opposite preferences onto their newspapers. Quite clearly, the price of objectivity is an increase in the level of criticism from both partisan camps. If sophisticated citizens put a high value on information that comes with a well recognized political bias, such dissatisfaction is perhaps more easily understood. What is less understandable is why, in the modern media age, we should have developed such primitive expectations regarding media effects in the first place.

A significant stride forward in media research occurred when scholars began focusing more intently on agenda-setting (Erbring *et al.* 1980; MacKuen 1981). Efforts aimed at understanding the political consequences of media coverage switched from media bias to media coverage, based on the insight that media influence is not simply a matter of telling people how to vote, but rather a matter of influencing citizen perceptions regarding the importance of various issues and problems. Stated quite simply, the argument suggests that issues covered by the media become the issues that citizens view as important, and hence the connection between the real objective political landscape of policy problems and issues, on the one hand, and people's perceptions of that landscape on the other, depends fundamentally on what the media chooses to report.

Erbring *et al.* (1980: 21) made a lasting contribution to the agenda-setting literature in their effort to demonstrate "audience-contingent media effects embedded in an issue specific micro model of salience." They show that citizen perceptions of national problems include an important localized component that is explained on the basis of varying media reports. Even if the variability of local conditions is taken into account, the coverage of various policy problems by the local media has a pronounced impact on what citizens think is important. At the same time, they show that receptivity to the media agenda is "audience contingent"—it is dependent on people's own concerns and preferences.

This last point is especially important because it reinforces the emergent trend in political sociology—the intersection between environmentally supplied information and the preferences of individual citizens. In the case of social communication, citizens attempt to construct patterns of social interaction that minimize their exposure to disagreeable political messages, and in other instances socially communicated messages are either disguised or misperceived in ways that exaggerate agreement (MacKuen *et al.* 1990). The exercise of citizen preference is more problematic in the case of the media for two reasons. First, only a very restricted range of alternative choices is typically available—very few citizens read any newspaper other than the locally supplied alternative. Second, even when more alternatives are available, such as the case of television news, they are typically quite homogeneous and uniform in the content of their coverage. Nevertheless, individual preference is still invoked in the *interpretation* of environmentally supplied information: people develop a combative stance toward the media by imputing an unfavorable political bias. And they respond selectively to media coverage—they are not equally likely to think that all issues covered by the media are equally important.

Other subtle forms of media influence have been pursued in recent studies by Iyengar and Kinder (1987) and Iyengar (1991). In a series of carefully and creatively constructed experiments, Iyengar and Kinder examine not only the consequences of agenda-setting by the media, but also the impact of media "priming." Citizens are limited in the amount of information they can employ in rendering political judgments, and hence they make choices based on a limited number of concerns. Correspondingly, the crucial determinant of party and candidate success is often the sets of criteria used by citizens to make their political decisions, and Iyengar and Kinder establish the potential of the media to "prime" citizen decision-making by establishing the basis of evaluation. Once again, the results support the importance of an intersection between environmentally supplied information and the political preferences of those who are exposed to the information source. In the words of Iyengar and Kinder (1987), "Priming is greatest when viewers are predisposed, by virtue of a well-developed, accessible theory, to see a connection between the president's responsibilities and the condition of the country." And for most citizens, this means that partisan preferences intersect with media reports to determine the extent of media priming effects.

In a sequel effort, Iyengar (1991) pursues the importance of media framing—media effects on the attribution of responsibility for policy problems. He argues that television's primary mode of communication—episodic coverage of particular events—encourages the attribution of responsibility for policy problems to individuals rather than to larger social and political forces. As before, we see the importance of the intersection between citizen preference and predisposition on the one hand, and environmentally supplied information on the other. In particular, partisanship tends to limit the effects of framing. In his own words, "Individuals' partisan affiliations provide them an important resource with which to resist the suggestions of news frames" (126).

The extent to which individual citizens play active or passive roles in the acquisition and interpretation of information from the media is open to alternative interpretations. Certainly Downs (1957) would ascribe an important role to individuals as they aggressively pursue the economies that derive from locating information sources that convey a compatible bias. In contrast, Iyengar and Kinder (1987) see individuals as the victims of a media that lies beyond their control—two of their chapter titles describe citizens as the victims of agenda-setting and priming on the part of the media. Regardless of the particular interpretation given to the role of citizens, media studies almost necessarily focus on the question of whether viewers respond to the media as purposeful, instrumentally motivated consumers of political information.

## B The politics of race

The emergent focus on this intersection between the individual and the environment is illustrated quite well with respect to several controversies in the literature on race and politics in the United States. In general, this literature aims to explain the source of racial conflict in politics. In particular, why and under what circumstances do white citizens demonstrate political preferences, choices, and behaviors that are hostile to black citizens?

The authors of this chapter have been involved in efforts aimed at connecting the politics of race to political and electoral change. Carmines and

Stimson (1989) argue that the evolution of race is responsible for a transformation of American politics. Beginning in the Congress, filtering through party organizations, and ultimately being demonstrated in elections and voter loyalties, the American political landscape has been reorganized in terms of race. Huckfeldt and Kohfeld (1989) argue that race has not only grown in importance, but its growth has led to a concomitant decline in the importance of class as an organizing principle in contemporary American politics. And thus the importance of class frequently has been eclipsed by the ascendancy of race as the primary line of cleavage in American elections.

These efforts share a common viewpoint in which race is understood as one among several possible lines of political conflict. And thus they might be subsumed under what Bobo (1983) and Glaser (1994) refer to as group conflict theories of racial hostility. The underlying logic is that, under particular social, institutional, and environmental circumstances, race becomes crucial as a line of cleavage in politics. Racial politics is not, in short, something that simply occurs in the heads of voters. Instead it is a product of time, place, and circumstance, as a particular setting gives rise to the realization of particularly defined interests and impulses. For Carmines and Stimson this means that the politics of race became ascendant as the result of political élites who chose a propitious moment to place the race issue on the political agenda. For Huckfeldt and Kohfeld, this means that race typically grows more important when the political dominance of whites is numerically threatened by the presence of blacks, and when politicians construct coalitional appeals that successfully exploit the racial divide.

The first line of resistance to arguments such as these is the assertion that race does not matter, or at least that the political importance of race is radically overstated. The most recent advocate of this position is Abramowitz (1994) who writes in response to Carmines and Stimson (1989). The general line of argument, which was offered earlier in the work of Wolfinger and Arseneau (1978), is that racial attitudes do a poor job in explaining partisanship differences or voting choices once the effects of other policy opinions are taken into account. Hence, people's attitudes toward civil rights are independent of partisan choice.

Several responses are available to the "race does not matter" critique. One response, rooted in cognitive political psychology, argues that for many voters a broad range of issue opinions are organized cognitively with respect to race, and hence race tends to permeate a diverse range of public policy concerns (Hamill, *et al.* 1985). A second response, rooted in the political sociology of race, argues that the "race does not matter" critique fails the macro-political reality check. If the political consequences of race do not matter, why do 90 percent of voting blacks regularly support Democratic candidates for the presidency, while a majority of voting whites have supported Republican presidential candidates in every election since 1964?

Even among those who accept the premise that race *does* matter, there is significant disagreement regarding the nature of racial effects in politics. Group conflict theorists argue that racial politics is the result of political competition between whites and blacks over scarce resources, and hence is rooted in reality-based assessments on the part of political actors. In contrast, the symbolic racism thesis argues that racial conflict is, in fact, divorced from reality—from the real circumstances and interests of the actors. Rather, racial conflict is rooted in prejudice and hostility toward blacks, and hence it is stimulated and manipulated on the basis of symbolic issues and appeals (Sears *et al.* 1979).

Our goal here is not to take sides in these debates—we occupy such an intellectually partisan role in other places! Rather our purpose is to demonstrate the roots of these alternative explanations with respect to enduring traditions in political behavior research. Group conflict theories of racial conflict lie in the mainstream of a tradition in political sociology that conceives of macro-political outcomes at the intersection between individual purpose and orientation on the one hand, and the political and social environment on the other. The other two arguments—"race does not matter" and "racial conflict is the same as prejudice"—are both rooted in individualistic conceptions of political behavior and traditions that seek the explanation of macro-political outcomes in terms of either wholly individualistic impulses or wholly internalized cultural norms. Notice that the group conflict thesis is essentially noncommittal with respect to the role played by prejudice in racial conflict. Advocates of the group conflict position might argue either that prejudice plays a crucial role in the dynamic of racial conflict, or that it plays a minor role. What the group conflict position *does* argue is that prejudice alone is not an adequate explanation of racial conflict—racial conflict might exist quite apart from prejudice and, alternatively, prejudice might exist quite apart from racial conflict.

## C Heuristically guided political decision-making

It is not news that most citizens have only limited knowledge and information about politics and public affairs. While a minority of citizens are well informed, a majority—and by many accounts, a large majority—lack information about basic facts, salient issues, and significant political figures. If most citizens lack such elementary political information, how can they act as effective political participants? Indeed, how can they formulate meaningful political opinions? And yet plainly most citizens do have political opinions and express them on a vast array of political topics. Moreover, although these attitudes obviously do not represent reasoned judgments logically derived from full-blown and well-articulated philosophical premises, neither are they capricious or whimsical. Most people seemingly know why they prefer one policy position over another, and this provides at least the potential for effective political action. How is this possible, given their lack of attention to politics and their paucity of information about public affairs?

Anthony Downs (1957) was the first analyst to give serious attention to the problem of political information in modern mass democracies. Reasoning that the costs of becoming informed about the details of political issues generally outweigh the relative benefits to be derived from voting on an informed basis, he argues that it is irrational for most citizens to become fully knowledgeable about public affairs. Downs (1957: 210) identifies three types of costs involved in becoming informed:

1) Procurement costs are the costs of gathering, selecting, and transmitting data.
2) Analysis costs are the costs of undertaking a factual analysis of data.
3) Evaluative costs are the costs of relating data or factual analysis to specific goals, i.e., of evaluating them.

Together, these costs constitute a formidable barrier to becoming political informed—too formidable, Downs argues, for most citizens rationally to invest the time, attention, and resources needed to become politically informed. Rather, rational citizens have strong incentives to develop methods of avoiding the substantial costs of information acquisition. They do so by developing a variety of short-cuts in the gathering and use of information. These shortcuts allow citizens to make political decisions and form political preferences without becoming fully informed about the content and details of political issues. In short, their use leads to minimally informed—but informed nonetheless—citizenry.

Since Downs wrote so insightfully about the rational basis of political ignorance, a large literature in cognitive and social psychology focusing on decision-making under conditions of complexity and uncertainty has emerged (for a comprehensive review, see Ottati and Wyer (1990)).

Beginning with Tversky and Kahneman's highly influential paper (1974), this work has disclosed that individuals use a wide variety of heuristics—that is, cognitive shortcuts that reduce complex problem-solving to more simple judgmental operations—to make many social decisions.

Kahneman and Tversky have identified four fundamental heuristic principles that allow individuals to focus on a limited set of information to make decisions rather than engaging in a detailed analysis of all available information. These principles are availability, representativeness, adjustment, and simulation. Briefly, availability refers to the salient features of a decision-making situation, so that individuals need not attend to all available or even relevant information. For example, in deciding whether to identify oneself as a liberal, citizens may focus only on the most prominent liberal spokesman such as Senator Kennedy or Jesse Jackson rather than considering all those political figures and organizations that fall under the liberal category. Representativeness focuses on the probability that an item or piece of information is part of the appropriate class of all information. Thus, representativeness bears on the typicality of information—whether it possesses the central or defining properties of a given class of information. A representative heuristic principle would be invoked, for example, if a voter relied on the race of the candidate in a congressional contest between an African-American and a white to estimate which candidate is the Democrat (or Republican). In a racially contested election, African Americans are far more likely to be Democrats than Republicans while whites are just as likely to be Republicans. The fact that this is not invariably the case only shows that no heuristic principle, including representativeness, can be employed without a certain probability of error.

The final two heuristic principles discussed by Kahneman and Tversky involve more cognitive processing than either the availability or representativeness principles. The adjustment principle refers to the changes that are made in decisions after making an initial, tentative response. The initial or anchoring response is adjusted as relevant new information is brought to bear on the decision. Thus, in estimating the political ideology of a candidate it might be useful to know that she/he is a Democrat and therefore, more likely to be a liberal based on this initial piece of information. But this initial estimate might very well be adjusted if it turns out the candidate opposes all abortions, supports the death penalty, and favors a strict balanced budget. In other words, the anchoring estimate is altered as more relevant information becomes available.

The final heuristic principle studied by Kahneman and Tversky involves still more cognitive processing by the decision-maker. The simulation heuristic comes into play when relevant information is lacking and the individual must try to anticipate the consequences of a given decision scenario. In deciding whether to support a given military project, the Secretary of Defense attempts to decide whether the project is feasible, effective, and worth the investment in terms of providing greater security to the nation. Notice that existing information is only of limited value in this situation because the crucial factor is the probable reaction of your potential adversaries—thus highlighting the substantial amount of cognitive processing involved in the application of the simulation heuristic.

Applications of heuristics in political psychology do not usually make use of the specific heuristic principles discussed by Kahneman and Tversky but rather specific political heuristics felt to be more relevant to political decision-making. Yet some of these heuristics can be usefully interpreted within the Kahneman–Tverksy framework. The use of stereotypes to make political judgments, for example, clearly involves the application of the representative heuristic. Poor people, for example, may be seen as lazy and irresponsible—those undeserving of governmental assistance. Alternatively, they may be viewed as hard-working and responsible but unable to find appropriate jobs; this perception is likely to foster greater support for governmental employment programs. Thus, stereotypical judgments are based on the essential or defining features of a group—their representativeness.

Brady and Sniderman (1985; also see Sniderman, *et al.* 1991) discuss what they term a likability heuristic. This heuristic enables citizens to make inferences about the issue orientations of various political groups. The authors reason that if citizens have an opinion on a given political issue and have affective evaluations of various political groups, then they will attribute their opinion to favored groups, and the opposite opinion to disfavored groups. This turns out to lead to fairly accurate estimates of the issue stands of specific political groups—even though many citizens lack detailed information of the groups' issue positions.

As Mondak (1994) has noted, Brady and Sniderman's likability heuristic can be viewed as an application of the adjustment heuristic. The individual's own issue position provides an anchoring for the estimated position of the particular group. Additional information—specifically, their affective evaluation of political groups—leads to the estimates of their issue positions. And this estimate is adjusted as new information becomes available.

Both Carmines and Kuklinski (1990) and Mondak (1993) discuss what may be termed source cues, a particularly important decision heuristic for politically inattentive and uninformed citizens. Their studies show that citizens, in formulating policy assessments and even in developing their own policy position, often rely on prominent political figures. Thus, attributing the defense buildup in the 1980s to President Reagan, affected citizens' evaluation of that policy but only among respondents who lacked detailed policy information (1993). Similarly, the experimental studies conducted by Carmines and Kuklinski (1990) indicate that citizens are more likely to form their own opinion on issues if salient political leaders (i.e., Senators Kennedy or Helms) are identified either as proponents or opponents of the issue proposal.

By using heuristic principles we see how citizens can make relatively informed judgments about politics on the basis of limited—even minimal—information. In the process, we also see how new developments in political psychology have helped re-establish the role of the citizenry as purposeful, instrumentally motivated actors in the political process.

## IV Conclusion: converging models of democratic citizens in political behavior

In answering the fundamental challenge of the democratic citizen embedded in the political economy tradition, political behavior research—and, indeed, political science—has come closer to defining a distinctively political model of citizenship. The strategic actor found in the political economy tradition has reappeared in the political sociology and political psychology traditions. In the process, a convergence has emerged among these three traditions in the resulting model of democratic citizenship that they propose. The core of this convergence lies in a model of citizens who act purposefully on the basis of their own goals, ambitions, and needs.

We are intentionally avoiding the "r" word, and many political economists will not be satisfied with the implementation of citizen purpose as it is typically employed in studies of political behavior. At the same time, a great many students of political behavior have come to accept something that approaches Popkin's description of the reasoning voter (1991: 7): "They think about who and what political parties stand for; they think about the meaning of political endorsements; they think about what government can and should do. And the performance of government, parties, and candidates affects their assessments and preferences." Moreover, and as Popkin points out, such reasoning is accomplished on the basis of "low information rationality"—short-cut methods of collecting and processing information about politics.

Perhaps the major controversy surrounding the conception of purposeful citizens concerns whether such a model of instrumental behavior has any real analytic consequence. That is, how much analytic purchase has been obtained by beginning with the assumption that citizens act instrumentally on the basis of their own purposes? According to Simon (1985), the assumption (or acknowledgment) of rationality yields very modest analytic reward—simply asserting that citizens are rational takes us very little distance toward a convincing political analysis. Rather, a series of auxiliary assumptions must be made in order to generate analytic utility from the assumption of instrumental citizen behavior, and it is at this point that the analysis truly begins. (See Lodge and Hamill 1986; Lodge *et al.* 1989.)

The auxiliary assumptions are important for a variety of reasons. First, they provide meaning to the model of purposeful citizens and instrumental behavior. One model of purposeful citizens generates analytic utility by suggesting that citizens update their assessments of parties and politicians by reviewing past experience (Fiorina 1981). Another model suggests that instrumental citizens reduce information costs by obtaining information from other citizens (Huckfeldt and Sprague 1994). Still another model suggests that citizens reduce their information costs through a series of analytic short-cuts, or heuristics (Sniderman *et al.* 1991). These analyses, and many more, are at least partially motivated by the assumption that citizens act in reasonable ways to realize their goals. But beyond that point, they may be radically dissimilar in the models of citizen decision-making that they put forward.

Moreover, while the concept of citizen purpose may lie beyond the reach of direct empirical scrutiny, the auxiliary assumptions are often subject to the bright light of empirical evaluation. Thus, the premise of citizen purpose does not mean much until we determine how citizen's purposes are realized. Therein lies, for example, the contribution of Fiorina's retrospective voting model: it articulates the consequences of purposeful citizens at the same time that it provides a basis for empirical evaluation.

Indeed, a great deal of contemporary research in political behavior is concerned with working out the consequences of citizen purpose. The work of Sniderman *et al.* (1991) provides an exemplar of this general research program: his focus on the manner in which citizens reason about their choices and his articulation of "reasoning chains" are some of the ways that the concept of citizen purpose is given meaning. In short, the citizenship challenge posed by political psychology has been answered, in large part, by turning the challenge on its head. Rather than focusing on how much citizens fail to know about politics, the new look examines the way in which citizens quite sensibly attempt to contain the costs of acquiring and processing political information.

We might also suggest that a "new look" pervades a great deal of political sociology research as well. Rather than focusing on the extent to which individuals are reflections of the environments and contexts and groups within which they are embedded, the new look examines the manner in which citizen purpose is realized or thwarted within particular configurations of time, place, and setting. Rather than seeing individual behavior as a straightforward product of exogenous determinants, citizen choices are seen as being interdependent with their surroundings. At the same time that preferences are informed by information taken from an environmentally determined supply, citizens also impress their preferences on the environment by way of their own information sampling criteria. In short, political information is obtained at the intersection between individually defined citizen purpose and environmentally determined information supply.

Perhaps unintentionally, students of political economy produced a citizenship challenge which ultimately provided part of the solution to companion challenges posed by political psychology and political sociology. By recognizing the costs as well as the benefits of citizenship, the evolving analysis of political behavior sees many old problems in a new light. In the attempt to come to grips with citizens who were neither independent nor well informed, a revised model of citizenship has emerged—a model of the citizen as a cost-conscious consumer and processor of political information who, while taking her duties seriously, has successfully reduced the impulse to be consumed by politics and political affairs. The end product is a more realistic view of citizenship and citizens in democratic politics—a view which understands citizens in the contexts of the real-world constraints imposed by their own inherent limitations as collectors and processors of political information. In many ways, this revised view of the citizen is not only more realistic but also more political, organized around a view of interests and information that is located in the characteristics not only of individuals, but also of the groups and settings to which they belong.

## References

Abramowitz, A. J. 1994. Issue evolution reconsidered: racial attitudes and partisanship in the US electorate. *American Journal of Political Science*, 38: 1–24.

Aldrich, J. H. 1993. Rational choice and turnout. *American Journal of Political Science*, 37: 246–78.

ALT, J. E. The impact of the Voting Rights Act on Black and White voter registration in the South. Pp. 351–77 in *Quiet Revolution in the South: The Impact of the Voting Rights Act 1965–1980*, ed. C. Davison and B. Grofman. Princeton, N.J.: Princeton University Press.
BECK, P. A. 1991. Voters' intermediation environments in the 1988 presidential contest. *Public Opinion Quarterly*, 55: 371–94.
BERGER, B. M. 1960. *Working-Class Suburb*. Berkeley: University of California Press.
BOBO, L. 1983. Whites' opposition to busing: symbolic racism or realistic group conflict? *Journal of Personality and Social Psychology*, 45: 1196–210.
BOUDON, R. 1986. *Theories of Social Change*. Berkeley: University of California Press.
BRADY, H. E., and SNIDERMAN, P. M. 1985. Attitude attribution: a group basis for political reasoning. *American Political Science Review*, 79: 1061–78.
BURT, R. S. 1987. Social contagion and innovation: cohesion versus structural equivalence. *American Journal of Sociology*, 92: 1287–335.
CALVERT, R. C. 1985. The value of biased information: a rational choice model of political advice. *Journal of Politics*, 4: 530–55.
CAMPBELL, A.; CONVERSE, P. E.; MILLER, W. E.; and STOKES, D. E. 1960. *The American Voter*. New York: Wiley.
CARMINES, E. G., and KUKLINSKI, J. H. 1990. Incentives, opportunities, and the logic of public opinion in American political representation. Pp. 240–68 in *Information and Democratic Process*, ed. J. A. Ferejohn and J. H. Kuklinski. Urbana: University of Illinois Press.
—— and STIMSON, J. A. 1989. *Issue Evolution: Race and the Transformation of American Politics*. Princeton, N.J.: Princeton University Press.
—— and HUCKFELDT, R. 1992. Party politics in the wake of the Voting Rights Act. Pp. 117–34 in *Controversies in Minority Voting*, ed. B. Grofman and C. Davidson. Washington, D.C.: Brookings Institution.
—— —— and MCCURLEY, C. 1995. Mobilization, countermobilization, and the politics of race. *Political Geography*, 14: 601–19.
CHONG, D. 1991. *Collective Action and the Civil Rights Movement*. Chicago: University of Chicago Press.
COLEMAN, J. S. 1964. *Introduction to Mathematical Sociology*. New York: Free Press.
—— KATZ, E.; and MENZEL, H. 1966. *Medical Innovation: A Diffusion Study*. Indianapolis, Ind.: Bobbs-Merrill.
DOWNS, A. 1957. *An Economic Theory of Democracy*. New York: Harper and Row.
ERBRING, L.; GOLDENBERG, E. N.; and MILLER, A. H. 1980. Front-page news and real-world cues: a new look at agenda setting by the media. *American Journal of Political Science*, 24: 16–49.
—— and YOUNG. A. A. 1979. Individuals and social structure: contextual effects as endogenous feedback. *Sociological Methods and Research*, 7: 396–430.
ERIKSON, R. S.; WRIGHT, G. C.; and MCIVER, J. P. 1993. *Statehouse Democracy: Public Opinion and Policy in the American States*. New York: Cambridge University Press.
EULAU, H. 1963. *The Behavioral Persuasion in Politics*. New York: Random House.
—— 1986. *Politics, Self, and Society*. New York: Cambridge University Press.
FEREJOHN, J. A., and KUKLINSKI, J. H., eds. 1990. *Information and Democratic Processes*. Urbana: University of Illinois Press.
FINIFTER, A. 1974. The friendship group as a protective environment for political deviants. *American Political Science Review*, 68: 607–25.
FIORINA, M. P. 1981. *Retrospective Voting in American National Elections*. New Haven, Conn.: Yale University Press.
—— 1990. Information and rationality in elections. Pp. 329–42 in *Information and Democratic Processes*, ed. J. A. Ferejohn and J. H. Kuklinski. Urbana: University of Illinois Press.

Gans, H. J. 1967. *The Levittowners*. New York: Vintage.

Giles, M. W., and Evans, A. 1985. External threat, perceived threat, and group identity. *Social Science Quarterly*, 66: 50–66.

Glaser, J. M. 1994. Back to the black belt: racial environment and white racial attitudes in the South. *Journal of Politics*, 56: 21–41.

Granovetter, M. 1973. The strength of weak ties. *American Journal of Sociology*, 78: 1360–80.

Grofman, B.; Handley, L.; and Glazer, A. 1993. Modeling the link between race and electoral change using aggregate data. Mimeo., University of California, Irvine.

Hamill, R.; Lodge, M.; and Blake, F. 1985. The breadth, depth, and utility of class, partisan, and ideological schemata. *American Journal of Political Science*, 29: 850–70.

Huckfeldt, R. 1983. Social contexts, social networks, and urban neighborhoods: environmental constraints on friendship choice. *American Journal of Sociology*, 89: 651–69.

—— 1984. Political loyalties and social class ties: the mechanisms of contextual influence. *American Journal of Political Science*, 28: 399–417.

—— 1986. *Politics in Context: Assimilation and Conflict in Urban Neighborhoods*. New York: Agathon.

—— and Kohfeld, C. 1989. *Race and the Decline of Class in American Politics*. Urbana: University of Illinois Press.

—— and Sprague, J. 1987. Networks in context: the social flow of political information. *American Political Science Review*, 93: 1197–216.

—— —— 1988. Choice, social structure, and political information: the informational coercion of minorities. *American Journal of Political Science*, 32: 467–82.

—— —— 1990. Social order and political chaos: the structural setting of political information. Pp. 23–58 in *Information and Democratic Processes*, ed. J. A. Ferejohn and J. H. Kuklinski. Urbana: University of Illinois Press.

—— —— 1991. Discussant effects on vote choice: intimacy, structure, and interdependence. *Journal of Politics*, 53: 122–58.

—— —— 1993. Citizens, contexts, and politics. Pp. 281–304 in *Political Science: The State of the Discipline II*, ed. A. Finifter. Washington, D.C.: American Political Science Association.

—— —— 1995. *Citizens, Politics, and Social Communication*. New York: Cambridge University Press.

Iyengar, S. 1990. Shortcuts to political knowledge: the role of selective attention and accessibility. Pp. 281–303 in *Information and Democratic Process*, ed. J. A. Ferejohn and J. H. Kuklinski. Urbana: University of Illinois Press.

—— 1991. *Is Anyone Responsible?* Chicago: University of Chicago Press.

—— and Kinder, D. R. 1987. *News that Matters*. Chicago: University of Chicago Press.

Jackman, R. W. 1993. Response to Aldrich's rational choice and turnout: rationality and political participation, *American Journal of Political Science*, 37: 279–90.

Jennings, M. K., and Niemi, R. G. 1974. *The Political Character of Adolescence*. Princeton, N.J.: Princeton University Press.

—— —— 1981. *Generations and Politics: A Panel Study of Young Adults and Their Parents*. Princeton, N.J.: Princeton University Press.

Katz, E. 1957. The two-step flow of political information: an up-to-date report on a hypothesis. *Public Opinion Quarterly*, 21: 61–78.

Key, V. O., Jr., and Munger, F. 1959. Social determinism and electoral decision: The case of Indiana. Pp. 281–99 in *American Voting Behavior*, ed. E. Burdick and A. J. Brodbeck. Glencoe, Ill.: Free Press.

——, Heard, H. 1949. *Southern Politics*. New York: Alfred A. Knopf.

KLAPPER, J. T. 1960. *The Effects of Mass Communication.* New York: Free Press.
KNOKE, D. 1990. *Political Networks.* New York: Cambridge University Press.
LANGTON, K. P., and RAPOPORT, R. 1975. Social structure, social context, and partisan mobilization: urban workers in Chile. *Comparative Political Studies,* 8: 318–44.
LAZARSFELD, P. F.; BERSELSON, B.; and GAUDET, H. 1944. *The People's Choice.* New York: Columbia University Press.
LODGE, M., and HAMILL, R. 1986. A partisan schema for political information processing. *American Political Science Review,* 80: 505–19.
—— McGRAW, K. M.; and STROH, P. 1989. An impression-driven model of candidate evaluation. *American Political Science Review,* 83: 399–419.
McPHEE, W. N.; SMITH, R. B.; and FERGUSON, J. 1963. A theory of informal social influence. Pp. 74–103 in *Formal Theories of Mass Behavior,* ed. W. N. McPhee. New York: Free Press.
MACKUEN, M. B. 1981. Social communication and the mass policy agenda. Pp. 19–146 in *More than News,* ed. M. MacKuen and S. L. Coombs. Beverly Hills, Calif.: Sage.
—— 1990. Speaking of politics: individual conversational choice, public opinion, and the prospects for deliberative democracy. Pp. 59–99 in *Information and Democratic Processes,* ed. J. A. Ferejohn and J. H. Kuklinski. Urbana: University of Illinois Press.
—— ERICKSON, R.; and STIMSON, J. 1989. Macropartisanship. *American Political Science Review,* 83: 1125–45.
MATTHEWS, D. R., and PROTHRO J. W. 1963. Social and economic factors and negro voter registration in the South. *American Political Science Review,* 57: 24–44.
MILLER, W. E. 1956. One-party politics and the voter. *American Political Science Review,* 50: 707–25.
MONDAK, J. J. 1993. Source cues and policy approval: the cognitive dynamics of public support for the Reagan agenda. *American Journal of Political Science* 37: 186–212.
—— 1994. Cognitive heuristics, heuristic processing, and efficiency in political decision making. Vol. iv, pp. 117–41 in *Research in Micropolitics,* ed. M. X. Delli Cappini, L. Huddy and R. Y. Shapiro. Greenwich, Conn.: JAI Press.
OLSON, M. 1965. *The Logic of Collective Action.* Cambridge, Mass.: Harvard University Press.
OSTROM, E. 1972. Metropolitan reform: propositions derived from two traditions. *Social Science Quarterly,* 53: 474–93.
—— 1990. *Governing the Commons.* New York: Cambridge University Press.
OTTATI, V. C., and WYER, R. S., Jr. 1990. The cognitive mediators of political choice: toward a comprehensive model of political information processing. Pp. 186–216 in *Information and Democratic Process,* ed. J. A. Ferejohn and J. H. Kuklinski. Urbana: University of Illinois Press.
POPKIN, S. L. 1991. *The Reasoning Voter: Communication and Persuasion in Presidential Campaigns.* Chicago: University of Chicago Press.
PRZEWORKSI, A. 1974. Contextual models of political behavior. *Political Methodology,* 1: 27–61.
—— 1985. Marxism and rational choice. *Politics and Society,* 14: 379–409.
—— and SPRAGUE, J. 1986. *Paper Stones.* Chicago: University of Chicago Press.
SEARS, D. O.; HENSLER, C. P.; and SPEER, L. K. 1979. Whites, opposition to busing, self-interest or symbolic politics? *American Political Science Review,* 73: 369–84.
SIMON, H. A. 1957. *Models of Man.* New York: Wiley.
—— 1985. Human nature in politics: the dialogue of psychology with political science, *American Political Science Review,* 79: 293–304.

Sniderman, P. M. 1993. The new look in public opinion research. Pp. 219–46 in *Political Science: The State of the Discipline II*, ed. A. Finifter. Washington, D.C.: American Political Science Association.

—— Brody, R. A.; and Tetlock, P. E. 1991. *Reasoning and Choice.* New York: Cambridge University Press.

Sprague, J., and Westefield, L. P. 1979. An interpretive reconstruction of some aggregate models of contextual effects. Presented at the annual meeting of the Southern Political Science Association, Gatlinburg, November 1–3.

Stimson, J. A. 1991. *Public Opinion in America: Moods Cycles and Swings.* Boulder, Colo.: Westview Press.

Tversky, A., and Kahneman, D. 1974. Judgment under uncertainty: heuristics and biases. *Science,* 185: 1124–31.

Tedin, K. L. 1974. The influence of parents on the political attitudes of adolescents. *American Political Science Review,* 68: 1579–92.

Tingsten, H. 1963. *Political Behavior: Studies in Election Statistics,* trans. V. Hammarling. Totowa, N.J.: Bedminster; originally published 1937.

Weber, M. 1966. On the concept of sociology and the meaning of social conduct. Pp. 29–58 in *Basic Concepts in Sociology,* trans. H. P. Secher. New York: Citadel Press; originally published 1925, ch. 1 of *Wirtschaft und Gesellschaft.*

Wright, G. C., Jr. 1976. Community structure and voting in the South. *Public Opinion Quarterly,* 40: 200–15.

Wolfinger, R. E., and Arsenau, R. B. 1978. Partisan change in the South, 1952–1976. Pp. 179–210 in *Political Parties: Development and Decay,* ed. L. Maisel and J. Cooper. Beverly Hills, Calif.: Sage.

Chapter 9

# Political Behavior: Reasoning Voters and Multi-party Systems

Franz Urban Pappi

STUDYING political behavior at the level of the general electorate has changed considerably from the early days, characterized by the predominance of the Michigan model in the 1960s (Campbell *et al.* 1960; 1966) and the juxtaposition of this social-psychological approach with the rational choice approach of Downs (1957). Beginning with revisionist arguments in favor of issue voting (Rusk 1987), and continuing through both traditional and rational choice ideas about retrospective voting (Fiorina 1981), a concept of the "reasoning voter" has emerged.

This concept functions as a bridge between political psychologists and "realist" versions of the rational choice approach—versions which try to predict actual voting behavior, instead of focusing exclusively on equilibrium conditions for the demand and supply of policy packages. Reasoning voters are approximately rational, trying to come to terms with a decision situation about which they are only vaguely informed. Judgmental heuristics are used to solve "Simon's puzzle" of how to decide rationally "with limited information and processing capacity" (Sniderman, Brody and Tetlock 1991: 18).

Within a rational choice framework, Popkin adopts the term "reasoning voter" to describe a similar situation in which "voters actually do reason about parties, candidates, and issues" (1991: 7), "investing" their vote in collective goods on the basis of "costly and imperfect information under conditions of uncertainty" (1991: 10). As contrasted with private investors, these "public investors" have less incentive to gather costly information. Hence Popkin characterizes this choice situation as one of low-cost rationality, where the premium is on gathering inexpensive information from friends and using informational and calculative short-cuts.

The general insight contained in the concept of the reasoning voter is not new. Since the original *homo economicus* was gradually turned into a utility maximizer, subjectively calculating the probabilities of outcomes, it was only a small step to realize that the acquisition of perfect information is often too expensive (a point already present in Downs (1957)). What is new is that cognitive psychology can now offer a set of experimentally tested results: about "schemata" for acquiring, processing and retrieving information; about dominant "frames" simplifying the goal structure; and about habits as quick decision rules in frequently recurring decision situations (Esser 1991).

These theories enable the cognitive psychologist to model individual decision-making. But in political science in general, and in voting research in particular, it is also necessary to consider the interface between the choices of individuals and the world of party politics and government policy. This chapter focuses on problems of preference formation and party choices of reasoning voters in multi-party systems. This is a necessary and separate task, in a field where the research agenda focuses heavily on the American two-party system and its presidential elections.

Rational choice is behavior which is consistent with one's preferences, the preferences themselves having to fulfill minimal conditions of consistency such as transitivity and so on. Thus, we will first consider problems of party preferences of voters in multi-party systems. Second, the forces influencing these preferences are taken into account. Building on the concept of the reasoning voter, we shall consider how variables like issue proximities, retrospective evaluations of government performance and party identification can be interpreted in a model of preference formation in multi-party systems. The final task will be to analyze the decision problem itself, taking into account future expectations about possible government coalitions, incentives for sophisticated voting and instrumental versus expressive or symbolic voting.

The overall model of this chapter is built on the assumption that, in multi-party systems, voters' party preference profiles are the crucial link between the factors influencing reasoning about parties and the final voting decision. In this sense, "party preference" has an analogous causal status as "comparative candidate evaluations" in the revised Michigan model (Markus and Converse 1979). From among possible influences on party preference, issue proximities and retrospective evaluations are identified as factors most proximate to the process of preference formation in the assumed funnel of causality. Issue proximities constitute a substantive prospective measure, while retrospective evaluations pertain to the past performance of the present government.[1]

The first and third sections of this chapter touch upon what Huckfeldt and Carmines (above: chap. 8) call the "economists' challenge" to sociological and social psychological approaches to the study of voting behavior. Purposeful, rational action has as its pre-condition clear preferences (Section I) and reasoning about the consequences of one's action (Section III). This latter task is complicated when the outcome which counts is the aggregate of all people's votes rather than the result of one individual's action alone. Section II's discussion of preference formation takes up political psychological concerns about the process of becoming informed, thus rendering possible the matching of personal goals with the party-political means of achieving those goals.

## I Party preferences in multi-party systems

In a two-party system, the relation between party preference and party choice is rather simple, once the voter has decided to participate in an upcoming election at all. Either a voter prefers party or candidate *A*, in which case s/he will choose this alternative; or else s/he prefers party or candidate *B* and will, therefore, choose that alternative; and indifference should lead to abstention. It would be irrational, in a two-party setting, to vote for a party which is not one's first choice.

The choice problem in a three-party system is not a straightforward extrapolation of the two-party case. There, it may be rational to vote strategically for one's second most preferred party if one expects one's most preferred party to lose and one's least preferred party otherwise to win. The strict preference order between three parties contains $3! = 6$ logical possibilities, and even more if ties of two or all three parties are contemplated. With three possible answers to every paired comparison between three parties (counting pairs for the same parties only once, not twice as ordered pairs), there exist 27 different possible preference orderings; among the 13 consistent individual rank orders, one profile has all three parties tied, six preference profiles have two-parties tied at either the first or last rank, and another six constitute strict rank orders. From three parties upward, it is no

[1] Future expectations are more difficult to integrate in models for multi-party systems, since the future government depends not only on election results but also on future coalition-building. Coalition expectations influence the final decision outcome, insofar as the voter has to reason about the government which will be a possible consequence of the various different ways of casting his or her vote which are compatible with his or her party preference profile.

longer self-evident that voters have consistent preferences (contrary to the assumptions built into frequently used measurement devices such as "feeling thermometers").

Ideally, one would hope that the preferences of individual voters are consistent enough for them to be expressed in a precise utility function which is not just the trivial outcome of measurement by fiat. Using paired comparisons between German parties, it can be shown that the minimal condition for rational voters—transitivity of strict preferences—is fulfilled for over 90 percent of the electorate (Pappi 1983). Brady and Ansolabehere (1989: 149) report similarly that only 2 to 10 percent of American respondents had intransitive candidate orderings over candidates in presidential primaries.[2]

In normal election studies, the data which one would need for the valid construction of utility functions over a set of parties are not available. But in many European election studies, "feeling thermometer" or "sympathy scalometer" questions are asked for all parties competing for seats in parliament; and it is possible to exploit this type of data for a construction of the party preference profiles of voters. Before discussing different types of measurements, though, we have first to clarify the theoretical status of the concept of "party preference."

I propose to treat "party preference" as the central intervening variable between the reasoning about and evaluations of parties, on the one hand, and voting behavior, on the other. The status of this variable is, as I have said, similar to that of "comparative candidate evaluations" in American models of the presidential vote. But to make it strictly equivalent parties would have to be presented as concrete options in the specific upcoming election. That is to say, one would have to ask voters about their party preferences concerning one specific election: for a federal election, for example, one would have to ask, "Which party do you prefer most in federal politics for the upcoming legislative period?"; and when two elections (a federal and a state election, for example) are held simultaneously, one would have to ask two election-specific party preference questions.

The alternative is to conceptualize parties as consistent players in different games and to assume that voters (especially in multi-party systems) develop general party preferences. In most democracies, the different government levels are not of equal importance, so that, empirically, parties are evaluated overwhelmingly with respect to their national role.

The degree of election- or situation-specificity is one dimension characterizing party preference. A second is the time dimension. Should "party

[2] They report that the transitivity axioms are also usually fulfilled for indifference judgments and for negative as well as positive preferences among candidates.

preference" just tap the momentary feelings towards parties, or should it measure a more permanent attitude? Theoretically, it seems that one's present party preference profile must be influenced by the profile of an earlier period: there has to be some continuity in these preferences. Alternatively, though, we might ask the question in a way which stresses the momentary character of preference, to the detriment of its permanent aspects. Thus on the "time" dimension, party preferences might represent either a more permanent or a more momentary attitude, just as on the first dimension party preferences might be either situation- or election-specific or not.

The concept of "party preference" is, of course, related to the concept of "party identification." In its original Michigan version, party identification is conceptualized as a permanent positive identification with a party as a general (not situation-specific) attitude object. For European multi-party systems, party identification questions have the shortcoming that a certain degree of identification with one party cannot be interpreted as the inverse of the identification with the other parties. An American-style seven-point scale (running from strong to weak Democrats through Democratic leaners and independents to strong Republicans) cannot be constructed for multi-party systems. Dutch authors have proposed an alternative measure of permanent party preference profiles which avoids concentration on the most-preferred party. The proposed question is supposed to measure general electoral utility of parties by asking respondents, for each party, how probable it is that they "will ever vote" for it (van der Eijk *et al.* 1986). Since "will ever vote" transcends the momentary aspect of preference, focusing on elections in general and not on any specific election, answers to this question can be interpreted equivalent to expressions of party identification in multi-party systems.

The more conventional "sympathy" ratings of parties stress more the momentary (though not, it is thought, election-specific) character of these attitudes. Monthly time series do show many ups and downs for the different parties. But these movements are normally not decomposed into their different components, such as the overall mean for the established parties, the variance of the respondents' ratings or the relative party ranks; and the latter may remain relatively stable, even if the mean level of sympathy or the size of the distances between parties changes. What may look on the surface like a clear candidate for a "momentary general party preference" may thus contain information on more permanent aspects of party preference, measuring an attitude somewhere in between a momentary and a permanent preference. This interpretation comes close to party identification as "a running tally of retrospective evaluations" (Fiorina 1981: 89) for every party a voter perceives as relevant, where s/he continually adds positive evaluations for each party and substracts negative ones depending on experiences during legislative periods.

Let us now briefly consider the other two possible conjunctions of the situation- or election-specificity and time dimensions. Pre-election polls sometimes ask situation-specific, momentary party rankings. The only combination of the two dimensions that might seem empirically empty is permanent election-specificity. But even there one can imagine voters who have different, but relatively stable, party preferences for different types or levels of elections. For example, they would never vote for a post-communist party at the national level, but they consider the respective local party as an option for a first or middle rank. Since voting research normally focuses on national elections, election-specificity is a neglected variable.

Different devices exist for assembling data on party preference profiles: paired comparisons, rankings or ratings. The latter two methods impose consistent profiles by their very design, whereas paired comparison data allow empirical tests of consistency. Conventional ranking or rating scales are frequently analyzed in search of a simple spatial representation of party preference profiles. In those studies, the results of factor analysis (or of unfolding techniques applied to preference or dominance data) are interpreted as the joint decision space of the electorate, summarizing the rankings or distances of the voters from the parties in the statistically best possible way. Numerous analyses of this type have been performed for multi-party systems, including Converse's early (1966) study for Finland and Nannestad's sophisticated recent (1994) study for Denmark. Unfolding analysis allows us to test hypotheses about the one-dimensionality of a party space. Factor analysis, in contrast, is more often applied for explanatory than for confirmatory purposes; and the interpretations of the resulting solutions do not always appear plausible, especially where more than one dimension is identified.

## II Preference formation

Among the many forces affecting party preference, those singled out here are those with the most immediate impact on reasoning voters, who must make up their minds using the available political information and without investing in expensive monitoring.

Citizens in modern democracies are continuously exposed to a steady flow of political information from the mass media, from discussions in their everyday lives, and so on. Not everyone is interested in politics all the time; some people may have difficulties inferring from news about domestic affairs any conclusions about the role of parties and the consequences for their party preferences. Informational short-cuts which could be relevant in this situation are: "schemata," to process and retrieve political information; "framing mechanisms," to simplify the evaluation of parties; and "retrospective evaluations" of government performance.

The concepts of schema and frame were developed by cognitive psychologists (Hastie 1986; Kahneman and Tversky 1982) and are frequently applied to voting and public opinion research (Sniderman 1993; Iyengar 1991). Schema theory has, however, aroused controversy as a way of understanding political attitudes and behavior (Kuklinski *et al.* 1991; Lodge, McGraw, Conover, Feldman and Miller 1991). The core of this controversy, as whenever a new concept is imported from one discipline into an established research field in another, is whether we gain fresh insights or whether we just retell old stories in a new jargon. Schemata—"the set of cognitions relevant to some concept" (Kuklinski *et al.* 1991: 1342)—help the individual in information processing. Memory structures, and information input, processing and retrieval are modeled by psychologists for individual persons. Political psychologists, though, are more interested in convergence among citizens in their definitions of social situations; and for that task we already had concepts such as belief systems (Converse 1964) and ideologies as political codes (Klingemann 1979).

Here, such concepts are discussed in relation to reasoning voters and their information-processing problems. Some of these older concepts were initially offered as "realist" conceptions of voters, counterposed to normative models of an informed and perfectly rational voter. What were originally conceived as deviations from from *homo economicus* or from the *homo politicus* of civics textbooks can nowadays be reinterpreted as the modal citizen, who uses informational short-cuts and judgmental heuristics in reasoning about politics and parties. These findings from cognitive psychology have led to new conceptualizations of rational choice (Lindenberg 1990) which are more realistic, but which should nevertheless allow formal modeling.

The minimally reasoning citizen of traditional voting research is the retrospective voter. Following the traditional reward-punishment theory of Key (1966), retrospective voters simplify their reasoning by evaluating their own well-being in the immediate past for which they hold the incumbent government responsible, rewarding it for positive developments and blaming it for deteriorations.[3]

[3] Alternatively, retrospective evaluations might be mediated rather than direct, the citizen relying not only on his or her own experience but also on judgments of others or of the mass media about the accomplishments of the incumbent government (Fiorina 1981).

But what is the impact of retrospective evaluations on party preferences in multi-party systems, where coalition governments are the rule and not the exception? How citizens allocate credit for overall government achievements among parties within a coalition is essentially an empirical question. In European voting studies, questions on party competence for different policy domains are sometimes asked, and these measures can be viewed as proxies of performance evaluations. The respondents are asked to rate or rank the importance or salience of each problem, and then they are asked to name which party they think would be best in handling the issue. Even without an explicit time frame, competence evaluations seem inevitably to be based on past experiences; and in the case of elections to the European Parliament, these experiences will more likely stem from the participation of parties in national governments than in their European roles.

Simple retrospective evaluations of government performance are based on the voters' experiences and seem not to be prone to rationalizations in the way that party competence judgments often are. Even scholars relying on party competence measurements concede that these may be only reflections of more permanent affective ties to parties (Küchler 1991: 101)—that is, a consequence and not a cause of party preference. The easier the evaluation task is made for respondents, the more they are tempted to report non-attitudes. This is a severe problem, especially for smaller parties, whose competence may be less visible when in government and almost impossible to evaluate when in opposition.

Rationalization is an easy option for respondents when they are just asked to pick the most competent party, without having to compare it specifically with other parties. Sniderman *et al.* (1990) have shown that informed and less informed American voters differ in the relative importance that "incumbent approval" and "comparative prospective evaluations of candidates" have for their votes. The more informed voters reason more, basing their final decision less on incumbent approval and more on a comparative candidate evaluation. They act as optimizers, whereas less informed voters just attempt "to decide whether the way things have been have been good enough," thereby acting as satisficers (Sniderman *et al.* 1990: 131). In parliamentary systems with coalition governments, satisficers are confronted with a more difficult task than in the American presidential system; but they may nevertheless choose, for example, between the larger coalition party (or the party of the prime minister) and the major opposition party. In Germany, only the Christian Democrats and the Social Democrats present candidates for the chancellorship; so that system approaches a two-party constellation, at least for the less informed who thus have a chance to avoid a complex comparison between all party options.[4]

Party competence questions have another disadvantage as probes of factors influencing the formation of party preferences: those questions are sometimes ambiguous, in utility terms. When a respondent is asked to evaluate the competence of a party with respect to a certain policy domain such as European unification, s/he might name the party which s/he perceives as the most able promoter of unification—even though s/he is actually against further unification. A party is properly deemed "competent" if it is able to pursue its goals effectively, whether or not the respondent agrees with these goals, and a respondent who opposes further European unification can nevertheless believe that a party in favor of such a policy is very competent (and is, therefore, an able opponent of his or her own goals). Weird as this possible interpretation may look, at the very least it shows that competence questions are not a very direct way to measure the utility component of party preference. Retrospective evaluations of government performance, and judgments about future policies of the parties which a voter is able to compare with his or her own policy preferences, are much more plausible ingredients of preference formation than competence ratings.

At the core of the rational voter model are the proximities of voters' "ideal points" to the perceived party positions on issue scales. Downs (1957) originally postulated that this voter will invest his or her vote in that party which promises the best returns in the next legislative period, and issue proximities are supposed to measure these prospective returns. Issue positions are linearly ordered sets of policy options, resulting in a one-dimensional scale on which voters have single-peaked, symmetric preference curves. The further a voter perceives a party's position to be (in either direction) from his or her own ideal point, the greater the utility loss s/he expects if that party were to pursue its policy in government.

The reasoning voter is here assumed to differ from the original rational voter in two major respects. S/he does not form her party preferences exclusively on the basis of issue proximities, and s/he does not act as a consequential investor in that part of his or her future well-being which is influenced by government. The latter aspect will be discussed in the next section. The first aspect was already dealt with, showing how retrospective

[4] But then it would be important to measure incumbent performance directly, instead of using questions built on the assumption that the respondents reason carefully about the comparative competence of the parties.

evaluations of government performance as well as issue proximities affect preference-formation.[5]

Issue voting raises many problems for both the informed and the less informed voters. The first group is not necessarily homogeneous: these voters are not all driven by the same problems, for which they look to the government for solutions. The latter group often simply will not know which policies different parties offer as solutions. The final result could be an electorate balkanized into many issue publics, with a large rump for which issue proximities are utterly meaningless. Multi-party systems should have even more uninformed voters than two-party systems, since it is presumably more cumbersome to gather and process data on many parties than it is on only two; and assuming a proliferation of one-issue parties in such settings, balkanization would more probably be even more characteristic of multi-party than of two-party systems.

Interestingly enough, these hypotheses are radically untrue to life in multi-party systems. A tentative explanation is given by a theory analogous to the "functional theory of party identification" (Shively 1979): citizens without direct information on parties' positions concerning many issues will apply informational short-cuts; and the more they are forced to do that by features of the political system, the better oriented citizens in multi-party systems become compared to those in two-party systems.

From Downs (1957) forward, ideology has been discussed as a possible information short-cut. This use of ideology should not be confused with ideological thinking in the sense of "political sophistication" (Converse 1964). In that latter sense, ideologues have the highest level of political conceptualization; they have opinions on many issues and organize their political beliefs, using abstract principles like "liberalism" or "conservativism" which are not idiosyncratic but rather which are shared by the élites; and their different attitudes are, therefore, consistent as shared "official" ideologies. The original estimate of the percentage of such ideologues in the American electorate was 2.5 percent, plus another 9 percent who where characterized as "near-ideologues" (Campbell *et al.* 1960: 249). Even though these numbers have increased, compared to the low levels within the American electorate in the late 1950s (see Abramson 1983: 273), ideological thinking has remained a minority affair and not a powerful device for "cognitive misers" in general to use in orienting themselves in the world of politics (Smith 1989).

Organizing one's own attitudes in terms of "official ideologies" which

[5] Other possible factors (such as candidate characteristics, and so on) cannot be excluded in principle. But in multi-party systems these presumably are more election-specific and less general types of impact on party preference formation.

function as constraints for individual belief systems is not at all necessary when citizens try to orient themselves politically with the help of ideological labels like "left" and "right." Of course, in order for them to function as orientation short-cuts at all, members of the public have to attach these labels to political parties. But even unsophisticated voters should be able to apply the left–right schema as a mere orientation device, assuming the mass media uses these labels more-or-less consistently in its discussion of political topics. Fuchs and Klingemann "view the left–right schema as a mechanism for the reduction of complexity, which serves primarily to provide an orientation function for individuals and a communication function for the political system" (1990: 205). They find that over 90 percent of the West German and Dutch electorates have at least a minimum understanding of the labels "left" and "right," whereas this figure is lower in the United States. Since the schema is closely linked in Europe to the cleavages characterizing party systems, it is able to fulfill its function as an information short-cut about political contents.

Many European researchers use the left–right scale in surveys, both in order to generate data on the perceived positions of parties and to ask the respondents where they place themselves on the same scale. As a result, one has all the necessary data for a simple one-dimensional spatial model, in which differences between ideological self-placement and perceived party positions are interpreted as utility-loss terms. This is the most straightforward operationalization of Downs's idea that many party systems are characterized by one overwhelming ideological dimension (an economic left–right scale) on which both parties and voters are located, giving general hints in a complex world about their policy preferences. This operationalization is appealing, since voters in European multi-party systems do indeed perceive the parties' positions on the left–right scale relatively correctly (for Germany, see Klingemann 1972 and Pappi 1983: 427; for Italy see Sani 1974). But in order to use these data as generalized issue proximities, one has in addition to make two assumptions:

1. the left–right dimension is the only major schema applied to parties, or at least there is not another schema which would result in a different linear order among the parties;
2. the respondents are able to use the same orientation device about themselves as about parties, which is to say, they also perceive their own policy preferences in general left–right terms.

Hinich and his collaborators (Hinich and Pollard 1981; Enelow and Hinich 1984; Hinich and Munger 1992) have developed a theory which, instead of being built on these assumptions, actually allows an empirical test of them. One can interpret this approach as a type of schema theory, which postulates that citizens in modern democracies do indeed use ideological labels attached to parties as information short-cuts about concrete policy positions of these parties. But contrary to Downs's original model, the citizens do not have to place themselves directly on the ideological dimensions—since their primary goals are their policy preferences which can (given certain further assumptions) be summarized as derived ideological positions.

Downs (1957: 132–3) had originally proposed a simpler model:

> each party takes stands on many issues, and each stand can be assigned a position on our left–right scale. Then the party's net position on this scale is a weighted average of the positions of all the particular policies it upholds. Furthermore, each citizen can apply different weights to the individual policies, since each policy effects some citizens more than others.

Hinich and Munger (1992: 9) say this model does not work. It cannot guarantee coherence in ideological messages, due to the different weights the voters assign to the policy positions. "If we allow for some overall coherence, we must still address the question of how voter preferences are expressed, for this expression determines which ideology parties try to associate themselves with in order to win elections." They solve this central puzzle with a new spatial theory of ideology in which "the ideological dimension(s) (where politics is debated and decided)" are related systematically "to the complex, n-dimensional policy space (which voters actually care about)." This goal is achieved by postulating that voters have knowledge of the ideological positions of parties on one or more ideological dimensions, and that voters use that knowledge to guess about the policy stands of these parties.

Thus, one has only to assume that citizens have a vague idea of how some ideological dimension (left–right, for example) is related to policy stands (on economic redistribution, for example). Suppose a voter has the idea that leftist parties are more in favor of redistribution. From that knowledge, s/he can derive the probable positions of the parties on this policy dimension, placing parties more toward the redistributive end of the scale the more leftist they are perceived to be. But the citizen does not necessarily locate himself or herself on the left–right ideological scale: instead s/he merely has a particular preference concerning redistributive issues.

Voters may differ in their perceptions of how much the policy positions of the parties differ with respect to the basic ideological dimension. Some voters perceive rather large differences between left and right parties with respect to redistributive issues; others may think that nowadays left and

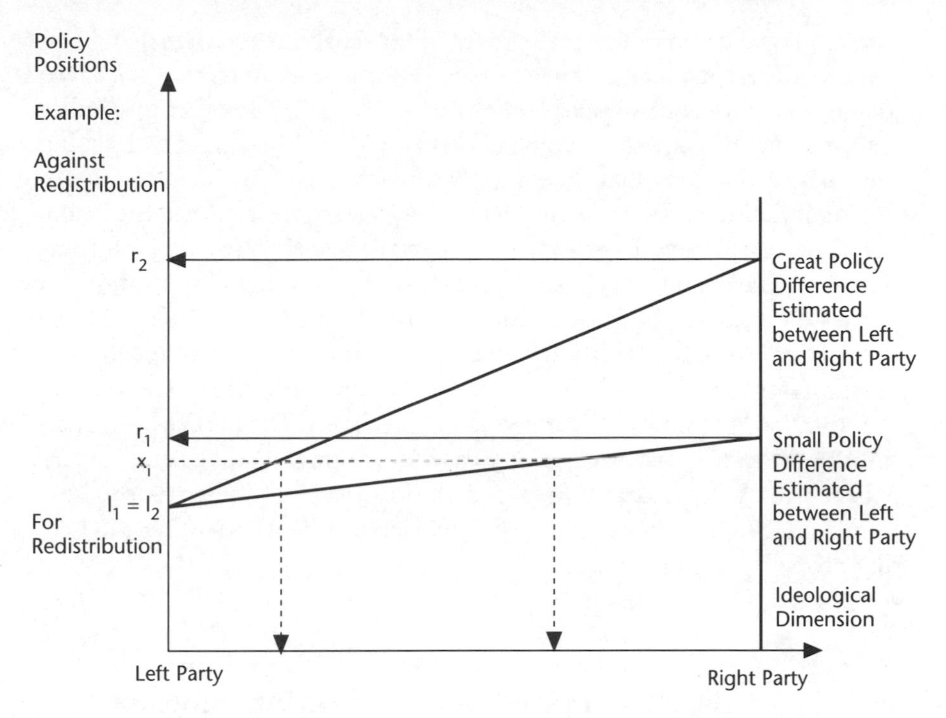


**Fig. 9.1** *The policy preference of voter* i ($x_i$) *and two possible estimates of the policy positions of a left* ($l_1$, $l_2$) *and right party* ($r_1$, $r_2$)

right parties do not differ very much concerning such issues. Depending on these estimates, voters with the same policy preference may end up at different derived positions on the underlying ideological dimension (see Figure 9.1).

Where campaigns focus on a single issue, this theory would just complicate the decision situation unnecessarily. The aim of the theory is to find a few ideological dimensions consolidating several different concrete policy issues at play in campaigns or in politics more generally. Even if the number of issues is quite large, it may well be that the underlying ideological space is only one-dimensional. Still, it is an empirical question whether a party system can be characterized by only one latent ideological dimension along which the parties differ, or whether more than one latent ideological dimension underlies the political position-taking in day-to-day politics.[6]

Empirically, there are two ways to apply Hinich's theory. One possibility

[6] Thus, the German liberal party is arguably perceived to occupy a middle position between the Christian Democrats and the Social Democrats only when attention is focused on law and order, abortion rights or new social issues: in economic policy terms, the same party is perceived to the right of the Christian Democrats (Laver and Hunt 1992; Pappi 1994).

is to employ data on policy preferences of voters and perceived party positions for different issues, using factor analytic methods to recover the latent ideological space (Enelow and Hinich 1984: 213–15). The other possibility is to measure directly the ideological knowledge of the voters concerning the parties. Assuming that there is only one left–right dimension underlying policy differences in a multi-party system, one can ask the usual left–right question with respect to the parties, and then use this information as an independent variable to predict the perceived policy positions of parties on specific policy dimensions (Hinich and Munger 1992: 23–5). Whichever estimation method one uses, one is able to derive ideological positions for the voters which are then used as the "derived ideal points" on the so-called "predictive" dimension or dimensions. The parties have positions in the same space, so it is very easy to compute ideological distances between the voters' ideal points and the positions of the parties. These terms can then be used as independent variables, along with others, to predict people's party preferences.

## III Party preferences and voting behavior

On the model here in view, then, retrospective evaluations of parties and issue proximities serve, firstly as predictors of party preference formation and, secondly and through that, as predictors of voting behavior. Clearly, "retrospective evaluations" and "issue proximities" are topics about which voters themselves reason—which is a major argument in favor of this model. But when it comes to actual voting behavior, it is an open question of whether voters have to reason anew come every fresh election. Could not party preference or party identification of an earlier period serve as a standing decision? At least a substantial subset of voters might have finished their reasoning some time ago, and their party preferences take only the rudimentary form that they confidently know which party they like most. Then it is possible to vote in general elections just out of habit: voting being a repeated action for most voters, participation and choice could be an almost automatic consequence of a stable attitude towards one favored party, outside the reach of conscious decision-making (Ronis *et al.* 1989).[7]

[7] On the other hand, though, habits can also be seen as a conscious decision between (on the one hand) a standard option with a certain known utility and (on the other hand) an alternative with additional outcomes whose evaluation incurs additional information costs: the utility of the additional option will be expected only with a given probability, and it will not be as certain an outcome as the standard option, so the new alternative will be selected only if its expected utility is rather high, compared to the search costs (Esser 1991: 67).

This "habit" concept is appealing, at least as applied to older cohorts with frequent experience of participating in democratic elections (Converse 1976). Some caveats should nonetheless be entered. Voting is indeed repeated behavior, but with much longer time intervals than in everyday behavior. Citizens are regularly confronted with news about parties but are relatively seldomly required to choose among them. In addition, voting behavior is not costless, so that a decision to participate at all in a certain election may not be self-evident, especially when bad news about one's favorite party abounds. Reflecting upon such caveats, we can conclude that a certain amount of reasoning beyond pure habit is needed.

Whatever the forces influencing preference formation, reasoning voters summarize their impressions from party politics in a preference profile. Even pure-habit voters merely prefer their standing party to all others, and are only less open to new party signals in the ongoing election campaign; so that their party preference of the earlier period carries over to the new election without short-term impacts. What remains to reason about are the election-specific conclusions our potential voters draw from their preferences.

Logically, the decision situation is clear enough: one has to decide whether to participate in the upcoming election, and which party to choose. In a two-party system, the most-preferred party will obviously be the party of one's choice, so the participation decision is the more difficult one. It gave rise to the voters' paradox—why vote at all, when the expected utility of one's own participation is very small or even negative, considering the participation costs. Those costs are certain, whereas the utility incomes of future governments have to be discounted by the small probability that the vote of the respective individual is decisive. Research shows that turnout at the aggregate level is positively correlated with the mean size of the party differential (utility term) and the closeness of the aggregate vote for the competing parties (probability term) (see Grofman below: chap. 29). Still, the overall impact of participation norms (civic duty) is very important at the individual level. Hence, many authors, even those starting from rational choice premises, end up taking participation norms as a major factor determining turnout (Barry 1978)—and as one which is independent of the instrumentally rational calculus of party choice.

Those norms serve an expressive function. The satisfaction comes from the act of voting as such, not from its consequences for future governments as in the case of instrumental action. "Public choice orthodoxy seems to assume that though non-instrumental considerations are relevant in getting voters to go to the polls in the first place, such considerations cease to bear once the voter slips behind the curtain to pull a lever or mark a card"

(Brennan and Lomasky 1993: 35). Brennan and Lomasky argue that both expressive and instrumental factors play a role in party choice, as others have argued before them (see Goodin and Roberts 1975). An example of expressive preferences are ethical principles which people uphold and see as symbolized by a party program, without asking about the consequences of the principles in political reality. In Weberian terms, one can characterize individual voters as both instrumentally rational (*zweckrational*) and *wertrational*, that is, oriented "to an absolute value . . . involving a conscious belief in the absolute value of some ethical, aesthetical, religious, or other form of behavior, entirely for its own sake and independently of any prospect of external success" (Weber 1965: 175). Since a single voter is almost never decisive in a general election, expressive considerations gain importance compared to the investment decision when a citizen invests his or her vote to maximize his or her returns from future governments.

In multi-party systems with coalition governments, citizens as investors of votes have to overcome even more obstacles than in two-party systems with alternating governments. Although having normally only the option of choosing which party to support, they are also interested in the future coalition. When the parties do not announce their coalition preferences before the election, or when election outcomes are quite uncertain, it is impossible to formulate a rational strategy aiming to secure a particular coalition government. Then voting degenerates to a reporting of first preferences. Suppose now, though, that party competition is polarized between government and opposition parties, and the government parties promise to continue their coalition: a rational voter whose most preferred party is coalition partner *A* and whose second preference is coalition partner *B* may then vote sophisticatedly, voting for *B* instead of *A* if s/he thinks s/he is more decisive for the electoral success of *B* than of *A*.[8] But in general, we expect that voting in multi-party systems is even more characterized by expressive considerations than in two-party systems, since the complexity of the decision situation discourages prospective instrumental voting. Parties advocating this or that as an absolute value, rather than focusing on practical success-oriented politics, flourish more than in a majoritarian democracy with alternating governments. And high turnout figures seem to confirm this type of *Gesinnungsdemokratie* which often degenerates to *Stimmungsdemokratie*.

[8] In Germany, coalition majorities often depend both on the relative advantage of the largest compared to the second-largest party and on the success of the junior partner gaining more than the 5 percent threshold of valid votes. In this situation, it makes sense to choose the party whose electoral prospects are less certain, thus maximizing the decisiveness of one's vote for the future preferred coalition government (Eckstein 1995).

Both instrumental and expressive considerations can influence the party preference formation of the reasoning voter. What is at issue is not the causes but the consequences of party preferences for actual behavior. Is it possible to show that the voter-as-consumer draws different conclusions from his or her given party preference profile than does the voter-as-investor? The latter has to calculate his or her expected-utility streams from different future governments, whereas the former needs incentives which reward the revealing of a preference in an election.

Guttman *et al.* (1994) use the distinction between non-voting due to indifference and due to alienation to differentiate voters-as-investors and voters-as-consumers.[9] Interpreting feeling thermometers as direct measures of utility, they construct for the American presidential election of 1976 a measure of indifference (the *difference* of those scale scores between Ford and Carter) and a measure of alienation (the *absolute level* of the preferred candidate's scale score). Analyzing the University of Michigan's 1972–1976 panel, they find the absolute utility level has a statistically significant effect on the probability of voting, but the utility difference term does not.[10] This sensitivity to absolute utility levels and insensitivity to utility difference suggests that voters are behaving as consumers rather than investors, and that expressive considerations are more determinative of their final act of voting.

The investor-voter in multi-party systems has to take into account the coalition possibilities, guessing probabilities of alternative coalitions on the basis of different election results. Naturally, investor-voters will reason about these possibilities during the election campaign. In between elections, however, behavior of government officials and parties is monitored, and consequences are drawn from good and bad news and experiences, in a way which reflects upon general rather than election-specific party preference; and these alternative election-specific coalition possibilities do not loom large in investor-voters' calculations.

During that preference-formation phase, voters-as-consumers do not necessarily differ from the voter-investors. The two diverge only as the upcoming election draws closer. The consumer-voter needs more impetus from the mass media or from peers to become involved; and the more

[9] Harking back to Hotelling's (1929) original example of ice cream vendors at the seashore, they argue that "[c]onsumers abstain from buying ice cream if the closest vendor is sufficiently distant. The positions of alternative vendors are irrelevant to the decision to buy ice cream [or, for voters-as-consumers, a party's program]. Voter-investors, in contrast, are interested in the victory of the favored party against its rival(s). Here, the difference between the positions of the favored party and its rival(s), and not the voters political distance to his favored party, is all-important" (Guttman *et al.*, 1994: 198).

[10] Rattinger and Krämer report similar results for Germany (1995), where alienation is known as *Politikverdrossenheit.*

biased the information is in one direction or the other, the higher the expressive value of his or her preference revelation will be, provided that s/he agrees with the partisan flavor of this information. The voter-as-investor, in contrast, will take account of coalition expectations and may contemplate sophisticated voting; and since voting for one's second preference depends on calculations of possible aggregate results, the election laws and the different aggregation devices embodied in them may here have a direct impact on individual behavior.

## IV Conclusion

Continental European democracies have developed parliamentary systems in which governments are normally formed by coalitions. Two-party systems are the exceptions. The modal category is a moderate pluralism—either in Sartori's (1976) original meaning of the term of three to five parties, or with a larger number of parties which are nevertheless not sharply polarized combined with anti-system parties which gain a substantial number of votes. On the one hand, these party systems make it very difficult for voters to anticipate future governments. But on the other hand, they facilitate political orientation by providing ideological signals about the positions of parties.

The reasoning voter in multi-party systems will, as I have shown, develop a "party preference profile" as a summary measure of his or her experiences and information about the performance and policy offers of the parties. Among the factors influencing preference formation are performance evaluations of governments and issue proximities. But this short list could be easily supplemented: by future expectations; by competence evaluations of party leaders and candidates for government offices; and so on. These factors are substantively linked to the utility that a citizen can associate with a party as a possible provider of collective goods.

Once citizens have formed consistent preferences for the parties, the major problem is then predicting their voting behavior, taking into account the election-specific coalition expectations as an additional factor influencing the causal path from party preference to voting behavior. But since the single voter is not decisive, public investors differ from private investors (Popkin 1991: 10) and expressive considerations gain an importance in their own right as instrumental rationality is downgraded for voters in mass electorates (Brennan and Lomasky 1993). It is not that voters are irrational but, rather, that the voting mechanism (for instance, proportional voting in parliamentary multi-party systems) is a less-than-perfect mechanism for revealing the electorate's policy preferences.

## References


ABRAMSON, P. R. 1983. *Political Attitudes in America*. San Francisco: Freeman.

BARRY, B. 1978. *Sociologists, Economists and Democracy*. Rev edn. Chicago: University of Chicago Press.

BRADY, H. E., and ANSOLABEHERE, S. 1989. The nature of utility functions in mass publics. *American Political Science Review*, 83: 143–64.

BRENNAN, G., and LOMASKY, L. 1993. *Democracy and Decision*. Cambridge: Cambridge University Press.

CAMPBELL, A.; CONVERSE, P. E.; MILLER, W. E.; STOKES, D. E. 1960. *The American Voter*. New York: Wiley.

—— —— —— —— 1966. *Elections and the Political Order*. New York: Wiley.

CONVERSE, P. E. 1964. The nature of belief systems in mass publics. Pp. 206–61 in *Ideology and Discontent*, ed. D. E. Apter. New York and London: The Free Press, Collier-Macmillan.

—— 1966. The problem of party distances in models of voting change. Pp. 175–207 in *The Electoral Process*, ed. M. K. Jennings. and H. Zeigler. Englewood Cliffs, N.J.: Prentice-Hall.

—— 1976. *The Dynamics of Party Support*. Beverly Hills, Calif.: Sage.

DOWNS, A. 1957. *An Economic Theory of Democracy*. New York: Harper and Row.

ECKSTEIN, G. 1995. *Rationale Wahl im Mehrparteiensystem. Die Bedeutung von Koalitionen im räumlichen Modell der Parteienkonkurrenz*. Frankfurt am Main: Peter Lang.

EIJK, C. VAN DER; NIEMÖLLER, B.; and TILLIE, J. N. 1986. The two faces of 'future vote': voter utility and party potential. Mimeo.: University of Amsterdam.

ENELOW, J. M., and HINICH, M. J. 1984. *The Spatial Theory of Voting*. Cambridge: Cambridge University Press.

ESSER, H., ed. 1991. *Alltagshandeln und Verstehen*. Tübingen: Mohr (Siebeck).

FIORINA, M. P. 1981. *Retrospective Voting in American National Elections*. New Haven, Conn.: Yale University Press.

FUCHS, D., and KLINGEMANN, H.-D. 1990. The left-right scheme: theoretical framework. Pp. 203–34 in *Continuities in Political Actions*, ed. M. K. Jennings, J. van Deth *et al.* Berlin: Walter de Gruyter.

GOODIN, R. E., and ROBERTS, K. W. S. 1975. The ethical voter. *American Political Science Review*, 69: 926–8.

GUTTMAN, J. M.; HILGER, N.; and SCHACHMUROVE, Y. 1994. Voting as investment vs. voting as consumption: new evidence. *Kyklos*, 47: 197–207.

HASTIE, R. 1986. A primer of information-processing theory for the political scientist. Pp. 11–39 in *Political Cognition*, ed. R. R. Lau and D. O. Sears. Hillsdale, N.J.: Lawrence Erlbaum Associates.

HINICH, M. J., and MUNGER, M. C. 1992. A spatial theory of ideology. *Journal of Theoretical Politics*, 4: 5–30.

—— and POLLARD, W. 1981. A new approach to the spatial theory of electoral competition. *American Journal of Political Science*, 25: 323–41.

HOTELLING, H. 1929. Stability in competition. *Economic Journal*, 39: 41–57.

Iyengar, S. 1991. *Is Anyone Responsible? How Television Frames Political Issues.* Chicago: University of Chicago Press.

Kahnemann, D., and Tversky, A. 1982. Risiko nach Maß—Psychologie der Entscheidungspräferenzen. *Spektrum der Wissenschaft*, 5: 89–98.

Key, V. O. 1966. *The Responsible Electorate: Rationality in Presidential Voting 1936–1960.* New York: Vintage.

Klingemann, H.-D. 1972. Testing the left-right continuum on a sample of German voters. *Comparative Political Studies*, 5: 93–106.

—— 1979. Measuring ideological conceptualizations. Pp. 215–54 in *Political Action*, ed. S. H. Barnes and M. Kaase. Beverly Hills, Calif.: Sage.

Küechler, M. 1991. Issues and voting in the European elections 1989. *European Journal of Political Research*, 19: 81–103.

Kuklinski, J. H.; Luskin, R. C.; and Bolland, J. 1991. Where is the schema? going beyond the "s" word in political psychology. *American Political Science Review*, 85: 1341–55.

Laver, M., and Hunt, B. W. 1992. *Policy and Party Competition.* London: Routledge.

Lindenberg, S. 1990. Rationalität und Kultur. Die verhaltenstheoretische Basis des Einflusses von Kultur auf Transaktionen. Pp. 249–87 in *Sozialstruktur und Kultur*, ed. H. Haferkamp. Frankfurt: Suhrkamp.

Lodge, M.; McGraw, K. M.; Conover, P. J.; Feldman, S.; and Miller, A. H. 1991. Where is the schema? critiques. *American Political Science Review*, 85: 1357–80.

Markus, G. B., and Converse, P. E. 1979. A dynamic simultaneous equation model of electoral choice. *American Political Science Review*, 73: 1055–70.

Nannestad, P. 1994. Dänisches Wahlverhalten 1971–1979: Ein Modell reaktiven Wählens und einige empirische Ergebnisse. Pp. 285–306 in *Parteien, Parlamente und Wahlen in Skandinavien*, ed. F. U. Pappi and H. Schmitt. Frankfurt: Campus.

Pappi, F. U. 1983. Die Links-Rechts-Dimension des deutschen Parteiensystems und die Parteipräferenz-Profile der Wählerschaft. Pp. 422–41 in *Wahlen und politisches System*, ed. M. Kaase and H.-D. Klingemann. Opladen: Westdeutscher Verlag.

—— 1994. Parteienwettbewerb im vereinten Deutschland. Pp. 219–48 in *Das Superwahljahr. Deutschland vor unkalkulierbaren Regierungsmehrheiten?*, ed. W. Bürklin and D. Roth. Köln: Bund-Verlag.

Popkin, S. L. 1991. *The Reasoning Voter.* Chicago: University of Chicago Press.

Rattinger, H., and Krämer, J. 1995. Wahlnorm und Wahlbeleiligung in der Bundesrepublik Deutschland. *Politische Vierteljahresschrift*, 36: 267–85.

Ronis, D. L.; Yates, J. F.; and Kirscht, J. P. 1989. Attitudes, decisions and habits as determinants of repeated behavior. Pp. 213–40 in *Attitude Structure and Function*, ed. R. P. Breckler, J. Steven and A. G. Greenwald. Hillsdale, N.J.: Lawrence Erlbaum Associates.

Rusk, J. G. 1987. Issues and voting. Pp. 95–141 in *Research in Micropolitics, ii. Voting Behavior II*, ed. S. Long. Greenwich, Conn.: JAI Press.

Sani, G. 1974. A test of the least distance model of voting choice, Italy 1972. *Comparative Political Studies*, 7: 193–208.

Sartori, G. 1976. *Parties and Party Systems.* Cambridge: Cambridge University Press.

Shively, W. P. 1979. The development of party identification among adults: explorations of a functional model. *American Political Science Review*, 73: 1039–54.

Smith, E. R. A. N. 1989. *The Unchanging American Voter.* Berkeley: University of California Press.

Sniderman, P. M. 1993. The new look in public opinion research. Pp. 219–44 in *Political Science: The State of the Discipline II*, ed. A. Finifter. Washington, D.C.: American Political Science Association.

—— Brody, R. A.; and Tetlock, P. E. 1991. *Reasoning and Choice.* New York: Cambridge University Press.

—— Glaser, J. M.; and Griffin, R. 1990. Information and electoral choice. Pp. 117–35 in *Information and Democratic Processes*, ed. J. A. Ferejohn and J. H. Kuklinski. Urbana: University of Illinois Press.

Weber, M. 1965. Social action and its types. Pp. 173–9 in *Theories of Society*, ed. T. Parsons, *et al.* New York: Free Press; one-volume edition.

Chapter 10

# Political Behavior: Institutional and Experiential Approaches

Patrick Dunleavy

MUCH about political behavior research ought simply be taken as read. The introduction of mass surveys into political science and political sociology has been an important advance. Methods for asking questions have advanced a bit, and techniques for analyzing survey data have advanced a lot, in the last 30 years. Debates about the foundations of political alignment between the Columbia school, the Michigan model and exponents of "issue voting" have been valuable. The extension of survey-based studies from the U.S., first to western Europe and thence to a wide range of other countries, and some small movement towards asking internationally standardized questions, have generated additional insights. Taking all that as granted, critics (Dunleavy 1989) still ask, "is there more to learn" about mass political behavior?

The question reflects a certain amount of disillusionment with political behavior studies, especially with electoral analysis, as the field has developed in recent years. Once it was apparently the authentic "big science" area of the discipline, with large and capital-intensive projects, elaborate and arcane technologies of its own, and the apparent promise of knowledge-accumulation on "normal science" lines. But the pace of advance in political behavior research has undeniably slowed in the last two decades. Fundamental debates which 1950s and 1960s authors were confident could be sorted out by better analytic techniques, larger sample sizes, or more refined survey instruments have instead seemed more and more immune to further empirical resolution. Patterns of causation involved in political alignments remain as disputed, on theoretical and value grounds, now as ever they were—both within the field of political behavior, and between its specialists and outsiders.

As in other fields:

> Although sophisticated statistical manipulation of research data can sometimes reduce the number of causal explanations compatible with the observed correlation, ultimately the choice between them is made in the light of what seems reasonable to the researcher, perhaps in the light of other research, or in terms of the obvious temporal order among the variables (Halfpenny 1984: 4).

Electoral analysis in particular, relies to an extraordinary degree on the indispensable context of understanding provided by common sense (Lindblom and Cohen 1979: 17) in ways which often vitiate the "scientific" status of key findings. For example, Nie, Verba and Petrocik carefully document a rise in issue voting among American voters in the 1970s, compared with the 1950s. But they then attribute this rise to the de-polarization of politics in the Eisenhower years, rather than to increasing sophistication amongst voters (Nie *et al.* 1979: 192)—even though they have no non-reactive measures of polarization, relying only on impressionistic evidence for this key proposition.

The current stagnation in political behavior research, coming after the strong hopes invested in it in its early years, has meant an internal decline in the behavioral optimism which initially strongly informed the field and an external loss of interest in its findings. Scholars specializing in political behavior have increasingly seemed divorced from theoretical debates elsewhere, concerned only with the routinized refinement or extension of existing work. And political scientists in other fields, such as public choice, the theory of the state, or the study of political parties and social movements, have increasingly discussed political behavior in terms divorced from survey-based empirical enquiry—as if its limits were now seen as too constraining. I first discuss the crisis of political behavior research as a fading of a previously strong modernist vision, and then examine some alternatives which show that there is indeed still more to learn about political behavior.

## I "Modernism" in political behavior research

At first sight, the suggestion that conventional electoral studies and survey-based work on political behavior is "modernist" may seem not very credible. Lyotard offers one of the most famous definitions of modernism, apparently light-years removed from the everyday concerns of this field:

> I will use the term modern to designate any science that legitimates itself with reference to a metadiscourse of [some] kind making explicit appeal to some grand narrative, such as the dialectics of Spirit, the hermaneutics of meaning, the emancipation of the rational or working subject, or the creation of wealth (Lyotard 1984: xxiii).

The casual observer will find few echoes in electoral analysis of Hegelianism, Marxian historical materialism, Habermassian critique or similarly "metaphysical" ideas.

Yet postwar political behavior research undeniably has its own grand narrative and shares many of the features which postmodernists have cited as undesirable stigmata of a modernist outlook. In this narrative the modern period is marked out as unique by the progress of democratization, which alone provides the raw materials on which most political behavior research is based. The whole field took off in the U.S., the earliest democracy if still one of the more equivocal ones in terms of voting participation rates (Vanhanen 1984: 115; 1990). And the first large-scale survey research came about at exactly the time when the stalled progress of democratization rebounded from the low point of the early 1940s, with a huge postwar growth in the number of countries with basically free elections and the proportion of the world population living there, as shown in Table 10.1.

**Table 10.1** The growth of liberal democracy, 1900 to 1995

| | *Liberal democracies* | | *World population* | |
|---|---|---|---|---|
| | *States* | *Population (millions)* | *Numbers (millions)* | *% living in democracies* |
| 1900 | 4 | 130 | 1,608 | 8 |
| 1910 | 9 | 163 | n.a. | n.a. |
| 1920 | 18 | 371 | 1,860 | 20 |
| 1930 | 15 | 346 | 2,008 | 18 |
| 1940 | 12 | 217 | 2,294 | 10 |
| 1950 | 26 | 869 | 2,516 | 34 |
| 1960 | 32 | 1,169 | 3,019 | 39 |
| 1970 | 31 | 1,263 | 3,693 | 34 |
| 1980 | 35 | 1,636 | 4,450 | 37 |
| 1990 | 39 | 2,070 | 5,246 | 39 |
| 1995 | 40 | 2,263 | 5,765 | 39 |

*Sources*: Sources for the democracy figures are research by Government Research Division, London School of Economics. World population data are drawn from Borrie (1970: 6) for 1900–40, and since 1950 from UN World Population Estimates and Projections. I thank John Hobcraft of the LSE Population Studies Department for help with these data.

American pluralist political sociologists who first developed survey-based research indubitably saw democratization as an integral element in modernization, the rationalization and secularization of society, and the spread of "advanced" ideas and culture. Their missionary zeal was reflected in subtle ways in the co-operative studies which they co-authored with indigenous European scholars or those in other countries. The opening sentence of the original edition of Butler and Stokes's *Political Change in Britain* (1969: 15) declared: "The possibility of rulers being constitutionally driven from office in a free election is relatively new in the history of government." Electoral analysis was to be the unique key to the foundational processes shaping the new political age.

The key modernist feature of mainstream election studies within this implied grand narrative was its strong totalizing approach. Right from the start, the Michigan school in the U.S. and in Britain attacked earlier surveys based in particular local communities, and instead espoused a single national survey inherent in which was a very high level of aggregation of de-contextualized data (Campbell *et al.* 1960; Butler and Stokes 1969/1974). Their work developed large cross-sectional surveys which could normally only be carried out by a single team commanding considerable capital resources, whose controllers became powerful academic oligarchs in danger of either dominating research in a proprietorial manner or rationing access to defining survey questions to like-minded exponents of the new orthodoxy. Once a series of consistent studies had been created, and over-time panel surveys had been set up, these researchers held data sets which could be and were analyzed as if they stood outside time. Many of these studies were explicitly premised on the view that the political context of the whole "modern" period has been so uniform as to render unnecessary any detailed cross-referencing between voters' responses to survey questions and their immediate political environment (see especially Heath *et al.* 1991).

The methodological assumptions underpinning this totalizing approach were simple in the extreme. Central to virtually all political behavior research, and still virtually unchallenged in 1990s studies, has been the search for *the single best decision algorithm* with which to characterize the alignments and behavior of an entire electorate. For any analytic issue, the search is for a restricted set of variables embedded in a single equation setting out their relative influence which captures more of the variance to be explained than any other competing variables and equation. In many studies, the nominal concern with "theoretical" understanding was quickly jettisoned, and the analysis of causal pathways was relegated to the margins: instead they ran, in effect, fairly simple-minded computer-tournaments between a pool of potential independent variables, and tournaments which were only lightly refereed by the analyst at that. Electoral research thus became a contest of regression or log-linear models, each of them offering a single summation of how voters as a whole have aligned between parties or across issues. With the passage of time analysts became more skilled at analysing residuals, and searching for interaction effects between variables. But what has never been seriously disputed is that a single "comprehensive" decision algorithm should be sought, as a means of answering any outstanding question.

To make matters worse, analysts often slip into discussing voters in the archetypal singular. They do so not just in their titles—*The Changing American Voter* (Nie *et al.* 1979) or *How Britain Votes* (Heath *et al.* 1985) or *The American Voter* (Campbell *et al.* 1960)—but also in their texts:

> The function of party identification is to enable the elector to cope with political information and to know which party to vote for (Harrop and Miller 1987: 134).
>
> [W]hile the social psychology of the voter may not have changed much . . . (Heath et al. 1991: 200)
>
> [T]he emphasis is once more on . . . the unchanging psychology of the voter (Crewe and Norris 1992: 19).

These apparently small slips in fact betoken a deeply influential process of reification and abstraction, inherent in the whole enterprise of aggregated survey data analysis, away from the experiences and choices of actual people.[1]

A key consequence of the totalizing approach has been the construction of analyses which exclude analytically inconvenient minorities. In the early days, those were chiefly kinds of voters too few in number to fit easily into cross-tabulations, or whose incorporation would prevent the use of analytic techniques requiring a dichotomous dependent variable. Catt (1996) demonstrates that the minorities which typically "disappeared" from key pieces of analysis have included, in the U.K.: everyone in Northern Ireland, where the party system is different from mainland Britain; everyone in northern Scotland, who are too expensive to poll; non-respondents to surveys; respondents who declare themselves as non-voters; and voters for all "minor" parties, such as the nationalist parties in Scotland and Wales and, for many years, even people choosing the Liberals or Liberal Democrats (who number up to a quarter of all voters since the 1980s). U.K. electoral

[1] A variant of this are comments in which voters as a whole are described as a single unit with anthropomorphic features. For example: "We begin . . . by considering whether the 1992 electorate was unusually hesitant . . . The idea that the electorate has grown more volatile . . . is often advanced" (Clifford and Heath 1994: 6).

studies represented voters as overwhelmingly concerned with the two main parties long after this had ceased to be true of real world elections. Other "minorities" were vaporized analytically by simple moves such as always referring to the archetypal singular voter as "he" and "him"; by categorizing the occupational class of married women from that of their husband (creating a kind of "household" class for women only); or by not exploring clearly relevant aspects of people's social locations. Thus mainstream election studies have been constantly surprised by, and slow to respond to, the growth of politically significant cleavages and issues (for example in Britain, public–private sector conflicts, ethnicity, sexual orientation and environmental issues).

A more fundamental corollary of the monistic tendencies of mainstream voting studies has been the complete underdevelopment of techniques for exploring whether and how different groups may in fact choose how to vote using quite distinct algorithms. No proposition is more centrally denied and left unexplored than this. Very few studies have systematically tried to *simulate* voters' decisions, to define a sequence of operations and criteria which might approximate the algorithms that actual voters employ in making choices. This neglect also has consequences for the ways in which researchers interpret seemingly "obvious" data. For example, although political scientists know well enough that voters can make choices only among available options over which they have little control, Catt (1989) notes how consistently analysts slip into equating all votes received by a party as "support"—glossing the blank fact of a ballot marked in a particular way as an indication of a positive ideological or emotional response to that party, and thereby masking or denying negative voting motivations. It is a small step from there to masking the doubts, cynicisms and misgivings of voters about political leaders and parties: "As long-established actors on the political stage *it is natural* that the parties should have become objects of mass loyalty or identification" (Butler and Stokes 1974: 34, my emphasis). Sheltering behind highly aggregated decontextualized data, and with a ready supply of such quick or biased deconstructions of voter behavior to hand, electoral analysis has added disappointingly little to our knowledge of the *diversity* of ways in which people think or act politically.

It was not always so. The early 1950s survey work had a radical edge to it. In debunking the previous normatively based models of rationalistic voters, and in showing that ordinary people operated with differently structured political information from political élites, this research did not widely decode politics in left–right terms and often seemed to explain voters' alignments in line with family influences or long-term identifications with parties. Indeed, it was in this phase that voting studies seemed least modernist in their explicit repudiation of "logocentric" models and recognition of voter "irrationality." At this stage, too, studies not based on survey work still had some influence on "mainstream" thinking—such as Robert Lane's brilliant essay *Political Ideology* (1962) which intensively explored the attitudes and reactions of just fifteen individuals against their fully contextualized social backgrounds and in a manner informed by a wealth of social and political theory.

But the initial cutting edge of survey research, its disconcerting impact in shifting conventional perceptions of societal processes, has long since disappeared under accumulated layers of orthodoxy-building. Political behavior research has progressively "rationalized" the survey-based image of voters, rebuilding *The Rational Public* (Page and Schapiro 1992) in a revisionist downgrading of what stable and efficacious democratization requires of citizens. And the initial diversity of political behavior methodologies has been replaced by a suffocating reliance on quantifiable survey data alone. Virtually all this information relies on pre-coded questions, devised by the analysts alone (rather than based on respondent's own conceptions), and often asking people to carry out completely artificial operations, such as ranking political candidates or parties on fifty-point thermometer scales simply in order to be able to use sophisticated analytic methods (Page and Jones 1979). The meaning of survey responses is very rarely checked against other kinds of data—there is no reinterviewing of respondents using different approaches, little cross-referencing of responses to differently phrased or slanted questions, and virtually no triangulation of quantitative and qualitative information. Single survey instruments dreamed up intuitively by analysts with a little piloting have predominated in published research, their meaning taken as unequivocal and inferred on a single "common-sense" basis (Bishop *et al.* 1978; Sullivan *et al.* 1978). The main exception, which typically takes analysts one step further from people's experiences, is the use of compound scales or indices, some pre-designed—some "emerging" iteratively as possible constructs out of the data analysis, and others computer-dredged from the response patterns by computers and rationalized *post hoc*, as with the "political principles" constructed from commercial poll data using factor analysis by Rose and McAllister (1986).

The primary internal challenge to mainstream political behavior studies since the 1970s has come from versions of rational choice theory. But their arrival has so far posed no threat at all to the older orthodoxy's totalizing approach and neglect of diversity. Originally based on a relatively complex and in principle falsifiable set of procedures, rational choice theory embodies very powerful but restrictive assumptions about actors' mind-sets, levels of information, and consistency—assumptions easily glossed in building more "realistic" applied variants. Adapted to testing within the survey methods and data analysis repertoire already developed (Fiorina 1981), simplified rational choice approaches have increasingly emerged as an extreme form of "logocentrism"—privileging an objectified model of formal rationality and ascribing it universally to all political actors (Hinich and Munger 1994). This approach too is totalizing, again lending itself to phrasing in terms of an archetypal singular "voter": issue voting accounts "assume that, on each issue, the voter compares his [*sic*] own stance with that of the parties. The voter then, it is claimed, aggregates his preferences on the different issues and votes for the party which offers him, on balance, the largest number of preferred policies" (Heath and McDonald 1988: 96).

From this level of discussion, it is impossible that there will emerge any empirically diversified view of the complex bases of alignments or of the multiple choice algorithms followed by citizens.

And issue voting models take further the democratization "grand narrative" left mostly implicit in older studies. They portray an electorate once held captive by habitual or traditional family, ethnic, interest group and social class influences now (becoming) liberated by acclimatization to liberal democracy, mass education and expanded mass media coverage to function as autonomous sources of intentionality. For example, Rose and McAllister (1986) argued that British elections had moved decisively from "closed class" to "open" competitions; Sarlvik and Crewe (1983) detected a "decade of dealignment" in British politics during the 1970s; and Himmelweit *et al.* (1985) claimed evidence of a new sophistication in voters' choices. In all these studies modernism's teleological streak was evident. The lure of a future (imputed) destination or a detected trend led researchers to gloss present realities. Deviations, inconsistencies, backsliding and contradictions in current data are all too likely to be easily masked for analysts making such strong commitments to purposive history.

Both "issue voting" approaches and mainstream "party identification" and "political sociology" accounts agree that a fully quantitative, highly aggregate, survey data approach is the appropriate "scientific" one for political behavior research. The primary point of mass surveys is always just to reconstruct an account of institutionally significant changes: this party's loss of support, that party's gain. Authors of both schools are aware of the restrictive limitations of survey methods, the inherent losses of understanding involved in reducing people's complex meanings to dots and dashes on computer disk. But the overwhelming response has been to carry on "perfecting" survey sampling methods, question and survey instrument design, and the methods of analysis deployed to make sense of the data.

There is now little difference between non-academic surveys (carried out for the mass media or party organizations) and the questionnaire design used by political scientists. Both political sociology and rational choice accounts in different ways have helped cement the close and symbiotic relationship which has developed between political behavior research and practitioners of political control and manipulation technologies. In the last thirty years, as it has become professionally reputable and commercially "applicable," the subfield has increasingly restructured the electoral realities it purports to independently describe. As with virtually all other branches of professional knowledge, a key consequence of the discipline's existence has been an expansion of the technical capacity to analyze, predict and shape social behavior—in this case, party behavior and electoral outcomes. Elections are now pictured by elected politicians, political party professionals, political and electoral consultants, large polling organizations, mass media commentators, advertising corporations and large numbers of voters themselves in ways which have been extensively reshaped by the mainstream and issue voting orthodoxies. The expanded "control" capability apparently conferred by political behavior research fits closely with an "engineering" model of scientific endeavor in which truth is associated with the experimental ability to successfully manipulate social forces and predict behavior (Camhis 1979).

The potential dangers in this "industrial" application of knowledge need little elaboration (Foucault 1980). The analytic neglect of minorities and concentration on quantitative modeling both find support in the adequacy of the existing tool-kit for control purposes, for successful campaigning. Political science, in this approach, is not an abstract quest for knowledge for its own sake but rather for an effective understanding of those forces (and presumptively only those forces) which structure the competition for political power (for a strong example, see Heath *et al.* 1994). In this fundamental sense most political behavior research has been system-biased, concerned with understanding those mass political phenomena which condition the transfer or retention of institutional power, and unconcerned with phenomena which seem ineffectual under current arrangements. To pick one key example, in countries with "plurality rule" or "list"-style proportional representation electoral systems, political scientists and opinion pollsters have amassed volumes of data charting voters' first preference alignments across parties over time; but they have almost ignored the structure of voters' second or subsequent preferences, because those preferences do not immediately determine which political élite gains control of government. One hardly has to agree with Foucault that the state is "a mythical abstraction whose importance is a lot more limited than many of us think" (quoted in Hoffman 1995: 162) to see the limitations inherent in such a narrowly institutional approach as that.

The final modernist feature of political behavior research has been its reliance on over-polarized antimonies and false dichotomies. It is a recurring feature of electoral studies in particular that complex theoretical positions are repeatedly expressed in a lowest-common-denominator form apparently required for empirical testing. In this distorted mirror of intellectual debate voters are either "rational" or expressive, issue attitudes are either "real" or non-attitudes, actors have "perfect" information or they do not, they operate in quite distinct "market" or "political" contexts (Brennan and Lomasky 1993), and so on. Given the defective nature of the questions asked, the inherent ambiguities of language, changes in the "framing" of questions by survey design, or external shifts in the political or policy environment, however, in many cases it is actually quite unclear how someone with a perfectly worked-out position should respond. Consider for example Heath and McDonald's (1988) panel study of some 900 British voters: they argue that, between 1983 and 1987, voters in their study were more "consistent" in their party identification (itself dichotomized as "Conservatives" and "all others," although in 1983 the effective number of parties was 2.9) than in their attitude towards nationalization–privatization. But suppose I favored the privatization of the expensive telephone system (implemented in 1983), but opposed that of the well-run gas industry (1985), and expected to make money from electricity privatization (projected in 1987). Am I then "inconsistent"? Surely only within a very strange or limiting view of the world. At this level of aggregation, it is simply impossible for analysts to make most of their everyday judgments about how to deconstruct people's responses stick *except* by forcing them into arbitrary and artificial frameworks of their own devising. In these circumstances, it is unclear whether Schumpeter's maligning of his fellow citizens is more applicable to voters or to researchers studying them.[2]

In summary, the modernist character of political behavior research can be detected in many places: in its implicit grand narrative of democratization; in the strong totalizing and scientistic emphasis of quantitative survey-based

[2] In Schumpeter's (1943/1987: 262) famous phrase, "The typical citizen drops down to a lower level of mental performance as soon as he enters the political field. He argues and analyzes in a way that he would readily recognize as infantile within the sphere of his real interests. He becomes a primitive again."

research; in the monism of research methods and marginalizing of social minorities; in the search for a single best decision algorithm; in the "engineering" conception of knowledge, in the system-biased character of knowledge development; and in the reliance on over-polarized antinomies.

## II Alternatives

In the social sciences it is always easier to formulate criticisms than to find new ways of proceeding. Pragmatically, for the limitations of modernist political behavior research to be taken seriously, it is important to show that they are correctable. If the only other option were to lapse into a vague postmodernist relativism and subjectivism, reiterating in a new guise the pre-behavioral criticisms of voting studies of the 1950s, then the *faute de mieux* option of sticking with what we have would no doubt prevail. However, as a result of the easier diffusion of processing power and the capability to store and retrieve large quantities of unstructured data, plus the accumulation of new quantitative and qualitative techniques, we seem to stand at an important turning point—a genuinely "modern" period—in the development of social science methods. The trick will be to maintain intellectual control of the new potentials. Here three avenues offer ways forward to a new style of systematic political behavior research: methodological pluralism, disaggregating information, and a shift from an institutional to an experiential focus.

### A Methodological pluralism

Methodological pluralism entails taking seriously the known limitations of pre-coded survey questions in uncovering people's complex meanings. Instead we should model political behavior in several different ways, seeking to triangulate the perspective offered by one approach with that conveyed by others. A first step entails abandoning reliance on single survey instruments (or on more complex scales) devised by analysts in isolation, and the adoption of a more forensic survey investigation style in which alternatively worded questions on the same issue seek to uncover groups of people with distinct ways of understanding or choosing. Expensive to undertake, this approach probably implies the end of "omnibus" surveys, themselves a key support for over totalizing analytic approaches.

A next stage involves integrating survey results with other forms of quantitative information. For instance, the potential insights from cross-referencing élite-level and mass-level data within a sophisticated multi-theoretical analysis have recently been impressively demonstrated by Iversen (1994). His work builds on the theoretical space opened up by Przeworski and Sprague (1986), who also showed how much could be derived from analyzing another kind of data, cross-national election results.[3] To take another example, Q methodology provides a useful way of constructing ideological maps by examining how small numbers of respondents feel in detail about a large number of statements, culled from actual political discourse rather than invented by the analyst (Dryzek and Berejekian 1993; Dryzek 1994). The approach could be supplemented by a more systematic way of selecting statements for inclusion, based on textual analysis of mass media databases (such as Profile or Lexis).

A further and more speculative stage would be to use the ability to search flexibly for patterns and associations in free-text databases so as to create an important supplement to mass surveys, or in the further future perhaps a replacement for them. In the 1950s, within the technology of the day, surveys seeking fixed responses to analysts' questions were the only way of gathering data systematically from a representative sample of 1,000 people in a way that could subsequently be analyzed. But in the 21st century it should be perfectly feasible to have 1,000 lightly structured and rather open-ended "conversations" between trained interviewers and respondents for a similar period of time, and to record what gets said verbatim. Storing and sifting this new kind of record using expanded information technologies, we could then analytically surface what influences people's alignments and attitudes by a process of *post-hoc* interrogation of their full text. Cross-referencing the connected texts provided by print and broadcast media databases with respondents' conversations could also begin to address the hitherto irresolvable problems of assessing the extent of citizens' autonomy and dependence in the political sphere. It should certainly allow a far more precise charting of the origins and circulation of ideas, deconstructions of meanings, and the lineages of propaganda and resistance in liberal democracies.

The current rapid growth of expertise in qualitative methods offers important additional diversity in approaches (Devine 1994). Before any mass survey finding could be taken seriously it might need to be supported by findings using other methods, such as: close textual analysis of taped interview transcripts for a subset of respondents in the main survey; a compilation of life-histories and self-completion questionnaires (possibly

[3] Also important in terms of theoretical space are the directional model of Rabinowitz and MacDonald (1989), and the preference-shaping model (Dunleavy 1991: chaps. 4 and 5).

free-form); detailed investigation at a variety of spatial levels of how respondents' experiences were shaped; and intensive studies designed to unpick the interrelationship between direct (hence more autonomous?) experiences and mediated experiences. Ethnographic studies of particular groups, and even participant observation studies of political behavior, have also been suggested (Devine 1994). If people's experiences can only be very partially and inadequately tapped by survey instruments, then it may simply be necessary to accept that and look for other available methods of gauging them. A loss of quantification and aggregation may well result from this switch in approach, but if so that reflects the way that people's experiences are, a situation that research must work around rather than seeking to deny.

## B Disaggregating information

Disaggregating information is an important supplement to methodological pluralism. It means getting away from decontextualized and aggregated data. Instead analysts would seek to start their investigations as close as possible to the understandings of a set of people themselves, structuring research around the units and levels at which their experiences are organized. The main dimensions involved will be set by the nature of the issues being investigated, but normally they must include social locations and territorial areas.

Picturing social locations accurately depends in part on social theory, and in part on how people themselves see social arrangements. A process of iteration between theoretical work and empirical results is crucial here. Most political behavior research classifies people in terms of conventional functional indices such as occupation, ethnicity, language-group, religion, housing, industry-type, etc. These "objective" measures need to be supplemented by other self-classifications of people. They may also involve constructing some overarching categories to capture otherwise diffused effects lost in issue-by-issue categorizations. For example, there is evidence that Americans extensively structure their thinking around what Lane (1991) terms "the market experience," learning common lessons from very diverse economic interactions which they apply across different contexts; and they perceive their interactions with government as qualitatively different in character from their private sector dealings (Lane 1983; 1986). In western Europe too, public–private sector conflicts in production and consumption contexts, have had strong implications for party competition and political behavior (Dunleavy 1986). Hence it may make more sense to reaggregate data around these "experienced" sectoral categories than to rely on more fragmented official classifications.

Disaggregating (and reaggregating) across geographical areas is also important. National-level data provide only a summary of regional and local situations—a summary which can be misleading, especially in cross-national analyses where each country (no matter how small or large) counts for one, and none for more than one. For example, Lijphart (1994) provides an authoritative account of the effects of electoral systems. Yet in his data set, countries such as Iceland and the U.S. (which is 1,000 times larger in population) feature as equivalent units; and, furthermore, the main type of data are averages of "deviation from proportionality" (DV) statistics for a given electoral system, averaged across many different elections. Probe Lijphart's figures only a little, however, and difficulties emerge. In Britain, voters in the component regions of the country *experience* levels of DV in the treatment of their choices by the electoral system that are on average one and a half times greater than the national figure, which is artificially lowered by biases towards the two main parties off-setting each other (Dunleavy and Margetts, 1993). Meanwhile in Spain, people in cities and regions experience the voting system as *more* proportional than the national DV figure suggests, since the overall Spanish score reflects malapportionment effects as well as disproportionality. Here, geographical disaggregation produces a picture of the world radically different from conventional institutional accounts.

## C Shifting from an institutional to an experiential focus

A shift from an institutional to an experiential focus entails a consistent effort to move away from system-biased ways of doing research. The concept of "experiences" as the focus of analysis may seem odd to modern eyes. From differing perspectives, both American pragmatists such as John Dewey (1922) and the British conservative philosopher Michael Oakeshott (1933) emphasized the grounded and contextualized quality of experiences, and the ways in which they accumulate into intuitions (which Dewey called "funded experiences"). These earlier insights were marginalized by postwar behavioral research, however. In studies of political behavior the orthodox assumptions have been that people formulate attitudes (in mainstream accounts) or preferences (in issue-voting models), which they subjectively recognize and then directly display in actions. This stance ignores the many ways in which situational constraints (such as interaction and inter-dependency effects) can prevent attitudes or preferences being expressed in behavior, and perhaps even recognized (Dowding 1991).

Scurrying across the complexities and contextualization of people's experiences can also lead analysts to draw conclusions which are unwarranted. Take a simple example: accounts of national elections generally assume that the alignments being charted are formulated in relation to that specific election. In fact, people might quite rationally choose whom to vote for on the basis of the overall success of that alignment across a variety of elections—after all, most countries have at a minimum national, state or regional, constituency, and local elections (Dunleavy and Margetts 1995). Again, collecting information on how many preferences people have across parties, and how positive or negative these evaluations are, may be critical for understanding in an academic (as opposed to a "control") sense the mainsprings of alignments or political attitudes. Even when we move on to supposedly "objective" data, such as that derived from electoral outcomes, the distinction between an institutional and an experiential approach remains important. The institutional way of calculating DV, for example, captures part of the picture—the ways in which electoral systems condition the transfer of seats (and hence access to state power) between political parties (Taagepera and Shugart 1989: 104–11). But it neglects the equally important question of how voting systems condition people's experiences of voting, by either recognizing or ignoring their choices (Dunleavy and Margetts 1994).

An experiential approach is also preferable because political science does not have a very good record in anticipating what will or will not prove to be institutionally important. Major political change almost invariably takes the form of previously "peripheral" or marginal phenomena suddenly assuming core importance, or growing steadily in overall influence despite a string of specific defeats. Most key social movements—the women's movement, civil rights in the U.S., environmentalism in the early 1970s and later 1980s, sexual politics, and contemporary far-right politics—have emerged into prominence largely unheralded by political behavior research focusing myopically only on those things which are already institutionally significant.

The *gestalt* shift entailed in moving from institutional to experiential measures is a subtle one. Its key element is a determination to map and chart those interactions, processes and linkages experienced as important by citizens themselves, whether or not they issue in immediately efficacious political consequences. Existing totalizing approaches have reached a plateau in their ability to explain. A more differentiated approach could help to illuminate why under one set of conditions diverse choice algorithms result in a given aggregate outcome, when a relatively small change of conditions can produce different overall results.

## III Conclusions

In political behavior research, as in other areas of modern thought, "we have paid a high enough price for the nostalgia of the whole and the one" (Lyotard 1984: 81), and suffered enough in intellectual self-understanding from "the coercion of a formal, unitary and scientific discourse" (Foucault 1980: 85). It does not follow, however, that we should simply lapse into relativistic critique, as postmodern theory in general might be taken to recommend. In other areas—such as architecture, planning, literature, and various parts of the social sciences—postmodernist criticisms have been taken on board as part of an overall process of constructive disciplinary self-renewal, resulting in significant changes of approach. For political behavior research the same potential now exists. And the diminishing returns to effort from existing approaches suggest that the painful costs of adjustment might be more easily borne.

## Acknowledgments

I would like to thank Helena Catt of the Department of Politics, Auckland University for stimulating many of the ideas here.

## References

Bishop, G.; Tuchfarber, J.; and Oldendick, R. 1978. The nagging question of question wording. *American Journal of Political Science*, 22: 250–69.

Borrie, W. D. 1970. *The Growth and Control of World Population*. London: Weidenfeld and Nicholson.

Brennan, G., and Lomasky, L. 1993. *Democracy and Decision*. Cambridge: Cambridge University Press.

Butler, D., and Stokes, D. 1969/1974. *Political Change in Britain*. 2nd edn. London: Macmillan.

Camhis, M. 1979. *Planning Theory and Philosophy*. London: Tavistock.

Campbell, A.; Converse, P.; Miller, W. E.; and Stokes, D. 1960. *The American Voter*. New York: Wiley.

Catt, H. 1989. What Do Voters Decide? Tactical Voting in British Politics. Ph.D. dissertation, Department of Government, London School of Economics and Political Science.

—— 1996.

Clifford, P., and Heath, A. 1994. The election campaign. Pp. 7–24 in Heath, Jowell and Curtice (1994).

Crewe, I., and Norris, P. 1992. In defence of British election studies. Pp. 3–25 in *British Elections and Parties Yearbook 1991*, ed. I. Crewe, P. Norris, D. Denver and D. Broughton. Hemel Hempsted: Harvester-Wheatsheaf.

Devine, F. 1994. Learning more about political behavior: beyond Dunleavy. Pp. 215–28 in *British Elections and Parties Yearbook 1994*, ed. D. Broughton *et al.* London: Frank Cass.

Dewey, J. 1922. *Human Nature and Conduct.* New York: Henry Holt.

Dowding, K. 1991. *Political Power and Rational Choice.* Aldershot, Hants.: Edward Elgar.

Dryzek, J. S. 1994. Australian discourses of democracy. *Australian Journal of Politics*, 29/2; 221–39.

—— and Berejekian, J. 1993. Reconstructive democratic theory. *American Political Science Review*, 87: 48–60.

Dunleavy, P. 1986. The growth of sectoral cleavages and the stabilization of state expenditures. *Environment and Planning D: Society and Space*, 4: 129–44.

—— 1989. Mass political behavior: is there more to learn? *Political Studies*, 38: 453–69.

—— 1991. *Democracy, Bureaucracy and Public Choice.* Hemel Hemsptead: Harvester-Wheatsheaf.

—— and Margetts, H. 1993. Disaggregating Indices of Democracy: Deviation from Proportionality and Relative Reduction in Parties. Paper to Workshop on "Measuring Democracy," the European Consortium for Political Research Joint Sessions, University of Leiden, 11–17 April.

—— —— 1994. Auditing democracy: the case for an experiential approach. Pp. 155–81 in *Defining and Measuring Democracy*, ed. D. Beetham. London: Sage.

—— —— 1995. The rational basis for belief in the democratic myth. Pp. 60–88 in *Preferences, Institutions and Rational Choice*, ed. K. Dowding and D. King. Oxford: Oxford University Press.

Fiorina, M. 1981. *Retrospective Voting in American National Elections.* New Haven, Conn.: Yale University Press.

Foucault, M. 1980. *Power/Knowledge*, trans. and ed. C. Gordon. New York: Pantheon.

Halfpenny, P. 1984. *Principles of Method.* London: Longman.

Harrop, M., and Miller, W. L. 1987. *Elections and Voters.* Basingstoke: Macmillan.

Heath, A., and McDonald, S. 1988. The demise of party identification theory? *Electoral Studies*, 7: 95–107.

—— Jowell, R.; and Curtice, J. 1985. *How Britain Votes.* Oxford: Pergamon.

—— —— —— Evans, G.; Field, J.; and Witherspoon, S. 1991. *Understanding Political Change: The British Voter, 1964–87.* Oxford: Pergamon.

—— —— ——, Taylor, B., eds. 1994. *Labour's Last Chance? The 1992 Election and Beyond.* Oxford: Pergamon.

Himmelweit, H.T.; Humphreys, P.; Jaeger, M.; and Katz, M. 1985. *How Voters Decide.* New York: Academic Press.

Hinich, M. J., and Munger, M. C. 1994. *Ideology and the Theory of Political Choice.* Ann Arbor: University of Michigan Press.

Hoffman, J. 1995. *Beyond the State.* Oxford: Polity.

Iversen, T. 1994. The logic of electoral politics: spatial, directional and mobilizational effects. *Comparative Political Studies*, 27: 155–89.

Lane, R. E. 1962. *Political Ideology: Why the American Common Man Believes What He Does.* New York: Free Press.

—— 1983. Procedural goods in a democracy: how one is treated versus what one gets. *Social Justice Research*, 2: 177–92.

—— 1986. Market justice, political justice. *American Political Science Review*, 80: 383–402.
—— 1991. *The Market Experience*. Cambridge: Cambridge University Press.
LIJPHART, A. 1994. *Electoral Systems and Party Systems*. Oxford: Oxford University Press.
LINDBLOM, C. E., and COHEN, D. 1979. *Useable Knowledge*. New Haven, Conn.: Yale University Press.
LYOTARD, J.-F. 1984. *The Postmodern Condition*. Minneapolis: University of Minnesota Press.
MARCH, J. 1988. *Decisions and Organizations*. Oxford: Blackwell.
NIE, N. S.; VERBA, S.; and PETROCIK, J. 1979. *The Changing American Voter*. Cambridge, Mass: Harvard University Press.
OAKESHOTT, M. 1933. *Experience and Its Modes*. Cambridge: Cambridge University Press.
PAGE, B. I., and JONES, B. 1979. Reciprocal effects of policy preferences, party loyalties, and the vote. *American Political Science Review*, 73: 1071–89.
—— and SCHAPIRO, R. Y. 1992. *The Rational Public: Fifty Years of Trends in Americans' Policy Preferences*. Chicago: University of Chicago Press.
PRZEWORSKI, A., and SPRAGUE, J. 1986. *Paper Stones: A History of Electoral Socialism*. Chicago: University of Chicago Press.
RABINOWITZ, G., and MACDONALD, S. E. 1989. A directional theory of issue voting. *American Political Science Review*, 83: 93–121.
ROSE, R., and MCALLISTER, I. 1986. *Voters Begin to Choose: From Closed Class to Open Elections in Britain*. London: Sage.
SARLVIK, B., and CREWE, I. 1983. *Decade of Dealignment*. Cambridge: Cambridge University Press.
SCHUMPETER, J. 1943/1987. *Capitalism, Socialism and Democracy*. 6th edn. London: Counterpoint; originally published 1943.
SULLIVAN, J. L.; PIERESON, J.; and MARCUS, G. 1978. Ideological constraint in the mass public: a methodological critique and some new findings. *American Journal of Political Science*, 22: 233–49.
TAAGEPERA, R., and SHUGART, M. S. 1989. *Seats and Votes*. New Haven, Conn.: Yale University Press.
VANHANEN, T. 1984. *The Emergence of Democracy: A Comparative Study of 119 States, 1850–1979*. Helsinki: Societas Scientiarum Fennica.
—— 1990. *The Process of Democratization: A Comparative Study of 147 States, 1980–88*. New York: Taylor and Francis.

Chapter 11

# Political Behavior, Old and New

Warren E. Miller

## I The excitement of new worlds to conquer

A historical review of political behavior in the United States should really encompass at least three epochs, the new, the old and the very old. As a member of the research community for almost fifty years, it seems appropriate for me to comment briefly on the old and the new, but I will leave the tracing of the longer lines of intellectual origin to the sociologists of knowledge and the philosophers of science. If one, therefore, bypasses the early work of Stuart Rice (1928), Samuel P. Hayes, Jr. (1932), or better known figures such as Merriam and Gosnell (1924), the era of the "Old" must be introduced with Lazarsfeld and Berelson. The Huckfeldt–Carmines chapter (above: chap. 8) provides a good statement on their work, as well as that of our role model in political science, V. O. Key, Jr. The only major omission from their review of the literature are the contributions provided by Stein Rokkan, including his well-known collaboration with Lipset. Rokkan, along with Lazarsfeld, saw politics and mass political behavior as manifestations of social structure and social experience. Against that Old World backdrop, the introduction of a micro-analytic social psychological perspective was more dramatic than many retrospective accounts reveal.

For many of us in the next generation, the world of political behavior was introduced in 1952 with the first national election study carried out by the Political Behavior Program of the University of Michigan's Survey Research Center. The 1952 study did not spring full-blown from imaginative minds (or deep pockets) at the University of Michigan. Under the leadership of Pendelton Herring, then President of the Social Science Research Council, the Council Committee on Political Behavior had been formed in

1949 and was, in fact, the sponsor of the 1952 study. In the same period the Ford Foundation created the Center for Advanced Study in the Behavioral Sciences in Palo Alto, symbolizing the emergence of the behavioral emphasis in the social sciences. Political behavior in general, and electoral behavior in particular, owes much to the private foundations, including Rockefeller, Carnegie, Russell Sage and Markle. These organizations provided a context of enthusiasm, as well as funding, throughout the decades of the 1950s and 1960s. The Ford Foundation, through the efforts of Foundation officers such as Peter de Janosi and Kalvin Silvert supported research activities and the development of institutional infrastructures associated with pioneering work in political behavior at numerous sites in South America as well as Europe. The National Opinion Research Center at Chicago and the Institute for Social Research at Michigan were new institutional homes for behavioral research.

Organized support for the behavioral approach to Political Science was centered in the SSRC, but the American Political Science Association provided disciplinary legitimacy, in part through the leadership provided by Evron Kirkpatrick, Executive Director of the then recently established Washington office. As one example, Kirkpatrick was directly responsible for the recognition of Political Science by the National Science Foundation. Kirkpatrick and others in the APSA leadership also were involved in making the International Political Science Association a supporting mechanism for cross-national collaboration. IPSA recognized "work groups" that provided legitimizing umbrellas covering like-minded research scholars whose research projects would often turn to UNESCO for institutional support.

It is useful to begin this essay with the many references to institutional support for political behavior in order to make the point that there were many new, highly relevant, organizational initiatives outside the halls of the academy proper. In their newness, the leaders of the initiatives conveyed a shared sense of optimism and challenge that encouraged members of the political science research community to explore new work ways. It also, unhappily, permits one to contrast the "Old" with its many sources of funding support with the "New," in which NSF has, by default, come close to the role of sole source supplier in the United States. However, to my mind, the contrast is less well defined by the resulting levels of financial support than by the diversity of organizations outside of government and academe that were ready to help those who saw in the new methods of data collection and data analysis the means of doing innovative research on traditional intellectual problems of the disciplines.

Which is not to say there was no institutional resistance in the halcyon days of the 1950s and 1960s. For too long the norm across political science departments in the United States was for each department to have one political behaviorist in captivity—one, but usually no more. Even at Michigan, the late George Belknap—a co-author with Angus Campbell of the pioneer article on Party Identification (and a University of Chicago Ph.D. under Avery Leiserson, a member of the SSRC Political Behavior Committee) was not accepted as a member of the department, and Samuel J. Eldersveld was the lone representative of the behavioral persuasion in the Michigan department until the late 1950s. Indeed, the era began with some likelihood that political behavior would become political sociology as it was on the eastern side of the Atlantic. Illustratively, James C. Davies, one of the colleagues working on the 1952 study, joined the American Psychological Association because of a fear that political science would never accept the social psychological perspective being emphasized at Michigan.

## II Multi-disciplinary origins

Put more positively, political behavior had its origins in the multidisciplinary world of post-World War II social science. For example, the leader of the quadrumvirate responsible for *The American Voter* (1960), Angus Campbell, was trained as an experimental psychologist at Stanford and was one of the pre-World War II initiators of social psychology as an academic discipline. The second in the alphabet of authors, Philip Converse, was an early product of the University of Michigan's post-World War II doctoral program in social psychology who brought a degree in sociology to his emerging specializations (after a baccalaureate in English literature). As the third of the group, I was a mutant political scientist whose academic work as a doctoral student in social science at Syracuse University consisted of one course in international relations and three sequences of course work in the methodologies and research methods of anthropology, sociology and social psychology. Donald Stokes was the only one with professional training as a political scientist out of Princeton and Yale; and even he came to Michigan on a pre-doctoral fellowship to study mathematical statistics in an era in which the slide rule and McBee sort cards were only slowly giving way to Hollerith cards and analog computers.

In the 1950s Michigan was joined by Yale University and the University of North Carolina as pre-eminent centers for the study of political behavior. In the case of UNC, the disciplinary environment was similar to

Michigan in that sociology, anthropology and social psychology were companion disciplines to political science, and research scholars benefitted from the UNC Institute for Social Research—a parallel to the Michigan ISR. The North Carolina focus in political behavior was somewhat more sociological than at Michigan, and their signal contribution, among many noteworthy works, was the Prothro and Matthews study, *Negroes and the New Southern Politics* (1966). The fieldwork was done by the Michigan Survey Research Center and the study design included four intensive community studies purposively selected from the primary sampling units that were the basis for their cross-sectional study of voting behavior in the South.

The most celebrated of the community studies of political behavior was directed by Robert Dahl at Yale (another member of the SSRC Political Behavior Committee) and published as *Who Governs: Democracy and Power in an American City* (1961). It complemented other political behavior research of the day as empirical work of a behavioral orientation clearly prompted by questions of primary concern to political science.

In 1962 the Michigan SRC carried out the fieldwork for a study of public attitudes toward and understanding of the United States Supreme Court. The principal investigators were Joseph Tanenhaus, Professor of Political Science at the University of Iowa, and Walter Murphy, Professor of Politics at Princeton. Their data collection reflected the excitement created in different corners of the discipline by the capacity to use the new methods and techniques of social science to create data tailored to the research needs of the research scholar. The generic contrast, to be found in sociology as well as political science, lay in the many differences between data collected in the course of society's social bookkeeping for administrative or bureaucratic purposes (birth rates, divorce rates, election statistics) and data collected expressly for use in social research.

It is true that *The American Voter*, as the successor to *The Voter Decides* (Campbell, Gurin and Miller 1954) with its early emphasis on predispositions (party, candidate, and issue orientations), made heavy use of attitudes and beliefs in attempting to understand individual level vote decisions. And the psychological nature of party identification apparently imprinted the emphasis on social psychology as the hallmark of the Michigan school. It should be noted, however, that politically relevant relationships between the individual citizen and the social group—both primary and secondary—were major themes in the Michigan data collections of 1952 and 1956, well beyond the innovative measurement of party identification. And, indeed, at least seven chapters in *The American Voter* were given over to "sociological" concerns.

It is pertinent to note how many of the themes introduced by Huckfeldt and Carmines under the heading of political sociology's challenge were, by yesterday's lights, in the same intellectual mode as *The American Voter.* At least in part because Ted Newcomb of Bennington Study fame was a prominent part of our social and intellectual environment, we started planning for what has become the flagship of studies of political socialization hard on the heels of the completion of *The American Voter.* The initial planning and negotiating with the Danforth Foundation, the ultimate funding source for the first 1965 data collections under the direction of Kent Jennings, were prompted by our shared conviction that crucial values and beliefs of voters are first formed in pre-adult years, and the contributions of family, school and peers to that formation should be systematically studied. The idea that there was something essentially sociological and therefore "different" about our projected panel study of political socialization and maturation did not occur to us. The urgency of the need to expand the age range coverage of our subjects and the need to introduce diachronic variations in their experience into our research design seemed obvious and completely in the intellectual mode of *The American Voter.* In like manner the data skillfully analyzed in Finifter's study of politics and the workplace (1974) were collected by Miller and Stokes in 1961 as that year's Detroit Area Study. To be sure, the Detroit Area Study was a research vehicle maintained by the Department of Sociology at Michigan but, as with the Jennings socialization study, being in Michigan made the workplace study seem a natural outgrowth of analyses pursued in Section IV (chaps. 12–18) of *The American Voter.*

Political Science *was* the intellectual and disciplinary home for the Miller–Stokes study of Representation in Congress. That study joined a growing list of innovative uses of new methodologies to investigate age-old questions of political behavior. The Wahlke–Eulau study of state legislators, published as *The Legislative System* (1962), had begun in 1955 (again with the sponsorship of the SSRC Committee on Political Behavior) and demonstrated the feasibility of carrying out systematic data collection from political élites. This capacity was exercised again by Eulau a decade later as he and Prewitt published *Labyrinths of Democracy: Adaptations, Linkages, Representations, and Policies in Urban Politics* (1973).

In 1974, near the end of the "Old" period, three closely interrelated studies were promoted by the Michigan Center for Political Studies. With support from the National Science Foundation and Russell Sage, the 1974 elections were the center of attention for interrelated studies of (1) the voters (2) the campaigns for election to the U.S. House of Representatives as reported by a national sample of candidates and campaign managers in contested races and (3) the media presentation to the voters of those campaigns. Some of the research of Erbring and MacKuen reported by Huckfeldt and Carmines exploited the data from that 1974 effort. Studies of campaign strategies as planned by campaign managers, as reported by editors and political reporters, and as viewed by voters had been synchronized because of the obvious functional interdependence of the three sets of actors. Incidentally, the three element design, this time centered in elections to the United States Senate, is currently being replicated by two of my Colleagues, Professors Kim Kahn and Patrick Kenney at Arizona State University. Both are Political Scientists and both are well trained in behavioral research approaches and methods that originated in a wide array of social science disciplines.

The list could be extended, but the point is not so much to review the research program originating with the Michigan Center for Political Studies as to illustrate how many different studies were seen by their principal investigators as tightly interconnected parts of a substantive, coherent whole that only coincidentally embraced components that might be identified with different disciplines and were made possible by a thoroughly inter-disciplinary set of research methods and techniques. Each study was shaped by principal investigators with somewhat different combinations of interests and skills and varied disciplinary training but with virtually no self-conscious interest in being "inter-disciplinary." The strengths of the studies depended on borrowing concepts, hypotheses and research methods and techniques from each of several academic disciplines and applying them as the internal logic of each component of an overall research design took shape. It may be useful to sort out the various disciplinary contributions—or origins—of "New" research on political behavior. Among the "Old" studies the interdisciplinary mix of insights and skills seemed thoroughly natural.

## III New aspects of political behavior research

The Old and the New differ in at least three respects. In part because so many of the "Old" data collections were simple cross-section snapshots of a large and complex set of "variables" derived from large and heterogeneous populations, methodological creativity was called upon to overcome limitations inherent in the data. Inadequacies were particularly obvious in data intended to represent a "process" such as involved in voters arriving at vote choices. This need for methodologies to overcome deficiencies in the data was complemented by an inventiveness that was soon its own spur, speeding new developments as courses on "research methods" proliferated and "political methodology" became the focus for professional organizational activity.

For whatever reasons, analytic sophistication, or complexity, now seems to be of the essence in much newly published work. The propensity to begin data analysis with simple bivariate or trivariate inquiries has disappeared in order to hasten on to "model" a new idea with many-termed multi-variate equations. Whether this, in general, is a mark of the superiority of the "New" over the Old is a matter for another set of chapters. It would appear, however, that being up-to-date in Method City is at least as important as testing a new substantive idea or theoretical insight. As a consequence, "New" research is less often inspired by the prospect of a new venture into heretofore unchartered domains of politics and government and more often a trek through essentially familiar terrain in a new analytic vehicle.

The evolution of new analytic perspectives and techniques has, of course, been made possible largely through the agency of the computer, a second source of transformations from the "Old" to the "New." As the data were collected for Almond and Verba's *Civic Culture,* Dahl's *Who Governs,* or *The American Voter,* data processing and analytic calculations were accomplished with the eighty-column Hollerith (IBM) card and mechanical card sorting devices—the counter sorter or the "101" with its complex wiring boards to produce more complex counting and sorting. The original data for the Elmira Study underlying Lazarsfeld and Berelson's seminal volume, *Voting* (1959), are still carried in the ICPSR archives on multiply-punched cards—with two variables often encoded in a single column dividing the twelve rows between them. Analyses for *The Voter Decides* were based on "analysis cards," in which a sub-set of variables pertinent to a single piece of analysis had been transferred from their original storage cards and brought together so they were physically located on the same analysis card.

Although this cumbersome technology was a vast improvement over hand recording on tabulating paper, it placed a high premium on data management, and documentation tailored to fit each stage, and it placed severe limits on the complexity of analytic techniques that could be used. The computer changed—and is still changing—all of that. All twenty-three studies in the American NES forty years series are now available, with complete detailed documentation, on a single compact disc that also carries front-end software to facilitate access to the data. The capacity of computer software for multivariate analyses will today accommodate data sets ranging across scores of variables for thousands of cases. These storage, retrieval and processing capabilities have encouraged the proliferation of both methodological and substantive interests and have made possible research that could scarcely have been imagined forty years ago.

The third crucial change that has occurred over the past thirty years has been in the social organization of access to data. The first years of the behavioral revolution were marked by a veritable explosion of empirical studies based on data collected from large and politically significant populations. In electoral behavior, there were studies of national electorates, and in legislative behavior, studies of second chambers in national governments. However, by virtue of traditional academic culture, and in the absence of any alternative, exploitation of each new data collection in the 1950s was largely restricted to the principal investigators and their graduate students.

In an effort to expand the numbers of scholars who could have access to the 1952 and 1956 Michigan election studies, the same SSRC Political Behavior Committee so essential to the funding of those studies sponsored and underwrote two summer training seminars at the Michigan Survey Research Center in 1954 and 1958. Participants included such latent luminaries as Robert Lane and Heinz Eulau. Converse, Miller, and Stokes were the instructors and each summer was devoted to teaching about and learning about codebooks and crosstabs and computation of Chi squares. The summers were exciting, new experiences in collegial learning and doing, and new experiences in the world of data processing.

The 1954 experience inspired Eulau to return to Ann Arbor in 1955 and launch his work on politics and social class. The modal sequel to the seminars was less rewarding. Scholars returned to their departments with a new awareness of the absence of the infrastructure necessary to extend or continue the Michigan experience. And this even in many of the best departments of political science. Out of an awareness of the difficulty experienced by most seminarians who wanted to continue their new work ways at home, the idea of the Inter-University Consortium for Political Research was born.

The idea had two components. The first was to establish the cultural norm of sharing data. The traditional norm gave to the creator or generator of data a virtually proprietary right to control access to the data. The new concept emphasized sharing data resources, particularly if they were initially created through public funding. The other principal component of the original idea of the Consortium was the goal of creating a new organization to facilitate individual research efforts, but to be supported by departments of political science. It would maintain data libraries (unfortunately improperly labelled "archives"), built on data contributed by principal investigators to be shared with colleagues. The organization would also provide the training needed for access to the data and for facilitating data analysis. It would provide the infrastructure needed for the exploitation of extensive data resources to be shared by individual research scholars otherwise separated from each other.

Today ICPR has become ICPSR, and through the early leadership of people such as Jean Blondel, Stein Rokkan, Erwin Scheuch and Rudolf Wildenmann there is a thriving ECPR. Bergen, Cologne, Essex, and others have established data archives that complement the largest of the data collections maintained in Ann Arbor by the ICPSR.

Most recently, yet another organizational innovation has appeared to promote the "Comparative Study of Electoral Systems." With the leadership of John Curtice from England, Sören Holmberg from Sweden, Hans-Dieter Klingemann from Germany, Steven Rosenstone from the United States and Jacques Thomassen from the Netherlands, plans are underway to make national election studies, around the world, vehicles for systematic cross-national comparative studies. The focus is sharply on electoral behavior, but the micro-analysis of individual behavior in each country will be enhanced with data on the contexts provided by social and economic factors relevant to each country.

## IV The impact of the new on the old

The ultimate consequences of the revolutions in methods, computer applications and human organization cannot be anticipated in any detail. There will certainly be an extraordinary increase in the sheer volume of research. Technical standards will almost certainly go up on all fronts—data collection, data analysis and theoretical explanations of political behavior in varied settings. However, given cultural and linguistic differences across communities of scholars, it is also likely that research technology will continue to outpace and overwhelm substantive discovery.

On the other hand, concerns such as those expressed in the previous chapter are not simply methodological concerns, and they are slowly being addressed, or redressed. The work of Huckfeldt and Sprague, among others, certainly speaks directly to the need for disaggregation and reaggregation at the level of individual experience. Among the "older" works that attended to similar concerns involving a quite different topic is the Verba–Nie volume on political participation. In general, I think it is as easy to underestimate the very real concerns of "Old" behavioral researchers over questions of heterogeneity among their respondents as it is to underestimate the cost of doing something about it. At the same time, it may be that there is a real problem in overestimating the extent or the consequence of experientially based heterogeneity. At least in the domain of trying to explain an elector's vote choices, the search for evidence that voters coming from different circumstances use different paradigms in their decision-making has not been rewarding. Noting that virtually every summary measure of a relationship represents some "average" among voters, including many for whom the summary is just not appropriate, is well and good—but what to do? I do not think the fault lies in a preoccupation with institutional definitions of meaningfulness so much as with the inadequacy of resources with which to design proper data collections. The "Old" did not do well in this domain, but not because the problems were not recognized.

Franz Pappi's concern is, in my view, another version of the Dunleavy concern. The particular application to the multi-party election has not drawn enough attention in the United States largely because it is so seldom that American voters have a real opportunity for strategic voting in Presidential elections. It is possible that a repetition of 1992 would change all of that. The chapter by Pappi (above: chap. 9) is crucially important because it complements the other preoccupations with American politics and American political science with a major contribution from European political culture. In particular, it brings the consideration of coalition formation into the analytic calculus.

On this score I can only say, tell us more. Most American analysts do not cope well when the voter has more than two alternatives. A third candidate is either ignored, as were Wallace in 1968 and Anderson in 1980, or given separate treatment, as with Perot in 1992. The notion that some voters may have a calculus based on strategic voting is often enunciated but seldom examined. At the same time, I would note the possibility of utilizing candidate "thermometer" data as an alternative to many examples of modeling rational choice theories. The problem can at least be reduced if analyses seek to adopt the voter's perspective and are, therefore, limited to the apparent contest, voter-by-voter, between the two most favored candidates. In 1992, analyses eliminating less preferred candidates provided persuasive explanations of the Perot vote without invoking concepts of strategic voting (Miller and Shanks 1996).

## V New definitions of citizenship

In as much as each line of inquiry into mass electoral behavior develops its own empirical generalizations united by their own logic, and progresses by the constant interplay of theory and evidence, it would take the unseen guiding hand of an Abigail Smith to produce *a* new empirical model of *the* democratic citizen. It seems to me that we are not in the process of developing a new model of *the* democratic citizen so much as we are learning about the conditions and circumstances under which varied and different categories of citizens respond politically in understandable and understandably different ways. Although much political philosophy and much élite political behavior rests on one or another conception of the average or typical or modal member of the rank and file citizenry, much of the research on large populations, such as national electorates, documents the heterogeneity of responses to common stimuli.

More generally, it seems to me that the challenge of democratic theory has, indeed, supplied "much of the agenda for political behavior research during the past 50 years," but only in the broadest sense. It is true that the felt need to understand better the voting behavior of an entire electorate is drawn from the conviction that elections are vital institutions in democratic politics. Much work on specific topics is on the agenda by virtue of institutional definitions of a problem. It is also true that *after the fact* there is often speculation about the meaning which a newly verified generalization has for this or that aspect of normative philosophy. However, it is also true that verification of empirical generalizations is seldom sought because of their potential implication for theories of democratic governance. Generalizations are usually re-examined because verification, or rejection, has implications for our theories about the causes of a given behavior.

What we have learned about mass electoral behavior certainly has meaning for theories of democratic politics. But as yet we are often uncertain as to what the various meanings ultimately will be. It may be that greater self-conscious attention to the assumptions that underlie democratic theory *would* channel programs of research to meet various challenges posed by the theory. Thus far the meandering course of research into mass electoral behavior has more often followed the decisions of scholars intent on better "understanding" some obvious feature of political topography than testing a normative construction.

## References

Almond, G., and Verba, S. 1963. *The Civic Culture.* Princeton, N.J.: Princeton University Press.

Campbell, A.; Converse, P. E.; Miller, W. E.; and Stokes, D. E. 1960. *The American Voter.* New York: Wiley.

—— Gurin, G.; and Miller, W. E. 1954. *The Voter Decides.* Evanston, Ill.: Row Peterson.

Dahl, R. A. 1961. *Who Governs?* New Haven, Conn.: Yale University Press.

Eldersveld, S. J. 1964. *Political Parties.* Skokie, Ill.: Rand McNalley.

Eulau, H., and Prewitt, K. 1973. *Labyrinths of Democracy.* Indianapolis, Ind.: Bobbs-Merrill.

Finifter, A. W. 1974. The friendship group as a protective environment for political deviants. *American Political Science Review,* 8: 607–25.

Hayes, S. P., Jr. 1932. Voter's attitudes towards men and issues. *Journal of Social Psychology,* 2: 164–82.

Huckfeldt, R., and Sprague, J. 1995. *Citizens, Politics, and Social Communications.* New York: Cambridge University Press.

Jennings, M. K., and Niemi, R. G. 1974. *The Political Character of Adolescence.* Princeton, N.J.: Princeton University Press.

—— —— 1981. *Generations and Politics.* Princeton, N.J.: Princeton University Press.

Lazarsfeld, P. F.; Berelson, B. R.; and Gaudet, H. 1948. *The Peoples's Choice: How the Voter Makes Up His Mind in a Presidential Campaign.* New York: Columbia University Press.

—— —— and McPhee, W. N. 1959. *Voting.* Chicago: University of Chicago Press.

Merriam, C. E., and Gosnell, H. F. 1924. *Non-Voting.* Chicago: University of Chicago Press.

Miller, W. E., and Shanks, J. M. 1996. *The New American Voter.* Cambridge, Mass.: Harvard University Press.

Prothro, J., and Matthews, R. 1966. *Negroes and the New Southern Politics.* Chapel Hill: University of North Carolina.

Rice, S. 1928. *Quantitative Methods in Politics.* New York: Knopf.

Rokkan, S., and Lipset, S. M., eds. 1967. *Party Systems and Voter Alignments.* New York: Free Press.

Tingsten, H. 1937. *Political Behavior: Studies in Elections Statistics.* Stockholm: Knopf. London: P. S. King.

Verba, S., and Nie, N. H. 1972. *Participation in America.* New York: Harper and Row.

Wahlke, J.; Eulau, H.; *et al.* 1962. *The Legislative System.* New York: Wiley.

Part IV

# Comparative Politics

Chapter 12

# Comparative Politics: An Overview

Peter Mair

## 1 Introduction: the discipline of comparative politics

EVER since Aristotle set out to examine differences in the structures of states and constitutions and sought to develop a classification of regime types, the notion of comparing political systems has lain at the heart of political science.[1] At the same time, however, while perennially concerned with such classic themes as the analysis of regimes, regime change, and democracy and its alternatives, comparative politics is not a discipline which can be defined strictly in terms of a single substantive field of study. Rather it is the emphasis on comparison itself, and on how and why political phenomena might be compared, which marks it out as a special area within political science. Indeed, precisely because there is no single substantive field of study in comparative politics, the relevance and value of treating it as a separate sub-discipline has often been disputed (see the discussion in Verba 1985; Dalton 1991; Keman 1993*a*).

The discipline of comparative politics is usually seen as being constituted by three related elements. The first, and most simple element is the study of foreign countries, often in isolation from one another. This is usually how comparative politics is defined for teaching purposes, especially in Anglo-American cultures, with different courses being offered on different countries, and with numerous textbooks being published about the individual countries which are incorporated in these courses. In practice, of course, however useful this approach may be in pedagogical terms, there is

[1] See Books II.b and IV.b of Aristotle's *The Politics*.

often little real comparison involved, except implicitly, with any research which might be included under this heading being directed primarily to the gathering of information about the individual country or countries concerned. Indeed, one of the problems associated with the distinctiveness, or lack of distinctiveness of comparative politics as a sub-discipline is that an American scholar working on, say, Italian politics is usually regarded by her national colleagues as a "comparativist," whereas an Italian scholar working on Italian politics is regarded by her national colleagues as a "non-comparativist." This, of course, makes nonsense of the definition.

The second element, which is therefore more relevant, is the systematic comparison between countries, with the intention of identifying, and eventually explaining, the differences or similarities between them with respect to the particular phenomenon which is being analyzed. Rather than placing a premium on the information which may be derived about these countries, therefore, the emphasis here is often on theory-building and theory-testing, with the countries themselves acting as cases. Such an approach clearly constitutes a major component of political science research more generally, and, indeed, has been the source of some of the most important landmark texts in the discipline as a whole (e.g. Almond and Coleman 1960; Almond and Verba 1965; Lipset and Rokkan 1967; Lijphart 1977).

The third element within comparative politics is focused on the method of research, and is concerned with developing rules and standards about how comparative research should be carried out, including the levels of analysis at which the comparative analysis operates, and the limits and possibilities of comparison itself. Precisely because the act of comparison is itself so instinctive to both scientific and popular cultures, this third element is sometimes assumed by researchers to be unproblematic and hence is neglected. And it is this neglect, in turn, which lies at the root of some of the most severe problems in the cumulation of research, on the one hand, and in theory-building and theory-testing, on the other hand.

Unusually, then, comparative politics is a discipline which is defined both by its substance (the study of foreign countries or a plurality of countries) and by its method (see Schmitter 1993: 171). At the same time, of course, this immediately undermines its distinctiveness as a field of study. In terms of its method, for example, comparative politics is hardly distinctive, in that the variety of approaches which have been developed are also applicable within all of the other social sciences. Indeed, some of the most important studies of the comparative method (e.g., Przeworski and Teune 1970; Smelser 1976; Ragin 1987) are directed to the social sciences as a whole rather than to political science *per se*. In terms of its substantive concerns, on the other hand, the fields of comparative politics seem hardly separable from those of political science *tout court*, in that any focus of inquiry can be approached either comparatively (using cross-national data) or not (using data from just one country). It is evident, for example, that many of the fields of study covered in the other chapters of this book are regularly subject to both comparative and non-comparative inquiries.[2] If comparative politics is distinctive, therefore, then it is really only in terms of the combination of substance and method, and to separate these out from one another necessitates dissolving comparative politics either into political science as a whole or into the social sciences more generally.

Given the impossibility of reviewing the broad span of developments in political science as whole, and, at the same time, the undesirability of focusing on methods of comparison alone, a topic which has already received quite a lot of attention in the recent literature (see, for example, Collier 1991; Keman 1993*b*; Bartolini 1993; Sartori and Morlino 1991), this chapter will deal instead with three principal themes, focusing in particular on the contrast between the ambition and approach of the "new comparative politics" of the late 1950s and 1960s, on the one hand, and that of the current generation of comparativists, on the other (for a valuable and more wide-ranging review, see Daalder 1993). The first of these three themes, which is discussed in Section II, concerns the scope of comparison, which is perhaps the principal source of difference between the earlier and later "schools" of comparative politics. Although much tends to be made of the contrasting approach to institutions adopted by each of these two generations of scholars, and of the supposed neglect and then "rediscovery" of institutions and the state as a major focus of inquiry, this can be misleading, in that the apparent absence of an institutional emphasis in the 1950s and 1960s owed more to the global ambitions (the scope of their inquiries) of that earlier generation, and hence to the very high level of abstraction at which they constructed their concepts, rather than to any theoretical downgrading of institutions *per se*. Concomitantly, the rediscovery of institutions in the 1980s and 1990s owes at least as much to the reduction in the scope of comparison, and hence to the adoption of a lower level of conceptual abstraction, as it does to any theoretical realignment in the discipline.

The second theme, which is discussed in Section III, concerns the actual topics and questions which are addressed in comparative political inquiries, and where quite a marked shift in focus can be discerned, with

[2] It is thus interesting to note that when she was preparing the second edition of *The State of the Discipline*, published in 1993, Ada Finifter was advised that "not only do we need more comparative chapters than there were in the first edition, but all the chapters should be comparative" (Finifter 1993: viii).

much more attention now being devoted to "outputs" rather than to "inputs," and to the outcomes of politics and the performance of government rather than to the determinants of politics and the demands on government (see also Rogowski 1993). This also relates to the changing scope of comparison, in that it clearly makes much more sense to ask whether politics matters—a question of outputs and outcomes—when the scope of comparison becomes restricted to just a small number of relatively similar cases. The third theme, which will be addressed here in Section IV, concerns some of the problems which are currently confronted in comparative research, with particular attention being devoted, on the one hand, to the role of countries as units of analysis, and, on the other, to the use and, indeed, virtual fetishization of indicators. The chapter will then conclude with a brief discussion in Section V of some present and future trends in comparative politics, focusing in particular on the renewed emphasis on context, as well as on in-depth case analysis.[3]

## II Scope

Writing in the early 1960s, in a most valuable and broadly based review of the past and present states of comparative politics, Harry Eckstein (1963: 22) noted that comparative politics could then be characterized by "a reawakened interest in large-scale comparisons, a relatively broad conception of the nature of politics and what is relevant to politics, and a growing emphasis upon solving middle-range theoretical problems concerning the determinants of certain kinds of political behavior and the requisites for certain kinds of political institutions." Eckstein's reference point here was to the early stages of what is often now considered to have been "the golden age" of comparative politics, when a series of major and path-breaking research programs were initiated by Gabriel Almond and his colleagues on the American Social Science Research Council's Committee on Comparative Politics (founded in 1954). And what is perhaps most striking in this characterization, and what was also perhaps the most important feature of the new approach developed by the Committee, was precisely the attention which was beginning to be devoted to "large-scale comparisons." Rejecting the then traditional and almost exclusive emphasis on the developed world, and on western Europe and the United States in particular, and rejecting also the use of a conceptual language which had been developed with such

[3] For an earlier version of some of this discussion, focusing in particular on the comparative method, see Mair 1995.

limited comparisons in mind, Almond and his colleagues sought to develop a theory and a methodology which could at one and the same time both encompass and compare political systems of whatever sort, be they primitive or advanced, democratic or non-democratic, western or non-western. As Almond (1970: 16) was later to emphasize in a subsequent review of work of the Committee and of the development of comparative politics in this period, their strategy had been intended to bring together scholars working on countries across the globe, and to persuade them that they were "members of a common discipline concerned with the same theoretical problems and having available to them the same research methodologies."

The broadening of concerns in a geographic or territorial sense was also necessarily accompanied by a broadening of the sense of politics itself, and, in particular, by a rejection of what was then perceived as the traditional and narrowly defined emphasis on the study of formal political institutions. Indeed, reading the work of the major comparativists of the 1950s and 1960s, one is constantly struck by an almost palpable frustration with the approach to the study of political institutions which had prevailed up to then. Two factors were particularly relevant here. In the first place, the traditional emphasis on institutions was seen to privilege the formal and legal aspects of politics at the expense of what might be termed politics "in practice," and to privilege the "official" story at the expense of what was increasingly believed to be an alternative and "real" story. Thus "realism" rather than "legalism" was to become the keyword for the new comparativists. Secondly, a broadening of the sense of politics was also required in order to incorporate a recognition of less formally structured agencies and processes which spread the scope of the political quite far beyond the formal institutions of government alone. This shift developed directly out of the new global ambitions of the discipline, with the rejection of legalism going hand in hand with the rejection of a primary focus on western polities. Moreover, not only did this new approach allow for a more nuanced analysis of non-western regimes, but it also encouraged the new generation of comparativists to pay attention to less formalized aspects of politics even within the study of the western regimes themselves. Thus students of western European politics were now encouraged to abandon their "formal and institutional bias" and to focus instead on "the political infrastructure, in particular on political parties, interest groups, and public opinion" (Almond 1970: 14).

Global ambitions, and the need to develop a more broadly defined conception of politics and the political system, had two important consequences. The first was simply the beginning of an extraordinarily fruitful research program in comparative politics, the sheer scale, coherence and ambition of which has since remained unrivalled,[4] and the recollection of which remains enshrined in an image of this period as being the "golden age" of the discipline. "Comparative politics is [now] and has been disappointing to some," noted Verba (1985: 29) in a pessimistic review, "but it is disappointing in comparison to past aspirations and hopes." Since that golden age, it is often felt, the discipline has gone into retreat, with scholars complaining, at least in conversation with Verba, about "division, fragmentation, and atomization in the field . . . [and the lack of] clear direction, leadership, or a commonly held and agreed-upon set of theoretical underpinnings" (1985: 28).[5] Second, conscious that "the challenge facing comparative politics [was] to elaborate a conceptual apparatus in keeping with the vastly extended global scale of its empirical investigations" (Rustow, 1957/1963: 65), there also emerged a new approach to the study of politics which was to be encapsulated within the now much criticized notions of "structural-functionalism." Prior to this, as noted above, comparative politics had been dominated by the study of established, clearly-defined, and economically advanced democratic systems, all of which were more or less characterized by an apparently sharp division between state and civil society, and by a conception of the state which viewed it as composed of specific (and comparable) institutions—executives, parliaments, bureaucracies, judiciaries, military forces, and so on—each playing its own specific role within the system. Global comparisons, by contrast, implied not only the inclusion of non-democratic regimes, but also very underdeveloped countries with so-called "primitive" political systems, in which it was not only difficult to establish the boundary between state and civil society, but in which it was also sometimes almost impossible to identify specific political institutions with a specific purpose.

Along with global ambition, therefore, came the abandonment of an emphasis on the formal institutions of government, and, indeed, the abandonment of an emphasis on the notion of the state itself, which was to become translated into the more abstract references to "the political system." As Almond (1990: 192) later noted, this new terminology enabled scholars to take account of the "extra-legal," "paralegal" and "social" institutions which were so crucial to the understanding of non-western politics, and, as Finer (1970: 5) suggested, was required in order "to encompass pre-state/non-state societies, as well as roles and offices which might not be

[4] See, for example, in addition to a lengthy series of important monographs: Almond and Coleman 1960; Binder *et al.* 1971; Coleman 1965; LaPalombara 1963; LaPalombara and Weiner 1966; Pye 1962; and Pye and Verba 1965.

[5] See also the review by Daalder (1993: 20), who speaks of the "exhilaration" associated with the "political development boom" in the 1960s.

seen to be overtly connected with the state." Moreover, this new language could also serve the interests of those students who remained concerned with western polities, since even here a new wave of scholarship had begun to "[discover] that governmental institutions in their actual practice deviated from their formal competences" and had begun to "[supplement] the purely legal approach with an observational or functional one. The problem now was not only what legal powers these agencies had, but what they actually did, how they were related to one another, and what roles they played in the making and execution of public policy" (Almond, Cole and Macridis 1955/1963: 53). Hence the emergence of structural-functionalism, in which certain quite abstractly defined functions were defined as being necessary in all societies, and in which the execution and performance of these functions could then be compared across a variety of different formal and informal structures.

Since then, of course, this then novel and path-breaking approach has itself been subject to extensive criticism and counter-reaction, with a new wave of scholarship emerging in the 1980s which stressed the need to return to the study of institutions and to restore primacy to an analysis of "the state." If the approach of Almond and his colleagues might be characterized as one which "identif[ied] the subject matter of political science as a kind of activity, behaviour, or, in a loose sense, function . . . no longer limited in any way by the variable historical structures and institutions though which political activities may express themselves" (Easton 1968: 283; see also Fabbrini 1988), then the new approach which began to be asserted in the 1980s was one in which context became crucial, and in which it was precisely the "variable historical structures and institutions" which were now seen to play a central role (Thelen and Steinmo 1992). In the first place, institutions, and the state itself, now increasingly came to be seen as relevant "actors" in their own right, in the sense that they, or those who occupied their offices, were seen to have their own autonomous interests, and were thus also part of "real" politics (e.g. Skocpol 1985; see also Mitchell 1991). Second, and perhaps most crucially, institutions were also seen to have a major determining effect on individual behavior, setting the parameters within which choices were made and through which preferences were both derived and expressed (March and Olsen 1984; Shepsle and Weingast 1987). Third, institutions, and institutional variations in particular, were also seen to have a major effect on outcomes, with the capacity of actors to realize their ends being at least partially determined by the institutional context in which they operated (e.g. Scharpf 1988; Lijphart 1994*a*).

From one reading, then, we appear to witness an almost cyclical process, in which institutions, and possibly even the state, are initially privileged as the basis on which political systems might be compared; in which these institutions are later relegated as a result of the prioritizing of "a realism that recognized the processual character of politics" (Almond 1990: 192); and in which they then acquire a new relevance as part of that real politics itself, and as the context which determines individual behavior and performance. From this reading, therefore, we see a series of paradigmatic shifts (Evans *et al.* 1985), which travel right to the heart of comparative political analysis itself. From another reading, however, the contrasts are much more muted. In a trenchant review of some of the early work of the neo-statists and new institutionalists, for example, Almond was at pains to emphasize the real continuities which existed across the different schools, arguing that there was little in this so-called new approach which was not already present, either implicitly or explicitly, in much of the earlier literature, and that its terms were essentially "indistinguishable from 'behavioral' or structural functionalist definitions" (Almond 1990: 215).

But while Almond may have been correct in claiming that the reality underpinning the new terminology is less novel than has been claimed, the conceptual language involved is certainly different, and it is here that the key to the contrast between the two approaches can be found. In brief, it is not a problem of whether Almond and his colleagues neglected the importance of the state and of institutions more generally, or of whether Skocpol and many of the new institutionalists have now redressed that imbalance; this is, in the main, a fairly futile debate. Rather, and returning to the main question, it is a problem of the scope of the comparisons involved. For while Almond and his colleagues were consciously developing a conceptual language which could address the need for global comparisons, even when the particular analysis was in practice restricted to just one case or to just a handful of cases, much of the work engaged in by the more recent comparativists is explicitly adapted for application to a more limited (and often quite unvaried) set of comparisons, be it limited to regions (western Europe, Latin America, etc.), or even, as in the case of Skocpol (1979), Hall (1986), or Scharpf (1988), to just a very small number of countries. The result is that while Almond and his colleagues were required to operate at a very high level of abstraction (see Sartori 1970), developing concepts which could travel to and be relevant for all possible cases, the more recent school of comparativists have contented themselves with a relatively middle-range or even low level of abstraction, in which the specificities of context become crucial determinants (see also below).

It is not therefore a problem of shifting paradigms, but rather a problem of shifting levels of abstraction, which, in turn, is induced by a shifting scope of comparison. In this sense, as was the case with the structural-functionalist "revolution" in the late 1950s and 1960s, the change is not so much a reflection of developments at the level of theory, but rather at the level of method. For once comparisons become more limited in scope, whether by restricting the focus to one region, or to a small number of cases, it becomes possible to bring into play a degree of conceptual specificity and intensiveness which is simply not feasible at the level of global, all-embracing comparisons. In other words, institutions and the state come back in not only because they are seen to be more important *per se*, but also because the lower levels of abstraction involved have allowed them to come back in, and have created the room for this type of grounded analysis. In the end, therefore, what is striking about the categories adopted by the structural functionalists is not the fact that they were more process-oriented, or that they were more society-centered, or whatever, which is in any case highly debatable (Almond 1990: 189–218); rather, what is striking about these categories is the enormously high level of abstraction which they required in order to allow them to travel from world to world, and in which institutional specificity was absorbed upward into the more abstract notions of role, structure and function. If institutions and the state have come back into prominence, therefore, it is at least partly because the scope of comparison has become more restricted,[6] and it is this which is perhaps the most striking development within comparative politics in the last two decades or so.

This narrowing of the scope for comparison can be seen in a variety of ways. In the first place, and most practically, it can be seen in the now virtual absence of comparative analyses with a global, or even cross-regional ambition. To be sure, a variety of contemporary textbooks on comparative politics (e.g. Blondel 1990; Hague *et al.* 1992), as well as a number of established courses,[7] do attempt to remain inclusive, and aim to develop a framework which can accommodate first-, second- and third-world systems. With very few exceptions, however, contemporary research in comparative politics tends to be restricted by region, or even to a very small number of cases,[8] notwithstanding the fact that there now remain few, if any, *terrae incognitae*. This orientation clearly stands in sharp contrast to at

[6] Although it may also be argued that distinct ideological impulses were involved: see, for example, Almond (1990: 189–218), Mitchell (1991), and Chilcote (1994: 121–76).

[7] See, for example, the discussion about "Teaching Comparative Politics for the Twenty-First Century," in *PS: Political Science and Politics*, 28 (1995): 78–89.

[8] One relatively recent exception is the study of government ministers by Blondel (1985), which, appropriately enough, is described on the jacket copy as "build[ing] a framework that is devoid of national reference." One possible counterweight to the tendency to limit the number and range of cases may yet be provided by the growing interest in democratization processes, which clearly have a cross-regional relevance (see also below).

least the ambitions which were originally expressed by the Committee on Comparative Politics in the 1950s, and to that earlier work which, even when restricted to just one or a handful of cases, persisted in applying concepts which were believed to be universally valid.

Second, there is an increasing tendency for the profession as a whole to become compartmentalized into more or less self-sufficient groups of, for example, Europeanists, Africanists, and Latin Americanists, with very little communication taking place across the boundaries of regional expertise. In part, this is simply a consequence of the pressures for increased specialization; in part, however, it is also a consequence of increased professionalization, with the critical mass of scholars in the different fields of expertise, and their associated journals, now having grown sufficiently to allow for self-sufficiency. In a somewhat different context, Almond (1990: 13–31) has already famously referred to the development of "separate tables" in political science, by which groups of scholars are divided on the basis of both ideology (left versus right) and method (soft versus hard). Perhaps more realistically, however, we can also conceive of the separate tables being constituted by regional specialists, with their separate European, Asian, Latin American, and African kitchens, and, even within these parameters, being increasingly further subdivided by academic specialisms, with the party people eating separately from the public policy people, and with the local government experts eating separately from those involved in electoral research. For not only has the growth of the discipline acted to cut regional specialists off from one another, but, even within the different regions, it has also tended to foster the self-sufficiency of specialist fields, each with its own narrow network and its own set of journals (or, to continue the analogy, with its own menu), accentuating the trend towards fragmentation which was already regretted by Verba in 1985 (see above, and also Keman 1993*a*; for a more sanguine view of the process, see Macridis and Brown 1986, and Dalton 1991).

Third, and perhaps most importantly, the methodological debate within comparative politics, and perhaps within the comparative social sciences more generally, has increasingly tended to stress the advantages of "small N" comparisons. Thus, for example, it is quite instructive to compare Lijphart's 1971 review of the comparative method, which devoted considerable attention to ways in which scholars could compensate for, or overcome, the problem of having to deal with just a small number of cases, with a similar and more recent review by Collier (1991), which devoted a lot of attention to the sheer advantages of small N comparisons.

From one perspective, this new attitude can be seen to gell with many of the sentiments expressed by much of the other recent writings on the comparative method, whether these be within political science, sociology, or history, or even within an attempted multi-disciplinary synthesis (e.g., Ragin 1987; 1991), and which lay considerable stress on "holistic" analysis and on the need for in-depth understanding of particular cases. From another perspective, however, and notwithstanding the shared desire to move away from global comparisons and universal categories, much of this contemporary work in comparative politics might better be seen as consisting of two distinct "schools" or approaches (see also Collier 1991: 24–6). On the one hand, there are those researchers who persist in attempting to derive generalizable conclusions or in attempting to apply generalizable models across a range of countries which, in contrast to the global ambitions of the first postwar generation of comparativists, is usually limited in terms of region or status. On the other hand, there are also those researchers who seem increasingly wary of multiple case comparison, even when limited to a relatively small N, and who stress the advantages of close, in-depth analyses of what is at most a small handful of countries, in which the advantages offered by looking at the whole picture are seen to outweigh the disadvantages suffered by limited applicability.[9] Despite their contrasts, however, there is a sense in which each approach can lay claim to offer the best option for the future. As Collier (1991) notes, for example, recent advances in quantitative techniques now appear to afford a much greater opportunity for statistical analyses across relatively small numbers of cases, and may lend the conclusions derived from such analyses a greater strength and authority.[10] Relatively in-depth qualitative case analyses, on the other hand, despite their obvious limitations, have the advantage of being more grounded, and, at least at first sight, can also prove more sensitive to the insights now being afforded by both the "new institutionalism" and the rational choice paradigm.[11] Indeed, the renewed interest in case studies in recent years, and the associated emphasis on understanding the full context in which political decisions are made, has most certainly been stimulated by the potential offered by these new insights (see also Section V, below).

[9] See, for example, Rhodes's (1994) discussion of state-building in the United Kingdom, which includes a trenchant defense of the capacity of the case-study to produce generalizable conclusions.

[10] As, for example, is the case with the very closely argued analysis by Scharpf (1988) concerning the capacity of governments to implement public policies.

[11] See, for example, Tsebelis's (1989: 119–234) application of his nested games approach in a close analysis of British Labour Party activists, Belgian consociationalism, and French electoral coalitions.

## III Questions

In many respects, the broad direction of the questions addressed by comparative political inquiry has remained largely unchanged through generations, and perhaps even through centuries. How might regimes be distinguished from one another? What accounts for regime stability, and what accounts for regime change? Which is the "best" form of government? The attention devoted to these "big" questions has tended, of course, to ebb and flow with different generations of scholarship, with interest being recently reawakened in the aftermath of the recent wave of democratization (see, for example, Diamond and Plattner 1993), and being reflected most obviously in the extraordinary volume of new literature on transitions to democracy and on constitutional engineering and institutional design.[12] Indeed, it is precisely this reawakened interest in democratization, and the search for general patterns and predictions, which may well restore a sense of global ambition to comparative politics, since it is really only in this context that students of developing countries are beginning to reopen lines of communication with those whose field has been largely restricted to the developed west, and that the expertise of students of the former "second world" is finally seen as being relevant to mainstream comparative politics.

But these are clearly the classic themes, the hardy perennials of comparative politics, and once we move beyond these it is possible to see quite important shifts in the sorts of questions which tend to be addressed. In a recent review of the state of comparative politics, for example, Rogowski (1993: 431) noted five trends from the 1980s which certainly appeared to suggest a new research agenda, and which included: "A far greater attention to the economic aspects of politics . . . Increased interest in the international context of domestic politics and institutions . . . An altered and sharpened focus on interest groups . . . A revival of interest in state structures and their performance . . . [and] Further work on nationalism and ethnic cleavages." This is, of course, just one list among potentially many, and even after the lapse of just a couple of years, one might be inclined to relegate the once pronounced concern with, say, interest groups, and give priority instead, say, to the burgeoning interest in transitions to democracy and in the working of democracy itself. Notwithstanding any such qualifi-

[12] For one very good recent example, see the stimulating and very thoughtful debate on the respective merits of presidential systems and parliamentary systems among Linz, Lijphart, Sartori, and Stepan and Skach in Linz and Valenzuela (1994: 3–136). Indeed, it might even be said that it is here that we see the practice of comparative politics at its very best, and hence also, this being a recent debate, the ideal counter-argument against the notion that, in some way or other, the discipline has now begun to pass its sell-by date.

cations, however, what is particularly striking about this list, and what would surely also be common to almost any other such list which might currently be prepared, is the attention devoted to the outputs, or even simply the outcomes, of political processes and political institutions, and hence the attention to politics as an independent rather than a dependent variable. In other words, what is striking here is the sheer extent of concern with the impact of politics rather than with the determinants of politics (see, for example, Weaver and Rockman 1993). This is where the increased interest in political economy comes into play, for example, as well as that in state structures and institutions, whether the latter be framed within a more traditional discourse (e.g. Lijphart 1994*a*) or within the terms of reference of the new institutionalism (e.g. Hall 1986; Evans *et al.* 1985).

Here too, then, it is also possible to discern a difference between the new generation of comparativists and that which blossomed in the late 1950s and 1960s. Nor is this simply coincidental, for, at least in part, it is the abandonment of the ambition towards global comparison and universalism which appears to have provided the space in which these new questions can become relevant. There are two steps involved here. In the first place, as noted above, a restriction of the scope of comparison has allowed more attention to be devoted to institutional specificities, and this in itself has helped make it possible to ask whether politics matters. Second, restricting the scope of comparison also means that it now makes more sense to ask whether politics matters than would have been the case in comparisons which attempted to embrace three different worlds, since, in the latter case, and inevitably so, differences in levels of economic development, or even political culture, would have been likely to appear much more relevant (e.g. Castles 1982). Indeed, once comparisons are restricted to relatively similar cases, such as, for example, the advanced industrial democracies, in which the levels of economic development, or the patterns of political culture, or the structures of society are relatively invariant, then the researcher is almost necessarily forced back onto the inevitably varying political structures and processes.[13] And precisely because the possible "determinants" of politics—at the level of economy, (contemporary) culture, or society—in these similar cases do vary so little, these varying political structures and processes then increasingly assume the status of an explanans rather than an explanandum, and thus help to draw attention to enquiries into outcomes and outputs. Whatever the reasons, however, it is certainly true that

[13] As well as, increasingly, onto the variance of historical traditions, where a more nuanced version of political culture and political traditions then also comes into play (see, for example, Castles 1989; 1993; Katzenstein 1984: 136–90; Putnam 1993). The danger here, however, is that an emphasis on the crucial role played by historical traditions may sometimes lead to essentially *ad hoc* explanations, if not to a degree of fatalism.

comparative political inquiries are now much more likely than before to ask about the difference which politics makes, rather than to ask what makes politics different. In other words, confronted with variation in institutional structures and political processes, contemporary scholars are now much more likely to want to assess the impact of this variation rather than, as before, and most notably in the late 1950s and 1960s, asking why these differences have emerged in the first place, and this clearly does indicate a major shift in the direction of comparative research.

Evidence of this shift can be seen partially in the variety of new trends noted by Rogowski (1993, see also above), as well, indeed, as in almost any reading of the contemporary literature (see, for example, Keman 1993*b*). It can also be seen, and perhaps more interestingly, in the trajectory of individual scholars and schools of research. Among individual scholars, for example, it is possible to cite the case of Arend Lijphart, who has for long been one of the foremost authorities in the discipline, and whose work has progressed over time from an inquiry into the conditions which gave rise to certain types of democracy to an inquiry into the consequences of certain types of democracy. Lijphart's first major work in the field of comparative politics concerned the elaboration of a typology of democratic regimes, in which the various types identified, and most notably consociational democracy, were defined on the basis of two crucial determining variables—the degree of conflict or co-operation among élites, on the one hand, and the degree of fragmentation or homogeneity in the political culture, on the other, with the latter being located firmly within a conception of social divisions and social pluralism (Lijphart 1968). What is most interesting in this particular context, however, is that as Lijphart's work developed, and as he attempted to modify and build on these initial ideas, the specifically social side of the equation became less and less important, such that in his highly influential depiction of two more generalized models of democracy (Lijphart 1984), the question of the social determinants of the political structures with which he was concerned was essentially relegated to the margins (Lijphart 1984, see also Bogaards 1994). Ten years later, in his most recent work in this field, the change in emphasis was even more evident, with the inquiry now having shifted into the question of the performance of the different types of democracy, and with the question of determinants being almost wholly ignored (Lijphart 1994*a*).

Similar shifts can also be seen among different schools of research, with the democratization literature offering perhaps the most obvious example of the way in which the explanans has moved from an emphasis on the "objective" social and economic conditions for democracy (e.g. Lipset 1959) to an emphasis on the importance of élite decision-making, on "voluntarism," and on the types of institutions and political structures involved. Whether democracy can emerge, therefore, and whether it can be sustained, is now seen to be much less dependent than before on levels of social and economic development and much more dependent on political choices (Rustow 1970), on "crafting" (Di Palma 1990), as well as on the outcomes of rational actions and information (e.g. Przeworski 1991). As Karl (1991: 163) puts it, "the manner in which theorists of comparative politics have sought to understand democracy in developing countries has changed as the once dominant search for prerequisites to democracy has given way to a more process-oriented emphasis on contingent choice" (see also Karl 1991, more generally, as well as Whitehead, chap. 14, below). In a similar sense, the question of the consolidation and sustainability of new democracies is now also seen to be much more closely associated with the actual specifics of the institutions involved (e.g. Linz and Valenzuela 1994). Here, then, as is also more generally the case in a variety of different fields of inquiry in comparative politics, the questions now revolve much more clearly around what politics does, rather than what makes politics the way it is, with the result that, more than two decades after an early but very powerful appeal for just such a shift (Sartori 1969), comparative inquiries are now finally more likely to emphasize a political sociology rather than simply a sociology of politics.

## IV Problems

At one level, work in comparative politics is often frustrating. The scholar devotes much time and effort in gathering comparable cross-national data, in ensuring that no relevant factor has been excluded from the analysis, and in building a general and preferably parsimonious model which can explain the phenomenon in question wherever and whenever it occurs, only then to be confronted at some conference or other with some national expert who complains that it's really not like that around here and who then goes on to offer a much more nuanced but essentially idiographic counter-explanation (what Hans Daalder refers to as the "Zanzibar ploy"). At another level, of course, work in comparative politics allows one to be happily irresponsible, in that it is always possible to pre-empt the Zanzibar ploy by prefacing one's broad theory with the caution that while the conclusions are not necessarily true for any particular country, they are nevertheless certainly true more generally. In both cases, however, the real difficulty is essentially the same: although country tends to form the unit of analysis and observation, the scholar must nonetheless work at one remove from country, and, regardless of whether the number of cases is limited or extensive, must translate a national experience into an operational category. And without wishing to enter into a discussion of the pros and cons of different comparative methods, this immediately presents those engaged in comparative politics with two particular problems.

The first of these problems was already alluded to by Rogowski (1993), and has frequently been highlighted in contemporary discussions of the discipline, and concerns the extent to which country continues to provide a meaningful unit of analysis. One aspect of this problem is the difficulty of identifying what is specific to national politics in an increasingly international environment. Insofar as comparative research does increasingly focus on outcomes and outputs, for example, then it is also increasingly likely to resort to explanations and determinants which lie outside the control of any one national state. To be sure, it is possible to construct a similar-cases research strategy in which precisely the same international environment is common to all the relevant cases, and in which it can then be taken as a given which will not explain any subsequent cross-national variation which might be found (see, for example, Scharpf 1988), but the opportunities for such a strategy are necessarily both limited and limiting (Mair 1995). In any case, to the extent that national institutions and national governments lose their capacity to mould their own national environments, then to that extent the study of comparative politics faces potentially severe problems.[14] A second aspect of this problem concerns the sheer validity of country as a unit of analysis, even regardless of any relevant international context. The difficulty here is posed by the simple fact that countries themselves change over time, and hence in addition to puzzling over cross-national variation, researchers also need to be conscious of cross-temporal variation, in which country *A* at time *X* might differ as markedly from country *A* at time *Y* as it does from country *B* at time *X* (Bartolini 1993). Indeed, this difficulty becomes particularly acute when research is focused on institutional structures, since it is usually at this level that significant changes can and do occur. In other words, if institutions do matter, how can those countries be analyzed in which these very institutions change? One possible solution to this problem which is emerging with increasing frequency is simply the dissolution of country into particular subsets of variables, with the recent study by Bartolini and Mair (1990), and most especially that by Lijphart (1994*b*), offering useful examples of

[14] And not just comparative politics: note Susan Strange's (1995: 55) provocative suggestion that with the waning of the state as the most important unit of analysis, "much of Western social science is obsolescent, if not yet quite out-of-date."

the gains which can be made by abandoning the notion of countries as single and indivisible cases and by the adoption of multiple observations from each country. The focus of Lijphart's recent study is electoral systems and their political consequences, and it is precisely these electoral systems, rather than countries as such, which constitute the relevant cases in the enquiry. Thus, for example, although France is one of the twenty-seven democracies included by Lijphart in his research, France as such does not constitute one of the relevant units of analysis; rather, the six different electoral formulae which France has adopted since 1945 constitute six of the total of seventy cases which are analyzed in the study (Lijphart 1994*b*). To be sure, this is far from a novel strategy, and a similar approach has long been adopted in comparative coalition research, for example. Nevertheless, it is an increasingly common strategy, and suggests a much greater willingness to experiment with alternative units of analysis and hence to make provision for cross-temporal variation (Bartolini 1993).

The second problem involved here is perhaps more acute, and involves the reliability of the various measures and indicators which are used in order to translate national experiences into comparable operational categories, a problem which has become even more pronounced as scholars have attempted to build into their analyses measures of variation in political institutions and political structures. Social and economic explanans have always proved relatively easy to operationalize, and in this sense the appeal of "objectivity" in the sociology of politics (Sartori 1969) has always been easy to appreciate, not least because of the apparent reliability of such sources of data as the World Bank, the OECD, the European Union, and even survey research. Once institutions begin to be measured and compared, however, reliability appears to falter, while at the same time hard data—in the sense of data which mean the same thing in every context—often prove unavailable. The result is an endless search for suitable "indicators," and even, at the extreme, the apparent fetishization of such indicators. One useful example of such an approach was the Lange–Garrett–Jackman–Hicks–Patterson debate which took in the pages of the *Journal of Politics* in the late 1980s concerning the relationship between leftwing strength, as measured by party and organizational (i.e. trade-union) variables, and economic growth, and which was subsequently cited in a review of recent developments in the comparative method (Collier 1991: 22), as "an exemplar of a methodologically sophisticated effort by several scholars to solve an important problem within the framework of a small-N quantitative analysis." The debate did certainly represent a very valuable and important contribution to comparative political research, and it was also certainly marked by a pronounced methodological and statistical sophistication, with much of the to-ing and fro-ing between the authors revolving precisely around different methodological approaches. That said, however, it was also striking to see how the initial question of whether economic growth can be associated with leftwing strength was eventually transformed into a problem of statistical technique and case-selection, and how the more fundamental problem of how exactly leftwing strength could be measured and operationalized was essentially ignored. In other words, while the methodology was debated, the indicators themselves were taken from granted. And when one goes back to that debate, and looks to see precisely how these crucial indicators were derived, then one is simply directed to an article from the early 1980s, in which the "the left" is "broadly defined to include Communist, Socialist, Social Democratic, and Labour parties, as well as several small parties that are to the left of centre on a Downsian ideological continuum," and in which leftist strength in government is indicated by the extent to which these parties control government, "as indicated by their control of portfolios in the cabinet," as well as by "the strength of governing leftist parties in parliament" (Cameron 1984: 159), while levels of trade union membership and the organizational unity of labor are based on data reported in the *Europa Yearbook* (Cameron 1984: 165).

Now, my point is not that these indicators are worthless; far from it—they might well be very solid, and could certainly have been the best that could be found at the time of the original study by Cameron. What must be emphasized, however, is that they are simply indicators; they are not, nor can they ever hope to be, the real thing. And hence if a long debate is to rage in a reputable journal concerning the very important substantive question of whether leftist strength can be associated with economic growth, surely one of the first questions that springs to mind should not be about statistical techniques, but should rather be about the accuracy and reliability of the indicators themselves. For if the indicators no longer offer the best indication of what is supposed to be the underlying reality, then no amount of statistical engineering will result in the cumulation of understanding. Is some notion of the "Downsian left of centre" the most appropriate dividing line to define left and right, or might some other measure not be tested? Is control over portfolios *per se* the best indicator of governmental influence, or might account not be taken of precisely which portfolios were involved? Might the level of membership in leftist trade unions not offer a more appropriate measure of leftwing strength than membership in trade unions *per se*, and did the *Europa Yearbook* really continue to remain the best source of hard, reliable, cross-national data for this crucial variable? In the end, of course, these indicators might well prove to have been the best possible indicators then available to the contributors to this busy debate; what is simply surprising is that nobody thought to check this out.

There are, of course, numerous other examples which might be cited in which potentially fallible or arbitrary indicators have been accorded an almost biblical status. The Castles–Mair (1984) data on the left–right placement of parties in a number of western democracies, for instance, are generally seen as quite authoritative, and continue to be frequently employed in studies which follow along similar lines to the work cited above. These data probably are authoritative; but it is also possible that they are not, and the picture which they draw, based on a relatively small number of expert opinions in one snapshot sample, should not perhaps be accorded the significance and weight which they normally receive, and should certainly not be automatically assumed to have a validity extending both long before, and long after, their actual application. The same might also be said of the various indicators which were initially developed by Arend Lijphart (1984) as a means of elaborating his influential distinction between majoritarian and consensus democracies, and which have subsequently been incorporated in a variety of different analyses; although these particular indicators may well offer one of the best means by which these two types of democracy might be distinguished, they are not necessarily the only option, and any application of Lijphart's indicators should certainly take account of the specific time period (1945–80) to which they apply, in that a different slice of time can lead to quite a different categorization of the cases (see, for example, Mair 1994). Robert Putnam's (1993) modern classic on Italian democracy is certainly far-reaching in both its argument and its implications, and has been highly praised for its capacity to link patterns in contemporary political culture to their early modern foundations; but even here, despite the intellectual breadth of the study, the key measure of institutional performance on which the analysis depends is based on just a small number of indicators, some of which derive from observations which were taken in the course of only one calendar year (Morlino 1995).

The real problem here, then, as is often the case in comparative political research more generally, is that the analysis of the relationship between variables is assumed to be more important than the quality and reliability of the variables themselves, a problem which has become even more acute as increased priority has been accorded to various institutional and political factors, and their operational indicators. It is also a very severe problem, for despite the evident increase in statistical and methodological sophistication of comparative political research in recent years, and despite the very obvious theoretical ambition, the actual data which are employed remain remarkably crude (see also Schmidt 1995). And since it is precisely this lack of solid comparable data which is encouraging the virtual fetishization of whatever indicators might be available, regardless of their potential fallibility, it must surely remain a priority for comparative political research to follow the advice laid down by Stein Rokkan on many different occasions, and to continue to stimulate the collection of systematically comparable data which can really "pin down numbers" (cited by Flora 1986: v–vi) on cross-national variation.

## V Conclusion: present and future trends

All studies in comparative politics share at least one attribute: a concern with countries, or macro-social units, as units of analysis or, at least, as units of observation (Ragin 1987; Keman 1993*a*). At the same time, comparative analysis will also often seek to arrive at generalizable propositions, which, in their most extreme form, would seek to explain phenomena whenever and wherever they occur. The inevitable result is a tension between an emphasis on country-specific factors, on the one hand, and universal relationships, on the other. But whereas the then new comparative politics of the 1950s and 1960s tended to place the emphasis on universal relationships, and thus global comparisons, the tendency within comparative research over the past decade or so has been to move away from general theory by emphasizing the relevance of context.

In part, this tendency reflects the renewed influence of historical inquiry in the social sciences, and especially the emergence of a "historical sociology" (Skocpol and Somers 1980; Abrams 1982) which tries to understand phenomena in the very broad or "holistic" context within which they occur (see also Thelen and Steinmo 1992, and Section II above). More general theories, by contrast, are seen to involve the artificial disaggregation of cases into collections of parts which can then be compared cross-nationally, and in which the original configuration of the aggregated "whole" is forgotten (see Ragin 1987: ix–x). Understanding the full picture as a whole and in depth is therefore seen to be preferable to a more general explanation of particular fragments of that picture. In part, however, this return to context is also the result of exhaustion and frustration. When the universe of comparative politics expanded in the late 1950s and 1960s, and when data on more and more countries became available to comparativists, there developed an inevitable tendency to compare as many cases as possible, and research tended to be driven by the elaboration of deductive models which could then be tested with as big an N as possible. Explanations were then enhanced through either an expansion of data sets, or through a refinement of the explanatory variables, or through a clearer specification of precisely what needed to be explained. Much of the development of coalition theories in the period from the 1960s to the 1980s, for example, can be seen in this way, with an ever more extensive range of countries being included as cases; with more variables being added to the models, such as policy, ideology, governing experience, and so on; and with more precise definitions of what actually constituted a "winning" coalition (see the reviews in Browne and Franklin 1986; Budge and Laver 1992). In a similar vein, much of the work which sought to assess the impact of "politics" on public policy outcomes (e.g. Castles 1982) developed by means of taking in as many cases as possible, and then by enhancing explanatory capacity through the constant refinement of the measures and definitions of "politics" (involving party ideology, party policy, institutional structures, structures of interest representation, and so on), on the one hand, and the measures and definitions of "outcomes" (levels of expenditures, policy styles, different policy sectors, and so on), on the other. In both fields of study, therefore, the goal remained one of explaining the relevant phenomenon in as general a manner as possible, while seeking to improve the capacity to explain by a constant modification of measurement tools.

Most recently, however, this strategy appears to have changed, not least because the capacity to enhance the amount of variance explained has more or less exhausted itself, with a further refinement of the various models now appearing to offer little in the way of explanatory gains. Coalition theorists, for example, now tend to place much more emphasis on inductive models (e.g. Pridham 1986), and are now much more concerned with understanding the broader national context within which each coalition game is played out, while those who are attempting to explain public policy outcomes are now tending to revert much more towards in-depth, case-sensitive, holistic studies. Francis Castles, for instance, who has pioneered much of the best comparative work in this latter area, has recently gone from developing broad, deductive models in which context played little or no role (Castles 1982), to more culturally specific studies in which distinct, but largely unquantifiable "traditions" (the English-speaking nations, or the Scandinavian nations) are accorded an important role (Castles and Merrill 1989; Castles 1993), as well as to more country-specific studies, in which the national context appears paramount (Castles 1989). The result has been a shying away from more generalized models and a renewed emphasis on the deeper understanding of particular cases or countries, where, often inductively, more qualitative and contextualized data can be assessed, and where account can be taken of specific institutional circumstances, or particular political cultures. Hence we see a new emphasis on more culturally specific studies (e.g., the English-speaking nations), and then nationally specific studies (e.g., the UK alone), and even institutionally specific studies (e.g., the UK under the Thatcher government). Hence also the recent and increasingly widespread appeal of the very disaggregated approaches which emerge within the "new institutionalism" (e.g., Tsebelis 1990; Ostrom 1991).

At the same time, however, it would be largely mistaken to read this recent shift as simply a return to the old emphasis on the study of individual countries which pre-dated the efforts of the 1954 Committee on Comparative Politics, in that there remains one major contrast between the earlier single-country approach and the present rediscovery of context, a contrast which has now begun to play a crucial role in the development of comparative political science as a whole. For whereas the earlier focus on single-country studies was developed at a time when political science itself was at a very early stage of development, and at a time when the centers of disciplinary excellence were concentrated in just a handful of departments in a small number of countries, the present concern with context has emerged following a massive expansion of the discipline in terms of both internationalization and professionalization (Daalder 1993). Formerly, for example, collections of national studies such as that represented by the pioneering Dahl "oppositions" volume (Dahl 1966) were quite exceptional, in that it was only rarely that scholars with expert knowledge on countries or cases could be found and brought together to discuss the application of similar hypotheses to their countries or cases. Nowadays, however, this sort of pooling of resources has become quite commonplace, and forms a core strategy within many cross-national (but usually regionally specific) research projects in a variety of different disciplines. This is particularly the case within comparative politics, where the development of common training methods and paradigms, together with the expansion of formalized international networks of scholars (such as the European Consortium for Political Research, ECPR) have insured that political science scholars, at least in the different regions, have now begun to speak what is essentially the same disciplinary language. As a result, it is now relatively easy, money permitting, to bring national experts together and then to cumulate their knowledge into a broad comparative understanding which is at the same time sensitive to the nuances of different contexts (see, for example, Pridham 1986; Budge *et al.* 1987; Castles 1989; Katz and Mair 1994; Laver and Shepsle 1994). And precisely because these local experts are being brought together, and then aggregated, as it were, it is proving possible, through the combination of in-depth and more generalized approaches, to build up plausible, convincing, and yet sufficiently nuanced comparative analyses. In other words, as a result of the international networks and cross-national collaboration which has been facilitated by the professionalization of political science as a whole, case-study analysis is now being adapted to generalizable theories and models, thus offering a strong potential for linkage between these two traditionally distinct approaches. This, then, is the current stage at which comparative political research finds itself: the bringing together of more case-sensitive, context-sensitive groups of studies which, through team effort, and through collaborative group effort, can genuinely advance comparative understanding, and can genuinely contribute to the development of comparative politics.[15] It is, to be sure, a form of comparison which is much more limited in scope than that envisaged by the Committee; perhaps paradoxically, however, and to return to Eckstein (1963: 22), it is also a mode of comparison which seems much better suited to "solving middle-range theoretical problems," even though, as suggested above, these problems are now more likely to concern the consequences of politics, rather than, as Eckstein saw it, its "determinants."

## References

Abrams, P. 1982. *Historical Sociology*. Ithaca: Cornell University Press.

Almond, G. A. 1970. *Political Development*. Boston: Little, Brown.

—— 1990. *A Discipline Divided*. London: Sage.

—— Cole, T.; and Macridis, R. C. 1955. A suggested research strategy in Western European politics and government. *American Political Science Review*, 49: 1042–9; reprinted in Eckstein and Apter 1963: 52–7.

[15] At the same time, however, this strategy also carries its own dangers. More specifically, the combined effect of both internationalization and professionalization now risks creating an essentially two-tier profession, in which there are the genuine comparativists, on the one hand, that is, those who initiate and design such cross-national projects, and who are then responsible for the cumulative interpretation; and the country experts, on the other hand, that is, those who, time after time, as participants in these projects, interpret their own country or case in the light of the frameworks set by the project initiators. This distinction, to be sure, need not be hard and fast, and those whose job it is to interpret country *X* for project *A* may later develop their own projects and recruit their own teams of experts. In practice, however, much depends on the research and training infrastructures within the different countries, such that those national political science professions in which there is a greater emphasis on the need for cross-national research, and those in which funding is available for such research, will tend to produce the project initiators; whereas those in which the focus is more nationally oriented will tend to produce the country experts. It is thus no accident that comparative politics, and even comparative European politics, is now disproportionately dominated by American scholars, who are the principal beneficiaries of the cross-nationally oriented American National Science Foundation.

ALMOND, G. A., and COLEMAN, J. S., eds. 1960. *The Politics of Developing Areas.* Princeton: Princeton, N.J.: University Press.

—— and POWELL, G. B. 1966. *Comparative Politics: A Developmental Approach.* Boston: Little, Brown.

—— and VERBA, S. 1965. *The Civic Culture* Boston: Little, Brown; originally published 1963.

ARISTOTLE. n.d. *The Politics,* trans. E. Barker. Oxford: Clarendon Press, 1946.

BARTOLINI, S. 1993. On time and comparative research. *Journal of Theoretical Politics,* 5: 131–67.

—— and MAIR, P. 1990. *Identity, Competition and Electoral Availability: The Stabilisation of European Electorates, 1885–1985.* Cambridge: Cambridge University Press.

BINDER, L.; COLEMAN, J; LAPALOMBARA, J; PYE, L.; VERBA, S.; and WEINER, M. 1971. *Crises and Sequences in Political Development.* Princeton: Princeton University Press.

BLONDEL, J. 1985. *Government Ministers in the Contemporary World.* London: Sage.

—— 1990. *Comparative Government.* Hemel Hempstead: Philip Allan.

BOGAARDS, M. 1994. 25 Jaar Pacificatiedemokratie: een Evaluatie. M.A. thesis, Department of Political Science, Leiden University.

BROWNE, E. C., and FRANKLIN, M. N. 1986. Editors' introduction: new directions in coalition research. *Legislative Studies Quarterly,* 11: 469–83.

BUDGE, I., and LAVER, M. J. 1992. Coalition theory, government policy and party policy. Pp. 1–14 in *Party Policy and Government Coalitions,* ed. Michael J. Laver and Ian Budge. Basingstoke: Macmillan.

—— ROBERTSON, D.; and HEARL, D., eds. 1987. *Ideology, Strategy, and Party Change.* Cambridge: Cambridge University Press.

CAMERON, D. R. 1984. Social democracy, corporatism, labour quiescence, and the representation of economic interests in advanced capitalist society. Pp. 143–78 in *Order and Conflict in Contemporary Capitalism,* ed. J. H. Goldthorpe. Oxford: Oxford University Press.

CASTLES, F. G., ed. 1982. *The Impact of Parties.* London: Sage.

—— ed. 1989. *The Comparative History of Public Policy.* Oxford: Polity Press.

—— 1993. On religion and public policy: does Catholicism make a difference? *European Journal of Political Research,* 25: 19–40.

—— and MAIR, P. 1984. Left–right political scales: some "expert" judgments. *European Journal of Political Research,* 12: 73–88.

—— and MERRILL, V. 1989. Towards a general model of public policy outcomes. *Journal of Theoretical Politics,* 1: 177–212.

CHILCOTE, R. H. 1994. *Theories of Comparative Politics.* 2nd edn. Boulder, Colo.: Westview.

COLEMAN, J., ed. 1965. *Education and Political Development.* Princeton, N.J.: Princeton University Press.

COLLIER, D. 1991. New perspectives on the comparative method. In Rustow and Ericksen 1991: 7–31.

DAALDER, H. 1993. The development of the study of comparative politics. In Keman 1993*b*: 11–30.

DAHL, R. A., ed. 1966. *Political Oppositions in Western Democracies.* New Haven, Conn.: Yale University Press.

DALTON, R. J. 1991. Comparative politics of the industrial democracies: from the golden age to island hopping. Vol. ii, pp. 15–43 in *Political Science,* ed. William J. Crotty. Evanston, Ill.: Northwestern University Press.

DIAMOND, L., and PLATTNER, M. F., eds. 1993. *The Global Resurgence of Democracy.* Baltimore, Md.: Johns Hopkins University Press.

Di Palma, G. 1990. *To Craft Democracies.* Berkeley: University of California Press.
Easton, D. 1968. Political science. Vol. xii, pp. 282–98 in *International Encyclopedia of the Social Sciences*, ed. D. L. Sills. London: Macmillan.
Eckstein, H. 1963. A perspective on comparative politics, past and present. In Eckstein and Apter 1963: 3–32.
—— and Apter, D. E., eds. 1963. *Comparative Politics: A Reader.* New York: Free Press.
Evans, P. B.; Rueschemeyer, D.; and Skocpol, T., eds. 1985. *Bringing the State Back In.* New York: Cambridge University Press.
Fabbrini, S. 1988. Return to the state: critiques. *American Political Science Review*, 82: 891–9.
Finer, S. E. 1970. Almond's concept of the political system. *Government and Opposition*, 5: 3–21.
Finifter, A. W. 1993. Preface. Pp. vii–x in *Political Science: The State of the Discipline II*, ed. A. W. Finifter. Washington D.C.: American Political Science Association.
Flora, P. 1986. Preface. Vol. i, pp. v–viii in *Growth to Limits: The Western European Welfare States Since World War II*, ed. P. Flora. Berlin: de Gruyter.
Hague, R.; Harrop, M.; and Breslin, S. 1992. *Comparative Government and Politics.* 3rd edn. Basingstoke: Macmillan.
Hall, P. 1986. *Governing the Economy: The Politics of State Intervention in Britain and France.* New York: Oxford University Press.
Karl, T. L. 1991. Dilemmas of democratization in Latin America. In Rustow and Ericksen 1991: 163–91.
Katz, R. S., and Mair, P., eds. 1994. *How Parties Organize.* London: Sage.
Katzenstein, P. 1984. *Small States in World Markets.* Ithaca, N.Y.: Cornell University Press.
Keman, H. 1993*a*. Comparative politics: a distinctive approach to political science? In Keman 1993*b*: 31–57.
—— ed. 1993*b* *Comparative Politics.* Amsterdam: Free University Press.
LaPalombara, J., ed. 1963. *Bureaucracies and Political Development.* Princeton, N.J.: Princeton University Press.
—— and Weiner, M., eds. 1966. *Political Parties and Political Development.* Princeton, N.J.: Princeton University Press.
Laver, M., and Shepsle, K. A., eds. 1994. *Cabinet Ministers and Parliamentary Government.* Cambridge: Cambridge University Press.
Lijphart, A. 1968. Typologies of democratic systems. *Comparative Political Studies*, 1: 3–44.
—— 1971. Comparative politics and the comparative method. *American Political Science Review*, 65: 682–93.
—— 1977. *Democracy in Plural Societies.* New Haven, Conn.: Yale University Press.
—— 1984. *Democracies.* New Haven, Conn.: Yale University Press.
—— 1994*a*. Democracies: forms, performance, and constitutional engineering. *European Journal of Political Research*, 25: 1–18.
—— 1994*b*. *Electoral Systems and Party Systems.* Oxford: Oxford University Press.
Linz, J. J., and Valenzuela, A., eds. 1994. *The Failure of Presidential Democracy* (Vol. i, *Comparative Perspectives*). Baltimore, Md.: Johns Hopkins University Press.
Lipset, S. M. 1959. Some social requisites of democracy: economic development and political legitimacy. *American Political Science Review*, 53: 69–105.
—— and Rokkan, S. 1967. Cleavage structures, party systems, and voter alignments: an introduction. Pp. 1–64 in *Party Systems and Voter Alignments*, ed. S. M. Lipset and S. Rokkan. New York: Free Press.

MACRIDIS, R. C., and BROWN, B. E. 1986. Comparative analysis: method and concepts. Pp. 1–22 in *Comparative Politics*, ed. R. C. Macridis and B. E. Brown. 6th edn. Chicago: Dorsey Press.

MAIR, P. 1994. The correlates of consensus democracy and the puzzle of Dutch politics. *West European Politics*, 17: 97–123.

—— 1995. Landenvergelijkend Onderzoek. Pp. 213–38 in *Leren van Onderzoek: Het Onderzoeksproces en Methodologische Problemen in de Sociale Wetenschappen*, ed. W. Hout and H. Pellikaan. Amsterdam: Boom.

MARCH, J. G., and OLSEN, J. P. 1984. The new institutionalism: organizational factors in political life. *American Political Science Review*, 78: 734–49.

MITCHELL, T. 1991. The limits of the state: beyond statist approaches and their critics. *American Political Science Review*, 85: 77–96.

MORLINO, L. 1995. Italy's civic divide. *Journal of Democracy*, 6: 173–7.

OSTROM, E. 1991. Rational choice theory and institutional analysis: towards complementarity. *American Political Science Review*, 85: 237–43.

PRIDHAM, G. 1986. An inductive theoretical framework for coalitional behaviour. Pp. 1–31 in *Coalitional Behaviour in Theory and Practice*, ed. G. Pridham. Cambridge: Cambridge University Press.

PRZEWORSKI, A. 1991. *Democracy and the Market*. Cambridge: Cambridge University Press.

—— and TEUNE, H. 1970. *The Logic of Comparative Social Inquiry*. New York: Wiley.

PYE, L., ed. 1962. *Communications and Political Development*. Princeton, N.J.: Princeton University Press.

—— and VERBA, S., eds. 1965. *Political Culture and Political Development*. Princeton, N.J.: Princeton University Press.

PUTNAM, R. D.; with LEONARDI, R. and NANETTI, R. Y. 1993. *Making Democracy Work*. Princeton, N.J.: Princeton University Press.

RAGIN, C. 1987. *The Comparative Method*. Berkeley: University of California Press.

—— ed. 1991. Special issue: issues and alternatives in comparative social research. *International Journal of Comparative Sociology*, 32/1–2.

RHODES, R. A. W. 1994. State-building without a bureaucracy: the case of the United Kingdom. Pp. 165–88 in *Developing Democracy*, ed. I. Budge and D. McKay. London: Sage.

ROGOWSKI, R. 1993. Comparative politics. Pp. 431–50 in *Political Science: The State of the Discipline II*, ed. A. W. Finifter. Washington D.C.: American Political Science Association.

RUSTOW, D. A. 1957. New horizons for comparative politics. *World Politics*, 9: 530–49; reprinted in Eckstein and Apter 1963: 57–66.

—— 1970. Transitions to democracy: towards a dynamic model. *Comparative Politics*, 2: 337–63.

—— and ERICKSEN, K. P., eds. 1991. *Comparative Political Dynamics*. New York: HarperCollins.

SARTORI, G. 1969. From the sociology of politics to political sociology. Pp. 65–100 in *Politics and the Social Sciences*, ed. S. M. Lipset. New York: Oxford University Press.

—— 1970. Concept misformation in comparative politics. *American Political Science Review*, 64: 1033–53.

—— and MORLINO, L., eds. 1991. *La Comparazione nelle Scienze Sociale*. Bologna: Il Mulino.

SCHARPF, F. 1988. *Crisis and Choice in European Social Democracy*. Ithaca, N.Y.: Cornell University Press.

SCHMIDT , M. G. 1995. The party-does-matter hypothesis and constitutional structures of

majoritarian and consensus democracy. Paper presented to the International Colloquium on Cross-National Studies of Public Policy in the 1980s and 1990s, University of Heidelberg, 19–20 June.
Schmitter, P. C. 1993. Comparative politics. Pp. 171–7 in *The Oxford Companion to Politics of the World*, J. Krieger. New York: Oxford University Press.
Shepsle, K. A., and Weingast, B. R. 1987. The institutional foundations of committee power. *American Political Science Review*, 81: 85–104.
Skocpol, T. 1979. *States and Social Revolutions*. Cambridge: Cambridge University Press.
—— 1985. Bringing the state back in: strategies of analysis in current research. In Evans *et al.* 1985: 1–45.
—— and Somers, M. 1980. The use of comparative history in macro-social inquiry. *Comparative Studies in Society and History*, 22: 174–97.
Smelser, N. 1976. *Comparative Methods in the Social Sciences*. Englewood Cliffs, N.J.: Prentice-Hall.
Strange, S. 1995. The defective state. *Daedalus* 124/2: 55–74.
Thelen, K., and Steinmo, S. 1992. Historical institutionalism in comparative politics. Pp. 1–32 in *Structuring Politics: Historical Institutionalism in Comparative Perspective*, ed. S. Steinmo, K. Thelen and F. Longstreth. Cambridge: Cambridge University Press.
Tsebelis, G. 1990. *Nested Games*. Berkeley: University of California Press.
Verba, S. 1985. Comparative politics: where have we been, where are we going? Pp. 26–38 in *New Directions in Comparative Politics*, ed. H. J. Wiarda. Boulder, Colo.: Westview Press.
Weaver, R. K., and Rockman, B. A. 1993. Assessing the effects of institutions. Pp. 1–41 in *Do Institutions Matter?*, ed. R. K. Weaver and B. A. Rockman. Washington D.C.: Brookings Institution.

Chapter 13

# Comparative Politics: Micro-behavioral Perspectives

Russell J. Dalton

DURING this century there have been three waves of democratic expansion that have included dramatic periods of theoretical and empirical development in the social sciences. The first occurred after the turn of the century, when Woodrow Wilson, Harold Gosnell, Walter Lippmann, and others re-examined the nature of politics in modern mass democracies. The second period followed the Second World War. It attempted to identify the requisites for stable and successful democracy and the factors that undermined the democratic process in interwar Europe. This period included scholars such as Barrington Moore, Hannah Arendt, Gabriel Almond, Raymond Aron, and Seymour Martin Lipset.

We are now living through a third period of democratic ferment that is producing a dramatic surge of academic research on the theme of democratization and the nature and conditions of democratic politics. The political systems of Central and Eastern Europe were transformed in an amazing process of regime change. Popular pressures moved ahead the democratization process in East Asia, ranging from the people power movement in the Philippines to the democratic reforms in South Korea and Taiwan. A wave of democratic elections has swept across Sub-Saharan Africa in the first half of the 1990s. These democratic transitions have created new freedoms for these publics, and new theoretical and political questions for social scientists. For the first time we are witnessing a transition from communism to democracy, and the nature and destination of this transition process is unclear. Similarly, the expansion of democracy to societies rooted in Confucian traditions raises questions about the cultural bases of democracy in these societies.

As these democratic transitions are occurring, new challenges to the democratic process have arisen within established democracies such as the United States and Western Europe. Most advanced industrialized nations face similar problems in dealing with a changing economic structure, new forces of cultural change, and a new relationship between the citizenry and the government. Cultural diversity and the pressures of ethnic fragmentation are now common problems for European states in the East and West. The political demands presented by environmentalists, the women's movement and other citizen groups are affecting virtually all advanced industrial societies. New and expanded patterns of political participation are a common phenomenon in most established democracies. Equally common are questions about the changing nature of electoral behavior and electoral choice in advanced industrial democracies. Everywhere, it seems, new questions about the nature of democracy are developing.

It is too soon to tell whether this period of political change will yield the type of theoretical and empirical advances in the social sciences that accompanied the two previous periods. Certainly, our scientific tools are more sophisticated than in earlier periods, and our knowledge about societies and politics is much greater. These new events provide distinctive opportunities to test our theories, expand the boundaries of knowledge, and to develop new theories. We normally observe political systems in a general state of equilibrium, when stability and incremental change dominate our findings. Now we have opportunities to examine questions of more fundamental change and adaptation that often go to the heart of our theoretical interests, but which we can seldom observe directly.

The democratization process itself is the subject of another chapter in the *New Handbook* (chap. 14). The present chapter reviews some of the recent major research advances in comparative political behavior. It is not possible to provide a comprehensive review of the field in a few pages (see Dalton and Wattenberg 1993; Klingemann and Fuchs 1995). Instead, we focus on a few major areas of research.[1] These areas were chosen for two reasons. First, they represent areas that I considered have made significant scientific advances in recent years. Second, although these examples are largely drawn from research on advanced industrial societies, these areas also have applicability to the process of democratic transition for emerging democracies. These are areas where our present knowledge can be expanded in the context of this global wave of democratization.

[1] The one additional area that should be included in this essay, if space permitted, is political participation. This area is important both because of the significant research advances that have been made in the past decade and because of the potential relevance of this area to the topics of developing democratic experience and expanding the institutional bounds of democratic politics (for reviews of this research see Dalton 1996; Rosenstone and Hansen 1993; Parry, Moyser and Day 1992; Verba *et al.* 1995).

## I Political culture and democratization

One of the most powerful social science concepts to emerge from the previous wave of democratization studies was the concept of political culture. Gabriel Almond and Sidney Verba's seminal 1963 study, *The Civic Culture*, contended that the institutions and patterns of action in a political system must be *congruent* with the political culture of the nation. Culturalist studies have been especially important in the study of democratization, as analysts tried to identify the cultural requisites of democracy (Almond and Verba 1963; 1980; Verba 1965; Baker, Dalton and Hildebrandt 1981).

Three kinds of culturalist studies have been most visible in the democracy literature. The first is the "civic culture" theory of Almond and Verba. Drawing upon evidence from five democratic societies, they held that a nation's political culture exerted an independent influence on social and political behavior. Culture sets norms for behavior that members of society acknowledge and generally follow, even if they personally do not share these values. This is by far the most influential approach; work done along its lines is voluminous and is not limited to democratic systems (see the reviews in Almond and Verba 1980). The second approach is the "authority-culture" theory (Eckstein 1966). Eckstein's work is especially relevant to present concerns because he is one of the few cultural theorists to discuss the dynamic aspects of culture and how culture can play a role in processes of political change (Eckstein 1988; 1990). Aaron Wildavsky developed a third distinct version of political culture analysis (Wildavsky 1987). Wildavsky drew upon Mary Douglas's grid-group approach to develop a typology of cultures based on four distinct life styles. These types were based on social relations and the values they exemplified.

Despite the heuristic and interpretive power of the concept of political culture, other scholars raised questions about the precision and predictive power of the concept. Max Kaase (1983) penned the saying that measuring political culture is like trying to nail jello to the wall. That is, the concept lacked precision and often became a subjective, stereotypic description of a nation rather than an empirically measurable concept. Some analysts saw political culture in virtually every feature of political life, others viewed culture as a residual category that explained what remained unexplainable by other means. Even more problematic was the uneven evidence of culture's causal effect.[2] Political culture studies often were based on a public opinion

[2] Another criticism questioned whether culture was a cause or effect of institutional arrangements (Barry 1970). I consider this a somewhat artificial distinction. Although the thrust of cultural theory emphasized its influence over institutional arrangements, the clear intent of Almond and Verba

survey of a single nation. In such a research design it was difficult to isolate the role of culture in influencing national patterns of political behavior.

Even before the current wave of democratic transitions, political culture studies had been enjoying a revival of academic interest. Drawing upon the 1981 World Values Study, Ronald Inglehart (1990: chap. 1) provided new evidence on the congruence between broad political attitudes and the democratic stability for a set of twenty-two nations.[3] Robert Putnam's (1993; cf. 1973) research on the development of regional governments in Italy provided even more impressive testimony in support of cultural theory. Putnam compared the performance of regional governments in Italy with an imaginative array of measures. He found that the cultural traditions of a region—roughly contrasting the co-operative political style of the North to the more hierarchic tradition of the South—was the most potent predictor of the performance of their respective governments. Even more telling, Putnam demonstrated that these cultural traditions show deep historical roots in earlier patterns of civic association. Putnam's very creative and systematic study of cultural influences has given rise to a general renaissance in cultural studies.

The recent global democratization wave renews the importance of questions about the congruence between culture and political system, and raises a new set of research questions for political culture research. Normally, political institutions and the basic principles of a regime are constant, thus it is difficult to study the interaction between institutional and cultural change. However, the recent shifts in regime form in a vast array of nations create new opportunities to study the congruence between cultural and institutional choices. To what extent did political change in Eastern Europe arise from the public's dissatisfaction with the old regimes; to what extent can the prospects for democracy in this region be judged by the conduciveness of their cultures to democratic politics? For instance, we can examine how citizens evaluate different political systems based on real experience, thus testing the link between political norms and institutional choices in a way that is not realistic in a single political context. More generally, current events revive past debates on the continuity of culture and the ways in which cultural norms can be transformed (Almond and Verba 1980). Eckstein's recent research on cultural change also suggests that political culture should be studied as an extension of other patterns of social relations and the general levels of "civic inclusion" existing in a society

was to draw attention to cultural patterns so that governments and élites could respond to these inheritances and in some cases remake the culture (Verba 1965).

[3] Inglehart's findings were subsequently criticized by Muller and Seligson (1994) on methodological grounds, and Inglehart has responded to these criticisms with new data from the 1990–91 World Values Study (Inglehart forthcoming).

(Eckstein 1988; 1990). The depth and breadth of cultural norms compatible with democracy may be an important factor in explaining the course of the political transitions now occurring worldwide.

Almost as soon as the Berlin Wall fell, survey researchers were moving eastward. We are quickly assembling a wealth of initial findings on the political attitudes of Russians and East Europeans, and this includes many studies of political culture. For instance, several groups of researchers have found surprisingly high levels of support for basic democratic principles in the former Soviet Union (Miller *et al.* 1993; Gibson, Duch and Tedin 1992; Finifter and Mickiewicz 1992). Furthermore, research from other Eastern European nations paints a roughly similar picture of broad public approval of democratic norms and procedures (McIntosh and MacIver 1992; Dalton 1994*b*; Weil 1993). Although one must worry about the depth of these responses, whether they reflect enduring cultural norms or the temporary response to a traumatic set of political events, the publics in most post-Soviet states began their experience with democracy by espousing greater support for democratic principles than was expected. Rather than the apathy or hostility that greeted democracy after transitions from right-wing authoritarian states, the cultural legacy of communism appears to be much different.

An equally rich series of studies are emerging for East Asia. Doh Shin and his colleagues are assembling an impressive mass of survey evidence on the development of democratic attitudes in South Korea (Shin, Chey and Kim 1989; Shin and Chey 1993). Despite the government's hesitant support for democracy, the cultural foundations of democracy are more extensively developed. Similar research has developed in Taiwan, where the transition to democracy has been accompanied by supportive attitudes among the public (Chu 1992). Perhaps the most exciting evidence comes from studies of the People's Republic of China. Even in this hostile environment, Andrew Nathan and Tianjian Shi (1993) find that the pre-Tiananmen Chinese public espoused surprising support for an array of democratic principles. One might question whether these opinions are sufficiently ingrained in these various publics to constitute enduring features of the political culture, but even these endorsements of democracy are a positive sign about the prospects for democracy.

In summary, social science has made great progress in the last ten years in developing the empirical evidence supporting the congruence thesis underlying the political cultural model, and in collecting new evidence of citizen beliefs in the emerging democracies of Europe and East Asia. Yet, these empirical successes have not yet been balanced by the type of theoretical innovation and creativity that marked the two previous democratization waves. Political scientists should do more than just collect new data on old questions of survey research—though replication is an important and valuable element of science. To move the field ahead, now is the time to ask additional questions. For instance, is there is but one "civic culture" that is congruent with the working of a democratic system. Experience would suggest that there are a variety of "democratic" cultures, as well as ways to define culture, which require mapping and further study (Flanagan 1978; Seligson and Booth 1993; Almond and Verba 1980). Equally important, our conceptualization of the elements of a political culture, and their interrelationships, has made relatively little progress since the *Civic Culture* study.

More generally, much of the new wave of empirical research in democratizing nations does not expand the theoretical bounds of political culture research. Tests of the contrasting models of Almond and Verba, Eckstein, and Wildavsky should find a rich medium in these new political experiences, and one should expect that new theoretical frameworks will emerge from these studies. Equally important, because the world is in flux, we now have the ability to examine cultural theory as a predictive tool. We can examine how the congruence between culture and institutions develops, because a host of nations are now in the process of political transition. Attempts to test theories of cultural change, or theories on the nonpolitical origins of political culture, are fertile fields for research during this unusual period of political change. There are a host of other questions that involve the creation of cultural norms and political identities, and the overlap between personal preferences and perceived social norms. One can see bits of progress here and there, but not the frontal assault on our theorizing about the world that came from earlier democratization waves.

It may be that the current pattern of research represents the achievement of a mature science; with well-developed instruments and research questions, further research becomes incremental rather than the creative theoretical work of earlier democratic waves. Still, I see the potential for theoretical and methodological creativity as the (so-far) missed opportunity of this democratization wave.

## II Value change and modernization

Another area where comparative political behavior has made major strides involves changes in public values. Early behavioral research examined the relationship between the development of an industrial society and the changing values of the public (e.g., Inkeles and Smith 1974). In the last two decades this research has been extended to the further processes of value change that accompany the development of advanced industrial, or post-industrial, society.

Ronald Inglehart's thesis of post-material value change (Inglehart 1977; 1990; Abramson and Inglehart 1995) has furnished the most important framework for studying the changes affecting mass publics in advanced industrial democracies. Inglehart's explanation of value change is based on two premises. First, he suggests that the public's basic value priorities are determined by a scarcity hypothesis: individuals place the greatest value on things that are in relatively short supply.[4] The second part of Inglehart's theory is a socialization hypothesis: individual value priorities reflect the conditions that prevailed during one's pre-adult years. The combination of both hypotheses produces a general model of value formation: an individual's basic value priorities are formed early in life in reaction to the socioeconomic conditions (personal and societal) of this period, and once formed, these values tend to endure in the face of later changes in life conditions.

Inglehart uses this model to argue that the socioeconomic forces transforming Western industrial societies are changing the relative scarcity of valued goals, and consequently the value priorities of Western publics. Older generations are still more likely to emphasize traditional "material" social goals, such as economic well-being, social security, law and order, religious values, and a strong national defense. Having grown up in an environment where these traditional goals seem relatively assured, younger generations of Westerners are shifting their attention toward "post-material" goals of self-expression, personal freedom, social equality, self-fulfillment, and maintaining the quality of life.

What is most significant about Inglehart's post-material thesis is its broad relevance to the study of advanced industrial societies. His concept of value change was immediately useful in explaining generational differences in public attitudes toward the Common Market. In addition, the underlying dimension of material vs. post-material values is related to the public's growing interest in environmental issues, women's rights, consumer protection and other quality of life issues.

The emergence of post-material values can be linked to democratization themes in two ways. First, post-material value orientations have partially redefined the nature of politics in advanced industrial societies. These

[4] In his earlier work, Inglehart attempted to generalize the scarcity hypothesis into a broader theoretical framework of Abraham Maslow's hierarchy of human values; the Maslovian framework has become less prominent in Inglehart's more recent writings.

interests have led to the formation of new citizen movements that have become active and vocal participants in the democratic process. The environmental movement and women's groups, for example, have pressed for their alternative political agenda (Dalton 1994*a*; Gelb 1989). Often these issues have placed them in conflict with established economic interests, such as business lobbies and labor unions. These new social movements also have been joined by New Left or Green parties that advocate their positions within the electoral and parliamentary arenas (Mueller-Rommel 1989). In short, these orientations have contributed to many of the political controversies that now divide the public and political groups in advanced industrial democracies.

In addition to alternative social goals, post-materialists pressed for changes in the style of democratic politics. Post-materialism contributed to the changing action repertoires of Western publics (Barnes and Kaase 1979; Jennings and van Deth 1990). Post-materialists are more likely to use unconventional forms of political action, such as protests and other forms of élite-challenging behavior. Similarly, citizen groups and Green parties called for an expansion of the democratic process to allow greater involvement of the public in policy-making and policy administration. Post-materialists favor citizen advisory groups, referendums, and other forms of direct democracy over the style of limited representative democracy practiced heretofore. In short, post-materialism has brought the principles of democracy into question—the so-called "Crisis of Democracy" literature (Klingemann and Fuchs 1995)—by creating a debate between representative democracy and participatory democracy.

The second implication of the post-material thesis is to highlight the contrasting values of those nations now undergoing democratic transition in Eastern Europe or East Asia. There have been attempts to expand the post-material concept to these societies (Inglehart and Siemenska 1990; Gibson and Duch 1994), but this seems a questionable research hypothesis. Inglehart formulated post-materialism as the consequence of the economic and political developments that these societies are now just beginning. We should therefore expect that the publics and élites in Eastern Europe and East Asia will place greater stress on the material goals that once dominated the politics of Western democracies. For instance, while the Dutch may be striving to become post-materialists, the Poles are hoping to achieve the materialist excesses to which the Dutch have become accustomed. In addition, the democratizing nations should place greater weight on developing institutionalized forms of representative democracy, and may want to avoid the participatory democracy advocated in advanced industrial societies. This suggests a significant divergence in the immediate goals of advanced industrial democracies and the emerging democracies.[5] Both sets of nations are becoming more democratic, but with different definitions of what democracy means and how it should function.

## III Electoral change

Elections are the central procedures of representation in modern democracies, and the past generation of research has yielded dramatic advances in our knowledge about how voters reach their decisions. One of the major themes in contemporary electoral research involves changes in the relative weight of the factors influencing voting decisions. Political choice in most Western democracies was traditionally structured by class, religious and other social divisions. Because individuals were often ill-prepared to deal with the complexities of politics, they relied on the political cues of external reference groups in reaching political decisions. In addition, social institutions such as the unions and churches were major political actors, influencing both political élites and their membership. Seymour Lipset and Stein Rokkan summarized this position in their famous conclusion: "the party systems of the 1960s reflect, with but few significant exceptions, the cleavage structures of the 1920s" (1967: 50). Early electoral research largely substantiated Lipset and Rokkan's claims.

As this theme of stable, cleavage-based voting became the conventional wisdom, dramatic changes began to affect these same party systems. The established parties were presented with new demands and new challenges, and the evidence of partisan change became obvious. Within a decade the dominant question changed from explaining the persistence of electoral politics to explaining electoral change (Dalton *et al.* 1984; Crewe and Denver 1985).

The growing emphasis on electoral change was linked to mounting evidence that the class and religious divisions were decreasing in influence. For instance, in the second edition of *Political Man*, Lipset demonstrated a decline in the level of class voting for several Western democracies (Lipset 1981: appendix). Collaborating research came from Australia (McAllister 1992), Britain (Franklin 1985), Germany (Baker, Dalton and Hildebrandt

[5] In his most recent work, Inglehart (forthcoming; Abramson and Inglehart 1995) argues for a general process of global value convergence. To an extent I agree, if one defines convergence in terms of support for liberal and democratic values as expressed in Western political traditions. However, the post-material model maintains that advanced industrial societies are evolving into a new form of social and political organization. This gap between advanced industrial democracies and emerging democracies is the divergence emphasized here.

1981), Japan (Watanuki 1991) and other advanced industrial democracies (Inglehart 1990; Lane and Ersson 1991). Thus one of the major findings from the last decade of electoral research holds that social positions no longer determine political positions as they did when social alignments were solidly frozen.[6]

Mark Franklin and his colleagues have compiled the most comprehensive evidence supporting this conclusion (Franklin *et al.* 1992). They tracked the ability of a set of social characteristics (including social class, education, income, religiosity, region, and gender) to explain partisan preferences. Across fourteen Western democracies, they found a consistent erosion in the voting impact of social structure. The rate and timing of this decline varied across nations, but the end-product was the same. They conclude with the new "conventional wisdom" of comparative electoral research: "One thing that has by now become quite apparent is that almost all of the countries we have studied show a decline . . . in the ability of social cleavages to structure individual voting choice" (Franklin *et al.* 1992: 385).

In many Western democracies the erosion of group cleavages as an influence on electoral choice was parallelled by a decline in the ability of partisan attachments (or partisan identifications) to explain political behavior. The strength of party attachment has weakened in several Western democracies over the past generation (see review in Dalton 1996). Similarly, there has been a decrease in party-line voting and the increase in partisan volatility, split-ticket voting, and other phenomena indicating that citizens are no longer voting according to a party line. Perot's strong showing in the 1992 American presidential election, the collapse of the Japanese party system, or Berlusconi's breakthrough in Italian politics provide graphic illustrations of how weakened party ties open up the potential for substantial electoral volatility.

The decline of long-term predispositions based on social position or partisanship should shift the basis of electoral behavior to short-term factors, such as candidate image and issue opinions. There is evidence that the new electoral order includes a shift toward candidate-centered politics. Martin Wattenberg (1991) has documented the growing importance of candidate image in Americans' electoral choices, and comparable data are available for other Western democracies (Bean and Mugham 1989). Furthermore, there are signs of a growing personalization of political campaigns in Western democracies: photo opportunities, personalized

[6] Of course, in the social sciences nothing goes undisputed. A variety of British and American scholars have questioned the evidence on the decline of class voting (e.g., Heath *et al.* 1991). I do not find their evidence convincing.

interviews, walkabouts, and even televised candidate debates are becoming standard electoral fare (Kaase 1994).

The decline in long-term influences on the vote has also increased the potential for issue voting. Mark Franklin (1985) showed that the decreasing influence of long-term forces on British voting decisions was counterbalanced by an increased impact of issues on the vote (also Baker *et al.* 1981: chap. 10; Eijk and Niemoeller 1983; Budge and Farlie 1983; Rose and McAllister 1986). Oddbjorn Knutsen (1987) and others linked the rise of these cross-cutting issue interests to the erosion of previous social cleavages.[7] In reviewing the evidence from their comparative study of voting behavior, Mark Franklin (1992: 400) supports this point, concluding: "if all the issues of importance to voters had been measured and given their due weight, then the rise of issue voting would have compensated more or less precisely for the decline in cleavage politics."

These changes in the sources of electoral choice, as well as related changes in patterns of political participation and the individual's relationship to the political system, lead to what we would call the "individualization of politics." This has involved a shift away from a style of electoral decision-making based on social group and/or party cues toward a more individualized and inwardly oriented style of political choice. Instead of depending upon party élites and reference groups, more citizens now try to deal with the complexities of politics and make their own political decisions. What is developing is an eclectic and egocentric pattern of citizen decision-making. Rather than socially structured and relatively homogeneous personal networks, contemporary publics are more likely to base their decisions on policy preferences, performance judgments, or candidate images.

The relationship between the individual and the media both contributes to these trends, and reinforces them (Semetko *et al.* 1991). The contemporary media provide voters with a greater variety of information sources, and potentially a more critical perspective of established political actors such as parties, labor unions, and industries. Access to a diverse media environment enables the public to become active *selectors* of information rather than passive *consumers* of political cues provided by others. In addition, the ability to see candidates and parliamentary leaders on television

[7] There is considerable debate on the content of this new issue voting. Some issues represent the continuation of past social conflicts, now without a group base. Other issues tap the new political controversies of advanced industrial societies. Yet another approach argues that such "position issues" have been overtaken by a new emphasis on "valence issues" that assess the performance of government on broadly accepted goals, such as judging parties on their ability to guide the economy or foreign policy. Thus the growth of issue voting has created new questions on what issues are important.

has caused more attention to be paid to the personal attributes of politicians, such as competence and integrity. The expansion of the 1992 American presidential campaign into new media forums illustrates this point, and similar developments can be observed in other Western democracies, albeit in more modest form, as new communications technologies change the patterns of information flow.

The individualization of politics also displays itself in the increasing heterogeneity of the public's issue interests. The post-material issues of environmentalism, women's rights, and life styles choices have been added to the already full agenda of advanced industrial democracies. In addition, schema theory argues that citizens are becoming fragmented into a variety of distinct *issue publics.* Rather than politics being structured by a group benefits framework, which often reflected socially derived cues, citizens now tend to focus on specific issues of immediate or personal importance.

For advanced industrial democracies, these developments have an uncertain potential for the nature of the democratic electoral process (Dalton 1996; Klingemann and Fuchs 1995). These changes can either improve or weaken the "quality" of the democratic process and the representation of the public's political interests. The nature of contemporary political beliefs means that public opinion is simultaneously becoming more fluid and less predictable. This uncertainty forces parties and candidates to become more sensitive to public opinion, at least the opinions of those who vote. Motivated issue voters are more likely to at least have their voices heard, even if they are not accepted. Furthermore, the ability of politicians to have unmediated communications with voters can strengthen the link between politicians and the people. To some extent, the individualization of electoral choice revives earlier images of the informed independent voter that we once found in classic democratic theory (Popkin 1991).

At the same time, there is a potential dark side to these new forces in electoral politics. The rise of single-issue politics handicaps a society's ability to deal with political issues that transcend specific interests, such as the U.S. budget deficit. Élites who cater to issue publics can leave the electorally inactive disenfranchised. Too great an interest in a single issue, or too much emphasis on recent performance, can produce a narrow definition of rationality that is as harmful to democracy as "frozen" social cleavages. In addition, direct unmediated contact between politicians and citizens opens the potential for demagoguery and political extremism. Both extreme right-wing and left-wing political movements probably benefit from this new political environment, at least in the short term.

For the emerging democracies, there is an apparent similarity to the portrait of voting choice we have just described. Emerging party systems are unlikely to be based on the stable-group based cleavages, especially when the democratic transition has occurred quite rapidly, as in Eastern Europe.[8] Similarly, new electorates are also unlikely to hold long-term party attachments that might guide their behavior. Thus, the patterns of electoral choice in many new democracies may be based on the same short-term factors—candidate images and issue positions—that have recently gained prominence in the electoral politics of advanced industrial democracies.

These similarities are only superficial, however. They do not reach below the surface of the electoral process. Advanced industrial democracies are experiencing an evolution in the patterns of electoral choice that flow from the breakdown of long-standing alignments and party attachments, the development of a more sophisticated electorate, and efforts to move beyond the restrictions of representative democracy. The new electoral forces in Western democracies are also developing within an electoral setting in which traditional group-based and partisan cues still exert a significant, albeit diminishing, influence.

The democratic party systems of Eastern Europe and East Asia face the task of developing the basic structure of electoral choice—the political frameworks that Lipset and Rokkan examined historically for the West. This presents the unique opportunity to study this process scientifically: to examine how new party attachments take root, the relationships between social groups and parties are formed, party images are created, these images are transmitted to new voters, and citizens learn the process of electoral choice and representative democracy. The venerable Lipset–Rokkan framework may provide a valuable starting point for this research, and the Michigan model of party identification may provide a framework for studying how new political identities may form. However, the creation of party systems in the world of global television, greater knowledge about electoral politics (from the élite and public levels), and fundamentally different electorates is unlikely to follow the pattern of Western Europe in the 1920s.

To answer these questions will require a dynamic perspective on these processes of partisan and electoral change. It is frankly too soon to determine how political scientists will respond to these challenges. There has already been an impressive development of the empirical base of research in these new democracies—a development that took decades in some

[8] The exception may be the party systems of Latin America and East Asia (Taiwan and South Korea) which might be able to integrate existing social cleavages because of the different nature of these democratic transitions (e.g., Remmer 1991; Chu 1992).

Western democracies. There are many encouraging signs and impressive empirical studies emanating from Eastern Europe and East Asia. The true test, however, is whether scholarship focuses on these broad questions, or simply becomes a replicant of earlier scholarship in the West.

## IV Conclusion

As political scientists, we have just lived through what are arguably the most significant political events of our lifetimes: the collapse of the Soviet empire and the global democratization wave. This is a conclusion that echoes across the chapters of this volume. These events touch the very core of many of our most basic questions about the nature of citizen politics and the working of the political process. Normally we study democratic systems that are roughly at equilibrium and speculate on how this equilibrium was created (or how it is changing in minor ways). Moreover, during the earlier waves of democratic transition the tools of empirical social science were not available to study political behavior directly. The current democratization wave thus provides a virtually unique opportunity to address questions on identity formation, the creation of political cultures (and possibly how cultural inheritances are changed), the establishment of an initial calculus of voting, and the dynamic processes linking political norms and behavior. The answers will not only explain what has occurred during this democratization wave, but may aid us in better understanding the basic principles of how citizens function within the democratic process.

### Acknowledgments

I would like to thank Harry Eckstein and Martin Wattenberg for our on-going discussions about political culture and electoral change which contributed to the ideas presented here.

### References

Abramson, P. I., and Inglehart, R. 1995. *Value Change in Global Perspective.* Ann Arbor: University of Michigan Press.

Almond, G., and Verba, S. 1963. *The Civic Culture.* Princeton, N.J.: Princeton University Press.

—— —— eds. 1980. *The Civic Culture Revisited.* Boston: Little, Brown.

Baker, K.; Dalton, R.; and Hildebrandt, K. 1981. *Germany Transformed.* Cambridge: Harvard University Press.
Barry, B. 1970. *Sociologists, Economists and Democracy.* London: Collier-Macmillan.
Barnes, S.; Kaase, M.; *et al.* 1979. *Political Action.* Beverly Hills, Calif.: Sage.
Bean, C., and Mughan, A. 1989. Leadership effects in parliamentary elections in Australia and Britain. *American Political Science Review,* 83: 1165–79.
Budge, I., and Farlie, D. 1983. *Explaining and Predicting Elections.* Boston: Allen and Unwin.
Chu, Y.-H. 1992. *Crafting Democracy in Taiwan.* Taipei: Insitute for National Policy Research.
Crewe, I., and Denver, D. T., eds. 1985. *Electoral Change in Western Democracies.* New York: St. Martin's.
Dalton, R. 1994*a*. *The Green Rainbow.* New Haven, Conn.: Yale University Press.
—— 1994*b*. Communists and democrats: attitudes toward democracy in the two Germanies. *British Journal of Political Science,* 24: 469–93.
—— 1996. *Citizen Politics in Western Democracies.* 2d edn. Chatham, N.J.: Chatham House.
—— and Wattenberg, M. 1993. The simple act of voting. Pp. 193–219 in *The State of the Discipline II,* ed. A. Finifter. Washington, D.C.: American Political Science Association.
—— Flanagan, S.; and Beck, P., eds. 1984. *Electoral Change in Advanced Industrial Democracies.* Princeton, N.J.: Princeton University Press.
Eckstein, H. 1966. *Division and Cohesion in Democracy.* Princeton, N.J.: Princeton University Press.
—— 1988. A culturalist theory of political change. *American Political Science Review,* 82: 789–804.
—— 1990. Political culture and political change. *American Political Science Review,* 84: 253–8.
Eijk, C. van der, and Niemoller, C. 1983. *Electoral Change in the Netherlands.* Amsterdam: C.T. Press.
Finifter, A., and Mickiewicz, E. 1992. Redefining the political system of the USSR. *American Political Science Review,* 86: 857–74.
Flanagan, S. 1978. The genesis of variant political cultures. Pp. 129–66 in *The Citizen and Politics,* ed. Sidney Verba and Lucian Pye. Stamford, Conn.: Greylock.
Franklin, M. 1985. *The Decline of Class Voting in Britain.* Oxford: Oxford University Press.
—— *et al.* 1992. *Electoral Change.* New York: Cambridge University Press.
Gelb, J. 1989. *Feminism and Politics.* Berkeley: University of California Press.
Gibson, J., and Duch, R. 1994. Postmaterialism and the emerging Soviet democracy. *Political Science Research,* 47: 5–39.
—— —— and Tedin, K. 1992. Democratic values and the transformation of the Soviet Union. *Journal of Politics,* 54: 329–71.
Heath, A., *et al.* 1991. *Understanding Political Change: The British Voter 1964–1987.* New York: Pergamon.
Huntington, S. P. 1991. *The Third Wave: Democratization in the Late Twentieth Century.* Norman: Oklahoma University Press.
Inglehart, R. 1977. *The Silent Revolution.* Princeton, N.J.: Princeton University Press.
—— 1990. *Culture Shift in Advanced Industrial Society.* Princeton, N.J.: Princeton University Press.
—— forthcoming. *Modernization and Postmoderization.* Ann Arbor: University of Michigan.
—— and Siemienska, R. 1990. A long-term trend toward democratization? Paper presented at the annual meetings of the American Political Science Association.
Inkeles, A., and Smith, D. 1974. *Becoming Modern: Individual Change in Six Developing Countries.* Cambridge, Mass.: Harvard University Press.

JENNINGS, M. K., and VAN DETH, J., eds. 1990. *Continuities in Political Action.* Berlin: de Gruyter.

KAASE, M. 1983. Sinn oder Unsinn des Konzepts "Politische Kultur" für die vergleichende Politikforschung. Pp. 144–72 in *Wahlen und politisches System*, ed. M. Kaase and H.-D. Klingemann. Opladen: Westdeutscher Verlag.

—— 1994. Is there personalization in politics? *International Political Science Review*, 15: 211–30.

KLINGEMANN, H.-D., and FUCHS, D., eds. 1995. *Citizens and the State.* Oxford: Oxford University Press.

KNUTSEN, O. 1987. The impact of structural and ideological cleavages on West European democracies. *British Journal of Political Science*, 18: 323–52.

LANE, J.-E., and ERSSON, S. 1991. *Politics and Society in Western Europe.* Beverly Hills, Calif.: Sage.

LIPSET, S. M. 1981. *Political Man.* Expanded edn. Baltimore, Md.: Johns Hopkins University Press.

—— and ROKKAN, S., eds. 1967. *Party Systems and Voter Alignments.* New York: Free Press.

MCALLISTER, I. 1992. *Political Behavior: Citizens, Parties and Elites in Australia.* Sydney: Longman Cheshire.

MCINTOSH, M. E., and MACIVER, M. A. 1992. Coping with freedom and uncertainty: public opinion in Hungary, Poland and Czechoslovakia, 1989–1992. *International Journal of Public Opinion Research*, 4: 375–91.

MILLER, A. H.; REISINGER, W. M.; and HESLI, V. L., eds. 1993. *Public Opinion and Regime Change: The New Politics of Post-Soviet Societies.* Boulder, Colo.: Westview.

MUELLER-ROMMEL, F., ed. 1989. *The New Politics in Western Europe.* Boulder, Colo.: Westview.

MULLER, E., and SELIGSON, M. 1994. Civic culture and democracy. *American Political Science Review*, 88: 635–52.

NATHAN, A., and SHI, T. 1993. Cultural requisites for democracy in China. *Daedalus*, 122: 95–124.

PARRY, G.; MOYSER, G.; and DAY, N. 1992. *Political Participation and Democracy in Britain.* Cambridge: Cambridge University Press.

POPKIN, S. 1991. *The Reasoning Voter.* Chicago: University of Chicago Press.

PUTNAM, R. 1973. *The Beliefs of Politicians.* New Haven, Conn.: Yale University Press.

—— 1993. *Making Democracy Work.* Princeton, N.J.: Princeton University Press.

REMMER, K. 1991. The political economy of elections in Latin America. *American Political Science Review*, 87: 393–407.

ROSE, R., and MCALLISTER, I. 1986. *Voters Begin to Choose: From Closed-class to Open Elections in Britain.* Beverly Hills, Calif.: Sage.

ROSENSTONE, S., and HANSEN, J. 1993. *Mobilization, Participation, and American Democracy.* New York: Macmillan.

SELIGSON, M., and BOOTH, J. 1993. Political culture and regime type: Nicaragua and Costa Rica. *Journal of Politics*, 55: 777–92.

SEMETKO, H., *et al.* 1991. *The Formation of Campaign Agendas.* Hillsdale, N.J.: Lawrence Erlbaum.

SHIN, D. C.; CHEY, M.; and KIM, W.-W. 1989. Cultural origins of public support for democracy in Korea. *Comparative Political Studies*, 22: 17–238.

—— —— 1993. The mass culture of democratization in Korea. *Korean Political Science Review*, 27: 137–60

VERBA, S. 1965. Germany. *Political Culture and Political Development*, ed. L. Pye and S. Verba. Princeton, N.J.: Princeton University Press.

VERBA, S., SCHLOZMAN, K., and BRADY, H. 1995. *Voice and Equality: Civic Voluntarism in American Politics.* Cambridge: Harvard University Press.

WATANUKI, J. 1991. Social structure and voting behavior. Pp. 49–83 in *The Japanese Voter*, ed. S. Flanagan *et al.* New Haven, Conn.: Yale University Press.

WATTENBERG, M. 1991. *The Rise of Candidate Centered Politics.* Cambridge, Mass.: Harvard University Press.

—— 1994. *The Decline of American Political Parties.* 4th edn. Cambridge, Mass.: Harvard University Press.

WEIL, F. 1993. The development of democratic attitudes in Eastern and Western Germany in a comparative perspective. Vol. i, pp. 195–225 in *Research on Democracy and Society*, ed. F. Weil *et al.* Greenwich, Conn.: JAI Press.

WILDAVSKY, A. 1987. Choosing preferences by constructing institutions. *American Political Science Review*, 81: 3–23.

Chapter 14

# Comparative Politics: Democratization Studies

Laurence Whitehead

## I Introduction

THIS chapter is about the comparison of national political processes, structures and systems. In particular, it draws on the work comparing the dynamics of authoritarian regimes, the processes by which such regimes may lose control, transitions from authoritarian rule, and the possible establishment and even consolidation of "democratic" regimes. There is a lengthy history of comparative political analysis on these themes as they concern Latin America, but of course these issues are of more than regional interest. They are of central importance for the study of contemporary politics in southern and east-central Europe and have also attracted growing interest in sub-Saharan Africa, in various parts of Asia and in all ex-communist countries. Synthetically, we can say that the comparative analysis of such "regime transitions" has been one of the major growth industries within the field of political science over the past decade. It has offered a way of organizing analysis of political processes in a wide variety of countries, sometimes new, usually poor and often somewhat unstable—countries which would not otherwise figure at the center of a political science discipline that is largely based in, and concerned with, the politics of old, rich and stable countries.

Given the prominence and extensiveness of democratization processes in the real world, a political science discipline which offered no systematic or well-grounded approaches to the interpretation of this reality would be abdicating from an essential task. But our chances of producing a strong predictive theory are slight. Despite a decade of work on transitions from authoritarian capitalist rule, our discipline was not well-placed to predict or even anticipate the wave of democratizations which swept the ex-Soviet bloc after 1989; nor can we now offer high probability predictions about whether or how the remaining Communist Party ruled countries (China, Cuba, Korea, Vietnam) will fall into line. Even the core terminology we use—"breakdown," "liberalization," "transition," "consolidation" and "democracy" itself—is inherently somewhat fuzzy. It is open to more than one definition and vulnerable to selective appropriation (think how the Reagan administration sought to apply these labels differentially in Central America). Indeed, given the extraordinary range of diverse situations in which it is applied, it necessarily encompasses a family of meanings, each of which is to some extent context-dependent.

## II Who is it for?

In order to clarify the place of the democratization literature in the general corpus of work on comparative politics, this chapter tackles three questions: "Who is it for?," "What is it like?" and "How is it done?" In the lead article in the inaugural issue of the journal *Comparative Politics*, Harold Lasswell noted that interest in acquiring comparative knowledge has tended to come in sudden bursts. Writing from the perspective of American political science, he singled out three instances, each partly linked to U.S. involvement in an international conflict. The Spanish American War stimulated the comparative study of world politics and colonialism; the First World War and the Bolshevik Revolution deepened territorial studies focused on a Europe-centered world; and, on a vastly greater scale, the Second World War, the onset of the Cold War and decolonization gave rise to crash programs of training and research. He inferred

> that an effective demand for more comparative knowledge depends on the shared expectation of political elites that they will be better off if they broadened the territorial scope of their political information. Further, some political scientists see an opportunity to improve the stock of knowledge available to the discipline and to raise their own value position in reference to colleagues (Lasswell 1968: 3).

Broadening out from the U.S. case, Lasswell added that civilized imperial powers—both in phases of expansion and of relative decline—in general encouraged some study of comparative government and that transnational organizations (e.g., engaged in foreign trade or missionary enterprises) also tended to support at least partially systematic comparative analysis.

Revolutionary-minded counter-élites were also likely to promote the comparative study of institutions. But, he wondered, would the bursts of interest occasioned by such élites in search of practical knowledge generate a permanent difference in the spatial and temporal dimensions of political knowledge, or "do most of us return to American government after a brief tour of a newer world?" (Lasswell 1968: 5). His conclusion was reassuring for the future of comparative politics, since he believed that the "core knowledge" of our discipline had been changed by the "behavioral revolution" in a direction which permanently expanded the demand for comparative analysis based on scientific methods of theory-formation and data collection. However, he acknowledged that this had produced disappointing results so far, and he argued that this was because "the comparative method as applied has been insufficiently contextual, inadequately problem-oriented, and unnecessarily restrictive in technique. In brief it has been insufficiently configurative" (Lasswell 1968: 6).

A quarter-century later some aspects of the Lasswell diagnosis now come to us as from a different world, but on two major themes an update is worth pursuing. These are "Who is comparative politics for?" and "How is it to be done?"— and of course the two are intimately linked. In order to assess whether this latest sudden "burst" of activism in the field of comparative politics is likely to prove ephemeral we need to consider the question of "who for?" as well as of "how?" Indeed, by reflecting on "who for?" we may also find some clues as to appropriate methods and suitable yardsticks for judging our performance. (For example, in contrast to Lasswell's imperial élites and revolutionary counter-élites, many new users of our services may be more forgiving of accounts which have low predictive power, although they may be more exacting as regards their context-sensitivity.)

The demand for intellectual guidance in understanding such contemporary processes as transitions from authoritarian rule and the consolidation of fragile democracies (inherently comparative topics) has recently been so strong and widespread that some political scientists were bound to respond.

This contemporary demand for a comparative politics of democratization arose from many sources. Early British interest revolved around hopes of exporting the "Westminster system" in the process of decolonization, rapidly fading as most such efforts proved a failure. But from the 1970s interest revived as new democracies arose, first in southern Europe, later in Latin America and parts of Asia during the 1980s, and eventually in the post-Soviet territories as well as in the formerly one-party states of sub-Saharan Africa. Undoubtedly ruling élites in the dominant western states took an early view on this subject, either because (like Kissinger) they feared it might disrupt their idea of world order or because (like Brandt or Carter) they hoped it might promote it. But on the whole western governments did not display much sophistication or seek fresh analytical insights until relatively late in the day.

By contrast, in the Spanish-speaking countries both activist and scholarly energies became focused on this topic well before its practical importance had been demonstrated. If such totemic regimes as the Salazar and Franco dictatorships could be replaced by more-or-less stable and functioning liberal democracies, then not only the ruling élites but the educated classes of Latin America more generally needed to know how that startling development was likely to impinge on their political circumstances. It was not so much "northern" political élites as "southern" political communities who demanded fresh insights into the comparative politics of democratization. More recently, emerging political communities in eastern Europe, and indeed in South Africa, have broadened and deepened the sources of demands still further. Of course "northern" demand also expanded rapidly, but much of this was derivative from these more urgent "southern" requirements.

Producing comparative politics for these new users is a different enterprise from servicing imperial-type élites. For the sake of brevity the contrast will be stated in a stylized form which inevitably overstates the reality. To start with the most obvious contrast, it makes a difference that the comparisons assumed some linguistic competence in Spanish as well as in English. Not only does this permit the coining of such illuminating neologisms as *democradura* (composite of "democracia" and "dictadura") and *dictablanda* (referring to hard democracy, soft dictatorship). More importantly it prompts the respectful exchange of ideas, and even some syncretism, between rival linguistic communities each with their own separate and rich traditions of political theory and experience. (For example, "the state," "civil society" and "the rule of law" are all key liberal categories with strong but somewhat distinctive resonances in each of the two communities.)

A second contrast is between imperial élites with access to large concentrations of state power, accustomed to projecting their wills externally while strongly sheltering the home population from outside disruption, and tentatively formed political communities with only partial access to what are, in any case, rather fragile instruments of state power and aware of their own vulnerability to impositions from without. The first type of user may demand scientifically-accredited findings with strong predictive potential which can assist in getting things done and which do not raise too many awkward normative dilemmas. Users of the second type, on the other hand, can be expected to have more tolerance for indeterminacy and less expectation of scientifically validated political prescriptions, both because such a stance reflects their own experiences of uncertainties and power limitations and also because they tend to associate methodological positivism with an amoral and perhaps undemocratic style of politics ("power politics"). A more "interpretative" rather than "explanatory" style of analysis may be associated with the participatory practice of "listening to the people."

A third (closely related) contrast is between the style of comparison congenial to a strong homogeneous and self-confident polity, few of whose citizens are likely seriously to entertain the possibility of an invidious verdict on their domestic arrangements, and a collection of weak, heterogeneous polities whose citizens can all too easily be assailed by feelings of insecurity, or possibly even of inferiority, especially when the comparisons made are too relentless and "objective." In the first case, existing political institutions and practices can be more-or-less taken for granted, or just marginally adjusted in the light of international comparisons. In the second case, supposed or actual lessons from elsewhere can produce a transformative impact, since all political arrangements (not just specifics, but also foundational principles) are provisional and no reasoned case for change can be reliably pre-screened out of the public debate.

## III What is it like?

Such contrasts help to explain two major respects in which the comparative democratization literature seems to differ from much earlier work on comparative politics: i) greater relativism and ii) more open engagement with normative dilemmas. In both cases we are dealing with differences of degree, and again for reasons of brevity what follows involves a certain overstatement.

On *relativism*, we should first notice that the transitions literature generally compares countries of more-or-less "equal" standing, which means that it cannot easily reduce the experience of any one case to a mere reflection or contrast with a leading case. In principle, all the separate national realities require equivalent consideration. This should mean that the work must be more genuinely comparative, in that if an argument is framed in terms perceived in country *B* as favoring country *A*, then it will elicit a counter-argument that requires equal consideration. (For example, the governments of Argentina, Chile and Uruguay can each make a case for the merits of their specific approach to civil–military relations—just as the limits of each country's argument can be detected by considering the strong points in the rivals' claims.) Genuinely comparative work of this kind requires relatively "dense" interpretations of each case. It penalizes frivolous or over-schematic treatment of the nuances of a particular situation; it obstructs the "shoe-horning" of intractable processes into inappropriately pre-fabricated explanatory boxes. "Dense" or contextually sophisticated interpretations of each case affect, in turn, processes of concept-formation and methods for evaluating relevant evidence. Whereas data-gathering and hypothesis-testing may seem rather straightforward and untroublesome procedures in a strong and secure political system, the same operations raise more doubts and create more ambiguities when conducted in a series of insecure and differentiated polities which are all hypersensitive about the implications of the comparative exercise for their own internal harmony.

This is all the more so because such comparisons have an unmistakably *normative engagement.* In his foreword to the pioneer *Transitions from Authoritarian Rule* study, Abraham Lowenthal was explicit about the "frank bias for democracy" of the underlying research project, which he characterized as an exercise in "thoughtful wishing" (in O'Donnell *et al.* 1986: x). Such conscious bridging of the fact-value distinction underlines the point that comparative work on democratization belongs somewhere towards the humanities end of the social science spectrum.

Typically, this normative engagement is not merely a value preference of the researcher or a prescriptive implication of his or her findings; it is built into the very fabric of the analysis. The best way to illustrate the intimacy of this engagement is by sketching a simplified narrative of a dialogue between theorists and practitioners. Faced by the Franco regime's history of intransigence, many Spanish democrats believed they should work for a "democratic rupture"; they believed, in other words, that democracy could and should be reconquered by mass mobilization from below. By contrast, theoretical reflection and comparative analysis gave rise to an opposed view, which suggested the greater efficacy of an alternative path to the same (but was it quite the same?) objective. On this view a transition to democracy could best be secured through an interactive process of élite negotiation, in the course of which important elements of the old regime would offer a fresh start and the anti-dictatorial opposition would be deradicalized. This proved both an effective prescription and a powerful instrument for the empowerment of some groups and the disempowerment of others. It also legitimated a particular set of "rules of the game," rules which became the foundational principles of Spanish democracy and precluded others. Its success in Spain contributed to its practical influence elsewhere as well as to its enhanced theoretical appeal. Disentangling fact from value in this narrative would be a thankless task.

Similarly, subsequent comparative work on democratization in South America argued (against the previously dominant dependency paradigm) for the autonomy of parties and for the primacy of domestic over international determinants of regime change. Again, in practical terms these arguments proved efficacious; and again, they contributed to the empowerment of certain types of actor and the disempowerment of others, and they may even have contributed to the legitimation of a new hegemonic discourse. Once again, therefore, the normative content of the analysis is inextricably interwoven with its analytic structure, which derives its authority as much from its identification with prevailing political processes as from its strictly scientific or methodological credentials. In the same way, we can also associate the appeal of a more critical counter-current of comparative studies (which focuses on societal democratization rather than the political regimes, and which values participation and therefore social redistribution above institutional reform) as much with its normative orientation as with its analytical clarity or its evidentiary basis.

The combination of relativism with normative engagement helps to protect this type of work from degenerating into a new ruling ideology. Since the diverse processes all demand equivalence of respect, and since for the most part their outcomes remain open-ended, the engagement is more with dilemmas than with clearly defined solutions.

Whether or not such relativism and such normative engagement are desirable features of general work in comparative politics is a separate question. In this particular area of work, the point to stress is that, while the demand for comparative interpretation is strong, the required performance yardstick in terms of scientific rigor is relatively low. Qualities such as "insight," "judgment" and "persuasiveness" (qualities which we tend to associate with a classical arts education) seem as much in demand as definitional precision, formal proof and strong predictive power (the traditional hallmarks of "scientific" work). Obviously, this stark polarity presents a false dilemma. In general our procedure should be to select methods which enable us to achieve objective confirmation of relevant general explanations. But the balance of methods available to us will vary according to the explanatory tasks in hand.

If this is so, explaining "democratization" may require the interpretative skills of the comparative historian at least as much as the logico-deductive clarity of, say, the game theorist. As Lasswell put it, "in order to discover the principal likenesses and differences to be studied, the entire context must be continually scanned . . . [and] the observational techniques must be multiple" (1968: 6, 14). I would add that the explanatory categories and concepts we use may have to be adjusted, "stretched" and even reinterpreted in the course of a dialogue between comparative analysis and the political processes under consideration. Such cognitive processes as scanning, synthesizing multiple sources and qualities of information, and adjusting our concepts in interaction with experience need not scandalize us, since most social and interpersonal judgments can be well formulated in this way. Such highly developed academic disciplines as history rely heavily on such procedures, and they are generally credited with producing impressive results (when best practice is observed).

Undoubtedly this kind of comparative work raises important questions of method, certain of which we are about to consider. Before that, we need to note that the main "users" of comparative work on democratization are not technical specialists but generalists struggling to direct and co-ordinate political communities which are inexperienced in self-government. By background they may be lawyers, journalists, community leaders or bureaucrats—all occupations which privilege the skills of persuasion and synthesis and which require daily management of social indeterminacy. As democratization proceeds, such social groups knit together into an increasingly self-confident and authoritative political stratum, eager to absorb the explanations of their predicament that can be derived from comparative political analysis. It thus emerges that the dominant style of comparative work dictated by the task of explaining democratization may well be particularly congenial to the new consumers of such work. Indeed, given not only their occupational skills but the lessons of political experience, which indicate to them that democratization is a complex and protracted process of social communication and persuasion, they would be likely to reject explanations couched in overly positivist or determinist form. Moreover, in the absence of securely predictable regularities, the demand for normatively grounded interpretations can be expected to increase.

## IV How is it done?

The rest of this chapter considers the questions of: i) predictability vs. contingency in comparative political analysis; ii) the methods appropriate for assessing and comparing complex dynamic political processes; iii) concept-formation and adjustment in this type of work; iv) some implications concerning the "objectivity" or "scientific status" of its results, and its relationship to democratic values. From this rather theoretical discussion the conclusion derives some pragmatic precepts which may be taken as suggestions for practitioners of this science-cum-art.

## A Predictability/Contingency

On *predictability/contingency*, without repeating well-known debates on political science methodology, suffice it to say that few of the processes we study are governed by laws so reliable that causation can be equated with prediction (or retrodiction). Moreover, as Mill put it 150 years ago, when studying the phenomena of politics and history, "Plurality of Causes exists in almost boundless excess and effects are, for the most part, inextricably interwoven with one another. To add to the embarrassment, most of the enquiries in political science relate to the production of effects of a most comprehensive description . . . results likely to be affected directly either in *plus* or in *minus* by nearly every fact which exists, or even which occurs, in human society" (Mill 1843/1973: 452).[1] Clearly, such observations apply with particular force to the study of democratization.

## B Dynamic processes

Additional problems arise when comparing and assessing dynamic political processes. After all, some contemporary social philosophy goes much further, asserting that the classical logician's model of a discrete cause acting to produce a given effect (or cluster of causes producing clusters of effects) needs sweeping reconsideration. The idea is that whatever explanatory value we can obtain will come from characterizing the specific way in which the two are bound together. In place of a "billiard-ball" model of causation, macro-political change involving the redefinition of actors' perceptions and identities might better be illuminated by reference to a reflexive model, or one in which dialogue or debate provides the engine of change. In this case "cause" and "effect" may be considered *internal* to the process of change and bound together by the interdependencies of persuasive argumentation, rather than acting externally and discretely on each other. Any historically grounded explanation of an abstract process such as "democratization" requires consideration of what the term *means* or

[1] In Book 6, Chapter 10, Section 3, Mill (1843/1973) went on to label attempts to discover laws of progress through which we could predict future events as "mostly chargeable with a fundamental misconception of the true method of social philosophy." Mill was clear however that, if political scientists used correct methods and did not overstate what they could achieve, the results would be a "real process of verification" (p. 917), despite the low predictive power of the results.

*signifies* to those involved, meanings which can only be constructed out of the intentions and understandings of the participants.[2] Since such explanations involve interactive processes of social persuasion, they are almost bound to invoke normative as well as empirical consideration.[3] Moreover, although dialogue and debate may be powerful processes for the identification and elimination of error, their results are always open to revision (rival subjectivities never disappear); in many cases the provisional outcomes are inconclusive (several coherent theses may remain in finely balanced contention after failed propositions have been rejected); and as we have all witnessed on some occasion or another, the strongest argument does not necessarily emerge unscathed (outcomes may be messy compromises). On this somewhat relativist model of social explanation, not only would we have to surrender strong predictability but the "value-free" standpoint of the analyst would also be put under challenge, and a strong tolerance for contingency would be required.

At this point, those most wedded to classical scientific paradigms may resist, feeling that unless they cling to the only coherent framework of explanation they know then all will be lost. The comparative study of democratization could be useful here, even to those who have no interest in it as a substantive field of enquiry, in offering some reassurance that it is possible to study important processes that are characterized by contingency, discursive open-endedness and the coexistence of competing subjectivities without abandoning the pursuit of explanatory cogency or neglecting good standards of verification. It could indicate that political science has something worthwhile to say about questions of the greatest public importance, something which can best be said by a discipline which is open to a degree of methodological pluralism and which values its antecedents in philosophy and the humanities, as well as pressing its ambitions to the status of "science."

Consider such illustrative generalizations as these: i) two of the surest ways for an authoritarian coalition to disintegrate, thus opening the way to a possible democratic transition, are via succession crisis or via defeat in an external military engagement; ii) the chances of democratization in any given country will be materially affected by the outcome of similar processes in the regionally dominant state; iii) where state formation has

[2] In the precise terminology of the logician Georg Henrik von Wright (1971: esp. chs. 3 and 4) such social explanations rely, not on Humean causation under covering laws, but on "motivational necessitation through practical inferences." Because they depend on the *intentions* of agents they are "teleological," that is to say, geared to expected future consequences, rather than necessitated by past antecedents ("caused," in the Humean sense).

[3] For the more general case that explanation is an inherently normative undertaking, see Garfinkel (1981: chs. 5 and 6).

already been completed (i.e. territorial boundaries are secure, and national identities are well-formed) democratic consolidation will be much easier than where state formation and democratization have to be attempted simultaneously. Although such propositions may sound rather ambiguous and loose, and although they yield only "fuzzy" and "qualitative" statements of probability, they do in fact provide *some* basis for prediction (as well as a broader framework for interpretation). Such generalizations are worth having, certainly when compared to the predicament of those who only have only one singular national experience to judge by.

This therefore brings us to much broader questions. What does the comparative democratization literature indicate about the most appropriate methods for assessing and comparing complex and dynamic macro-political processes? One key contribution was made in 1970 by Dankwart Rustow, who introduced the notion of a "dynamic model," in which the conditions favoring the *initiation* of a democratic transition might be quite different from those required for its subsequent *consolidation* (Rustow, 1970).[4] The two sets of conditions were sketched in a very general way. Subsequent work has refined the two types of conditions and has attempted to deal with the question of how they might be linked—again, relying partly on further theoretical reflection concerning the nature of democratization processes in general and partly on induction, taking into account a larger number of instances and more in-depth work on specific cases.

For the purposes of this chapter, three methodological points should be stressed. a) The presentation of a "dynamic model" was a significant advance on previous work, which had tended merely to list "prerequisites" or "social correlates" of democracy. Under this new approach, the emphasis shifted from supposed trigger conditions to the dynamics of interaction once the process had started. b) On their own neither the theoretical reflections nor the comparative case material was very instructive; but studied

[4] This article has been widely used, but it is not always remembered that Rustow used the central argument, in a more general way, in his earlier contribution to the inaugural issue of the same journal. There he noted, "the distinct (indeed often the opposite) conditions necessary for the birth and for the survival of a given type of regime. For example, postcolonial states originate in militant anticolonialism but survive through careful management of their economic and foreign policy problems. Military regimes typically result from secret plotting and armed revolt but endure as they obtain a wider basis of support in alliance with civilian bureaucrats or a political party. Charismatic leaders, according to Weber, establish their credentials by performing seeming miracles, but preserve their legitimacy through routinization and bureaucratization. An absolute monarchy is best sustained by unquestioned acceptance of tradition and heredity but evidently cannot be newly founded on the same principle. Democracy arises through conflict and compromise but survives by virtue of growing consensus" (Rustow 1968: 51). (Note the initial reference to "necessary conditions" jars with the examples which contain tendency statements and refer implicitly to the logic of strategic interactions *within* regimes, rather than to necessary conditions *external* to them).

together they opened a fruitful line of enquiry and eventually led to improved comparative research. c) Although the complex processes being compared were inherently diverse, with major differences in timing, process, and outcome, nevertheless they were viewed holistically (not disaggregated into component parts in the search for closer resemblances); and it was shown that in this field diversities of process could nevertheless be handled within a single comparative framework.[5]

As democratization studies have expanded, many variants of this general approach have been attempted. Some authors have attached prime importance to "modeling" the strategies of key actors as they interact and compete for power in the course of democratization; various specific experiences have thereby been grouped and classified to fit the categories suggested by game theory (for a clear example see Colomer 1991). An alternative perspective has directed attention to interaction effects (learning processes) *between* concurrent processes (e.g. Spanish political actors may have perceived their alternatives and shaped their strategies in response to apparently analogous events in Portugal). Alfred Stepan developed a typology of alternative "paths" to redemocratization, directing attention to various types of international context (in O'Donnell *et al.* 1986)—an approach I have subsequently extended, by for example distinguishing between the consequences of various types of "imposition" of democracy and by contrasting them with democracy through decolonization and through "convergence" (Whitehead 1996).

One point to note here is that, despite the diversity of paths that have been suggested and the evident differences of starting point and initiating conditions, a case can be made that the outcomes—at least so far—have been relatively homogeneous and can be reduced to just a small number of overlapping alternatives. Perhaps in reaction to this, some of the most recent attempts to probe behind these apparent similarities of outcome have shifted away from considering internal differentiations within democratization processes, redirecting attention toward the way such processes may be interacting with simultaneous macro-social changes in other areas (the "decline of the state-centric matrix" or the shift from state-led to market-oriented economics or the development of "civil society"). (Cavarozzi (1992) provides an illustration.)

All these various exercises in comparison involve a two-way iterative process. A simplified dynamic model (or set of middle-range generaliza-

[5] Of course, comparative history has long been pursued along these lines. At a more theoretical level, Weber developed the notion of "ideal types" for the same reasons. The claim made here is not that Rustow invented a new method (indeed, the contrast between conditions for regime creation and for regime maintenance can be traced back at least to Aristotle's *Politics*), merely that he pioneered its application to the comparative study of democratization.

tions), derived from initial theory plus induction, directs attention to certain strategic variables which are proposed as motors of change. By some process of selection or judgment, a range of (usually ongoing) cases are identified as possible exemplars of the model. Narratives of these cases are then compiled with attention directed particularly (though not exclusively) to the strategic variables suggested by the initial model or framework. But since these are narratives of complex macro-political processes characterized by uncertainty and value conflict, no unambiguous confirmation or rejection of the model is to be expected from these narratives. Rather, they are likely to raise doubts, or to suggest some respecification of the original model and generalizations. In principle this two-way process of adjustment and interrogation can continue through several cycles, particularly since many of the narratives unfold in unexpected directions and place pressures on the encompassing framework which tend to keep it loose and open-ended.

How satisfactory are such methods? From a purist standpoint they are obviously untidy. The best defense for them is that they are "appropriate," given the intractable (though vitally important) nature of the realities they purport to explain. Instead of prescribing a *single* method to be pursued with exclusivist rigour, it may be more appropriate to recommend the application of *multiple* overlapping methods, each of which can sharpen our understanding of some part or other of the phenomena under consideration, but none of which on its own has yet established sole dominance. The tension between complex narrative and coherent explanatory framework is obviously troubling, but it could also be a creative tension: more might be lost than gained by resolving this conflict unilaterally in favor of either one side or the other. There are intellectually respectable antecedents for such a methodological approach.[6] It may be a virtue of the comparative democratization debate that it forces into the open the case for procedures that are also followed, though in a less avowed manner, in other areas of our discipline.

[6] J. S. Mill (1843/1973: 917) recognized the importance of thorough case studies in comparative politics, noting that since its "empirical laws" (or interpretative framework) must be based on but a few instances "if therefore, even one or two of these few instances be insufficiently known . . . and are therefore not adequately compared with other instances, nothing is more probable than that a wrong empirical law will emerge instead of the right one."

Compare Geoffrey Hawthorn's recent volume on how to think about counterfactuals in social explanation. "The causal connections or runnings-on that we have been able to detect . . . turned out either to be phrased at a level that is so general as to be insufficiently informative and not address our interests in explanation; or to be so conditional as not to be general; or, when they have generated testable propositions, to be false. Because the answers to questions about social change . . . have to be hedged with so many conditions, any account of any particular change, if it is to respect the conditionality of the instance, has itself to be relatively particular and accordingly complex. And the more complex it becomes . . . the more it suggests alternatives" (1991: 160–1).

## C Concept-formation

One particular implication of this approach has attracted scholarly attention, perhaps because it engages with some much broader debates in contemporary philosophy. This concerns *concept-formation and adjustment*: the way in which the concepts (some prefer to say the "categories") which structure the interpretative framework are generated, defined, related to each other and adjusted in the course of the enquiry.[7] Thus Collier and Mahon embark on a potentially far-reaching critique of what they call "classical categorisation," as contained in Sartori's well-known 1970 *APSR* article. They invoke Wittgenstein as the source for their approach to "family resemblances," and Lakoff for the notion of "radial categories" (Collier and Mahon 1993: 852).[8] Such "decentering" of the meaning of core concepts could easily lead them into culturally relativist or postmodern terrain, but they quickly pull back, concluding that Sartori's article "deservedly remains a benchmark for scholars of comparative politics," although "some caution and refinement are in order." The refinements they propose create "an opportunity for broader and more flexible application by increasing the category's extension. Yet this very flexibility can lead to major scholarly disputes about whether the category fits the cases under study."

This cautiously innovative article raises much bigger issues than it confronts. In practical terms, the prudent conclusion is that something like Weberian "ideal-type" analysis still provides an effective way to combine imaginative, analogical—even intuitive—methods of theory construction with continued commitment to "objective" verification and impersonal procedures of replication. On an alternative view, the key categories deployed in the democratization literature are so intensely normative and judgmental that their emergence and adjustment must be an inherently evaluative process. For example, to label a democracy "consolidated" is not merely to categorize but also to judge. It is both normative and empowering. If analysts have discretion to select categories that are not mere reflections of given reality, and if their method requires iteration between their concepts and highly contested specific political processes, can they still claim to be practising a neutral "science" to the exclusion of a more subjective "art"?

[7] After acknowledging that "broad comparison is difficult, that political and social reality is heterogenous, that applying a category in a given context requires detailed knowledge of that context, and that it is easy to misapply categories," Sartori (1970) provided useful guidance on the conceptual "travelling" and "stretching" that might be required to accommodate new cases.

[8] Cf. Gallie (1956), who proposes a subjectivist view of the meaning of "democracy."

## D Objectivity and scientific status

What, then, is the scientific status of these democratization studies? Before this question creates too much *angst*, we should pause to recall that even in the natural sciences well-founded doubts have emerged about the traditional view of how theories emerge and are verified. The central theme in Isaac Newton's thought was the establishment of the dominance of spirit over matter by demonstrating the activity of God in the ordinary operation of nature. Most scientific theories are to some extent undetermined by the evidence mustered to "prove" them. Despite the fact that scientists are typically interacting cognitive agents, with a variety of aims and pursuing a variety of cognitive strategies, their methods, unsentimentally analyzed, still permit rational scientific advance (Kitcher 1993). Thus, especially when the object of study is not nature but politics, theory construction is unlikely to be fruitful unless it can in some way take into account the "subjective set" of the agents for whom it is intended.

A political scientist who was bereft of all social knowledge would lack all requisites for organizing and interpreting the evidence. But if language, history and knowledge of context are constitutive elements in the process of theory construction, then successful work may require some *combination* of rigor and attunedness. Indeed a degree of methodological pluralism and a diversity of sources of recruitment may be vital requirements for progress in political science in general as well as, more obviously, for comparative politics in particular. When the task involves comparing complex dynamic processes occurring in more than one country, then the training demands for social attunedness become that much more exacting. (This is the core case for area studies.) One major element in the understanding of complex processes characterized by high contingency is to compare them with plausible interpretations of what *might* otherwise have happened. Yet the selection of appropriate counterfactuals is a most demanding activity. It requires extensive knowledge of the context and of appropriate analogies or apparently comparable cases; and it requires skilled judgment. In short it requires both science and art.

Finally, if the object of study is democratization, it will not be possible to generate an adequate interpretative framework or properly to evaluate the instantiating cases without a good familiarity with the democratic norms, principles or values which are required in order to structure the emerging political order. Altogether then, the comparative study of democratization requires a degree of normative involvement (attunedness, familiarity with context, capacity to relate to democratic aspirations) which, far from serving as a substitute for analytical clarity and respect for the evidence, constitutes an indispensable complement to such attributes.

## V Conclusion

It is clear that there are multiple styles of comparative analysis—enumeration, sampling, paired comparison, case studies, modeling and so on—some of which are highly positivist, ranging through a spectrum to the strongly normative and even subjective. All are, in one way or another, implicitly or explicitly theory-laden. A pluralist approach leads to the conclusion that several rival theoretical perspectives may have some explanatory power, and therefore that multiple comparative styles embodying alternative theoretical standpoints may be worth attempting. The performance of different theories and alternative styles of enquiry are likely to depend heavily upon the nature of the explanatory task. Thus, a highly positivist and predictive theory may be appropriate for modeling and comparing the properties of alternative decision rules under proportional representation; whereas the revolutionary and perhaps anti-democratic potential of various strands of Islamic fundamentalism can perhaps best be illuminated by a much more intuitive, historically informed and culturally grounded style of comparative analysis.

Comparative work on democratization requires the thorough and careful evaluation of a large range of contextual factors. As Mill stressed long ago, such evaluations require persons who are "competently skilled" (not just trained in logical analysis). Only such persons are "capable of preparing the materials for historical generalisation, by analysing the facts of history, or even by observing the social phenomena of his own time. No other will be aware of the comparative importance of different facts, nor consequently know what facts to look for or to observe; still less will he be capable of estimating the evidence of facts which, as is the case with most, cannot be ascertained by direct observation or learnt from testing, but must be inferred" (Mill 1843/1973: 917).

Among the skills required, we should not underestimate the complications of translating key terms from one language to another (or even from one country to another within the same language community). Comparativists such as Joseph La Palombara have long been uncomfortably aware of the fact that such simple (even apparently naturalistic) political terms as "party" carry very different connotations in different settings—a point underscored by democratization studies in the ex-communist countries, where in contrast to most of South America the term has been associated with anti-democratic systems of rule. In fact, comparing regime changes in whole political systems requires familiarity not just with the special connotations attaching to individual terms but also the more general political idioms in which these terms are embedded.

Quentin Skinner's defense of context-sensitive interpretations of political theory can be extended to this type of comparative political analysis more generally.

> We can hope to attain a certain kind of objectivity in appraising rival systems of thought. We can hope to attain a greater degree of understanding, and thereby a larger tolerance, for various elements of cultural diversity. And above all, we can hope to acquire a perspective from which to view our own form of life in a more self-critical way, enlarging our present horizons instead of fortifying local prejudices (Skinner, in Tully, 1988: 287).

This requires the acquisition of highly refined judgmental skills. It should not be equated with the abandonment of all standards of objectivity (as the first sentence quoted makes clear) although it is likely to stimulate a heightened awareness of the many ways in which complex political analysis is almost inevitably shaped by the standpoint of the analyst. Whilst such self-awareness creates some difficulties, it also pays some dividends. It can help the analyst to understand the clashes of power and perception which drive regime change, without becoming uncritically wedded to the viewpoint of any one protagonist. Since democratization involves the forging of common identities and perceptions through a process of interaction and persuasion, it is highly desirable for the analyst to apprehend the various competing positions without being captured by any of them. Indeed, it is difficult to see how one could construct a credible general account of a democratization process without including some consideration of the values of compromise and tolerance that provided the *differentia specifica* needed to justify the costs and risks of embarking upon a regime change.

It has been the contention of this chapter that the practice of comparative politics is to some extent an "art" (i.e., a matter of judgment and persuasion, as well as of formal proof). Insightful and convincing interpretations of major real-world political phenomena may be generated to some extent by inductive as well as by deductive reasoning, with the help of (properly disciplined) intuition as much as with imitations of experimental technique. This is especially true in the field of democratization studies where, perhaps more than elsewhere, Rawls's recent dictum holds—"many of our most important judgments are made under conditions where it is not to be expected that conscientious persons with full powers of reason, even after free discussion, will arrive at the same conclusion . . . these burdens of judgment are of first significance for a democratic idea of toleration" (Rawls 1993: 29; see also Bohman 1991).

Fortunately for the discipline of political science there is a noble pedigree for the kind of "practical reason" and normative engagement, based on the arts of social judgment and persuasion, which has tended to characterize the best work in the field of comparative democratization. By tradition Aristotle is supposed to have written the constitutions of 158 Greek polities (although only that of Athens survived) before undertaking the comparative analysis in the *Politics*. Machiavelli tried to persuade the Italian city states to save themselves by organizing citizen militias. Madison wrote the *Federalist Papers* out of a "normative engagement" with the consolidation of a fragile U.S. constitutional republic, selecting his comparative arguments accordingly. Tocqueville studied American democracy as part of a program to consolidate a liberal order in France, and after 1848 he entered government there in an attempt to steer the new democracy to safety. Weber was a key adviser to the drafters of the Weimar Constitution, making every effort to use his prestige as a social scientist and his knowledge as a comparativist to assist in stabilizing the transition to a democratic republic in post-Wilhelmine Germany.

Given such antecedents the current generation of students of comparative democratization need not feel too insecure if their work leads them into areas which are more normative, subjective and prescriptive than now characterizes other branches of political science. Even shifts in emphasis from causal explanation to understanding, from proof to judgment, from demonstration to persuasion, can be licensed by the nature of their subject-matter.

## References

Bohman, J. 1991. *The New Philosophy of Social Science*. Oxford: Polity.

Cavarozzi, M. 1992. Beyond transitions to democracy in Latin America. *Journal of Latin American Studies*, 24: 665–84.

Collier, D., and Mahon, J. Jr. 1993. "Conceptual stretching" revisited: adapting categories in comparative analysis. *American Political Science Review*, 87: 845–55.

Colomer, J. M. 1991. Transitions by agreement: modeling the Spanish way. *American Political Science Review*, 85: 1283–302.

Gallie, W. B. 1956. Essentially contested concepts. *Proceedings of the Aristotelian Society*, 56: 167–98.

Garfinkel, A. 1981. *Forms of Explanation*. New Haven, Conn.: Yale University Press.

Hawthorn, G. 1991. *Plausible Worlds*. Cambridge: Cambridge University Press.

Kitcher, P. 1993. *The Advancement of Science*. Oxford: Oxford University Press.

Lasswell, H. D. 1968. The future of the comparative method. *Comparative Politics*, 1: 3–18.

Mill, J. S. 1843. *A System of Logic.* Vol. vii, in *Collected Works*, ed. J. M. Robson. Toronto: University of Toronto Press, 1973.

O'Donnell, G.; Schmitter, P.; and Whitehead, L., eds. 1986. *Transitions from Authoritarian Rule: Prospects for Democracy.* 4 vols. Baltimore, Md.: Johns Hopkins University Press.

Rawls, J. 1993. *Political Liberalism.* New York: Columbia University Press.

Rustow, D. A. 1968. Modernization and comparative politics. *Comparative Politics*, 1: 37–51.

—— 1970. Transitions to democracy: toward a dynamic model. *Comparative Politics*, 2: 337–63.

Sartori, G. 1970. Concept misformation in comparative politics. *American Political Science Review*, 64: 1033–53.

Tully, J., ed. 1988. *Meaning and Context: Quentin Skinner and his Critics.* Oxford: Polity.

Von Wright, G. H. 1971. *Explanation and Understanding.* London: Routledge and Kegan Paul.

Whitehead, L., ed. 1996. *International Dimensions of Democratization.* Oxford: Oxford University Press.

Chapter 15

# Comparative Politics, Old and New

David E. Apter

## I Introduction

FROM the start, comparing has been a particular way of connecting ideas derived from political philosophy and theory to empirical events and phenomena. The primary emphasis is on power. The purpose is to determine what difference differences make between the ways power can be deployed—not power in general, of course, but as organized in political systems and generated at national and sub-national levels. Interpreting the significance of differences in the uses and allocations of power by different political systems is the common enterprise underlying various alternative approaches to comparative politics.

Before discussing how comparative politics has evolved, some clarifying definitions are in order. When we speak of political "system" we mean that its components are interdependent, a change in one involving changes in others. Political systems, at a minimum, have as a primary responsibility (one might call it their original function) the maintenance of order over defined jurisdictions, for which they have a monopology of coercive force. Sovereign jurisdictions we call the state (Poggi 1990). "Government" is the chief instrumentality through which the political system works. "Civil society" refers to those networks of society (such as voluntary organizations, non-governmental organization, private educational and religious facilities, etc.) which are outside of government or state control but perform public functions (schools, etc.). How it intervenes, and the way its power is delimited defines the type or character of the state (democratic, authoritarian, etc.). "Democracy," following Schumpeter (1947: 269), can be defined as "that institutional arrangement for arriving at political decisions in which individuals acquire the power to decide by means of a competitive struggle for the people's vote." To the degree to which government intervenes in civil society we speak of the "strong state" (Birnbaum 1982), that is, one where government accepts a high level of responsibilities for the welfare of its citizens. Where these responsibilities are fulfilled by bodies outside the state we speak of a "strong civil society" (Badie and Birnbaum 1983). There is, however, no clear or even necessary correspondence between government intervention and social benefit.

Strong or weak, democratic or authoritarian, political systems are important to the extent that they are "configuring," that is, to the extent that they establish laws and orders effectively governing political conduct. The fit between prescribed and actual political behavior varies extensively in time and place, however. As citizens of the state or individuals and groups in civil society change and elude prescribed behavior, by legal and legislative means or by means of confrontational actions (reflecting a variety of circumstances), the result is changes in values and beliefs, alterations in principles of justice, or modificatiions in the pursuit of highly valued goals. Hence, included among the critical concerns of comparative politics are how well different types of political system are indeed "configuring," how such types can be established and maintained, and how perceived discrepancies between prescribed and actual political behavior can be mediated. Defining the good political system and ensuring a good fit between such a system and actual political processes is central within the broader range of comparative concerns. Insofar as there is widespread consensus that democracy is the best available political system, most comparative political inquiry shows a concern with democracy: how to realize it, sustain it, adapt and improve it, and how to deal with threats to its survival both from within and without.

Comparisons of political systems and how they work tend to be made on the basis of states which are their concrete surrogates. Most comparison of political systems is by countries, institutions within (sub-systems), and case. A variety of strategies is available: functional, multivariate, phenomenological, and so on. Any chosen strategy of research will depend on the general approach followed, the nature of questions posed or hypotheses being tested. In this respect comparative politics, insofar as it goes beyond mere description, can be said to be the empirical side of political philosophy or political theory.[1] Among the more characteristic concerns have been the exploration of differences between political systems in relation to conflict or compromise, power and accountability, efficacy and justice.

[1] The link between them can thus hardly be separated from comparative methods, which, however, would require more treatment than we have space for here.

Concrete political systems "types" include a wide variety of alternatives, from "tribes," to the polis, to states, monarchical and republican, democratic and authoritarian, presidential and parliamentary. Within each, there is also wide variety in how factions and coalitions form and re-form, interests are pursued, and, depending on constitutional structures, linkages are sustained between civil society and the state (whether in terms of kinship, ecclesiastical bodies, political movements, political parties or electoral systems).

Among the variety of comparative approaches, three will be singled out here for discussion: institutionalism, developmentalism, and neo-institutionalism. The first approach tends to focus on the specific workings of political systems *per se*: presidential and parliamentary, unitary and federal, parties and voting, committees and elections. The second approach incorporates broad theories of societal change. The third approach combines both. Institutionalism constitutes the bedrock of comparative politics. It remains foundational.[2] Even most recent texts remain "institutionalist."[3] That is, they describe how the political system of a state works by detailing the structure and functioning of government and its practices. What came to be called the "new" comparative politics—developmentalism (political and economic)—placed more emphasis on societal change rather than on techniques of governance, and in so doing drew considerably from other social science disciplines. In turn, "neo-institutionalism" not only brought the state back in but modified the preoccupations of the developmentalists in a direction of greater operationalism more tailored to the way political systems and states work.

## II Institutionalism

Institutionalism was more or less the exclusive approach in comparative politics, up to and considerably after World War II. Its original emphasis was on law and the constitution, on how government and the state, sovereignty, jurisdictions, legal and legislative instruments evolved in their different forms. Of significance were varying distributions of power and how these manifested themselves in relations between nation and state, central and local government, administration and bureaucracy, legal and constitu-

[2] For good examples of standard comparative texts following in the tradition of institutionalism see Friedrich (1968) and Finer (1949).

[3] Compare, for example, the categories in an "old" institutionalist text like Herman Finer's *Theory and Practice of Modern Government* (1949) with the latest edition of William Safran's *The French Polity* (1995). The categories in both are much the same.

tional practices and principles. Such evolution began in antiquity when ideas of political system were first articulated (Bryce 1921), with democracy as a teleological outcome. However, if institutionalism emphasized the uniquely western character of democracy, it also proclaimed its universality. Democracy meant differentiated civil government, legislatures and courts, executive powers and local government, municipalities. Comparative politics involved the detailed examination of how these instrumentalities worked, including a strong emphasis on reform (expanding the suffrage, the problem of oligarchy, reducing the dangers to established order by such doctrines as anarchism, socialism, and communism)—not least of all in a context of growing social upheaval, world wars, depression and totalitarianism.[4]

In these terms comparative politics is virtually coterminous with the origins of political science. One might say that the relation between political philosophy and comparative politics has been reciprocal. Each has contributed to the other in terms of the analysis of power as well as perfectible ideals of justice. Classical concerns were with the best state as an embodiment of reason, wisdom, and rationality, and how well it nurtured the civic virtue of citizens.[5]

Institutionalism, deriving its original examples from both republican and imperial Rome, might be said to have evolved out of antique concerns plus enlightenment doctrines of natural and positive law. Law represented an organic relationship between superordinate and subordinate magistrates and jurisdictions. Scholars of comparative institutions were mainly lawyers. They examined for example Justinian's *Institutes*, the contributions of the Commentators and Glossators (not to speak of the Code of

[4] Whether country-by-country, function-by-function, or instrumentality-by-instrumentality, their primary preoccupation was with states and governments, constitutions and their amendment, rights and their guarantees, unitary and federal systems, centralization and decentralization, regionalism and localism, questions of majority and minority representation, cabinet government or cabinet dictatorship, multi-party versus two-party systems, constituencies, electorates, first and second chambers, legislative committees, and electorates, procedures, the readings of bills and their debate, voting, and closure, the role of committees, and increasingly, the role of public opinion and the press.

[5] As has been pointed out elsewhere in this volume (Almond above: chap. 2), the original typologies of political system in Plato's *Republic* or the *Laws*, or Aristotle's *Politics*, drew inspiration from concrete comparisons between Sparta, Athens, Persia and other states, and ascribed differences between people (classes and "races"). So, for example, within the polis barbarians were distinguished from Greeks, slaves from citizens, aristocrats from plebeians—with such distinctions prescriptively validating the concentration and dispersal of power according to typologies of political systems based on the one, the few, and the many. These political-system types were better or worse according to how well they sponsored virtue, prudence, moderation, prowess in war, individual and civic discipline within the good state, as well as specifying the circumstances under which the good state might decay. Indeed, moral improvement and the prevention of decay was a principal concern in comparative politics from Plato and Aristotle on, and according to which one could compare typologies not only for "best" systems but most feasible alternatives, including in that the state which provided the best nurture for its citizens.

Hammurabi, Gaius' *Institutes*, Salic and Germanic law, and so on). For some Roman law was a source of inspiration. Others were influenced by social compact theories which focused attention on legitimacy in terms of representation, the relationship between the individual and the community, the citizen and the state defined the nature of constitutionalism. In these terms political philosophy and law became the foundations for the institutional study of comparative politics (Strauss 1959).

A third ingredient, as the above discussion implies, was history. Here too the emphasis was on the evolution of the state out of the polis and the origins of conciliarism, but in terms of specific benchmark events, struggles between church and state, between ecclesiastical and secular authority, over kingship and feudal barons, and the civil wars and revolutions which transformed the matter of individualism and social compact theories of authority from abstract principles to matters of life and death (Gough 1957).[6]

These intimate and intricate connections between political philosophy, law, and history took the form of two different but overlapping traditions, Continental and Anglo-Saxon. For the discipline of comparative politics, it was the latter which became more important. With a pedigree which can be traced to Bracton in the 13th century, it includes such figures as Blackstone, Anson, Stubbs, Dicey, Vinogradoff and Maitland. Institutionalism, then, has a history of constitutionalism marked by the transfer of general and specific powers from monarchs to assemblies, by means of rights represented in charters with democracy a function of parliamentary supremacy. In turn institutionalism included the examination of procedures and instruments by means of which freedom could be made to serve as a precondition for obligation.[7] In short, if comparative politics was about the evolution of democracy, democracy was considered an instrument for the moral perfection of man "to which his own nature moves" (Barker 1946), the evidence of such "movement" being the great democratic revolu-

[6] The institutionalist paradigm really took shape however during the Enlightenment period. The earlier emphasis on categories of people and their differing "natures" was transformed to a universe of individuals whose differences were relatively minor. It mattered, of course, whether one's view of man in a state of nature as compared to the civic community was or was not benign. For Hobbes, for whom it was not, there was no question of democracy. But most other theorists emphasized more benign properties such as Adam's Smith's propensity to truck, barter and exchange (i.e. identifying a universe of interests), the problem being how best to reconcile individual liberty with community rules. For Rousseau this was the general will; for Locke, the exercise of civil responsibility, a matter of parliamentary representation and sovereignty.

[7] The Continental tradition of institutionalism was also concerned with social contract theory. It retained a more robust connection to the natural rights tradition as embodied in Roman law, ecclesiastical conciliarism, in a context of evolving nationalism. The latter took the form of an evolutionary historicism, a teleology, i.e. the ineluctable emergence of democratic institutions out of specific conflicts, such as between papal conciliarism and monarchy, the medieval corporation and the secular state (Gierke 1950).

tions—the English, the American and the French, the latter exhibiting two powerful and competing alternatives, liberal constitutionalism of 1789 and radical Jacobinism of 1792 (Furet and Ozouf 1989).

How to realize the ideas of these revolutions constitutionally was one way in which history-as-events became embodied in modern principles of government. If each revolution is represented as a system of government best suited to man's nature, what were the most appropriate institutional arrangements for each? What, in each case, would maximize the configuring power of democratic and libertarian constitutions?

Above all institutionalism was concerned with democracy as a system of order with open ends. That emphasized the centrality of choice. If order was one priority, choice was another. Both became standards for evaluating governments. Comparing in these terms the governments of England, the U.S. and France after their revolutions: British parliamentarianism was represented as the model parliamentary system because of its superior stability, the American was the model presidential system because of its choice (and localism), while the French was the unstable version of the first. In this sense governments and states could be judged by their distance from the first two, with on the whole the first being preferred to the second.[8]

In these terms institutionalism was concerned with defining those political arrangements best able to square the circle—between order and choice, individual and community, citizen rights and obligations, according to accountability and consent, executive and legislative authority, electoral arrangements, the jurisdictions of courts and magistrates, and the relative virtues of unwritten versus written constitutions (a debate still going on in England), the virtues of unitary versus federal systems, parliamentary versus presidential systems, the workings of cabinet government (Jennings, 1936/1947), the role of a Privy Council and the significance of its absence, the transformation of imperial household establishments into administrative organizations (Robson 1956), the evolution of local government, the procedural rules of parliamentary behavior (Campion 1950), judicial review, the role of magistrates, committees and committee systems (Wheare 1955), electoral systems (Mackenzie 1958; Lakeman and Lambert 1959) and, above all, political parties (Ostrogorski 1964; Michels 1915/1958; Duverger 1954).

A formidable array of figures personify such concerns, including Carl Schmitt in Germany, Ivor Jennings, Ernest Barker and Harold Laski in England, Leon Deguit and Andre Siegfried in France, Carl Friedrich and

[8] England was the prototype of the stable, unitary, parliamentary democracy; France, the unstable one; and America an example of the virtues of federalism and localism. A number of American scholars, including Woodrow Wilson, favored a parliamentary system for the United States.

Herman Finer in the United States, to name only a few. What they had in common was not only an extraordinary empirical knowledge of how such institutions worked, including specialized instruments like political parties or parliamentary committees, but a common knowledge of classical, medieval and social contract history and law.

Institutionalists did not only study the workings of democracies or authoritarian alternatives in configural terms. They recognized that institutions "work" only insofar as they embody the values, norms, and principles of democracy itself. Hence institutionalism was never simply about mechanisms of governing but was also about how democratic principles were "institutionalized." That suggested that only some societies were "fit" for democracy, while others would become so when they had evolved accordingly. Hence, for example, tutelary colonialism as an appropriate way to nurture and encourage democracy by means of legislative devolution, recapitulating metropolitan experience in colonial territories (Hancock 1940; Wight 1957).

One might say that institutionalism was and remains the centerpiece of comparative politics. Reformist and prescriptive it evolved first in an age of European nationalism when the central problem was how to secure and make viable the connections between nations divided by language, culture, religion, and local nationalisms.[9] Economic factors came to play an increasingly important role as what Arendt (1963) called the "social question" became more and more preoccupying, trade unions became better organized and, together with political movements of many varieties, pressed for greater political participation, greater equality, a redefinition of equity and challenged liberal principles with socialist and other ideological alternatives. Institutionalism had to address the question of how government could deal with unemployment, the business cycle, negative social conditions, the emergence of class politics, political movements, and protest movements extra-institutional in methods if not in principles. And the more institutionalism became concerned with political economy, the more attention it paid to fiscal and monetary institutions and policies in a context of Keynesianism, especially as protection against radicalized party politics. Challenges to the principle of private property from leftist parties using Marxist or socialist theories, not to speak of the spread of socialist and communist parties in Europe with their claims to social as well as civil rights, raised the question not only of totalitarian alternatives like communism or fascism but also of whether parliamentary socialism was a likely next step in the evolution of democracy (Schumpeter 1947), the social

[9] German scholars, in particular, were preoccupied with how to elevate a shared and common national jurisdiction and citizenship.

welfare state and social or "industrial democracy" (Clegg 1951; Panitch 1976) coming to be seen as an alternative to totalitarianism, and a means of preventing citizens from voting democracy out and totalitarianism in.[10] That of course turned attention both to political parties and voting patterns, as well as to the potential attractions of single-party bureaucratic and authoritarian rule in different totalitarian systems (Friedrich and Brzezinski 1962).

Perhaps institutionalists had too much confidence in the configuring power of democratic political systems. They were unable to deal "theoretically" with the indisputable and marked discrepancies between institutionalist theory and practice, when it came to establishing democratic constitutions in newly independent countries after the First and Second World Wars (Huntington 1992).[11] Institutionalists regarded the mostly unanticipated emergence of totalitarian governments in Russia and Italy, and the failure of the Weimar constitution and the rise of Nazism as deviant forms of political behavior. Moreover, as radical Marxist communist parties and other extremist groups grew in strength, especially in Europe, and began challenging not only the way democracy worked but democracy itself, it became obvious that more attention had to be paid to psychological, economic, social and organizational factors in ways outside the conventions of institutional analysis. If even the best democratic constitution (Weimar) could not guarantee that democracy would work, there was also a plethora of examples of countries with good constitutions and bad governments (the Soviet constitution of 1936).[12] Institutionalism was inadequate to the test imposed by constitutional engineering. It assumed that countries without democracy were frustrated democracies waiting only to be liberated. Nor was the record better where democracy was made a condition of transition to independence after the Second World War (Huntington 1992).

[10] "The welfare state is the institutional outcome of the assumption by a society of legal and therefore formal and explicit responsibility for the basic well-being of all of its members" (Girvetz 1968: 512). Examples include the New Deal in the U.S., Beveridge in the U.K., the Popular Front in France, and the emergence of the social democratic state, as in Sweden.

[11] With the exception of such cases where colonial territories evolved towards dominion status under the Statute of Westminster of 1931 and within the British Commonwealth, Canada, New Zealand, Australia, South Africa, and, a striking departure, India, most efforts to establish democratic institutions in a hot-house way have been unsuccessful.

[12] The first widespread efforts at democracy by means of "institutionalist social engineering" were in countries emerging from the break-up of the Austro-Hungarian, Turkish, and Russian empires. With few exceptions, such as Czechoslovakia and Finland, such efforts failed (Headlam-Morley 1929). The second, including the decolonization process after the Second World War, has hardly been a success. Today the efforts to establish democracy in Eastern Europe and Russia remain more hope than realization.

## III The "new" comparative politics

The "new" comparative politics, with its emphasis on growth and development, was part of the more general optimism of the period after World War II. But if the premise and promise of development represented the good, the evil was communism and the Cold War. In the west, every move to the left was a gain for the Soviet Union. Every move towards democracy was a gain for the United States and its allies. The result of such Manicheanism was that, no matter how virtuous the policies undertaken to promote the first, they were to some degree morally diluted (if not contaminated) by the pervasiveness of the second. That gave developmentalism a certain ambiguity, quickly exploited by so-called "Third World" countries. Such ambiguity extended to efforts at institution-building, not so much in terms of Europe (in the post-war recovery period including the Marshall Plan), but "decolonization" in colonial territories.[13] Even more ambiguity characterized the developmentalism practiced by the U.S. in Latin America, under the Alliance for Progress, which to many was simply "neo-imperialism"—with much of the "Third World," with a self-proclaimed but ambiguously practiced "neutralism" between the "first" world (the west) and the "second" (the U.S.S.R. and other socialist countries), rejecting democracy in favor of one-party states and personal rule, with more or less explicit genuflections to socialism (vaguely defined). Indeed, so morally clouded were the politics of development that the main metropolitan countries allowed themselves to be almost as much manipulated as manipulating.

In effect, the political problem was how to combine decolonization with devolution of powers democratically, by redirecting nationalism—that is, by changing its venue away from the state towards it—within a context of "new nations." Colonialism then became tutelary rather than hegemonic. By so doing, it was hoped, democratic institutions would become the instruments of the state-in-becoming, a positive, developmental state. By the same token, this would prevent "stage-skipping"—the communist alternative of the one-party state, "skipping" the "bourgeois phase" and proceeding directly to socialism. What was at stake were two very different conceptions of "underlying reality." For the first, the market plus democracy (a double market, economic and political), would constitute a moving equilibrium given expertise and outside aid. For the other, such a stage was by its very nature neo-imperialist, hegemonic, substituting economic control for political. In this sense devolution for one was the substitute for

[13] It suited would-be dictators to charge that democratic constitutions which marked the transition to independence in many formerly colonial countries represented a neo-colonial inheritance.

revolution while revolution for the other was the alternative to devolution (Algeria becoming one of the examples of the latter for France, Vietnam for the U.S., etc.).

Competing pulls between left and right also had consequences in western countries. In Europe the equivalent of devolution was social welfarism and social democracy, not least of all (as in France or Italy) where there were large, legal and well-organized and financed communist parties. It spawned a huge literature on worker participation (the "Yugoslav model") and participatory democracy (Pateman 1970). Modest doses of socialism became appropriate modifiers of liberal capitalism. A great deal of comparative analysis was devoted to the evolution and the problems of the social welfare state (Offe 1984).

Such differences manifested themselves in two alternative approaches to developmentalism: modernization theories and dependency theories. Modernization theorists included a very diverse and loosely clustered group of comparative studies specialists such as Gabriel Almond, Samuel Huntington, David E. Apter, Lucian Pye, Myron Weiner, Leonard Binder, Edward Shils and Talcott Parsons, as well as a variety of others, some of whom combined case materials with broadly analytical books on comparative development. If they shared an ancestral figure, it was Max Weber. Dependency theory—whose putative ancestor was Marx—was even more diverse, including economists like Paul Baran and Andre Gundar Frank, historians like Perry Anderson and Eric Hobsbawm and political scientists like Gavin Kitching, Colin Leys and Benedict Anderson.

For a good many of the first group, "decolonization plus growth plus democratization" appeared to be a legitimacy formula for independence, especially under the patronage of tutelary colonialism (Shils 1962). For members of the second group, that was a strategy of hegemony and domination, and one had to attack such a formula in principle as well as practice. As a result scholars using much the same material and data drawn from the same country or samples could come to quite opposite conclusions, Kenya being a good example (cf. Leys 1974; Kitching 1980; and Bienen 1974).

Whatever the effects of such politicization on comparative politics as a field, the result was to make the comparative politics less Euro-centered and more concerned with how to build democracy in countries to which it was not indigenous. There was less faith in the configuring powers of constitution and government, and more in the need for a simultaneous and mutual process of institution-building from the bottom up and the top down. The developmental state had to assume responsibilities for sponsoring and stimulating development and, in effect, controlling the consequences (Apter 1965). Within the broad framework of development theory, there was the explicit assumption that, sooner or later, development would eventually result in replication of the same key social and cultural values and institutions as those in industrial societies—especially since it was assumed that with growth there would come a division of labor, the evolution of a middle class, private as well as public enterprise and so on. Successful development would sweep away "traditional" parochialisms and "primordialisms" (Geertz 1963) and establish pre-conditions for democracy. In turn, democracy would optimize the conditions for development. So, as the state was better able to benefit from, mediate, and control the consequences of growth, growth would generate new opportunities within societies, making for stable transitions.

That all required more understanding of little known cultures and practices. Where previous institutionalists dealt with political economy in connection with unemployment, fiscal policy, controling the business cycle and so on, the new emphasis was on continuities between the "great transformation" from pre-industrial to industrial societies in the west and its recapitulation within what was increasingly called the "Third World" (Polanyi 1944).[14] The analytical emphasis shifted away from state to societal structures—as well as to how best to introduce the values and cultural principles of democracy, how to socialize and motivate people in terms of these values, or how best to internalize them. In these terms one could examine problems such as how to ensure that nationalism, the driving and mobilizing force for independence and autonomy, would come to incorporate democratic norms and political values.

In general, then, developmentalism led to the comparison of societies with widely different social and political institutions and cultural practices. The central hypotheses were drawn from how "modern" institutions evolved in the west: the shift from theocratic to secular; from status to contract; pre-capitalist to capitalist; static to evolutionary notions of societal change, organic and mechanical solidarity; traditional to legal rational authority; *Gesellschaft* and *Gemeinschaft*; and for those of a more radical persuasion, transformation from pre-capitalism to bourgeois democracy and the prospects of socialism. These large-scale distinctions, refined in field studies, formed the basis for comparisons centering on the problems of social change and how these favored or undermined democratic potentialities. Controlling and rectifying social strains incurred in the process came to define the primary role of the state, with politics being seen as a matter of maintaining political balance, stability and viability. Where such

[14] It would be hard to overestimate the impact of this work on a whole generation of comparativists.

strains could not be mediated and governments fail to become institutionalized, the propensities grow for authoritarian regimes and "praetorianism" (Huntington 1968).

It would be wrong to say that the more the "new comparative politics" emphasized social change, the less concerned it was with specific political institutions. But in its attempts to apply, in the form of hypotheses, what had been learned from the transition from pre-industrial to industrial society in the west, it attached as much importance to society as the state, with power being generated by diverse sources, not all of them conventionally political.[15] Indeed, what came to be called the "tradition vs. modernity" distinction attempted to derive salient values and norms which, internalized and socialized, would make for successful transitions to both modernity and democracy (as well as identifying those which were less receptive or more resistant). For this comparative theorists could draw on a virtual pantheon of social historians, historical sociologists and anthropologists—Max Weber, Émile Durkheim, Ferdinand Töennies, George Simmel, Vilfredo Pareto, George Ostrogorski, Roberto Michels, Robert Redfield, B. Malinowski, A. R. Radcliffe-Brown, E. E. Evans-Pritchard, Claude Levi-Strauss and so on—posing questions of the connections between belief and social practices.[16]

Emphasis on the institutionalization, internalization and socialization of norms drew particularly on learning theory imported from social psychology and on value theory imported from political anthropology. How different cultures and ethnic groups responded to innovation was another central concern, incorporating theories of Erik H. Erikson (1968) on identity, David McClellend (1961) on "achievement motivation," and John Dollard (1939) on frustration–aggression theory; these foci were represented in a wide variety of case materials, from the very comparisons between "traditionalism" versus "modernity" (Eisenstadt 1973; Rudolph and Rudolph 1967) to theories of political violence (Gurr 1971), conditions of political integration (Geertz 1963) and analyses of ethnic conflict (Horowitz 1985).[17]

[15] Emphasizing qualitative rather than quantitative methods and functional frameworks social change theorists were oriented towards the problem of how to "equilibrate" norms appropriate to development and democracy, internalize them in the form of appropriate behavior, socialized in terms of roles and role networks, which in turn would reinforce and institutionalize norms. It is lack of "fit" between these that produces "strains," the rectification of which constitutes the "political" problem.

[16] The political interest in cultures might be said to begin with national character studies. See for example, Inkeles (1972).

[17] One should also note the significance of "psychological" emphases applied to the analysis of political violence (Gurr) and more psychoanalytical approaches by Ivo K. Feierabend and Rosalind L. Feierabend.

A good many studies of modernization were strongly influenced by sociologists, perhaps the most influential being Talcott Parsons. But the systematic comparison of societies as well as state systems, and in terms of political outcomes, was reflected in the work of many others, including Seymour Martin Lipset, Philip Selznick, Daniel Bell, Arthur Kornhauser, Philip Converse, Ralf Dahrendorf, Morris Janowitz, Edward Shils and Alain Touraine. Among their concerns were problems of ethnicity, primordialism, and the need to understand a society's "central values" and the variable responses of political cultures to change (Apter 1963/1971).[18]

Political economy, which for institutionalists was a matter of financial institutions, the role of treasury and central banks and of course the problems of the business cycle or the significance of unemployment for the evolution of democracy (Schumpeter 1947: 47), shifted to "development." Major figures of a liberal persuasion and influential in comparative politics using market theory included W. W. Rostow, W. Arthur Lewis (1957) and Albert Hirschman—the first concerned with what might be called "the American century," the second with Africa and the Caribbean, the last with Latin America.

It was in terms of "alternative" political economy theories that comparativists using "modernization" theory became separated from those who were to become "dependency" theorists. The latter represented both critical theories of capitalism and imperialism, and offered alternative prescriptions for socialism to be realized from above, through the one-party state, thus "skipping" a phase of bourgeois democracy. Such concerns were best represented by Paul Baran's *The Political Economy of Growth* (1962), which influenced several generations of *dependistas* in Latin America and contributed heavily to what became a corpus of radical developmental comparisons, from case materials (Leys 1974) to the more comparative studies of Frank, Cardozo, Suret Canale and Amin, drawing not least of all on work of Althusser, E. P. Thompson and Poulantzas but also incorporating work of many others.

If development theory, whether in the form of modernization studies or dependency theory, was caught up in Cold War conflicts as these manifested themselves in the Third World, such conflicts were also reflected in differences between comparative methodologies like functionalism versus dialectical analysis.[19] The first favored equilibrium theory in a context of

[18] The concern here was with "political institutional transfer," for example how and to what extent it would be possible to "institutionalize" western parliamentary structures in an African setting. See Apter (1963/1971*b*).

[19] Modernization and dependency theory became mutually adversarial. The first emphasized institution-building in a context of economic growth. The other emphasized the contradictions of growth under capitalism, with its neo-colonial past, and pointed out "necessary" neo-imperialist political

liberal capitalism as the foundation of democracy. The other favored conflict theory *en route* to socialism.

Depending upon which perspective one took, nationalism could take various forms: absolutist (Anderson 1986); as the vehicle for integration (Apter 1963/1971*b*; Coleman 1958); praetorianism (Huntington 1968); mobilizing support by creating a national discourse (Anderson 1991); a force for transformation, using political parties and the party-state as the instrument (Gellner 1983; Hobsbawm 1990); a disintegrative force (Migdal 1988); or, in their different contexts, virtually all of these things (Almond, Flanagan and Mundt 1973).

Such matters were incorporated in broad comparisons as well as case studies which examined in depth large themes of change, development, hegemony, power.[20] Out of these very diverse components and persuasions came a wide variety of broad comparisons and political ethnographies. They dealt with comparisons within and between Third World countries, one-party states more generally, authoritarianism and the problems posed by reinforcing social cleavages for stable democratic rule. Virtually all aspects of society were examined for the implications for political life, including the effects and consequences of education and educational systems, the role and place of élites, civic culture and its socialization in civic communities (Almond and Coleman 1960; Coleman 1965; Almond and Verba 1963).[21] A crucial emphasis on all sides was the matter of ideology, particularly nationalism as an alternative to or in conjunction with radicalization. Nationalism became the basis for examination of legitimacy, party mobilization, mass movements, populism and leadership, particularly as these related to authoritarianism and the rejection of democracy (Ionescu and Gellner 1969; Linz and Stepan 1978; O'Donnell 1973).

One of the more general criticisms of both modernization and dependency theory (that is, of developmentalism generally) was that politics seemed to be reduced to reflexes of economics or to societal processes. If the developmentalists criticized the institutionalists for their inability to deal in a satisfactory theoretical way with discrepancies between the

consequences. Each constituted a "critical theory" of the other. Each sponsored an extensive program of comparative and case materials. Both have been applied to developmental principles of political economy both within industrial countries (metropoles), and third world countries (peripheries). In hindsight, modernization studies gave too little emphasis to the state as an actor in itself while dependency theories treated it as an agency of hegemonic classes and powers.

[20] Within the confines of the single case, comparison tended to be diachronic, i.e. showing internal changes over time. Broader comparisons tended to be synchronic.

[21] Indeed, a genuine corpus of materials emerged—including the work of LaPalombara, Weiner, Pye, Coleman and Binder on bureaucracy, the penetration of western institutions in non-western settings, and a host of similar issues—mainly under the auspices of the Committee on Comparative Politics of the Social Science Research Council.

configuring power of the state and the complexities of social life which confounded the best laid constitutions, they also sinned in the opposite direction with their broad theories (Tilly 1984).

## IV Neo-institutionalism

What we will call "neo-institutionalism" combines older institutionalist concerns with developmentalism. Restoring "political system" to center-stage, it combines an interest in what are now called "less developed countries" with interest in Europe. Neo-institutionalism can be said to have evolved out of a general concern with pluralist democracy (Dahl 1982; Dogan 1988). It incorporates political behavior, including voting behavior and the analysis of changing fortunes of political parties and the significance of these changes for the state (Lipset and Rokkan 1967; Rokkan 1970) and problems of élites and democratization (Linz and Stepan 1978). Particularly concerned with social welfare and social democratic alternatives to authoritarianism, neo-institutionalists shifted away from the old institutionalist preoccupation with the Great Depression and towards the generalization of the social welfare state, of which Scandinavian and Dutch experiments with social democracy as well as Labour Party Britain represented significant examples.[22] Everywhere in Europe, too, there was political movement towards greater intervention of the state on behalf of its more disadvantaged citizens.

The comparative emphasis was on political parties, how they work, how coalitions form, public attitudes change, and the role of élites, bureaucracies, and politicians within different types of regimes. Where developmentalism stressed the need for growth as a way of contributing to democracy, neo-institutionalism examines the way governments confronting the negative consequences of growth, including environmental and pollution problems and the absorption of immigrants, where marginalization of industrial workers and polarization between a functional élite and a growing underclass of the functionally superfluous exacerbates tensions and promotes extremism. It includes, too, explanations for the reversal of the social welfare and social democratic state, and a return to the liberal state which was the main concern of insitutionalists—not least in terms of questions of governability under conditions in which the most efficacious policies are not politically feasible, and the most politically feasible policies are

[22] The latter a result of a general sense of obligation by a grateful government to their returning veterans and citizens after the Second World War.

not efficacious (Leca and Papini 1985). Finally, such matters are being evaluated against the backdrop and fallout from the implosion of the Soviet Union. If the the end of the Cold War has provided a third round of democratic opportunity, so too there are new opportunities for religious and ethnic sectarianism and fundamentalism—neither of which were anticipated or fitted with "social change" theories.[23]

Where developmentalists and neo-institutionalists come closest together is in their concern with "transitions" to democracy. The latter employ somewhat different strategies for analyzing that problem. The most conventional is the broad comparison based on historical cases, using class and state formation within what might be called a "post-Weberian" framework. Early examples include the work of Reinhard Bendix (1964) and Barrington Moore (1966). Later examples include Skocpol (1979) and O'Donnell, Schmitter and Whitehead (1986)—the first three using comparisons from France and England, India and Japan, Russia and China, both in terms of classes and in terms of the role of bureaucracy and the state; and the latter using Latin American examples. All draw general inferences about state formation in terms of democracy and totalitarianism.

Other analysts have emphasized the link between industrial capitalism and parliamentary democracy, the critical historical role of labor (Rueschemeyer, Stephens and Stephens 1992), and the significance of social protest and anti-state activism generally (Tilly 1978; Tarrow 1994).[24] Here one might argue that if capitalism appears to be a necessary condition for democracy it is certainly not a sufficient one (Lipset 1994).

There has also been a renewed emphasis on statistical studies comparing such factors as education, growth rates and urbanization. Inkeles and Smith's *Becoming Modern* (1974)—using a multiplicity of variables to measure hierarchy, stratification, stability, and so on, in six countries—was perhaps a forerunner. And, in a more opposite direction, comparative political economy is now being applied to particular cases like France. Among the concerns growing out of such studies is the concentration and dispersal of power in parliamentary regimes, and the centrality of electoral systems and voting behavior. Variations on such themes include the possibilities of "consociationalism"—with its emphasis on how to establish viable democratic institutions in the face of deep-seated social cleavages—originally developed in a case study of cleavage politics in Uganda (Apter

[23] Eastern Europe has to some extent replaced the Third World as the focus for "constitutional engineering," as well as the transformation of former socialist systems to capitalism. By the same token there has been growing interest in the breakdown in democratic regimes. See Linz and Stepan (1978).

[24] The latter, a growing field, includes comparative work on social movements in Poland, Chile, France and elsewhere by Alain Touraine, Charles Tilly and Sidney Tarrow.

1961) and was extended first to the Netherlands and then to a variety of other contexts from Austria to South Africa by Lijphart (1977; 1984) and others, using "grand coalition" theories and mutual veto mechanisms to establish tendencies towards or against democracy. Still another emphasis is on the interplay of political sub-systems, how opportunities are created for "negotiated agreements" (Di Palma 1990).

Another important strand in neo-institutionalism is the use of rational choice theory, which is more and more frequently being applied to the question of democracy in terms of what might be called the "double market," the intersection between the economic marketplace and the political—an approach pioneered by Downs (1957) and Olson (1965; 1982) and developed in a variety of contexts by Hechter, Bates, Laitin, Rosenbluth and others. For Przeworski (1991) the crucial element in the survival of democratic regimes lies in their capacity to generate incentives such that political groups that lose still have more to gain from competing within a democratic framework than they do from overturning it. In contrast to both the old institutionalists and modernity theorists (like Huntington, for example), this would assume that it is not necessary to believe in democracy in order to support it. What counts more for Przeworski is whether economic needs are being met, and the degree to which reforms result in unemployment, poverty and reduced inequality. With these changing concerns, not only institutions of government have become central again, but so too have the problems in western social welfare and social democratic states, including how to pay for compensatory programs and entitlements—raising questions about the proper role and scope of government and the limits of state intervention.

Nor is political culture forgotten. Using Italy as a case study in a context of regional politics, Putnam (1993) argues quite convincingly that it is the presence or absence of civic traditions, the civic community, which constitutes the key variable. His approach combines some of the work of the modernization theorists with a concern with particular institutional arrangements, relying on both analytical and quantitative forms of analysis within the configurative tradition of institutionalism.

Finally, political economy has combined with institutional comparisons in Europe, including responses by political parties to changes in the economy, European integration, and of course the disappearance of socialism not only in Eastern Europe and Russia but also the decline of socialism and social democracy in the west. Among the present concerns are how to pay for the social welfare and social democratic state, the impact on party alignments of the decline of the left generally, and such specific concerns as the transformation of the British Labour Party towards an acceptance of market principles as against nationalization, the denationalization phase of Mitterrandism and the fractionalization of the Socialist Party in France, and so on. There is debate between how much democracy is a function of its procedures and efficacy, and how much on prior cultural traditions or culture shifts (see, e.g., Inglehart 1990; Abramson and Inglehart 1995).

These more specific concerns fit into larger comparisons between, for example, Scandinavian and other social democratic countries like the Netherlands or France, "strong" or interventionist states which have high social overhead costs and elaborate welfare programs. Among recent and significant treatment of such matters one can include work by Peter Hall (1986), John Zysman (1983) and Peter Katzenstein (1978).[25]

Neo-institutionalism, then, is less constitutional than the old, and more prone to economic analysis insofar as it deals with fiscal and monetary policy, banks, markets and globalization. But it is also concerned with locating changes in the legislative process, shifts in long established party politics (such as the impact of Mitterrandism, Thatcherism or Reaganism on the principles and practices of government), not to speak of new social formations, coalitions, and so on, as these impinge on the state. Like the old, it is concerned with the state as an instrumentality in its own right, with its own tendencies and needs, and, as a configuring power, how it determines the nature of civil society. In general one can say that neo-institutionalism is more connected to social and political theory, and less to political philosophy, than its predecessor, and also more engaged in political economy.

There is renewed attention paid to the importance of legal structures, the significance of their presence or absence in, say, Russia or China—not to speak of the specific instrumentalities by means of which representative institutions derive their legitimacy from the consent of the governed. Above all, neo-institutionalism brings us back to the eternal question of the significance of proportionality in political systems, the original question of Plato as well as Rousseau, who was explicit about the need for government as a system of mutual proportionalities between wealth and power, rulers and ruled.[26]

## V An evaluation

This very brief overview of some of the newer tendencies in comparative analysis cannot, of course, do justice to the varieties of comparative politics

[25] See also Sen (1984), Lipton (1980) and Lindblom (1977).

[26] See the discussion of proportionality in Masters (1968: 340–50).

being undertaken today. What should also be understood is that with each change in the analytical focus of comparison, different intellectual pedigrees are invoked; and with each turn of the methodological screw comes a shift in comparative methods and operational strategies (quantitative and statistical, stochastic processes, path analysis, network analysis, as well as functionalism, structuralism, coalitional and vector analysis, social ecology, and so on) (Golembiewski, Welsh, and Crotty 1968/1969). There have also been good collections on these matters: early ones like Eckstein and Apter's *Comparative Politics* (1963) and cyclopedic efforts to cover the range of the field such as with Grawitz and Leca's *Traité de science politique* (1985), as well as more modest efforts such as Badie's *Le développement politique* (1980), Dogan and Pelassy, *How to Compare Nations* (1984), Wiarda's *New Directions in Comparative Politics* (1985), Andrain's *Comparative Political Systems* (1994). As originally suggested, these different styles of analysis have been accompanied by a steady interest in the use of methods, the appropriate units for comparison, what theoretical principles and ideas to use for hypotheses, what techniques will provide covariance, and what constitutes the basis of valid explanation. Issues range from "the N problem" versus the case study to the virtues and deficiencies of grand theories (Skinner 1985), and what Tilly (1984), attacking theories of the latter sort, has called "big structures, large processes, and huge comparisons." Whatever the emphasis, the newer comparative political analysis has tended to employ a variety of empirical methods, functional, analytical, quantitative, statistical, as against descriptive comparisons (country-by-country, institution-by-institution).

There is always a problem with how best to incorporate theoretical questions and hypotheses in case materials so that they do not simply illustrate what is already known (a reinforcement effect), or simply add to details without substantively increasing general knowledge (a trivialization problem). The advantage of case studies is their depth, their preoccupation with internal characteristics of social and political life. The problem is how to strike the right balance. Few case studies involving detailed description of politics have had much impact on comparative politics, except for illustrative purposes. Those who do fieldwork are often parochialized by area or country studies and, because detailed knowledge tends to make generalization difficult, anti-theoretical. This is not always the case: field studies like those of Geertz, Coleman, Apter, Ashford, LeRoy Ladurie, Furet, Lewin, Tucker, Scalapino, in contexts as varied as Indonesia and Morocco, Africa, Japan, China, France, Russia (to take some more or less at random), all bring broad theory to bear in specific situations. Major themes are embodied in case studies such as Coleman's *Nigeria, Background to Nationalism*

(1958), Apter's *Ghana in Transition* (1963/1971*b*) and *Political Kingdom in Uganda (1961/1996)*, Kitching's *Class and Economic Change in Kenya* (1980), Stepan's *The State and Society* (1977), Fagen's *The Transformation of Political Culture in Cuba* (1969), Schmitter's *Interest Conflict and Political Change in Brazil* (1971) and Friedman, Pickowicz and Selden's, *Chinese Village, Socialist State* (1991), to mention only a few. None of these studies is simply an exercise in detailed knowledge, nor simply applies what is already known theoretically to particular countries. All have contributed to the body of theoretical knowledge to provide both a richness to comparative politics and, on occasion, a more phenomenological understanding as with Geertz's (1973) emphasis on "reading" politics as a social text. Moreover, a significant proportion of the studies which had major impact on comparative politics by means of case materials are done not by area specialists but by outsiders—sometimes causing great debate between comparativists and area studies specialists.[27]

However, there is nothing like a good case study to reveal the shortcomings of overgeneralized comparative theories, one which deals with the interconnections between sub-systems, introduces valences and variables which a national or central governmental perspective may obscure. It can serve as an antidote as well to rational choice theories which push the level of rationality to the level of the political system as a whole system, when a variety of other rationalities may be involved in sub-systems and sub-sets which prejudice the center but make sense to those involved.

The need for close analysis of the case varies according to the questions being posed, of course. Much depends on both the requirements of knowledge in depth where such knowledge is available, as with China, or as with Japan where it is difficult to work without knowledge of language, history, culture, art and so on. Such knowledge may be lacking in countries such as those in Africa, where there were few written materials prior to colonialism except perhaps in Arabic, and where the recuperation of the past may require the use of oral history. But one of the best reasons for case work, in addition to these more obvious requirements, is that a good case can temper broadly comparative theories which tend to become obvious and overkill rather quickly. Moreover, comparative theories are too often "surprised" by events which their theories not only could not predict but insulated them against, the implosion of the Soviet Union being a good example.

If one applies a tough standard like predictability to the study of comparative politics it is clearly no better or worse than any of the other

[27] One only has to think of the controversies aroused by Theda Skocpol's *States and Social Revolutions* (1979).

sub-fields of political science, or the social sciences more generally. There are simply too many variables, and it is difficult to know which are the most salient. How much does democracy depend on "pre-requisites" of culture, or education, or specific civic élites? How much will it depend on perceived negative experiences with authoritarian rule? None of these questions can be answered in any decisive way. Nor can one establish some minimal level of social conditions. Concluding his overview of the comparative analysis of democracy, Lipset (1994) argues that while it is possible to draw conclusions from the experiences and characteristics of democratizing countries by correlations between democracy and economic growth and changes in stratification, there are too many other significant relationships for these to be conclusive. More generally we can agree that "given the multivariate nature of whatever causal nexus is suggested, it is inevitable that any given variable or policy will be associated with contradictory outcomes."

If so, what can be said in favor of comparative political analysis? For one thing it sensitizes observers to the differences between their own societies and others, and some of the consequences of difference. It makes one alive to the complexities and multiplicities of interaction between norms, values, institutions and social structures, and the varied forms of political behavior which, even when they appear similar to our own, might nevertheless mean quite other things to those who engage in it.

For the big questions—developmental change, changing notions of equity and justice, the proportions and balances between equity and allocation, choice and order—one can make projections, anticipate, become aware of consequences (Apter 1971*a*). One can distinguish how the same behavior in one setting leads to different outcomes in another. For example, risk-taking, which is essential to entrepreneurial innovation, can also produce and feed off violence (Apter 1996). One can also anticipate significant problems (problems of single-issue politics, parochial forms of nationalism, localism, sectarianism, and the revival of ethnic, religious, racial, and other boundaries) leading to less tolerance rather than more, and with it dangers of a negative rather than a positive pluralism. In these terms, the decline of the left has left a "space" for primordial revivalism in which democracy as understood is a last not a first consideration. Another critical question is how democratic political systems will deal with connections between innovation and growth, on the one hand, and marginalities (economic, social, ethnic, religious), on the other. Finally, one can ask whether or not there can be an "excess" of democracy, which will overload its capacities and result in too much fine tuning of moral sensibilities. In the name of democracy, interests can become elevated to the level of rights, reducing the prospects of negotiable solutions and generating hostility and mutual antagonism, less rather than more tolerance, and fewer rather than more political options.

Of course even if democracy is a universal system, there remains the question of how best to adapt it to the varieties of circumstance, old and new, which it will have to confront—not least of all extra-territorial associations, regionalism, globalism, and a variety of functional and political associations, private and public, which may alter the character of sovereignty and cast doubt on the sanctity of territorial boundaries. But despite the need for adaptive variation, what does suggest itself, tentatively, is a controversial conclusion. Examining these different approaches—institutional, developmental, neo-institutional—their different emphases and strategies of research and the large corpus of empirical studies, one is forced to conclude that there seems to be a relatively limited and specific ensemble of institutions which enable democracy to work in any meaningful sense. Despite "experiments" to the contrary, there are only a limited number of structural possibilities for the democratic state. No dramatically new alternative democratic formula has replaced what socialists once derided as "bourgeois" democracy. Nor has some formula for a culturally-specific democratic design, uniquely fitted to the particularities of a single country, emerged in any strong sense of the term. Democracy may have "vernacular" forms, but by and large these are not very satisfactory in dealing with problems of contemporary political life.

## References


Abramson, P. R., and Inglehart, R. 1995. *Value Change in Global Perspective.* Ann Arbor: University of Michigan Press.

Almond, G., and Coleman, J. S. 1960. *The Politics of the Developing Areas.* Princeton, N.J.: Princeton University Press.

—— and Verba, S. 1963. *The Civic Culture.* Princeton, N.J.: Princeton University Press.

—— Flanagan, S. C.; and Mundt, R., eds. 1973. *Crisis, Choice, and Change.* Boston: Little, Brown.

Althusser, L. 1969. *For Marx.* London: Penguin.

Amin, S. 1980. *Class and Nation, Historically and in the Current Crisis.* New York: Monthly Review Press.

Anderson, B. 1991. *Imagined Communities.* London: Verso.

Anderson, P. 1986. *Lineages of the Absolutist State.* London: Verso.

Andrain, C. F. 1994. *Comparative Political System.* Armonk, N.Y.: M. E. Sharpe.

Apter, D. E. 1961/1997. *The Political Kingdom in Uganda.* Rev. edn. 1968. Princeton, N.J.: Princeton University Press.

Apter, D. E. 1963/1971*b*. *Ghana in Transition*. Rev. edn. Princeton, N.J.: Princeton University Press; originally published 1963.
—— 1965. *The Politics of Modernization*. Chicago: University of Chicago Press.
—— 1971*a*. *Choice and the Politics of Allocation*. New Haven, Conn.: Yale University Press.
—— ed. 1996. *The Legitimization of Violence*. London: Macmillan.
—— and Saich, T. 1994. *Revolutionary Discourse in Mao's Republic*. Cambridge, Mass.: Harvard University Press.
Arendt, H. 1963. *On Revolution*. New York: Viking.
Badie, B. 1980. *Le développement politique*. Paris: Economica.
—— and Birnbaum, P. 1983. *The Sociology of the State*. Chicago: University of Chicago Press.
Barker, E. 1946. *The Politics of Aristotle*. Oxford: Clarendon Press.
Bendix, R. 1977. *Nation-Building and Citizenship*. Berkeley: University of California Press.
Bienen, H. 1974. *Kenya: The Politics of Participation and Control*. Princeton, N.J.: Princeton University Press.
Birnbaum, P. 1982. *La logique de l'Etat*. Paris: Fayard.
Bryce, J. 1921. *Modern Democracies*. New York: Macmillan.
Campion, G. 1950. *An Introduction to the Procedure of the House of Commons*. London: Macmillan.
Cardozo, F. H., and Falleto, E. 1979. *Dependency and Development in Latin America*. Berkeley: University of California Press.
Clegg, H. 1951. *Industrial Democracy and Nationalization*. Oxford: Blackwell.
Coleman, J. S. 1958. *Nigeria: Background to Nationalism*. Berkeley: University of California Press.
Di Palma, G. 1970. *Apathy and Participation*. New York: Free Press.
—— 1990. *To Craft Democracies*. Berkeley: University of California Press.
Dogan, M., and Pelassy, D. 1984. *How to Compare Nations*. Chatham, N.J.: Chatham House.
Dollard, J. 1939. *Frustration and Aggression*. New Haven, Conn.: Yale University Press.
Downs, A. 1957. *An Economic Theory of Democracy*. New York: Harper.
Duverger, M. 1954. *Political Parties*. New York: Wiley.
Eckstein, H., and Apter, D. E., eds. 1963. *Comparative Politics*. New York: Free Press.
Eisenstadt, S. N. 1973. *Tradition, Change, and Modernity*. New York: Wiley.
Erikson, E. H. 1968. *Identity*. New York: Norton.
Fagen, R. 1969. *The Transformation of Political Culture in Cuba*. Stanford, Calif.: Stanford University Press.
Finer, H. 1949. *Theory and Practice of Modern Government*. New York: Holt.
Frank, A. G. 1966/1969. *Capitalism and Underdevelopment in Latin America*. New York: Monthly Review Press.
Friedman, E.; Pickowicz, P. G.; and Selden, M. 1991. *Chinese Village, Socialist State*. New Haven, Conn.: Yale University Press.
Friedrich, C. J. 1968. *Constitutional Government and Democracy*. Waltham, Mass.: Blaisdell.
—— and Brzezinski, Z. K. 1962. *Totalitarian Dictatorship and Autocracy*. New York: Praeger.
Furet, F., and Ozouf, M. 1989. *A Critical Dictionary of the French Revolution*. Cambridge, Mass.: Harvard University Press.
Geertz, C., ed. 1963. *Old Societies and New States*. New York: Free Press.
—— 1973. *The Interpretation of Cultures*. New York: Basic.
Gellner, E. 1983. *Nations and Nationalism*. Ithaca, N.Y.: Cornell University Press.

GIERKE, O. V. 1950. *Natural Law and the Theory of Society, 1500–1800.* Cambridge: Cambridge University Press.

GIRVETZ, H. K. 1968. Welfare state. Vol. xvi, pp. 512–21 in *International Encyclopedia of the Social Sciences*, ed. D. L Sills. New York: Macmillan.

GOLEMBIEWSKI, R. T.; WELSH, W. A.; and CROTTY, W. A. 1968/1969. *A Methodological Primer for Social Scientists.* Chicago: Rand McNally.

GOUGH, J. W. 1957. *The Social Contract.* Oxford: Clarendon Press.

GRAWITZ, M., and LECA, J., eds. 1985. *Traité de science politique.* 4 vols. Paris: Presses Universitaires de France.

GURR, T. R. 1970. *Why Men Rebel.* Princeton, N.J.: Princeton University Press.

HALL, P. 1986. *Governing the Economy.* New York: Oxford University Press.

HANCOCK, W. K. 1937–42. *Survey of Commonwealth Affairs.* 2 vols. New York: Oxford University Press.

HEADLAM-MORLEY, A. 1929. *The New Constitutions of Europe.* Oxford: Oxford University Press.

HOBSBAWM, E. J. 1990. *Nations and Nationalism Since 1780.* Cambridge: Cambridge University Press.

HOROWITZ, D. L. 1985. *Ethnic Groups in Conflict.* Berkeley: University of California Press.

HUNTINGTON, S. 1968. *Political Order in Changing Societies.* New Haven, Conn.: Yale University Press.

—— 1993. *The Third Wave: Democratization in the Late Twentieth Century.* Norman: University of Oklahoma Press.

INGLEHART, R. 1990. *Culture Shift in Advanced Societies.* Princeton, N.J.: Princeton University Press.

INKELES, A. 1972. National character and modern political systems. In *Psychological Anthropology*, ed. F. L. K. Hsu. Cambridge, Mass.: Schenkman.

—— and SMITH, D. H. 1974. *Becoming Modern.* Cambridge, Mass.: Harvard University Press.

IONESCU, G., and GELLNER, E., eds. 1969. *Populism.* London: Weidenfeld and Nicolson.

JENNINGS, W. I. 1947. *Cabinet Government.* Cambridge: Cambridge University Press.

KATZENSTEIN, P. 1978. *Between Power and Plenty.* Madison: University of Wisconsin Press.

KITCHING, G. 1980. *Class and Economic Change in Kenya.* New Haven, Conn.: Yale University Press.

LAKEMAN, E., and LAMBERT, J. D. 1959. *Voting in Democracies.* London: Faber and Faber.

LECA, J., and PAPINI, L. 1985. *Les démocraties sont-elles gouvernables?* Paris: Economica.

LEWIS, W. A. 1957. *The Theory of Economic Growth.* London: George Allen and Unwin.

LEYS, C. 1974. *Underdevelopment in Kenya.* Berkeley: University of California Press.

LIJPHART, A. 1977. *Democracy in Plural Societies.* New Haven, Conn.: Yale University Press.

—— 1984. *Democracies: Patterns of Majoritarian and Consensus Government in Twenty-One Countries.* New Haven, Conn.: Yale University Press.

LINDBLOM, C. E. 1977. *Politics and Markets.* New York: Basic.

LINZ, J., and STEPAN, A., eds. 1978. *The Breakdown of Democratic Regimes.* Baltimore, Md.: Johns Hopkins University Press.

LIPSET, S. M. 1994. The social requisites of democracy revisited. *American Sociological Review*, 59: 1–22.

—— and ROKKAN, S., eds. 1967. *Party Systems and Voter Alignments.* New York: Free Press.

LIPTON, M. 1980. *Why Poor People Stay Poor.* Cambridge, Mass.: Harvard University Press.

MCCLELLAND, D. C. 1961. *The Achieving Society.* Princeton, N.J.: Van Nostrand.

MACKENZIE, W. J. M. 1958. *Free Elections.* London: Allen and Unwin.

MASTERS, R. 1968. *Rousseau.* Princeton, N.J.: Princeton University Press.

MICHELS, R. 1958. *Political Parties.* New York: Free Press.
MIGDAL, J. S. 1988. *Strong Societies and Weak States.* Princeton, N.J.: Princeton University Press.
MOORE, B. 1966. *The Social Origins of Dictatorship and Democracy.* Boston: Beacon Press.
O'DONNELL, G. 1973. *Modernization and Bureaucratic Authoritarianism.* Berkeley, Calif.: Institute of International Studies.
—— SCHMITTER, P. C.; and WHITEHEAD, L. 1986. *Transitions From Authoritarian Rule: Comparative Perspectives.* Baltimore, Md.: Johns Hopkins University Press.
OFFE, C. 1984. *Contradictions of the Welfare State.* Cambridge, Mass.: MIT Press.
OLSON, M. 1965. *The Logic of Collective Action.* Cambridge, Mass.: Harvard University Press.
—— 1982. *The Rise and Decline of Nations.* New Haven, Conn.: Yale University Press.
OSTROGORSKI, M. 1964. *Democracy and the Organization of Political Parties.* New York: Doubleday/Anchor.
PANITCH, L. 1976. *Social Democracy and Industrial Militancy.* Cambridge: Cambridge University Press.
PATEMAN, C. 1970. *Participation and Democratic Theory.* Cambridge: Cambridge University Press.
POGGI, G. 1990. *The State.* Stanford, Calif.: Stanford University Press.
POLANYI, K. 1944. *The Great Transformation.* Boston: Beacon Press.
POULANTZAS, N. 1973. *Political Power and Social Classes.* London: New Left Books.
PRZEWORSKI, A. 1991. *Democracy and the Market.* Cambridge: Cambridge University Press.
PUTNAM, R. 1993. *Making Democracy Work: Civic Traditions in Modern Italy.* Princeton, N.J.: Princeton University Press.
—— with LEONARDI, R. and MANETTI, R. Y. 1993. *Making Democracy Work: Civic Traditions in Modern Italy.* Princeton, N.J.: Princeton University Press.
PYE, L. 1968. *The Spirit of Chinese Politics.* Cambridge: MIT Press.
—— and PYE, M. 1985. *Asian Power and Politics: The Cultural Dimensions of Politics.* Cambridge, Mass.: Harvard University Press.
ROBSON, W. A., ed. 1956. *The Civil Service.* London: Hogarth Press.
ROKKAN, S. 1970. *Citizens, Elections, Parties.* Oslo: Universitetsforlaget.
RUDOLPH, L. I., and RUDOLPH, S. H. 1967. *The Modernity of Tradition.* Chicago: University of Chicago Press.
RUESCHEMEYER, D.; STEPHENS, E. H.; and STEPHENS, J. D. 1992. *Capitalist Development and Democracy.* Chicago: University of Chicago Press.
SCHMITTER, P. C. 1971. *Interest Conflict and Political Change in Brazil.* Stanford, Calif.: Stanford University Press.
SCHUMPETER, J. 1947. *Capitalism, Socialism, and Democracy.* New York: Harper.
SAFRAN, W. 1995. *The French Polity.* White Plains, N.Y.: Longman.
SEN, A. 1984. *Resources, Values and Development.* Cambridge, Mass.: Harvard University Press.
SHILS, E. 1962. *Political Development in New States.* s'Gravenhage: Mouton.
SKINNER, Q., ed. 1985. *The Return of Grand Theory in the Human Sciences.* Cambridge: Cambridge University Press.
SKOCPOL, T. 1979. *States and Social Revolutions.* Cambridge: Cambridge University Press.
STEPAN, A. 1977. *The State and Society.* Princeton, N.J.: Princeton University Press.
STRAUSS, L. 1959. *What is Political Philosophy?* New York: Free Press.
SURET-CANALE, J. 1971. *French Colonialism in Tropical Africa.* London: Hurst.
TARROW, S. 1994. *Power in Movement.* Cambridge: Cambridge University Press.
THOMPSON, E. P. 1963/1965. *The Making of the English Working Class.* London: Gollancz.

Tilly, C. 1978. *From Mobilization to Revolution.* New York: Random House.
—— 1984. *Big Structures, Large Processes, Huge Comparisons.* New York: Russell Sage Foundation.
Touraine, A. 1965. *Sociologie de l'action.* Paris: Editions du Seuil.
Wheare, K. C. 1955. *Government by Committee.* Oxford: Clarendon Press.
Wiarda, H. J. 1985. *New Directions in Comparative Politics.* Boulder, Colo.: Westview Press.
Wight, M. 1946. *The Development of the Legislative Council 1606–1945.* London: Faber.
Zysman, J. 1983. *Governments, Markets, and Growth.* Ithaca, N.Y.: Cornell University Press.

Part V

# International Relations

Chapter 16

# International Relations: An Overview

Kjell Goldmann

THERE are two main approaches to the task of reviewing a discipline: by surveying its findings or by examining its foundations. Neither will be attempted here. The former approach is open only to those who are well-read across the entire discipline, which few can claim to be in the case of the multifarious discipline of international relations (IR).[1] The latter is what Kenneth Waltz set out to do in his contribution to *A Handbook of Political Science* (1975) which was then elaborated into *Theory of International Politics* (1979); I have no illusion of being able to offer anything comparable.

What I propose to do is more limited. Four contentious issues will be reviewed: (1) the role of the state in present world politics; (2) the reason for doing research about matters such as this; (3) the significance of purpose and meaning in IR; and (4) whether theories of IR are to be seen as tentative conjectures or as instruments of power. These are issues of concern to many of us IR scholars but which we debate in a confusing way.

One reason for the confusion, I submit, is the prevailing conception of the field of IR as an arena in which a small number of "schools" or "approaches" fight out their differences. If all the problems of a discipline are subsumed under an all-encompassing opposition between "realism" and one or two other "schools" or "approaches," they easily become obscured—as if by political parties during an election campaign. It should be useful to review such issues, while avoiding the presumption of an all-out bi-polar or tri-polar confrontation.

[1] It has long been debated whether IR is a separate discipline, a subfield of political science, or a multidisciplinary field of study (an "inter-discipline," as Olson and Onuf suggest (1985: 4)). I will not go into this matter, which some find interesting. It is for the sake of simplicity that IR will be called a discipline rather than a sub- or inter-discipline in this chapter.

In what follows I shall distinguish between differences over substance, over values and over methods. Students of IR, like other social scientists, may differ: (1) over the substantive features of the subject matter they are investigating; (2) over the values that ought to be promoted by research; and (3) over the research methods that ought to be used. To make these distinctions is not to suggest that substance, value and method are unrelated to each other; there is no nostalgia here for the illusions of value-free research and theory-free substance. The reason for making these distinctions is pragmatic rather than epistemological: the appropriate way of managing conflict between academics depends on the nature of their differences.

Thus if the difference is over substance (if it concerns, say, the validity of deterrence theory, or the relative merits of interdependence and institutions for peace-building) further analysis is the obvious way of proceeding. If on the other hand the difference is over value (if it concerns, say, whether deterrence is a justifiable way of avoiding war, or whether social justice rather than the avoidance of war should be our focus) peaceful coexistence would seem to be the appropriate academic response; political disagreement between scholars cannot be resolved by academic analysis and should not be resolved by political dictat. Differences over method (whether for example deterrence theory is a hypothesis to be tested or an ideology to be deconstructed, or whether it is possible or impossible to gain future-oriented knowledge about peace-building) are the difficult ones, since an academic community needs to be both open-minded and principled on this score. A pre-condition for meaningful debate about matters such as these, at any rate, is that we be clear whether we are debating substance, value or method. This has not always been the case in the discipline of IR.

In the growing literature about IR as an academic discipline, there is agreement that there has been a shift from "grand" to "middle-range" theory in the last quarter-century, with Waltz's *Theory of International Politics* as a lonely exception. The conventional wisdom of the discipline is that the pursuit of "grand" theory has proven a "chimera" (Hoffmann 1977: 52) and that the generalizing aspirations of the 1960s "now seem untenable and even pretentious" (Olson and Onuf 1985: 13). Another trend, it is often suggested, is the emergence of international political economy as a subfield on a par with the study of international peace and security and the conditions for international order. A third, oft-mentioned feature of the discipline is persistent U.S. dominance: Stanley Hoffmann's frequently quoted characterization of IR is as an "American social science" (Hoffmann 1977).

The most remarkable feature of the discipline's self-image, however, has been mentioned already: the image of a field characterized by fundamental cleavages giving rise to recurrent "debates" of central importance. The first debate, according to the standard view, was between the original "idealism" of the discipline and the challenge of "realism." What participants pretentiously called the "second grand debate" was conducted between advocates of a "scientific" approach and of a "traditional" approach. Then, we are told, a "third debate" ensued. One party to it was "neo-realism": realism, it turned out, "was not dead; it had merely gone underground" (Banks 1984: 13). It is less clear who its adversary was: "globalists" (Maghroori and Ramberg 1982); "pluralists" and "structuralists" (Banks 1984: 15; Hollis and Smith 1990: 38–40); "globalists" and "neo-marxists" (Holsti 1985), "post-positivism" (Lapid 1989); the "transactional paradigm" and the "globalist paradigm" (Knutsen 1992: 235–6); "world society approaches" and "structural approaches" (Olson and Groom 1991); "interdependence" and "globalism" (Wæver 1992: 19–21); or maybe "Critical International Relations Theory" (George 1989, Brown 1993: 12).

*Theory of International Politics* was a main object of the "third debate." Waltz's work is seen in the literature as a realist counter-attack against "bureaucratic politics," "interdependence" and other subnational and transnational concerns (Wæver 1992: 20; Brown 1993: 5). The counter-counter-attack has been intense. There is by now a large literature in which *Theory of International Politics* is criticized from all conceivable points of view. The social-scientific equivalent of political correctness has become to dissociate oneself from *Theory of International Politics.* Criticisms range from substance to method to value. Neo-realism has been criticized for being wrong in its account of what it sets out to account for, for being based on untenable epistemology, for being politically conservative. It is intimated that these things naturally go together. Peculiarly, there seems to be agreement that "realist orthodoxy"—with its emphasis on the international system, nation-states and problems of peace and security—has maintained its hold on the discipline in spite of the massive criticisms leveled against it for years.[2]

The debate over realist theory has helped to specify theoretical issues and propositions of great importance. It is joined, in this spirit, by David Sanders in his contribution to this volume. It has been less helpful in defining the field as one of opposing overarching "schools" which between them exhaust our menu for choice. The conventional image of IR has, in effect, become one of an oligopolistic competition between U.S.-led conglomerates offering packages which have to be accepted or rejected *in toto,* rather

[2] For content analysis data pertaining to this matter see Goldmann 1995: 246.

than one of a marketplace in which large numbers of independent producers from all corners of the world offer a diversity of products. There is controversy, furthermore, about what each of the packages contains. The consumer's task has been made even more difficult by the tendency to advertise the packages with flashy concepts.

The rest of this chapter, therefore, is written in the spirit of a consumer's guide. The object is to make it easier to see what the disagreements in the discipline are and are not about. It goes without saying that major issues such as those raised here can be surveyed only in broad outline within the compass of this chapter.

## I The significance of the nation-state

One issue concerns the nation-state. No concept is more important for the discipline of international relations. The state, in the sense of nation-state, is a defining feature of "international." If the state did not exist, neither would IR. "Without the concept [of state] to fall back on, scholars would have to abandon the claim that there is something unique about the 'international' or 'interstate' realm" (Ferguson and Mansbach 1989: 2). And yet mainstream IR research is commonly criticized for being "state-centric."

It could be argued that the concept of nation-state is overly ambiguous, controversial, or normative (Ferguson and Mansbach 1989: 41–80)—too much so to define an academic discipline. This, however, is rarely what critics of the discipline's "state-centrism" have in mind; they rarely question the meaningfulness of the notion of "international" relations. What they criticize is, rather, a particular assumption which they believe is dominant in the discipline—namely, that "nation-states or their decision makers are the most important actors for understanding international relations" (Vasquez 1983: 18).

One's view of the validity of this assumption has been one of the litmus tests in sorting scholars into contending camps. The controversy may, however, be about at least three different things.

### A The state as object of support

On one interpretation, the issue is whom to support and whom to oppose. When the significance of "non-state actors" such as international organizations and transnational popular movements is emphasized, this sometimes seems to mean that phenomena such as these *ought* to be considered legitimate international actors, alongside states and governments—that is, some non-state actors *ought* to be important. "State-centrism," in this sense, is a defense of the nation-state against internationalist challenges. At the same time, when the significance of non-state actors like multinational corporations (MNCs) is stressed, the point is sometimes apparently to emphasize that the activities of MNCs are cause for as much concern as are those of states and governments—that we should devote attention not only to the "direct" violence of states, but also to the "structural" violence exercised by multinationals (on structural violence see Galtung (1969)). And when the concept of national interest is rejected, this sometimes is on the ground that the interests with which we *ought* to identify are those of individuals or social classes or transnational groupings or maybe even systems of ideas, rather than those of states.

Thus the controversy over "state-centrism" is in part one over political legitimacy and rightful concern. This matter is one of political values rather than of social-scientific analysis. Two of the foremost critics of the state concept in IR theory actually urge us to regard this concept as a frankly normative one (Ferguson and Mansbach 1989: 87). Evidently they are prepared to embrace the conclusion that the very definition of "international relations" as a field of study—and its inclusion in a handbook of political science—are political acts.

### B The state as object of study

On a second interpretation, the issue is about what to seek to explain and understand. In this sense if no other, mainstream IR research is unquestionably "state-centric." To cite a programmatic statement reprinted in a widely-used text from a quarter-century ago, "the substantive core of international relations is the interaction of governments of sovereign states" (Platig 1969: 16). So it remains. If the discipline has an established research program, this is it.

This emphasis may be criticized on the grounds that the actions and interactions of nation-states and governments are unimportant, that the "real groups or entities in politics" are classes or transnational coalitions (Vasquez 1983: 214), and therefore that it is a waste of time to examine why states and governments act and interact the way they do. Or, less radically, states and governments are increasingly constrained by interdependence and institutions, war is becoming increasingly dysfunctional, national security concerns are getting obsolete—in short, the "pillars of the

Westphahlian temple" are "decaying" (Maghroori and Ramberg 1982: 16–17; Zacher 1992). The issue of "state-centrism," thus interpreted, is whether developments such as these reduce our ability to explain and understand whatever it is that concerns us (peace and security, political economy) by focusing on state or government actions and interactions.

This issue cannot but be empirical. Three lines of research are apt to shed light on the validity of assumptions that the actions and interactions of states and governments are becoming unimportant.

- First is the examination of the extent to which international outcomes can be explained and understood without taking nation-states and governments into account.
- Second is the enquiry into the way in which the influence, independence and identity of nation-states have in fact been affected by the growth in international interdependence and institution-building. Integration in Western Europe, it has been paradoxically argued, has been a condition for the strengthening of state power rather than a cause of its demise (Milward 1992).
- Third is the exploration of the extent to which the independence, influence and identity of one's own state have, in fact, ceased to be regarded as top-priority issues in politics.

The answers may well vary across subfields, particularly between international-peace-and-security and international-economic-policy: the lack of distinctions between issue-areas (Vasquez 1983: 214) has in the past confounded the debate over the significance of the state.

Of those three lines of research, the third is especially important because of the crucial role played by the urge for national independence in mainstream theory. Authors do not always explain why they regard the state system as a constant. Morgenthau (1985: 535), however, is an exception. World government is impossible, he argues, because of people's attitudes:

> [T]he overwhelming majority would put what they regard as the welfare of their own nation above everything else, the interests of a world state included. . . . [T]he peoples of the world are not prepared to . . . force the nation from its throne and put the political organization of humanity on it. They are willing and able to sacrifice and die so that national governments may be kept standing.

According to this line of thought, "state-centrism" in research simply reflects the fact that national independence is regarded by *peoples* (not just by their governments) as a fundamental human need or value—an idea shaped and reinforced by political developments over centuries and proclaimed as a right in solemn, twentieth-century international declarations. Whether attitudes actually are now turning in new directions is, perhaps, the most basic issue raised by debates over "state-centrism."

## C The state as explanatory factor

On a third interpretation, the accusation of "state-centrism" means that mainstream research errs in its assumption that the actions and interactions of states and governments can be fully explained at the state and inter-state levels. This criticism, importantly different from the last, is a distinctly peculiar one. Whereas the discipline of IR remains unquestionably state-centric in its choice of *explananda*, that is not true so far as its *explanantes* are concerned. Even the 1960s text cited earlier lists a large number of intra-state and non-state factors with which students of IR need to be concerned (Platig 1969: 17–19).

True, Waltz's *Theory of International Politics* is limited to the level of the inter-state system. But this is in explicit opposition to trends in the discipline. It is strange that this unusual work has come to epitomize mainstream research. There, almost every conceivable level of analysis has been embraced. Standard collections in comparative foreign policy—such as East, Salmore and Hermann (1978) and Hermann, Kegley and Rosenau (1987)—are virtual inventories of the many-folded search in which the discipline has been engaged. If anything, the variety exceeds that in other parts of political science, thanks to the inclusion of individual-level phenomena such as perception alongside situational variables such as crisis (both reflecting common notions of what distinguishes international from domestic politics). What can be said about this eclecticism is that we know little about the relative significance of factors at different levels. What cannot be said is that it is "state-centric," in this third sense.[3]

## D Summary: "state-centrism" as empirical issue

The issue of "state-centrism" is, thus, in part political and in part a non-issue. There remains the challenge of investigating how the state and attitudes toward the state are affected by the new international relations that are emerging. Scholars have long pointed out that politics is becoming transnational and that nation-states are becoming penetrated, but there is

[3] For quantitative evidence bearing on this issue see Goldmann 1995: 248.

more to be done. Post-Cold War Europe, with its paradoxical combination of revived nationalism and advancing internationalism, has much to offer empirically minded scholars on this score. The European Union in particular is emerging as a major object of study for IR scholars with an interest in the foundations of their field.

## II The purpose of research and the end-of-the-Cold-War problem

Differences about the purpose of IR research, implicit in the "debates," could usefully be made explicit. The proper focus and method of research depends on the reason for conducting it. An orientation useful for one purpose may be unsuitable for another. Three purposes of IR research are outlined in the present section, not to typologize authors but to focus attention on the way in which orientation is contingent on purpose.

The problem that the end of the Cold War poses for the discipline will be used as illustration. Few IR scholars anticipated the changes that took place in world politics in the second half of the 1980s. It is difficult to quarrel with Gaddis's (1992: 53) conclusion that "none of the three general approaches to theory that have evolved since 1945"—the behavioural, the structural and the evolutionary, in Gaddis's view—"came anywhere close to anticipating how the Cold War would end." Indeed, IR scholarship from the mid–1950s to the mid–1980s may be characterized as the accumulation of explanations of the Cold War's persistence. Propositions ranged from international anarchy with its attending security dilemmas, to the built-in propensity of arms races to escalate, to the machinations of military-industrial interests, to the propensity of men and women to reject discrepant information. The sum was a powerful theory of international non-change. There was little basis in this rich literature to expect the East-West confrontation to end—and to do so peacefully and in a short time. Hence the "end-of-the-Cold-War problem."

The appropriate response, I shall argue, is either concern or excitement or indifference, depending on one's research purpose. A threefold typology of research purposes is suggested in Figure 16.1.[4] The point, to reiterate, is not to put labels on authors—there has been more than enough of

[4] The typology is reminiscent of one originally introduced by Habermas, who distinguishes between empirical-analytical science, historical-hermeneutical science, and critical theory (Brown 1992: 201–2). My typology, however, relates merely to the purpose of conducting research and not to broader differences between orientations.

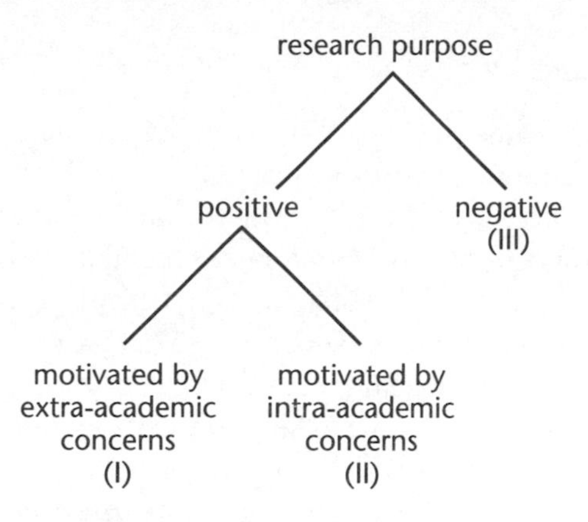


Fig. 16.1 *Three purposes of research*

"schools" and "approaches" in IR—but to pave the way for an assessment of research orientations as means to ends. Controversy over means is confusing, if contenders are driven by unacknowledged differences over ends.

A distinction is first made between positive and negative research. In the terminology adopted here, the purpose in the former case is to add positively to existing knowledge; and in the latter case, it is to examine the limitations of what others take to be knowledge. Positive research is further subdivided into research motivated by extra- and intra-academic concerns, respectively. The purpose in the former case is to produce social effects; the purpose in the latter case is to improve our comprehension of ourselves and our existence. A similar distinction has not been made with regard to negative research for reasons that will be explained below.

Terminology is a problem here, since obvious labels may have unintended overtones. What I will call type-I research is similar to what some might call "problem-solving" or maybe "positivism." Type III includes, among other things, what is known as "post-positivism" and "critical theory." It has seemed best to avoid labels such as these.

The three types of purpose are not mutually exclusive in the sense that a scholar must choose between them. On the contrary, it has been commonly taken for granted that intra-academic comprehension and extra-academic applicability go hand in hand, and that negative criticism is a condition for positive comprehension. I hope to show, however, that the tension between them is increasingly obvious in the field of IR and hence that research orientations are not equally suitable regardless of purpose.

## A Type-I research

Type-I objectives would seem to have dominated the scholarly study of international relations. Those preferring to call themselves "peace researchers" have been particularly explicit (or pretentious) on this score. Johan Galtung, for one, has compared peace research with medical science and has argued that the object is to do away with the quack-doctoring of international relations (Galtung 1964: 4; 1967: 13–20). David Singer has maintained that continued acquiescence with pre-scientific analysis qualifies as "the ultimate in war crimes" (Singer 1986: 59). Such Enlightenment zeal has been common in IR, whether the object of study has been decision making, deterrence, crisis management, peace-making or international order. Alexander George, an exceptionally respected political scientist, follows Galtung in drawing a parallel between IR research and the medical profession (George 1993: 17–18).

Type-I objectives have not been associated with a particular methodology. So-called "Quantitative International Politics" (QIP) was launched under its banner a generation ago, and statistical data analysis as well as mathematical modeling have become standard procedure in type-I research about peace and war (e.g. Russett 1972; Hoole and Zinnes 1976). At the same time, when QIP was criticized for invalid abstraction, this came not least from scholars with extra-scientific concerns, and that in part was what the "second debate" was about. Type-I research can thus take "humanistic" as well as "scientific" forms.

What is inherent in a type-I orientation is something else: the need to be concerned with the future—to anticipate future developments to which we may have to adapt, to assess the outcomes of courses of action that we may choose to pursue, or at least to contribute to the intelligent consideration of such matters. Few seem to have believed in the feasibility of strict prediction. But future-oriented knowledge in a looser sense has long been sought by IR scholars using a variety of methods.

The end of the Cold War is problematic from this perspective. It will not do to argue that there has been no ambition to predict particular events. We must be able to anticipate particular events (at least major events like the end of the Cold War) with reasonable certainty in order for our knowledge to be policy relevant. Nor will it do to demonstrate that existing theory did in fact imply that the Cold War would come to some sort of end at some future point in time. For policy-makers and citizens looking ahead in 1983 or 1984, at the peak of what was termed the "new" Cold War, crucial issues were not just whether but when and how the Cold War would come to an end. Sooner-or-later theory is not good enough for type-I purposes.

There is no getting away from the fact that the difficulty of anticipating the end of the Cold War on the basis of IR theory is cause for concern in a type-I perspective. Different conclusions may be drawn from this insight.

- One is that, in spite of this spectacular failure, it remains sensible to take the findings of IR scholars into account in policy-making, but we should refrain from claims to the effect that we can provide strong, research-based advice on international affairs. What needs to be modified is our way of participating in public debate—our extra-academic rhetoric, so to speak (see George (1993) for the elaboration of a similar view).
- A second possibility is to make distinctions between more and less predictable international relations. History, according to this thought, is a better guide to the future in some respects than in others. Scholars are beginning to distinguish between the North and the South on this score. Matters are changing, they argue, but only among the developed countries; traditional patterns are likely to continue, but only in the developing world (Holsti 1991; Jervis 1992).
- A third possibility is to conclude that a type-I orientation remains appropriate only as a long-term goal. Presently existing IR theory is too weak to be of more than marginal use for policy-making. Matters will become different in the long term, but type-II is a more appropriate orientation for the time being.
- There is a fourth possibility: to conclude that type-I research has proven impossible in principle because of the inherent unpredictability of international relations. The end of the Cold War may be taken as confirmation of the fact that the hope of solving international problems with the help of IR research is illusory.

This is an important debate for today's IR scholars, but only insofar as they assume that the purpose of their efforts is of type I. The matter is different on other assumptions, as will now be argued.

## B **Type-II research**

What is problematic from the type-I point of view is an exciting challenge from a type-II perspective. Some of the theory from which the end of the Cold War could not have been predicted had a respectable conceptual and empirical basis. Obviously, the better a theory, the more exciting discrepant evidence. The issue raised in this perspective is not whether we should stop doing what we have been doing but how to improve our theory in the light of new experience.

One example must suffice. It has been common for scholars pondering the end of the Cold War to emphasize the need to take political beliefs and values as well as domestic-political structures and processes into consideration, in addition to (or in place of) the factors traditionally emphasized in IR theory. Even though the IR tradition is eclectic on this score, theory has been dominated by structures of various kinds, ranging from the international system to states to bureaucracies to cognitions. Neither the substantive content of foreign policy ideas nor the domestic politics by which competing ideas lead on to foreign policy actions have attracted commensurate attention—an American trait, according to Hoffmann (1977: 58). The end-of-the-Cold-War experience emphasizes the need to bring politics back in, so to speak—politics in the dual sense of political ideas and domestic political process.

This will complicate theorizing. There is a dilemma between parsimony and comprehensiveness (Allan 1992). The trade-off is a long-standing bone of contention in type-II research. But there should be room for both simplifiers and complexifiers among those seeking comprehension for its own sake. The matter is more intriguing so far as type-I research is concerned. Scholars should recognize, it has been argued, that "too strict a pursuit of the scientific criterion of parsimony in their efforts to theorize is inappropriate for developing useful policy-relevant theory and knowledge" (George 1993: 140). It is difficult, however, to see how policy could be made and citizens' opinions formed except on the basis of models singling out the main features of a situation or a problem. Whereas it may be possible to take more-or-less everything into account in an academic effort to explain the past, this may not be a realistic objective for policy-makers and citizens who are pondering the future.

The obvious solution to the end-of-the-Cold-War problem in type-II research—to include political ideas and domestic political processes in models of international politics—thus may be problematic within a type-I orientation. Politics in the sense of ideas and dynamic processes would seem to be inherently less predictable than structures; if these are the kind of variables that must be taken into account in a valid theory of international relations, this is a serious problem in a type-I perspective. Conversely, it may be a source of inspiration in a type-II perspective, since most of us find it more rewarding to plunge into the substance of the play and the performance of the actors than into the structure of the theatre.

## C Type-III research

Emancipation from the constraints implied in what is taken by others to be knowledge is the essence of type-III research. Its focus is on conceptions of necessities and impossibilities. To demonstrate that the necessary may be avoided and that the impossible may be actualized is what type-III research aims to do.

The distinction between the extra- and the intra-academic is difficult to maintain in the case of negative research. Logically, a difference can be made between extra-academic emancipation and intra-academic skepticism similar to the one between affirmative and skeptical postmodernism (Rosenau 1992). It is controversial, however, whether radically negative research can be a means to extra-academic ends; it has been argued that "by refusing any task more positive than that of undermining existing theory, [this approach] effectively leaves the world as it is" (Brown 1992: 218). Since it does not matter for the purposes of this chapter whether the reason for conducting negative research is extra- or intra-academic, it has seemed unnecessary to join the debate about the political implications of postmodernism by attempting to distinguish between the two. What distinguishes both from positive research is that effective criticism suffices for goal-attainment and that is the important point in the present context.

Academic wrecking operations may be limited to specific propositions, based on traditional epistemology and intended to show that what is widely accepted is ill-supported in fact, thus clearing the way for research with a positive purpose. This is research on the border between types I or II and type III. It is something else that justifies the retention of type III as a separate category, to be distinguished from types I and II. What is counted as knowledge in research of types I and II, some IR scholars argue, are "arbitrary cultural constructs." This is the thesis of scholars who "resist knowing in [the modernist] sense." Their emphasis is on "ambiguity, uncertainty and the ceaseless questioning of identity"; their objective is "to interrogate limits, to explore how they are imposed, to demonstrate their arbitrariness and to think *other*-wise" (Ashley and Walker 1990: 262–4, emphasis in original). Their task is to "destabilise the assumptions [IR theorists] . . . take for granted and then to show how other ways of thinking might be opened up" (Walker 1992: 23). This, of course, is the application to IR of widespread views about the arbitrariness and oppressiveness of knowledge claims.

Now, what has made the end of the Cold War an issue among students of IR is the tension between theory and evidence. If one rejects the very idea that the validity of theory is contingent on its agreement with data, as some

IR scholars do explicitly and others implicitly (for example, Walker 1992: 19), the appropriate attitude toward the datum of the end of the Cold War is one of indifference. What is from other points of view a piece of discrepant evidence, and hence a reason either for concern or for excitement, is irrelevant if "truth is something we can—and must—do without in our thinking" (Brown 1992: 204). If data-based theory is an illusion, the occurrence of a datum is a non-event.

## D **Summary: three professional roles**

Three professional roles of IR scholars have now been outlined: the constructive citizen; the detached analyst; and the systematic skeptic. One's assessment of the discipline hinges on one's role-conception. The "debates" have proceeded as if all students of IR had the same purpose, or as if their purposes were compatible. Neither presumption is valid. Some type-III research assumes that a type-I or II orientation is based on illusions. Similarly, the search for future-oriented knowledge inherent in type-I research may lead to results that are superficial from the point of view of type-II analysis, while the more profound comprehension sought in a type-II perspective may do little to solve the problems central to type-I research.

Paradoxically, the most controversial of the roles is the one which IR scholars have traditionally adopted, the role of expert citizen. It is from this point of view that the performance of the discipline—as exemplified by the end-of-the-Cold-War problem—is a disappointment. The choice, however, is not "between detailed, high-confidence prescriptions for action and nothing at all," as Alexander George puts it (1993: 17). One of the larger challenges that lie ahead is precisely to go beyond George's pioneering efforts in considering the methodological requirements of type-I research and to consider how they differ from those of research of type II.

At any rate, rational debate within the discipline would benefit if role-conceptions were made explicit. Today's encounters between type-I professors and type-III graduate students would be more rewarding, or at least less confusing, if it were clear that the meaningfulness of "positivist" research and the relevance of Foucault depend on what we set out to do.

## III Purpose, meaning and institutions

Mentalism—the view that phenomena are explicable only in terms of creative and interpretative minds—was challenging structuralism in the study of international relations by the first half of the 1990s. Purpose and meaning were emphasized, just as in other parts of political science (see Giandomenico Majone chap. 26 below). The "subjective" was stressed rather than the "objective." The implications of this challenge—the current variation on a perennial theme—will be considered in the present section with regard to the analysis of foreign policy and of international institutions.

### A The analysis of foreign policy

There is nothing new in an interest in the subjective side of international relations. It is a myth that mainstream IR theory has conceived of states as soulless creatures responding "automatically" or "mechanically" to external stimuli. Even in a work like *Theory of International Politics*, Waltz explains why structure determines behavior in terms of mental processes (rational calculation and socialization). What is hypothesized in such reasoning is not that men and women are machines but rather (*a*) that they act rationally to further their objectives, on the basis of their definition of the situation, and (*b*) that their definition of the situation accords with reality, as assessed by outside observers, to an extent sufficient to give structural theory some degree of explanatory power. This is the logic of systemic theory such as Waltz's as well as of many other works in which structures, whether international or domestic, are presumed to determine state action and inter-state interaction. It is also the logic of deterrence theory, arms race theory from Richardson (1960) on, and other theories of escalatory and de-escalatory processes.

Mental phenomena, furthermore, have loomed large in studies of foreign policy decision-making and bargaining. In particular, much has been published about perceptions and cognitive processes. The psychological approach to international relations is well-established. Why, then, the feeling that the subjective side has been improperly left out?

One reason is dissatisfaction with the explanatory power of non-ideational theory. It is said that structural features such as Waltz's explain little unless actors' perceptions of meaning are taken into account (Wendt 1992) and that much in foreign policy cannot be explained except by reference to ideas (Goldstein and Keohane 1993). The end of the Cold War has been taken as evidence that ideas must be taken into account in IR analysis, as we have seen.

The shift in emphasis from structure to ideas in fact may reflect a wish to substitute "understanding" for "explanation" in the study of IR. "Explanation" is generally presumed to deal with causes, and causes are generally presumed to be "external" to the actor whose behavior is to be explained. "Understanding," in contrast, is taken to deal with meaning, and meaning can only be understood from "within." According to Hollis and Smith, there are "two plausible stories to tell, one from the outside about the human part of the natural world and the other from inside a separate social realm. One seeks to explain, the other to understand . . . [and] combining the two stories is not as easy as it at first seems" (Hollis and Smith 1990: 6–7).

Many, however, recommend a combination. Views vary as to whether foreign policy actions are in fact properly "understood" if exogenous variables remain unaccounted for: some seem satisfied with an interpretation couched in terms of the preferences and perceptions of actors, whereas others insist upon the further step of asking why preferences and perceptions are as they are. Views also vary as to whether every link in the chain between cause and effect must be investigated in order for an "explanation" to be convincing: no cliché is more familiar to international relationists than the assertion that the black boxes of decision-making must be opened up in order to make proper explanation possible. Thus, one of the divisions in the discipline, often confounded with other differences, concerns what constitutes a satisfactory analysis of action and especially whether and how to mix the "external" and the "internal" in the analysis of foreign policy. By the mid–1990s the trend was to emphasize the "inside" and de-emphasize the "outside."

A second reason to urge more attention be paid to purpose in the analysis of international relations may be the way in which the substance of politics has been marginalized. Mainstream analysis examines whether foreign policies are "adaptive" (Rosenau 1970) and why states "realign" (Holsti 1982), rather than the substantive contents of the adaptations and the realignments. Many of us have attempted to find out what makes governments "co-operate" or "defect," rather than to examine what governments do in fact do when they co-operate and defect. The explananda of mainstream theory, furthermore, have been structures (international, organizational, cognitive) and processes, rather than purposes.[5] The purposes of

[5] Consider, for example, the index of a major volume such as *New Directions in the Study of Foreign Policy* (Hermann, Kegley and Rosenau 1987): it contains several entries under topics such as "cognitive processes," "decision-making," "information processing" and "national attributes"; but it includes no references to "goal," "purpose" or "program" and a single reference each to "national interest" and "ideology" (both of which are briefly mentioned in a paper about political economy).

foreign policy, to the extent that they have been mentioned at all, have been assumed rather than investigated ("power," "security," etc.). Against this background, Holsti's 1991 work—analysing the substantive motives states have had for waging war and the prevailing attitudes toward war in various epochs—constitutes one noteworthy effort among several to bring more of substantive policy analysis into the theoretical analysis of foreign policy.

More may be at stake, though. To argue that we should take into account actors' substantive purposes and their own interpretations of reality is different from denying that any external reality exists independently of the perceptions of actors. Now, however, the discipline of IR is witnessing the appearance of mentalism in that literal sense, according to which all there is to study is how actors believe things are or say things are. According to this view, actors' environments exist only in their beliefs or in their texts. The constraints to which they are submitted are neither structural nor material, but merely perceptual or textual. Issues such as these have been discernible in recent "debates" and are becoming increasingly topical, especially in a type-II perspective.

## B International relations as institutions

Institutions may be seen as generalized purposes and meanings. International relations may be seen as institutionalized, in the sense that "much behavior is recognized by participants as reflecting established rules, norms and conventions" (Keohane 1989: 1). Thus an emphasis on purpose and meaning in analysing action, and a focus on institutions in analysing systems, are two sides of the same coin. In both cases, there is a stress on the subjective, on seeing things "from within."

When the "new institutionalism" was launched in political science its aims were, among other things: (1) to emphasize the autonomous role of politics in society; (2) to focus on "the ways in which political life is organized around the development of meaning through symbols, rituals and ceremonies" rather than on decision-making and the allocation of resources; and (3) to see political actors as responding to obligations and duties rather than to calculated self-interest (March and Olsen 1984: 735). Neither the first nor the second point seems to have impressed students of IR, but the third point has been all the more important.

As regards the autonomy of politics at the international level, the tone was historically set by Morgenthau, who conceived of politics in terms of "the concept of interest defined in terms of power." This concept, he wrote, "sets politics as an autonomous sphere of action and understanding apart from other spheres, such as economics . . ., ethics, aesthetics, or religion." It alone makes a theory of politics possible, "for without it we could not distinguish between political and nonpolitical facts" (Morgenthau 1985: 5). When institutionalism made a breakthrough in IR in the mid–1980s, however, it appealed to those critical of what was conceived of as realist orthodoxy. To join Morgenthau in emphasizing the autonomy of politics would have been an anathema. Apart from Morgenthau and maybe Waltz, students of IR appear to have been unconcerned with the distinction between the political and the non-political, as is evident from their inclination to include a variety of economic, psychological and other "non-political" variables in their research.

As regards the second point, students of IR have certainly been aware of the role of "symbols, rituals and ceremonies" in inter-state relations, ranging from the formalities of diplomatic negotiation to the power demonstrations of tacit bargaining. They appear, however, to have been prone to interpret these phenomena in instrumental terms, as methods that actors use to further their objectives (to make others accept their demands as legitimate, to signal commitments and intentions, etc.). The increasing interest in purpose and meaning does not seem to include an increased concern with the non-instrumental side of politics.

The remaining point concerns obligations and duties rather than calculated self-interest as the propellant of political action. Even if it has been difficult to determine what precisely the "debates" have been about, the role of self-interest has often been suggested as the crux of the matter. "Liberal institutionalism" (Keohane 1989) is maybe the most recent incarnation of the rejection of the notion that "interest defined in terms of power" is the essence of international politics.

Institutionalism may be attractive to students of IR in a fourth way, as well. One advantage of institutions is that they can change. According to non-realists, static necessity rules in the harsh world of realism; institutionalism, in contrast, assumes that action is determined by man-made institutions, not by laws of nature. What is more, institutions transform interests (Wendt 1992). Institution-building has been a standard ingredient of peace proposals since the Middle Ages, and present-day institutionalism promises to show that this is a realistic rather than an idealistic notion.

In part, the renewed interest in institutions has taken the form of the study of international regimes, construed as "implicit or explicit principles, norms, rules and decision-making procedures around which actors' expectations converge in a given area of international relations" (Krasner 1983: 186). Regimes differ from institutions in general by being "more specialized arrangements that pertain to well-defined activities, resources, or geographical areas" (Young 1989: 13). A large literature exists about regime formation and regime change, which is one dimension of the phenomenon of "governance without government" (Rosenau and Czempiel 1992) that has interested many students of IR.

The challenge of the institutional approach is well-known. It has to explain both compliance and change. Moreover, it has to do both at the same time.

## 1 Compliance

Keohane suggests that we distinguish between institution as a "general pattern or categorization of activity" and as a "particular human-constructed arrangement, formally or informally organized" (Keohane 1988: 383). Sovereign statehood is an instance of the former, the United Nations of the latter. Compliance is more problematic in the latter case than in the former: problematic though it may be whether particular institutions can be counted on to determine action, few students of IR would deny that the institution of sovereign statehood has an impact on international politics. Indeed, the necessities presumed to obtain in realist theory about anarchic politics may be seen as institutions, in the sense of general patterns that have evolved over long periods (Wendt 1992). Whatever the issue between realists and others, it is not whether patterns such as these are important.

When it comes to the sources of compliance, we may distinguish between a "rationalistic" and a "reflective" approach. In the former, institutions are assumed to affect actors' calculation of cost (Keohane 1988: 166). The logic is identical with that of structural theory: institutions are features of actors' environments that affect what is rational and hence what will likely be. The distinguishing feature of the reflective approach is the assumption that preferences and perceptions are not fixed but are affected by institutions. The assumption of fixed preferences and perceptions, it is argued, prevents us from understanding "how interests change as a result of changes in belief systems," thus obscuring rather than illuminating "the sources of states' policy preferences" (Keohane 1988: 390–1). Logically, this too is analogous with some forms of structuralism: features of the environment (in this case, institutions) are presumed to affect actors' preferences and perceptions, and hence their conceptions of rationality, and hence their actions. Institutions and structures are indistinguishable from this point of view. Or, better, institutions *are* social structures in the sense of patterns that exist regardless of single actors and actions.

Where do we go from here? What is an insitutionalist research program like? One possibility is to set out to gain generalized knowledge about the factors determining the impact of various kinds of institutions on actors' preferences and perceptions under varying circumstances; this in effect has been suggested by Wendt (1992) and embarked upon by Young (1979). Another possibility is to emphasize the indeterminate nature of "changes taking place in consciousness" (Keohane 1988: 391) and to focus on ex post investigation of the impact of institutions on perceptions and preferences in single cases.

The former should appeal to type-I researchers, who would see that the more we know about the conditions for compliance, the more useful data about institutions are for future-oriented purposes. The latter should be attractive from the point of view of type-II research, for whom the institutionalist perspective promises to enrich our understanding of what may be called the role of political culture in IR. A particular challenge is to explore the interaction between political cultures at the international, national and subnational levels.

## 2 Institutional change

If compliance is problematic mainly as regards particular institutions, general institutions are problematic as regards change. Whereas particular institutions may be changed by agreement among those concerned, general institutions are largely contingent on the practices of individual actors, and compliance means that they tend to remain as before. Institutionalists argue that non-institutionalists are incapable of accounting for change, but they face problems of their own in this regard.

The common-sense solution is to assume that actions and rules are interdependent. Action is affected but not determined by rules; it strengthens or weakens rules, depending upon compliance. This thought, elaborated by Dessler (1989) in relation to the agent–structure debate in the form of what he calls a "transformational model," is in truth not much different from the old notion that international law is both an influence on and a function of state practices. This notion in turn may be seen as an instance of the view that action results from previous action and affects the conditions for future action—in other words, feedback and learning. What may be new in recent scholarship of a constructivist orientation is a focus on the role of interaction in "constructing" identities and interests (Wendt 1994), a notion that has often been left implicit in IR analysis.

So far as single institutions are concerned, the issue is obviously the degree to which rules influence action and the degree to which action influences rules. An institution may be strong in the sense of being a powerful influence on actors and in the sense of being invulnerable to discrepant action. Some institutions may be both influential and invulnerable (the institution of state sovereignty, perhaps); others may bc influential but vulnerable (a newly established international regime, perhaps) or invulnerable but non-influential (the UN Charter prohibitions of the use of force, perhaps); still others may be weak on both accounts. The realist position translated into these terms is that some institutions (state sovereignty, the rules of "power politics") are so strong in terms of both influence and invulnerability that it is "realistic" to treat them as facts of international life. Radical internationalism would seem to imply that the institutions of state sovereignty and "power politics" are indeed dangerously influential but at the same time vulnerable to change by discrepant action. Moderate internationalism implies that even if the institutions of state sovereignty and "power politics" are both influential and invulnerable, other institutions can be set up as counterweights that will grow influential and invulnerable over time (Goldmann 1994).

The strength of international institutions, in the dual sense of influence and invulnerability, deserves to be a major focus of research. What this means depends on whether one's purpose is of type I or type II. In the latter case, the task is to examine the interplay of rules and actions in sufficient detail to provide explanation and understanding. Carlsnaes, who has outlined a program for such research, says that the effort will be "painstaking" but that it will "undoubtedly give a 'truer' and more 'realistic' picture of the complex, dynamic and determinative empirical mechanisms and processes preceding and underlying a given foreign policy decision"; this, he suggests, indicates in a nutshell "both the costs and the benefits of incorporating the agency-structure issue into foreign policy analysis" (Carlsnaes 1992: 266). Viewed from a perspective of type I, the need is more simply to investigate the predictive utility of such information about institutions as can be acquired with a lesser effort: the rules of international regimes, the substantive contents of foreign policy doctrines, the conditions for institutional change by deliberate action, and such like.

This, however, brings us back to the question of how to analyze action. Regardless of whether we think that satisfactory analysis makes it necessary to examine the actor's perceptions and preferences in each individual instance, or whether we rest content with postulating them, an analysis in terms of institutions will require us to operate at two levels of subjectivity: the shared subjectivity of institutions; and the individual subjectivity of actors. What brings about action is assumed to be the actor's subjective preferences together with his subjective perception of the subjective perceptions and preferences of others—all of this being observable in a fashion that is at least quasi-objective. I suspect that this paradoxical position is widely held in the discipline of IR, a fact overshadowed by our propensity to engage in polarizing "debates."

## IV Theory: conjecture or discourse?

When Stanley Hoffmann edited a text called *Contemporary Theory in International Relations* three and a half decades ago, he made a distinction between "theory as a set of answers" and "theory as a set of questions." By the former he meant "a set of explanatory hypotheses which purport to reveal the rules of the game of international politics." By the latter he meant "efforts to devise a right way of studying the phenomena of world affairs" by focusing on "the most important variables, even though no attempt is made at the outset to present any laws governing the behavior of those variables"; the claim is merely "to put into the hands of the researchers an adequate tool for their studies" (Hoffmann 1960: 29).

To make some such distinction between claims that theorists make or between functions that theories perform has recently become even more important. IR academics seem to disagree over what sort of thing a theory is. What is to some a set of conjectures à la Popper (that is, assumptions that are accepted provisionally, pending efforts at refutation) is to others a discourse oppressively compeling us to adopt a particular mode of thought. Similarly, what is to some an ideal type with which empirical observations may be compared for the purpose of identifying interesting deviations is to others a truth-claim. Only by postulating some such difference over what theory is can we comprehend the intensity of the attacks on the mainstream by "dissidents in exile" (Ashley and Walker 1990).

The debate about *Theory of International Politics* is indicative of such a difference. Waltz's work, on one interpretation, is a contribution to an ongoing search for an understanding of what goes on in world politics. Those taking Waltz's account of some features of international politics as their point of departure will find that it has limitations, such as the difficulty of pinpointing the explanatory power claimed for it. They will also find it easy to point to empirical evidence that is difficult to reconcile with it. If conjectures and refutations are the name of the game, however, bold conjectures such as Waltz's are what is needed. Without them, there would be little worth refuting.

While some critics of *Theory of International Politics* have rested content with indicating explanatory variables that have been left out of Waltz's conception of systemic theory, others have gone further. Waltz has thus been criticized for having written a book intent on explaining persistent patterns in the present international system rather than one about something else, namely, the history of this system (Ruggie 1986; Cox 1986). Even more indicative of what is at stake is Ashley's (1986: 258) assertion that what emerges from Waltz's book is an "ideology that anticipates, legitimizes and orients a totalitarian project of global proportions: the rationalization of global politics." Ashley (1986: 255) pointedly prefaces his criticism with a quotation from Bourdieu claiming that "the specifically symbolic power to impose the principles of the construction of reality . . . is a major dimension of political power." The vehemence of the attacks on *Theory of International Politics* suggests that it has been seen more widely as the exertion of discursive power rather than as a provisional statement to be exposed to efforts at refutation.

The discussion about the use of game theory in IR analyis provides another indication of a cleavage over what theory is and is not. Critics have gone beyond posing the question of whether there is sufficient similarity between international relations and standard games such as the Prisoners' Dilemma in order for a theory of the latter to improve our understanding of the former. Game-theoretical analysis is also said to address the wrong problem, since it is concerned with the implications of given preferences and perceptions without asking how preferences and perceptions are formed (Jervis 1988: 322–9). More fundamentally, to "model human actions as individually oriented, instrumentally rational actions" is a form of "microeconomic imperialism" whose "ideological character" should be highlighted rather than obscured; the use of game theory, it has been argued, implies that conflict "should" be regarded as driven by interests and not by passion; people do not accept the model of "perspicacious, self-interested decision-makers" as representing "how they should act in all circumstances" (Hurwitz 1989: 115–16).

Feminist critiques of IR theory, which Ann Tickner considers in her contribution to this book (below: chap. 18), may be a third indication of disagreement over what theory is. The claim is not merely that mainstream theory is limited in its ability to explain what it sets out to explain because of its being gendered. The claim is also that mainstream theory actively prescribes male-oriented concerns. Again the view is of theory as oppressingly normative rather than conjectural and analytic.

Walker (1992: 5–6) is clear about what a normative view of theory entails. IR theories, he asserts, are interesting mainly as "expressions of the limits of the contemporary political imagination." Attempts to "think otherwise about political possibilities are constrained by categories and assumptions that contemporary political analysis is encouraged to take for granted." IR theory may be read as "a constitutive practice whose effects can be traced in the remotest interstices of everyday life."

Maybe it should not simply be assumed that, say, the use of game theory by some academics has such profound implications. Granted, however, that established theory is normative in the sense of defining the central puzzles and legitimate concerns of a discipline and that this is not devoid of political implications, there is a dilemma here, since many if not most IR scholars take it for granted that "theory as a set of questions" is indispensable in research. The dilemma is obvious; that is not the issue. The issue is whether we should set out to refute theories by empirical observation and conceptual reflection or by examining whether their implications accord with our preferences—whether, to put it bluntly, our proper professional role is that of a truth-seeker or an ideology-producer. This is one of the matters that the debate in the discipline is about.

## V Conclusion

Students of IR disagree about what ought to be their main concern (peace and security, political economy, the situation of women, etc.). They disagree over whether to favor the integration, internationalization and transnationalization of states and politics or the maintenance of national independence and identity. They differ, furthermore, as to whether their proper role is that of an activist, an analyst, or a critic—and, it seems, over the legitimate role of political persuasion in social-scientific writing. Students of IR are thus like other academics, disagreeing over matters that are political rather than academic.

They also differ over matters of substance, most fundamentally about the states system as a political institution (its impact on action, its vulnerability to change) as well as over the relative significance of various factors at various levels in accounting for foreign policy, including in particular the autonomous role of ideas. Differences over substance invigorate research, but only if there is agreement on the fundamentals of method. If there has been an appreciable consensus on these matters in the discipline, it is being challenged on three fronts: (*a*) what a proper analysis of action is like; (*b*) whether knowledge is an "arbitrary cultural construct" or whether there can be objective standards; and (*c*) whether theory is a constructive guide for research or an oppressive constraint on thinking. These challenges, rather than divisions over value and substance, make international relations today a fragmented discipline.

## Acknowledgments

I am indebted to Lisbeth Aggestam, Jens Bartelson, Jan Hallenberg, Robert Keohane, Alexa Robertson, Jacob Westberg and the editors for their comments on earlier versions.

## References

The multiplicity of traditions in IR thinking is outlined in T. Nardin and D. R. Mapel, eds., *Traditions of International Ethics* (1992). A recent, book-length survey of present-day research, including work in languages other than English, is A. J. R. Groom and M. Light, eds., *Contemporary International Relations: A Guide to Theory* (1994). A much-quoted work about the problem of international order is H. Bull, *The Anarchical Society* (1977). Influential contributions to the debate about neo-realism, including much of Waltz's *Theory of International Politics*, may be found in R. O. Keohane, ed., *Neo-realism and Its Critics* (1986); see also B. Buzan, C. Jones and R. Little, *The Logic of Anarchy: Neo-realism to Structural Realism* (1993). The vast scientifically oriented literature about peace and war includes, for example, B. Bueno de Mesquita, *The War Trap* (1981) and J. A. Vasquez, *The War Puzzle* (1993). For an example of the political-economy orientation see D. Baldwin, *Economic Statecraft* (1985).

Allan, P. 1992. The end of the Cold War: the end of international relations theory? Pp. 226–41 in *The End of the Cold War*, eds. P. Allan and K. Goldmann. Dordrecht: Martinus Nijhoff.

Ashley, R. K. 1986. The poverty of neo-realism. In Keohane 1986: 255–300.

—— and Walker, R. B. J., eds. 1990. Speaking the language of exile: dissidence in international studies. *International Studies Quarterly* (special issue), 34: 257–416.

Baldwin, D. 1985. *Economic Statecraft*. Princeton, N.J.: Princeton University Press.

Banks, M., ed. 1984. *Conflict in World Society*. Brighton: Wheatsheaf Books.

Brown, C. 1992. *International Relations Theory*. London: Harvester Wheatsheaf.

Brown, R. 1993. Introduction: toward a new synthesis of international relations. Pp. 1–20 in *From Cold War to Collapse*, ed. M. Bowker and R. Brown. Cambridge: Cambridge University Press.

Bueno de Mesquita, B. 1981. *The War Trap*. New Haven, Conn.: Yale University Press.

Bull, H. 1977. *The Anarchical Society*. London: Macmillan.

Buzan, B.; Jones, C.; and Little, R. 1993. *The Logic of Anarchy*. New York: Columbia University Press.

Carlsnaes, W. 1992. The agency-structure problem in foreign policy analysis. *International Studies Quarterly*, 36: 245–70.

Cox, R. W. 1986. Social forces, states and world orders: beyond international relations theory. In Keohane 1986: 204–54.

Dessler, D. 1989. What's at stake in the agent-structure debate? *International Organization*, 43: 441–73.

East, M. A.; Salmore, S.; and Hermann, C. F., eds. 1978. *Why Nations Act*. Beverly Hills, Calif.: Sage.

Ferguson, Y. H., and Mansbach, R. W. 1989. *The State, Conceptual Chaos and the Future of International Relations Theory*. Boulder, Colo.: Lynne Rienner.

Gaddis, J. L. 1992. International relations theory and the end of the Cold War. *International Security*, 17: 5–58.

Galtung, J. 1964. An editorial. *Journal of Peace Research*, 1: 1–4.

Galtung, J. 1967. *Fredsforskning.* Stockholm: Prisma.
—— 1969. Violence, peace and peace research. *Journal of Peace Research,* 6: 167–91.
George, A. 1993. *Bridging the Gap.* Washington, D.C.: United States Institute of Peace.
George, J. 1989. International relations and the search for thinking space: another view of the third debate. *International Studies Quarterly,* 33: 269–79.
Goldmann, K. 1994. *The Logic of Internationalism.* London: Routledge.
—— 1995. Im Westen nichts Neues: seven international relations journals in 1972 and 1992. *European Journal of International Relations,* 1: 245–58.
Goldstein, J., and Keohane, R. O., eds. 1993. *Ideas and Foreign Policy.* Ithaca, N.Y.: Cornell University Press.
Groom, A. J. R., and Light, M., eds. 1994. *Contemporary International Relations.* London: Pinter.
Hermann, C. F.; Kegley, C. W. Jr.; and Rosenau, J. N., eds. 1987. *New Directions in the Study of Foreign Policy.* Boston: Allen and Unwin.
Hoffmann, S. 1960. *Contemporary Theory in International Relations.* Englewood Cliffs, N.J.: Prentice-Hall.
—— 1977. An American social science: international relations. *Daedalus,* 106: 41–60.
Hollis, M., and Smith, S. 1990. *Explanation and Understanding in International Relations.* Oxford: Clarendon Press.
Holsti, K. J. 1982. *Why Nations Realign.* London: Allen and Unwin.
—— 1985. *The Dividing Discipline.* Boston: Allen and Unwin.
—— 1991. *Peace and War: Armed Conflicts and International Order 1648–1989.* Cambridge: Cambridge University Press.
Hoole, F. W., and Zinnes, D. A., eds. 1976. *Quantitative International Politics.* New York: Praeger.
Hurwitz, R. 1989. Strategic and social fictions in the prisoner's dilemma. Pp. 113–34 in *International/Intertextual Relations,* ed. J. Der Derian and M. J. Shapiro. Lexington: Lexington Books.
Jervis, R. 1988. Realism, game theory and co-operation. *World Politics,* 40: 317–49.
—— 1992. The future of world politics: will it ressemble the past? *International Security,* 16: 39–73.
Keohane, R. O., ed. 1986. *Neo-realism and Its Critics.* New York: Columbia University Press.
—— 1988. International institutions: two approaches. *International Studies Quarterly,* 32: 379–96.
—— 1989. *International Institutions and State Power.* Boulder, Colo.: Westview Press.
Knutsen, T. L. 1992. *A History of International Relations Theory.* Manchester: Manchester University Press.
Krasner, S., ed. 1983. *International Regimes.* Ithaca, N.Y.: Cornell University Press.
Lapid, Y. 1989. The third debate: on the prospects of international theory in a post-positivist era. *International Studies Quarterly,* 33: 235–61.
Maghroori, R., and Ramberg, B., eds. 1982. *Globalism Versus Realism.* Boulder, Colo.: Westview Press.
March, J. G., and Olsen, J. P. 1984. The new institutionalism: organizational factors in political life. *American Political Science Review,* 78: 734–49.
Milward, A. S. 1992. *The European Rescue of the Nation-State.* London: Routledge.
Morgenthau, H. J. 1985. *Politics Among Nations.* 6th edn., rev. by K. W. Thompson. New York: Knopf.
Nardin, T., and Mapel, D. R., eds. 1992. *Traditions of International Ethics.* Cambridge: Cambridge University Press.

Olson, W. C., and Groom, A. J. R. 1991. *International Relations Then and Now*. London: HarperCollins Academic.
—— and Onuf, N. 1985. The growth of a discipline: reviewed. Pp. 1–28 in S. Smith, ed., *International Relations*. Oxford: Basil Blackwell.
Platig, R. 1969. International Relations as a Field of Inquiry. Pp. 6–19 in *International Politics and Foreign Policy*, ed. J. N. Rosenau. 2nd edn. New York: Free Press.
Richardson, L. F. 1960. *Arms and Insecurity*. Pittsburgh: Boxwood.
Rosenau, J. N. 1970. *The Adaptation of National Societies*. New York: McCaleb-Seiler.
—— and Czempiel, E. O., eds. 1992. *Governance without Government*. Cambridge: Cambridge University Press.
Rosenau, P. M. 1992. *Postmodernism and the Social Sciences*. Princeton, N.J.: Princeton University Press.
Ruggie, J. G. 1986. Continuity and transformation in the world polity: toward a realist synthesis. In Keohane 1986: 131–57.
Russett, B. M., ed. 1972. *Peace, War, and Numbers*. Beverly Hills, Calif.: Sage.
Singer, J. D. 1986. Research, policy, and the correlates of war. Pp. 44–59 in *Studies of War and Peace*, ed. Ø. Østerud. Oslo: Norwegian University Press.
Vasquez, J. A. 1983. *The Power of Power Politics*. London: Frances Pinter.
—— 1993. *The War Puzzle*. Cambridge: Cambridge University Press.
Wæver, O. 1992. *Introduktion til Studiet af International Politik*. Copenhagen: Forlaget Politiske Studier.
Walker, R. B. J. 1992. *Inside/Outside*. Cambridge: Cambridge University Press.
Waltz, K. N. 1975. Theory of international relations. Vol. viii, pp. 1–85 in *Handbook of Political Science*, ed. F. I. Greenstein and N. W. Polsby. Reading, Mass.: Addison-Wesley.
—— 1979. *Theory of International Politics*. Reading, Mass.: Addison-Wesley.
Wendt, A. 1992. Anarchy is what states make of it: the social construction of power politics. *International Organization*, 46: 391–425.
—— 1994. Collective identity formation and the international state. *American Political Sceince Review*, 88: 384–96.
Young, O. R. 1979. *Compliance and Public Authority*. Baltimore, Md.: Johns Hopkins University Press.
—— 1989. *International Co-operation*. Ithaca, N.Y.: Cornell University Press.
Zacher, M. W. 1992. The decaying pillars of the Westphalian temple: implications for international order and governance. In Rosenau and Czempiel 1992: 58–101.

Chapter 17

# International Relations: Neo-realism and Neo-liberalism

David Sanders

THERE has been a dramatic increase in the diversity and range of theorizing about international relations over the past two decades. Not only has the analytical rigor of "orthodox" theoretical approaches been strengthened, but the introduction of additional perspectives has brought new theories, epistemologies and even ontologies to bear on traditional questions about inter- and intra-state behavior. This chapter focuses on the two main strands of the "orthodoxy": neo-realism and neo-liberalism. In particular, it seeks to assess how far the increased analytical rigor injected by game theory into the neo-realist/neo-liberal debate has contributed to our ability to "explain" or "understand" the behavior of state and non-state actors in the global system.

I approach this task, first, by reviewing the logic which led scholars to import game theoretic language and models into the analysis of international relations in the first place. I then identify a limited number of analytic weaknesses that stem from this importation. Finally, I conduct a "thought experiment" which attempts to specify what neo-realism and neo-liberalism might look like if their efforts to constitute versions of rational choice theory were substantially downgraded. I describe the combined result as "concessional realism"[1]—a simple but flexible set of propositions about nation-state behavior in the contemporary international system. The research program suggested by "concessional realism" is rather different from that engendered by the current neo-realist–neo-liberal debate. It implies a much more direct focus on the problems of categorizing and identifying national and transnational "interests." It also implies a much

[1] This description derives from Spegele (1983).

stronger focus both on the problem of "Hobbesian fear" (Butterfield 1958) and on theory-guided empirical research which examines a wider range of "real" instances of foreign policy decision-making.

## 1 Origins: traditional realism, neo-realism and neo-liberalism

The current concern of neo-realist and neo-liberal scholars with game theoretic formulations originated partly with Waltz's efforts (Waltz 1979) to convert "traditional realism" into a "neo-realist" or "structural" theory. Traditional realism was both a simple decision-making theory and a proto-structural theory about outcomes in the international system (Morgenthau 1967; Carr 1946). In decision-making terms, it offered an unambiguous, if simplistic, analysis of foreign policy calculation. State strategy was aimed fundamentally at maximizing the state's interests and was underpinned by three "Hobbesian" motives: achieving and maintaining the state's security; satisfying the economic demands of politically significant sections of the domestic population; and enhancing the state's international prestige. The paramount need for security was best achieved by maximizing the state's power capabilities. Traditional realism took on the character of a proto-structural theory in two senses. First, the condition of international anarchy (which derived from the absence of a Leviathan-like world government) was seen as the determining structural factor that lead decision-makers to adopt "safety first" strategies of realpolitik in order to protect and maximize the interests of their respective nation-states. Second, the character and outcomes of the interactions between different states were determined by the overall pattern of national interests: friendship and co-operation between states were considered to derive fundamentally from convergences of their respective national interests; enmity and confrontation from conditions of interest-divergence.

Waltz's central claim was that any analysis of international politics which confined itself merely to the attributes of the (nation-state) units or to the interactions between units was fundamentally "reductionist" and therefore inadequate (Waltz 1979: 18–37). On the contrary, what was required was a thoroughgoing analysis of international structure and its consequences both for nation-state behavior and for the outcome of nation-state interactions. Notwithstanding Waltz's critique of the Hobson–Lenin thesis, what he attempted to develop was precisely what Marxism always claimed to provide: a structural explanation of state behavior.

Waltz developed the notion of structural explanation in two ways. First, in his exposition of balance of power theory, Waltz (1979: 126) attempted to provide a structural explanation of the dominant alliance strategy (the avoidance of power preponderance) that states pursue. As in traditional realism, a pivotal role was accorded to the notion that under anarchic conditions there is no security for the junior partner(s) in a winning coalition. Second, Waltz developed a structural explanation of system outcomes. Defining structure as a set of constraining conditions which produce a gap between intention and outcome, Waltz (1979: 89–93, 119–22) drew a powerful analogy between balance of power theory and the theory of perfect competition. Under perfect competition, because of the structure of the market in which there are no barriers to entry and perfect information, firms which aim to maximize profit end up minimizing it (i.e. earning a normal profit) because more firms enter the market if greater than normal profits are being made. Waltz argued that structure exerts a similar set of effects in international political systems, where a balance of power (the outcome) emerges "fortuitously" as a result of each state independently pursuing its own self-interest (the intention).

The idea that outcomes occur which are neither intended nor desired by any of the actors involved had, of course, been a familiar theme in game theory since the 1950s. Indeed, Prisoner's Dilemma had long been recognized as a possible restatement of the Hobbesian security problem which was central to traditional realism (Axelrod 1970; Snyder 1971). It was therefore quite natural that the "unintended and undesired outcome" principle should have been taken up by a new generation of neo-realists who sought to develop it further both in terms of other sorts of game and in terms of iterated games. It was perfectly possible for international actors to prefer mutual co-operation, but the structure of the situation in which they found themselves produced an outcome of mutual defection. Equally naturally, neo-realism's neo-liberal opponents engaged in a similar exercise—though with the objective of showing that the structural constraints on co-operation implied by these games were far weaker than neo-realists supposed. In essence, Waltz's efforts to transform realism into a scientific, structural theory led directly to international relations theorists placing much more emphasis on game-theoretic approaches. The logic was simple. International relations theory should aspire to the status of structural theory; game-theoretic models described the structure of the situation in which nation-state decision-making takes place: it was obvious that the two should be combined to produce a new and more sophisticated theoretical apparatus—a task which both neo-realists and neo-liberals undertook with considerable vigor and enormous skill.

## II Limitations of the game theoretic approach to international relations theory

The neo-realist/neo-liberal debate has been criticized from a wide range of different positions, ranging from post-Marxian critical theory to feminism. Much of this criticism has been epistemological. Neo-realists and neo-liberals are variously accused of failing to recognize that their theories merely serve to justify an existing power structure (Peterson 1992); of reifying the concept of causality (Ashley 1986); of underestimating the importance of political discourses (Enloe 1994; Campbell 1992); and of failing to understand the centrality of subjective meanings, rules and rule-following behavior (Hollis and Smith 1990). I am not concerned to dwell on such critiques here. Rather, I seek to provide a critique that accepts the basic "neo-positivist" epistemology of the neo-realists and neo-liberals (Keohane 1993: 297).

Given that it is not possible here to review all of the limitations of game theoretic approaches to understanding state behavior, I focus on two key problems: their failure to provide a convincing analysis of the notion of "interests" and their ineffective specification of the notion of "structural constraints."

### A Problems relating to the role of national interests

In traditional realist theory, national interests played a pivotal role. In the discourse of foreign policy debate, "the national interest"—even in an age of proliferating international institutions and regimes—still appears to pre-occupy the private calculations and public utterances of a wide range of national leaders. The continuing centrality of interests (though they admittedly do not refer explicitly to national interests) is also acknowledged by Axelrod and Keohane (1993: 88): "Perceptions define interests . . . [To] understand the degree of mutuality of interests (or to enhance this mutuality), we must understand the process by which interests are perceived and preferences determined." Axelrod and Keohane (1993: 88) go on to specify the way in which rational choice theorists approach this problem: "One way to understand this process is to see it as involving a change in payoffs, so that a game such as Prisoners' Dilemma becomes more or less conflictual . . ."

I have no quarrel whatsoever with Axelrod and Keohane's assertion that "we must understand the process by which interests are perceived." As I understand it, the game theoretic perspective assumes that a state's "interests" are maximized if its payoff gains are maximized (Axelrod and Keohane 1993: 88–91), while its payoff gains can themselves be regarded as "gains in capabilities" (Baldwin 1993*b*: 16–17). In this sense "understanding the process by which interests are perceived" seems to consist in showing how different preference orderings can arise and how they correspond to different varieties of payoff structure: a given actor's interests are considered to vary according to the character of the particular game that it is playing at any one time. Unfortunately, this is more a definition of interests—defined in terms of payoff structures—than an account of the "process" whereby interests are perceived and determined.

A theory of interests—for that is what Axelrod and Keohane are in effect calling for[2]—cannot be based on the notion that, if we have understood the principles of gain maximization, then we have also understood the way in which nation-states seek first to specify, and then to protect and promote, their interests. We may be happy to assume that it is "in a state's interests" to maximize its gains. But this is not equivalent to saying that the principle of gain-maximizing reveals all that needs to be known about interest protection and promotion.

Interests need to be considered in their historical context. Before we can speculate intelligently either about the formation of interests or about the consequences of states having conflicting or overlapping interests, we need to know what states' leaders perceive their respective states' interests to be. This can only be achieved through extensive, laborious and difficult empirical study. To be sure, some sort of "interest typology" might prove useful in this context. It might make sense, for example, to differentiate between a nation-state's economic-ecological interests (which could consist in maximizing the long-term economic and ecological welfare of its population) and its political-security interests (which could consist in maximizing the state's ability to respond rapidly and effectively to any external threats or challenges that might impinge upon it in the future[3]). I am obviously not in a position here to offer a theory of nation-state (or sub- or supra-state) interests. Such a theory is nonetheless an essential prerequisite of a satisfactory explanation both of nation-state behavior and of the outcomes of inter-state interactions. It is also likely that a satisfactory theory of nation-state interests can only be developed, as it were, from the "bottom-up"; from a considered analysis of a large amount of (currently uncollected)

[2] Indeed, Keohane (1993: 294) explicitly suggests that "without a theory of interests . . . no theory of international relations is possible."

[3] I take it that this was what Morgenthau (1967: 5–6) had in mind when he wrote that statesmen "think and act in terms of interests defined as power."

empirical evidence relating to policy-makers' interest-perceptions. What international relations theory really needs is a new Quincy Wright: someone who has command of an enormous range of substantive case studies of interests and who possesses the theoretical insight to offer a simplifying synthesis of what s/he observes. I would certainly not want to advocate a purely inductivist approach to the analysis of international relations—that way lies mindless empiricism. Equally, as Hanson (1958) observed, there is a place for both inductive and deductive theorizing in the process of systematic enquiry. In the analysis of nation-state interests, at least for the time being, we desperately need a little less theoretically based deduction and a little more empirically based induction.

## B The ineffective specification of "structural constraints"

As noted earlier, Waltz defines structure as a set of constraining conditions that produces a gap between intention and outcome. He goes on to suggest that the two main structural characteristics of an international system are the particular distribution of capabilities that it exhibits and the (anarchic) ordering principle upon which it is based.[4] Accepting this definition, it is legitimate to ask how far the methods adopted by game theorists are capable of capturing the way in which international structure might be considered to constrain the outcomes of state interactions.

In terms of the importance of the distribution of capabilities, such methods have achieved a considerable amount. For example, Snidal's recent (1993) work on the importance of relative gains under varying conditions of system polarity—which shows that the impact of relative gains declines as the number of actors increases—continues a long line of studies that have very effectively analyzed the connections between polarity, the decision calculus of major actors and system outcomes.

The record is less convincing, however, in relation to game theorists' efforts to analyze the structural consequences of the anarchic character of the international system. The key difficulty in this context is whether the options defined even by multidimensional, multi-choice payoff matrices bear sufficient resemblance to the "real" choices of "real" policy-makers for the structure of the matrix to be regarded as an analogue for the (anarchic) structure of the international system. This question is particularly problematic with regard to the question of uncertainty. Uncertainty is one of the

[4] Waltz's (1979: 93–7) third component of structure as applied in the domestic context, the functional differentiation of units, is of no relevance in international political systems: nation-states vary in capabilities, not in function; they are "like units."

critical features of an anarchic structure that the payoff matrix approach seeks to encapsulate. The core of the problem in this regard is that payoff matrices radically understate the degree of uncertainty that actually faces policy-makers in the international system. A *kxk* matrix specifies uncertainty in terms of the *k* alternative strategies available to an opponent.[5] Yet, even in a *k*-actor situation in the "real world," the uncertainties are much greater than this *k*-choice position would imply. In deciding between competing policy options, decision-makers in "real" situations are typically confronted with the following uncertainties:

- How will different factions inside the opposing country react?
- Will this course of action strengthen the position of more or less friendly factions in the opposing country?
- Will this course of action be seen as a precedent for the future or will it be seen as *sui generis*?
- How will public opinion in the opposing country react?
- Will this course of action increase the intransigence of the opposing decision-makers or will it make them more amenable to compromise in the future?
- How will domestic public opinion react?
- How will different domestic groups or parties react?
- How will the decision-maker's own party react? Will it strengthen the leadership's hand or that of opposing factions within the party?
- Will this option have collateral effects on third parties?
- How will these effects, if they occur, affect this country's relationship with the third parties in the future?
- And, most important of all, how likely is it that this course of action will actually lead to the outcome that is desired?

The answers to most of these questions are very rarely known in the "real world" of international politics. Yet the uncertainty about all of them constitutes a crucial element of the decision calculus of the foreign policy-maker that needs to be explicitly taken into account if "real" choices, and hence "real" outcomes, are to be explained. Neither payoff matrices nor the sort of "win-set" models envisaged by Putnam (1988) in his two-level theory of games begin to capture the "real" uncertainties that confront the "real" policy-maker. Yet it is precisely these "real" uncertainties that

[5] The greatest degree of uncertainty occurs, of course, when an actor does not know what payoffs it will receive given either its own and/or its opponent's actions. This sort of situation has been handled extensively by game theorists. See, for example, Rasmussen 1989.

structure—that is, act as structural constraints upon—both the choices that are made and the outcomes that obtain in the "real world." The question must remain open therefore as to whether game-theoretic perspectives adequately reflect a central aspect of the anarchic international structure which they were expressly intended to encapsulate.

## III Towards a game theory-free concessional realism

Although the foregoing discussion is in no sense offered as a thoroughgoing critique of game-theoretic approaches to international relations, it does have some implications for the way in which the neo-realist/neo-liberal debate might be developed. The failure of the game theory approach to take adequate account of either interests or uncertainty suggests that it might be worthwhile considering what neo-realism and neo-liberalism would look like if they were shorn of their game-theoretic superstructure. The "thought experiment" that follows offers precisely such a portrayal. In deference to the increasing recognition that neo-realist and neo-liberal propositions need to be subjected to more explicit empirical scrutiny (Keohane 1993), the "experiment" emphasizes the importance of direct empirical testing. The experiment also seeks to maintain the view adopted by diplomatic historians that realism constitutes both a theory of decision-making and a simple structural theory about the origins and outcomes of nation-state behavior.

In offering revised statements of neo-realist and neo-liberal theories—which are presented in Appendices 17A and 17B—I follow Lakatos (1970) in assuming that all theories have both (1) a largely non-negotiable and non-falsifiable "core" and (2) a set of testable propositions that are to a greater or lesser degree derived from that core. In terms of the testable propositions, I distinguish between (i) statements that can be empirically evaluated by a detailed examination of the decision calculus of individual policy-makers (the "decision-making level"); and (ii) statements that require empirical testing in the context of the transnational pattern of interests or co-operation (the "structural level").

Readers will doubtless decide for themselves as to the individual and collective adequacy of the theoretical statements outlined in Appendix Tables A17.A and A17.B. Four features of the tables merit particular attention.

First, it is clear that the content of the tables could not by any stretch of the imagination be described as sophisticated. ("Simplistic stuff" is one self-criticism that occurs to me.) In comparison with the recent works of Keohane, Snidal, Grieco, Powell, Baldwin and many others, there is a huge loss of abstraction and theoretical rigor. My response to this potential criticism is to ask both myself and the reader whether this loss actually matters significantly. If we work to the agenda defined by Tables A17.A and A17.B, we will certainly know less about the theoretical world. However, I am not convinced that working to the agenda of these would necessarily tell us any less about the world of observation—which is the world we want to understand and explain anyway—than more sophisticated, rigorous and abstract analyses. The strength of the propositions embodied in Tables A17.A and A17.B lies partly in their parsimony and partly in the unambiguous empirical research agenda that they imply: a series of hypotheses that can be tested individually and directly by examining a wide range of historical case studies.

The second key feature of the tables concerns the concept of interests. The statement of concessional realism accords some prominence to perceptions of national interests. However, it is evident that a considerable amount of empirical work needs to be undertaken in order to determine what decision-makers actually perceive their states' interests to be.[6] This investigation also implies the need for an analysis of how perceptions of interests are shaped by country-specific historical forces and by institutional practices and commitments at both the national and international level. This in turn implies an emphasis on domestic political structures and processes that is not normally allowed for in neo-realist theory. It is an emphasis, however, whose importance has certainly been recognized by neo-liberals and institutionalists.[7] Given the obvious importance of domestic considerations in the study of foreign policy, it seems appropriate that a putative theory of international relations should make due allowance for the possible operation of domestic factors in the genesis of nation-state behavior.

A third feature of Tables A17.A and A17.B that merits attention concerns the concept of structure. It could be argued that the propositions as they stand significantly underplay the role of international structure and that the only hint of an analysis of structural effects, in Waltz's sense, occurs in the non-falsifiable core of realism (in Table A17.A, proposition 2*b*). There is certainly nothing comparable to the sort of structural effects that Waltz describes in his discussion of the analogy between perfect competition and

[6] Given the increasing role of contemporary international institutions, the notion of "decision-makers' perceptions" should probably be extended to include the leaders of major transnational institutions.

[7] See, for example: Axelrod and Keohane 1993: 101–2; Elman 1995.

balance of power theory. In response to this criticism, I would argue that the overall pattern of perceived interests—which can only be specified by first considering states' perceptions of their interests—is just as much a structural characteristic as Waltz's "distribution of capabilities." In this sense, in looking at overlapping and contradicting patterns of interest, the model outlined in Table A17.A does take some account of possible structural effects. If this still represents a weak analysis of structural effects in Waltz's sense, then so be it. I can only observe in mitigation that, as far as I am aware, no one else (Waltz included) has managed to produce an account of the effects of international structure that would bear comparison with the way in which structural effects operate under conditions of perfect competition.

My final comment about the theoretical restatements presented in Tables A17.A and A17.B concerns the alleged tensions between neo-realism and neo-liberalism. Although neo-realism and neo-liberalism accept some of the same preliminary assumptions (notably that the international system is an anarchic one composed of self-regarding, interest-maximizing states), it is generally accepted that the two theories do generate different expectations about the prospects for co-operation between nation-states—with the neo-liberal position, obviously, being the more optimistic. It is also widely acknowledged that neo-realism seems to provide a better model for analyzing the military-security problems that states confront, while neo-liberalism offers a more useful account of their relations in the political-economic sphere (Grieco 1993*a*: 131).

The question that obviously arises in this context is why this pattern of differential success should be observed. Grieco gets the closest to providing an answer when he discusses his *k* coefficient—which measures a state's "sensitivity to gaps in payoffs" in its dealings with another state. Grieco (1993*b*: 323) suggests that, "In general, *k* is likely to increase as a state transits from relationships in what Karl Deutsch terms a 'pluralistic security community' to those approximating a state of war . . . [The] level of *k*, for example, will be lower if a state's partner is a long-term ally rather than a long-term adversary."

What Grieco is implying here is that the character of the relationship between any pair of states will depend upon specific historical circumstances. If nations have established a reasonable degree of mutual understanding and trust through a long period of co-operation, they are less likely to worry about relative gains (they are insensitive to "gaps in payoffs") and are therefore more likely to co-operate with each other in the future than are nations which, for whatever historical reasons, are deeply suspicious of each other. To historians, of course, this is self-evident. Yet to international relations theorists, somehow, its significance seems to be underestimated. As a formal theorist, Grieco quite rightly incorporates his *k* coefficient into his formal model of the decision calculus of the state under anarchic conditions. In so doing, however, he misjudges the explanatory importance of historical circumstance. It might well be the case that states worry less about relative gains when they are dealing with long-term friends and allies. The crucial point, however, is that when "real" states deal with long-term friends and allies the whole character of their interaction is fundamentally different from the sort of interaction that occurs when they deal with states for whom they retain a sense of Hobbesian fear. Grieco's *k* coefficient—though it is an extremely clever device—simply doesn't go far enough. It cannot be stressed enough that states discriminate in the ways that they relate to other states. Just as an individual relates differently to each of her/his various friends and acquaintances, so a state will behave and calculate differently towards the various states with which it has contact. And precisely because states discriminate, they make different sorts of calculation about the costs and benefits of co-operation depending upon whom they are dealing with. Nation-states may or may not be sufficiently alike in the problems that they confront to merit Waltz's description of them as "like units." "Real"—discriminating—decision-makers most assuredly do not see all other states as "like units": they invariably regard some actors as "friends" and others as either real or potential "enemies."

What all of this suggests is that the attempt to develop a single decision calculus (based on either absolute or relative gains) in order to analyze nation-state decision-making is unlikely to capture the sheer inconsistency of decision-making in the "real" world. At a minimum, we need to develop two different models of nation-state decision-making—one for situations where historical circumstances mean that a condition of Hobbesian fear still exists between the parties involved; and one for situations where Hobbesian fear has, for whatever historical reasons, been transcended. If we think about it, this is precisely what neo-realism and neo-liberalism have, in their separate ways, already provided us with. Where Hobbesian fear exists, military-security issues predominate and the neo-realist model proves broadly satisfactory. Where Hobbesian fear has been transcended, political-economic issues predominate and the neo-liberal model works smoothly. If this characterization fails to explain why Hobbesian fear is sometimes transcended and sometimes not, I am unconcerned: these are matters of historical contingency rather than questions of theory.[8]

[8] Consider, for example, the case of postwar Anglo-American relations. It would be ridiculous to suppose that an abstract theory could explain why U.S.–U.K. relations were so close in the 1960s that

If the plausibility of this argument is accepted, the implications for the propositions specified in Tables A17.A and A17.B are clear. Proposition 8 of Table A17.A suggests that:

> The boundaries of anarchy/Hobbesian fear can be redrawn (through changing alliances and bloc memberships) but never eliminated without Leviathan. Where the boundaries of Hobbesian fear have been eliminated between countries, political and economic co-operation between them will be more readily achieved.

One way of thinking about neo-liberal theory is to regard it as a set of sub-hypotheses that fall within the confines of proposition 8. Where Hobbesian fear has been eliminated, the neo-liberal propositions summarized in Table A17.B would come into operation; where it has not, realism would still hold sway.[9] The resultant combination could be described as a "concessional realist" model. Just as realism, in transmuting itself into neo-realism, made an epistemological "concession" to positivism (Spegele 1983), so neo-realism now needs to make a substantive concession to neo-liberalism by recognizing that, in specified circumstances, the neo-liberals simply tell a more plausible story. It should also be stressed that the idea of neo-liberalism as a special case of concessional realism is not intended to constitute a "demotion" of neo-liberalism, merely a convenient location of it. Neo-liberals should be reassured that such a location would not imply a diminution of the theoretical power of their analysis. The removal of Hobbesian fear does not mean the end of anarchy. It merely means that the anarchic structure in which states must operate is "mature" as opposed to "immature" (Buzan 1991). There is still no Leviathan to see fair play and there is still considerable uncertainty both about the aims and future behavior of other actors and about the outcomes of co-operation or confrontation. As with all compromises, this proposed fusion of neo-realist and neo-liberal thinking will probably appeal to neither. We shall see.

the Americans could trust the British sufficiently to sell them nuclear weapons. (Indeed Waltz (1979: 121) counsels against expecting a theory to "tell us why state *X* made a certain move last Tuesday.") The U.S. might have been uncertain about what foreign policies Britain would pursue in a wide range of contexts. What it was certain about, however, was that (1) the British would never use the weapons against the U.S. and (2) that the British were sufficiently responsible not to use them without close consultation with Washington. The U.S. decision could, of course, be analyzed in terms of either relative or absolute gains calculations, but this would not explain how the extraordinary degree of trust that existed had been built up over the previous 25 years. The nature of the calculation that the U.S. government made was dependent on the character of the relationship that existed between Washington and London; and that character was in turn determined by contingent historical circumstances.

[9] The extent to which neo-liberal propositions were also relevant under conditions of Hobbesian fear could still be investigated empirically.

## Acknowledgments

I am grateful to Hugh Ward, Eleni Moraiti, Bob Goodin and participants at the XVIth World Congress of the International Political Science Association in Berlin for their helpful comments.

## References

Ashley, R. 1986. The poverty of neo-realism. In Keohane 1986: 255–300.

Axelrod, R. 1970. *Conflict of Interest.* Chicago: Markham.

—— and Keohane, R. O. 1993. Achieving co-operation under anarchy: strategies and institutions. In Baldwin 1993*a*: 85–115.

Baldwin, D. A., ed. 1993*a*. *Neo-realism and Neo-liberalism.* New York: Columbia University Press.

—— 1993*b*. Neo-liberalism, neo-realism and world politics. In Baldwin 1993*a*: 3–28.

Bull, H. 1977. *The Anarchical Society.* London: Macmillan.

Butterfield, H. 1958. *Christianity and History.* London: Collins.

Buzan, B. 1991. *People, States and Fear.* 2nd edn. Boulder: Lynne Rienner.

Campbell, D. 1992. *Writing Security.* Manchester: Manchester University Press.

Carr, E. H. 1946. *The Twenty Years' Crisis, 1919–1939.* London: Macmillan.

Enloe, C. 1994. Margins, silences and bottom rungs: how to overcome the estimation of power in the study of international politics. Paper presented to the conference on "After Positivism: Theory and Method in World Politics," University of Wales, Aberystwyth, July 1–3 1994.

Galtung, J. 1964. A structural theory of aggression. *Journal of Peace Research*, 1: 95–119.

Grieco, J. M. 1993*a*. Anarchy and the limits of co-operation: a realist critique of the newest liberal institutionalism. In Baldwin 1993*a*: 116–40.

—— 1993*b*. Understanding the problem of international co-operation: the limits of neo-liberal institutionalism and the future of realist theory. In Baldwin 1993*a*: 301–38.

Hanson, N. R. 1958. *Patterns of Discovery.* Cambridge: Cambridge University Press.

Hobbes, T. 1651. *Leviathan*, ed. J. Plamenatz. London: Fontana, 1962.

Hollis, M., and Smith, S. 1990. *Explanation and Understanding in International Relations.* Oxford: Clarendon Press.

Krasner, S. D., ed. 1983. *International Regimes.* Ithaca, N.Y.: Cornell University Press.

Keohane, R. O., ed. 1986. *Neo-realism and Its Critics.* New York: Columbia University Press.

—— 1993. Institutional theory and the realist challenge after the cold war. In Baldwin 1993*a*: 269–300.

Krasner, S. D. 1993. Global communications and national power: life on the pareto frontier. In Baldwin 1993*a*: 234–49.

Lakatos, I. 1970. Falsification and the methodology of scientific research programmes. Pp. 91–196 in *Criticism and the Growth of Knowledge*, ed. I. Lakatos and A. E. Musgrave. Cambridge: Cambridge University Press.

Morgenthau, H. J. 1967. *Politics Among Nations.* 4th edn. New York: Knopf.

Peterson, V. S., ed. 1992. *Feminist Revisions of International Relations Theory.* London: Lynne Reiner.

Powell, R. 1993. Absolute and relative gains in international relations theory. In Baldwin 1993*a*: 209–33.

Putnam, R. D. 1988. Diplomatic and domestic politics: the logic of two-level games. *International Organisation*, 42: 427–61.

Rasmussen, E. 1989. *Gains and Information.* Oxford: Basil Blackwell.
Snidal, D. 1993. Relative gains and the pattern of international co-operation. In Baldwin 1993*a*: 170–208.
Snyder, G. H. 1971. Prisoners' Dilemma and Chicken models in international relations. *International Studies Quarterly,* 15: 66–103.
Spegele, R. D. 1983. Alarums and excursions: the state of the discipline in international relations. *Politickon* (South Africa), 10: 51–72.
Waltz, K. N. 1979. *Theory of International Politics.* Reading, Mass.: Addison-Wesley
Ward, H. 1993. Game theory and the politics of the global commons. *Journal of Conflict Resolution,* 37: 203–35.

# Appendix 17A

# Concessional Realism

**Table A17.A Propositions summarizing concessional realism**

## A The non-falsifiable core of concessional realism

1 The international system is an anarchic, self-help one. In the absence of a single effective central authority, the nation-state has to rely primarily—though not necessarily exclusively—on its own efforts to protect the economic and security interests of its citizens.

2 The Hobbesian state of nature model, with certain modifications, provides a useful analogue for the international system. Specifically:

*a*) It is not the internal characteristics of the state that determine its external behavior though its internal characteristics may affect the way in which its leaders define national interests. The crucial determining factor in external behavior is the external environment in which the state finds itself. This environment engenders a high level of uncertainty for state decision-makers and fosters a sense of "Hobbesian fear" among them.

*b*) If one state behaves in a consistently belligerent fashion in an anarchic system, then all potentially affected states must either follow a realpolitik strategy in return or else contrive the protection of another state that does. A realpolitik strategy can be defined as follows: *A* will increase its (usually military) power capabilities in order (i) to increase its ability to use force (or the threat of force) against *B* and (ii) to minimize the chances that *B* will use force (or the threat of it) against *A*.

c) As a result of (*a*) and (*b*) states are in a continuous state of preparedness to use force for either offensive or defensive purposes.
d) For a variety of historical reasons, states can redraw the boundaries of Hobbesian fear and as a result combine themselves into more or less cohesive blocs. In security matters, the calculations and behavior of these blocs will be analogous to the calculations and behavior of states.
e) The external policies of states or blocs are driven by three main motives: security ("safety"), gain ("satisfying appetites"), and glory/prestige ("reputation").

3 Under anarchy, the state's dominant behavior strategy (disagreed) is to maximize its own power capabilities (Bull 1977) or to avoid the development of power preponderance elsewhere (Waltz) or to avoid relationships in which it is obliged to play a subordinate role (Galtung 1964) or to prevent others from making advances in their relative capabilities (Grieco).

4 Under anarchy, the pursuit of this dominant strategy produces a tendency towards pre-emptive expansionism or political imperialism among the states which possess the physical capability to undertake it. The forms which this imperialism take will vary according to historical context. Nation-states will use whatever political and/or economic mechanisms are convenient to sustain and strengthen their existing position in the international system. Sometimes imperialism will consist in the development of international institutions that operate disproportionately to reinforce the interests of the sponsoring state(s).

## B Concessional realist propositions capable of being tested at the individual, decision-making level

5 Given the continuing threat to the state or bloc from its external environment, the most important motive underlying external policy is always the attainment of national (or bloc) security; economic gain calculations will always be subordinated to security calculations if there is any tension between the two.

6 Under anarchy, the intentions of potential antagonists are inferred from their capabilities, not from their express statements or promises. Monitoring of opponents' capabilities is a permanent feature of international politics.

7 Grieco's binding thesis: "If states share a common interest and undertake negotiations on rules constituting a collaborative arrangement, then the weaker but still influential partners will seek to insure that the rules so constructed will provide for effective voice opportunities for them and will thereby prevent or at least ameliorate their domination by stronger partners" (Grieco 1993*b*: 331).

8 The boundaries of anarchy/Hobbesian fear can be redrawn (through changing alliances and bloc memberships) but never eliminated without Leviathan. Where the boundaries of Hobbesian fear have been eliminated between countries, political and economic co-operation between them will be more readily achieved.

9 Moral principles have no real force in international politics when nations' vital interests are at stake except as *ex post* justifications for actions based on other, interest-maximizing calculations.

10 International law has no real constraining force in international politics when nations perceive that their "vital interests" are at stake. Legal principles, like morals principles, are espoused merely to justify decisions already taken on the basis of interest-maximizing criteria.

## C Concessional realist propositions capable of being tested at the structural level

11 Co-operation between or among nation-states derives from decision-makers perceiving that there is a convergence of economic and/or security interests between or among the states involved. Confrontation or discord between nation-states—especially including overt violent conflict—is the consequence of decision-makers perceiving that there is a serious divergence of interests between the states involved.

12 If the decision-makers in competing states perceive that the security interests of their respective states are intensely divergent, no amount of co-operation between the states involved can improve the relations between them. Co-operation in "areas that don't matter" cannot contribute to the transformation of the relations between states: co-operation has no autonomy independently of the logic of the pattern of perceived interests.

13 In the security sphere, if power capabilities are relatively evenly distributed between or among potential antagonists, then a balance of power can be said to exist between or among them. Such a balance, in turn, produces a position of mutual deterrence in which each party calculates that it would incur more overall cost than overall benefit if it were to engage in aggression. If no such balance exists, then either coercive diplomacy or war (or both) will occur.

14 Outcomes in international politics are determined entirely by the pattern of political and economic power relations. The parties with the strongest array of relevant capabilities will always prevail in any conflict or negotiation.

# Appendix 17B

## Neo-liberalism

Table A17.B Propositions summarizing neo-liberalism

## A The non-falsifiable core of neo-liberalism

1 The international system is an anarchic, self-help one. The states that comprise it are self-regarding interest-maximizers.

2 The effects of anarchy in the contemporary international system are strongly mitigated by the relatively high levels of interdependence shared by many, if not all, nation-states. Interdependence consists in (*a*) economic interpenetration in terms of international trade and financial flows; (*b*) nation-states' collective interest in avoiding a major nuclear war; and (*c*) nation-states' collective interest in avoiding ecological catastrophe.

3 States co-operate with one another when they recognize that they share a mutual interest in so doing and recognize that reciprocity will produce a positive gain for each of them.

4 States sometimes recognize that the long-term costs of confrontation with another state or group of states is so great that it is better to contrive ways of co-operating with that state or group of states; this may help to break down the sense of mistrust that is felt between the states' leaderships.

## B Neo-liberal propositions capable of being tested at the individual, decision-making level

5 In interactions with other states, the fundamental goal of the state is to maximize the absolute gains that it makes.

6 Productive co-operation between states is easier to contrive among states if fewer states are involved in the co-operation effort.

7 Actors try to change the context in which they interact—they seek to reduce the sense of Hobbesian fear and mistrust that they feel towards other actors—in order to increase the chances of co-operation. This leads to the creation of international institutions and regimes and prompts states to attempt to gain acceptance for new norms of international behavior.

8 Over time, international institutions can influence the ways in which national decision-makers perceive the interests of their respective states. In particular, institutions can facilitate further self-interested co-operation by reducing uncertainty; by increasing the quality, evenness and volume of information; and by establishing and reinforcing a state's reputation for reliability.

9 The leaderships of states which enjoy an extensive network of transnational ties are more likely to develop common perceptions and expectations which in turn will facilitate further co-operation. This process is also likely to be fostered by the growth of "epistemic communities."

10 International regimes ("sets of implicit or explicit principles, norms, rules and decision-making procedures around which actors' expectations converge in a given area of international relations") facilitate co-operation and reinforce reciprocity by "delegitimizing defection" (Krasner 1983: 3).

## C Neo-liberal propositions capable of being tested at the structural level

11 International institutions are capable of developing sufficient political resources to enable them to engage in supranational enforcement, thereby overcoming

the constraining effects of anarchy. Equally, states will not always abide by their international commitments.

12 Institutional inertia can lead international institutions to persist after the perceived convergence of interests that created them has disappeared.

Chapter 18

# International Relations: Post-Positivist and Feminist Perspectives

J. Ann Tickner

In the *Handbook of Political Science*, published in 1975, Richard Smoke claimed that "the existing theory . . . of the field [is] probably not capable of coping with a world which is changing so rapidly and so dangerously, both in its military technology and in the patterns of its international politics" (Greenstein and Polsby 1975: 339). In spite of these prescient warnings, it is doubtful whether any of the authors in the "International Politics" volume of the *Handbook* could have predicted the extent of the changes that have taken place since its publication both in the international system and in the discipline of international relations.[1] As we grope toward a better understanding of this confusing and changing world, the optimism about the possibility of theoretical progress, which many of these authors expressed, has largely dissipated. After a brief outline of the contents of the *Handbook*, I will elaborate on the erosion of this theoretical consensus. Having reviewed some post-positivist critiques of mainstream theory, I will suggest some ways of facilitating conversations across epistemological and theoretical divides. In conclusion, I will examine how some recent feminist perspectives are making contributions to this reconstructive project. The intention of this paper is to highlight critical and feminist approaches rather than provide a comprehensive overview of the field.

[1] The naming of the volume, "International Politics" rather than "International Relations" is instructive and evidences the belief in the possibility of constructing *political* theories to explain international relations.

## I International relations theory in the handbook

Except for two chapters, one on international law (Lipson) and one on interdependence and integration (Keohane and Nye), all the authors in the *Handbook* employed realist, state-centric assumptions and focused on traditional issues of war and peace described, for the most part, in terms of conflictual relations between the great powers. Claiming that "the agonizing Vietnam experience cast a pall over the whole topic of limited war and over researchers' interest in exploring it further," Richard Smoke's chapter on national security, centered on the evolution of superpower nuclear strategy (Greenstein and Polsby 1975: 321). While George Quester avowed that we should be "distressed by gross inequalities of income" in the world, his chapter was concerned with economic issues only insofar as they impacted on the potential for military conflict (Greenstein and Polsby 1975: 237). The fact that the *Handbook* contained only American authors spoke to the prevailing U.S. hegemony, both in the international system and in the discipline that described it.

For the most part, the authors of the *Handbook* were positive about the potential for theoretical progress. Neo-realist Kenneth Waltz was optimistic about the possibility of constructing systems-level theory capable of generating testable hypotheses (Greenstein and Polsby 1975: 15). His own structural explanation of state behavior has centered on systems-level balance of power theory modeled on microeconomics (Waltz 1979). In her *Handbook* chapter, Dina Zinnes also expressed optimism that the discipline was capable of assembling a cumulative body of knowledge with which to develop a scientific study of international politics. In contrast to Waltz's abstract structural analysis, Zinnes emphasized empirically based research possibilities; her approach focused on systematic data-gathering and the search for patterns and relationships among variables which could lead to the testing of hypotheses about the foreign conflict behavior of states (Greenstein and Polsby 1975: 92–9).

## II An eroding consensus

Subsequent ferment in the discipline suggests that this optimism about the potential of "scientific" theories of international politics was short-lived. Only ten years after the publication of the *Handbook*, K. J. Holsti, a historically and empirically oriented scholar, asserted that international theory was in disarray: what he saw as a three-centuries-long intellectual consensus that had provided hypothetical answers to critical questions about international politics was breaking down (Holsti 1985: 1). The incorporation into the discipline of new issues and actors challenged the hegemony of state-centric analysis and its focus on issues of war and peace apparent in most of the chapters in the *Handbook*. According to Holsti, the lack of consensus about subject-matter and how to theorize about it was throwing the field into a state of theoretical confusion.

Holsti's pessimistic analysis of the state of international theory was partially confirmed by other writers in the 1970s and early 1980s, although some scholars saw the splits among traditions in a more positive light. In his *Anarchical Society*, Hedley Bull, the leading theorist of the British-Australian school, reviewed the Grotian, Hobbesian and Kantian traditions of international relations. Bull drew inspiration for his key concept, the anarchical but social order of inter-state relations, out of a moderate synthesis of these traditions (Bull 1977). In a similar trichotomization, Hayward Alker and Thomas Biersteker noted the continuing presence in international theory of both realists and idealists, scientists and traditionalists, as well as the impact of Cold War allegiances or orientations. They hoped for cumulative knowledge resulting from dialogues across such groupings (Alker and Biersteker 1984).

Changes in the international system, partially responsible for this disciplinary disarray, were clearly evident in the 1970s. The OPEC cartel, shocks to the international monetary system and early signs of global recession, together with a détente between the United States and the Soviet Union, moved economic issues to the top of the international relations' agenda. With the benefits of liberal interdependence in doubt, the 1970s also spawned the birth of neo-realist approaches to international political economy and the rebirth of nationalist ones (Gilpin 1987). Combining realist assumptions with liberal predispositions, certain scholars began to analyze economic relations between states not only in terms of conflict but also in terms of their potential for building international institutions or regimes (Krasner 1982; Keohane 1984). These scholars used rational choice theory and micro-economic models to explain states' foreign economic policy behavior.

Recognition of the serious plight of the South in this turbulent global economy, together with its demands for a New International Economic Order, brought some recognition by other western scholars of the Latin American dependency approach and world-systems theory. These scholars began to focus, not on a world of autonomous state actors, but on a class-based global economy with boundaries of inequality between core and periphery created and maintained by the structural condition of uneven political and economic development (Galtung 1980; Wallerstein 1976).

With the nuclear stalemate at the center of the system, certain national security analysts began to turn their attention to what Smoke described as the previously neglected topic of limited war. "Low intensity" conflict and wars of intervention demanded different types of analysis from conflict between powerful states at the center of the system (Klare 1992). Few of these wars have been international in the sense of trans-boundary incursions: scholars have begun to look for their origins, not in the structure of the international system, but in ethnic and religious identities, or failures of state-building exacerbated by outside intervention (Jackson and Rosberg 1982).

The end of the Cold War, the lack of military rivalry between the major powers, and the continuation of conflict in the periphery have stimulated a lively debate over the emerging structure of the system. Holding to his previous structural analysis, Kenneth Waltz (1993) has predicted a shift to a multipolar world in which Germany and Japan will rearm with nuclear weapons. Seeing an eternal pattern of conflict in an anarchic world of competitive states, Waltz continues to affirm his earlier assumptions: he is not sanguine about current optimism as to the likelihood of peace at the core of the system and instead predicts an emerging multipolar competitive balance of power.

While Waltz clings to his structural balance of power model, a quite different neo-Kantian literature has emerged which claims that war at the core of the system is unlikely because democracies rarely fight each other. Testing both normative and structural models for their relative explanatory power, Bruce Russett (1993) finds a strong correlation between democratic governments and the lack of war between them. Similarly, in what they term "the zone of peace," Max Singer and Aaron Wildavsky (1993) see a fundamentally different world order in which conflict will be confined to the peripheries of the international system. While they admit that the "zones of turmoil" contain 85 percent of the world's population, Singer and Wildavsky believe their troubles are an inevitable developmental stage through which they must proceed before becoming prosperous and democratic. In these literatures, an emerging North–South dichotomy has begun to replace an East–West one. Boundaries between states of similar power and capabilities have been replaced by boundaries between rich and poor and strong and weak.

## III The current state of international theory

While they acknowledge that revolutionary changes associated with the end of the Cold War have turned the world upside down, Yale Ferguson and Richard Mansbach claim that "there may today be less anarchy in world politics than in theories about it" (1991: 363). Given the recent upheavals in the international system, many theorists believe the potential for the type of theoretical development that the authors of the *Handbook* predicted has dissipated even further than when Holsti was writing in the mid-1980s. In a severe indictment of the discipline, John Lewis Gaddis has argued that, since no major international relations theory, behavioral, structural or evolutionary, was able to predict the end of the Cold War, these theories were either artifacts of the Cold War, in which case they lacked the universal applicability they claimed, or else they were simply wrong (Gaddis 1992: 53). Claiming that anticipating the future is a principle objective of all the major theoretical approaches to international relations, Gaddis attributes this failure to inappropriate attempts by theorists to gain "scientific" legitimacy. In their quest for objectivity and predictability, what Yosef Lapid has called, "a frustrating worshipful relationship to the natural sciences," (1989: 246) international theorists are, according to Gaddis, employing traditional methods of the physical and natural sciences at a time when natural scientists are turning to new theories that can accommodate indeterminacy, irregularity, and unpredictability.

Gaddis's skepticism about the potential for constructing one objective and predictive science of international relations is shared by many contemporary scholars. Voices of dissent are proliferating; having moved beyond the introduction of new issues and actors, certain scholars have begun to question the epistemological foundations of the discipline.[2] Proclaiming a "third debate" which is challenging the positivist consensus of the early 1970s, Yosef Lapid sees "the demise of the empiricist, positivist promise" (Lapid 1989: 236).[3] "Scientific" claims about "truth," "rationality," "objectivity," and "reality" are coming under renewed critical reflection as a shift from a relatively exclusive focus on mechanistic, causal explanations to a greater interest in historically contingent interpretative theories is taking place.[4] Faced with what they describe as a world of multiple realities, post-

[2] Many of these scholars are cited in works referred to in the Further Readings section at the end of this chapter.

[3] Lapid identifies the first debate as realism versus idealism and the second as the history versus science debate of the 1950s and 1960s.

[4] For an analysis that compares these two theoretical traditions see Hollis and Smith (1991). Robert Keohane (1988) recognized these two traditions, which he labelled rationalist and reflective, in his presidential address to the International Studies Association.

positivists are attempting to "deconstruct" the traditional discipline, and the relations it studies, by re-examining its fundamental assumptions. Many of them see the discipline as too emeshed in the security interests of the United States to be able to offer any meaningful understanding of global political life. Denying the positivist claim that it is possible to speak about an objectified world from the position of a detached neutral observer, Jim George (1994) portrays realist theory as political practice which makes and remakes the world in its own image.

According to Rob Walker (1993), less tainted explanations are unlikely to emerge without a more sustained reconsideration of the fundamental theoretical and philosophical assumptions that have informed conventional international relations theory. Binary oppositions such as inside–outside, identity–difference, and space–time, embodied in the principle of state sovereignty and its fixing of territorial boundaries, have disallowed the possibility of defining a world politics which can encompass the multiplicity of local and global spaces and identities in which our lives are emeshed.

Certain post-postivists are applauding the proliferation of these new voices and approaches to international theory; constituting what Lapid terms the celebratory response, these scholars are questioning the assumption that convergence of belief is necessary for maturity in science and exploring the possibility that a diversity of viewpoints may be compatible with scientific rationality and objectivity. Lapid believes that we should not be seeking to discover some universal scientific method nor the attainment of some objectively valid truth about world politics. Rather, we should be "promoting a more reflexive intellectual environment in which debate, criticism, and novelty can freely circulate" (Lapid 1989: 250). However, the possibility of constructive debate is highly contested. The divide between those who believe that the entire project of positivist social science is fundamentally suspect and those who claim that methodological pluralism is inhibiting the cumulation of scientifically based knowledge is large. I will now review some recent attempts to bridge this divide.

## IV **Prospects?**

If the optimism about theory building of the 1970s has proved illusory, what directions should we take as we seek a better understanding of our complex world? Claiming that "today, many schools of international theory contend but few genuinely communicate," Yale Ferguson and Richard

Mansbach (1991: 365) suggest that the tasks at hand are to find points of convergence between mainstream and dissident literatures and to construct research agendas that do not abandon empirical analysis entirely. They urge social scientists to recognize that, in their disciplines, the ultimate truth will always be unknowable. "Theory reflects the dominant political and social norms of its time and place and [that] international relations research unavoidably reflects the norms of scholarly communities . . . in which they are emeshed" (Ferguson and Mansbach 1991: 366).

While recognizing the subjective dimension of international theory, Ferguson and Mansbach believe there is still room for a less rigid empiricism than positivists have allowed and one which is informed by a greater self-consciousness. Critical of the "scientific" approach's search for parsimony in the interest of gaining explanatory power, they suggest that we can only "broaden the search for understanding by painting our pictures with greater detail" and by overcoming the ahistoricism and ethnocentricism of traditional theory (Ferguson and Mansbach 1991: 369). Empiricism must allow for human intentionality and be grounded in an understanding that human beings are motivated by identities and values.

Since "interstate" and "intrastate" political arenas are inextricably linked, theories of international relations must begin to challenge the boundaries of traditional theory that have kept these two arenas separate. If human agency is to be brought into international theory as critics of structural approaches demand, there is going to be a blurring of the micro–macro boundary, the intellectual boundary between domestic and international politics so crucial to neo-realist theory. Problematizing this boundary demands a closer examination of the state. Since it is individuals not states who are actors, "we must go behind the normative/legal facade to see who or what is actually 'behaving' and the factors that influence behavior" (Ferguson and Mansbach 1991: 370). If we are to reach a better understanding of our complex world, we must be willing to construct theories that can move freely across levels and boundaries and incorporate the multiplicity of actors which, today, constitute the realm of international politics.

## V Feminist contributions

Echoing several of these themes, Thomas Biersteker (1989: 264) has argued that international theory is in need of a non-degenerative, critical pluralism that can "provoke critical scrutiny of dominant discourses, empower marginalized populations and perspectives, and provide a basis for alternative conceptualizations." Feminism, one of the newest approaches to international theory, is making some important contributions to these goals. While feminist approaches cannot provide comprehensive solutions to all the field's needs, I believe they can make some important contributions to the critical and constructive pluralism these scholars envision.[5]

One of the central goals of feminism in all disciplines has been to uncover voices that have not previously been heard. Since women speak from the margins of international politics, they can offer a perspective outside the state-centric focus of traditional, western international relations. Since feminist approaches are based on the experiences of women whose lives have generally been ignored in theory construction, they can contribute to broadening the empirical base upon which we build our theories. Feminist perspectives on international relations include a growing body of literature from outside the west, literatures which can help to overcome the ethnocentricity of the discipline (Sen and Grown 1987; *Alternatives* 1993).

## A New facts

In his critical assessment of the theoretical pluralism of the third debate, K. J. Holsti argues for theoretical progress based on the additive and dialectical accumulation of knowledge for "[I]ntellectual progress exists largely in keeping up with the facts" (1989: 257). Claiming that the purpose of theoretical activity is to enhance our understanding of the world of international politics, Holsti maintains that theories must remain grounded in reality. The implication of his assessment is that most critical theorists do not live up to such standards.

While many contemporary feminists would agree that theory must remain grounded in reality, they would point to multiple international realities and their associated "facts." Their reconceptualizations and explanations are often focused on different sets of experiences than those that concern mainstream theorists. Contending that human knowledge has always been equated with knowledge about men, generally privileged men in dominant cultures, feminists question the possibility of using such knowledge and its associated sets of facts, to arrive at any adequate or comprehensive understanding of the "realities" that women and men, separately and together, experience.

[5] I shall be drawing on a variety of feminist approaches including empiricist, standpoint and postmodern.

By offering some "new facts," I shall suggest some different ways of looking at world politics which have the potential for generating new theoretically oriented research agendas. Building a more comprehensive picture of reality, which includes knowledge based on women's lives, is not an easy task, for data on women is sparse. Clearly much more data-gathering on the domestic and international dimensions of women's lives—data not based on presupposed distinctions between these realms—is needed.

The 1993 United Nations *Human Development Report* finds that, although some countries do better than others, no country treats its women as well as its men. In industrial countries, gender discrimination is mainly in employment with women often getting less than two-thirds of the employment opportunities and about half the earnings of men. In the developing countries, women receive less adequate health care, nutritional support and education than men; women make up two-thirds of the world's illiterate population (United Nations 1993: 16–17). Women working for minimal wages provide a "docile" labor force for multinational corporations in "export-processing" zones. Home-based work and domestic service, where women are often forced to emigrate to support their families, are also sources of under-remunerated and often exploited labor (Enloe 1990; Peterson and Runyan 1993).

The National Organization for Women estimated, in their 1990 Resolution on Women in Combat, that 80–90 percent of casualties due to conflict since World War II have been civilians, the majority of them women and children. Although the empirical bases for this estimate need further development, the magnitude of female vulnerability thus evidenced, and the possibilities for its amelioration, demand further investigation. Women are particularly vulnerable to rape in times of war, an issue which has only recently begun to be recognized. Data on war, generated by international relations scholars, has tended to concentrate on battle deaths during periods of open hostilities. Looking at the long-term effects of war, we find that women are particularly vulnerable to dislocation and negative economic consequences. Women and children constitute 80 percent of the global refugee population, a phenomenon usually attributable to intergroup violence or military conflict (Seager and Olson 1986: 27). Refugee women are among the poorest of the world's poor. This type of evidence demonstrates that women suffer disproportionately from both physical and structural violence in all societies.

Feminists believe such gross, pervasive, but differing disparities in welfare require investigation. However, they also claim that these persistent inequalities cannot be explained by incorporating these data into existing theoretical frameworks. Only by looking at these issues with gender-sensitive lenses, which also take into account ethnic, racial and class structures, can gender hierarchies which perpetuate these pervasive, but historically variable patterns be understood.[6] By introducing this gendered analysis, feminist perspectives have opened up large areas for future comparative and international research virtually untouched by previous comparative and international theorists.

If these "new facts" find women in disproportionately high numbers at the bottom of the socio-economic scale in all societies and disproportionately vulnerable to military "protectors," similar "facts" also demonstrate women's near absence in the realm of foreign policy-making. Women make up less than 5 percent of the world's heads of state (United Nations 1991: 6). In the United States in 1987, women constituted less than 5 percent of the senior Foreign Service ranks and, in the same year, less than 4 percent of the executive positions in the Department of Defense were held by women (Tickner 1992: 1; McGlen and Sarkees 1993). When critics call attention to exceptions to these generalizations, leaders like Indira Ghandi, Golda Meier or Margaret Thatcher, feminists point to hierarchically gendered structures which have propelled these "masculine gendered" women to the top, and demand a refocusing of investigations on the power and changeability of such structures (Peterson and Runyan 1993). Like post-positivist approaches more generally, they call for the need to ground knowledge in history and culture while paying special attention to the various ways in which gender has been constructed and deployed in terms that denote unequal power relations between women and men (Enloe 1993; Sylvester 1994).

## B Enhancing objectivity

These new "facts" challenge claims of universality and gender neutrality made by traditional theorists. Feminists are also searching for a redefinition of the meaning of objectivity as part of an attempt to construct an epistemology which, while it acknowledges the impossibility of arriving at one universal truth, can still lead to better, more widely shared understandings of the world.

As Sandra Harding (1990: 141–3) asserts, women's lives in any culture, society, or historical period have been devalued and neglected as

[6] I am defining gender as a social construction denoting unequal relations between men and women. The meaning of masculine and feminine varies across time and cultures. In the West, certain characteristics, such as power, autonomy, rationality, and public are generally perceived as masculine, while their opposites, weakness, dependence, emotion and private are stereotypically associated with femininity.

starting-points for all types of scientific research. Harding argues that the perspective or standpoint of the outsider or the excluded, is likely to produce more objective knowledge than that of dominant groups whose ways of thinking fit all too closely with dominant institutions and conceptual schemes. Harding suggests that, since marginalized groups have less interest in maintaining the status quo, their knowledge is less likely to be distorted by the desire to legitimate existing structures of power, a distortion noted also by post-positivist critics of realist international relations theory. Correcting distortions and enhancing objectivity require overcoming the excessive reliance on the lives of privileged individuals (usually men), in social science research. It also requires a critical examination of the potentially biasing beliefs of those who are considered to be inside the "scientific community."[7] The "new facts" presented above were ones suggested by such a feminist perspective.

Acknowledging the impossibility of representing all the world's multiple realities in terms of one universal truth, feminists nonetheless seek theories which offer us what Donna Haraway (1988: 580) has called a "reliable account of things, . . . an earth-wide network of connections, including the ability partially to translate knowledges among very different—and power-differentiated—communities." Since she believes that what is claimed to be universal or "objective" knowledge is, in reality, largely knowledge of privileged men, Haraway argues for what she calls "embodied objectivity" or "situated knowledge." For Haraway, situated knowledge does not mean relativism, but rather shared conversations leading to "better accounts of the world." Situated knowledge "allows us to become answerable for what we learn how to see" (Haraway 1988: 583).

How can these feminist conceptions of objectivity help us to construct a better account of international relations? If women's voices and women's experiences in the world, as well as those of other marginalized groups, have not been taken seriously, a feminist perspective on international theory must begin by questioning claims of universality. With its focus on the "high politics" of national security, conventional international theory has focused on issues that are associated with the experiences of privileged men. Harding's claim that members of dominant groups are given legitimacy as speakers and historical agents (Harding 1990: 142) would seem particularly applicable to international theory. We are socialized into believing that war and power politics are spheres of activity with which men have a special affinity and expertise and that their voices in describing, explaining and prescribing for a conflict-prone world are, therefore, likely

[7] For one such analysis which focuses on the discourse of US defense intellectuals see Cohn (1987).

to be more authentic. This rigid boundary separation between international and domestic politics, evident in international theory, has led to an exclusion of spheres of activity with which women have traditionally been associated. Investigating these spheres of activities, as well as the realm of high politics, from the perspective of the outsider, can shed some new light on the state and the boundaries within which our traditional understanding of international relations has been framed.

## C New conceptual frameworks

In considering how we might use this new knowledge to critically examine claims of objectivity made for explanations of states' behavior in the international system, I shall conclude by pointing toward some new, more complex conceptual frameworks, within which knowledge about international relations can be more adequately cumulated.

Feminists are suggesting that international theories which claim to offer objective and universal explanations of state behavior have been constructed out of the behavior of men. Traditional realism has built its theories of state behavior on models of classical writers, such as Machiavelli, Hobbes and Rousseau, who openly acknowledged that their depiction of human behavior was based on the behavior of men (Grant 1991).[8] Neorealism has depicted states rather differently, as abstract unitary actors whose actions are explained through laws universalized across time and place: states appear to act according to some higher rationality that is presented as independent of human agency. Yet, in reality, these depictions of state behavior rely on gendered models. Using rational choice and game theoretic models, such explanations draw on the instrumentally rational behavior of individuals in the marketplace, behavior that, in the modern West, has been more typical of men than women. Feminists suggest that these partial models tend to prioritize certain aspects of state behavior associated with conflict and draw our attention away from other activities in which states are also engaged, such as interdependent economic activities and building co-operation and community in other domains.

Donna Haraway (1989: 4) claims that all scientific theories, including the natural sciences, are embedded in particular kinds of stories, or what she calls "fictions of science." Conventional international theory has also constructed fictional stories about the boundaries between "domestic" and international politics. These myths, based on Hobbes's depiction of the

[8] For a more general critique of realism's interpretations of these foundational texts see Walker (1993).

state of nature, have framed our understanding of a dangerous "anarchic" world outside where aggressive behavior goes unsanctioned due to the lack of enforceable laws. State boundaries between anarchy on the outside and an orderly domestic space on the inside must be protected through the use of force if necessary.

Feminists challenge this depiction of the international environment. They argue that, as a model of human behavior, Hobbes's depiction of individuals in the state of nature is partial at best; if life were to go on for more than one generation women must have been involved in activities such as reproduction and child rearing (Tickner 1992: 46). State of nature myths in international relations also encourage unfavorable depictions of those on the "outside" which can lead to misperception and objectification. Feminist analysis allows us to see how often these unfavorable images are constructed through an association with a devalued "femininity" when characteristics, such as irrational or emotional, often associated with women, are used to portray those on the outside of the state. As David Campbell notes (1992: 238), this discourse that is used to secure the identity of those on the inside, through association of danger with those on the outside, is frequently framed in gendered terms.

In attempting to construct less ethnocentric images than those which have typically informed traditional explanations of state behavior, we might draw on Haraway's notion of "situated knowledge" which, she believes, can produce a more reliable understanding of the world by allowing our objects of investigation their independent integrity. For Haraway (1988: 592), situated knowledge requires that the object of knowledge be pictured as an actor or an agent rather than as a resource. For international theory to produce this kind of knowledge, it must become more sensitive to cultural differences and less committed to artificial boundary distinctions. While acknowledging the impossibility of producing one true story or one true set of facts, international relations scholars must continue to search for mutually shared understandings of our complex world. To move further toward this goal could be the basis for a realistic optimism about the future of the discipline of international relations.

## Acknowledgments

I should like to thank participants in the PIPES seminar at the University of Chicago, especially Jennifer Mitzen and Monica Duffy Toft, for their comments on an earlier draft of this paper. Hayward Alker also provided many useful insights.

## References

FURTHER READINGS

*Post-positivist Perspectives.* In addition to works cited in the text, a good introduction to a variety of authors in this fast-growing field can be found in Der Derian (1995), which is organized in terms of some common issues addressed by realists, from both the English and American schools, liberals, post-positivists, constructivists, postmoderns, and feminists. An earlier volume, Der Derian and Shapiro (1989), contains a variety of post-structural and postmodern approaches. For collections of post-positivist writings which emphasize culture and identity see Lapid and Kratochwil (1995) and Shapiro and Alker (1996).

*Feminist Perspectives.* Feminist approaches which critique mainstream international relations theories and epistemologies can be found in edited volumes by Grant and Newland (1991) and Peterson (1992). Single-authored texts focused on international theory include Sylvester (1994) and Tickner (1992). Peterson and Runyan (1993) and Enloe (1990; 1993) each examine and attempt to reconstitute world politics through gender-sensitive lenses. Whitworth (1994) and Stienstra (1994) both apply a feminist/neo-Gramscian framework to the traditional cultural assumptions embedded in the practices of international organizations.

ALKER, H., and BIERSTEKER, T. 1984. The dialectics of world order: notes for a future archaeologist of international savoir faire. *International Studies Quarterly*, 28: 121–42.

*ALTERNATIVES.* 1993. Special Issue: Feminists Write International Relations. *Alternatives*, 18/1.

BIERSTEKER, T. 1989. Critical reflections on post-positivism in international relations. *International Studies Quarterly*, 33: 263–7.

BULL, H. 1977. *The Anarchical Society: A Study of Order in World Politics.* New York: Columbia University Press.

CAMPBELL, D. 1992. *Writing Security: United States Foreign Policy and the Politics of Identity.* Minneapolis: University of Minnesota Press.

COHN, C. 1987. Sex and death in the rational world of defense intellectuals. *Signs*, 12: 687–718.

DER DERIAN, J., ed. 1995. *International Theory: Critical Investigations.* New York: New York University Press

—— and SHAPIRO, M., eds. 1989. *International/Intertextual Relations: Postmodern Readings of World Politics.* Lexington, Mass.: Lexington Books.

ENLOE, C. 1990. *Bananas, Beaches, and Bases: Making Feminist Sense of International Politics.* Berkeley: University of California Press.

—— 1993. *The Morning After: Sexual Politics at the End of the Cold War.* Berkeley: University of California Press.

FERGUSON, Y., and MANSBACH, R. 1991. Between celebration and despair: constructive suggestions for future international theory. *International Studies Quarterly*, 35: 363–86.

GADDIS, J. L. 1992. International relations theory and the end of the cold war. *International Security*, 17: 5–58.

GALTUNG, J. 1980. *The True Worlds: A Transnational Perspective.* New York: Free Press.

GEORGE, J. 1994. *Discourses of Global Politics: A Critical (Re)Introduction to International Relations.* Boulder, Colo.: Lynne Rienner.

GILPIN, R. 1987. *The Political Economy of International Relations.* Princeton, N.J.: Princeton University Press.

GRANT, R. 1991. The sources of gender bias in international relations theory. In Grant and Newland. 1991: 8–25.

Grant, R., and Newland, K., eds. 1991. *Gender and International Relations*. Bloomington: Indiana University Press.
Greenstein, F., and Polsby, N., eds. 1975. *International Politics*. Vol. viii of *Handbook of Political Science*. Reading, Mass.: Addison-Wesley.
Haraway, D. 1988. Situated knowledges: the science question in feminism and the privilege of partial perspective. *Feminist Studies*, 14: 575–99.
—— 1989. *Primate Visions: Gender, Race, and Nature in the World of Modern Science*. New York: Routledge.
Harding, S. 1990. Starting thought from women's lives: eight sources for maximizing objectivity. *Journal of Social Philosophy*, 21: 141–9.
Hollis, M., and Smith, S. 1990. *Explaining and Understanding International Relations*. Oxford: Oxford University Press.
Holsti, K. J. 1985. *The Dividing Discipline: Hegemony and Diversity in International Theory*. Boston: Allen and Unwin.
—— 1989. Mirror, mirror on the wall, which are the fairest theories of all? *International Studies Quarterly*, 33: 255–61.
Jackson, R., and Rosberg, C. 1982. Why Africa's weak states persist: the empirical and the juridical in statehood. *World Politics*, 35: 1–24.
Keohane, R. 1984. *After Hegemony: Co-operation and Discord in the World Political Economy*. Princeton, N.J.: Princeton University Press.
—— 1988. International institutions: two approaches. *International Studies Quarterly*, 32: 379–96.
Klare, M. 1992. The development of low-intensity-conflict doctrine. Pp. 37–53 in *Intervention Into the 1990s: U.S. Foreign Policy in the Third World*, ed. P. Schraeder. Boulder, Colo.: Lynne Rienner.
Krasner, S. 1982. Structural causes and regime consequences. *International Organization*, 36: 185–206.
Lapid, Y. 1989. The third debate: on the prospects of international theory in a post-positivist era. *International Studies Quarterly*, 33: 235–54.
—— and Kratochwil, F., eds. 1995. *The Return of Culture and Identity in International Relations*. Boulder, Colo.: Lynne Rienner.
McGlen, N., and Sarkees, M. R. 1993. *Women in Foreign Policy: The Insiders*. New York: Routledge.
Peterson, V. S. 1992. *Gendered States: Feminist (Re)Visions of International Relations Theory*. Boulder, Colo.: Lynne Rienner.
—— and Runyan, A. S. 1993. *Global Gender Issues*. Boulder, Colo.: Westview Press.
Russett, B. 1993. *Grasping the Democratic Peace: Principles for a Post-Cold War World*. Princeton, N.J.: Princeton University Press.
Seager, J., and Olson, A. 1986. *Women in the World: An International Atlas*. London: Pan Books.
Sen, G., and Grown, C. 1987. *Development, Crises, and Alternative Visions: Third World Women's Perspectives*. New York: Monthly Review Press.
Shapiro, M., and Alker, H. 1996. *Challenging Boundaries: Global Flows, Territorial Identities*. Minneapolis: University of Minnesota Press.
Singer, M., and Wildavsky, A. 1993. *The Real World Order: Zones of Peace/Zones of Turmoil*. Chatham, N.J.: Chatham House Publishers.
Stienstra, D. 1994. *Women's Movements and International Organizations*. New York: St. Martin's Press.
Sylvester, C. 1994. *Feminist Theory and International Relations in a Postmodern Era*. Cambridge: Cambridge University Press.

Tickner, J. A. 1992. *Gender in International Relations: Feminist Perspectives on Achieving Global Security*. New York: Columbia University Press.
United Nations. 1993. *Human Development Report*. New York: Oxford University Press.
—— 1991. *The World's Women: Trends and Statistics 1970–1990*. New York: United Nations.
Walker, R. B. J. 1993. *Inside/Outside: International Relations as Political Theory*. Cambridge: Cambridge University Press.
Wallerstein, I. 1976. *The Modern World System*. New York: Academic Press.
Waltz, K. 1979. *Theory of International Politics*. Reading, Mass.: Addison-Wesley.
—— 1993. The emerging structure of international politics. *International Security*, 18: 44–79.
Whitworth, S. 1994. *Feminism and International Relations: Towards a Political Economy of Gender in Interstate and Non-Governmental Institutions*. New York: St. Martin's Press.

Chapter 19

# International Relations, Old and New

Robert O. Keohane

INTERNATIONAL relations at the end of the 20th century is in a state of ferment and confusion. The Cold War has disoriented the Cold Warriors and many students of Soviet–U.S. relations, while the internationalization of the world economy has blurred boundaries between domestic and international politics, and made it harder to treat these two subjects as separate spheres. Controversies in the discipline have mixed enlightenment and confusion in unfortunately unequal proportions. We are at sea; but in a fascinating ocean with new creatures to observe, interacting in new ways.

Perhaps a short chapter can only irritate and befuddle; I hope not. In the space available, I will focus on what I see as central methodological, conceptual and theoretical issues in the study of international relations. Section I focuses on the objectives and methods that seem most productive for understanding this subject. Section II briefly examines Realism, which for half a century was the dominant approach to the study of international relations in the western world. Realism today encounters serious anomalies as a result of transnational relations, the peaceful behavior of democracies toward one another and the growing importance of international institutions. Section III highlights the role of assumptions about choice and necessity, while Section IV lists some research programs in international relations that I regard as promising. The chapter concludes with some suggestions for productive lines of future research.

## I Objectives and methods

In the terms set by Professor Goldmann's contribution to this *Handbook* (chap. 16: above), I am a "Type II" student of international relations, seeking understanding of how states relate to each other and to transnational actors and forces. Negative criticism (Type III) is both necessary and easy, given the lamentable state of our knowledge, but it quickly reaches the point of diminishing returns. As Professor Goldmann points out, we have been poor at predicting the future, and our causal models are weak, hence inferences from the academic study of international relations to policy (Type I) cannot be taken too seriously. Our essential task is *understanding*.

Understanding does not mean prediction. Professor Goldmann discusses John Gaddis's (1992) criticism that international relations theory failed to predict the end of the Cold War. Gaddis is right, of course, about the inadequacy of international relations theory. We do not understand well the extent to which international pressures contributed to the collapse of the Soviet Union, nor do we have strong theories that help us anticipate reactions to this seismic change in world politics. However, the criticism that we failed to predict the demise of the Soviet Union misunderstands what students of international relations can do. The end of the Cold War was a very complex phenomenon, with sources deep in the Soviet Union's social, political, and economic system, as well as in international relations; and it was affected by choices made by Mikhail Gorbachev, which another leader might not have made. It was a classic "conjunctural" event in Hirschman's terms (Hirschman 1970). Claiming that political scientists should have predicted the end of the Cold War is like demanding that if there had been scientists 65 million years ago, they should have been able to predict that the earth would soon collide with a comet or asteroid, and that this collision would lead to the extinction of the dinosaurs!

However, the fact that we lack theories that would enable us to understand the effects of the end of the Cold War on world politics certainly should make us humble. In this respect we compare unfavorably with scientists explaining the *effects* of a cosmic collision. Our present confusion also reveals the emptiness of claims that we possess a "policy science" to which policy-makers refuse to listen "at their peril." We don't have recipes for foreign policy success. Our manifest inability to predict complex events, whether minor or earthshaking, should also teach us that although it is useful to seek to develop and test conditional generalizations, the accumulation of such generalizations is unlikely to lead to successful prediction of events that result from the conjuncture of multiple causal paths. All of our generalizations will be conditional and probabilistic, and will be associated with large margins of uncertainty. Significant events are the combinatorial result of so many forces that even if we had probabilistic generalizations of much higher quality than those we have now, we would not be able to make successful predictions. The model of science as increasingly successful prediction based on increasing numbers of validated generalizations is not appropriate for our field.

This acknowledgment, however, does not legitimize an escape from serious attempts at both descriptive and causal inference, requiring a combination of theory and empirical work. We seek intersubjective knowledge—theory linked to evidence, which will be persuasive to students of international relations with a wide variety of prior views on the nature of the field. This does not mean that we believe that we can ever be totally "objective"—of course, our purposes, limitations and biases affect our work. But there is an ideal of objective knowledge toward which we can strive. And we can apply scientific methods to help us get there: attempts to make inferences according to publicly known rules of inference; clear, stated hypotheses; public specification of evidence; estimates of the degree of uncertainty associated with our hypotheses; efforts to search for *disconfirming* evidence. There has been, recently, a lot of Quixotian jousting at the windmill of naïve positivism—but much less serious attempt to demonstrate a *superior alternative* to the sophisticated falsificationism of Popper (1968) and Lakatos (1970).

If we really want to understand international relations, we should seek both contextually sensitive description and descriptive inference, and conditional, probabilistic causal propositions. We will not find "necessary and sufficient conditions" under which deterministic laws would hold. We should seek to specify hypotheses and look for their observable implications *at any level of analysis.* This is an important point, often obscured by the misleading "levels of analysis" argument. Systemic propositions have implications at the level of foreign policy decision-making and can be tested at that level, as well as on the basis of outcomes at the systems level. For instance, the propositions that states balance against power (Waltz 1979), that they balance against threat (Walt 1987), or that they seek to "hide" from conflict (Schroeder 1994), imply different discussions within foreign ministries, and different policy initiatives, as well as different international outcomes.

If we should not be seeking the will-o'-the-wisp of prediction—or feeling guilty because we cannot predict wars and the ends of cold wars—what should we be doing? We need to make sense of historically unique events that result from conjunctures of other complex phenomena, and from human choices: we cannot fully explain them, but we can describe the causal mechanisms that produced them. We also need to examine their consequences in a way that goes beyond narrative description to explanation, examining the effects of these major causes in the light of our understanding of the constraints affecting international politics.

One way to do this is suggested by the theory of the demise of the dinosaurs developed by Luis and Walter Alvarez in the late 1970s. The Alvarez hypothesis was that a cosmic collision 65 million years ago threw tremendous quantities of dust into the air, reducing global temperature and thus killing the dinosaurs. Put this way, the Alvarez team had just made a speculation, like many in political science. But they went further; they asked themselves what else would be true if they were right, and deduced that a layer of iridium, not previously noticed, would be found at the appropriate place in the earth's crust. When it was, their theory was dramatically supported—not proved, since there remain alternative processes that could have created this result. They did not predict the collision; but given the collision, they explained why the dinosaurs died (King, Keohane, Verba 1994: 11–12).

Another exemplar of this method is found in mystery fiction: I will call it the "Marple–Dagliesh Method" after two famous detectives created by Agatha Christie and P. D. James, respectively. Detectives do not explain murder generically, but a unique murder. They rely on forensic science, but no amount of science ever caught a murderer by itself. They are both theoreticians and empiricists: they begin with some evidence ("clues"), formulate tentative hypotheses that are consistent with the evidence at hand and scientific generalizations, look for evidence that will test those hypotheses, find new evidence, reformulate their hypotheses, and so on, until the villain is found. No predictive "policy science of murder" exists, yet we believe that detectives Marple and Dagliesh do good work when they deduce who the murderer was. Students of international relations would do well to imitate these detectives by carefully observing and describing events, specifying the causal mechanisms that could have led to these results, and testing hypotheses with evidence. The test of our work will not be whether it yields accurate predictions of complex events—it won't—but whether it enhances our ability to anticipate patterns of action.

## II The rise and fall of realism

In the study of world politics, theory typically follows practice, rather than *vice versa.* Adolf Hitler's accession to power in Germany and the subsequent crises and war that resulted, led to the revival of Realism as a school of thought, emphasizing the role of national interests and of power and criticizing the naïve idealism of much of the interwar period (Carr 1946, Morgenthau 1948). During the Cold War, Realism was the dominant approach, and it had great value. It served as an antidote to ideological thinking (Waltz 1959) and as a source of caution, emphasizing the principle of avoiding over-extension by keeping ends aligned with the means to achieve them (Lippmann 1943). It should not be forgotten that major Realist thinkers such as Kennan, Lippmann, Morgenthau and Waltz compiled a distinguished record of early opposition to the war in Vietnam. Furthermore, Kenneth Waltz's systematization of Realism as "neo-realism," or "structural realism," provided both a firmer basis for understanding world politics from a Realist standpoint and a clearer point of reference for its critics.

By the time Waltz systematized Realism, however, its fit with the world was already beginning to erode: as Hegel observed, the owl of Minerva only flies at dusk. The internationalization of the world economy had led, among the major industrialized democracies, to a pattern of "Complex Interdependence" (Keohane and Nye 1977), characterized by multiple issues in world politics, multiple actors (not just states), and the inefficacy of force on many issues. Complex Interdependence is an ideal-type, but where its features are approximated, Realist theory is misleading.

The anomalies that Complex Interdependence illuminates have three dimensions. First, the state, while still the most important type of actor in world politics, no longer plays as dominant a role as in the past: transnational, as opposed to interstate, relations have increased in importance. Transnational forms of communication, from short-wave broadcasting to communications satellites to the internet, have relaxed the state's hold on information. Direct foreign investment means that multinational corporations have a major presence around the world. Exclusive notions of sovereignty as control over population within a well-defined territory have been challenged by concern over the implications on state action or inaction on issues ranging from human rights to environmental protection.

Secondly, in contrast to the assumptions of Realism, liberal democracies behave differently in foreign policy than do non-democracies. Depending on one's definition, liberal democracies either do not fight each other at all or at least very much less than one should expect (Doyle 1983; Russett 1993; Owen 1994). In the realm of law, the courts of liberal democracies enter into dialogues with one another, interpreting each other's decisions on the basis of broadly shared legal principles; they treat judgments by courts of autocracies very differently (Burley 1993). The demonstrated ability of modern liberal democracies to enter into co-operative international institutions, it can be argued, owes something to the transparency of their internal procedures and, consequentially, the increased credibility of their promises (Keohane 1984: 258–9; Cowhey 1993). Such variation between the behavior of democracies and non-democracies produces difficulties, to say the least, for a theory that views states as "like units," constrained by the structure of the international system (Waltz 1979), and that interprets the behavior of the United States and the Soviet Union during the Cold War as showing "striking similarities" (Waltz 1993: 45). Waltz's analysis overlooks the fact that the core alliances of the United States and the Soviet Union were fundamentally different—one voluntary, based on incentives and common values (NATO); the other coerced, based on agreements with regimes kept in power by military force (Warsaw Pact). NATO has held together, despite Waltz's predictions; the Warsaw Pact quickly collapsed.

The third anomaly concerns the major role played in contemporary world politics by international institutions: "persistent and connected sets of rules (formal and informal) that prescribe behavioral roles, constrain activity, and shape expectations" (Keohane 1989: 3). Realism has proved unable to explain the expansion in number and scope of international institutions, which was dramatic in the years immediately after World War II and which has continued since. Governments continue to invest substantial material and symbolic resources in maintaining and extending such institutions. In the mid-1990s, the European Union is the most highly institutionalized international organization in history: not a state, but a coherent entity with executive, legislative, and judicial organs, and an increasing body of binding law. NATO is by some measure the strongest and most durable alliance in history—a security institution of considerable note. The World Trade Organization, incorporating the General Agreement on Tariffs and Trade (GATT), supersedes in comprehension and specificity any previous global set of rules for the governance of major transactions. The International Monetary Fund and the World Bank are controversial—some might say that John Maynard Keynes's fear that they would grow up as "brats [who] will grow up politicians" (Gardner 1980: 266) has come true—but they certainly are large and powerful bureaucracies—more so than any globally operating bureaucracies previously known to the human race. The fact that states support the growth of these international institutions suggests that they must perform some functions that serve state interests.

In view of the discrepancy between Realism and reality, it is not surprising that Realism has taken some hard knocks recently. Realism has underemphasized domestic politics and international institutions, and it has failed to develop a theory of change. It has also overemphasized the role of the state as an actor. Yet the world has not become benign: the vicious wars in Bosnia and central Asia have reaffirmed Realist insights about the capabilities of human beings for violence, and the continued weakness of international controls on groups bent on attacking their neighbors. Whatever "anarchy" means, the lack of common government means that power-seeking actors can use force against others—to which the only effective counterweight may be force. Such conflict can be self-perpetuating out of fear and power. Although under auspicious circumstances international institutions that promote co-operation can be built they are difficult to construct, and fragile. Harmony is not the natural condition of the human race. But Bosnia also *refutes* the most pessimistic realists' "back to the future" scenarios (Mearsheimer 1990). There has been no return to 1914 alignments and no general war in Europe. As noted above, liberal democracies fight one another very rarely, although we do not have a good theory of why this is so. We observe a complex mixture of interaction, ranging from deadly warfare to institution-building and extensive co-operation, creating security for an enormous outpouring of foreign investment. The puzzle outgrows any theory that we have.

## III Choice and necessity

The ultimate premise of Realist thinking is that state action is based on necessity: the anarchy of world politics leads to security dilemmas that can only be resolved through self-help. Choice is severely constrained, hence variation in social systems, in ideas, or in leaders' qualities is, in the last analysis, not very important. Material capabilities affect what is necessary, and therefore essentially determine action. This is what is known as a "structural" theory.

In contrast, liberal and institutionalist theory focuses on variation rather than constancy of state behavior—variation that it attributes not merely to variation in material capabilities but also to variation in political processes and particularly in the character of human institutions, domestic and international. When state élites do not foresee self-interested benefits from co-operation, liberal institutionalists anticipate neither co-operation nor the institutions that facilitate co-operation. When states can jointly benefit from co-operation, however, we expect governments to attempt to construct such institutions. Institutions can provide information, reduce transaction costs, make commitments more credible, establish focal points for co-ordination, and in general facilitate the operation of reciprocity. Their construction may be blocked, not only by conflicting interests but by difficulties in making credible commitments, finding or constructing focal points, or in maintaining the conditions for reciprocity to operate. By seeking to specify the conditions under which institutions can have an impact and co-operation can occur, institutionalist theory tries to show under what conditions Realist propositions are valid.

In the liberal internationalist view, domestic and international institutions interact, reinforcing one another. Democratic domestic politics and the growth of international society are strongly associated with the growth of international institutions; international society and institutions create "space" for domestically derived preferences. I use Hedley Bull's definition of international society: "International society exists when a group of states, conscious of certain common interests and common values, form a society in the sense that they conceive themselves to be bound by a common set of rules in their relations with one another, and share in the working of common institutions" (Bull 1977: 12).

What institutions do can be understood by thinking about strategic action in game-theoretic terms. In game theory, strategic behavior is conditional in part on one's payoffs—the "structure" of the game, which reflect power as well as preferences. But behavior also depends on the information one has: for instance, the conflictual outcome in Prisoners' Dilemma depends on a lack of credible communication between the players. Indeed, *the major lesson of modern game theory is how sensitive behavior is to different information conditions.* Since world politics involves strategic interaction, any good theory of world politics needs to treat information as a variable. We have to add a strategic perspective to structural theory.

Many situations in international relations are like Prisoners' Dilemma: co-operation yields joint gains, but without credible commitments by others, each actor must plan for the worst case, yielding bad outcomes. In other words, the equilibria of Prisoners' Dilemma are suboptimal. Other situations resemble Assurance games, which have Pareto-optimal equilibria, but which are vulnerable to disruption by uncertainty or misinformation. Many other situations are games of co-ordination, with multiple equilibria, all superior to the status quo but with different distributional implications for the actors involved (Krasner 1991; Martin 1992*b*; Morrow 1994). These situations are different in important ways, but they have one feature in common: where potential joint gains from co-operation exist, *information creates the capacity to co-operate.* Interests as interpreted by actors depend on information as well as on underlying payoffs. What it is good for us to do depends on our expectations, and others' expectations; on the credibility of our commitments. Institutions can provide information, and they can facilitate linkages among issue areas (Martin 1992*a*), both of which may alter states' interests by increasing the payoffs to co-operation. So institutions matter, even if they cannot enforce rules from above, *because they change actors' conceptions of their interests.*

Societies that are insulated from the threat of immediate attack, which do not live on the security "knife-edge," can exercise some influence over how their governments define self-interest, and hence over how they should act. Unlike the stylized states of Realist theory, they are not forced to accept the "necessity" of accepting worst-case analysis and seeking to pre-empt possible opponents. In the 19th century, such opportunities were provided by geography to island nations such as Great Britain and those remote from Eurasia, such as the United States, Canada, and Australia. In the 20th, such opportunities are increasingly provided by human-constructed institutions and by the spread—more gradual and halting than some enthusiasts imagine—of democracy. Democracies create international institutions; such institutions make it easier for countries so inclined to co-operate with one another, in a "virtuous circle." If world politics were a pure self-help system, with no margin for survival, domestic politics would not matter. It matters in part because there is room for policy variation in the shadow of international institutions and international society.

Realism and institutionalism both emphasize the important of reciprocity in international relations. Realism emphasizes its negative side: retaliation, giving "evil for evil." Surely arms races and spirals toward war can sometimes be described in these terms. Reciprocity—"tit for tat"—can lead to feuds, and when it is difficult to ascertain whether others have co-operated, the consequences can be terrible (Downs *et al.* 1986; Signorino 1995). Institutionalism emphasizes that reciprocity in one form or another can provide the basis for co-operation (Axelrod 1984; Ostrom 1990; Keohane and Ostrom 1994). Institutions make regulation by reciprocity feasible (Ostrom 1990: 90). Reciprocity can operate within societies, as well as between them. Interests faced with retaliation may press for more co-operative policies: exporters may oppose protection more vigorously (Gilligan 1993), multinational firms in import-competing industries may not demand protection for themselves for fear of incurring the wrath of governments in whose jurisdictions they operate (Milner 1987).

Neither institutionalism nor liberalism is an "alternative" to Realism. Democracies and institutions only thrive under certain conditions: vicious circles of demagoguery and war are as evident in contemporary world politics as virtuous circles of democracy and co-operation. The question is not one of replacing Realism entirely but of making its propositions conditional, and of bringing insights from liberalism and institutionalism to bear on situations where the conditions for democracy and international institutions are favorable.

## IV The new international relations

We have a huge agenda. We need to understand both continuity and change; how states deal with one another but also how they relate to non-state actors and by which principles they operate internally; how lack of common government induces severe conflict, in which reciprocity plays a role, but also how institutions based on reciprocity can ameliorate such conflict. It is no wonder that we have trouble. But we can hardly avoid any of these tasks. All the more important that we not lose ourselves either in layers of narcissistic self-criticism or premature policy advice!

To make progress, we will have to pursue promising research programs. Research programs begin with theoretical insights or arguments—not just with a set of puzzles. Each of these programs selects particular *explanatory variables* to emphasize, which explains both their frequent disagreements and their potential complementarity. There are many potential research programs, and creativity will generate more. So my list is not presumed to be exhaustive, only illustrative of some promising directions. I have chosen four: the extension of Realist analysis, liberal institutionalism, the mutual effects of domestic and international politics, and the analysis of subjectivity.

1) *Realism with modifications.* A Realism formulated non-dogmatically, as a set of premises within which students devise and test conditional generalizations, continues to have great potential for understanding many areas of world politics. For instance, its focus on relative gains may illuminate areas of international relations, if expressed clearly and in a way consistent with expected utility theory (Powell 1991; Bueno de Mesquite and Lalman 1992). Indeed, relative gains in this form are relevant not only to hostile behavior on issues of war and peace but also to strategic trade policy in economics (Baldwin 1993). But the search for universal laws of world politics is misguided. We need to ask: *under what conditions* do Realism's hypotheses fit reality? We need to avoid pseudo-laws of a deterministic kind, but to seek conditional generalizations, statements of possibility, and contextually sensitive explanation relying on valid causal inference. Realism may alert

us, as detectives, to possibilities and to causal processes found in history, but which cannot simply be assumed to apply in contemporary cases.

2) *Liberal institutionalism.* This research direction was discussed above, and in Professor Goldmann's contribution (chap. 16: above). The core issue for students of international institutions is how to understand the interaction of rules and actions, which involves explaining both conformity to rules and change in them (Kratochwil 1989). For liberals, the central issue has been how liberal democracies behave differently, in international relations, from non-democracies. Liberal institutionalists, combining both perspectives, need to seek to understand better how democratic politics and international institutions relate to one another: to what extent are they mutually reinforcing, as claimed above? Liberal institutionalism, like Realism, may benefit from increasing formalization, in the form of rational choice theory, as long as the historical and comparative contexts of action are kept in mind and propositions generated by theory are tested empirically, rather than simply being applied to "toy models" invented by the investigator.

3) *The impact of international structures and processes on domestic politics.* This is what Peter Gourevitch (1978; 1986) has called "the second image reversed." For example, what are the effects of internationalization of the world economy on domestic electoral institutions (Cameron 1978; Rogowski 1987)? How are increases in international trade and investment likely to alter domestic political cleavages (Rogowski 1989; Frieden 1991)? How are changes in global communications affecting politics in various countries? Comparative work is crucial on these issues, since international-level changes are refracted through established domestic institutions and practices, such that uniform results should hardly be expected.

4) *The role of subjectivity and the limits of materialistic analysis.* Nationalism, in both its liberal and xenophobic varieties, can hardly be understood without attention to the role of myths and ideas—both intrinsically and how they are instrumentally manipulated for strategic purposes. As Alexander Wendt (1992) has emphasized, to a considerable extent societies construct their own identities. This social construction is not a mystical process but can be analyzed, using the tools of game theory as well as of anthropology (Geertz 1973; Bates and Weingast 1995).

One way to approach issues of subjectivity is to begin with a structural analysis incorporating the principle of rationality, then to relax various assumptions, having to do with unique equilibria, common knowledge of cause–effect relationships, or the irrelevance of differences in principled beliefs and worldviews (Goldstein and Keohane 1993). In this way, one can move step-by-step toward more realistic interpretations of subjectivity in international relations, while retaining the clear explanatory framework provided by a combination of structural and strategic theory. However, a significant drawback of this incremental approach is that it does tend to "privilege" the materialist-rationalist framework, requiring justifications for deviation from its assumptions, without symmetrically demanding that materialist rationalism justify the interests and identities of the actors with which it begins its analysis. It will be a challenge for international relations theory over the next few years to devise ways to understand how the formation of actors' identities and interests is affected by how they think, without simply falling into circular pseudo-explanations or quasi-philosophical jargon. To guard against such degeneration, students of subjectivity need to address real research problems. A set of claims divorced from real international relations problems is not a research program.

## V Conclusion

This chapter has attempted to make three principal points. First, students of international relations need to transcend the limitations of Realism, examining how domestic politics and international relations interact, and exploring the role of information and institutions, without discarding the core insights of Realism that have persisted in western thought for centuries. We need to question assumptions about structure and necessity, while seeking to explain variations in choices made, and exploring how reciprocity works.

Second, we need to connect our theories, or elements of them, to one another in appropriate ways, to solve puzzles. In the wonderful image of Professor Goldmann, international relations should not be viewed as "an oligopolistic competition between U.S.-led conglomerates offering packages that have to be accepted or rejected *in toto*." We should try to rein in our tendency to wage paradigmatic warfare; and journal editors in our field should resist the temptation to make their pages more "exciting" by stimulating such counterproductive jousting.

Finally, we need continually to link theory with empirical work on real international problems—not expecting general laws but linking social science with history, using strategic, game-theoretic analysis with sensitivity to context, seeking to understand specific conjunctures in the light of general principles. All of our theories are in danger of becoming dogmatic and arid if they are not linked to actual experience. Theory and empirical analysis are not separate domains, much less watertight compartments; in progressive research programs, they must be closely connected.

## Acknowledgments

The author is grateful to Lisa L. Martin for comments on the talk prepared for the International Political Science Association meetings in Berlin, August, 1994, that formed the basis for this chapter.

## References

FURTHER READINGS

The most important statement of classical Realism remains Hans J. Morgenthau, *Politics Among Nations* (1948). Kenneth N. Waltz, *Theory of International Politics* (1979) is the major neo-realist work. The concept of Complex Interdependence is discussed in Robert O. Keohane and Joseph S. Nye, Jr., *Power and Interdependence* (1977); that of international society is put forward in Hedley Bull, *The Anarchical Society* (1977). Institutionalist analysis is represented by Robert O. Keohane, *After Hegemony* (1984); Kenneth A. Oye, ed., *Co-operation Under Anarchy* (1986); and Lisa L. Martin, *Coercive Co-operation* (1992). On the democratic peace see Bruce M. Russett, *Grasping the Democratic Peace* (1993). Reciprocity is discussed by Robert Axelrod, *The Evolution of Co-operation* (1984), and by Elinor Ostrom, *Governing the Commons* (1990). A review of recent controversies can be found in Baldwin, ed., *Neo-realism and Neo-liberalism* (1993).

AXELROD, R. 1984. *The Evolution of Co-operation.* New York: Basic Books.

BALDWIN, D. A. ed., 1993. *Neo-realism and Neo-liberalism: The Contemporary Debate.* New York: Columbia University Press.

BATES, R., and WEINGAST, B. 1995. A new comparative politics: integrating rational choice and interpretivist perspectives. Working Paper, Center for International Affairs, Harvard University.

BUENO DE MESQUITA, B., and LALMAN, D. 1992. *War and Reason: Domestic and International Imperatives.* New Haven: Yale University Press.

BULL, H. 1977. *The Anarchical Society: A Study of World Order in World Politics.* New York: Columbia University Press.

BURLEY, A.-M. S. 1993. International law and international relations theory: a dual agenda. *American Journal of International Law*, 87: 205–39.

CAMERON, D. A. 1978. The expansion of the public economy. *American Political Science Review* 72: 1243–61.

CARR, E. H. 1946. *The Twenty Years' Crisis: 1919–1939.* 2nd edn. London: Macmillan; originally published 1939.

COWHEY, P. F. 1993. Domestic institutions and the credibility of international commitments: Japan and the United States. *International Organization*, 47: 299–326.

DOYLE, M. 1983. Kant, liberal legacies and foreign affairs, Part I. *Philosophy and Public Affairs*, 12: 205–35.

Downs, G.; Rocke, D. M.; and Siverson, R. M. 1986. Arms races and co-operation. In Oye 1986: 118–46.

Frieden, J. A. 1991. Invested interests: the politics of national economic policies in a world of global finance. *International Organization*, 45: 425–52.

Gaddis, J. L. 1992. International relations theory and the end of the Cold War. *International Security*, 58: 5–58.

Gardner, R. N. 1980. *Sterling-Dollar Diplomacy in Current Perspective: The Origins and the Prospects of our International Economic Order*. Rev. edn. New York: Columbia University Press; originally published 1956.

Geertz, C. 1973. *The Interpretation of Cultures*. New York: Basic Books.

Gilligan, M. 1993. *Conditional Trade Policy and the Demand for Liberalization: U.S. Trade Policy Since the Civil War*. Ph.D. dissertation, Department of Government, Harvard University.

Goldstein, J., and Keohane, R. O. 1993. *Ideas and Foreign Policy: Beliefs, Institutions and Political Change*. Ithaca: Cornell University Press.

Gourevitch, P. A. 1978. The second image reversed. *International Organization*, 32: 881–912.

—— 1986. *Politics in Hard Times: Comparative Responses to International Economic Crises*. Ithaca: Cornell University Press.

Hirschman, A. O. 1970. The search for paradigms as a hindrance to understanding. *World Politics*, 22: 329–43.

Keohane, R. O. 1984. *After Hegemony: Co-operation and Discord in the World Political Economy*. Princeton: Princeton University Press.

—— 1989. *International Institutions and State Power: Essays in International Relations Theory*. Boulder, Colo.: Westview.

—— and Nye, Joseph S. Jr. 1977. *Power and Interdependence: World Politics in Transition*. Boston: Little, Brown.

—— and Ostrom, E., eds. 1994. *Local Commons and Global Interdependence: Heterogeneity and Co-operation in Two Domains*. London: Sage.

King, G.; Keohane, R. O.; and Verba, S. 1994. *Designing Social Inquiry: Scientific Inference in Qualitative Research*. Princeton: Princeton University Press.

Krasner, S. D. 1991. Global communications and national power: life on the Pareto frontier. *World Politics*, 43: 336–66.

Kratochwil, F. V. 1989. *Rules, Norms and Decisions: On the Conditions of Practical and Legal Reasoning in International Relations and Domestic Affairs*. Cambridge: Cambridge University Press.

Lakatos, I. 1970. Falsification and the methodology of scientific research programs. Pp. 91–196 in I. Lakatos and A. Musgrave, eds., *Criticism and the Growth of Knowledge*. Cambridge: Cambridge University Press.

Lippmann, W. 1943. *U.S. Foreign Policy: Shield of the Republic*. Boston: Little, Brown.

Martin, L. L. 1992*a*. *Coercive Co-operation: Explaining Multilateral Economic Sanctions*. Princeton: Princeton University Press.

—— 1992*b*. Interests, power and multilateralism. *International Organization* 46: 765–92.

Mearsheimer, J. 1990. Back to the future: instability in Europe after the Cold War. *International Security*, 15: 5–56.

Milner, H. V. 1987. *Resisting Protectionism: Global Industries and the Politics of International Trade*. Princeton: Princeton University Press.

Morgenthau, H. J. 1948. *Politics Among Nations*. New York: Alfred O. Knopf.

Morrow, J. D. 1994. Modelling the forms of international co-operation: distribution versus information. *International Organization* 48: 387–424.

Ostrom, E. 1990. *Governing the Commons: The Evolution of Institutions for Collective Action.* Cambridge: Cambridge University Press.

Owen, J. M. 1994. How liberalism produces democratic peace. *International Security,* 19: 87–125.

Oye, K. A., ed. 1986. *Co-operation Under Anarchy.* Princeton: Princeton University Press.

Popper, K. A. 1968. *The Logic of Scientific Discovery.* New York: Harper and Row.

Powell, R. 1991. Absolute and relative gains in international relations theory. *American Political Science Review,* 85: 701–26. Reprinted in Baldwin 1993: 209–33.

Rogowski, R. 1987. Trade and the variety of democratic institutions. *International Organization,* 41: 203–24.

—— 1989. *Commerce and Coalitions: How Trade Affects Domestic Political Alignments.* Princeton: Princeton University Press.

Russett, B. M. 1993. *Grasping the Democratic Peace: Principles for a Post-Cold War World.* Princeton: Princeton University Press.

Schroeder, P. 1994. Historical reality vs. neo-realist theory. *International Security,* 19: 108–48.

Signorino, C. 1995. Simulating international cooperation under uncertainty: the effects of symmetric and asymmetric noise. Working Paper, Center for International Affairs, Harvard University.

Walt, S. M. 1987. *The Origins of Alliances.* Ithaca: Cornell University Press.

Waltz, K. N. 1959. *Man, the State, and War.* New York: Columbia University Press.

—— 1979. *Theory of International Politics.* Reading, Mass: Addison-Wesley.

—— 1993. The emerging structure of international politics. *International Security,* 18: 44–79.

Wendt, A. 1992. Anarchy is what states make of it: the social construction of power politics. *International Organization,* 64: 391–426.

Part VI

# Political Theory

Chapter 20

# Political Theory: An Overview

Iris Marion Young

POLITICAL theorists in the last quarter-century have been primary custodians of a conception of the political as a participatory and rational activity of citizenship. This contrasts with the more usual concept of politics assumed by popular opinion, journalism, and much social science. On this latter view, politics is a competition among élites for votes and influence, and citizens are primarily consumers and spectators. Hannah Arendt's work remains a milestone in twentieth century political theory because she gave us an inspiring vision of the political as active participation in public life, which many political theorists continue to guard and preserve.

In that vision the political is the most noble expression of human life, because the most free and original. Politics as collective public life consists in people moving out from their private needs and sufferings to create a public world where each can appear before the others in his or her specificity. Together in public they create and recreate through contingent words and deeds the laws and institutions that frame their collective life, regulate their ever-recurring conflicts and disagreements, and spin the narratives of their history. Social life is fraught with vicious power competition, conflict, deprivation, and violence, which always threaten to destroy political space. But political action sometimes revives, and through a remembrance of the ideal of the ancient *polis,* we maintain the vision of human freedom and nobility as participatory public action (Arendt 1958).

Arendt distinguished this concept of the political from the social, a modern structure of collective life which she believed increasingly eclipsed the political. Modern economic forces and mass movements collude to create a realm of need, production and consumption outside the household.

Government institutions increasingly define their task as managing, containing and attending to this ever-expanding social realm—through education, public health policy, policing, public administration, and welfare. People's lives may be well cared for as a consequence, and government more or less efficient in its administration, but according to Arendt in the modern state genuine public life sinks into a swamp of social need (Canovan 1992: chap. 4).

Although they often wish to preserve her vision of the political, contemporary theorists have largely rejected Arendt's separation of the political from the social, and her backward-looking pessimism about the emergence of mass social movements of the oppressed and disenfranchised. The more common judgment is that social justice is a condition of freedom and equality, and thus that the social must be a main focus of the political (Pitkin 1981; Bernstein 1986).

In her political theory of welfare state discourse, Nancy Fraser recasts Arendt's concept of the social and suggests that much contemporary citizen activism in public life should be conceptualized as politicizing the social (Fraser 1989). In this chapter I take up this suggestion, and construct an account of political theory in the last two decades through the theme of the politicization of the social. The story I tell is of course a construction, from my own point of view, which emphasizes some aspects of political theorizing in the last twenty-five years, and de-emphasizes others.

The theme of the politicization of the social, for example, will lead me to say little about the massive literature in recent political theory which takes some aspect of the historical canon of political theory as its subject. Much of this scholarship has been influenced by or influenced contemporary concern for social justice and participatory democracy, however. J. G. A. Pocock's *The Machiavellian Moment*, for example, has influenced contemporary civic republicanism. James Miller's reading of Rousseau, to take another example (Miller 1984), is influenced by contemporary discussions of participatory democracy.

Likewise, this chapter will make little reference to recent political theory that uses techniques of rational choice theory (other chapters will be devoted to this subject), although much of this literature expands and illuminates the issues of social justice and welfare which I do discuss in the first section. Nor will I discuss the interesting and important work on the history of law and policy done by political theorists. Most of my attention in this chapter, finally, will be on English language political theory, though I will refer to some German and French writers.

The theme of politicizing the social aptly organizes the great body of recent political theory, providing several illuminating perspectives from which to see these theories. In one way or another, the theoretical trends I discuss either reflect on the conditions of social justice, or express and systematize the politics of recent social movements, or theorize flows of power in institutions outside as well as inside the states, or inquire about the social bases of political unity. My account divides recent political theory into six sub-topics, each of which expresses a different form of politicizing the social: social justice and welfare rights theory; democratic theory; feminist political theory; postmodernism; new social movements and civil society; and the liberalism-communitarianism debate. Recognizing that many works in recent political theory overlap these categories, I nevertheless try to locate most works in one of them.

## I Social justice and welfare rights theory

In 1979 Brian Barry could look back on two decades of political theory and find the first nearly barren and the second producing bumper crops. With him I will locate the publication of John Rawls's *A Theory of Justice* (1971) as the turning-point. It is no accident that the decade of the 1960s intervened between the barren field of political theory and the appearance of this groundbreaking book. Despite its rhetoric of timelessness, *A Theory of Justice* must be read as a product of the decade that preceded it. Would civil disobedience occupy a central chapter in a basic theory of justice today?

*A Theory of Justice* mapped a theoretical terrain from the thicket of demands and responses to the Black civil rights movement and journalistic attention to poverty: social justice. Rawls insists in *A Theory of Justice*, on the priority of his first principle of equal liberty. Most of the vast literature responding to this book in the last twenty-five years, however, has paid more attention to his second principle, which prescribes equal opportunity in awarding positions and says that social and economic inequalities should benefit the least advantaged. Whether Rawls intended so or not, most interpreted *A Theory of Justice* as recommending an activist and interventionist role for government, not only to promote liberties, but to bring about greater social and economic equality.

Hitherto principled political commitment to social equality and distributive economic justice were most associated with socialist politics. Insofar as commitment to such principles had made their way into public policy in liberal democratic societies, many understood this as a result of the relative success of socialist and labor movements in producing certain compromises and concessions from the dominant economic powers (Piven and

Cloward 1982; Offe 1984). *A Theory of Justice* presented norms of social and economic equality within a framework that claimed direct lineage with the liberal tradition.

A major issue of political conflict in the last two decades, as well as earlier, is about whether a liberal democratic state should legitimately aim to ameliorate social problems and economic deprivation through public policy. If Rawls supplied the philosophical framework for the side of the debate that advocates public policies directed at improving the relative distributions of the least advantaged, Robert Nozick's *Anarchy, State and Utopia* (1974) supplied a framework for the other. Nozick argues against the use of what he calls "patterned" principles of justice, that is, principles which require public actors to aim at bringing about particular patterns of distribution. Instead he advocates an unpatterned principle, which specifies the procedures through which holdings are legitimately acquired. In Nozick's theory, any pattern of distribution is just which arises from free transfer of initially legitimate holdings. Nozick finds that conformity with patterned principles requires interfering with consensual economic interaction wherever these would yield outcomes deviating from the desired patterns, and this interference with free exchange makes them wrong.

Nozick's is a theory that gives primacy to liberty over any attempts to undermine distributive inequality; Rawls seeks to construct a theory in which commitment to liberty and equality are compatible. Many articles and collections of essays in the next decade debate the issue of whether a commitment to more egalitarian patterns of distributive justice is compatible with liberty (Arthur and Shaw 1978; Kipnis and Meyers 1985).

Several political theorists continue the Rawlsian project of demonstrating that liberty is not only compatible with greater social equality, but requires it. Amy Gutmann (1980) adds participatory democracy to the values that egalitarian liberalism must promote. Bruce Ackerman (1980) offers a liberal egalitarian conception of social justice that relies on a method of neutral dialogue rather than an imagined social contract. With this dialogue method he re-examines some reasons for questioning a utilitarian theory of justice, and directly refutes entitlement theory's claim that a commitment to collective property and state regulation of distribution is incompatible with liberalism. Contemporary normative arguments for welfare rights, or a welfare liberal conception of justice, similarly aim to systematize a social democratic political program consistent with liberal values and explicitly refuting more libertarian interpretations of those values (Wellman 1982; Goodin 1988; Sterba 1988).

In his idea of social justice as the promotion of people's capabilities, Amartya Sen aims to show the falsity of an opposition between equality and freedom. Equal moral respect for persons implies for development ethics that their capabilities should be nurtured. The most coherent meaning of freedom consists in this nurturing and exercise of capabilities. While Sen's development ethic is egalitarian in the sense of arguing for major redistributive resource allocation to those deprived of opportunities to develop and exercise capabilities, he argues against any simple notion of equal rights, equal liberties, or the equal distribution of goods, because such notions fail to take account of the variety of human needs and situations (Sen 1985; 1992).

Kai Nielsen (1985) argues for the compatibility of liberty and equality in a more explicitly Marxist and socialist vein, devoting a large chunk of his argument to a refutation of Nozick. Even some Marxist-inspired interpretations of justice aim to make an anti-exploitation social and economic theory compatible with a Rawlsian normative theory (Peffer 1990; Reiman 1990). Others insist, however, that different social class positions generate different pictures of society, and different, incompatible, conceptions of justice (e.g., Miller 1976). Thus Milton Fisk (1989) argues that liberal egalitarianism is a contradictory normative theory responding to the contradictory social formation of welfare capitalism, and that both are the outcome of an uneasy class compromise. I believe that there is considerable truth in the claim that both the liberal democratic welfare state and a normative theory that attempts to reconcile the liberal tradition with a commitment to radical egalitarianism are fraught with tensions. Perhaps the promised fourth volume of Brian Barry's *Treatise on Social Justice* will further clarify the requirements of just economic distribution.

A long tradition of normative political theory about international relations concentrates largely on issues of war and peace and the adjudication of conflict between states. Perhaps directly inspired by the social divisions that arose around the war in Vietnam, the period we are considering continues that tradition. Michael Walzer's *Just and Unjust Wars* (1977) is notable both for its reference to the tradition of just war theory, and its original and creative analyses of distant and recent historical events, including the Vietnam War. More original to contemporary political theory, however, is application of questions of social justice, welfare and distribution to international relations. In *Political Theory and International Relations* (1979), Charles Beitz argued that Rawls's principles of justice can be used as the basis for evaluating and criticizing the distributive inequality between developed societies of the North and less developed societies of the South. More recently Thomas Pogge has developed a careful and persuasive extension of a Rawlsian approach to problems of global justice (Pogge 1989, pt. 3). Political theorizing about social and economic inequality across national boundaries remains underdeveloped. Some important work has begun, however, on immigration issues and international justice (Barry and Goodin 1992; Whalen 1988); environment and international justice (Goodin 1990); hunger and obligation to distant peoples (Shue 1980; O'Neill 1986).

## II Democratic theory

Literature on social justice and welfare politicizes the social by asking whether government ought explicitly to try to ameliorate social oppression and inequality. But Arendt's critique of such expanded attention to the social as conceiving public life as social housekeeping might apply to much of this literature. With some exceptions, this literature tends to conceive citizens as rights-bearers and receivers of state action, rather than as active participants in public decision-making.

Nurtured by social movement calls for participatory democracy in the 1960s and 1970s, in the last two decades normative theorizing has flowered that takes speech and citizen participation as central. Carole Pateman's still widely cited *Participation and Democratic Theory* (1970), set much of the agenda for contemporary participatory democratic theory. That work criticized a plebescite and intergroup pluralist conception of democracy, and rearticulated an ideal of democracy as involving active discussion and decision-making by citizens. It argued that social equality is a condition of democratic participation, and that democratic participation helps develop and preserve social equality. This means that the sites of democratic participation must include social institutions beyond the state in which people's actions are directly involved, particularly the workplace.

C. B. Macpherson articulated a framework for critique of the passivity and utilitarianism of dominant conceptions of liberal democracy, and for an alternative more active conception of democracy. It is a measure of how much intellectual discourse has changed in the last twenty years that today reflection on conceptions of human nature seems quaint. Yet Macpherson's analysis of political theories according to whether they assume the nature of human beings as primarily acquisitive consumers of goods or primarily as developers and exercisers of capacities remains a useful way to orient democratic political theory. The perspective of possessive individualism will inevitably regard the political process as a competition for scarce goods, where the competitors' desire for accumulation knows no limits. If one redefines the human good as the development and exercising of capacities, however, then democratic theory takes a wholly different turn. Distributive justice becomes only a means to the wider good of positive freedom, which is itself a social good because realized in co-operation with others. Freedom is the opportunity to develop and exercise one's capacities, and actively engaged citizen democracy is both a condition and expression of such freedom (Macpherson 1973; 1978; cf. Carens 1993). Macpherson's interest in capacities is similar to Sen's mentioned earlier, and is similarly motivated by a conviction that the meaning of freedom should be deepened in political theory and practice.

Several recent political theorists take as a basic value such an expanded notion of freedom, as an absence of domination and positive capacity for self-realization and self-determination. Equality can be best understood as compatible with freedom in this sense, rather than in the narrower, usually property-based sense of freedom as liberty from interference. Thus one aspect of contemporary democratic theory concerns articulation of the conditions of genuine democratic citizenship. People who are deeply deprived cannot be expected to exercise the virtues of democratic participation, and are seriously vulnerable to threats and coercion in the political process. Too often wealth or property function as what Michael Walzer (1982) calls "dominant" goods: inequalities in these economic relations will generate inequalities in opportunity, power, influence, and the abilities to set one's own ends. So serious commitment to democracy presupposes social measures that limit the degree of class inequality and guarantee that all citizens have their needs met (Bay 1981; Green 1985; Cunningham 1987; Cohen and Rogers 1983). Most of those who theorize this relation of social and political equality to democracy concentrate on issues of class. Influenced by feminist analysis, however, a few notice the need to address issues of the gender division of labor to support political equality and participation (Green 1985; Walzer 1982; Mansbridge 1991).

Participatory approaches to democratic theory hold that democracy is a hollow set of institutions if they only allow citizens to vote on representatives to far away political institutions and protect those citizens from government abuse. A fuller democracy in principle means that people can act as citizens in all the major institutions which require their energy and obedience. As I will discuss in a later section, this conclusion has opened both contemporary political practice and theory to interest in civic associations outside both state and corporate life as the most promising sites of expanded democratic practice. Following Pateman's lead, however, contemporary democratic theory has also shown a renewed interest in workplace democracy. Through practices of workplace democracy, several writers argue, citizens can both begin to realize the social and economic equality that they find a condition for democratic participation in the wider polity, and at the same time live the value of creative self-governance in one of the most regular and immediate aspects of modern life (Schweickart 1980; Dahl 1985; Gould 1988). The relative impotence of political theory in setting the agenda of political debate may be revealed by the fact that such thoroughly articulated arguments have little influence on discussion of workplace practices.

At the beginning of the period I am reviewing, the theory of political democracy was largely identified with a theory of interest group pluralism. Inspired by contemporary participatory democratic experiments and institutions, important critiques of this interest group pluralism emerged with well developed alternative conceptions of democracy based on active discussion. In *Beyond Adversary Democracy,* Jane Mansbridge (1980) argued that conceptualizing the democratic process as the competition among interests is too narrow, and she offered a model of "unitary" democracy as one in which participants aim at arriving at a common good through discussion. Wisely, she also argued that unitary democracy has limits, and suggested that both adversary and unitary democracy are necessary in a robustly democratic polity.

Benjamin Barber took up the impulse of this classification and critique, but argued in *Strong Democracy* (1984) that an ideal of unitary democracy is too conformist and collectivist. He proposed a model of strong democracy instead, as a participatory model in which citizens form together a public commitment to a common good but where social pluralities of interest and commitment remain. It is not clear to me, however, that Barber's and Mansbridge's models are all that different.

Following these important texts, recent years have seen an explosion of theorizing about democracy as a discussion-based form of practical reason. Ideals and practices of democratic decision-making that emphasize reasoned discussion have received important further development and refinement (Cohen 1989; Spragens 1990; Sunstein 1988; Michelman 1986; Dryzek 1990; Habermas 1992; Fishkin 1991; Bohman 1996). Though I consider this a very important trend in contemporary political theory, as currently articulated the notion of deliberative democracy has at least two problems. On the whole the models too much assume the need for a unity of citizens as either a starting-point or goal of deliberation (Young 1996). Theories of deliberative democracy, moreover, have for the most part not grappled with the facts of modern mass democracy that nurtured the theory of interest group pluralism. In particular, theorists of participatory and deliberative democracy either ignore issues of representation or roundly reject representation as compatible with democracy (Hirst 1990).

Representation remains sorely undertheorized. In recent years a few theorists have theorized representation in the context of strong democratic theory (Burnheim 1985; Beitz 1989; Bobbio 1984; Grady 1993), but much more work needs to be done. Future work theorizing structures of representation in a large-scale strong democracy would do well to build on the recent *magnum opus* of the patriarch of liberal pluralism himself, Robert Dahl (1989).

## III Feminist political theory

One of the most original and far-reaching developments of political theory in the past quarter-century is work in feminist political theory. Feminist theorists politicize the social by questioning a dichotomy of public and private and thereby proposing that family relations, sexuality, and the gendered relations of street, school and workplace are properly political relations.

It is impossible to do justice to the varied accounts of feminist political theorists in this short space. One theme that runs through much of this work is a deconstruction of the dichotomy of public and private presupposed by both traditional and contemporary political thought. The public realm of politics can be so rational, noble, and universal only because the messy content of the body, meeting its needs, providing for production, caretaking, and attending to birth and death, are taken care of elsewhere. Male heads of households derive their power to make wars, laws and philosophy from the fact that others work for them in private, and it is no surprise that they would model nobility on their own experience. But a modern reflexive political theory should recognize that the glory of the public is dialectically entwined with the exploitation and repression of the private, and the people restricted to that sphere so they can take care of people's needs. However the analysis proceeds, feminist political theory concludes that 20th-century politics requires a basic rethinking of this distinction and its meaning for politics (Okin 1979; Clark and Lange 1979; Elshtain 1981; Nicholson 1986; Young 1987; Landes 1988; Shanley and Pateman 1991).

Much feminist political theory analyzes the masculinism of a universal reason that abhors embodiment and honors the desire to kill and risk life (Hartsock 1983; Brown 1988). Beginning with the ancients, courage tops the list of citizen virtues, which promotes the soldier as the paradigm citizen. Machiavelli is celebrated as the father both of modern *realpolitik* and republicanism because he fashions the account of political man so clearly relying on images of risk, danger, winning, and the competition of sport and battle. Hannah Pitkin's brilliant study of Machiavelli relies on feminist psychoanalysis as well as critiques of the public–private dichotomy to expose the grounds of this masculinist citizen in a psychic opposition between self and other (Pitkin 1984).

Many feminist critics focus on the idea of the social contract to uncover different assumptions about human nature, action and evaluation that exhibit masculine experience and develop a one-sided account of the possibilities of political life and political change. Several have focused on the assumptions of individualism, atomistic autonomy and independence that structure the image of the rational citizen in modern political thought. Carole Pateman (1988) argues that the idea of the individual assumed by social contract theory is actually male, because that concept of individual assumes an independence from bodily caretaking that can only obtain if someone else is doing it for one. Other feminist critics argue that the concept of the rational autonomous individual of social contract theory carries an image of the person as self-originating, without birth and dependence. If the original dependence of all human beings on others were to replace this assumption of self-generation, then the entire edifice that constructs social relations as effects of voluntary bargains would collapse. Some writers have explored alternative starting-points for a conception of society and the political, which begins with premises of connectedness and interdependence rather than autonomy and independence (Held 1987).

Feminists have subjected many of the important terms of political discourse to searching analysis, including power (Hartsock 1983), authority (Jones 1993), political obligation (Hirschmann 1992), citizenship (Dietz 1985; Stiehm 1984; Bock and James 1992), privacy (Allen 1988), democracy (Phillips 1991) and justice (Okin 1989).

The questions and conclusions in this conceptual literature are extremely diverse, but the arguments tend to cluster around two projects. First, feminist analysis argues that theories of justice, power, obligation, and so on reflect male gendered experience, and must be revised if they are to include female gendered experience. Often the criticism takes the form of arguing that the generality that political theorists claim for their concepts and theories cannot in fact be general because the theories do not notice the facts of gender difference and take these into account in formulating their theories. Thus Susan Okin argues, for example, that Walzer's arguments about justice become inconsistent when the facts of male domination within communities are taken into account.

Second, these conceptual analyses often claim that political theory tends to disembody these central political concepts. Thus Nancy Hartsock argues, for example, that dominant theories of power repress the relation childhood experience of vulnerability, and presume a rigid self-other dichotomy that reduces power to competition and control. Thinking power in terms of embodiment would draw the attention of political theorists to power as power-to and not simply power-over (cf. Wartenberg 1990). Joan Tronto reflects on issues of power in the context of care-giving and its implications for politics and policy (Tronto 1993). Much feminist discussion about the concept of equality, to take another example, has questioned whether equal respect for women should imply identical treatment to men, because women experience pregnancy and childbirth, and suffer other vulnerabilities because of sexist society (Scott 1988; Bacchi 1991).

Feminist arguments about individualism, the public–private dichotomy, contract theory, and the implicit bias in Western ideas of reason and universality have influenced some work of male political theorists concerned with contemporary issues (e.g. Green 1985; Smith 1989). Much political theory continues using these assumptions, apparently feeling obliged neither to revise their approaches in light of feminist critiques nor to give arguments against them.

## IV Postmodernism

I bring postmodernism under the theme of politicizing the social in at least two respects. First, like some of the feminist writers I have mentioned, many postmodern political theorists concern themselves with the movement and flow of power throughout society, and with how social power conditions and is conditioned by political institutions and conflicts. Second, many postmodern thinkers insist that political actors should be understood as not-necessarily-coherent products of social processes, rather than as unanalyzed origins of conflict and co-operation.

The work of Michel Foucault stands perhaps most directly as a towering contribution to political theory, at the same time as this work challenges many traditional assumptions. Foucault thinks that political theory and discourse continue to assume a paradigm of power derived from pre-modern experience, and that since the eighteenth century a new structuration of power has operated. The old paradigm conceives power as sovereignty: the repressive force of a rule's decree of what is allowed and forbidden. The new regime of power operates less through command and more through disciplinary norms. In this modern regime the king and his agents do not reach out from a center to control unruly subjects through fear. Instead, ruling institutions percolate up from the ground, in the outlying capillaries of society, which discipline bodies to conform with norms of reason, order and good taste. Power proliferates and becomes productive in the disciplinary institutions that organize and manage people in a complex division of labor: hospitals and clinics, schools, prisons, welfare organizations, police departments (Foucault 1979; 1980; Burchell *et al.* 1991).

Political theory has yet fully to absorb and evaluate their picture of power as the productive and proliferating process of disciplinary institutions. William Connolly (1987) has taken a leading role in showing Foucault's ideas as a challenge to political theory's uncritical reliance on Enlightenment ideas. He argues that norms are always double-sided and ambiguous, and that we should resist the bureaucratic impulse to discipline ambiguity. A few political theorists have examined the concept of power in light of Foucault's work (Smart 1983; Wartenberg 1990; Spivak 1992; Honneth 1991). More engagement with Foucault's ideas will require rethinking the concepts of state, law, authority, obligation, freedom, and rights, as well.

The critical force of Foucault's analysis is apparent. But this theorizing cries out for normative ideals of freedom and justice by means of which to evaluate institutions and practices. Several political theorists make important arguments that Foucault's theorizing is implicitly contradictory because he refuses to articulate such positive ideals (Taylor 1984; Fraser 1989, chap. 1; Habermas 1990, chaps. 9 and 10).

Several other French writers associated with postmodernism have been important for political theory, including Lacan, Derrida, Lyotard, Baudrillard, and Kristeva. I will discuss only a few other themes that arise for political theory out of the work of these writers.

Postmodern thinkers have questioned an assumption that unified individual subjects are the units of society and political action. Subjectivity is a product of language and interaction, not its origin, and subjects are as internally plural and contradictory as the social field in which they live. This ontological thesis raises serious questions for political theory about the meaning of moral and political agency. Interpreting Merleau-Ponty along with some of the others I have mentioned above, Fred Dallmayr (1981) offers a vision of political process where a desire to control dissipates.

Several writers take up the Derridian critique of a metaphysics of presence (Derrida 1974) to argue that a desire for certainty and clear regulatory principles in politics results in repressing and oppressing otherness, both in other people and in oneself (White 1991). In *Identity\Difference* William Connolly (1991) gives a twist to this thesis by claiming that such unifying politics produces a resentment too quick to blame and not open enough to ambiguity. Bonnie Honig (1993) applies these sorts of arguments to the texts of political theorists such as Kant, Rawls and Sandel; she argues that their desire for a unifying theoretical center in political theory oppressively expels subjects who deviate from their models of rational citizen and community.

I find the most important consequence of postmodern critiques of identifying thinking to lie in a reinterpretation of democratic pluralism. Democratic politics is a field of shifting identities and groups that find affinities and contest with one another (Yeatman 1994). Ernesto Laclau and Chantal Mouffe's 1985 book, *Hegemony and Socialist Strategy*, has been influential along these lines. They argue that the Marxist concept of the revolutionary agency of the working class is a metaphysical fiction inappropriate to the contemporary period of proliferating radical social movements defined by multiple identities and interests. Along lines similar to Lyotard's (1984) analysis of the myth of the People as the subject of historical narrative, Claude Lefort argues that modern politics, especially in relatively free and democratic societies, can rest no legitimacy on a unified "will of the people." On the contrary, modern democracy is precisely the process of contesting claims ungrounded in any unified subject (Lefort 1986). Radical democratic politics should be understood as the coalescing of plural social movements in civil society, to deepen democratic practice both in the state and society (cf. Mouffe 1993). I heartily endorse a political theory that appreciates social heterogeneity and is suspicious of efforts to unify (see Young 1990). Much of this writing, however, seems either to identify normative standards of justice and freedom as themselves suspicious, or not to refer to issues of freedom and justice at all. The task for a political theory sensitive to the repressive implications of identifying logic and exclusionary normalization is to develop methods of appealing to justice less subject to these criticisms.

## V New social movements and civil society

In the last twenty years movements have proliferated whose style and demands go beyond claims for rights or welfare: environmentalism, peace movements, group based movements of national resistance and cultural pride, feminism, gay and lesbian liberation. Most paradigmatically, the aim of many of these movements is to politicize the social, to bring many habits of everyday social interaction and culture into reflective question and contestation. Some recent political theory conceptualizes the political styles and implications of these sorts of movements, often called "new social movements" (Melucci 1989; Boggs 1986; Moores and Sears 1992).

These movements are called "new" for at least two reasons. First, on the whole their issues do not primarily concern inclusion in basic citizenship rights nor the enlargement of economic rights. Their issues are more specifically social—respect and self-determination for cultural difference, responsibility and pluralism in everyday lifestyle, reflection on power in social interaction, participation in decisions in social and economic, as well as political, institutions. Secondly, the form of organization of these movements does not replicate the mass movement form of political party or union, a unified bureaucracy seeking power through resource mobilization. Instead, these new social movements tend to be networks of more local groups, each with their own principles and style, that nevertheless act in concert *en masse* in some protest actions.

Some important political theory reflects systematically on the normative political principles embodied in some of these movements. The environmental movement, for example, provides substance for reflection on basic normative issues of value, social rationality, and democratic participation (Sagoff 1988; Goodin 1992; Dryzek 1987).

Despite the importance of anti-racist social movements emphasizing self-determination, full inclusion, cultural pluralism, and reparation for past injustice, political theorists have paid surprisingly little attention to issues of race and racism. Debates about affirmative action have been the most common context for reflecting on issues of racism among American political theorists (Goldman 1979; Bowie 1988; Ezorsky 1991). This focus, while important, is rather narrow. Judith Shklar's (1991) account of the meaning of American citizenship in the shadow of the legacy of slavery gives a more profound place to racist ideology and fears for understanding political discourse. Bernard Boxill (1984) offers a thorough philosophical grounding for many claims of African Americans about social justice, including affirmative action, self-respect, and reparations. Bill Lawson and Howard McGary (1992) conceptualize a political theory of freedom by reflecting on slave narratives. Drawing on the critical reactions of Black feminists and feminists of color to dominant feminist discourse, Elizabeth Spelman (1988) critically questions the use of an undifferentiated category of gender in social and political theory. Along similar lines, Nancy Caraway (1991) synthesizes much of the writing of feminist women of color as they dialogue with the ethnocentrism of much feminist political and social theory.

Andrew Sharp's carefully argued book, *Justice and the Maori* (1990) stands as a model of theorizing of indigenous people's movement in the context of advanced industrial society. Some political theorists in Australia and Canada have begun to take up the challenge of normative theorizing about indigenous people's issues (Carens 1993; Kymlicka 1993; Wilson and Yeatman 1995). Though there is significant work on indigenous people's issues by legal theorists in the United States (e.g., Williams 1990), I see few signs of a discussion among U.S. philosophers and political theorists about normative questions concerning indigenous peoples.

Some recent theorizing about the role of state and bureaucracy in advanced industrial societies helps set a context for understanding the new social movements. Foucault's theory of normalization produced by bureaucratized human and social services holds that such operations of disciplinary social power also create their own resistances. Along somewhat different lines, Claus Offe (1984) gives an account of the modern welfare state as having depoliticizied processes of social control and public spending. The state has become an arena where officials conduct their real business more or less behind closed doors, and experts administer policies with a technical know-how that does not bring normative ends into view. Social movements politicize some of this activity from a position outside state institutions.

In his concept of the "colonization of the lifeworld," Jürgen Habermas (1987) offers a theoretical context for conceptualizing the meaning of new social movements. State and corporate institutions have developed their own technical rationality which has become "uncoupled" from the everyday life context of meaningful cultural interaction, and then these state and corporate imperatives turn around and constrain or distort the everyday lifeworld. Many new social movements can be interpreted as a reaction to this colonization, an attempt to open greater space for collective choice about normative and aesthetic ends, and to limit the influence of systemic imperatives of power and profit.

If it is true that state activity is largely technicized, then state institutions cannot function as the site of deliberative politics in advanced capitalist society. Rather, politics, in the sense of people meeting together to discuss their collective problems, raise critical claims about action, and act together to alter their circumstances, happens more in critical public spheres outside the state and directed at its actions. Habermas's major work of the early 1960s *The Structural Transformation of the Public Sphere* (English edition 1989), has received new attention by political theorists interested in participatory politics and critical normative discourse in late 20th-century society (Calhoun 1991).

For the purpose of such theorizing, a concept of civil society as the locus of free and deliberative politics has been emerging. The concept of civil society was used in the opposition movements of Eastern Europe throughout the 1980s, and this usage has influenced some of these theoretical developments. The concept has also been influential in opposition movements in South Africa and Latin America.

Leading proponents of the theory of civil society include John Keane (1984; 1988), Jean Cohen (1983) and Andrew Arato (Arato and Cohen 1992). Civil society consists in voluntary associative activity—the array of civic associations, non-profit service organizations, and so on that are only loosely connected to state and corporate economy. Activities of civil society do require a strong liberal state that protects the liberties of speech, association, and assembly. But the activities of civil society are more directly participatory than the way citizens relate to state decision-making apparatus.

Both Cohen and Arato and Keane thus look to civil society as the arena for deepening and radicalizing democracy. The public spheres of civil society can and should be enlarged by pushing back the bureaucratized functions of the state and structuring more areas of social life in terms of voluntary participatory organizations. These civic organizations can also serve as the stage from which to launch criticism of state policy and action.

The theory of civil society adds an important dimension to our understanding of politics as public action and participation. It also seems to occlude some concerns that appear more salient, however, when the focus is on state policy, namely concerns about economic inequality. The concept of civil society is ambiguous, moreover, about the relation of civil society to the economy. Not all theorizing about civil society and political theory distinguishes between economy and civil society as Cohen and Arato do. Some theorists identify the freedom of civil society with freedom of the market (see, e.g., Kukathas and Lovell 1991). Then the theory of civil society emerges as a new form of anti-state liberalism. Since all civil society theorizing agrees that modern welfare state bureaucracies tend to be undemocratic and dominative, the question arises of how a commitment to the active promotion of social justice can be made compatible with this view of politics and democracy.

## VI Liberalism and communitarianism

With his 1982 *Liberalism and the Limits of Justice*, Michael Sandel launched the current of contemporary political theory known as communitarianism, which can also be interpreted as politicizing the social. Communitarians claims to anchor political values like justice, rights, and freedom, in particular social and cultural contexts. Thus it conceptualizes the social as prior to and constitutive of the political.

Sandel argued that Rawls's theory of justice wrongly presupposes a moral self prior to social relationships guided by principles of justice, a self "unencumbered" by particular culture and commitments into which he or she is thrown. Principles of justice generated from such an abstract notion of self can serve only to regulate public relations among strangers in the most formalistic way. For a robust political theory of social union, Sandel suggested, justice must be supplemented by recognition of particular community bonds and commitments that constitute selves.

In *After Virtue*, Alistair MacIntyre (1981) levelled a more historically oriented challenge to liberalism. The economic and ideological changes of modern society create a modern dilemma of relativism. Religious and moral questions—questions about the good, the just, the virtuous—have become matters of private conscience or contesting political opinion. Liberalism is a system of formal adjudication among such competing and incommensurate opinions among which there is no means of deciding some are right and others wrong. In this modern world-view moral agents are released onto the landscape as disconnected, commodified and often cynical atoms. The late modern malaise can best be treated by looking for living communities of shared values and virtues that can serve as contemporary analogues of medieval guild communities, and other traditional self-ruling communities bound by common commitment to particular excellences.

In response to these and other communitarian critiques of liberalism's claim to transcend and bracket particular cultural contexts, some writers claimed that communitarianism implies an unacceptable relativism. If culture shapes norms, and there is no means of gaining a reflective distance for evaluating those norms and articulating principles of liberal reason, then we cannot evaluate morally and politically different social contexts. Communitarians responded that liberals aim at a dangerous and abstract universality.

By the mid 1980s the so-called liberalism–communitarian debate was flooding the pages of journals and books in political theory. But the debate was both too abstract and founded on a false dichotomy. Despite the fact that the aim of communitarians was to situate moral and political norms in the particular social contexts of full-blooded agents, they rarely discussed any particular communities (cf. Wallach 1987). It was difficult to find, moreover, any communitarian who would reject liberal values of equal respect, freedom of action, speech and association, or tolerance (cf. Gutmann 1985). Few self-proclaimed liberals, on the other hand, were ready to deny the power of particular cultural commitment in individual lives, though they might disagree with communitarians about the normative significance of these facts.

The liberalism–communitarianism debate did expose how much contemporary liberal political theory abstracts from social group affiliation and commitment to consider individuals only as individuals. It thus posed an important challenge of whether and how liberal theory ought to include recognition of particular contexts of social and cultural group difference. Will Kymlicka's *Liberalism, Community and Culture* (1989) represents a turning-point in this debate. Unlike many writings in this discussion, Kymlicka is not abstract about community and culture, but rather discusses the particular cultural and political situations of native peoples in relation to the liberal state of Canada. Staunchly adhering to the values of modern political liberalism, Kymlicka argues that these are not only compatible with, but require, the constitution of cultural rights that may sometimes imply special rights for endangered or oppressed cultural minorities. The key to his argument that such cultural rights follow from liberalism is his construction of individual rights as including an individual right to cultural membership, and thus to the maintenance of the culture of which one is a member.

Another Canadian contributor to this more contextualized discussion of cultural rights, Charles Taylor, is less certain that a principle of cultural recognition is compatible with at least some versions of liberalism (Taylor 1992). If we understand liberalism to require a universality to the statement of rights, such that laws and rules should apply equally to all in the same way, then politically recognizing and maintaining particular cultures sits uneasily with liberalism. The recognition and preservation of minority cultures may require special treatment and special rights for which there are good moral arguments, but arguments beyond the liberal individualist tradition (cf. Young 1989; 1990, especially chap. 6).

A different set of recent works has also aimed to produce a reconciliation between the stances of liberalism and communitarianism that were posited in the early 1980s. Liberalism has typically been interpreted as neutral among values and equally accepting of ways of life as long as their activities leave one another alone. Communitarianism, on the other hand, especially in MacIntyre's version, takes the good, as the ends of action, and virtue, as the disposition to bring about these good ends, as the moral commitment that liberalism has abandoned to relativism. Some writers have rejected the characterization of liberalism as neutral among ends and virtues, and have argued that liberalism itself implies particular cultural values, normative ends, and behavioral virtues (Macedo 1990; Galston 1991).

It is fitting to end my story of two decades of political theory by referring to the same writer with whom I began: John Rawls. The arguments of *Political Liberalism* (1993) are, to a significant degree, attempts to respond to the liberalism–communitarianism debate and the social context of multiculturalism in liberal society. Rawls moves in the opposite direction from Kymlicka and some of the other writers who aim to reconcile the values of political liberalism with public recognition for particular cultural norms and ways of life. Freedom and respect for particular "comprehensive doctrines," as he calls them, instead requires that they all agree on a set of principles guiding the interaction of distinct communities, but transcending them all. Multiculturalism is possible in a liberal society only if we re-draw a fairly clear border between what is properly public, the business of the constitutional and legal rules governing the whole society, and what is private, in the sense of matters of individual and community conscience and commitment.

Although the overlapping consensus that Rawls believes can emerge from the willingness of different cultures and communities of conscience to set fair terms of co-operation retains attention to social and economic inequality as well as liberty, I find *Political Liberalism* to constitute a retreat from the social. Rawls thinks that conflicts and ambiguities produced by the being together of concrete communities, about issues of sexuality, family, video content, religious dress in public, and countless other issues, are best handled by re-establishing a legal and political discourse that only admits into its realm issues already framed in terms of generalizable norms. The problem of political conflict in the late 20th century appears to be that particularist value claims of social groups have too much entered public discourse, and we would do better to distinguish those claims that can properly be adjudicated through public reason from those simple differences that are social or private.

While Rawls continues to say that the difference principle is important, in this later work he emphasizes the procedural mechanisms for arriving at and maintaining committed consensus on civil rights and liberties. Proposals to redistribute wealth and income so as to maximize the expectations of the least advantaged are much more controversial today than they were twenty years ago, even as the ranks of the least advantaged have been swelling. Being less advantaged overlaps significantly, moreover, with social positioning in terms of race, gender, ethnicity, and culture. Thus political claims about family values or recognition for cultural minorities have much to do with claims of social justice. Even where less tied to issues of economic disadvantage, the "politics of identity," whereby groups make claims for public recognition of the specificity of their cultural values, are not going away. For all these reasons the current temptation of political theory to retreat from the social threatens to make it even more irrelevant to politics than usual. Fortunately, there are signs that many political theorists will continue engaging with these fiercely difficult political issues of the late 20th century.

## Acknowledgments

I am grateful to Joseph Carens, Robert Goodin, Molly Shanley, Rogers Smith and Andrew Valls for helpful comments on earlier versions of this chapter.

## References

Ackerman, B. 1980. *Social Justice in the Liberal State.* New Haven, Conn.: Yale University Press.

Allen, A. 1988. *Uneasy Access: Privacy for Women in a Free Society.* Totowa, N.J.: Rowman and Littlefield.

Arato, A., and Cohen, J. 1992. *Civil Society and Political Theory.* Cambridge, Mass.: MIT Press.

Arendt, H. 1958. *The Human Condition.* Chicago: University of Chicago Press.

Arthur, J., and Shaw, W., eds. 1978. *Justice and Economic Distribution.* Englewood Cliffs, N.J.: Prentice-Hall.

Bacchi, C. 1991. *Same Difference.* Boston: Allen and Unwin.

Barber, B. 1984. *Strong Democracy.* Berkeley: University of California Press.

Barry, B. 1979. The strange death of political philosophy. Pp. 11–23 in *Democracy, Power and Justice.* Oxford: Clarendon Press, 1989.

—— 1989. *A Treatise on Social Justice.* Vol. i. *Theories of Social Justice.* Berkeley: University of California Press.

—— and Goodin, R. E., eds. 1992. *Free Movement.* University Park, Pa.: Penn State University Press.

Bay, C. 1981. *Strategies of Political Emancipation.* Notre Dame, Ind.: University of Notre Dame Press.

Beitz, C. 1979. *Political Theory and International Relations.* Princeton: Princeton University Press.

—— 1989. *Political Equality.* Princeton, N.J.: Princeton University Press.

Bernstein, R. 1986. Rethinking the social and the political. Pp. 238–59 in R. Bernstein, *Philosophical Profiles.* Philadelphia, Pa.: University of Pennsylvania Press.

Bobbio, N. 1984. *The Future of Democracy.* Minneapolis: University of Minnesota Press.

Bock, G., and James, S., eds. 1992. *Beyond Equality and Difference.* London: Routledge.

Boggs, C. 1986. *Social Movements and Political Power.* Philadelphia, Pa.: Temple University Press.

Bohman, J. 1996. *Public Deliberation.* Cambridge, Mass.: MIT Press.

Bowie, N., ed. 1988. *Equal Opportunity.* Boulder, Colo.: Westview Press.
Boxill, B. 1984. *Blacks and Social Justice.* Totowa, N.J.: Rowman and Allenheld.
Brown, W. 1988. *Manhood and Politics.* Totowa, N.J.: Rowman and Littlefield.
Burchell, G.; Gordon, C.; and Miller, P., eds. 1991. *The Foucault Effect.* Chicago: University of Chicago Press.
Burnheim, J. 1985. *Is Democracy Possible?* Oxford: Polity Press.
Calhoun, C. ed. 1991. *Habermas and the Public Sphere.* Cambridge, Mass.: MIT Press.
Canovan, M. 1992. *Hannah Arendt: A Reinterpretation of Her Political Thought.* Cambridge: Cambridge University Press.
Caraway, N. 1991. *Segregated Sisterhood.* Nashville: University of Tennesee Press.
Carens, J. H. 1993. Citizenship and aboriginal self-government. Ottawa, Ont.: Royal Commission on Aboriginal Peoples.
—— ed. 1993. *Democracy and Possessive Individualism.* Albany: State University of New York Press.
Clark, L., and Lange, L. 1979. *The Sexism of Social and Political Theory.* Toronto: University of Toronto Press.
Cohen, J. 1983. *Class and Civil Society.* Amherst: University of Massachusetts Press.
—— 1989. Deliberation and democratic legitimacy. Pp. 17–34 in *The Good Polity*, ed. A. Hamlin and P. Pettit. Oxford: Blackwell.
—— and Rogers, J. 1983. *On Democracy.* New York: Penguin.
Connolly, W. E. 1987. *Politics and Ambiguity.* Madison: University of Wisconsin Press.
—— 1991. *Identity\Difference.* Ithaca, N.Y.: Cornell University Press.
Cunningham, F. 1987. *Democratic Theory and Socialism.* Cambridge: Cambridge University Press.
Dahl, R. A. 1985. *A Preface to Economic Democracy.* Berkeley: University of California Press.
—— 1989. *Democracy and Its Critics.* New Haven, Conn.: Yale University Press.
Dallmayr, F. 1981. *Twilight of Subjectivity.* Amherst: University of Massachusetts Press.
Derrida, J. 1974. *Of Grammatology*, trans. G. C. Spivak. Baltimore, Md.: Johns Hopkins University Press.
Dietz, M. 1985. Citizenship with a feminist face: the problem with maternal thinking. *Political Theory*, 13: 19–37.
Dryzek, J. 1987. *Rational Ecology.* Oxford: Basil Blackwell.
—— 1990. *Discursive Democracy.* Cambridge: Cambridge University Press.
Elshtain, J. B. 1981. *Public Man, Private Woman.* Princeton, N.J.: Princeton University Press.
Ezorsky, G. 1991. *Racism and Justice.* Ithaca, N.Y.: Cornell University Press.
Fishkin, J. S. 1991. *Deliberative Democracy.* New Haven, Conn.: Yale University Press, 1991.
Fisk, M. 1989. *The State and Justice.* Cambridge: Cambridge University Press.
Foucault, M. 1979. *Discipline and Punish*, trans. A. Sheridan. New York: Vintage Books.
—— 1980. *Power/Knowledge*, trans. and ed. C. Gordon, L. Marshall, J. Mephan and K. Soper. New York: Pantheon Books.
Fraser, N. 1989. *Unruly Practices.* Minneapolis: University of Minnesota Press.
Galston, W. 1991. *Liberal Purposes.* Cambridge: Cambridge University Press.
Goldman, A. 1979. *Justice and Reverse Discrimination.* Princeton, N.J.: Princeton University Press.
Goodin, R. E. 1988. *Reasons for Welfare.* Princeton, N.J.: Princeton University Press.
—— 1990. International ethics and the environmental crisis. *Ethics and International Affairs*, 4: 91–105.

GOODIN, R. E. 1992. *Green Political Theory.* Oxford: Polity Press.
GOULD, C. 1988. *Rethinking Democracy.* Cambridge: Cambridge University Press.
GRADY, R. C. 1993. *Restoring Real Representation.* Urbana: University of Illinois Press.
GREEN, P. 1985. *Retrieving Democracy.* Totowa, N.J.: Rowman and Allenheld.
GUTMANN, A. 1980. *Liberal Equality.* Cambridge: Cambridge University Press.
—— 1985. Communitarian critics of liberalism. *Philosophy and Public Affairs,* 14: 308–22.
HABERMAS, J. 1987. *The Theory of Communicative Action,* trans. T. McCarthy. Boston: Beacon Press.
—— 1989. *The Structural Transformation of the Public Sphere,* trans. T. Burger with F. Lawrence. Cambridge: MIT Press; originally published 1962.
—— 1990. *The Philosophical Discourse of Modernity.* Cambridge, Mass.: MIT Press.
—— 1992. *Faktizität und Geltung.* Frankfurt: Suhrkamp Verlag. English translation: *Between Facts and Norms,* trans. W. Rehg. Cambridge, Mass.: MIT Press, forthcoming.
HARTSOCK, N. 1983. *Money, Sex and Power.* New York: Longman.
HELD, V. 1987. Non-contractual society. Pp. 111–38 in *Science, Morality and Feminist Theory,* ed. M. Hanen and K. Nielsen. Calgary: University of Calgary Press.
HIRSCHMANN, N. 1992. *Rethinking Obligation.* Ithaca, N.Y.: Cornell University Press.
HIRST, P. 1990. *Representative Democracy and its Limits.* Oxford: Polity Press.
HONIG, B. 1993. *Political Theory and the Displacement of Politics.* Ithaca, N.Y.: Cornell University Press.
HONNETH, A. 1991. *The Critique of Power.* Cambridge, Mass.: MIT Press.
JONES, K. 1993. *Compassionate Authority.* New York: Routledge.
KEANE, J. 1984. *Public Life in Late Capitalism.* Cambridge: Cambridge University Press.
—— 1988. *Democracy and Civil Society.* London: Verso.
KIPNIS, K., and MEYERS, D. T., eds. 1985. *Economic Justice: Private Rights and Public Responsibilities.* Totowa, N.J.: Rowman and Allenheld.
KUKATHAS, C., and LOVELL, D. W. 1991. The significance of civil society. Pp. 18–40 in *The Transition from Socialism,* ed. C. Kukathas, D. W. Lovell and W. Maley. Sydney: Longman Cheshire.
KYMLICKA, W. 1989. *Liberalism, Community and Culture.* Oxford: Clarendon Press.
—— 1993. Group representation in Canadian politics. Paper presented to IRPP project on Communities, the Charter and Interest Advocacy, Ottawa, Ont.
LACLAU, E., and MOUFFE, C. 1985. *Hegemony and Socialist Strategy.* London: Verso.
LANDES, J. 1988. *Women and the Public Sphere in the Age of the French Revolution.* Ithaca, N.Y.: Cornell University Press.
LAWSON, B., and MCGARY, H. 1992. *Between Slavery and Freedom.* Bloomington: Indiana University Press.
LEFORT, C. 1986. *The Political Forms of Modern Society.* Cambridge: Cambridge University Press.
LYOTARD, J.-F. 1984. *The Postmodern Condition.* Minneapolis: University of Minnesota Press.
MACEDO, S. 1990. *Liberal Virtues.* Cambridge, Mass.: Harvard University Press.
MACINTYRE, A. 1981. *After Virtue.* Notre Dame, Ind.: Notre Dame University Press.
MACPHERSON, C. B. 1973. *Democratic Theory.* Oxford: Clarendon Press.
—— 1978. *The Life and Times of Liberal Democracy.* Oxford: Oxford University Press.
MANSBRIDGE, J. J. 1980. *Beyond Adversary Democracy.* New York: Basic Books.
—— 1991. Feminism and democratic community. Pp. 339–95 in *Nomos* xxxv: *Democratic Community,* ed. J. W. Chapman and I. Shapiro. New York: New York University Press.
MELUCCI, A. 1989. *Nomads of the Present.* London: Hutchinson Radius.
MICHELMAN, F. 1986. Traces of self-government. *Harvard Law Review,* 100: 1–311.

Miller, D. 1976. *Social Justice.* Oxford: Clarendon Press.
Miller, J. 1984. *Rousseau: Dreamer of Democracy.* New Haven, Conn.: Yale University Press.
Mooers, C., and Sears, A. 1992. The new social movements and the withering away of state theory. Pp. 52–68 in *Organizing Dissent,* ed. W. K. Carroll. Toronto: Garamond Press.
Mouffe, C. 1993. *The Return of the Political.* London: Verso.
Nicholson, L. 1986. *Gender and History.* New York: Columbia University Press.
Nielsen, K. 1985. *Equality and Liberty.* Totowa, N.J.: Rowman and Allanheld.
Nozick, R. 1974. *Anarchy, State and Utopia.* New York: Basic Books.
O'Neill, O. 1986. *Faces of Hunger.* London: Allen and Unwin.
Offe, C. 1984. *Contradictions of the Welfare State.* Cambridge, Mass.: MIT Press.
Okin, S. M. 1979. *Women in Western Political Thought.* Princeton, N.J.: Princeton University Press.
—— 1989. *Justice, Gender and the Family.* New York: Basic Books.
Pateman, C. 1970. *Participation and Democratic Theory.* Cambridge: Cambridge University Press.
—— 1988. *The Sexual Contract.* Stanford, Calif.: Stanford University Press.
Peffer, R. 1990. *Marxism, Morality, and Social Justice.* Princeton, N.J.: Princeton University Press.
Phillips, A. 1991. *Engendering Democracy.* University Park, Penn.: Penn State Press.
Pitkin, H. F. 1981. Justice: on relating public and private. *Political Theory,* 9: 327–52.
—— 1984. *Fortune is a Woman.* Berkeley: University of California Press.
Piven, F. F., and Cloward, R. 1982. *The New Class War.* New York: Pantheon.
Pocock, J. G. A. 1975. *The Machiavellian Moment.* Princeton, N.J.: Princeton University Press.
Pogge, T. W. 1989. *Realizing Rawls.* Ithaca, N.Y.: Cornell University Press.
Rawls, J. 1971. *A Theory of Justice.* Cambridge, Mass.: Harvard University Press.
—— 1993. *Political Liberalism.* New York: Columbia University Press.
Reiman, J. 1990. *Justice and Modern Moral Theory.* New Haven, Conn.: Yale University Press.
Sagoff, M. 1988. *The Economy of the Earth.* Cambridge: Cambridge University Press.
Sandel, M. J. 1982. *Liberalism and the Limits of Justice.* Cambridge: Cambridge University Press.
Schweickart, D. 1980. *Capitalism or Worker Control?* New York: Praeger.
Scott, J. W. 1988. Deconstructing equality versus difference: or, the uses of poststructuralist theory for feminism. *Feminist Studies,* 14: 33–50.
Sen, A. K. 1985. *Commodities and Capabilities.* Amsterdam: North Holland.
—— 1992. *Inequality Reconsidered.* Oxford: Oxford University Press.
Shanley, M. L., and Pateman, C., eds. 1991. *Feminist Interpretations and Political Theory.* University Park, Pa.: Penn State Press.
Sharp, A. 1990. *Justice and the Maori.* Auckland: Oxford University Press.
Shklar, J. 1991. *American Citizenship.* Cambridge, Mass.: Harvard University Press.
Shue, H. 1980. *Basic Rights.* Princeton, N.J.: Princeton University Press.
Smart, B. 1983. *Foucault, Marxism and Critique.* London: Routledge and Kegan Paul.
Smith, R. 1989. "One united people:" second-class female citizenship and the American quest for community. *Yale Journal of Law and Humanities,* 1/2: 229–93.
Spelman, E. 1988. *Inessential Woman.* Boston: Beacon Press.
Spivak, G. C. 1992. More on power/knowledge. Pp. 149–73 in *Rethinking Power,* ed. T. E. Wartenberg. Albany: State University of New York Press.
Spragens, T. 1990. *Reason and Democracy.* Durham, N.C.: Duke University Press.

STERBA, J. 1988. *How to Make People Just.* Totowa, N.J.: Rowman and Littlefield.
STIEHM, J., ed. 1984. *Women's Views of the Political World of Men.* Dobbs Ferry, N.Y.: Transnational Publishers.
SUNSTEIN, C. 1988. Beyond the republican revival. *Yale Law Journal,* 97: 1539–90.
TAYLOR, C. 1984. Foucault on freedom and truth. *Political Theory,* 12: 152–83.
—— 1992. *Multiculturalism and the "Politics of Recognition."* Princeton, N.J.: Princeton University Press.
TRONTO, J. 1993. *Moral Boundaries.* New York: Routledge.
WALLACH, J. 1987. Liberalism, communitarians and the tasks of political theory. *Political Theory,* 15: 581–611.
WALZER, M. 1977. *Just and Unjust Wars.* New York: Basic Books.
—— 1982. *Spheres of Justice.* New York: Basic Books.
WARTENBERG, T. 1990. *Forms of Power.* Philadelphia, Pa.: Temple University Press.
WELLMAN, C. 1982. *Welfare Rights.* Totowa, N.J.: Rowman and Allenheld.
WHALEN, F. 1988. Citizenship and freedom of movement: an open admissions policy? Pp. 3–39 in *Open Borders? Closed Societies?*, ed. M. Gibney. New York: Greenwood Press.
WHITE, S. 1991. *Political Theory and Postmodernism.* Cambridge: Cambridge University Press.
WILLIAMS, R. A. 1990. *The American Indian in Western Legal Thought.* Oxford: Oxford University Press.
WILSON, M., and YEATMAN, A., eds. 1995. *Justice and Identity.* Wellington, N.Z.: Bridget Williams.
YEATMAN, A. 1994. *Postmodern Revisionings of the Political.* New York: Routledge.
YOUNG, I. M. 1987. Impartiality and the civic public: some implications of feminist critiques of moral and political theory. Pp. 56–76 in *Feminism as Critique,* ed. S. Benhabib and D. Cornell. Minneapolis: University of Minnesota Press.
—— 1989. Polity and group difference: a critique of the ideal of universal citizenship. Pp. 117–42 in *Feminism and Political Theory,* ed. C. Sunstein. Chicago: University of Chicago Press.
—— 1990. *Justice and the Politics of Difference.* Princeton, N.J.: Princeton University Press.
—— Forthcoming. Communication and the other: beyond deliberative democracy. In *Democracy and Difference,* ed. S Benhabib. Princeton, N.J.: Princeton University Press.

Chapter 21

# Political Theory: Traditions in Political Philosophy

Bhikhu Parekh

## I Background

IN much of the discussion of post-Second World War political philosophy, it is often argued:

1) that the 1950s and 1960s marked the decline or even the death of political philosophy, and the 1970s and 1980s its resurgence;
2) that the resurgence was caused, or at least stimulated, by a sharp rise in the level of political and ideological struggle brought about by such factors as the Vietnam War, the Civil Rights movement in the United States, the disintegration of the post-war consensus, and the emergence of the New Left; and
3) that Rawls's *A Theory of Justice* (hereafter *TJ*) symbolized the rebirth of political philosophy. (See, e.g., Barry 1991; Miller 1990; Held 1991.)

Propositions (2) and (3) presuppose that (1) is true. If (1) were shown to be false, we would not need (2) to explain it. And as for (3), we would not then see *TJ* as a historical benchmark, although of course it would still remain a major work in post-Second World War political philosophy. Although the images of death and resurrection have a deep emotional appeal to those shaped by Christianity, there is little evidence to support (1). Furthermore if the Vietnam War and other events were able to breathe new life into a dead or dying discipline, it would be extremely odd if, other things being equal, such infinitely more cataclysmic events as the Second World War, the Nazi and Communist tyrannies, and the Holocaust could not throw up important works in political philosophy. If (2) is true, (1) cannot be true.

Contrary to the general impression, the 1950s and 1960s were quite rich in political philosophy. In his long and seminal introduction to Hobbes's *Leviathan*, written in 1946, and in *Rationalism in Politics* (1962), Michael Oakeshott outlined a novel conception of the nature of political philosophy, challenged the dominant rationalism of western thought that was in his view largely responsible for the recent tragedies, and offered a highly original statement of conservatism that disengaged it from its traditional associations with religion, historicism, moralism, nationalism and social hierarchy. The 1950s and 1960s also saw the publication of Hannah Arendt's major writings, which have attracted more book-length studies than any other contemporary writer. She problematized the concept of human nature, criticized the apolitical character of traditional political philosophy, and argued that its assumptions and concepts needed to be radically revised if we were to make sense of the Stalinist and Nazi totalitarianism. She theorized many of these insights in her *The Human Condition* (1958), one of the seminal books on political philosophy in the twentieth century, as well as in such less impressive works as *Between Past and Future* (1961) and *On Revolution* (1963). During this period Isaiah Berlin published several important essays, the two most influential being "Two concepts of liberty" (1958) and "Does political theory still exist?" (1962). Relative to its size, the first essay has spawned more critical literature than any other contemporary work including Rawls's *TJ*. In these and other essays, Berlin challenged the moral monism of much of the traditional political philosophy including the liberal, stressed the incommensurability and irreducible plurality of moral values, and outlined a distinctly modern and highly influential form of liberalism.

The two decades also saw such other creative and influential political philosophers as Karl Popper, Leo Strauss, Eric Voegelin, C. B. Macpherson, F. A. Hayek, R. G. Collingwood, and George Santayana, Even Rawls's *TJ* was, as he himself said, largely an elaboration of the seminal ideas he had developed in the articles written between 1951 and 1963. The 1950s and 1960s also saw excellent pioneering works on Hobbes, Locke, J. S. Mill, Kant and others, which were often not histories of political thought as some commentators misdescribed them, but philosophical engagements with the ideas of past thinkers and thus essays in historically mediated political philosophy. During this period systematic attempt was made to construct a Marxist political philosophy by such writers as Althusser, Sartre, Habermas and Marcuse. Since Marx had dismissed philosophy as intellectual "onanism" and politics as a largely parasitic and epiphenomenal activity, and since his thought therefore had limited theoretical resources for constructing a political philosophy, the attempts by these and other writers to develop a Marxist political philosophy were truly remarkable. The growth of Marxist political philosophy stimulated its critique, and led to some of the best critical works on Marx during this period.

The political philosophy of the 1950s and 1960s was distinguished by several features, of which I shall note three. First, they were the decades of prima donnas or gurus. Hardly any of the major figures engaged in a critical dialogue with others or even referred to them. They did, of course, read and privately comment on each other's writings, and in some cases carried on an extensive private correspondence, but they said little about each other in their published writings or even met at professional conferences. The entire corpus of Arendt's writings contains only a couple of references to Oakeshott, and hardly any to Popper and Berlin. Others were no different. Each guru had his followers, some more than others, and established a distinct school whose members sympathetically developed the master's thought.

Secondly, although different writers during this period were preoccupied with different questions, they were acutely aware that their discipline had come under severe criticism from such diverse sources as logical positivism, linguistic philosophy, sociology of knowledge, behaviouralism, and Existentialism, and such historically orientated thinkers as Collingwood and Croce. In the Anglo-Saxon world it was widely argued that all inquiries were either empirical or normative, that traditional political philosophy was of the latter kind, and that since values could not be objectively defended, all normative inquiries including political philosophy were basically no more than personal preferences illegitimately claiming universal validity. Even those writers who rejected such a positivist view argued that since contemporary western societies were broadly agreed on moral values, political philosophy had no public role and was unnecessary. In short, political philosophy was either impossible, or unnecessary, or both.

Although the major political philosophers of the 1950s and 1960s responded to these and other criticisms in their own different ways, some taking them more seriously than others, they were all agreed that political philosophy was both a possible and necessary form of inquiry. It was universal in its scope, critical in its orientation, and aimed to give a rational account of political life by grounding it in basic human capacities and needs (Berlin), human nature (Strauss, Voegelin and Marcuse), human rationality (Popper), the human condition (Arendt), or human predicament, and later, human agency (Oakeshott). For them its irreplaceable contribution consisted in highlighting the fundamental features of human life in general and political life in particular, exposing bad arguments, attacking seductive but inherently unrealizable ideological projects, standing guard over the integrity of the public realm, and clarifying the prevailing form of political discourse. Almost all of them thought that political philosophy was primarily concerned to understand rather than to prescribe, that it operated at a level which prevented it from recommending specific institutions and policies, and that it could never become a *practical* philosophy. Many of them also thought that political philosophy had come to grief in the preceding decades because its heavy dependence on general philosophy made it a hostage to the latter's changing fashions. In their own different ways they therefore sought to establish its autonomy, and argued that political philosophy was not an *applied* philosophy, an extension to political life of general philosophical doctrines developed independently of it, but a relatively self-contained mode of inquiry with its own distinct categories and form of investigation.

Thirdly, political philosophers writing in the 1950s and 1960s had lived through the horrors of the Second and in some cases even the First World War, the rise of Fascist, Nazi and Communist totalitarianism, and the concentration camps, and were deeply troubled by the latent barbaric tendencies of European civilization. They traced the roots of these tendencies to rationalism (Oakeshott), historicism (Popper), moral monism (Berlin), the rise of the *animal laborans* (Arendt), relativism (Strauss), Gnosticism (Voegelin) and capitalism (Marcuse and other Marxists). Although fiercely critical of contemporary communism, most political philosophers of this period also took a critical view of liberal democracy. Even the non-Marxists, who championed what might loosely be called liberal values, worried about liberalism's asocial view of the individual, ahistorical view of rationality, preoccupation with material affluence, moral subjectivism, alliance with capitalism, and instrumental approach to politics. They were also critical of democracy and were disturbed by the ease with which the masses had been mobilized by the Fascists and the Nazis. They deeply cherished free society, but rightly refused to equate it with liberal democracy. In order to emphasize their distance from the latter, they called their preferred society "free," "open," "libertarian," "rational," "civil," "political or politically constituted community," "government by discussion," and so on.

Since political philosophy was flourishing in the 1950s and 1960s, the question arises as to why it was declared dead or in terminal decline during this period. Ignorance of the range of writings, the positivist dismissal of them as not "really" philosophy, behavioralist triumphalism, the naïve belief that a philosophical engagement with past thinkers was "history of ideas" and not political philosophy, the unfounded view that the problems they addressed were "dated" and irrelevant to modern times, etc. played a part. Paradoxically another important factor was the dominant normative conception of political philosophy, attacked by some but deeply cherished by others. As the remarks of its contemporary detractors show, many of them expected political philosophy to set "new political goals," to provide the modern age with a "coherent conception of its needs," to "prescribe" how we should live (Easton 1953; Laslett 1956). As we saw, most political philosophers of the 1950s and 1960s did not share this view and regarded political philosophy primarily as a contemplative, reflective and explanatory inquiry concerned to understand rather than to prescribe. Since their writings did not conform to their critics' narrow standards of what constituted "true" political philosophy, the latter predictably pronounced the discipline dead.

## II Recent political philosophy

The early 1970s saw the emergence of four new publications, all American and indicative of the fact that the owl of Minerva had now moved its nest from Europe to the United States. *Interpretation*, a journal with Straussian leanings, was launched in 1970, followed a year later by Rawls's *A Theory of Justice* and the launching of the multidisciplinary journal *Philosophy and Public Affairs*, and three years later by the narrowly academic journal *Political Theory*. Rawls's *TJ* built on the philosophical labors of his older contemporaries, and both continued and broke with their style of political philosophy.

Since they had seen off logical positivism, linguistic philosophy, behaviorism, historicism, and so on, he did not have to worry about these. Like them he too viewed political philosophy as critical in nature, universal in scope, and quasi-foundational in its orientation. In several other respects, however, he departed from the political philosophy of the 1950s and 1960s. For his predecessors political philosophy was primarily concerned to understand political life; for him it was primarily normative and a form of *practical* philosophy. They thought that it elucidated the fundamental features of human life including the basic human capacities and needs, and could not go below a certain level of generality; for him it was fully equipped not only to offer a theory of man but also to lay down a structure of desirable institutions, policies and practices. Although Rawls himself did not put it this way, his political philosopher was a law-giver, someone capable of devising an entire social structure on the basis of universally acceptable minimum principles, a view for which the writers in the 1950s and 1960s had little sympathy. Unlike them, again, Rawls made justice the master-concept of politics and gave it an unusually broad meaning. Since Rawls's philosophical ambition was different from theirs, he detached political philosophy from logic, rhetoric, ontology, and the history of Western civilization with which it had earlier been closely connected, and aligned it to such disciplines as economics, psychology, the study of political institutions, and social policy.

Rawls's *TJ* offered no new vision of man, no new insights into human nature, no novel analysis of the tensions and ambiguities of modernity, and lacked the historical and cultural depth of Arendt, Oakeshott, Voegelin, and others. The vision of society that he offered largely reaffirmed the post-war consensus, ironically at a time when it was coming under severe criticism from libertarian, Marxist, religious and other quarters. In spite of all this *TJ* was a work of considerable historical and philosophical importance. He showed how to construct a moral and political theory that was both philosophically satisfactory and true to our moral intuitions, and integrated critical theoretical reflection with the lived reality of practical life. Using such suggestive devices as the original position and reflective equilibrium, he articulated with great clarity the inner structure of a highly influential and mainly liberal-rationalist form of moral and political reasoning. He integrated such diverse disciplines as epistemology, moral philosophy, moral psychology, political theory and economics, and developed a multidisciplinary perspective on such complex concepts as justice, liberty and equality. Rawls not only built long-neglected bridges between political philosophy and other social sciences but also stressed the centrality of the former, thereby giving its practitioners a sense of pride and importance they had long craved for. All this, as well as the fact that Rawls's conventional moral and political vision was intellectually accessible and morally congenial to many a liberal academic, made it one of the most influential though not one of the most profound books of our times.

The post-1970s saw many new developments in political philosophy. The nature and scope of political philosophy became a subject of much direct and indirect debate and gave rise to four distinct views. First, since Rawls was an important figure at least in the United States, several writers accepted his view that political philosophy was a branch of moral philosophy, that the latter was essentially normative, and that the task of political philosophy was not only to develop general principles for evaluating the social structure but also to design appropriate institutions, procedures and policies. They applied his view of political philosophy to the analyses of justice, equality, international relations, etc., and although they sometimes arrived at different conclusions, their theories had similar logical structures (Ackerman 1980; Barry 1989; Beitz 1979).

Second, the older view of political philosophy that had much support in the Western tradition of political thought and that had been reaffirmed by Oakeshott, Arendt, Berlin, Voegelin and others, informed the work of such writers as Charles Taylor (1985; 1990), Alisdair MacIntyre (1981; 1988), and William Connolly (1988). For them political philosophy was primarily a contemplative and reflective inquiry concerned to understand human existence in general and the modern world in particular. It was neither a branch of moral philosophy nor normative in its orientation, although of course it had a strong moral dimension. It aimed to explore what human beings were like, what they had made of themselves in history, the nature of modernity, the distinguishing features of modern self-consciousness, and so on, and to use that understanding to illuminate both the specificity of contemporary political life and the range of choices open to us.

Third, some political philosophers such as Michael Walzer argued that political philosophy was embedded in the way of life of a specific community, that it was primarily concerned to articulate the latter's self-understanding, and that it was necessarily municipal in its scope and interpretive in its orientation (Walzer 1983; 1987). Finally, some others, such as Richard Rorty (1989), derived their inspiration from post-structuralist and especially postmodernist writers, challenging both the traditional distinction between the theoretical and other forms of thinking and the primacy of the former. For them theoretical thought not only did not enjoy a privileged access to truth, but often stood in its way. Its categories were too rigid, frozen and bipolar, and its obsession with logical consistency and system-building too unrealistic, to do justice to the ambiguities, contradictions and tensions of human life in general and political life in particular. In their view political philosophy needed to be tentative, exploratory, conversational, open-ended, ironic, sensitive to the ambiguity of life, and closer to the intuitive and untheorized thinking of literary writers and artists.

The past twenty years have also thrown up a remarkable body of literature exposing the sexist, racist, statist, élitist, nationalist and other biases of traditional political philosophy, including that of the 1950s and 1960s. Although these writings were sometimes polemical, insufficiently rigorous, and lacked a constructive impulse, they persuasively showed that the biases were not inadvertent and easily eliminable but deeply embedded in the very structure of traditional political philosophy, infecting its questions, answers and methods of investigation as well as its conceptions of rationality, basic human capacities, needs, moral reasoning and the good life. Feminist critiques have so far been the most impressive (Benhabib and Cornell 1987; Phillips 1991). Similar critiques from racial and other perspectives have only just begun. And there is as yet no systematic attempt to show how the imperialist experience shaped the basic assumptions and categories of much of the post-16th century political philosophy (Parekh 1994*a*). When these critiques are fully worked out and integrated, their cumulative impact is bound to entail a radical reappraisal of the nature and history of the discipline.

So far as substantive questions are concerned, several new ones have been placed on the agenda and many old ones explored from new angles. A few examples will illustrate the point. Some of the new questions arose as a reaction to Rawls's *TJ*. Since he made justice the master-concept of politics, many writers wondered whether that did not ignore or gave a distorted account of important aspects and areas of political life (Sandel 1982; Heller 1987; MacIntyre 1981; Nozick 1974). Since, contrary to Rawls's intention, his theory of justice was widely perceived to be biased towards liberalism, the question arose as to whether the state could ever be neutral between different visions of the good life and whether liberalism was a purely procedural and morally neutral device or whether it represented a substantive conception of the good (Raz 1986; Dworkin 1977; Galston 1991). The increasing concern about the quality of collective life encouraged work on the nature of the political community, participatory democracy, education for citizenship, virtues of citizens, etc. (Barber 1984; Gutmann 1987; Macedo 1990). Thanks to the hitherto marginalized groups' demands for public recognition and the cultural plurality of modern society, the cohesion of the polity has become a subject of acute debate, leading to such questions as the nature of national identity, the political role of education, the range of permissible cultural diversity, and the best way to combine the demands of national unity with those of cultural diversity (Kymlicka 1989; Parekh 1994*b*; Miller 1995). Political obligation, largely ignored by the earlier writers, became a subject of much debate from the mid-1960s onwards, and is explored from such novel angles as whether it can ever be moral in nature, how it relates to ethnic, communal and other obligations, how it is derived, and whether one can ever incur it in a polity that does not make adequate institutional provisions for active participation (Simmons 1979; Parekh 1993; Pateman 1985). The global context of moral and political obligation has also acquired salience, and considerable attention is paid to such questions as whether and what obligations we have to people in other countries, the moral significance of national boundaries, whether we have a humanitarian duty to intervene in the internal affairs of strife-torn countries, and the nature of international justice (Barry 1991; Beitz 1979; Held 1991). The environmental crisis has raised long-neglected questions about man's relations to nature and other animals, the nature and limits of private property, appropriate models of economic growth, and the limited capacity of contemporary political ideologies to deal with such questions (Singer 1993; Goodin 1992).

The debate between liberals and communitarians, both in general and in the context of many of the questions mentioned above, dominated the 1980s especially in the United States, with Europeans as interested but somewhat puzzled observers. The debate raised important questions about the nature of the self, the relation between the self and society, society and the state, politics and culture, and between personal and collective identity, the nature of the good life, the role of political philosophy, and the nature and grounds of morality. Unfortunately the debate was often marred by false polarizations and inadequately formulated fundamental questions. Although communitarians stressed the ontological, epistemological and moral importance of community, hardly any of them developed a systematic theory of community analyzing what it means and entails, whether and how it is possible in modern society, how it can be reconciled with deep cultural differences and the liberal stress on individual autonomy, whether it entails greater restrictions on the freedom of speech and expression than are acceptable in a liberal society, and whether political community requires moral community as its necessary basis. Several communitarians equated the community with the nation-state, and unwittingly endorsed an insidious form of nationalism and collectivism. Again, they talked about the radically situated self, but did not adequately explain if the concept made ontological sense, whether it did not presuppose a highly cohesive community that does not in fact exist, and how such a self could morally and emotionally reach out to the rest of mankind and treat the latter impartially and sympathetically. For their part, liberals freely talked about personal autonomy, choice, critical self-reflection and so on, without fully exploring the internal logic and cultural limits of these ideas, the nature, degrees and social pre-conditions of autonomy, and whether and why it should be held up as a universally valid ideal (Benhabib 1989).

Three broad features of recent political philosophy deserve particular attention. First, liberalism has become the dominant voice today, not only in the sense that Conservative, Marxist, religious and other voices are relatively subdued and that most political philosophers are of liberal persuasion, but also, and more importantly, in the sense that liberalism has now acquired unparalleled philosophical hegemony. It is more or less the absolute standard of moral and political evaluation today, all societies being divided into liberal and non-liberal, and the latter viewed as illiberal. Not surprisingly everyone is anxious to appear as a liberal, and legitimizes even his radical departures from liberalism in liberal terms. For example, Charles Taylor is reluctant to admit that Quebec's concern to preserve its distinct way of life with its consequent curtailment of some individual rights is a perfectly legitimate attempt to set up one type of good society which, though non-liberal, is not at all illiberal and oppressive. Instead he insists that it represents a different kind of liberalism (Gutmann 1992, but see Taylor 1994, where he sees through the whole exercise).

Liberal hegemony has had several unfortunate consquences. It has narrowed the range of philosophical and political alternatives, restricted our philosophical vocabulary, and deprived liberalism of an authentic and uncaricatured "other." Furthermore, it has turned liberalism into a meta-language, enjoying the privileged status of being both a language like others and the arbiter of how other languages should be spoken, both a currency and the measure of all currencies. The way in which this distorts the self-understandings of non-liberal systems of thought is too obvious to need spelling out. What is even worse, while the hegemonic liberalism has incorporated the moral, political and cultural insights of other ways of thought and become richer, it also runs the risk of forfeiting its identity and coherence and becoming an ideological Esperanto. One further consequence of the liberal hegemony is that, unlike the political philosophers of the 1950s and 1960s, we are increasingly losing the capacity to affirm our commitment to freedom and individuality while remaining critical of the prevailing structure of liberal democracy. Oakeshott and Popper could champion "civil" or "open" society and yet criticize the liberal society. It is doubtful if we can do this today without inviting incomprehension or charge of bad faith.

Secondly, the 1970s and 1980s have seen the decline of the age of the gurus in the Anglo-Saxon world though not on the continent of Europe, where they continue to flourish and set up their competing camps to which their Anglo-Saxon pilgrims periodically repair for their spiritual sustenance. Political philosophers today take full account of each other's works, and engage in a critical dialogue with them. No one is considered "big" enough to be treated with awe, or spared criticism out of a misplaced sense of philosophical or personal loyalty. This becomes clear if we compare the way Rawls has been discussed with the way Oakeshott, Strauss, Voegelin, Popper and others were discussed during their times. Although accorded the respect due to a creative thinker, Rawls has often been subjected to vigorous even savage criticism, and is seen as first among equals rather than as a guru or the founder of a school. His response to his critics is also quite different from that of the earlier political philosophers to theirs. He has painstakingly answered their criticisms, admitted his errors, and modified his views. Indeed in his two major works, he thanks more people than was done by almost all post-war political philosophers put together, and there is hardly an important idea of his which he does not magnanimously trace to others.

Thanks to the changes in the intellectual climate, ideas today are depersonalized, abstracted from their originators, discussed in their own terms, and treated as a public property. There is therefore a genuine sense of community among political philosophers based on their shared interest in a common body of thought. We know what the major controversies in the discipline are, what issues remain unexplored, and where the growth areas lie. As a result the history of the political philosophy of the 1970s and 1980s cannot be written in the same way as that of the 1950s and 1960s. The latter was dominated by individuals with few direct debates between them. Although a commentator can set up such debates, they necessarily involve a good deal of abstraction and artificiality and run the risk of violating the integrity of their subject-matter. The history of the political philosphy of this period is therefore inescapably *thinker-centered*, and not surprisingly that is how it has generally been written. By contrast the history of the succeeding two decades is *thought-centered*, and predictably it is generally written in terms of criss-crossing controversies.

Finally, recent political philosophy remains as parochial as its counterpart two decades earlier. It has shown little interest in the political experiences, problems and debates of the non-Western world. The latter continues to be treated as if its destiny lay in uncritically reproducing the historical experiences of the West, and its problems and aspirations are analyzed almost entirely in western terms. The ignorance of the non-Western world has several unfortunate consequences. Western political philosophy lacks both adequate protection against ethnocentric biases and a valuable source of critical self-consciousness. It is also unable to appreciate the full range of the different visions of the good life, and to develop culturally sensitive categories of thought and moral principles indispensable for dealing with an increasingly interdependent world. Since the West today exercises considerable political, economic and cultural power over the world, its distorted understanding of the latter encourages misguided policies and causes much avoidable moral and political havoc.

## III New challenges

In the light of our discussion, it is clear that for the first time in nearly a century, political philosophy is in a reasonably good health. It has survived some of the fiercest attacks, and built up an impressive tradition of inquiry that is hospitable to new experiential material and disciplinary alliances. Although the triumphalist 1970s and 1980s displayed unjustified arrogance towards their predecessors in the 1950s and 1960s, and although some of the latter in turn have sometimes taken an uncharitable view of the achievements of their successors, there is now a better appreciation of the coherence and continuity of political philosophy since 1945. If the discipline is to continue its progress, it should be ready to meet new challenges and suitably to reappraise its theoretical tools. Of the many challenges facing it today and likely to grow with time, two deserve particular attention. There are also several others, such as the increasing dissolution of the nation-state into both larger and narrower units, changes in the nature and content of the political, both the repressive and the emancipatory potential of the increasing plea for state intervention in social issues that have hitherto belonged to the private realm, and the restructuring of civil society, all of which affect the subject-matter of political philosophy as it has been defined for the past four centuries, but I will ignore these and related challenges.

First, as we saw earlier, contemporary political philosophy is characterized by at least four different views on its nature and scope. Some of these are deeply mistaken and in need of reconsideration. Political philosophy can never be merely municipal and interpretive, the former because one cannot philosophize about political life without some conception of what it is to be human and thereby introducing an inescapable universal dimension, the latter because a society's moral and political structure is never homogeneous and therefore every interpretation of it necessarily involves criticism and choice which, if they are not to be based on the political philosopher's personal preferences with all their attendant difficulties, require clearly stated and defended moral and political principles. A community's self-understanding is not out there waiting to be discovered and elucidated; it is necessarily constructed from a specific standpoint. It is striking that the thought of Michael Walzer, the ablest current advocate of the municipal and interpretive conception of political philosophy, is undergirded by a host of universalist and prescriptive claims (Carens 1995; Barry 1991). As for the postmodernist view of political philosophy, especially the version familiarized by Rorty, it is underpinned by a municipal and interpretive conception of political philosophy, and open to the same objections. Since it cannot rise above the prevailing form of communal self-understanding, this view of political philosophy also lacks the capacity to probe the latter's ambiguities, tensions and partiality, and contains a deep positivist bias.

This leaves us with two remaining views on the nature of political philosophy, namely political philosophy as a contemplative and reflective and as a moral and prescriptive form of inquiry. Each has its merits, but neither is adequate. Politics is concerned with how we should live as a community, and has an inescapable prescriptive dimension. However how we should live depends on who we are, what choices are open to us, what our current predicament is, etc., and cannot be decided without a patient and probing theoretical reflection on our traditions, character, history and social structure. A well-considered view of political philosophy therefore needs to emphasize both its contemplative and critical, reflective and prescriptive, dimensions.

The second challenge facing political philosophy today has to do with the problems arising out of the considerable cultural diversity of modern society. Many a past political philosopher largely and rightly assumed a culturally homogeneous society in which such explanatory and normative principles as they developed could be confidently applied to all or at least to the bulk of its citizens. For example, they assumed that whatever ground of political obligation they advanced, be it consent, fairness, gratitude, common good or self-realization, applied to all citizens alike and with more or less the same moral force. Today we can no longer make such an assumption. Some sections of citizens, such as the religious fundamentalists, moral sceptics and philosophical anarchists, do not accept the legitimacy of the established structure of authority, and can only be expected to obey it on prudential grounds. Even those who acknowledge the moral obligation to obey the law define its basis differently, depending on the central values of their cultural tradition. In the individualist moral tradition consent is a central value, and a plausible ground of political obligation. This is not the case with other cultural traditions, which stress such values as gratitude, love of God, communal solidarity and loyalty to ancestors, and define and ground political obligation differently. It is therefore doubtful whether there can be a uniform basis of political obligation in such a culturally plural society as ours. A well-considered theory of political obligation, as of legitimacy and authority, will necessarily have to be thin and formal, leaving sufficient moral spaces to be filled in differently by different moral traditions.

Cultural pluralism also requires reconsideration of the traditional understanding of such crucial concepts as equality, fairness, justice, social cohesion, political unity and freedom. Contrary to the standard liberal assumption, there are several different ways of treating people equally, organizing the just society and creating a united polity, and freedom can be defined in several ways of which the culturally specific and class-bound idea of negative liberty is only one and not the most coherent. This raises the question as to how we can arbitrate and decide between different interpretations of these concepts. Take the concept of equality. The standard difference-blind liberal view of it has its obvious strengths. However since human beings are culturally embedded, and since their differences mediate the consequences of our treatment of them, such a view of equality can easily lead to grave inequality and injustice. Broadly speaking equality refers to impartial application of a rule, and justice to its content. Since no rules are ever culturally neutral, albeit some less than others, they are bound to discriminate in favor of those whose ways of life and thought they reflect.

However once we take differences into account, all kinds of difficulties begin to arise. How can we insure that we are according equal treatment to those we treat differently? What differences should we take into account? How can we prevent differences from becoming rigid and frozen once they are institutionalized and embodied in legal categories? And how can we create social cohesion and a shared collective identity among such differentially treated citizens? Whatever view we take, we face difficult philosophical and moral problems. The problems are not new, for we do take differences into account as when we distinguish between the needs of men and women, young and old, able and disabled. A culturally plural society accentuates these problems and makes it particularly difficult to decide how to compare men and women who are individuated differently in different cultures, how to decide what differences are relevant, how to interpret and respond to them, and how we can be sure that two individuals belonging to different cultures are equal in relevant respects.

Since the Western tradition of political philosophy is largely predicated on the assumption of cultural homogeneity, it has considerable difficulty coping with these and related questions. A culturally plural society calls for a multiculturally grounded political philosophy, one that can build bridges between cultures, translate the categories of one culture into those of another, and skilfully and patiently evolve culturally sensitive and internally differentiated interpretations of universal categories and principles. We have only just begun to appreciate both the importance and the difficulties of such a political philosophy. Its importance is strikingly evident in the fact that Rawls had to revise his *TJ* within only a few years of its publication and to follow it up with *Political Liberalism* (1993), which rests on quite different philosophical assumptions. The difficulties involved in developing a multiculturally grounded political philosophy are equally strikingly evident in the fact that, in spite of its determined attempt to the contrary, *Political Liberalism* retains a strong monocultural orientation, and its political conception of justice, its view of public reason, its definition of the individual and its mode of ethical and philosophical reasoning carry little conviction with those not already committed to Rawlsian liberalism.

## Acknowledgments

I am grateful to Joseph Carens, Robert Goodin and Noel O'Sullivan for their helpful comments on this paper.

## References

David Held's *Political Theory Today* (1991) is a useful collection of articles on different aspects of contemporary political theory written by distinguished writers. Will Kymlicka, *Contemporary Political Philosophy* (1990), offers an excellent account of the major schools of thought in contemporary political philosophy, and Raymond Plant, *Modern Political Thought* (1991), a valuable analysis of the major debates. Finally, Stephen Mulhall and Adam Swift, *Liberals and Communitarians* (1992), is a good summary and critique of that important contemporary debate, organized in terms of the ideas of the principal individuals involved.

Ackerman, B. 1980. *Social Justice in the Liberal State.* New Haven, Conn.: Yale University Press.
Arendt, H. 1958. *The Human Condition.* Chicago: University of Chicago Press.
—— 1961. *Between Past and Future.* London: Faber and Faber.
—— 1963. *On Revolution.* London: Faber and Faber.
Barber, B. 1984. *Strong Democracy.* Berkeley: University of California Press.
Barry, B. 1989. *Theories of Justice.* Berkeley: University of California Press.
—— 1991. *Democracy and Power.* 2 vols. Oxford: Clarendon Press.
Beitz, C. R. 1979. *Political Theory and International Relations.* Princeton, N.J.: Princeton University Press.
Benhabib, S. 1989. *Situating the Self.* Cambridge: Polity Press.
—— and Cornell, D., eds. 1987. *Feminism as Critique.* Oxford: Polity Press.
Berlin, I. 1958. Two concepts of liberty. Reprinted pp. 118–72 in I. Berlin, *Four Essays on Liberty.* Oxford: Oxford University Press, 1969.
—— 1962. Does political theory still exist? Pp. 1–33 in *Philosophy, Politics and Society,* 2nd series, ed. P. Laslett and W. G. Runciman. Oxford: Blackwell.
Carens, J. 1995. Complex justice, cultural difference and political community. Pp. 45–66 in *Pluralism, Justice and Equality,* ed. D. Miller and M. Walzer. Oxford: Clarendon Press.
Connolly, W. E. 1988. *Political Theory and Modernity.* Oxford: Blackwell.
Dworkin, R. 1977. *Taking Rights Seriously.* London: Duckworth.
Easton, D. 1953. *The Political System.* New York: Knopf.
Galston, W. A. 1991. *Liberal Purposes.* Cambridge: Cambridge University Press.
Goodin, R. E. 1992. *Green Political Theory.* Oxford: Polity Press.
Gutmann, A. 1987. *Democratic Education.* Princeton, N.J.: Princeton University Press.
—— ed. 1992. *Multiculturalism.* Princeton, N.J.: Princeton University Press.
Held, D., ed. 1991. *Political Theory Today.* Oxford: Polity Press.
Heller, A. 1987. *Beyond Justice.* Oxford: Blackwell.

Kymlicka, W. 1989. *Liberalism, Community and Culture.* Oxford: Clarendon Press.
—— 1990. *Contemporary Political Philosophy.* Oxford: Clarendon Press.
Laslett, P. 1956. Introduction. Pp. vii–xv in *Philosophy, Politics and Society*, 1st series, ed. P. Laslett. Oxford: Blackwell.
MacIntyre, A. 1981. *After Virtue.* London: Duckworth.
—— 1988. *Whose Justice? Which Rationality?* London: Duckworth
Macedo, S. 1990. *Liberal Virtues.* Oxford: Clarendon Press.
Miller, D. 1990. The resurgence of political theory. *Political Studies*, 38: 421–37.
—— 1995. *On Nationality.* Oxford: Clarendon Press.
Mulhall, S., and Swift, A. 1992. *Liberals and Communitarians.* Oxford: Blackwell.
Nozick, R. 1974. *Anarchy, State and Utopia.* New York: Basic Books.
Oakeshott, M. 1946. Introduction. Pp. v–lxvii in T. Hobbes, *Leviathan*, ed. M. Oakeshott. Oxford: Blackwell.
—— 1962. *Rationalism in Politics.* London: Methuen.
Parekh, B. 1993. A misconceived discourse on political obligation. *Political Studies*, 41: 236–51.
—— 1994*a*. Decolonising liberalism. Pp. 105–26 in *The End of "Isms"?*, ed. A. Shtromas. Oxford: Blackwell.
—— 1994*b*. Discourses on national identity. *Political Studies*, 42: 492–504.
Pateman, C. 1985. *The Problem of Political Obligation.* Oxford: Polity Press.
Phillips, A. 1991. *Engendering Democracy.* Oxford: Polity Press.
Plant, R. 1991. *Modern Political Thought.* Oxford: Blackwell.
Rawls, J. 1972. *A Theory of Justice.* Oxford: Clarendon Press.
—— 1993. *Political Liberalism.* New York: Columbia University Press.
Raz, J. 1986. *The Morality of Freedom.* Oxford: Clarendon Press.
Rorty, R. 1989. *Contingency, Irony and Solidarity.* Cambridge: Cambridge University Press.
Sandel, M. 1982. *Liberalism and the Limits of Justice.* Cambridge: Cambridge University Press.
Simmons, A. J. 1979. *Moral Principles and Political Obligation.* Princeton, N.J.: Princeton University Press.
Singer, P. 1993. *Practical Ethics.* 2nd edn. Cambridge: Cambridge University Press.
Taylor, C. 1985. *Philosophical Papers.* 2 vols. Cambridge: Cambridge University Press.
—— 1990. *Sources of the Self.* Cambridge: Cambridge University Press.
—— 1994. Can liberalism be communitarian? *Critical Review*, 8: 257–63.
Walzer, M. 1983. *Spheres of Justice.* New York: Basic Books.
—— 1987. *Interpretation and Social Criticism.* Cambridge, Mass.: Harvard University Press.

Chapter 22

# Political Theory: Empirical Political Theory

Klaus von Beyme

EMPIRICAL political science divides into two mainstreams. The Weberian tradition is interested primarily in a reconstruction of social reality in a historical perspective and works *ex post facto* with typologies and ideal types. The Durkheimian tradition, deeply affected by French positivism after the fashion of Comte, takes as its motto *savoir pour prévoir* and is interested primarily in modeling reality by isolating dependent and independent variables.

This Durkheimian style of empirical political theory, in particular, supposes that models "should be tested primarily by the accuracy of their predictions rather than by the reality of their assumptions" (Downs 1957: 21). It, accordingly, has been particularly embarrassed by political science's failure to predict any major political events since 1945. The student rebellions of the 1960s, the rise of new fundamentalism, the collapse of communism, the peaceful revolution of 1989—all came as a surprise to political scientists.

Political science takes little comfort, either, in new tendencies in the natural sciences. Abandoning the old Baconian optimism that science does battle against ideology and superstition in the service of truth and utility, natural scientists influenced by autopoietic systems theory and chaos scenarios have given up on the idea of predicting major events on the macro level (Maturana 1985). Many social scientists have belatedly come to think similarly that macro-theoretical predictions are little more than informed guesswork. The evolution of events can be reconstructed only *ex post facto*, and the task of theory is to keep open various options (Luhmann 1981: 157).

Political scientists face further systematic distortion of theory-building peculiar to their own field. While the positivistic mainstream endlessly echoes Max Weber's plea for value-free science, developments in the history and philosophy of science undermine many tacit assumptions of that model. In its mature stage, social theory is increasingly subjected to "social and political imperatives" of society. Broader social aims and interests, more than any "internal logic of truth finding," have been shown to be a predominant impetus behind scientific research (Barnes 1992). In spite of continuous de-ideologization of theory-building in the 1980s, social science theories inevitably start from the social conditions embedded in structuring political discourses (Wagner *et al.* 1991: 77; Wagner and Wittrock 1993).

## I A chronology of shifting paradigms

Since the Second World War, there have been major shifts in the importance and focus of political theory driven largely from within the discipline itself. With the rise of the behavioral persuasion, itself disinterested in the great questions, came in the 1950s and 1960s a perceived decline in normative political theory (Miller 1990). Repudiating earlier metaphysical theories of the state, individual and groups became the starting-point of analysis. Theories such as Bentley's (1949) were revived. Positive political theory tried to restrict itself to conceptual analysis, and normative revolts remained isolated. In the late 1960s, however, there came a revival of "grand" political theory, frequently in ideologized form. But since the late 1970s the great debates between "positivists" and "Marxists" have been exhausted and in their place policy analysis has emerged as the middle level of a new theory-building exercise embracing both empirical and normative elements (von Beyme 1992: 248 ff.).

Major shifts in political theory have also arisen from extra-scientific political factors as well as from internal developments within the discipline. In most countries there was, over the postwar period, a decline of faith in the steering capacities of the political center. The 1960s and early 1970s were, especially in Europe, a time of planning illusions and Keynesian trust in anti-cyclical steering of the economy. In the 1970s, however, mainstream political science turned away from implementing grand ideological visions and toward empirical political studies, satisfying itself with typologies of policy cycles and political and societal actors. Theory-building in that period concentrated on variations in steering via social co-operation:

- consociationalism (Lijphart 1977);
- neo-corporatism (Schmitter 1981);
- societal co-operation (Willke 1983);
- generalized political exchange (Marin 1990);
- private interest group government (Streeck and Schmitter 1985), growing out of liberal-corporatist ideas, in part in resistance to attempts at "bringing the state back in" (Skocpol 1979); and
- models of the state as the steering center of society, as in "political cybernetics" (Deutsch 1966) and the "active society" (Etzioni 1968).

The 1980s experienced a new type of social and political actor, the new social movements. Schmitter dubbed this counterforce "syndicalism": although a misnomer outside Latin countries, he still believed neo-corporatism to represent the best protection against unruliness and ungovernability (Schmitter 1981). In the 1980s the ecological problem, in particular, entered the agenda of political theory. Most Anglo-Saxon political scientists hesitated to construct more than partial theories on the basis of new social movements (Goodin 1992). However, philosophically minded European thinkers such as Beck (1986/1992) hypothesized a "risk society" with quite different dynamics than classical industrial society. While resisting the temptation of the autopoietic bandwagon, Beck insists (independently of the Frankfurt School) on the necessity of completing modernity by adding a new non-technocratic and non-rationalistic component. In the 1980s, Habermas himself abandoned his late-Hegelian historicist project of reconstructing ever more typologies of crises; and while his work on discourse (1987) holds out feint hope that new movements would succeed in the defending the life-world (*Lebenswelt*) against the "system" with its alienating forces (bureaucratization, commercialization, justicialization), that hope seems resurgent in his more recent philosophy of the legal state (Habermas 1992).

Marxism withered away long before the collapse of "real socialism." Brilliant intellectuals such as Przeworski and Elster, building bridges from a variant of democratic socialism to rational choice theory, were dubbed "rational choice Marxists." Meanwhile, America experienced an unknown politicization of theory-building in the name of "political correctness" and "affirmative action" for underprivileged racial, ethnic and gender groups in society; even in the days of the "Caucus" within the American Political Science Association, a new political and normative thrust had never had such an impact as in the 1980s (Ricci 1984: 188–90). Despite the growing interest in green issues and new social movements, Europe had apparently exhausted its desire for politicization in the 1960s and 1970s. All this time

Marxism dominated the debate and set the agenda, even for "bourgeois" thought, foregrounding issues of emancipation and participation in a neo-Rousseauean wave of radicalism. This basic conflict had contributed to the internationalization of the debate. As soon as that latent intellectual civil war in political theory disappeared, a new trend towards regionalization of paradigms became apparent.

The 1980s also saw a decline of neo-conservative thought, parallel to the withering away of the leftists' paradigms which had provoked that conservative backlash. Neo-liberalism became the predominant conservative mood in many countries. The liberals, the main targets of the communitarians, "liberalized" in turn with fading counter-forces from the socialist camp and Marxist thought. Liberalism was able to turn back towards the ideal of a civil society, which became a basic consensus of enlightened democracies. They continued to emphasize the notion *l'homme* more than the participating *citoyen.* But they became more tolerant in turn toward political participation on the level of subsystems of the social system in terms of groups and new social movements. The peaceful velvet revolution in Eastern Europe showed the liberals that not all collective participatory democratic activity is bound to end up in a new authoritarian statehood (Cohen and Arato 1992).

## II The geography of paradigm shifts

From American surveys, one would infer there is a substantial uniformity within political theory worldwide. Galston's (1993) APSA overview registers hardly any European contributions, apart from Habermas and a few French postmodernists. Despite the artificial uniformity of the debate as presented from the American perspective (which hardly takes cognizance of a foreign book unless it is translated), there is actually a growing diversification in political theory on the macro-normative level, whereas on the level of partial theory relevant for empirical studies uniformity is growing.

In the late 1970s and in the 1980s new divergences of major national cultures also had an impact on social and political theory. Galtung (1983) half-seriously offered a typology of intellectual styles which accumulated further evidence in its favor throughout the decade:

- The French style, which is preoccupied with language and art in social theory and which retains a stubborn institutionalism absent until recently from the non-Francophone mainstream, influenced empirical political theory primarily via postmodern thought (Lilla 1994).

- The Teutonic style, which Galtung had in 1983 lumped together with Marxism and thus associated with all then-socialist countries, has now abandoned the Marxist track and turned to the political right. But autopoietic theory, especially Luhmann (1984) and the Bielefeld school (Willke 1983), is as abstract and as far from operationalization as Teutonic reasoning should be. Despite tendencies toward orthodoxy, Luhmann (1984), like most postmodernists, shows no dogmatic zeal and usually merely ridicules theoretical adversaries as being "old European ontologists." The German debate is highly influenced by Bielefeld agnosticism towards any possibility of political steering, much less changing the world. Actors' theories are ridiculed, and in many respects political theory in Germany has come close to abandoning the actor's perspective altogether. Nevertheless, political science as a whole will cling to the possiblity of tracing actors and their impact in the political process as a fundamental premise for certain types of study. A constructivist theory of science will facilitate this kind of "philosophy as if."
- There has never been a single Anglo-Saxon style of theorizing, as Galtung suggests. There were of course certain similarities between Britain and the USA, which were normally summarized under the rubric "pragmatic." But pragmatism as a philosophical dogma had less influence in Britain than in America. Positive political theory as an axiomatic, deductive type of theorizing (Riker and Ordeshook, 1973: xi) had few followers in Britain.

## III Theory and method: levels of theoretical analysis

Political theory is typically done with scant attention to methodological issues. A division of labor has grown up, according to which theoreticians are absolved of responsibility for operationalizing their propositions and empiricists, in turn, are absolved of responsibility for confronting theoretical issues and are allowed instead to treat methodological questions merely as matters of research technique. But theory without methodological framework is sterile, and only very abstract approaches (such as dialectical criticism or autopoietic systems theory) literally identify theory and method. The complete identification of theory and method is as detrimental to empirical work as is the complete separation of it. A balance—so far more common in sociology than political science—is necessary.

Approaches to theory-building can best be seen as a matrix which differentiates between macro-level and micro-level theories, on the one hand, and systems-based theories and actor-based theories on the other. This matrix appears as Figure 22.1. Hardly anyone works at the polar extremes. Only theory-building which operates on a very abstract level (such as Luhmann's) completely scorns actor-based theories, and by the same token approaches which start strictly with the individual (such as the behavioralists') need to introduce certain collective notions at a higher level of reasoning.

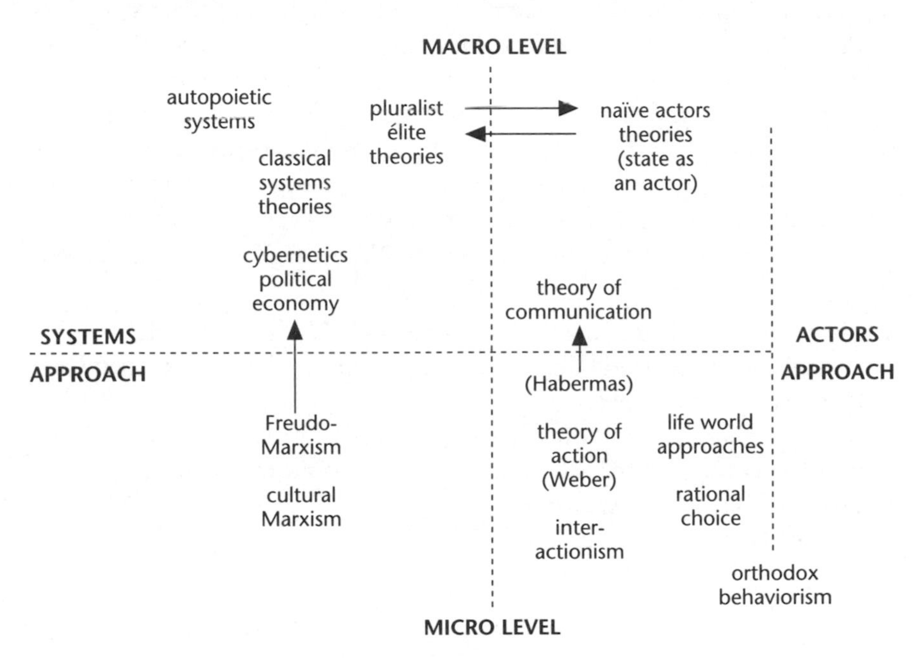


**Fig. 22.1** *Levels and approaches of theory building*

Between those polar extremes of autopoietic systems theory and orthodox behavioralism are many possible theoretical starting-points. There is, however, a tendency toward *rapprochement* between the extremist positions. By way of abstraction and induction, individualistic approaches can end up with models as abstract as certain systems theories. Systems theories, in turn, can be deductively differentiated into so many subfields and typologies of actors that they end up operating virtually at the micro-individualistic level of reasoning.

Political science thus gave up the classical modern notion of "one discipline, one method, one preferred unit of analysis." Mainstream political science lost faith in any fixed hierarchy of objects for research and methods of analysis. Political science theory-building predominantly takes place between the extreme poles of individualistic actor-based theories (such as interactionism and ethnomethodological approaches) and abstract system-based evolution. The meso-level, situated in between those micro and macro orientations, is essential for most questions of political science. Some notion of a collective actor is frequently adopted as an analytical device, even though every political scientist knows that institutions are not a unified entity of literally that sort.

The great recent success story in political science theory-building involves mathematical approaches, parallelling more mature social sciences such as economics. Rational choice modeling, in particular, conquered many U.S. departments and is now spreading through Europe. The APSA's account of the state of the discipline of political theory testifies to this triumph: in 1983, political theory consisted largely in a historical account of past empirical and normative political theories (Gunnell 1983), with William Riker (1983) explicating some of his favorite coalition games as a mere aside; ten years later, "formal rational choice" became a movement (Lalman *et al.* 1993: 77) and Riker was mentioned as an "early contributor." There seems now to be nothing more between "formal rational choice theory" and "normative political philosophy."

How do we explain this astonishing success story? There are several reasons for it.

- The neo-positivist claim for deductive political theory is easiest to implement with formal models.
- Rational choice approaches can be applied to any behavior, from the most egoistic rationality to the most altruistic behavior of Mother Teresa, who also maximizes her strategy of helping the deprived.
- Political science, concentrated on the meso-level between spheres of macro and micro theories, needs to assume that an actor-based approach is feasible. The actor of a rational choice approach is a construction which avoids questions about the real unity of a person.
- Rational choice encourages quantification and cumulative political science.
- Rational choice approaches were a counter-balance against the dominance of behavioral studies in earlier decades. It was easily combined with a multi-level analysis (especially in studies on the European Union) and with an enlightened neo-institutionalism, which spread in the 1980s (Scharpf 1989).

These advances of rational choice models point toward the stabilization of political science as a discipline. The development of theories in a discipline can hardly be the outcome of isolated individual predilections and insights. Only those theoretical approaches which comport with the internal rules of a scientific discipline, admitting of progressive elaboration by many hands, can form the basis for a theoretical mainstream.

## IV Politial trends and their impact on theory-building in the 1990s

A major shock to empirical political theory in the recent period came with the collapse of communism. That event not only forced revisions in theoretical explanations previously offered for developments in communist countries but also forced adaptations within the theoretical self-understandings of the victorious democracies themselves.

Most theories of modernization and the transition to democracy were modeled on developments in Southern Europe and South America in the 1970s (O'Donnell and Schmitter 1986). Many of those developments were not comparable to the 1989 revolution, however. With those unpredecentedly simultaneous transformations of both economic and political systems, old assumptions of modernization theorists about economic prerequisites being essential for the success of political democratization were set on their heads (Karl and Schmitter 1991). The unique character of peaceful revolutions in 1989 even led to a testing of chaos theories taken from recent developments in biology and physics, although most of the applications of those theories remained merely metaphorical (cf. Marks 1992).

The First, Second and Third Worlds grew more similar after 1989. The decline of communism discredited theories such as Barrington Moore's (1966) of alternative roads to modernity. The authoritarian road to modernity followed by half-industrialized countries such as Italy and Germany collapsed in 1945. The totalitarian road to modernity primarily followed by predominantly agrarian countries such as China and Russia ended in 1989. Most transitional systems are democratizing, but it is unlikely that the final product of this process will be fully fledged democracy anytime in the forseeable future. More likely there will be a proliferation of "anocracies," an admixture of anarchy and authoritarianism (Gurr 1991). Empirical political theorists have to confront the possibility of a certain degreee of backsliding among the consolidated democracies. The typological sequence of transitional societies—liberalization, democratization and consolidation—was difficult to find in Eastern Europe, useful though this typology may have been elsewhere in the 1970s.

The great transformation in the early 1990s has been interpreted in terms of a crisis of modernization propeling us into a postmodern world (Baumann 1990). Many former Marxists have turned to some anarchical variation of themes of postmodernity and patchworks of minorities. It is unlikely, however, that a clear evolution from modernity to postmodernity will take place. Most reasonable postmodernists accept postmodernism only as a mere stage of modernity which implements its basic principles in a more consequential and systematic way than classical modernity. Insofar as it is not simply equated with post-materialism or with certain processes of differentiation and individualization which may lead to further decline of the old class social stratification and towards a development of life styles (Beck 1986/1992), postmodernity is a set of theoretical assumptions rather than a clearly discernible new structure of society.

Some theoreticians in Europe (Beck, 1993: 158) used the revolutions of 1989 as proof that system-based theories without actors were wrong. In some respects, however, those "candle revolutions"—without revolutionary élites, ideologies, or mass organization—resemble more the "evolution without subjects" hailed in Luhmann's theories. Certainly postmodern elements were present in those transformation processes, but a new scarcity makes it unlikely that a post-materialist and postmodern lifestyle will soon develop in Eastern Europe. On the contrary, even postmodernism in the West was shaken by events in Eastern Europe.

The claim that "Communism = perversion of modernity; post-communism = enlightened postmodernity" is hardly tenable. Communism was already a hybrid of hyper-modernist megalomanic exaggerations of modernity, on the one hand, combined with pre-modern traits (implementing rationally planned systems through personalistic techniques of corruption, personal contacts and informal groups) on the other. In the West, postmodern theories emphasized a world of games. Lyotard's (1979) claim "let us quietly play" created an artificial world of games among parts of the Western intelligensia. This was possible only as long as it was protected by the Iron Curtain from major conflicts and intrusions; 1989 did away with this protection, and most of the postmodern problems are done away with by the more serious problems of survival.

Underlying the postmodernism debate is the search for a new balance between unity and plurality. The more successfully certain principles universalize, the more urgently components of plurality in modern societies emphasize their right to exist (Marquard 1987). This claim is normally realized by new social movements, but it has proven to be too early to see modernity on the road to a "society of movements" (Neidhardt and Rucht 1993). Empirical research suggests, rather, that the movements are vital for the first stages of the policy process—agenda-setting and policy formation—but that decision-making, implementation and evaluation are predominantly done by the traditional institutions and organized political forces, such as interest groups and parties.

It is also not by chance that the most recent normative debate between liberals and communitarians has now crossed the Atlantic and entered the vacuum left by the now-defunct Marxist ideological debates. A new normative minimal consensus is developing. The paradox of the early 1990s is this. Empirically minded European political scientists are looking for new analytical tools in America. But what they find there is broad skepticism toward the old positivistic, behavioralistic paradigms. They are also discovering a new message, rather normative in character. Furthermore, it is a message which they are now ready to accept, since the older social-democratic consensus in the North European countries has withered away. The pragmatic left in Europe—deeply affected by the erosion of communism, even though they did not share its views—needed a new normative orientation.

Though many superficial observers tend to think of the development of paradigms in terms of cumulative progress, we are increasingly realizing that there are Kuhnian revolutions but not in the same sense as in the natural sciences (Kuhn 1970). There are revivals of old positions. Neo-Aristotelianism is not as dead as the pre-Copernican vision of the world. In political theory, we see a series of small innovations rather than big revolutions. Most of them are not created by established mainstream thinkers but by theoreticians who stand apart from mono-disciplinary research in the spirit of "creative marginality" (Dogan and Pahre 1990: 182 ff.).

## References


Barnes, B. 1992. *T. S. Kuhn and Social Science.* New York: Columbia University Press.

Baumann, Z. 1990. From pillar to post. *Marxism Today.* Pp. 20–5.

Beck, U. 1986. *Die Risikogesellschaft.* Frankfurt: Suhrkamp. English edn: *Risk Society*, trans. M. Ritter. London: Sage, 1992.

—— 1993. *Die Entdeckung des Politischen.* Frankfurt: Suhrkamp.

Bentley, Arthur. 1949. *The Process of Government.* New edn. Bloomington, Ind.: Principia Press.

Beyme, K. von. 1992. *Die politischen Theorien der Gegenwart.* 7th edn. Opladen: Westdeutscher Verlag.

Cohen, J. L., and Arato, A. 1992. *Civil Society and Political Theory.* Cambridge, Mass.: MIT Press.

Deutsch, K. W. 1966. *The Nerves of Government*. New York: Free Press.

Dogan, M., and Pahre, R. 1990. *Creative Marginality: Innovation at the Intersections of Social Sciences*. Boulder, Colo.: Westview.

Downs, A. 1957. *An Economic Theory of Democracy*. New York: Harper.

Etzioni, A. 1968. *The Active Society*. New York: Free Press.

Galston, W. 1993. Political theory in the 1980s: perplexity admist diversity. Pp. 27–53 in *Political Science: The State of the Discipline II*, ed. A. Finifter. Washington, D.C.: American Political Science Association.

Galtung, J. 1983. Struktur, Kultur und intellektueller Stil. *Leviathan*, 11/3: 303–38.

Goodin, R. E. 1992. *Green Political Theory*. Oxford: Polity.

Gunnell, J. G. 1983. Political theory: the evaluation of a subfield. Pp. 303–38 in *Political Science: The State of the Discipline*, ed. A. Finifter. Washington, D.C.: American Political Science Association.

Gurr, T. R. 1991. The transformation of the Western state: the growth of democracy, autocracy and state power since 1800. In A. Inkeles, ed., *On Measuring Democracy: Its Consequences and Concomitants*. New Brunswick, N.J.: Transaction Publishers.

Habermas, J. 1987. *The Philosophical Discourse of Modernity*, trans. F. Lawrence. Cambridge, Mass.: MIT Press.

—— 1992. *Faktizität und Geltung. Beiträge zur Diskurstheorie des Rechts und des demokratischen Rechtsstaats*. Frankfurt: Suhrkamp.

Karl, T., and Schmitter, P. 1991. Modes of transition in South Africa, Southern Europe and Eastern Europe. *International Social Science Journal*, 43: 269–84.

Kuhn, T. S. 1970. *The Structure of Scientific Revolutions*. 2nd edn. Chicago: University of Chicago Press.

Lalman, D. *et al.* 1993. Formal rational choice theory: a cumulative science of politics. Pp. 77–103 in *Political Science: The State of the Discipline II*, ed. A. Finifter. Washington, D.C.: American Political Science Association.

Lijphart, A. 1977. *Democracy in Plural Societies*. New Haven, Conn.: Yale University Press.

Lilla, M. 1994. *New French Thought: Political Philosophy*. Princeton, N.J.: Princeton University Press.

Luhmann, N. 1981. *Politische Theorie im Wohlfahrtsstaat*. Munich: Olzog.

—— 1984. *Soziale Systeme*. Frankfurt: Suhrkamp.

Lyotard, J.-F. 1979. *La Condition Postmoderne*. Paris: Minuit. English edn: *The Postmodern Condition*, trans. G. Bennington and B. Masumi. Minneapolis: University of Minnesota Press, 1984.

Marks, G. 1992. Rational sources of chaos in democratic transition. *American Behavioral Scientist*, 35: 397–421.

Marin, B., ed. 1990. *Generalized Political Exchange*. Boulder, Colo.: Westview.

Marquard, O., ed. 1987. *Einheit und Vielfalt*. Hamburg: Meiner.

Maturana, H. R. 1985. *Erkennen. Die Organisation und Verkörperung von Wirklichkeit*. 2nd edn. Brunswick: Vieweg.

Miller, D. 1990. The resurgence of political theory. *Political Studies*, 38: 421–37.

Moore, B. 1966. *Social Origins of Dictatorship and Democracy*. Boston: Beacon Press.

Neidhardt, F., and Rucht, D. 1993. Auf dem Weg in die Bewegungsgesellschaft. *Soziale Welt*, 44/3: 305–26.

O'Donnell, G., and Schmitter, P., eds. 1986. *Transitions from Authoritarian Rule*. Baltimore, Md.: Johns Hopkins University Press.

Ricci, D. M. 1984. *The Tragedy of Political Science: Politics, Scholarship and Democracy*. New Haven, Conn.: Yale University Press.

Riker, W. H. 1983. Political theory and the art of heresthetics. Pp. 47–68 in *Political Science:*

*The State of the Discipline*, ed. A. Finifter. Washington, D.C.: American Political Science Association.

—— and Ordeshook, P. C. 1973. *An Introduction to Positive Political Theory*. Englewood Cliffs, N.J.: Prentice-Hall.

Rucht, D., ed. 1991. *Research on Social Movements*. Frankfurt/Boulder, Colo.: Campus/Westview.

Scharpf, F. W. 1989. Politische Steuerung und politische Institutionen. *Politische Vierteljahresschrift*, 30/1: 10–21.

Schmitter, P. 1981. Interest intermediation and regime governability in contemporary Western Europe and North America. Pp. 287–330 in *Organizing Interests in Western Europe*, ed. S. Berger. Cambridge: Cambridge University Press.

Schwarzenbach, S. A. 1991. Rawls, Hegel and communitarianism. *Political Theory*, 9: 539–71.

Senghaas, D. 1982. *Von Europa lernen*. Frankfurt: Suhrkamp.

Skocpol, T. 1979. Bringing the state back in: strategies of analysis in current research. Pp. 3–37 in *Bringing the State Back In*, ed. P.R. Evans, D. R. Rueschemeyer and T. Skocpol. New York: Cambridge University Press.

Streeck, W., and Schmitter, P., eds. 1985. *Private Interest Government*. London: Sage.

Wagner, P., and Wittrock, B. 1993. Social sciences and societal developments: the missing perspective. Mimeo., WZB, Berlin.

—— Wittrock, B.; and Whitley, R., eds. 1991. *Discourses on Society: The Shaping of the Social Science Disciplines*. Dordrecht: Kluwer.

Willke, H. 1983. *Entzauberung des Staates. Überlegungen zu einer sozietalen Steuerungstheorie*. Köningstein: Athenäum.

Chapter 23

# Political Theory, Old and New

Brian Barry

I SHALL take up three issues here each of which is common to two of the three chapters under discussion. Young and von Beyme both assert the importance of rational choice theory. But neither says much about it and that little seems to me in need of qualification. I shall devote the first section to this. In the second section I shall take sides on a question that divides Young and Parekh: the position of John Rawls in relation to earlier writers in the postwar period. Finally, in the third section I shall explain why I reject a proposition that Young and Parekh agree on, that contemporary mainstream liberal political philosophy is in need of radical reconstruction because of its inability to come to terms with cultural diversity.

## I Rational choice theory: successes and limitations

### A Normative applications

Iris Young lays emphasis on the contributions of rational choice theory to normative questions, while von Beyme is enthusiastic about prospects for "the stabilization of political science as a discipline" (von Beyme above: 525) by the rational choice approach. I shall follow this division, taking up normative applications here and then moving on to positive applications.

In accordance with my theme of "old and new" in political theory, let me begin with old wine in new bottles: the reworking of the classics in the light of rational choice theory. The two major figures whose political theories lend themselves most to fruitful re-examination in rational choice terms are Thomas Hobbes and David Hume. For both men make arguments that can be reconstructed formally, using game-theoretical apparatus.[1]

A standard objection to Hobbes's analysis is that, if human beings were so destructive as to need a state to contain their predatory tendencies, they could scarcely be constrained by a sovereign (see Hampton 1986: 63–74 for a discussion). The strength of game theory is to emphasize that "human nature" has many latent possibilities: which of them are realized in any given situation depends largely on the incentive structure that it exhibits. Thus, peaceful coexistence may degenerate with remarkable rapidity into violent conflict if fueled by fear, especially fear of pre-emptive attack. The logic of this was set out plainly by Hobbes (1651/1991: chap. 13), and can be seen at work in contemporary breakdowns of civil peace such as those in the former Yugoslavia. (See Hardin (1995: 142–82) for an analysis of group conflict on these lines.)

The same kind of apparatus can be deployed to explore fundamental questions in political theory without tying them to particular past thinkers. Thus, the justification of coercive authority is commonly argued to be that it is necessary to enforce co-operation. But co-operation may quite readily arise spontaneously, in spite of conflicting interests, where two parties interact repeatedly. Even in the unpromising setting of the First World War, informal truces grew up in sectors of the Western Front (Axelrod, 1984: 73–87).

The conditions for multilateral co-operation are more demanding, and have been explored theoretically by Michael Taylor (1987*a*), who has also looked at the ways in which societies without centralized coercive institutions manage conflict (Taylor 1982). Another line of empirical inquiry stimulated by this issue is the development of co-operation (e.g. on the exploitation of fisheries or allocation of water resources) in a state but outside the state's regulatory apparatus (Ostrom 1990). Contrary to Garrett Hardin's (1968) fable of the "tragedy of the commons," common resources have not normally been left to be exploited according to the logic of individual self-interest but have been subject to locally enforced limits on use.[2]

A closely related question (to a large extent an alternative way of approaching the same question) concerns the origin and maintenance of norms. Thus, since the soldiers on the Western Front had to appear to be fighting to satisfy their superiors, informal truces took the form of shelling the same targets each day at fixed times (Axelrod 1984: 86–7). We might

[1] On the Hobbesian contract, see Kavka 1986 and Hampton 1986; on Hume's "contract by convention" see Hardin 1982; for an assimilation of the two, see Taylor 1987*a*: 125–63.

[2] See Ostrom 1990: 224 n. 3 for the manorial system in Europe.

speak of a co-operative norm developing here: inadvertent breaches of it led to apologies for having "violated a situation of trust" (Axelrod 1984: 85). Similarly, the management of what Ostrom (1990: 13) calls "common-pool resources" entails the evolution of norms governing their exploitation.

Although Elster (1989) has declared norms to be inexplicable in rational choice terms, this is a trivial implication of his definitions, according to which normatively controlled action must be done without regard to consequences whereas rationality entails concern for consequences. However, many norms (such as those just cited) serve to co-ordinate mutually advantageous behavior and it is rational (in a means-end sense) for each person to observe them on condition that (enough) others do.[3]

"Rational choice theory" is usually taken to embrace social choice theory (the aggregation of preferences) as well as game theory (the analysis of strategic interactions). Social choice theory grew out of a problem in welfare economics: that of defining a "social welfare function" that would represent the judgments of a certain person about the relation between the welfare of a number of individuals and overall "social welfare" (Bergson 1938; 1954). Arrow (1963) turned the problem into a constitutional one about aggregating the judgments of a number of individuals about the relative preferability of alternative states of affairs. (See Arrow 1967 for an explicit acknowledgment of this move.) He then proved his famous "impossibility theorem" that there is no formula for aggregating sets of internally consistent individual preferences that can be guaranteed to create a consistent "social preference ordering" of states of affairs.

Although Arrow's result has been hailed as central to the concerns of political philosophers, it is not at all apparent why anybody should show any interest in a social welfare function as defined by Arrow (see Little 1952). The social states to be ranked would include, for example, "a complete description of the amount of each commodity in the hands of an individual" (Arrow 1963: 17). Yet in a liberal society there is no occasion for forming a collective view about the consumption-bundle of each individual or for forming a collective view about the use people make of their legitimate rights (Sugden 1989). Since there is no application for a collectively constituted view of a "best state of affairs," we do not need to tie ourselves into knots trying to make the "social welfare function" come out so that the "socially chosen" outcome is *x* or *y* depending on the decision taken by the holder of a right to exercise that right by doing *x* or *y*. In fact, there is

[3] See Hardin 1995: 108–14 for an analysis on these lines and a critique of Elster; also, on "the emergence of norms," Ullman-Margalit 1977. For an analysis of extralegal co-operative norms among ranchers in Shasta County, California, see Ellickson 1991.

no reason why we should be concerned with the deliverances of a "social welfare function" at all (Barry 1991*c*).

Despite the sophisticated literature generated by the "Arrow problem," I believe that the payoff to political philosophy lies in two rather simple ideas about voting. The first, which is at the core of Arrow's proof but was already established by Condorcet (1785), is that any vote among three or more alternatives is liable to yield a cycle (see further Black 1958: 159–80). Thus, it may be that a majority prefers $x$ to $y$, $y$ to $z$, and $z$ to $x$. The second, which flows out of this phenomenon, is that all voting schemes are open to manipulation: that is to say, wherever there are three or more alternatives, it may be possible for at least one voter to obtain a more preferred outcome by not voting in accordance with his "true" preferences. (The general result was proved in Gibbard (1973); a classic analysis of strategic voting is Farquharson (1969).) Between them these two results destroy any simple-minded idea that voting can establish an unambiguous "will of the people." How far this is taken to create a problem for the rationale of democratic institutions depends on what one regards that rationale as being (cf. Barry 1991*a* and Riker 1982).

## B Positive applications

I do not want to put a lot of weight on the normative–positive distinction. For conclusions about how things are may well have normative implications. An example from the earlier discussion is that research suggesting the viability (under certain conditions) of collective management of common-pool resources will tend to undermine the idea that the only alternatives are state regulation or a regime of private property.[4] Similarly, the "community power" debate was ostensibly an empirical question about "who governs" in American cities. But what was at stake was an evaluative conclusion about the adequacy of American political institutions. Roughly, were representative institutions a façade behind which the important decisions were taken by a (mostly business) élite, or was the day-to-day conflict of political actors not merely the appearance of widely dispersed power but an indication of the reality of it?

The "pluralists" gained the methodological high ground by proclaiming the axiom that only interests expressed in action should be counted (Dahl 1961; Polsby 1980). Their bemused opponents (Bachrach and Baratz 1970; Gaventa 1980) felt obliged to explain inaction by invoking "non-events and

[4] See Ostrom 1990: 8–21 for an explicit acknowledgment of this motivation.

non-decisions which, not surprisingly, were never found" (Dowding *et al.* 1995: 271) or "real interests" not recognized by the (non)agents (Lukes 1974). Yet a rational choice analysis would have revealed the folly of applying a "revealed preference" analysis to areas where problems of collective action were rife.[5]

The "logic of collective action" (Olson 1965/1971) tells us that, where the benefits of a good cannot be confined to those who contribute to its production, it is liable to be supplied at suboptimal levels or not at all because each person has an opportunity to "free ride" on the efforts of others. A public policy is an archetypal public good: if a steel town introduces a pollution-abatement ordinance (Crenson 1971), the benefits of cleaner air and the possible costs of reduced employment fall alike on those who worked for it, those who worked against it, and those who did nothing.

The implication is that political activity aimed at changing general public policy (as against aiming for some individually advantageous decision) is subject to the "logic of collective action." I may be quite clear that some reform would be in my interest, but still rationally conclude that it is not in my interest to incur a cost to help bring it about. The inference from my inaction that I would not value the outcome is as invalid as would be the conclusion that the two prisoners in a one-shot prisoner's dilemma who act rationally by confessing (Poundstone 1992) must prefer a longer to a shorter jail sentence because that was the outcome of their joint choices.

More broadly, I think that the most useful contribution of rational choice analysis is to pose the question: why are there apparently highly significant exceptions to the "logic of collective action?" Olson demonstrated that we cannot explain individual action in pursuit of a public good as a self-interested response to the benefit to be expected from contributing. This leads to a research program explaining why collective action does nevertheless occur. How, for example, can we explain the sometimes considerable costs incurred by civil rights activists in the 1960s in the American South? (see Chong 1991). And why does anybody outside the circle of those who expect to reap particularistic benefits take part in overthrowing a dictatorship? (see Taylor 1987*b*; Hardin 1995: 39–42).

The rational choice research program has undoubted achievements to its credit. At the same time, I am less enthusiastic about its past and less optimistic about its future as an all-encompassing explanatory scheme in political science than von Beyme. What he puts forward as a merit seems to be a drawback. In accounting for the "success story" of rational choice theory he adduces as one advantage that "rational choice approaches can be

[5] The point is made clearly in Dowding (1991: 84–114) and Dowding *et al.* (1995).

applied to any behavior, from the most egoistic rationality to the most altruistic behavior of Mother Teresa, who also maximizes her strategy of helping the deprived" (von Beyme above: 525).

There are two related contexts in which a rational choice analysis does not require for determinacy any assumption about the motives of agents. If we want to work out the strategies open to committee members to get as far as possible up their preference orderings, it makes no difference whether these orderings derive from egoism, altruism, a sense of duty, or any other motive. Similarly, if we assume that in multi-party systems parties wish to implement as much of their policy as possible, we can delimit the range of possible coalitions without having to ask why parties have the policies that they do have (Laver and Shepsle 1996). But what gives the analysis its bite is the assumption that each person on the committee or party in the legislature is concerned to get the best possible outcome, defined by its preferences over outcomes.

In less structured contexts, allowing preferences to take any value makes for a theory empty of application. In practice (contrary to what von Beyme suggests) rational choice explanations are usually based on an assumption of self-interested behavior. This incurs two opposite problems. First, only sheer dogma could sustain the view that political actors are never motivated by pursuit of the public good as they see it (Lewin 1991; Barry 1991*b*). Yet at the same time, many different specific objectives might be attributed to actors as flowing from self-interested motives. Thus, generality is lost but specificity is not attained.

There is a standing temptation to impute preferences to political actors in whatever way is required to make the model square with the observed outcome. Explanation of this kind is easy but unpersuasive, since a different model with different imputed preferences might perform equally well (Green and Shapiro 1994). The implication is that far more attention needs to be devoted to establishing, in attempts to explain specific events, what the beliefs and preferences of the actors actually were (Dowding 1995).

How might this be done? We can look at diaries or other contemporary documents where available.[6] Or we might simply ask people. For example, Schlozman *et al.* (1995) asked political participants what motivated them and arrived at conclusions that varied across different forms of activity in a plausible way. A more indirect strategy is to accept the burden of showing that an imputed preference-structure explains behavior in a whole range of situations in which the actor is involved and not only the situation in which

[6] See Chong 1991 for an imaginative use of such evidence in the context of the civil rights movement.

those preferences have to be ascribed to the actor to make the "predictions" of the model fit what actually happened.

## II The role of Rawls

Although they were written independently, Parekh's account of the history of political theory since 1945 reads like a direct contradiction of Young's. According to Young (above: 481), Rawls's *A Theory of Justice* (1971) can be located as the turning point between the barren and the fruitful years. Parekh (above: 503) denies that "the 1950s and 1960s marked the decline or even the death of political philosophy" and (as an implication of that denial) does not accept that *A Theory of Justice* "symbolized the rebirth of political philosophy." Since I am cited by both Young and Parekh as a supporter of the view that the former enunciates and the latter opposes, it seems appropriate that I should explain briefly what seems to me true in the claim that Rawls relaunched political philosophy.

When Peter Laslett made the much-quoted assertion that "political philosophy is dead," he explained that what he had in mind was the absence of contributions to a line of writers in English that he took to run "from Hobbes to Bosanquet" (Laslett 1956: vii). Richard Tuck has recently written on similar lines of "the absence of major works of political philosophy, of a more or less familiar kind, between Sidgwick and Rawls" (Tuck 1993: 72). Whichever late Victorian we take as the last in the line, nothing that Parekh says seems to me to impugn the claim that nobody until Rawls produced anything that represented a continuation of the canon of political thought, as traditionally conceived.

Parekh's description of the writers whose work he celebrates as gurus, and his remark that they generated disciples rather than critics, suggest that they might most aptly be thought of as purveyors of secular religion. And, indeed, it might be said of their writings that "it is with the mysteries of our Religion, as with wholesome pills for the sick, which swallowed whole, have the vertue to cure; but chewed, are for the most part cast up again without effect" (Hobbes 1651/1991: 256). There was not much of a structure of argument to get your teeth stuck into. Either you found the vision of life attractive or not; either way there was little point in trying to take it apart.[7]

What Rawls reintroduced with *A Theory of Justice* was political philosophy that could be chewed in the same way as the canonical books could be (and

[7] The exception is Isaiah Berlin's (1979) "Two Concepts of Liberty," originally published (as his Inaugural Lecture at Oxford) in 1958. This was indeed a work of analysis rather than oracular literature, as Parekh (above: 504) correctly observes. However, its impact was limited by its being presented in the guise of "conceptual analysis," even though there was a theory struggling to get out.

continued to be throughout the ascendency of the gurus). Parekh notes the vitality of the subject today but attributes this simply to "changes in the intellectual climate" (above: 513). This misses the point that it is precisely the virtues of *A Theory of Justice*—above all its systematic argumentative quality—that created that climate. Perhaps in a counterfactual world in which *A Theory of Justice* did not exist, something else would have taken its place; but there is no doubt that in the actual world it is Rawls who deserves the credit.

Parekh suggests that Rawls owed it to the gurus that he was able to put forward a theory that was "critical in nature, universal in scope, and quasi-foundational in orientation" (above: 503). I can see no basis for this claim. The index to *A Theory of Justice* does not contain entries for Arendt, Oakeshott, Popper, Strauss or Voegelin. The entries that are large enough to have subheads are for Aristotle, Bentham, Edgeworth, Hume, Kant, Mill and Sidgwick. The obvious inference is that Rawls saw himself as engaged in the same kind of activity as them, and carried on where they left off. More specifically, Rawls orientated himself to the two major liberal political theories of the past two centuries: Kantianism and utilitarianism. Rawls's critics in turn have tended to be inspired (if at some distance) by Hegel (Brown 1992). The post-Rawlsian debate has connected directly to the canon, bypassing the gurus.

## III Liberalism and the politics of difference

Readers of the chapters by Young and Parekh should put them in context by appreciating that their authors are both well-known as critics of liberalism, primarily on the ground that it is unable to accommodate diversity in beliefs and ways of life. I shall offer a defense of liberalism against this charge, thus I hope doing something to redress the balance.

The basic idea of liberalism is to create a set of rights under which people are treated equally in certain respects, and then to leave them to deploy these rights (alone or in association with others) in the pursuit of their own ends. In the past two hundred years, western societies have been transformed in accordance with the precept of equal treatment. Slavery has been abolished in the West Indies and Latin America as well as in the USA, where (belatedly) equal civil rights have been established. In the 1850s, the role of women in society, the polity and the economy was everywhere mediated through a male. A woman moved on marriage from the protection (and in effect guardianship) of her father to that of a husband—and the fate of those without such protection (the "old maids") was scarcely more eligible. The legal position of women has been changed unrecognizably since then in all western societies. The rationale for such transformations has been the liberal individualist principle, extended beyond its original application to religion so as to say that the state should relate to whites and blacks, or to men and women, through a uniform set of laws.

The liberal agenda is by no means completed. Anti-discrimination provisions can be strengthened. (Since discrimination is an inherently statistical notion, there is nothing contrary to the principle of liberal individualism in using measures that compare the demographics of the qualified applicant pool and the successful candidates for jobs.) The principle of "equal pay for equal work" can be pressed much further by comparing the skill and effort involved in work done predominently or exclusively by men and work done predominently or exclusively by women. Equal opportunity in education is still far from achievement. Above all, the kind of economic security that is necessary to give value to other rights is conspicuously lacking. It would be most effectively provided by an unconditional subsistence income (Parker 1989; Brittan 1995; Van Parijs 1995).

There has been a tendency in recent years (exemplified by Young and Parekh) for those in academia who are concerned for the position of women and minorities to turn their backs on the liberal agenda and argue instead for the politicization of group identities and for the abandonment of the liberal ideal of equal treatment under common laws. The chief objection made to liberalism is that, while pretending to be tolerant to diverse beliefs and ways of life, it is actually highly restrictive. The simplest refutation is that liberal societies quite clearly do not discriminate against, let alone prohibit, organizations that comprehensively violate liberal principles. Thus, with the exception of some liberal Protestant and Jewish congregations, it may be said that Christians, Mormons, Jews and Muslims belong to organizations that are undemocratic, draw their ministers from members of only one sex, and have doctrines that are more or less offensive to liberal tenets of sex equality. Moreover, religious (and other) groups are free to set up their own schools, even if they are avowedly devoted to indoctrination, so long as they meet some very weak requirements of minimal effectiveness in teaching standard school subjects.

It is, of course, true that (unless specially exempted) members of religious groups are required to comply with the law of the land. But this is simply an implication of the basic liberal idea that rights and resources are the currency in which all claims are adjudicated. Thus, whatever is a fair allocation of money is to be seen as a fair allocation of the opportunity to use a share of the society's resources for the pursuit of one's ends. People cannot legitimately make claims for additional economic resources on the ground that they require vintage claret and plovers' eggs to achieve levels of satisfaction that others achieve with less expensive comestibles (Dworkin 1981). Similarly, the cost of "expensive tastes" based on beliefs has to be borne by the person with the beliefs. (For a good discussion of the issues, see Jones (1994).) Those who have a conscientious objection to killing other people will find the military closed as a career; committed vegetarians will probably not want to work in slaughterhouses or butchers shops; those whose religious beliefs forbid trading on Fridays would be well advised not to take up shopkeeping; and so on. Motorcycle helmet laws bear harshly on those who get a thrill from riding bareheaded, and also mean that devout Sikhs have to find an alternative means of transport. Drug laws interfere with the pleasures of recreational drug users and prevent those for whom hallucenogenic mushrooms form part of their religious ritual from following it in that respect. Everyone finds some laws hard to keep: Porsche owners claim difficulty in keeping to speed limits; paedophiles have to sacrifice activities that may be central to their lives; and those whose cultural traditions dictate the genital mutilation of their daughters have to desist under penalty of law.

There are indeed some exemptions made to accommodate religious believers: an example is the provision for a waiver of humane slaughtering laws to permit kosher butchery.[8] But the rationale of any such exemptions (in as far as they are not a pure response to political pressure) is that they are pieces of indulgence based on a utilitarian balancing of costs and benefits. What must be emphasized is that they are not instantiations of some alleged deep principle of equal treatment according to which there is something *prima facie* illegitimate about a law that has a differential impact on different people according to their beliefs, cultural traditions, or personal proclivities. For all laws have a differential impact, and are bound to. Living together under a common set of laws entails acceptance of that. If there are good and sufficient reasons of a general kind for having a law prohibiting conduct of a certain kind, the reasons for someone's wanting to violate it—whether derived from religion, cultural traditions, or a strong personal impulse—are immaterial.

Any state, liberal or not, can delegate law-making powers to groups and permit them to act collectively in ways that contravene its own basic principles. Thus, it is hoped that the Chinese government will allow Hong Kong

[8] In citing this example (see Singer 1991: 153–6), I do not wish to suggest that it is justified. If people insist that any animals they eat must have been fully conscious at the time of death, this is an "expensive taste" that should not be satisfied at the expense of the suffering it entails. The alternatives are to give up meat or reinterpret the scriptures. Singer (1991: 154) notes "that even Orthodox rabbis are not unanimous in supporting the prohibition of stunning prior to killing: in Sweden, Norway and Switzerland, for example, the rabbis have accepted legislation requiring the stunning with no exemptions for ritual slaughter. Many Moslems have also accepted stunning prior to slaughter."

to be a liberal sub-polity after it resumes sovereignty over the colony in 1997. (The official formula for this is "one country, two nations.") Similarly, a liberal state could treat certain groups within it as in effect independent "nations" whose autonomy included a waiver of (some) liberal constitutional constraints.

If the idea of an independent "nation" within a liberal state were taken seriously, the limits on what it could do might plausibly be taken to be the limits within which a sovereign state could act without attracting valid external criticism for trampling on human rights. Examples that have been canvassed in relation to native American peoples in Canada and the US are that decision-making might take some traditional form incompatible with representative government based on universal suffrage; that the integral relation between religion and culture might be recognized by limiting rights to religious freedom; and that traditional usages with regard to property and other rights might be maintained, despite their being in conflict with equal treatment of the sexes.

Given the appalling record of European relations with the indigenous inhabitants of the New World, from the first contacts onward, it is hard not to be sympathetic to the idea that the much-violated treaties between independent contracting parties should finally be given some meaning in this way (Tully 1995). Yet the standard feminist complaint about the sentimentalizing of "community" clearly has particular force here. Can women who claim that they are disadvantaged by these arrangements reasonably be expected to accept the reply that this is the way their indigenous culture does things (or did the last time anybody can remember)?

Contrary to what is sometimes claimed (Kukathas 1992; Kymlicka 1995), there is no liberal principle that calls for deviations from liberalism to be permitted. Rather, what we have here is a clash between the value of collective autonomy and that of liberalism. The only other case in North America that might be conceived in similar terms is that of the Old Order Amish. They cannot claim a history of usurpation and oppression as can the native Americans. But they do not ask for special legislative powers; only for exemptions from a number of normal state-mandated requirements.[9] And they have, ever since arriving in the New World, made extraordinary efforts to insulate themselves. If the conclusion is drawn that the Old Order Amish should be treated as a group outside American society, what must be emphasized is again that this does not carry over to other groups. (As Amy Gutmann points out in a fine discussion (1995) the Amish cases have not been employed as precedents in the US courts, and rightly so given their peculiarity.)

[9] See Kraybill 1993 for an account of Amish conflicts with the state.

The reason for making an issue of the special situation of native Americans and (perhaps) the Amish is that there is a tendency to use the Amish as a springboard for wider claims about group rights (see for example Galston 1995), and to throw in native Americans with women, members of racial minorities, and the disabled as groups with legitimate claims to special treatment *qua* groups (see for example Young 1990: *passim*). This provides an entirely unwarranted support for claims that will not by themselves withstand scrutiny.

The issues raised here are ones about the extent to which the application of liberal principles should override group autonomy. Oddly enough, those who criticize liberalism for not being tolerant enough of diversity also criticize it for positing an inviolable division between the public and the private which licenses violence and exploitation (see for example Young 1990). A whole school of feminist theorists now regards this as such a commonplace that no argument is needed—merely a citation of some other feminist theorist. Yet it is manifestly false. Liberal polities have shown themselves prepared to intervene comprehensively in the affairs of private firms in pursuit of the health and safety of workers, trade-union recognition, protection of consumers and the environment, and anti-discrimination both among employers and customers. Nor are families a sacrosanct "private sphere." Incest has always been a crime, and there is no exemption from charges of battery on the ground that the victim is a spouse. Marital rape is now a crime in many jurisdictions, and there would be nothing inconsistent with liberal principles in prohibiting corporal punishment of children by parents.

It is true that many women are unwilling to complain to the police about an abusive husband (or one who beats or sexually abuses her children), and that housework and childcare duties tend to be unequally divided even when both spouses have full-time jobs outside the house. This largely reflects the unequal bargaining powers of women, who are (in the absence of a well-paid job) "only one man away from poverty" and are therefore unwilling to move out of abusive and exploitative marriages. The liberal solution lies in building up the resources of women so as to increase their independence. This could be done by setting child allowances at a level genuinely adequate to cover the cost of raising a child and (as earlier proposed) providing every adult with an unconditional income at subsistence level. Notice that this is a universal measure, applying to all alike, and is neutral between different household formations.[10]

The charge that liberalism stifles diversity sometimes takes the form of suggesting that liberal individualism is necessarily wedded to an "assimilationist"

[10] See Young 1995 for an argument in favor of such neutrality. Waldron 1993 is a good discussion of the relation between liberalism and neutrality.

ideal. Individual liberals may hope for a society in which, for example, gender is no more significant than eye colour (Wasserstrom 1980). But the framework of a liberal society leaves people free to associate with whomever they like, and to cultivate whatever distinctive ways of life they choose. Thus, there is absolutely nothing to prevent women from "drawing on images of Amazonian grandeur, recovering and revaluing traditional women's arts like quilting and weaving, or inventing new rituals based on medieval witchcraft" (Young 1990: 162). What is, however, true is that a woman who finds that a busy round of quilting, weaving and witchcraft leaves little time or inclination for a high-flying executive career cannot reasonably complain about not achieving one. "Assuming that justice ultimately means equality for women," that is "that all positions of high status, income and decision-making power ought to be distributed in comparable numbers to women and men" (Young 1990: 29) is valid only against a matching assumption of an equally distributed determination to achieve those positions.

The larger issue raised by the charge of coercive assimilationism is how far equal treatment requires employers (and workmates) to accommodate characteristics of employees that they regard as undesirable. A troubling phenomenon that this may bear on is the poor record of young black males in the USA in getting and keeping jobs—a record that is by no means fully accounted for by factoring in formal educational qualifications, and makes an increasingly striking contrast with the labour market experience of young black women. One explanation that has been offered (Jencks 1993: 128–9) is that there is a tendency for young black men to be perceived as having an "in your face" attitude that makes for difficult relations with superiors and co-workers in the organization. Supposing for the sake of argument that this is so, what follows?

Young apparently believes that the notion of a job is almost entirely socially constructed (Young 1990: chap. 7). This commits her to opposing the notion of meritocracy and to holding that co-workers can legitimately take part in drawing up job specifications. Even she may therefore have difficulty in resisting the conclusion that it is not an objectional form of discrimination to count courtesy and co-operativeness among the requirements of holding a job in the mainstream economy.

Liberals will be inclined to fear that, in seeking to eliminate the notion of objectively-definable qualifications for a job that can be derived from the nature of the job itself, Young is removing the best protection there is against the operation of free-floating prejudice in hiring and firing decisions. If the criteria of suitability for a job are up for grabs, why should not hairstyle, taste in personal adornment, sexual orientation, gender or race be potentially relevant? At the same time, however, liberals cannot rule out *a priori* the possibility that the cultural traits of some group may without unfairness disadvantage its members' employment prospects. The question for a liberal is whether or not the traits in question are genuinely related to job performance.

Employers must discriminate among job candidates if they are to take any decision at all. Discrimination becomes a pejorative term only when the criteria are irrelevant (eg racial discrimination). Liberals cannot afford the post-modern luxury of saying that relevance is in the eye of the beholder. The liberal conception of fairness depends on the possibility of reasoned argument about the appropriate criteria of relevance. The arguments may not in every case be conclusive and in the end the courts may well have to be called in to give an answer. But there is no way of avoiding the question.

The same line of thought applies to language. A country in which English is the primary vehicle of economic and political transactions does not need to take any official interest in the languages its inhabitants speak at home or in social gatherings. But at the same time it is under no moral obligation, on liberal premises, to prevent immigrants or their descendants who are not fluent in English from being restricted to menial jobs, disadvantaged in dealing with public officials and politically marginalized. Those who choose to migrate should accept that part of the deal is to adapt to the extent required to get on in the new society on its terms. Those who are not prepared to do so cannot reasonably complain if they fail to reap the benefits that attracted them in the first place (Kymlicka 1995).

The general theorem is that equality of opportunity plus cultural diversity will probably produce inequality of outcomes. Equality of outcomes requires either cultural uniformity or departing from equal opportunity so as to impose equal success rates for all groups. Young (1990: chap. 7) is somewhat drawn to the last but this would be ruled out by liberal individualist premises. A culturally diverse society cannot be conceived as one in which everyone is trying equally hard to achieve the same goals. The notion of equal opportunity has to be conceptualized in a way that takes account of that fact.

Even after all gratuitous barriers (including subtle ones) have been removed, it may well simply be that some ways of life and their associated values will lead to diminished motivation or capacity for high occupational achievement. A liberal will have to say that that is the unavoidable implication of diversity. But a liberal can also say (and should, in my view, say) that occupational achievement would in a just society make far less difference to people's standard of living (in the broadest sense) than it does now in countries such as Britain and the United States. In a just society, therefore, it would be much less important that some ways of life and values are (to a greater or lesser degree) incompatible with high occupational achievement, as conventionally measured.

To conclude: Parekh suggests that liberals have a great deal to learn from the rest of the world. I agree in this sense: we can learn what to avoid. Only a minority of states in the world provide their citizens with material necessities or with security against violence against each other or by the government. Although there are many reasons for this, an important one is the widespread violation of the basic liberal individualistic principle. On both of the criteria advanced, almost every African state is a worse place to live now than it was at the time of decolonialization. A large part of the responsibility for that decline lies in policies designed to further the interests of one group at the expense of others. I thus return to the theme of group conflict taken up in Section I (see Hardin 1995).

What we should be led to appreciate by extending our gaze over space and time is the improbability of liberal individualist arrangements in the world. They developed out of the divisions within Christianity in Western Europe, and have taken root to only a very limited extent in other countries except those of western European settlement. It is clear that the combination of private pluralism and public universality they require is extraordinarily hard to maintain. I believe that it is as precious as it is fragile—not because it is ours (a vacuous reason that traditionalists and postmodernists unite in giving) but because it is right. Young and Parekh propose to toss it away in pursuit of a politics of group identity and political recognition of cultural difference. I believe that they are profoundly mistaken.

## References

An entertaining and accessible introduction to game theory through the history and applications of the Prisoner's Dilemma game is Poundstone (1992). Booth, James and Meadwell (1993) and Dowding and King, eds. (1995) are useful in illustrating a variety of applications of rational choice theory. A stimulating book employing the rational choice approach in a variety of contexts (including Marxist historiography) is Carling (1991). Two normatively orientated books using a rational choice approach—the first sympathetic to utilitarianism, the second to rights thinking—are Hardin (1988) and Sugden (1986). Two books covering rational choice and social choice are Hargreaves-Heap *et al.* (1992) and Barry and Hardin (1982), the second an anthology with introductory material. The best critique of rational choice explanations in politics (though it is confined to examples in the United States) is Green and Shapiro (1994), though readers should perhaps ask themselves how many explanations utilizing alternative approaches within political science meet their exacting standards.

The *locus classicus* for the kind of liberal theory attacked by Young and Parekh is Rawls (1971), the best brief summary of which is Mulhall and Swift (1992: 1–33). An alternative and more accessible statement of broadly the same position as that taken by Rawls, with a defense against some criticisms that have emerged in the last twenty years, is Barry (1995). The criticisms discussed in this chapter may be found in Young (1990). There is an able response to these and some other recent anti-liberal arguments in Fullinwider (1995) and in Moon (1993), though Moon goes through a good deal of handwringing and agonizing before arriving at forthrightly liberal conclusions (see Macedo (1995) for a criticism of this tendency in Moon). A useful collection illustrating the debate between liberals and communitarians is Avineri and de-Shalit (1992). Mulhall and Swift (1992) covers much the same ground in a lucid critical survey. Kymlicka (1990) is a more comprehensive introductory text which includes a number of the themes taken up here.

ARROW, K. J. 1963. *Social Choice and Individual Values.* 2nd edn. New Haven, Conn.: Yale University Press; originally published 1951.

—— 1967. Values and collective decision making. Pp. 215–32 in *Philosophy, Politics and Society,* 3rd series, ed. P. Laslett and W. G. Runciman. Oxford: Blackwell.

AVINERI, S., and DE-SHALIT, A., eds. 1992. *Communitarianism and Individualism.* Oxford: Oxford University Press.

AXELROD, R. 1984. *The Evolution of Cooperation.* New York: Basic.

BACHRACH, P., and BARATZ, M. 1970. *Power and Poverty: Theory and Practice.* New York: Oxford University Press.

BARRY, B. 1991*a*. Is democracy special? Vol. i, pp. 24–60 in B. Barry, *Democracy and Power: Essays in Political Theory.* Oxford: Clarendon Press.

—— 1991*b*. Does democracy cause inflation? The political ideas of some economists. Vol. i, pp. 61–99 in B. Barry, *Democracy and Power: Essays in Political Theory.* Oxford: Clarendon Press.

—— 1991*c*. Lady Chatterley's Lover and Doctor Fischer's bomb party: liberalism, pareto optimality and the problem of objectionable preferences. Vol. ii, pp. 78–109 in B. Barry, *Liberty and Justice: Essays in Political Theory.* Oxford: Clarendon Press.

—— 1995. *Justice as Impartiality.* Oxford: Clarendon Press.

—— and HARDIN R., eds. 1982. *Rational Man and Irrational Society?* Beverly Hills, Calif.: Sage.

BERGSON, A. 1938. A reformulation of certain aspects of welfare economics. *Quarterly Journal of Economics,* 52: 310–34.

—— 1954. On the concept of social welfare. *Quarterly Journal of Economics,* 68: 233–52.

BERLIN, I. 1979. Two concepts of liberty. Pp. 118–72 in I. Berlin, *Four Essays on Liberty.* Oxford: Oxford University Press; originally published 1958.

BLACK, D. 1958. *The Theory of Committees and Elections.* Cambridge: Cambridge University Press.

BOOTH, W.J.; James P.; and Meadwell, H., eds. 1993. *Politics and Rationality.* Cambridge: Cambridge University Press.

BRITTAN, S. 1995. *Capitalism with a Human Face.* Cheltenham: Edward Elgar.

BROWN, C. 1992. *International Relations Theory.* Hemel Hempstead: Harvester-Wheatsheaf.

CARLING, A. 1991. *Social Division.* London: Verso.

CHONG, D. 1991. *Collective Action and the Civil Rights Movement.* Chicago: University of Chicago Press.

CONDORCET, M. J. A. N. Caritat, Marquis de Condorcet. 1785. *Essai sur l'Application de l'Analyse a la Probabilité des Decisions Rendues a la Pluralité de Voix.* Paris.

CRENSON, M. 1971. *The Unpolitics of Air Pollution.* Baltimore, Md.: Johns Hopkins University Press.

DAHL, R. A. 1961. *Who Governs?* New Haven, Conn.: Yale University Press.

DOWDING, K. 1991. *Rational Choice and Political Power.* Aldershot: Edward Elgar.

—— 1995. Interpreting formal coalition theory. In Dowding and King 1995: 42–59.

—— and KING, D. S., eds. 1995. *Preferences, Institutions and Rational Choice.* Oxford: Clarendon Press.

—— Dunleavy, P.; King, D.; and Margetts H. 1995. Rational choice and community power structures. *Political Studies,* 43: 265–77.

Dworkin, R. 1981. What is equality? Part i. Equality of welfare. *Philosophy and Public Affairs,* 10: 185–246.

Ellickson, R. C. 1991. *Order without Law: How Neighbors Settle Disputes.* Cambridge, Mass.: Harvard University Press.

Elster, J. 1989. *The Cement of Society.* Cambridge: Cambridge University Press.

Farquharson, R. 1969. *Theory of Voting.* New Haven, Conn.: Yale University Press.

Fullinwider, R. K. 1995. Citizenship, individualism, and democratic politics. *Ethics,* 105: 497–515.

Galston, W. 1995. Two concepts of liberalism. *Ethics,* 105: 516–34.

Gaventa, J. 1980. *Power and Powerlessness.* Oxford: Clarendon Press.

Gibbard, A. 1973. Manipulation of voting schemes: a general result. *Econometrica,* 41: 587–601.

Green, D. P., and Shapiro, I. 1994. *Pathologies of Rational Choice Theory.* New Haven, Conn.: Yale University Press.

Gutmann, A. 1995. Civic education and social diversity. *Ethics,* 105: 557–79.

Hampton, J. 1986. *Hobbes and the Social Contract Tradition.* Cambridge: Cambridge University Press.

Hardin, G. 1968. The tragedy of the commons. *Science,* 162: 1243–8.

Hardin, R. 1982. *Collective Action.* Baltimore, Md.: Johns Hopkins University Press.

—— 1988. *Morality within the Limits of Reason.* Chicago: University of Chicago Press.

—— 1995. *One for All: The Logic of Group Conflict.* Princeton, N.J.: Princeton University Press.

Hargreaves-Heap, S.; Hollis, M.; Sugden, R.; and Weale, A. 1992. *The Theory of Choice.* Oxford: Blackwell.

Hobbes, T. 1991. *Leviathan,* ed. R. Tuck. Cambridge: Cambridge University Press; originally published 1651.

Jencks, C. 1993. *Rethinking Social Policy: Race, Poverty and the Underclass.* New York: HarperCollins; originally published in 1992 by Harvard University Press.

Jones, P. 1994. Bearing the consequences of belief. *Journal of Political Philosophy,* 2: 24–43.

Kavka, G. S. 1986. *Hobbesian Moral and Political Theory.* Princeton, N.J.: Princeton University Press.

Kraybill, D. B., ed. 1993. *The Amish and the State.* Baltimore, Md.: Johns Hopkins University Press.

Kukathas, C. 1992. Are there any cultural rights? *Political Theory,* 20: 105–39.

Kymlicka, W. 1990. *Contemporary Political Philosophy.* Oxford: Oxford University Press.

—— 1995. *Multicultural Citizenship.* Oxford: Clarendon Press.

Laslett, P., ed. 1956. *Philosophy, Politics and Society,* 1st series. Oxford: Blackwell.

Laver, M., and Shepsle, K. 1996. *Making and Breaking Governments.* Cambridge: Cambridge University Press.

Lewin, L. 1991. *Self-Interest and Public Interest in Western Politics.* Oxford: Oxford University Press.

Little, I. M. D. 1952. Social choice and individual values. *Journal of Political Economy,* 60: 422–32.

Lukes, S. M. 1974. *Power: A Radical View.* London: Macmillan.

Macedo, S. 1995. Review of Moon 1993. *Political Theory,* 23: 389–93.

Moon, D. 1993. *Constructing Community.* Princeton, N.J.: Princeton University Press.

Mulhall, S., and Swift, A. 1992. *Liberals and Communitarians.* Oxford: Blackwell.

Olson, M. Jr. 1971. *The Logic of Collective Action.* 2nd edn. Cambridge, Mass.: Harvard University Press; originally published 1965.

OSTROM, E. 1990. *Governing the Commons.* Cambridge: Cambridge University Press.
PARKER, H. 1989. *Instead of the Dole.* London: Routledge.
POLSBY, N. W. 1980. *Community Power and Political Theory.* 2nd edn. New Haven, Conn.: Yale University Press.
POUNDSTONE, W. 1992. *Prisoner's Dilemma.* New York: Doubleday.
RAWLS, J. 1971. *A Theory of Justice.* Cambridge, Mass.: Harvard University Press.
RIKER, W. 1982. *Liberalism against Populism: A Confrontation between the Theory of Democracy and the Theory of Social Choice.* San Francisco: Freeman.
SCHLOZMAN, K. L.; VERBA, S.; and BRADY, H. E. 1995. Participation's not a paradox: the view from American activists. *British Journal of Political Science,* 25: 1–36.
SINGER, P. 1991. *Animal Liberation.* 2nd edn. London: HarperCollins; originally published 1975.
SUGDEN, R. 1986. *The Economics of Rights, Cooperation and Welfare.* Oxford: Blackwell.
—— 1989. Maximizing social welfare: is it the government's business? Pp. 69–86 in *The Good Polity,* ed. A. Hamlin and P. Pettit. Oxford: Blackwell.
TAYLOR, M. 1982. *Community, Anarchy and Liberty.* Cambridge: Cambridge University Press.
—— 1987*a*. *The Possibility of cooperation.* Cambridge: Cambridge University Press.
—— 1987*b*. Rationality and revolutionary collective action. Pp. 63–97 in *Rationality and Revolution,* ed. M. Taylor. Cambridge: Cambridge University Press.
TUCK, R. 1993. History. Pp. 72–89 in *A Companion to Contemporary Political Philosophy,* ed. R. Goodin and P. Pettit. Oxford: Blackwell.
TULLY, J. 1995. Cultural demands for constitutional recognition. *Journal of Political Philosophy,* 3: 111–32.
ULLMAN-MARGALIT, E. 1977. *The Emergence of Norms.* Oxford: Clarendon Press.
VAN PARIJS, P. 1995. *Real Freedom for All.* Oxford: Clarendon Press.
WALDRON, J. 1993. Legislation and moral neutrality. Pp. 143–67 in *Liberal Rights.* Cambridge: Cambridge University Press.
WASSERSTROM, R. 1980. On racism and sexism. Pp. 11–50 in *Philosophy and Social Issues.* Notre Dame, Ind.: Notre Dame University Press.
YOUNG, I. M. 1990. *Justice and the Politics of Difference.* Princeton, N.J.: Princeton University Press.
—— 1995. Mothers, citizenship, and independence: a critique of pure family values. *Ethics,* 105: 535–56.

Part **VII**

# Public Policy and Administration

Chapter 24

# Public Policy and Administration: An Overview

Barbara J. Nelson

CONSIDER the following political situations:

- After 27 years in jail and a tumultuous and uncertain political campaign, Nelson Mandela was elected President of South Africa. In his presidential address, as in his campaign, he called for a non-racial government in his country. In a speech full of hope and reconciliation, Mandela (1994: 338) asked the people of the world to stand by South Africa as it tackled "the challenges of building peace, prosperity, non-sexism, nonracialism, and democracy." One important part of this drama was invisible to the thousands who heard Mandela's speech. To get to this moment South Africa had to abandon its infamous system of three parliaments—for whites, coloreds and Asians (black Africans had no representation in parliament)—and replace it with a single all-South-African parliament. What was the process of constitution writing in the transition to democracy (Horowitz 1991)?
- In the last decades of the twentieth century, privately held wealth (assets less debts) has become increasingly concentrated in the United States. In 1970, the richest 1 percent of the population controlled about 25 percent of the wealth. In 1989, the richest 1 percent of the population controlled about 40 percent of the wealth (Bradsher 1995). During the same time, first welfare benefits and then social insurance benefits were cut through a combination of restrictive eligibility criteria, funding caps, taxation of benefits, and the elimination of programs. The rich got richer, and the poor, if they did not get poorer, certainly lived a more precarious existence. What combination of tax, investment, and benefit policies brought these changes about? What do increasing

disparities in wealth (and to a lesser extent income) mean for democratic governance?

- It was a hot, breezeless summer day when the Italian town of Seveso experienced one of the worst toxic waste accidents in Europe. A cloud of chemicals rained down on the town from ICMESA, a factory that produced pharmaceutical chemicals and industrial scents (Reich 1991). But when the clean-up was officially over not all the hazardous wastes, including the PCBs, were accounted for. Six years later they turned up in a slaughter house in southern France (Hilz 1992). How did the toxic waste get there? A few years earlier the U.S. military illegally exported military effluvia labeled as "cleaning fluid" to Zimbabwe. The perpetrators went to jail for fraud. But fraud is rarely the major issue. Should the more open international market in toxic waste be regulated and if so how and by whom? Within countries, where should toxic waste be stored and how, if at all, should it be transported?

These are the kinds of concerns that draw scholars, teachers, and public officials to the field of public policy. How does the field help us understand the origins and consequences of these situations? How does the accumulated knowledge and the depth of practice improve our responses to problems as crucial, yet as different, as these? This chapter analyzes the state of the discipline in public policy in partial response to these questions.

Academic fields within disciplines are like provinces within a country, the result of imperfectly remembered battles that define not only a territory and its borders, but also a turn of mind. From that perspective public policy is quite unlike the core fields, the inner Cabinet so to speak, of the discipline—quite unlike the fields organized around countries or political institutions. Instead of place or governmental function, public policy is distinguished by four intellectual imperatives: an interest in the *whole pattern of political systems and their processes*, a belief that *the consequences of governmental actions are important*, a struggle to produce *useful as well as theoretically and empirically sound knowledge*, and a conviction that *democracy matters.*

These imperatives are often contradictory, ambiguous, and unattainable. But they arose from a particular intellectual and political history and they have given shape to the conflicts as well as the contributions of the field called public policy. As such they structure the analysis and narrative of this chapter. The field is assessed in two complementary ways. The first section defines the policy process and examines the history of the policy enterprise. The second section analyzes the conflicts and findings of the literature arising from the four imperatives. These two approaches provide the perspective available through binocular vision of the same events, institutions, and ideas.

## I Organizing knowledge: definitions, structure and history

The field of public policy is American in intellectual origin in ways that are important to understanding its trajectory and contributions. This Americanism is best understood as a view of policy that assumes stable democracy and the persistence of independent platforms outside government from which scholars can analyze and criticize the directions of public actions. These assumptions exemplify American exceptionalism because few other countries have had either the stability or the separation of advice and responsibility experienced by the United States. Before World War II, élite British universities trained men for governance through exposure to the classical thinkers of the Western European tradition. During the same period, the study of political science in France and Germany focused on the proper administration of the state (Stein 1995). Thus, in Europe, the problem-solving tradition in political discourse focused primarily on state action. The same state-focused orientation was true in Japanese political science after World War II, even though the study of political science was considerably influenced by American definitions of the discipline (Inoguchi 1995). During the Cold War, the power of coercive and non-participatory communist states severely limited the development of an analytically independent political science applied to social problems (Ágh 1995; Wiatr 1995). In contrast, in Latin America, independent research units often provided advice and criticism directed at pressing problems, though their visibility—indeed their viability—waxed and waned with the rise and fall of democratic regimes (Sigal 1995). But nowhere was there quite the same combination of stable government and reliance on outside advice for problem solving as in the United States, and the field of public policy is bounded by these two often unstated assumptions. One result of these assumptions is the systematic lack of attention to normative issues including the struggle to build democratic institutions and the dilemmas of redistribution, equality, and liberty.

Historians of the development of political science in the United States have noted that political science has, by and large, defined government as a potentially good instrument of human creation, either as a check against the *demos*—the unruly masses—or as a positive force for progress.[1] Raymond Seidelman and Edward Harpham (1985) described the history of political science as the creation of a third approach to the politics and philosophy of the American experiment, an approach that was distinct from the two founding traditions. The first tradition was peopled by institutionalists like James Madison who were skeptical of human nature and thus established institutions to prevent people's baser passions from undermining governance. The second tradition was peopled by radical democrats like Thomas Paine who believed that the popular virtue cultivated in an active political community was the best safeguard for a system of good governance. Painting with a broad brush, Seidelman and Harpman (1985: 7) argued that since the progressive era "political scientists have sought a national State manned by trained experts and supported by responsible and virtuous popular democratic majorities." The field of public policy has deep roots in this "third tradition" with all of its inherent contradictions over concentrated or diffuse political power, the proper places of professionalism and popular participation, and the value of science compared to the art of politics in public problem-solving.

## A Definitions: the window to history

These contradictions are evident in the competing definitions of public policy. Interestingly, there are very few definitions of public policy *as a field* within the discipline. Lawrence Mead (1995: 1) captured the scope and sense of the field when he wrote that public policy is an "approach to the study of politics that analyzes government in the light of major public issues." Most authors move straight to the question of defining "public policy" and the "policy process." James Anderson (1990: 5) offered a representative definition when he wrote that a policy is "a purposive course of action followed by an actor or set of actors in dealing with a matter of concern. Public policies," he continued, "are those developed by governmental bodies and officials."[2]

Definitions of the policy process are more varied. Some closely link public policy with all governmental action. In this they are heirs to the *Staatswissenschaft* definitions of political science as a whole (Somit and

[1] For histories and critiques of the discipline of political science see: Crick 1959; Somit and Tanenhaus 1967; Ricci 1984; Janos 1986; Nelson 1989; Almond 1990; Farr and Seidelman 1993; Lowi 1992; 1993; Ross 1993; and Easton, Gunnell and Stein 1995; and Farr, Dryzek and Leonard 1995.

[2] Similarly, Larry Lynn (1980: 10) defined policy as "a specific set of government actions that will, by design or otherwise, produce a particular class of effects." Anderson's definition of public policy included the notion of government choosing not to act, a topic examined in more detail by Bachrach and Baratz (1963).

Tanenhaus 1967: 8). For instance, B. Guy Peters (1986: 6) wrote that "public policy is the sum of the activities of governments, whether acting directly or through agents, as it has an influence on the lives of citizens."[3]

The more prevalent definitions do not take such an encompassing view. Instead, they characterize the policy process as a set of problem-solving activities. Some definitions map *individual* problem-solving processes onto organizational settings. John Dewey (1910/1978) offered the first of these definitions, dividing public decision-making into five steps that moved from a sense of perplexity, to problem definition, to the formulation of alternative solutions, to the considerations of their implications, to experimentation with the preferred choice. Likewise, Harold Lasswell's (1971) formulation of the policy process described it as a creative decision process comprised of recommendation, prescription, invocation, application and termination.

Others in the problem-solving school draw their inspiration from *systems* theory with definitions based on inputs, transformations, and outputs (Easton 1965). For example, Garry Brewer and Peter deLeon (1983: 9, 17–21) wrote that "policy is society's most important decisions, actions backed by widespread approval and/or threat of sanctions." For them, policy is a system-level process that proceeds in six stages: initiation, estimation, selection, implementation, evaluation, and termination.[4]

Viewed outside of the narratives that animate them, these process definitions seem mechanistic in ways their authors were always at pains to disavow. But Deborah Stone (1988: 7) offered a larger critique of the stages-of-problem-solving definitions of the policy process. She criticized the "rationality project" implicit in these definitions. Speaking as much to economists engaged in policy analysis as to political scientists, Stone argued that the sequential model of the policy process

> parallels the cognitive steps of the rational model of decision making . . . The production model fails to capture what I [Stone] see as the essence of policy making in political communities: the struggle over ideas . . . Policy making . . . is a constant struggle over the criteria for classification, the boundaries of categories, and the definition of ideals that guide the way people behave.

So too, communities of ideas underpin the model of policy-making presented by Paul Sabatier and Hank Jenkins-Smith (1993) in their advocacy

[3] Likewise, Thomas Dye (1984: 1) characterized policy as "whatever governments choose to do or not to do."

[4] James Anderson (1990: 35) defined policy-making as a "sequential pattern of activity in which a number of analytically but not necessarily temporally distinct categories can be distinguished. These include problem identification and agenda setting, policy formulation, policy adoption, policy implementation, and policy evaluation."

coalition approach, where the creation of shared views and contention with other belief communities are foundational parts of the policy process. Increasingly this was the view of Aaron Wildavsky (1987: xv–xxi), who avoided defining "policy analysis" but worried about the polarization of policy élites—in effect the lack of commonalities in beliefs and structures of mutual accommodation among those charged with forging America's political future.

For all their differences, however, the three definitional traditions share one epistemologically important characteristic. All the definitions emphasize a holistic view of policy-making, a belief that the whole is greater than the sum of its parts, that individuals, institutions, interactions, and ideology all matter, even if there is notable disagreement about the proportional importance of each. If, as the old joke goes, existentialism means that *either* nothing is connected *or* everything is connected, then the vast majority of political scientists in the public policy field belong to the "everything is connected" persuasion. In the language of the structure–agency debate, most political scientists specializing in public policy see causation at and beyond the individual level.

The major dissent from this view is evident in the work of "olympian" public choice specialists, to use Herbert Simon's term in *Reason in Human Affairs* (1983). Simon and later Vincent Ostrom (1989) delineated the varieties of public choice schools. The work of scholars like William Riker and Peter Ordeshook (1973)—which assumed high levels of selfishness, complete information, a capacity to rank unambiguously all alternatives in a consistent manner, and maximization of expected utility—is less useful in solving institutionally based questions than in answering market-based questions. But the work of Elinor Ostrom (1986) and others relaxed the informational assumptions (and hence the ranking and utility assumptions) bringing questions about the relationship between individual behavior and organizational action to the fore. By and large, however, the public choice school is identified more with an interest in solutions to policy problems than with definitions of public policy as a process or a field.

## B Structure of the field

The history of the field of public policy is more the history of a discourse than of a conventional discipline or field comprised of ideas plus institutions, journals, and control of key resources. In fact the lack of these more material trappings of a field is a remarkable finding. The public policy section of the American Political Science Association was established in 1983, as part of the vertical integration of the discipline. The Policy Studies Organization, an independent group of political scientists, was established in 1971 but its lack of mechanisms for leadership succession restricted its success as a vehicle for field development.

The fragmentation of publishing in the field is emblematic of its multiple origins and the lack, for good as much as for ill, of an authoritative arbiter of ideas or approaches. For policy specialists in political science with strong disciplinary identification there is nothing equivalent to the *Public Administration Review.* The major political science journals occasionally publish policy articles and the *American Political Science Review* reports journal submissions in the field of "American Politics and Policy." But the name of this reporting category is representative of larger questions of boundaries.[5]

Several general policy journals exist (*Policy Sciences, Policy Studies Journal, Policy Studies Review* and the *Journal of Policy Analysis and Management*), but none is sponsored by an official organ of the American Political Science Association, although political scientists regularly publish in them and are often prominent in the editorial process. The lack of general journals for political scientists specializing in public policy is made up for by dozens of multidisciplinary journals on policy subjects, like *The Journal of Health Politics, Policy and Law,* which draw political scientists with substantive expertise to them. Similarly several journals focus on specific stages of the policy process, such as *Evaluation Quarterly* and the *Evaluation Studies Review Annual.*

The exchange of ideas that shape élite sensibilities occurs in yet another venue. The popular journals of cultural arbitration like *The Atlantic, The New Yorker, The Nation, Commentary, Policy Review, The Public Interest* and *Tikkun* frame the ideas of the reading élite. In analyzing the failure of health insurance reform in the United States from 1992 to 1994, Theda Skocpol (1996) demonstrated the importance of the ideas in William Kristol's policy memos as they played out in the Heritage Foundation's *Policy Review.*

Education for the public interest further fragments the field, because political science as a discipline does not routinely educate students for problem-solving. In describing the social organization of science, Don Price (1965) distinguished between the "scientific estate" whose purpose is to discover knowledge and the "professional estate" whose purpose is to

[5] A quick look at any article or book points up the multiplicity of fields a piece of scholarship might contribute to. For instance, is David Price's (1978) classic article "Policy Making in Congressional Committees: The Impact of Environmental Factors" in the legislative field because it is about Congress or in the policy field because it theorizes about the relation between policy content and political processes?

apply knowledge to solve practical social problems. In the main, liberal arts departments of political science teach about public policy at the undergraduate and graduate levels, but they do not provide education for action. One part of education for action—education for public problem-solving—has been professionalized since the 1920s and mainly occurs outside political science departments in public administration programs and policy schools (Stokes 1994; Barzelay and Kaboolian 1990; Crecine 1982).

It is evident that the field of public policy is subject to tremendous centrifugal forces. Lacking unifying institutions, the academic discourse is kept together by common training in political science, a concern for the four imperatives (holism, consequences, usefulness, and democracy), and for many people, the development of substantive areas of expertise. The strands of this academic discourse can be traced through the history of political science, but following the threads requires great care. There is more than a little temptation, especially in a short rendition, to cut out the knots and retie the threads in order to smooth over the weaving. In reality, the discursive history of the field is often tangled and ideas re-emerge, as much as they are passed on, from generation to generation.

## C History: to know the world and change it

That said, the most important thread in the history of public policy as a field is the *struggle to know the world and change it at the same time.* Scholars like John Dewey, Charles Merriam, Harold Lasswell, Herbert Simon, Charles Lindblom and Aaron Wildavsky located themselves in these struggles, even if their positions were different and their times dissimilar. The members of the field endeavor to examine and explain politics in ways that have at least a bounded generalizability (Chowdhury and Nelson 1994). But they also believe in applied knowledge and the links between experience and theory.

To look for bounded generalizations and to seek knowledge for and in application—these seem reasonably laudable goals in themselves. But they were not the only goals of political science, indeed they were often not the major goal. At least as powerful was the imperative for a pure science of politics, one that looked for universal truths and which was uncertain if not hostile toward applied knowledge.

The history of the discourse of public policy as a field is located in the contests between these approaches. *At stake were views about the primacy of science over other ways of knowing and the role of judgment in public problem-solving.* Those who supported scientism (such as William Bennett

Monroe and G. E. G. Catlin) believed that an empiricism akin to that used in physics was the best device for learning political truths. The applications that emerged from these truths were either self-revealing or someone else's job. In contrast, those with a more policy turn of mind, as well as many normative philosophers, thought that empirical methods were useful for some questions, but the range of tools available for policy analysis was much greater than an imperfectly transferred version of the scientific method. These tools included the importance of governmental experience, civic activism, and for some participation in the popular politics of social movement change. These experiences not only added to a scholar's (and a citizen's) ways of knowing about political life, but provided the settings if not the skills for honing political judgment.

Dorothy Ross (1991; 1993) reported that in the period of their establishment all of the social and behavioral sciences confronted the difficulties of simultaneously knowing the world and changing it. In the founding decades of the discipline, during the period when scientism had not taken hold fully, political scientists struggled to find links between the methods of the physical sciences and the sciences of society without giving up their roots in history and moral philosophy. Woodrow Wilson (1887/1993: 40) found little everyday difficulty connecting history, philosophical inquiry, comparative political functionalism, and the science of administration. In fact, Wilson's interest in an administrative science based on the "methods of the counting house" was based on an ethical concern about the corruption of public bureaucracies as well as the belief in the methods of business.

The separation of administration from "policy" (meant as politics in Wilson's usage) and the belief in governmental experience as a teacher of big truths was the first stance on public policy in the discipline. Its normative base has received too little attention in the histories of the discipline. To Wilson, Frank Goodnow (the first president of the American Political Science Association) and many others, developing the professional administrative capacities of the modern state was important to the nation as well as the study of political science. At every level, public bureaucracies were popularly viewed as having sunk into venality. Thus the separation of politics from administration was an invention that potentially protected the integrity of analysis and advice, as well as implementation.

But the forces of society and of discipline-building brought other views about policy into the discipline. Progressive reform made social scientists important players in the national conflicts over how the government and economy should be related. The growth of universities (and other independent groups capable of social analysis) changed the personal and social positions of political scientists, who had a base outside of government from which to establish knowledge for and about the state, a base that was consistent with the overall outsider critique of progressivism in ways not fully established when Wilson wrote his famous address "The Study of Administration" in 1887. University-based scientism, well rooted by 1920, created a changed set of standards for success in the interpretation-and-advice game.[6] Creating generalizable knowledge for its own sake became ever-more important and its importance was validated by an increasingly integrated system of higher education. By the second decade of the 20th century, political scientists as an interest group and as a profession had two sets of constituents—universities and governments—whose aims were often inconsistent.

What kind of scientism characterized political science in the inter-war period? Political scientists were more sophisticated than Baconian empiricists but remained metaphysical realists. They espoused an increasingly refined observational approach which was the heir to Baconism, mixed in with the hope for Newtonian generalizability as witnessed by the discussions on "economic man" and "political man." Those in the strict empirical wings of the discipline supported a philosophy of knowledge based in metaphysical realism (not a term they would have used to describe themselves), which argues that there is a real structure to the world, knowable with increasing accuracy through science but, known or not known, factually and truly present (Nussbaum 1992: 206; on contemporary issues of *mediated* metaphysical realism see Putnam, 1978).[7]

Charles Merriam charted one path through these forces that gave shape to the field of public policy. In his "The Present State of the Study of Politics" (1921) and *New Aspects of Politics* (1925) he sets an agenda for a political science that had human behavior at the individual level at its center, but whose method was more than mere statistical collation, and whose application took into account how group membership affected individuals and what individual behavior and belief meant to a democracy. While chroniclers of the discipline (Somit and Tanenhaus 1967: 113) concluded that Merriam "pointed the way to what would eventually become behavioralism," it was Merriam's practical aspects, his problem-solving emphasis, his stance apart from science and careerism that makes him important in the discourse of public policy. While works like *Non-Voting, Causes and Methods* by Merriam and Harold Gosnell (1924) portrayed the scientific method, though difficult to apply, as mostly neutral, there was also passion

[6] Scientism also provided a shield for advice that might be politically unpopular during the nativism and anti-communism of the post-World War I period.

[7] Metaphysical realism is a position that religious and moral philosophers can also hold. The reality of the world is established by God and perfectly knowable through the Almighty, but religious disciplines can help individuals know God's laws in the physical world.

and practicality in their conclusions. As such, Merriam escaped much of the scorn heaped on others who espoused scientism, leaving a legacy of "science and usefulness" that still shapes the field of public policy today.

Merriam is a central figure in the history of policy as a field for other reasons as well. Though very much committed to science in the service of humanity, he also strengthened the field of public administration by his service on the Brownlow Committee whose recommendations (1937) attempted to make the federal civil service more professional and managerially competent. Indeed, Donald Kettl (1993: 410) wrote of this period that the public administrators who went to Washington to help run the war effort had been trained in a " 'scientific management' approach to management that tried to elbow the politicians aside." Merriam also strove for an active populace, and his *Civic Education in the United States* (1934: 125) argued for an understanding of "propaganda, mass organization and manipulation, symbolism, and civic education itself" as well as the "prestige of rulers, custom, and . . . [the] deliberation . . . of representative bodies" if citizens were to enjoy and shape a democracy whose ultimate power was with the people.[8]

Perhaps more than any other figure in the history of the field, Merriam cherished goodwill toward knowledge from many sources, from a high science that was not the slavish imitator of methods in the natural sciences, from the knowledge learned from and applied to practice on real problems, and from a commitment to civic engagement in the name of sturdy democracy. He did not focus much on the social movement sources of public policy. Dorothy Ross reported that during this period sociology, and even economics, were stronger sources of this research than political science. It was not until the social movement revival of the 1960s that policy scholars in political science made the important distinction between policy processes with social movement rather than interest group roots (Nelson 1984; Evans and Nelson 1989; Mansbridge 1984).

Harold Lasswell, Merriam's student at the University of Chicago, and a supporter of a psychological approach to political science is another defining thinker in the history of the discourse. He was pragmatic like Dewey and convinced that science was a good servant like Merriam. His major contribution was to put the researcher as a person at the center of the policy

[8] Merriam also had a hearty respect for professionals—psychiatrists, social scientists, doctors, and social workers—who would help bring the benefits of scientific public policy to citizens. He straddles a position between belief in strong popular participation and the belief in the ameliorative and educational role of the well-educated in guiding the population. His notions of informed participation set a standard that was, and is, unrealistically high. But the stakes were high, too. In *Civic Education*, Merriam also feels the specter of an international economic depression, as well as international fascism and communism.

process in much the way Merriam put the individual as subject at the center of his work. He coined the phrase the "policy sciences" in the 1940s, made it a part of the discipline in his book *The Policy Sciences* (1951) written with Daniel Lerner, and presented a full-blown exegesis of its intellectual and organizational foundations in 1971 in his *A Pre-View of the Policy Sciences.*[9] In the same year, Yehezkel Dror published *The Design for Policy Sciences* which discussed when and how a policy sciences approach could be useful in solving real problems.

In a somewhat Habermasian way, Lasswell believed in the democratic importance of the discursive aspects of public life. He believed that good political decisions (judgments) were cultivated by discussion and experience—mostly among informed decision publics—and good judgment was as important as multidisciplinary, highly contextualized policy analysis in his approach to research. Robert Dahl and Charles Lindblom (1953/1976) jointly and severally took a more inclusive view of where the conversation that guides democracy ought to be located. In particular, Lindblom's work (1959; 1965) emphasized the immediacy and personalism of interactions among politicians and the importance of not underestimating everyday understandings of problems. Both Lasswell and Lindblom resisted the disillusionment with popular beliefs that Walter Lippmann's writings (1922; 1925) captured for progressives and which the era of screech radio and confessional television has reintroduced into American political debate.

Interestingly, where Lasswell emphasized public conversation, Dror emphasized governmental experience as the practice that guarded against a politically untethered scientism. In a now famous chart, Dror (1971: 19–23) listed policy subjects—agriculture, defense, and the like—and areas where behavioral sciences were "useless," "helpful," and "very helpful." In sum, his judgment was that decisions about metapolicy—like the shape of the new constitution in South Africa or aggregate levels of taxation—were always political. The policy sciences were most useful for middle-range problems, like demonstrating the consequences of particular parliamentary designs or estimating the economic impact of specific tax proposals. Dror's book is in large part a cautionary tale about letting technique be the vehicle for revealing public values.

Herbert Simon's work takes the opposite tack. *Administrative Behavior* (1947) represented an attempt to find out what is fundamental about humans as decision-makers. Simon wanted a science of human decision-

[9] Robert Lineberry (1982) argues that the policy sciences have a dual origin, in the democratic humanism of Harold Lasswell, and in the technical usefulness of operations theory, as brought to the war effort by Tjalling Koopmans. This point is made in deLeon and Overman (1989: 412).

making in organizations. His pathbreaking research found that individuals made circumscribed searches for information and that their decisions were based on bounded rationality. Douglas Torgerson (1995: 241) describes Simon's work as the search for ways to overcome the limits of human decision-making. "[T]he whole thrust of [Simon's] work is to find a way to coordinate the limited rationality of individuals into the organizational rationality of a more comprehensive decision making system . . . Simon stresses the potential to develop programmed decision procedures that can replace nonprogrammed ones." The emphasis on finding better political and policy solutions through technical policy analysis has many adherents. Among economists, Edith Stokey and Richard Zeckhauser's (1978) *A Primer for Policy Analysis* followed the trail that Simon blazed. Among political scientists, Robert Heineman, William T. Bluhm, Steven Peterson, and Edward Kearny were heirs to this approach in *The World of the Policy Analyst* (1990).

If Simon and others present the virtues of technical policy analysis, David Kirp (1992) gave voice to the democratic distrust that often accompanies them. Kirp noted that policy analysis on issues of welfare reform has often been quite solid, leading to solutions that, if they could get political support, would improve the lot of poor women and might eventually decrease public dependency. Long-term, intermediate solutions to unpopular problems have no political constituency, however. "Republicans," Kirp observed with prescience in early 1992, "would retail welfare, making it Willie Horton, the flag, and the ACLU all rolled into one" (1992: 695). By implication, Kirp takes a Madisonian view of the populace. Technical policy analysis is a regrettably weak but necessary dike against the flood waters of popular opinion stirred up by demagogues.

The rise of management approaches to public policy questions and, to a lesser extent, a renewed interest in civic education stand in contrast to the technocratic school that seeks to improve politics with the facts. In opposition to Woodrow Wilson's view of public administration, the public management approach seeks to understand the inseparable relation between politicians and bureaucrats. Aaron Wildavsky focused on this dilemma in *Speaking Truth to Power* (1987). Wildavsky's book, originally published in 1979, responded in part to what he saw were failures of policy analysis in Vietnam and the Great Society programs by investigating the sociology of knowledge and power. His view was that no amount of analytical torque could rachet the conflict of interests out of politics.

If public management is distinct from public administration, it is also distinct from management in the private sector. Barry Bozeman and Jeffrey Straussman (1991: 5) captured this difference by stating that, unlike private management, public management bears the "imprint of public authority." Almost all the research in the public management arena investigates how public authority manifests itself in governmental organizations. Hal Rainey (1991) and Larry Lynn (1987) combined an interest in organizational theory with lessons from public settings. They paid close attention to the importance of maintaining the support of constituencies, many of whom make incompatible demands on public organizations. The growing attention to leadership in public management also emphasizes constituencies, especially the creation of constituencies, as well as the importance of the ability to scan the environment, act quickly, and modify actions as goals are pursued (Doig and Hargrove 1990; Behn 1991). In research on the adoption of innovations, Martin Levin and M. Bryna Sanger (1994: 29–30) demonstrated that the "ready, fire, aim" approach Peters and Waterman (1982) found in successful private sector entrepreneurs applied to the public sector innovators as well.

The public management literature has not been sufficiently attentive to the problems of running large public agencies where change and continuity must coexist. Interestingly, some of the best literature on this problem focuses on the armed services. Judith Stiehm (1981; 1986) examined the origins and implementation of the policies to open the Air Force Academy to women and to make women eligible for all non-combatant positions in the Army. General H. Norman Schwarzkopf (1992) was director of personnel management for the Army when many of these policies were implemented. His autobiography described the training that the armed services—and no other public "agencies"—provide to senior staff in order to run just these kinds of organizations.

The small group of scholars interested in civic involvement in politics and policy-making take a stance against both the policy analysis approach with its emphasis on semi-automatic decision-making and the public management approach with its emphasis on professionalism. In *Strong Democracy* (1984), Benjamin Barber sought to create a better civic order through meaningful civic involvement. Unlike Charles Merriam (1934) and Thomas Reed (1930), their predecessors in the 1930s, current exponents of civic enhancement like Barber, Robert Bellah and colleagues (1985) and Sara Evans and Harry Boyte (1986) did not look to pre-collegiate or community education to make this connection between individuals, government, and the content of public life. Theirs was more a skills-for-conflict-in-a-democracy approach. For example, Harry Boyte (1992) was concerned with the over-moralization of political conflict, in part a legacy of the rise of social movement activism around the world.

The most recent development in the field of public policy is the atten tion given to the social movement sources of policy demands and policy content. Here is a subject area and wellspring of practical knowledge not much recognized in the histories of the field. Writing about Western Europe, Seymour Martin Lipset and Stein Rokkan (1967) noted that deep social conflicts gave rise to clusters of issue preferences around which European parties were formed. These social faultlines were "frozen" at the historical moment when parliamentary systems based on "universal" (often manhood) suffrage were established. In both older and newer democracies these old faultlines have begun to splinter. Among other reasons, this dealignment has come about because new constituents and new issues have not found an adequate home in the parties established at the invention of modern parliaments and because other forms of interest intermediation do not bring these conflicts to authoritative and acceptable resolution.[10]

Barbara Nelson (1984; 1995) recognized that the New Deal arrangement of economic issues within U.S. parties failed to provide a structure for introducing what were then understood to be private, and moral issues, into the policy stream. In *Making an Issue of Child Abuse*, she wrote that supporters of governmental intervention against child abuse had to create a new kind of issue, one that in the 1960s and 1970s required the "public use of private deviance." Similarly, Kristin Luker (1984) analyzed the invisibility and non-partisan nature of the early conflicts over abortion in *Abortion and the Politics of Motherhood.* The issue construction of class and regional parties in the United States did not easily absorb the public expressions of the moral economy (Scott 1976). Social movement research was frequently based on ethnographic, legal, and institutional methods (which in political science are more legitimate if conducted on élites—Supreme Court justices are a good example) and therefore brought into high relief the sociology of knowledge conflicts that have bedeviled the field.

This research also illuminated the dilemmas of serving anti-statist constituents through policy initiatives. *Why We Lost the ERA* by Jane Mansbridge (1984) and *Protest is Not Enough* by Rufus Browning, Dale Rogers Marshall, and David Tabb (1984) analyzed how social movement organizations as opposed to more conventional interest groups try to influence policy. Both books demonstrated the failure of direct action to have specific and lasting policy results, although direct action can set the parameters of the debate.

The destabilization of old policy coalitions plus the addition of policy demands from social movements are expressions of recent political

[10] See, for example: Chowdhury and Nelson 1994; Offe 1985; Dalton Flannigan and Beck 1984; Schmitter 1981.

influences on policy research. Policy research has always been influenced by the political climate. The internationalization of the world's economy, the lessening power of the nation-state, the movement away from hegemonic conflict between the superpowers, the trend toward small government in the face of a world population of over five billion, the reduction of public support for research, the rise of think-tanks, the creaking conflict between disciplines with a single epistemological base (methodological individualism in neo-classical economics) and multiple epistemological bases (most of the other social sciences)—all of these will contour the directions of the field. One of the next important political cleavages to be examined by policy scholars is likely to be "place." In politics and analysis the old *local-state-national triumvirate increasingly competes with two other spatial divisions: rural-suburban-urban and local-regional-international.*

## II What do we know? Research based on the four imperatives

What has the field of public policy contributed to the enduring body of knowledge in political science and to the repertoires of public officials, managers, citizens, and activists? Fields with unity of method and well-defined scope have a greater likelihood of creating cumulative knowledge and high theory than fields like public policy with its history of diffuse methods and subjects. So too, public policy has lacked a tradition of intellectual criticism that questions the deep structures of government and the state, a tradition that has helped the field of comparative politics develop normative and empirical theory in the face of similar problems of scope and methods. But within each of the four imperatives there is a richness of research that is both self-consciously about policy and contributes to theory and practice in other fields as well.

One measure of the diffuse quality of the field of public policy is the cottage industry in review articles and handbooks, not only summarizing the field as a whole but also reviewing its many parts and related enterprises. Most give an item by item recounting of the major writings in the field. Three are especially helpful. Peter deLeon and E. Sam Overman (1989) have written an excellent intellectual account of one part of the field in their "A History of the Policy Sciences." Donald Kettl's (1993) "Public Administration: The State of the Field" charted the relationship between the fractious siblings "political science" and "public administration." Douglas Torgerson (1995) presented a concise history of ideas about the relationship between science and practice in "Policy Analysis and Public Life: The Restoration of *Phronesis*?" Mindful of the bounty of all these resources, this assessment synthesizes the findings in the four intellectual imperatives that structure the field.

## A Holism

As a field, public policy embraces modeling the whole, but public policy is not alone in this approach. Modeling the whole is an honorable and widespread tradition in the social sciences. Most of the time this type of theorizing presents an explanation of social relations in society. Certainly interest group theories, pluralism, élite theories, and class theories have this goal in common: they seek to explain the big levers of social organization. The holism of the policy field is distinctive because the research has more concrete and circumscribed aims—to develop a single, or even several, general theories of *governmental processes,* and to a lesser extent, to embed these theories of governmental processes into larger understandings of the relations between state and society. Hugh Heclo (1972: 87) summarized the importance and the difficulty of a policy approach when he wrote that the aim of policy theory was to "become more truthful to the complexity of events."

The holistic imperative is expressed in two bodies of research, one which looks for theories of governmental functions and the other which looks for typologies of public problems from which patterns of governmental actions can be deduced. The functional approach to modeling the whole is best seen in efforts to understand the policy cycle as a complete system (May and Wildavsky 1978; Sharkansky 1970; Ranney 1968). This research stands against older traditions in political science, including legal formalism and old-style institutionalism, both of which lacked a dynamic, person-focused approach to governmental action. It is ironic, then, that the most significant contributions from research on the policy cycle have been insights into the workings of the separate stages of the policy process: agenda-setting, decision-making, implementation, evaluation, and termination.

The second body of research proceeds from Theodore Lowi's injunction that "policies determine politics" (1972: 229; cf. Lowi 1964). By this Lowi meant that different types of policies embody different types of relationships between individuals, groups, and the state and thus are characterized by different politics. In this approach the important analytical task is to determine the taxonomies of problem types and the causal theories that underpin them. This research tradition pays more attention to the links between governmental institutions and non-governmental actors than does the policy cycle approach.

One of the frontiers for policy research will be integrating the research on the policy cycle with the research on types of issues. To see the possibilities for this integration requires a whirlwind review of both the policy-stages literature and the issue-types literature, a review undertaken not to summarize the literature but to point out its directions.

## 1 The policy cycle

Policy-making rarely looks like the textbook discussions of the policy cycle. Sometimes a solution goes looking for a problem, like President Richard Nixon's need to have a foreign policy success shaping the timing of his diplomatic overtures to China (Kissinger 1979; Orfield 1975). Similarly, the content of policies is not merely determined in the decision-making phase. Rather, policy content is negotiated over and over again, in problem definition, legislation, regulation, and court decisions, and again in the decisions made by street-level bureaucrats. But even acknowledging the porous nature of the policy process, the stages of the policy process often have specific characteristics.

This is especially true during agenda-setting, the political process whereby conditions are transformed into problems.[11] The social and institutional processes that create publicly acknowledged problems worthy of governmental response vary from institution to institution. In *Agendas, Alternatives, and Public Policy*, John Kingdon (1984) demonstrated the importance of changing issue coalitions and issue enterpreneurship for Congressional agenda-setting. Nelson (1984) showed how agendas were set intergovernmentally, with Congress often being the last actor, and how the actual and symbolic content of issues shape the type of response institutions make.

But once recognized, the response to a problem is not self-evident. Most of the research by political scientists on policy decision-making tackles the problems of individual latitude and leadership within organizational settings (Neustadt 1960). Some of the theoretically richest work has looked at foreign policy decision-making. Graham Allison (1971) initiated a new era of research into policy decision-making with *The Essence of Decision.* The book is like the famous Japanese film "Rashomon" which tells four versions of the tale of a husband and wife attacked by a robber as they travel on an empty road through a forest. The husband dies and his body is found by a

[11] See, for example: McClain 1993; Cobb and Elder 1983; Polsby 1984; Downs 1972.

wood-cutter. Each version has a different character as the protagonist, and a different interpretation and causation of events. Similarly, Allison explained the events and outcome of the Cuban missile crisis from the perspectives of rational actor theory, administrative operating systems theory, and bureaucratic politics theory.[12] In ways he may not have intended, Allison's research demonstrated not only that different theoretical approaches produced different explanations of events, but also the importance of acknowledging the standpoint of one's theory (Harding 1987).

Historian Alexander George (1980) looked at the larger organization of foreign policy advice used by presidents Roosevelt, Truman, Eisenhower, Kennedy, Nixon and Carter. He assessed how cognitive style, sense of efficacy and competence, and orientation toward conflict shaped the structure of consultation. George showed that these dimensions clustered into three different structures of policy advice—formalistic, collegial, and competitive—with arguably different outcomes flowing from each structure. Not surprisingly, Franklin Roosevelt established a competitive organizational structure for policy advice, while Truman, Eisenhower and Nixon used a formalistic approach. Only Kennedy, of the presidents studied, employed a collegial approach, although Carter employed a mixed collegial and formalistic style.

The effects of theoretical assumptions and organizational form on decision-making have been investigated in domestic policy as well. Michael Thompson (1984: 336) used this approach to understand the environmental policy-making emerging from oppositional issue communities. He described decision-making for environmental policy as "examples not of decision-making under uncertainty but decision-making under contradictory certainties." Thompson's research is part of a tradition that examines how specific laws are adopted, with an emphasis on explaining how the content of legislation is affected by issue communities and by the expectations of long-term relationships. Political science has a venerable tradition of research on how contending groups of interests shape policy. Theodore Marmor (1973) recounted the story of the passage of Medicare. Paul Light (1985) investigated how the backbone of national social policy—the Social Security Act—has been revised as the coalitions that supported it have changed. Paul Peterson (1985) discussed how reform politics and professionalization created the coalitions necessary for expanded revenues and services for public schools.

The implementation literature arose from a dissatisfaction with a

[12] Allison's work has been criticized as promoting the bureaucratic politics model by not paying sufficient attention to instances when the other models predicted best. See Bendor and Hammond 1992.

decision focus in the policy field and from a recognition that, to quote the subtitle of Jeffrey Pressman and Aaron Wildavsky's book (1973), in a large, federal system "great expectations in Washington are dashed in Oakland." They argued that the complexity of joint action made it unlikely that good intentions promulgated by the Potomac would ever be adequately fulfilled in a beleaguered city by the Pacific. Another school of implementation research responded to the futility implied by Pressman and Wildavsky. Richard Elmore's article "Backward Mapping" (1982) showed the importance of planning implementation from the point of view of the deliverers. In the same vein, Michael Lipsky (1982) examined the range of discretion available to street-level bureaucrats, both as a question of implementation and of democratic practice. Eugene Bardach (1977), Robert Nakamura (1987), and Daniel Mazmanian and Paul Sabaier (1989) gave advice about how to develop robust implementation systems. But even these authors warned against the great stumbling-block in the path of implementation in a federal system, that losers at the legislative level would try to be winners in implementation. Donald Kettl (1983) described this as the great barrier reef of federalism.

Another group of scholars felt that the implementation dilemma was only resolvable by having smaller government, and thus less to implement. James Q. Wilson (1989; 1990) argued for smaller government, closer and thus more visible to the citizenry, and deregulated from centralized control. Wilson and others examined what happened to implementation when government decreased in size and was located closer to individuals. Interestingly, Wilson emphasized the gains in effective leadership that might occur from such a devolution. Emanuel Savas (1987) examined contracting for services, finding that the market exerted a good discipline on the prices of goods and services formerly provided by government. Steven Smith and Michael Lipsky (1993) were less optimistic when they analyzed the consequences of using non-profit agencies to provide public services. Originally designed to expand service and remove bureaucratic barriers, Smith and Lipsky (1993: 215) found that contracting for social services developed into a system "with extensive government intervention into the affairs of nonprofit agencies, a shift of many nonprofit agencies from the informal to formal care systems, greater homogeneity of services within particular service categories, a diminished role of the board of directors in agency governance, and destabilization among nonprofit agencies."

The shifting context of implementation as well as the lack of explicit goals makes formal evaluation of policies and programs very difficult, even though informal assessment and monitoring occur rather routinely. Military policies in Vietnam have been the subject of intense evaluation.

Both John Steinbruner (1974) and Robert McNamara (1995) concluded that in peace and war the planning and implementation of military procurement and deployment policy rested too much on faulty statistical analysis and too little on political analysis and an understanding of human decision processes under conditions of uncertainty and change. At the program level, the Great Society initiatives have been routinely evaluated, sometimes as in the case of the Seattle–Denver income maintenance experiments before the experiments were concluded.

A good deal of the methodological work on evaluation developed at the intersection of public policy and sociology. The texts by Carol Weiss (1972), Peter Rossi and Howard Freeman (1993) and Michael Patton (1992) have taught a generation of political scientists how to do program evaluations. In addition, Thomas Cook (1990) initiated research into meta-analysis of policy findings. Initially used in epidemiology to make large data sets out of a series of small ones, meta-analysis has developed into an evaluation technique that systematically examines the trends in findings from a series of studies.

In the rationalist view of the policy process, evaluations would contribute to the improvement of worthy programs and the discontinuation of unworthy ones. In point of fact, until recently, few governmental functions ever ended in the United States, leading Herbert Kaufman (1976) to ask *Are Government Organizations Immortal?* He found that of the agencies existing 1923, 94 percent had direct descendants in 1974. Some were transformed, like the horse cavalry into armored tank units (Katzenbach 1958). Others, like the Children's Bureau, went through cycles of prominence and quiescence. (Steiner 1976, Nelson 1984). William H. Starbuck and Paul S. Nystrom (1981) argued that scholars who approached Kaufman's data from a business school perspective would interpret his findings differently. Research on industrial firms counts mergers as producing only one organization. Likewise, when a firm goes out of existence and a new, similar one is founded, the second firm is not considered a lineal descendent of the first. The criteria for longevity are thus quite important in this research, as is a nuanced understanding of adaptability in systems characterized by political authority rather than market criteria.

Most of the research on termination is not called by this name, however. Rather it focuses on reducing the size of government and limiting the funding for or use of existing programs. The 1960s were not only a period of expansion of categorical grants but of increasing sophistication on the part of states in extracting funds from the federal government (Derthick 1975). In response, the political mechanisms used for termination and downsizing included the development of block grants in the 1970s, the decline in the real value of many welfare benefits in the 1980s, a second round of block grants in the 1990s, and balanced budget efforts that will stretch well into the 21st century. Block grants, which consolidate categorical grants, were first instituted in the Nixon administration to reduce cumbersome regulations arising from multiple levels of control and to lower overall funding levels. The Reagan block grant initiative was much more successful in this regard. George Peterson and his associates (1986: 7) found that "funding for the [Reagan] block grants, as enacted by Congress, and with all fiscal 1982 supplements, fell by approximately 8 percent from the 1981 categorical grant levels." Similarly, the Reagan tax cut of 1981 set the stage for the budget woes of the 1990s. Allen Schick (1994) noted that the federal tax burden as a share of GDP remained fairly constant from 1960 to 1990, at about 18.3 percent. But federal expenditures are about 3 percent higher than that. "The inadequacy of the revenue base received repeated legislative attention during the Reagan and Bush presidencies," wrote Schick (1995: 6). "Between 1982 and 1990 Congress enacted a dozen tax increases that augmented fiscal year 1990 revenue by $250 billion. But these measures were preceded by a huge tax reduction in 1981 that subtracted more than $300 billion from fiscal 1990 revenue." As a result President Clinton inherited a budget with less discretionary spending than President Reagan had at his inauguration. The resulting fight over what to cut in order to balance the budget is the newest face of program termination.

## 2 Issue typologies

The scholarship on issue typologies exists side by side with the work on the policy cycle. The focus of the issue typologies literature is not patterns of actions during stages of the policy cycle, but clusters of distinct issues each implying their own way to arrange the political relations between individuals, groups, and the state. The problem with this research is not that the typologies lack the power to explain political actions, but that there are a goodly number of them, each developed to deal with a different question of governance. The research that shows the relative power of each typology is just beginning (Ingram and Schneider 1993; Schneider and Ingram 1993).

In *The American Voter*, Campbell, Converse, Miller and Stokes (1960) discussed position and valence issues, defining position issues as those where individuals held conflicting views on what the government should do, and valence issues, a subset of the symbolic issues later posited by Murray Edelman (1988), as those uncontroversial issues which provided the glue that kept the polity from irreparable fissures. The problem with valence issues—motherhood, patriotism, and apple pie—is that they rest on social agreement about their unifying quality, an agreement that can unravel as new understandings of the problem emerge or new interests develop. The emergence of the issues of the moral economy into the lacunae of New Deal party politics is an example of such a change (Nelson 1995).

Between 1964 and 1972, Theodore Lowi elaborated a powerful typology of the issues addressed by Congress. He proposed that issues were arrayed by the proximity of governmental coercion (based on individual or environmental application) and the likelihood of coercion (remote or immediate). His extraordinarily durable typology distinguishes among distributive, regulatory, redistributive, and constituent policies. Robert Salisbury's (1968) work confirmed Lowi's unfolding typology, showing that policies permitting entities like professional organizations to be self-regulated had a different political logic than regulatory policies controled by government.

Lowi suggested that pluralist democracy only occurred when distributive policies were avoided because they encouraged mutual non-interference rather than democratic debate. Similarly, competition was most acute in regulatory policies and hence these were the policies, almost ironically, that promoted interest group formation and something close to popular participation. James Q. Wilson (1973; 1989; 1990) found the normative conclusions of Lowi's work wanting. He noted that a contest over issue positions may be good for democracy, but not for politicians who want to avoid clear winners and losers. Instead he proposed an issue typology that varied the concentration of costs and benefits and related his typology to the likelihood of groups forming. It is difficult, for example, for groups to emerge when costs are concentrated but benefits widely dispersed. The use of symbols and the creation of support in the populace as well as among those affected are among the methods to overcome the barriers to political action inherent in issues with these characteristics.

Helen Ingram and Anne Schneider (1993) were also dissatisfied with Lowi's typology. They proposed a schema that recognizes the increasing visibility of end-users of policies, an especially important consideration in the media age. The targets of public policy cluster into four ideal types: the advantaged, contenders, dependents, and deviants. These types vary by a number of characteristics including popular and élite beliefs about whether the targets are beneficial to society or a burden on it, what type of change will be made in the person by the policy, and what principle of justification supports each type of governmental action. Ingram and Schneider implied that because policy targets are socially constructed, they can be reconstructed as well. An example is the recent public debate about women on welfare that recategorizes them from dependents into deviants (Mink 1990; Reed 1992).

Scholars working in the public choice school cut the pie a different way. They compare market goods, those that are divisible, and collective goods, those that are indivisible. Some researchers like Mancur Olson (1965) and Terry Moe (1980) explained how people are induced to produce goods in which others share even when many consumers neither produce nor directly pay for them. Others, like Elinor Ostrom examined the flip-side of the collective goods problem, overuse of common assets. Her *Governing the Commons* (1990) was a landmark book in this vein. Its importance was that it demonstrated that the same problem—the tragedy of the commons—has several logics of response built into it, depending on the school of thought that analyzes it. Garrett Hardin (1968) introduced the memorable phrase "to symbolize the degradation of the environment that can be expected whenever many individuals use a scarce resource in common" (Ostrom 1990: 2). Ostrom contended that neither the state solution of total regulation nor the market solution of full privatization were the only ways to solve the overuse of common assets. She demonstrated formally and by policy examples a co-operative solution backed up with immediate, user-level sanctions. Her work is both a notable addition to the question of overuse of scarce goods needed by all, and a warning against the likelihood of finding an issue typology each of whose cells are characterized by just one political process.

How could the typology literature and policy cycle literature be integrated? Some initial steps are clear. First, the typology literature needs to be analyzed to determine the extent which the typologies overlap or are distinct. Second, the political process assumptions and consequences need to be clarified. Third, the institutional limitations of each typology need to be assessed. Most of the typologies focus on legislative activity, activity that is concentrated in the agenda-setting and decision-making stages. Is there a typology of issues around implementation that focuses on the discretion of street-level bureaucrats? Do decision-making and implementation in bureaucracies and courts imply a different set of issue typologies? Given the scope of the task, modeling the whole may still remain beyond the immediate grasp of the field.

## B Consequences

The research in the field of public policy has also emphasized the consequences of governmental actions for people. A large body of research seeks to answer the question "What happens to which people and why?" This intellectual imperative has the scope and the limits of the field in general.

The "what happens" question is usually defined in behavioral not normative terms. Ed Koch, the former mayor of New York, was famous for walking down the streets of the city asking his constituents "How am I doing?" He did not stop to engage them in conversation, however. The consequential imperative in the policy field often leads to the same emphasis on snapshot evaluations. But without this pragmatic emphasis, normative inquiries would be difficult to pursue due to insufficient information about the outcomes and impacts of governmental actions.

The imperative to consider consequences is best seen in the research on the antecedents of social spending. In 1963 Richard Dawson and James Robinson wanted to know why states varied in their social expenditures. Richard Hofferbert and David Cingranelli (below: chap. 25) reviewed the research oak that grew from the Dawson and Robinson acorn. They found that early research indicated that politics (as defined by party competition and simultaneous control of the legislature and the executive) did not matter as much as economics (as defined by income and tax levels) in explaining social expenditures.

Two partially overlapping bodies of research developed from these initial findings. The first sought to refine the models of social spending.[13] The authors working on these models had a dual motivation: their experience in studying government led them to believe that politics were not irrelevant to the distribution of governmental goods and bads, and they were unwilling to assume that there was so little connection between popular will, elected representation, and governmental activity in a democracy. They found that politics do matter, that politics are often the way that economic interests are manifested. Specifically, the cumulative weight of this research showed that depoliticized governmental structures (like city manager government) decoupled social policies from economic forces, that professionalism in legislatures increased social spending, that the content of party platforms predicted the kind of social policies parties adopted, that the nature of right-wing political opposition was more important in determining social expenditures than left-wing control, and that patterns of interest intermediation (such as corporatist versus liberal) helped to set the level of social spending. Governmental outputs in the area of social expenditures are thus a function of the deeply intertwined political and economic institutions and processes. Questions like "Are politics or economics more important" misrepresent the *nature* of the forces that influence governmental actions.

The second body of research, growing in part out of the interest in social

[13] See, for example: Lineberry and Fowler 1967; Cameron 1978; Castles 1982; Hicks and Swank 1992; and Klingemann, Hofferbert and Budge 1994.

spending, examined the origins of the welfare state, searching for the patterns of governmental response to the need to balance work, income, and family care in modern capitalist economies.[14] This research escaped from the short time-frame of the scholarship that focused on levels of spending, and in many ways it is emblematic of new directions in the policy field, directions where policy concerns are part of a larger set of questions about the relationship between state and society. Considered as a whole, this research showed that the overall level of economic wealth, the mix between industrial and other forms of production, the forms of land tenure, the extensiveness of the franchise and the staging of competition over new entrants to the electorate, the neutrality of bureaucracy, the existence of left parties and the specific nature of their constituencies, the strength of parliamentary oppositions, the prior existence and the total cost of public elementary schools, and the cultural assumptions about separate spheres for women all systematically affected the timing and the content of benefit programs. Moreover, adoption of social programs occurred in an environment where countries, provinces, and states learned from one another. Early and late adopters did not necessarily have the same characteristics (Walker 1969; Gray 1973; and Eyestone 1977). Indeed, in cross-national comparisons, late adopters of social programs often did so at *lower* levels of economic development than early adopters (Collier and Messick 1975).

The consequential imperative, especially the research on the variation and distribution of the products of governmental policies, has parallels in other substantive areas. Cross-national research by Henderson (1991) and Poe and Tate (1994) demonstrated that democratic countries engage in fewer egregious human rights violations than non-democratic ones. But the activities of democratic countries are less clearly laudable in other spheres. For example, industrial democracies tend to export their pollution, by directly selling their garbage and by encouraging the development of dirty industries in less industrialized countries that are "pollution rights abundant," to use Paul Krugman and Maurice Obstfeld's phrase (1994: 75). Such countries are "willing to tolerate an unusually high quantity of pollution relative to their supplies of other factors." Economically, the net gain to poor countries may be positive because pollution becomes a factor that can be embedded and traded in the goods of otherwise factor-poor countries. However, as Krugman and Obstfeld also noted "all this sounds economically very rational but it may sound equally highly immoral as a proposal that rich countries should 'export their pollution' to the third world."

Similar problems of the distribution of toxic waste sites exist within the

[14] See, for example: Tilly 1978; Flora and Heidenheimer 1981; Hage, Hanneman and Gargan 1989; Tarrow 1989; Esping-Andersen 1990; Nelson 1990; and Skocpol 1992.

United States. Case study research by Robert Bullard (1990) has shown, for instance, that toxic waste facilities have been disproportionately sited near minority communities (see also Commission for Racial Justice 1987; Shabecoff 1993). Most were found in the South where a combination of economic boosterism, low job mobility, and lack of Black political organization in rural areas made African-American communities particularly vulnerable. The Reverend Mac Legerton, who organized the people of Robeson County, North Carolina to resist the siting of two waste dumps, summed up the problem this way: "It is the same waste management equation that is being used all over the country . . . You take a poor, rural county, add a high minority population with historical racial, political, and economic divisions, and you have the most vulnerable community for siting of massive waste treatment facilities" (Shabecoff 1993: 240).

Not all the research on consequences looks at distribution, however. Some is driven by an interest in designing better responses (Bobrow and Dryzek 1987; deLeon 1988; Kelly 1988). For instance, Daniel Mazmanian and David Morell (1994) looked at a variety of methods for siting hazardous waste, including the concentrated storage approach so abhorred in Robeson County. In reviewing the literature, they found that single facility siting can be successful when a plant's operating procedures were open to community review. But they also found that treating waste management like a public utility with a guaranteed return on investment and non-competition—a model used in Europe—was also successful, as was equity-based regional siting that allocated all the waste produced regionally to many locations so that no community could escape from the responsibilities of waste management.

## C Useful knowledge

The impulse to design better systems for government is also part of the third imperative of the policy field: to produce useful knowledge. This imperative recognizes the social responsibilities of social scientists. Many of the central figures of the policy field—Merriam, Lasswell and Wildavsky, to name just a few—embraced this belief in their lives and writings. They supported a wide variety of ways for professionally educated political scientists to be useful, including serving on governmental commissions, acting as public intellectuals, testifying before public bodies, holding public office, and advising elected and appointed officials. Writing in 1939, Robert Lynd (1939/1967: 2–3) captured the importance of such work, noting that "the scholar-scientist is in an acute danger of being caught, in the words of one of Auden's poems, 'Lecturing on navigation while the ship is going down.' " His was not an abstract concern: "Nazi power-politics has stripped the social sciences in Germany of their intellectual freedom, while professors-in-uniform in Italy have been forced to betray their heritage by solemnly declaring the Italian population to be of Aryan origin. This is a critical time for social science" (Lynd 1939/1967: 1).

The drive to be useful and to see usefulness as one of the prime responsibilities of the academically-engaged social scientist predates Lynd's concerns. The intellectual origins of this imperative are clearly seen in the work of pragmatists like William James and John Dewey. James's famous aphorism of 1907 captures the active nature of his pragmatism: "The truth of an idea is not a stagnant property inherent in it. Truth happens to an idea. It becomes true, is made true by events" (1907/1975: 97). This quote is neither relativist nor anti-scientific, although it is sometimes portrayed as such. Rather it is a statement about the social construction of inquiry and the relationship between inquiry, evidence, and beliefs.

Notwithstanding the long history of these ideas in the social sciences, useful knowledge has an uncertain status in academic endeavors, including in the policy field in political science. In practice if not in principle, American universities have made the search for basic knowledge "unconstrained by the premature thought of practical use" a touchstone of their success (Stokes 1996). How did theory-driven basic research "win" over application? In part the question is posed too starkly. Theory won as the most important criterion for academic success in no small measure because it is useful. Those who study the creation of new knowledge know that it comes from many sources, trying to solve practical problems as well as chipping away at the limits of current theories. But there are also historical and social reasons for the primacy of basic knowledge in the academy. The Catholic monastic tradition from universities bequeathed a basic-truth approach to inquiry. C. P. Snow (1959/1988), Robert Lynd (1939/1967), and Irving Louis Horowitz and James E. Katz (1975) saw the class implications of this legacy. Writing from different political persuasions, Snow and Lynd argued that theoretical knowledge was (supposedly) disinterested, an aristocratic and pre-industrial view, where useful knowledge helped the bourgeoisie maintain its new found wealth and access to political power. Where Snow and Lynd worried that applied knowledge did not get enough attention within universities, Horowitz and Katz reported that applied social sciences might be too powerful in political life. In bourgeois democracy the results from applied social research could become the handmaiden of entrenched interests. Thus it was not that all theory "won" as the primary measure of academic success, but that a certain type of theory flourished in universities—that derived from curiosity rather than necessity, sheltered from political interference, and buoyed by independent peer review.

It is clear that there are strong social forces mitigating against political scientists interested in policy who want to be directly useful. Virtually everyone who has written about this imperative begins with a recognition that political and scientific life have different rules, procedures, ends, and rhythms. Unambiguous findings, rare enough in themselves, are not automatically persuasive. Henry Kissinger (1979: 39) wrote that "before I served as a consultant to Kennedy, I had believed, like most academicians, that the process of decision-making was largely intellectual and that all one had to do was to walk into the President's office and convince him of the correctness of one's views. This perspective I soon realized is as dangerously immature as it is widely held." Likewise, Charles E. Lindblom and David K. Cohen (1979: 1) opined that "in public policy-making, many suppliers and users of social research are dissatisfied, the former because they are not listened to, the later because they do not hear much they want to listen to."

It is easiest for political scientists to be useful when they are either occasional visitors or full-time participants in the political arena. In the 1970s an estimated 10,000 people testified before Congress each year, many of them social scientists. In the 1980s, there were approximately 10,000 federal civil servants with PhDs in social science disciplines. For those people who work for any branch of government and then go to work as lobbyists, it is the substantive and procedural knowledge gained in government service, not the connections, that most helps them do their jobs (Salisbury and Johnson 1989).

The hard part of being useful is making a sustained difference on the basis of scholarly research. Public officials are uncertain about the practical and scientific value of the social sciences, an uncertainty reflected by their equivocal place in the National Science Foundation. Excluded from the original legislation, the social and behavioral sciences face an uncertain future in an era where significant cuts in federal spending are expected. Furthermore, James Smith (1991) and William Lunch (1987) have argued that ideas not evidence are the mainstay of political decision-making in an increasingly nationalized and media-driven political system. Universities are no longer thought of as the sole source of analysis done outside government. They compete with think-tanks, which have become increasingly identified with particular belief-systems. Paul Light (1993) found that the capacity of the federal government to provide internal analysis was diminishing as well, with the balance of resources going toward *post hoc* analysis rather than formative evaluation.

As early as 1949, Robert Merton called for research on how to make applied social research more effective. Richard Nathan's *Social Science in Government: Uses and Misuses* (1989) provided a theoretically informed blueprint for accomplishing Merton's aims. Nathan argued that policy-makers had to play their part by being genuinely interested in the questions being asked, sufficiently uncertain about the answers that they want information, and willing to wait for the results. Under these circumstances, Nathan argued the case for demonstration projects with random assignment of participants, and evaluation studies with careful attention to institutional variables.

The prescriptive nature of this literature applies to research methods as well as research design. Some of the writing is directed at making scholars better at drawing political conclusions. T. Alexander Smith (1989) looked at how short-term time horizons affected policy-making. Richard Neustadt and Ernest May (1986) argued that it was important for policy analysts to be able to "think in time," that is understand how the characteristics of political eras affect patterns of political decision-making. Stuart Nagel (1990) and Richard Rose (1993) argued for the development of skills that encourage researchers and public officials to know what is contextual and what is transferable in the lessons of scholarship and politics. Rose proposed seven hypotheses about promoting wide-scale adoption of the lessons learned from programmatic experience or academic research. Lessons are more transferable when they are based on a limited number of unique elements; when programs can be delivered by more than one kind of institution; when diffusion occurs in a system of relatively equal resources; and when innovations have simple structures of cause and effect, require a small scale of change, are interdependent on other authorizing institutions, and embody widely held values.

But not all analysts see the value of getting experts and expert knowledge into policy-making. Just as David Kirp warned against depending too much on the populace for good policy, other scholars have warned against the power of experts in policy-making. Deborah Stone (1993) and Andrew Polsky (1991) saw the expansion of clinical expertise in the application of policies as a way to control individuals and promote larger normative agendas. Stone used the regulation of drinking during pregnancy as a case in point. She discussed the implications of the fetal alcohol syndrome, a clinical diagnosis, to public policy. She argued that the restrictions against maternal drinking during pregnancy, which occasionally included incarceration, were often more stringent than the penalties for drunkenness that causes great harm, including the death of children by drunk drivers. Stone (1993: 61) concluded that "some people would hold mothers to a much higher standard of care than other members of society," a kind of expertise gone awry.

## D Democracy matters

Stone's concern about how much and what kind of governmental intrusion citizens experience is linked to the fourth imperative: democracy matters. Indeed, all of the other imperatives—holism, the importance of the consequences of governmental actions, and the drive for useful knowledge—contribute to the democratic humanism that Lasswell felt best described the policy endeavor. Lasswell did not merely subscribe to Winston Churchill's quip that democracy was the worst form of government until one considered all others, but urged policy scientists to examine the ways that policies promote human dignity. In *A Pre-view of Policy Sciences* (1971: 42–3), he argued that it was important to *specify* how human dignity could be enhanced by attention to welfare, affection, respect, power, wealth, enlightenment, skill and rectitude.

Lasswell took a hopeful view of human nature, public participation, and political judgment. But others, equally committed to democracy, were more cautious. David Garson (1981) wrote that there are two schools of policy research, "empirico-analytic" and "neo-pluralist." While economists tend to belong to the first and political scientists belong to the second, the categories are by no means mutually exclusive. Peter deLeon and E. Sam Overman (1989) recognized the methodological and epistemological differences between the schools, with the empirico-analysts being associated with rigor and the neo-pluralists being associated with relevance. Although it is easy to overstate the differences between the two schools, they do have different implications for democracy. The analysts believe knowledge leads directly to wisdom, and as such they emphasize the role of professional training and research in the policy process. The democrats believe that there is a crucial stage of political judgment between knowledge and wisdom, and as such they value the political arts as well as the other activities. In observing the rigor versus relevance debate, deLeon and Overman (1989: 434) worried about the loss of attention to political judgment in policy-making in the United States: "Instead of being the final arbiter of policy issues the policy sciences have come to contribute to a polity of rational ideologies."

But for all of its importance and controversy, the democratic imperative is also the most diffuse, not only connecting to the other imperatives but to studies of state formation and to public ethics. Slowly the field is attending to the place of government within the state and society, and to the ways that state formation and cultural constructs set the stage for policy-making (Evans, Rueschemeyer and Skocpol 1985). Much of this research derives from the remarkable number of regime transformations in the last quarter-century, especially transitions from military and state socialist regimes to democracies. For instance, Guillermo O'Donnell (1973) and Sonia Alverez (1990) discussed the legacy of militarism and authoritarianism within bureaucracies on later efforts at democratic policy-making in Latin America. Similarly, John Rohr (1986) examined the legitimacy of the administrative state in the United States, asking thorny questions about how bureaucrats resolve competing claims from the politicians who tell them what to do and the public who consumes their goods and services.

Questions of personhood and citizenship have come to the fore through this attention to the state, and not merely in situations like South Africa, where the new constitution needed to give formal representation to the vast majority of the population who are black. The male model of work from which social benefits are derived and thus the uncertain legitimacy of public support for those who take care of children has been a major arena of scholarship on citizenship in North America and Western Europe (Pateman 1986). The legal personhood of the fetus is being debated worldwide. In some cases, like the United States, tolerance of religious differences and therefore the pluralist compact were at the core of the debate (Nelson and Carver 1994). In others, like Argentina, the conflict was over religiosity versus secularism (Feijoo 1995). Perhaps the most dramatic clash occurred during the reunification of Germany, where conflicts over abortion between East Germans and West Germans were so strong that they delayed formal constitutional integration. At the policy level, the clash was not between the godless communists who wanted abortion on demand and the pious westerners who wanted constitutional recognition for fetal personhood. Rather, the conflict was about the quality of the German abortion law, both how it was connected to Germany's Nazi past and its democratic present. Abortion is illegal in Germany, except when a woman legitimately invokes one of four compelling reasons (called indications). Jeremiah Reimer (1993: 168) wrote that "East Germans . . . were by no means eager to give up their more liberal abortion law. Polls also showed that most West Germans favored either a liberal *Fristenlösung* ['term' or 'periodic' model, roughly comparable to the trimester standard of *Roe v. Wade*] or complete decriminalization of abortion; only a minority agreed with the *Indikationslösung* [indications] of 1975 or wanted something more severe." If anything, the debate was about whether the majority position against fetal personhood in the first trimester could be recognized as the basis of the law, but everyone thought that the stakes for the fetus, for women, and for society were very high.

The connection between the state, citizenship, personhood and democracy is also evident in the attention to ethics in policy-making. Some like Joseph Carens (1995), Joan Tronto (1993) and Amy Gutmann (1987) connected the long tradition of moral philosophy, especially about the nature of the claims of membership, to pressing policy problems. Others, like Henry Aaron, Thomas Mann and Timothy Taylor (1994) and Amy Gutmann and Dennis Thompson (1984), examined the ways the social values of public life—including equality, liberty, compliance, freedom, and opportunity—conflict in practice. The volume edited by Aaron, Mann, and Taylor is especially interesting because it confronts the divide between economists and political scientists over the place of values in policy-making. In that volume, Jane Mansbridge (1994: 148–9) wrote that "we underestimate the prevalence of altruism if we count as altruistic only behavior that is demonstrably opposed to self-interest . . . Without what the framers called virtue (and I here call public spirit) . . . elections would not be possible, regulatory compliance would dwindle, and Congress would self-destruct." In another vein, Terry Cooper (1994) and Louis Gawthrop (1984) helped public employees think about the actions they might take when faced with conflict of interests, whistle-blowing, and public agencies characterized by shortages ranging from inadequate welfare payments to insufficient staff.

## III Conclusions

The uncertainties of making policy in an ethical manner provide a good way to conclude the chapter. These ethical uncertainties embody the twin aspirations of Americans toward government and its activities. Americans want good government and fear bad government, a duality rooted in American history. The challenge of democratic life is to provide opportunities for the first and safeguards against the second. The analytical and practical attention to holism, consequences, and useful knowledge in the field of public policy are proven aids in accomplishing both tasks. To become better at these tasks, policy scholars must give more attention to the relationship between government, the state, and society and to the normative issues that American exceptionalism hides. This research will improve the field's contribution to understanding the world and changing it, its central goal.

## Acknowledgments

A number of people contributed generously to this chapter through their comments or conversation. For their contributions, I would like to thank Philip M. Burgess, Kathryn A. Carver, Sara M. Evans, Robert Kudrle, Catherine E. Rudder, and Donald E. Stokes. For research assistance I would like to thank Rachel Garlin and Leila Hussein. My thanks also go to Deborah C. Miller and Laurie Krooss for assistance in producing the manuscript.

## References

Aaron, H. J.; Mann, T. E.; and Taylor, T., eds. 1994. *Values and Public Policy.* Washington, D.C.: Brookings Institution.

Ágh, A. 1995. The emergence of the "Science of Democracy" and its impact on the democratic transition in Hungary. In Easton, Gunnel and Stein 1995: 197–216.

Allison, G. 1971. *Essence of Decision: Explaining the Cuban Missile Crises.* Boston: Little, Brown.

Almond, G. A. 1990. *A Discipline Divided: Schools and Sects in Political Science.* Newbury Park, Calif.: Sage.

Alverez, S. E. 1990. *Engendering Democracy in Brazil: Women's Movements in Transition Politics.* Princeton, N.J.: Princeton University Press.

Anderson, J. E. 1990. *Public Policymaking.* Boston: Houghton Mifflin.

Bachrach, P., and Baratz, M. S. 1963. Decisions and non-decisions: an analytic framework. *American Political Science Review,* 57: 632–42.

Barber, B. 1984. *Strong Democracy.* Berkeley: University of California Press.

Bardach, E. C. 1977. *The Implementation Game.* Cambridge, Mass.: MIT Press.

Barzelay, M., and Kaboolian, L. 1990. Structural metaphors and public management education. *Journal of Policy Analysis and Management,* 9: 599–610.

Behn, R. D. 1991. *Leadership Counts: Lessons for Public Managers from the Massachusetts Welfare, Training, and Employment Program.* Cambridge, Mass.: Harvard University Press.

Bellah, R. N.; Madsen, R.; Sullivan, W. M.; Swidler, A.; and Tipton, S. M. 1985. *Habits of the Heart: Individualism and Commitment in American Life.* Berkeley: University of California Press.

Bendor, J., and Hammond, T. H. 1992. Rethinking Allison's models. *American Political Science Review,* 86: 301–22.

Berger, S., ed. 1981. *Organizing Interests in Western Europe.* Cambridge: Cambridge University Press.

Bobrow, D., and Dryzek, J. 1987. *Policy Analysis by Design.* Pittsburgh: University of Pittsburgh Press.

Bozeman, B., and Straussman, J. D. 1991. *Public Management Strategies.* San Francisco: Jossey Bass.

Boyte, H. C. 1992. The pragmatic ends of popular politics. Pp. 340–55 in *Habermas and the Public Sphere,* ed. C. Calhoun. Cambridge, Mass.: MIT Press.

Bradsher, K. 1995. Gap in wealth in U.S. called widest in west. *New York Times.* April 17: A1, D4.

Braybrooke, D., and Lindblom, C. E. 1963. *A Strategy of Decision.* New York: Free Press.

Browning, R. P.; Marshall, D. R.; and Tabb, D. H. 1984. *Protest Is Not Enough: The Struggle of Blacks and Hispanics for Equality in Urban Politics.* Berkeley: University of California Press.

Brownlow Committee (The President's Committee on Administrative Management). 1937. *Report of the President's Committee: Administrative Management in the Government of the United States.* Washington, D.C.: Government Printing Office.

Brewer, G. D., and deLeon, P. 1983. *Foundations of Policy Analysis.* Homewood Ill.: Dorsey Press.

Bullard, R. D. 1990. *Dumping in Dixie: Race, Class and Environmental Quality.* Boulder, Colo.: Westview Press.

Cameron, D. R. 1978. The expansion of the public economy: a comparative analysis. *American Political Science Review,* 72: 1243–61.

Campbell, A.; Converse, P.; Miller, W.; and Stokes, D. 1960. *The American Voter.* New York: Wiley.

Carens, J., ed. 1995. *Is Quebec Nationalism Just?* Montreal: McGill-Queens University Press.

Castles, F. G. 1982. The impact of parties on public expenditures. Pp. 21–96 in *The Impact of Parties: Politics and Policies in Democratic Capitalist States,* ed. F. G. Castles. Beverly Hills, Calif.: Sage.

Chowdhury, N., and Nelson, B. J. 1994. Research design and practice: methodological issues in feminist comparative research. Pp. 25–41 in *Women and Politics Worldwide,* ed. B. J. Nelson and N. Chowdhury. New Haven, Conn.: Yale University Press.

Cobb, R. W., and Elder, C. D. 1983. *Participation in American Politics: The Dynamics of Agenda-Building.* 2nd edn. Baltimore, Md.: Johns Hopkins University Press.

Collier, D., and Messick, R. 1975. Prerequisites versus diffusion: testing alternative explanations of social security adoption. *American Political Science Review,* 69: 1296–315.

Commission for Racial Justice. 1987. *Toxic Waste and Race in the United States: A National Report on the Racial and Socioeconomic Characteristics of Communities with Hazardous Waste Sites.* New York: United Church of Christ.

Cook, T. 1992. *Meta-Analysis for Explanation.* New York: Russell Sage.

Cooper, T. L., ed. 1994. *Handbook of Administrative Ethics.* New York: Marcel Drekker.

Crecine, J. P. ed. 1982. *The New Educational Programs in Public Policy.* Greenwich, Conn.: JAI Press.

Crick, B. R. 1959. *The American Science of Politics: Its Origins and Conditions.* Berkeley, Calif.: University of California Press.

Dahl, R. A., and Lindblom, C. E. 1976. *Politics, Economics, and Welfare.* Chicago: University of Chicago Press; originally published in 1953.

Dalton, R. J.; Flanagan, S. C.; and Beck, P. A. 1984. Electoral change in advanced industrial democracies. Pp. 3–22 in *Electoral Change in Advanced Industrial Democracies: Realignment or Dealignment?,* ed. R. J. Dalton, S. C. Flanagan, and P. A. Beck. Princeton, N.J.: Princeton University Press.

Dawson, R., and Robinson, J. 1963. Interparty competition, economic variables, and welfare policies in the American states. *Journal of Politics,* 23: 265–89.

deLeon, P. 1988. *Advice and Consent.* New York: Russell Sage.

—— and Overman, E. S. 1989. A history of the policy sciences. In Rabin, Hildreth and Miller 1989: 405–42.

Derthick, M. 1975. *Uncontrollable Spending for Social Service Grants.* Washington, D.C.: Brookings Institution.

Dewey, J. 1978. *How We Think.* Vol. vi, pp. 177–356 in *John Dewey: The Middle Works,* ed. J. B. Boydston. Carbondale: Southern Illinois University Press; originally published in 1910.

Doig, J. W., and Hargrove, E. C., eds. 1990. *Leadership and Innovation.* Abridged edn. Baltimore, Md.: John Hopkins University Press.

Downs, A. 1972. Up and down with ecology—the issue attention cycle. *Public Interest*, 32: 38–50.

Dror, Y. 1971. *Design for Policy Sciences*. New York: American Elsevier.

Dye, T. D. 1984. *Understanding Public Policy*. 5th edn. Englewood Cliffs, N.J.: Prentice-Hall.

Easton, D. 1965. *A Systems Analysis of Political Life*. Chicago: University of Chicago Press.

—— Gunnell, J. G.; and Stein, M. B., eds. 1995. *Regime and Discipline: Democracy and the Development of Political Science*. Ann Arbor: University of Michigan Press.

Edelman, M. J. 1988. *Constructing the Political Spectacle*. Chicago: University of Chicago Press.

Elmore, R. 1982. Backward mapping: implementation research and policy decisions. Pp. 18–35 in *Studying Implementation: Methodological and Administrative Issues*, ed. W. Williams. Chatham, N.J.: Chatham House Publishing.

Esping-Andersen, G. 1990. *The Three Worlds of Welfare Capitalism*. Cambridge: Polity Press.

Evans, P. B; Rueschemeyer, D.; and Skocpol, T., eds. 1985. *Bringing the State Back In*. New York: Cambridge University Press.

Evans, S. M., and Boyte, H. C. 1986. *Free Spaces*. New York: Harper and Row.

—— and Nelson, B. J. 1989. *Wage Justice*. Chicago: University of Chicago Press.

Eyestone, R. 1977. Confusion, diffusion, and innovation. *American Political Science Review*, 71: 441–7.

Farr, J., and Seidelman, R., eds. 1993. *Discipline and History*. Ann Arbor: University of Michigan Press.

—— Dryzek, J. S.; and Leonard, S. T., eds. 1995. *Political Science in History*. New York: Cambridge University Press.

Feijoo, del Carmen, M. 1995. The new Argentine constitution: gender issues and consolidation democracy. Paper delivered at Radcliffe College, Cambridge, Mass.

Flora, P., and Heidenheimer, A., eds. 1981. *The Development of Welfare States in Europe and North America*. New Brunswick, N.J.: Transaction Books.

Garson, G. D. 1981. From policy science to policy analysis: a quarter century of progress? *Policy Studies Journal*. 535–45.

Gawthrop, L. 1984. *Public Sector Management, Systems, and Ethics*. Bloomington: Indiana University Press.

George, A. L. 1980. *Presidential Decisionmaking in Foreign Policy: The Effective Use of Information and Advice*. Boulder, Colo.: Westview Press.

Gordon, L., ed. 1990. *Women, the State, and Welfare*. Madison: University of Wisconsin Press.

Gray, V. 1973. Innovation in the states: a diffusion study. *American Political Science Review*, 67: 1173–85.

Gutmann, A. 1987. *Democratic Education*. Princeton, N.J.: Princeton University Press.

—— and Thompson, D. eds. 1984. *Ethics and Politics: Cases and Comments*. Chicago: Nelson-Hall.

Hage, J.; Hanneman, R.; and Gargan, E. T. 1989. *State Responsiveness and State Activism: An Examination of the Social Forces and State Strategies that Explain the Rise in Social Expenditures in Britain, France, Germany, and Italy; 1870–1968*. London: Unwin Hyman.

Hardin, G. 1968. The tragedy of the commons. *Science*, 162: 1243–8.

Harding, S., ed. 1987. *Feminism and Methodology*. Bloomington: Indiana University Press.

Heclo, H. H. 1972. Review article: policy analysis. *British Journal of Political Science*, 2: 83–108.

HEINEMAN, R. A.; BLUHM, W. T.; PETERSON, S. A.; and KEARNY, E. N. 1990. *The World of the Policy Analyst.* Chatham, N.J.: Chatham House Publishers.

HENDERSON, C. 1991. Conditions affecting the use of political repression. *Journal of Conflict Resolution,* 35: 120–42.

HICKS, A., and SWANK, D. H. 1992. Politics, institutions, and welfare spending in industrialized democracies. *American Political Science Review,* 86: 658–74.

HILZ, C. 1992. *The International Toxic Waste Trade.* New York: Van Nostrand Reinhold.

HOROWITZ, D. L. 1991. *A Democratic South Africa? Constitutional Engineering in a Divided Society.* Berkeley: University of California Press.

HOROWITZ, I., and KATZ, J. 1975. *Social Science and Public Policy in the United States.* New York: Praeger.

INGRAM, H., and SCHNEIDER, A. 1993. Constructing citizenship: the subtle messages of policy design. Pp. 68–94 in *Public Policy for Democracy,* ed. H. Ingram and S. R. Smith. Washington, D.C.: Brookings Institution.

INOGUCHI, I. 1995. Democracy and the development of political science in Japan. In Easton, Gunnell, and Stein 1995: 269–94.

JAMES, W. 1975. *Pragmatism.* Cambridge, Mass.: Harvard University Press; originally published in 1907.

JANOS, A. C. 1986. *Politics and Paradigms: Changing Theories of Social Change in Social Science.* Stanford, Calif.: Stanford University Press.

KATZENBACH, E. L. Jr. 1958. The horse cavalry in the twentieth century: a study in policy response. *Public Policy,* 8: 120–49.

KAUFMAN, H. 1976. *Are Government Organizations Immortal?* Washington, D.C.: Brookings Institution.

KELLY, R. M., ed. 1988. *Promoting Productivity.* London: Macmillan.

KETTL, D. F. 1983. *The Regulation of American Federalism.* Baton Rouge: Louisiana State University Press.

—— 1993. Public administration: the state of the field. Pp. 407–28. in *Political Science: The State of the Discipline, II.* ed. A. W. Finifter. Washington, D.C.: American Political Science Association.

KINGDON, J. 1984. *Agendas, Alternatives, and Public Policies.* New York: HarperCollins.

KIRP, D. L. 1992. The end of policy analysis: with apologies to Daniel (*The End of Ideology*) Bell and Francis (*The End of History*) Fukuyama. *Journal of Policy Analysis and Management,* 11: 693–6.

KISSINGER, H. A. 1979. *The White House Years.* New York: Little, Brown.

KLINGEMANN, H.-D.; HOFFERBERT, R. I.; and BUDGE, I. 1994. *Parties, Policies, and Democracy.* Boulder, Colo: Westview.

KRUGMAN, P. R., and OBSTFELD, M. 1994. *International Economics: Theory and Policy.* 3rd edn. New York: HarperCollins.

LASSWELL, H. D. 1971. *A Pre-view of Policy Sciences.* New York: American Elsevier.

LERNER, D., and LASSWELL, H. D., eds. 1951. *The Policy Sciences.* Stanford, Calif.: Stanford University Press.

LEVIN, M. A., and SANGER, M. B. 1994. *Making Government Work: How Entrepreneurial Executives Turn Bright Ideas Into Real Results.* San Francisco: Jossey-Bass.

LIGHT, P. C. 1985. *Artful Work: The Politics of Social Security Reform.* New York: Random House.

—— 1993. *Monitoring Government: Inspectors General and the Search for Accountability.* Washington, D.C.: Brookings Institution.

LINDBLOM, C. E. 1959. The science of muddling through. *Public Administration Review,* 19: 79–88.

Lindblom, C. E. 1965. *The Intelligence of Democracy.* New York: Free Press.
—— and Cohen, D. K. 1979. *Usable Knowledge.* New Haven, Conn.: Yale University Press.
Lineberry, R. L. 1982. Policy analysis, policy sciences and political science. Paper presented to the American Political Science Association, Chicago, IL.
—— and Fowler, E. P. 1967. Reformism and public policies in American cities. *American Political Science Review,* 61: 701–17.
Lippmann, W. 1922. *Public Opinion.* New York: Harcourt, Brace.
—— 1925. *The Phantom Public.* New York: Harcourt, Brace.
Lipset, S. M., and Rokkan, S. 1967. Cleavage structures, party systems, and voter alignments: An introduction. Pp. 1–64 in *Party Systems and Voter Alignments,* ed. S. M. Lipset and S. Rokkan. New York: Free Press.
Lipsky, M. 1982. *Street Level Bureaucracy.* New York: Russell Sage.
Lowi, T. J. 1964. American business, public policy, case-studies and political theory. *World Politics,* 16: 677–715.
—— 1972. Four systems of policy, politics, and choice. *Public Administration Review,* 32: 298–310.
—— 1979. *The End of Liberalism: The Second Republic in the United States.* 2nd edn. New York: Norton.
Luker, K. 1984. *Abortion and the Politics of Motherhood.* Berkeley: University of California Press.
Lunch, W. M. 1987. *The Nationalization of American Politics.* Berkeley: University of California Press.
Lynd, R. S. 1967. *Knowledge for What? The Place of Social Science in American Culture.* Princeton, N.J.: Princeton University Press; originally published in 1939.
Lynn, L. E., Jr. 1980. *Designing Public Policy: A Casebook on the Role of Policy Analysis.* Santa Monica, Calif.: Goodyear.
—— 1987. *Managing Public Policy.* Boston: Little, Brown.
McClain, P. D., ed. 1993. *Minority Group Influence.* Westport, Conn.: Greenwood.
McNamara, R. 1995. *In Retrospect: The Tragedy and Lessons of Vietnam.* New York: Times Books.
Mandela, N. 1994. Inaugural address. P. 338 in *Facts on File.* New York: Facts on File.
Mansbridge, J. 1984. *Why We Lost the ERA.* Chicago: University of Chicago Press.
—— 1994. Public spirit in political systems. In Aaron, Mann, and Taylor 1994: 146–72.
Marmor, T. R. 1973. *The Politics of Medicare.* Chicago: Aldine.
May, J., and Wildavsky, A., eds. 1978. *The Policy Cycle.* Beverly Hills, Calif.: Sage.
Mazmanian, D., and Morell, D. 1994. The "NIMBY" syndrome: facility siting and the failure of democratic discourse. Pp. 233–49 in *Environmental Policy in the 1990,* ed. N. J. Vig and M. E. Kraft. 2nd edn. Washington, D.C.: Congressional Quarterly Press.
—— and Sabatier, P. 1989. *Implementation and Public Policy.* Lanham, Md.: University Press of America.
Mead, L. M. 1995. Public policy: vision, potential, limits. *Policy Currents,* February: 1–4.
Merriam, C. E. 1921. The present state of the study of politics. *American Political Science Review,* 15: 173–85.
—— 1925. *New Aspects of Politics.* Chicago: University of Chicago Press.
—— 1934. *Civic Education in the United States.* New York: Scribner's.
—— and Gosnell, H. F. 1924. *Non-Voting, Causes and Methods of Control.* Chicago: University of Chicago Press.
Merton, R. K. 1949. The role of applied social science in the formation of policy: a research memorandum. *Philosophy of Science,* 16: 161–81.

Mink, G. 1990. The lady and the tramp: gender, race and the origins of the American welfare state. In Gordon 1990: 92–122.
Moe, T. M. 1980. *The Organization of Interests.* Chicago: University of Chicago Press.
Nagel, S. S. 1990. Policy theory and policy studies. *Policy Studies Journal*, 18: 1046–57.
Nakamura, R. 1987. The textbook policy process and implementation research. *Policy Studies Review*, 7: 142–54.
Nathan, R. P. 1989. *Social Science in Government.* New York: Basic.
Nelson, B. J. 1984. *Making an Issue of Child Abuse.* Chicago: University of Chicago Press.
—— 1989. Women and knowledge in political science: texts, histories, and epistemologies. *Women and Politics*, 9: 1–25.
—— 1990. The origins of the two-channel welfare state: Workmen's Compensation and Mothers' Aid. In Gordon 1990: 123–52.
—— 1995. How issues of family violence have changed the political landscape: "the moral economy" and political parties in the United States. Paper delivered to the Family Violence Seminar, Children's Hospital and Harvard University, Boston, Mass.
—— and Carver, K. A. 1994. Many voices but few vehicles: The consequences for women of weak political infrastructure in the United States. In Nelson and Chowdhury 1994: 737–57.
—— and Chowdhury, N., eds. 1994. *Women and Politics Worldwide.* New Haven, Conn.: Yale University Press.
Neustadt, R. E. 1960. *Presidential Power.* New York: Wiley.
—— and May, E. R. 1986. *Thinking in Time: The Uses of History for Decision Makers.* New York: Free Press.
Nussbaum, M. 1992. Human functioning and social justice: in defense of Aristotelian Essentialism. *Political Theory*, 20: 202–46.
O'Donnell, G. 1973. *Modernization and Bureaucratic-Authoritarianism: Studies in South American Politics.* Berkeley: University of California Press.
Offe, C. 1985. New social movements: challenging the boundaries of institutional politics. *Social Research*, 52: 823–68.
Olson, M. 1965. *The Logic of Collective Action.* Cambridge, Mass.: Harvard University Press.
Orfield, G. 1975. *Congressional Power.* New York: Harcourt, Brace, Jovanovich.
Ostrom, E. 1986. Multiorganizational arrangements and coordination: an application of institutional analysis. Pp. 495–510 in *Guidance, Control and Evaluation in the Public Sector*, ed. F.-X. Kaufmann, G. Majone, and V. Ostrom. Berlin and New York: Walter de Gruyter.
—— 1990. *Governing the Commons.* New York: Cambridge University Press.
Ostrom, V. 1989. Some developments in the study of market choice, public choice, and institutional choice. In Rabin, Hildreth and Miller 1989: 861–82.
Pateman, C. 1986. The patriarchal welfare state. Pp. 231–60 in *Democracy and the Welfare State*, ed. A. Guttman. Princeton, N.J.: Princeton University Press.
Patton, M. 1990. *Qualitative Evaluation and Research Methods.* Newbury Park, Calif.: Sage.
Peters, B. G. 1986. *American Public Policy.* 2nd edn. Chatham, N.J.: Chatham House.
Peters, T., and Waterman, R. 1982. *In Search of Excellence.* New York: Harper-Collins.
Peterson, G. E.; Bovbjerg, R. R.; Davis, B. A.; Davis, W. G.; Durman, E. C.; and Gullo, T. A. 1986. *The Reagan Block Grants: What Have We Learned?* Washington, D.C.: Urban Institute.
Peterson, P. E. 1985. *The Politics of School Reform, 1870–1940.* Chicago: University of Chicago Press.
Poe, S. C., and Tate, C. N. 1994. Repression of personal integrity in the 1980s: a global analysis. *American Political Science Review*, 88: 853–72.

Polsby, N. W. 1984. *Political Innovation in America.* New Haven, Conn.: Yale University Press.

Polsky, A. J. 1991. *The Rise of the Therapeutic State.* Princeton, N.J.: Princeton University Press.

Pressman, J. L., and Wildavsky, A. 1973. *Implementation.* Berkeley: University of California Press.

Price, D. E. 1978. Policy making in congressional committees: the impact of environmental factors. *American Political Science Review,* 72: 548–74.

Price, D. K. 1965. *The Scientific Estate.* Cambridge, Mass.: Harvard University Press.

Putnam, H. 1978. *Meaning and the Moral Sciences.* Boston: Routledge and Kegan Paul.

Rabin, J.; Hildreth, W. B.; and Miller, G. J., eds. 1989. *The Handbook of Public Administration.* New York: Marcell Dekker.

Rainey, H. 1991. *Understanding and Managing Public Organizations.* San Francisco: Jossey-Bass.

Ranney, A., ed. 1968. *Political Science and Public Policy.* Chicago: Markham.

Reed, A. 1992. The underclass as myth and symbol: the poverty of discourse about poverty. *Radical America,* 24: 329–57.

Reed, T. 1930. Report of the committee on policy. *American Political Science Review,* 24: Appendix.

Reich, M. R. 1991. *Toxic Politics:.* Ithaca, N.Y.: Cornell University Press.

Ricci, D. M. 1984. *The Tragedy of Political Science: Politics, Scholarship and Democracy.* New Haven, Conn.: Yale University Press.

Riemer, J. M. 1993. Reproduction and reunification: the politics of abortion in united Germany. Pp. 167–88 in *From Bundesrepublik to Deutschland: German Politics after Unification,* ed. M. G. Huelshoff, A. S. Markovits and S. Reich. Ann Arbor: University of Michigan Press.

Riker, W., and Ordeshook, P. 1973. *An Introduction to Positive Political Theory.* Englewood Cliffs, N.J.: Prentice-Hall.

Rohr, J. A. 1986. *To Run a Constitution: The Legitimacy of the Administrative State.* Lawrence: University of Kansas Press.

Rose, R. 1993. *Lesson-drawing in Public Policy.* Chatham, N.J.: Chatham House.

Ross, D. 1991. *The Origins of American Social Science.* New York: Cambridge University Press.

—— 1993. The development of the social sciences. In Farr and Seidelman 1993: 81–112.

Rossi, P., and Freeman, H. 1993. *Evaluation.* Newbery Park, Calif.: Sage.

Sabatier, P. A., and Jenkins-Smith, H. C. 1993. *Policy Change and Learning: An Advocacy Coalition Approach.* Boulder, Colo.: Westview.

Salisbury, R. H. 1968. The analysis of public policy: A search for theories and roles. In Ranney 1968: 151–75.

—— and Johnson, P.; with Heinz, J. P., Laumann, E. O., and Nelson, R. L. 1989. Who you know versus what you know: the uses of government experience for Washington lobbyists. *American Journal of Political Science,* 33: 175–95.

Savas, E. S. 1987. *Privatization.* Chatham, N.J.: Chatham House.

Schick, A. 1994. *The Federal Budget: Politics, Policy, and Process.* Washington, D.C.: Brookings Institution.

Schmitter, P. C. 1981. Interest intermediation and regime governability in contemporary Western Europe and North America. In Berger 1981: 285–327.

Schneider, A., and Ingram, H. 1993. Social construction of target populations: implications for politics and policy. *American Political Science Review,* 87: 334–47.

Schwartzkopf, H. N. 1992. *It Doesn't Take A Hero.* New York: Bantam.

SCOTT, J. C. 1976. *The Moral Economy of the Peasant.* New Haven, Conn.: Yale University Press.

SEIDELMAN, R., with HARPHAM, E. J. 1985. *Disenchanted Realists: Political Science and the American Crisis, 1884–1984.* Albany: State University of New York Press.

SHABECOFF, P. 1993. *A Fierce Green Fire: The American Environmental Movement.* New York: Hill and Wang.

SHARKANSKY, I., ed. 1970. *Policy Analysis in Political Science.* Chicago: Markham.

SIGAL, S. 1995. Some institutional and political determinants of political science in Argentina. Pp. 229–48 in Easton, Gunnell, and Stein, 1995.

SIMON, H. 1947. *Administrative Behavior.* New York: Free Press.

—— 1983. *Reason in Human Affairs.* Stanford, Calif.: Stanford University Press.

SKOCPOL, T. 1992. *Protecting Soldiers and Mothers: The Political Origins of Social Policy in America.* Cambridge, Mass.: Harvard University Press.

—— 1996. *Boomerang: The Health Security Effort and the Anti-Government Turn in American Politics.* New York: Norton.

SMITH, J. A. 1991. *The Idea Brokers.* New York: Free Press.

SMITH, T. A. 1989. *Time and Public Policy.* Knoxville: University of Tennessee Press.

SMITH, S. R., and LIPSKY, M. 1993. *Non-Profits for Hire: The Welfare State in the Age of Contracting.* Cambridge, Mass.: Harvard University Press.

SNOW, C. P. 1988. *The Two Cultures and a Second Look.* Cambridge: Cambridge University Press; *The Two Cultures* originally published in 1959.

SOMIT, A., and TANENHAUS, J. 1967. *The Development of American Political Science.* Boston: Allyn and Bacon.

STARBUCK, W. H., and NYSTOM, P. C. 1981. Designing and understanding organizations. Pp. ix–xxii in *Handbook of Organizational Design*, ed. P. C. Nystrom and W. H. Starbuck. New York: Oxford University Press.

STEIN, M. B. 1995. Major factors in the emergence of political science as a discipline in western democracies: a comparative analysis of the United States, Britain, France, and Germany. In Easton, Gunnell and Stein 1995: 169–96.

STEINBRUNER, J. D. 1974. *The Cybernetic Theory of Decision.* Princeton, N.J.: Princeton University Press.

STEINER, G. Y. 1976. *The Children's Cause.* Washington, D.C.: Brookings Institution.

STIEHM, J. H. 1981. *Bring Me Men and Women: Mandated Change at the U.S. Air Force Academy.* Berkeley: University of California Press.

—— 1989. *Arms and the Enlisted Woman.* Philadelphia: Temple University Press.

STOKES, D. E. 1994. The changing world of the public executive. Paper delivered to the Conference on Public Affairs and Management in the Twenty-First Century, School of Public Affairs, Baruch College, City University of New York.

—— 1996. *Pasteur's Quadrant: Basic Science and Technological Innovation.* Washington, D.C.: Brookings Institution.

STOKEY, E., and ZECKHAUSER, R. 1978. *A Primer for Policy Analysis.* New York: Norton.

STONE, D. A. 1988. *Policy Paradoxes and Political Reason.* Glenview: Scott, Foresman and Company.

—— 1993. Clinical authority in the construction of citizenship. Pp. 45–67 in *Public Policy for Democracy*, ed. H. Ingram and S. R. Smith. Washington, D.C.: Brookings Institution.

TARROW, S. 1989. *Democracy and Disorder: Protest and Politics in Italy, 1965–1975.* New York: Oxford University Press.

THOMPSON, M. 1984. Among the energy tribes: a cultural framework for the analysis and design of energy policy. *Policy Sciences*, 17: 321–39.

TILLY, C. 1978. *From Mobilization to Revolution.* Reading, Mass.: Addison-Wesley.

Torgerson, D. 1995. Policy analysis and public life: the restoration of *Phronesis*? In Farr, Dryzek and Leonard 1995: 225–52.

Tronto, J. 1993. *Moral Boundaries: A Political Argument for an Ethic of Care.* New York: Routledge.

Walker, J. L. 1969. The diffusion of innovations among the American states. *American Political Science Review*, 63: 880–99.

Weiss, C. H. 1972. *Evaluation Research: Methods of Assessing Program Effectiveness.* Englewood Cliffs, N.J.: Prentice-Hall.

Wiatr, J. J. 1995. The impact of democraticization on political science in Poland. In Easton, Gunnell and Stein 1995: 217–28.

Wildavsky. A. 1987. *Speaking Truth to Power: The Art and Craft of Policy Analysis.* 2nd edn. New Brunswick, N.J.: Transaction; originally published in 1979.

Wilson, J. Q. 1973. *Political Organizations.* New York: Basic.

—— 1989. *Bureaucracy.* New York: Basic.

—— 1990. Judicial democracy versus American democracy. *PS: Political Science and Politics*, 23: 570–2.

Wilson, W. 1993. The study of administration. In Farr and Seidelman 1993: 33–48; originally published in 1887.

Chapter 25

# Public Policy and Administration: Comparative Policy Analysis

Richard I. Hofferbert
David Louis Cingranelli

## I Introduction

COMPARATIVE policy analysis is a field of study concerned with variations in the products of governmental activity over time and across different jurisdictions. The field has been guided by a relatively focused set of questions. To what extent are differences in the policies governments produce shaped by the social and economic contexts within which decisions are made? For example, are rich countries more likely to provide social benefits to their citizens than are less affluent countries? To what extent are policy dissimilarities systematically related to differences in governmental institutions and/or political conditions? Are leftist governments more likely than their rightist opponents to enact redistributive welfare policies? Did the modern welfare state get its impetus from the internal dynamics of the industrialization process, from the broadening and deepening of liberal democracy, or from some intricate combination of both?

The core of this research agenda was originally set by a series of studies in the 1960s.[1] The units of analysis were varied: American states, English county boroughs, cities in various countries, nation-states. Case studies, which had generally dominated policy analysis to that time, were eschewed in favor of aggregate, cross-jurisdictional statistical analyses, using one or another variation on relatively basic regression analysis correlating economic and political indicators with policy indicators (usually operationalized with expenditure data). The flowering of the field of comparative

[1] See, especially, Dawson and Robinson 1963; Cutright 1965; Dye 1966; Hofferbert 1966; Sharkansky and Hofferbert 1969. For a summary and critique of the first generation of comparative policy analysis, see Hofferbert 1972.

policy analysis, however, followed in the 1970s and onward, with advances in research design, methods of inquiry, and rather more sophisticated formulation of research questions. The result, more often than not, reinforced the cliché that "politics doesn't matter," especially when contrasted with the apparent influence of economic conditions on policy.

At stake in the questions posed by comparative policy analyses are concerns that lie at the core of democratic theory. Do variations in the form of democratic practices, such as modes of representation (for example, district population equality, proportional versus plurality elections), political party structure and performance, differences in constitutional detail, or outcomes of elections, relate systematically to the products of governmental action? By changing the partisan hue of those who govern, for example, can voters reasonably expect policy to be changed in more or less predictable directions? Can changes in institutional arrangements likewise redirect policy outputs? Or is the policy relevance of differences in either political performance or governmental institutions rendered largely trivial or moot by the overwhelming constraint imposed by economic resources or social conditions? Is the modern welfare state an exercise in "regulating the poor" (Piven and Cloward 1971) or is it the logical fruition of democratic politics (Schumpeter 1942)?

A wide range of policies has been studied, with a high degree of consistency in research questions and statistical techniques. The result has been that the field offers an excellent example of cumulative research in political science. It would be unreasonable to attempt here a comprehensive review and critique of the entire set of comparative policy studies. Rather, we shall concentrate on examples of research that illustrate each of two alternative foci within the field. And, in turn, we shall focus on a key analytical problem characteristic of each of those foci. First, we shall use the example of research on cross-national variations in welfare policy as an example of the *political economy* focus. Second, we shall use the example of research on party election programs as an example of the focus on *democratic theory.*

The political economy focus of comparative policy research seeks a comprehensive causal explanation that shows the relative strength of forces operative on a domain of policy. That is, it seeks a fully specified model of why and how policies vary over time and space. Causal inferences are paramount for this undertaking. The analytical problem to be explored is that of modeling the interaction between economic and political conditions. The focus is on the dependent variable, i.e., on the soundness of the explanation of policy variance.

The democratic theory focus of the field is not directly concerned with causation so much as with the quality of signals provided to voters by the competitive political process. This focus asks: How valid are the predictors provided to voters as to the policy consequences of their choices among parties and candidates in the contest for public office? The analytical problem to be explored with this research is that of the conflict between *causation* and *signaling*. The focus is on the independent variables, and in particular on how good various aspects of the political context are as signals of policy variance.

## II Comparative political economy: the case of welfare policy

### A Major studies

The body of comparative welfare state studies stands out as a good example of cumulative research in political science, enriched by the multi-national representation of contributing scholars. Phillips Cutright (1965), in a very early and influential study, set the challenge in contrasting the economic well-being of nations with their level of political democracy as determinants of national social security programs. The apparent impact of the latter was slight compared to the former, adding strength to the "politics doesn't matter" cliché.

In 1973, Anthony King published a two-article series that returned briefly to the case-analytic mode in order to examine the size and scope of the public sector in five advanced democracies. He explored carefully a set of well-formulated hypotheses regarding the relevance to differences in policy of such features as social power concentration, interest group strength, separated powers, federalism, and ideology. Concentrating particularly on the widely noted American limited government exceptionalism, King downplayed the importance of social and institutional conditions as explanations for national variations in public policy. He emphasized rather the significance of long-standing ideological differences among the peoples of different countries.

The concern for history and values is taken up most vigorously by Peter Flora and his associates in the Historical Indicators of Western European Democracy (HIWED) project (Flora and Heidenheimer 1981; Flora 1986). This project is indeed a landmark in the field. Flora and his associates assembled extensive time-series data, particularly on social insurance programs, in fifteen European countries. The data collection, reaching well back into the 19th century, is matched by narrative and careful classification of external conditions and political system attributes over time. Furthermore, the project, conducted during the 1970s, served as a training-ground for scholars who subsequently became leaders in the field (see, for example, Alber 1982).

The analyses produced by the HIWED project, and subsequently by its participants, are as rich in substantive content as would be expected of an undertaking of such scope. However, the very richness thwarts any effort at crisp summary. The work does not add up, however, to a ringing endorsement of the claim that political circumstances are irrelevant for policy. First, the assembled works demonstrate that economic resources are far from a thorough explanation of the evolution of policy choices. And second, through rich narrative as well as careful application of historical statistics, the HIWED research shows how leadership, long-standing social cleavages, and alternative institutional arrangements that may have made small differences at the time of policy inception have nonetheless cumulated to quite striking differences today. These points are well illustrated for specific cases, if not always proved for large sets of countries (e.g., Kuhnle 1981). The work represented in the publications of Flora and his colleagues adds the dimension of time and a richer field of explanatory elements to the intriguing list offered earlier by Harold Wilensky (1975). The result, however, was not a clear, crisp, quantifiable model that could satisfy the goal of full explanation sought by what we have labeled the *political economy* focus.

In the early 1980s, Francis G. Castles contributed a significant insight into the manner in which political differences might operate in shaping and molding the welfare state (Castles 1982). In addition to a careful statistical analysis of OECD countries' experience, his essay presents a quite readable critique and sharpening of theoretical reflection up to that time. In particular, he points out that for differing partisan composition of governments to yield consequential differences in output, those partisan configurations probably have to rest on significant divisions of class or ideology in the electorate itself. That is, it probably takes a certain sharpness of social cleavage, reflected in competing demands and in corresponding inter-party differences, for policy to change with changes in party.

Castles's 1982 essay is most often cited for his demonstration that, particularly in the context of European coalition governments, the extent of right-wing party control is consequential for variance in certain major policies central to the welfare state. Prior data analysis practices had focused on summary measures of left-wing control, for good reason, given the theory that it is working-class mobilization via leftist parties that has spurred the welfare state onward. The correlation of partisan composition and policy outputs, however, came much more sharply into focus when the former was measured as a percentage of the right-wing party in government (Castles 1982: 73). That is, the resistive power of the right seems more consistent than the initiating power of the left.

Castles' findings also indicated the differential relevance of explanatory variables for different policy domains in different time periods. Castles' work emphasized the role of politics in the shaping of public policy, especially in the realm of social welfare. It was, however, his theoretical reflections as much as his findings that made a lasting imprint on the agenda of future inquiry.

However, the political economy goal of a comprehensive theory, statable in a single equation, has proved elusive. That point is eloquently made by Gösta Esping-Andersen (1990), who draws together the various strands of comparative research on the welfare state. In the process, he also provides a useful synopsis of the main lines of political economy thinking from Adam Smith to the 1980s, as well as a quite comprehensive bibliography of comparative research (1990: chap. 1). His most cited contribution is the distinction between three types of welfare states: liberal (e.g. U.S., Canada, Australia); corporatist (also called 'conservative', including, e.g. Austria, France, Germany and Italy); and social democratic (Scandinavia).

His argument is that these types are distinct and that variance within them will be explained by models that are different across them.[2] In spite of a yeomanlike effort at theoretical integration and measurement, Esping-Andersen is forced into country-specific narratives to account not only for deviations, but also for the cases that fit each facet of his theoretical reflections. Yet it is his and other political economy theorizing of the 1980s that contain the most edifying efforts at theoretical synopsis of the field as it has evolved. (In addition to Esping-Anderson, see particularly Evans, Reuschmeyer and Skocpol 1985.)

Whereas Esping-Andersen, in his most vigorously argued theoretical reflection, deftly by-passes the barriers to quantitative analysis, Alexander Hicks and Duane Swank (1992) confront those barriers head-on, in a piece which must qualify as a fine example of the state of the art of comparative policy analysis in the 1990s. Under the label of "theory," they list a broad range of competing hypotheses about the evolution of the welfare state, returning to center-stage a concern for the political composition of governments. Building on the "new institutionalism" (Evans, Rueschemeyer and Skocpol 1985) Hicks

[2] Esping-Andersen's 1990 work, in this regard, built on an earlier piece by himself (1985) and by Walter Korpi (1983; 1989) demonstrating that different kinds of welfare states spend their money on different sorts of policies, and that partisan control has different apparent consequences for welfare budgets in different settings.

and Swank present a comprehensive operational model, tested using a pooled time-series, cross-sectional design. Public policy differences are indicated by share of national income (across eighteen countries over twenty-three years, with an impressive array of statistical controls appropriate for pooled analysis) devoted to social welfare programs (comparable to Wilensky's 1975 measure).

Among their independent variables are some interactive terms between party control and other elements in the model. Though they do not engage in the theoretical reflection necessary for a more nearly isomorphic statistical representation of the policy process, as elaborated below, we believe that this use of interactive terms presages an important and welcome shift in theorizing about the respective roles of politics and economics in shaping public policies.

In terms of substantive findings, it is a commentary on 30 years of inquiry that Hicks and Swank open their essay with the statement: "To date, no consensus has been reached on the controversy . . . over how politics matters for welfare spending in industrialized democracies" (1992: 658). Controling for their host of competing explanations, they offer evidence of a clear association of the partisan composition of governments and variations in policy outputs (see also Blais, Blake and Dion 1993). In particular, they note the apparent accommodation of policies of governments to the strength of opposition and the direction from which it comes. Thus leftist governments faced with a unified conservative opposition accommodate by being more centrist than their counterparts not so visibly restrained from without. The reverse also seems to hold for rightist governments faced with varying degrees of leftist opposition.

## B Key analytical problem: economics vs. politics

Unfortunately for theoretical integration, studies in which political variables have been found to correlate with policy are more scattered and inconsistent than the findings of economic influence. Do these findings really mean that economic factors such as wealth cause public policy outcomes, but political factors such as the degree of democracy do not? We think not. Scientific curiosity leads us to want to know how things fit together. Moral motivation wants to find things (independent variables) that can be changed or preserved in order to influence a desirable set of consequent circumstances (dependent variables). Neither scientific curiosity nor the goals of reform are served by mistaking associations for core causal processes.

To understand why people stop at an intersection with a traffic-light system, it is usually enough to know whether or not the light is red or green. The color of the light is the independent variable. The stopping or going is the dependent variable. But does the red light cause the cars to stop? Or is it the registration of the red on the driver's retina (a very important consideration for the color-blind)? Or is it the city ordinance passed by the city council which causes drivers to stop? Or is it the instruction manual which all drivers study? Or was the cause the pressure from the foot that pushed the pedal that activated the diskpad that pushed against the disk that . . . ?

The selection of causal elements will depend very much on the interests of the selector. One scholar's error term is another's causal cornucopia. Traffic engineers, automobile designers, sociologists, ophthalmologists, or students of law obedience will each use a different equation, leaving the items of interest to others in the error term. None will be wrong in so doing. Causation is a tricky concept, not made simpler by any configuration of statistical gadgetry. Just because the concatenation of *X*s in an equation shows it to be so, it would be foolish to tell the automobile designer that variations in the brake system are irrelevant so long as driver training and law enforcement are effective. Likewise, it would be foolish to tell the writers of constitutions or the leaders of political parties to abandon their work in the face of socioeconomic constraints.

Causal language confuses as often as it helps understanding. Such has been the history of the debate over whether politics or economics is more important as a determinant of policy. Certainly any research design that pits aspects of the political system or policy process against economic conditions risks misleading results. As early as 1979, Jeff Stonecash wrote a series of essays convincingly arguing for a set of statistical procedures to minimize this fundamental epistemological problem (Stonecash 1979; Stonecash and Hayes 1980).

Stonecash (1979) argues that the basic politics vs. economics problem is not "the selection of policy areas or the types of dependent variables used, but the theory specification employed." His argument is that the customary regression equations are not isomorphic of the processes sought to be identified through the statistics employed. And, we would argue, the increasing sophistication of more advanced regression techniques applied in subsequent years most commonly does not deal with this problem. The political system, argues Stonecash (1979: 464), "plays an intervening role, perhaps facilitating demand conversion. Politics does not simply 'cause' policy, independent of preferences and wealth."

Thus, politics is not viewed as having direct effects on policy, but as affecting the relationship between demands (with socioeconomic conditions as the stimulant and proxy measure) and policies. Politics is a necessary condition for this relationship to exist; in other words, without the political process, the relationship between demands and policies cannot exist. "Variation in the political processes across governments produces variations in the conversion of inputs into outputs, which is to say that the rule of transformation depends on the nature of the political process" (Stonecash 1979: 465).

The road to a solution, Stonecash effectively argues, is found in the careful use of interactive terms. Thus, let us assume, not unreasonably, that it is only relatively rich, industrial societies that produce demands for unemployment insurance. Let us also assume, however, that it takes a social democratic government to enact or enhance those benefits. The equation to capture this relationship is not:

$$\text{Insurance} = a + b_1(\text{econ}) + b_2(\text{socdem}) + e \qquad (1)$$

but rather:

$$\text{Insurance} = a + b_1(\text{econ}) + b_2(\text{socdem} \times \text{econ}) + e \qquad (2)$$

In equation 2, the term $b_2(\text{socdem} \times \text{econ})$ tells us the relative contribution of social democratic governments to what would otherwise have been produced.

The use of interactive terms does indeed present certain initial complexities of interpretation (e.g. built-in multicollinearity, the correct reading of standard errors, etc.), but the gain in getting the statistical specification to reflect more closely the real world processes justifies the added complexity.

There is no statistical manipulation, however, that will overcome the major barrier to a fully specified model of policy differences across any finite set of jurisdictions such as nation-states. Human creativity and caprice being what it is, it is nearly inevitable that the number of plausible explanatory variables will exceed the available number of cases. Gadgets for inflating *N*, such as pooled cross-time/cross-unit analysis, while serving certain limited purposes well, hardly solve the long-standing, seemingly intractable epistemological problems. On the other hand, more careful matching of statistical techniques to commonsense expectations about how the parts of the political process work together (such as fitting equations to the assumption that certain objective conditions are antecedent to rather than competitive with political conditions) will certainly advance useful understanding of why some jurisdictions produce one kind of policy and others produce another.

Until some such accommodation becomes commonplace, the state of the art will remain more art than life. So long as we continue to use the so-called *Casablanca model*, whereby we "round up the usual suspects" and place them on the right side of the equation, the goals of both political economy and democratic theory will continue to be ill-served.

## III Comparative democratic theory: the case of party election programs

### A Major studies

It is not unreasonable that so much attention has been given by comparative policy analysts to the role of parties in the policy process. Parties are, according to most modern democratic theory, the vehicles by which issues are articulated for electoral review. Parties are the organizing instruments for electoral contests, which may be fought over competing packages of such articulated issues. And, after the election, parties are the instruments responsible for the conduct of government, centrally located in the legislative process. As such, they should be highly dynamic, presenting and implementing an agenda that changes with the changing preferences or needs of the population.

Until recently, however, the party terms in the various policy equations were static. Implicit in the concern over parties in the policy process is a concern for their accountability, that is, the fulfillment of the party mandate, whereby the promises in election programs ("platforms," in American parlance) get reflected in the policies adopted by the parties that enter government after the election. Yet party has generally been operationalized not in terms of substantive stances in the election or elsewhere, but rather in terms of percentage or turnover between parties in parliamentary seats and/or government portfolios. Thus, the percentage of leftist or rightist party control of government is correlated with policy products, on the assumption that leftist or rightist parties always lean in one direction. The Swedish Social Democrats and the Italian Socialists, however, are not necessarily the same. Further, the German CDU of 1950 is hardly the CDU of 1995. Available information, however, has not, until recently, allowed for a more refined conceptualization of party variance.

It is obvious that the universe of policy discourse from which the parties articulate bundles of issues to present as formal programs to the electorate varies over time and place, just as the social conditions that give rise to political demands must certainly vary. Housing policy would probably not top the agenda in a well-housed country. Thus, while there may be a continuing, if fluid, distance between parties that sustains their identity, it is also reasonable to assume that each party regularly adjusts its issue scope and emphasis, even if rarely making any 180° turns.

The massive data collection effort initiated by the Manifesto Research Group (MRG) of the European Consortium for Political Research, and sustained by the Science Center–Berlin (Wissenschaftszentrum–Berlin: WZB), has made it possible to capture just such changes in the content and emphases of the parties over time, countries, and policy domains (Klingemann, Hofferbert and Budge 1994). The project coded each sentence in each pre-election published program of the more or less permanent parties into one of fifty-four unique thematic categories. The coding frame was devised over several years' consultation among scholars from several participating countries. To standardize for program length, percentages were calculated to reflect the proportion of the total number of sentences in the document devoted to each of the fifty-four thematic areas. The data archive holdings cover nearly all of the OECD countries' elections since World War II, plus the programs of the parties in the post-communist regimes of Europe.

The mandate and the conditions for accountability across ten democracies over the post-war period are the primary concerns of Hans-Dieter Klingemann, Richard I. Hofferbert, and Ian Budge's *Parties, Policies, and Democracy* (1994). It examines the congruence between party election program emphases and policy priorities across numerous domains in each of ten countries. Party emphases are operationalized as the percentage of election programs devoted to specific themes. Policy priority is measured by the percentage of expenditures devoted to areas matched to the election program themes. The results of their within-country time-series analyses are somewhat scattered, but a short list does stand out:

- Party election programs are remarkably stronger as predictors of post-election policy priorities than nearly all former writings on political parties would have led one to expect.
- Parties out of government as well as those in government often see their programmatic concerns reflected in policy priorities, but the programs of the winners indeed predict policy priority more often and far better than do those of the losers.
- The relative capacity of parties to deliver on their programmatic projections does not seem to vary with the institutional features (e.g., majority versus coalition; unified versus separated powers) often cited in constitutional theory as affecting accountability across otherwise democratic systems (e.g., Powell 1990).

The message is that politics matters very much, but institutional or constitutional variation within a set of otherwise comparably democratic countries, does not matter very much. This contrast should not detract from the positive findings in the Klingemann, Hofferbert and Budge research. In addition to demonstrating the vitality of election programs as predictors of subsequent policy behavior, this study also demonstrates very strong relationships—over time, across policies, in all countries—between party turnover and policy differences, when the latter are measured as priorities (i.e., percentages of outlays) rather than by absolute level of expenditure or ratio of expenditure to GDP or other indicators of system capacity or demand. Thus, Klingemann, Hofferbert, and Budge confidently conclude that both standing party differences as well as more transient emphases in election programs are quite congruent with post-election policy variations. It makes a difference for policy who wins and who loses. And there is ample evidence available at election time in the past record of the parties plus their current program stances for the voter to make an informed decision. Parties seem to work generally the way democratic theory says they should, largely regardless of institutional impediments that might be thought to dilute accountability.

To suggest that parties offer programs that are later reflected in policy is one important element in the chain of democratic forces. But there is another link in the chain that has been shown by current research to be rather strong, at least in one country. An assessment of the state of the art of comparative policy analysis must take note of the work by Robert Erickson, Gerald Wright, and John McIver (1993) on the congruence of public opinion and policy in the American states.

V. O. Key, Jr., once wrote: "Unless mass views have some place in the shaping of policy, all the talk about democracy is nonsense" (1961). Translated into a conditional hypothesis for comparative research, the statement would be: In democracies there is a positive relationship between popular preferences and policy outputs. Until the late 1980s, and even then in only a limited fashion, there was a virtually insurmountable barrier to a systematic test of this hypothesis. There had been no statistically reliable samples conducted within enough jurisdictions or time points sufficient to test the *public preferences* → *policy* linkage.

Unfortunately, even in a single country, such as the U.S., it is not statistically reasonable to report the subsets of a nationwide sample by subnational jurisdictions, such as states. In the case of American state studies, for example, not only would the individual states that happened to have residents showing up in a national sample contain statistically inadequate numbers of respondents, but customarily states are not used as sampling units in national surveys, and thus they have not been an appropriate subsetting criterion. Cross-national surveys in which comparable questions about policy preferences have been asked (e.g., Barnes and Kaase *et al.* 1979; Inglehart 1990), do not include sufficient numbers of countries to allow for statistically meaningful tests of the relationship between aggregated policy preferences and governmental outputs. Erikson, Wright, and McIver's work has overcome the sampling barrier for at least a cross-sectional/cross-state analysis of the *opinion* → *policy* linkage so central to democratic theory. In the course doing so, they have shifted some of the burden of proof back to those who claim that the policy processes of modern democracies lie beyond the effective reach of the very instruments that distinguish those systems.

The commercial world came to the rescue via the massive number of respondents (66,000) assembled by the Columbia Broadcasting System (CBS)/New York Times polls, conducted for day-by-day reporting of presidential elections. Based on samples reliable at the state level and combined with the very imaginative exploitation by Erikson, Wright, and McIver, these data form the basis for one of the most important breakthroughs in comparative policy analysis in the field's third decade.

The authors constructed, first a very general index of state *policy liberalism* (i.e., leftism), including indicators of state outputs for education, health and welfare, consumer protection, criminal justice, gambling regulation, women's rights, and tax progressivity. They then aggregated by state the results of the CBS/NY Times polls in response to the simple question (fitted to the peculiar American conception of the left-right scale): "Do you consider yourself a liberal, a moderate, or a conservative?"

Across forty-seven of the fifty states the simple correlation between policy liberalism and opinion liberalism was a stunning 0.82 (Erickson, Wright and McIver 1993: 78). A variety of other more sophisticated and complex analyses neither added to nor detracted from this essential message. At the most general level, that which the governments of the states produce would seem to be well synchronized with what the people want.

## B Key analytical problem: causality vs. signaling

Do the party programs cause congruent policy priorities? Does public opinion cause policy variation? Fortunately, democratic theory, in contrast to the goal of political economy, does not require that we establish causation, as such. The demonstration of certain associations between politics and policy is sufficient. Much of the effort to assess the correlations with policy of so-called political conditions (e.g., inter-party competitiveness, left/right control, content of party programs) rests upon the reliability of variance in such conditions as predictors of alternatives between which a voter may choose. Thus, the voter faced with a choice between parties or between party programmatic stances wants to know if those differences are likely to predict a difference in governmental behavior. Can the information be used with confidence? Will policies be different if the Tories or Labour wins? Will policies be different if the SPD election program emphasizes human services and the CDU emphasizes infrastructural development? Statistical associations from such differences in the past give a clue as to how useful the political information is at present. The analogy to the traffic signal is especially apt here. Rational behavior is conditioned on confidence in the reliability of the signal. If there is not a "dime's worth of difference" between the parties, then the voter's time spent in checking the record or listening to debate is wasted. And the democratic process is thereby decremented.

That decrement, however, is not demonstrated convincingly by a multiple regression equation pitting economic against *political* conditions. Parties are supposed to adapt a dynamic world to democratic processes. They sort through "causes" and present alternatives for dealing with them. Presumably the parties do not build their records or present their programs *in spite* of objective conditions, but rather *because of* such conditions. Suppose the labor market is increasingly uncertain, and the leftist party promises to cushion the uncertainty; or the transport system is in widespread disrepair and the rightist party promises to fix it. How much sense would it make to hold constant statistically the objective conditions (labor market, transportation conditions) in order check the extent to which party pronouncements have led to policy differences?

From the standpoint of democratic theory, the test is: are the party pronouncements, and thus the information available to the voter, accurate signals for future government action? Voters may differ on their assessment of the appropriate priority to attach to one or the other family of problems. They may foil the correlation between the policy problem and the policy enacted without violating the presumptions of democratic theory. But even if the correlation between the policy problem and the policy output is high, without some competitive signaling between competing parties, the requirements of democratic theory, that governments do what people choose, are not met. Democracy requires information and the possibility for voters to choose in a way that is substantively consequential. Are policies different when the choice goes one way as contrasted to when it goes another.[3]

[3] For a striking demonstration of tight congruence over time in a single country (U.S.) of objective conditions, public opinion, and party positions, see Shi 1995.

And when coefficients of various types affirm that such is indeed the case, it is theoretically inappropriate then to introduce "controls" that seem to wipe out these coefficients. Suppose that the correlation across twenty-five countries between percentage of the population between ages 5 and 20 correlates 0.85 with percentage of total budgetary outlays for education. Suppose that control of government by leftist parties adds only a statistically insignificant amount. Should we conclude from this finding that the parties are eternally and universally stupid? Although it is unlikely that parties always advocate spending money on things, even when there is no need or demand, common statistical procedures cast unwarranted doubt on the politics → policy signaling capacity. The bias in favor of the null hypothesis is, in such a case, a bias against a central tenet of democratic theory: that political parties are responsive to public needs.

Most of the statistical models employed in comparative policy analysis have run precisely such a risk of discarding a true hypothesis. The true hypothesis is most likely that, between their standing differences and their current program positions, the parties in the electoral process send out pretty accurate signals of what they will do in government. Another hypothesis, which is yet to be widely tested, is that policy variation probably tracks public preferences pretty well. In the drive to find *true causation*, many comparative policy studies have not only forgotten the old lessons of epistemology, but they have failed adequately to credit the signaling capacity of the political processes of democratic systems. And that is a mistake not only of statistical specification, but also of theory. This may well be a case where "simpler is better," from a technical standpoint.

## IV Conclusion

One might ask whether it is a condemnation or a compliment to the field that, as Hicks and Swank claim, 30 years of inquiry has produced no consensus on "how politics matters" for certain important areas of public policy. It is convenient to present our assessment under three headings: theory, substance, and research technology.

Any field in which a list of the same four or five questions can stimulate capable scholars to do good research for over three decades—any such list, in and of itself—has a pretty good implicit theory. But if one has a very ambitious conception of theory, taken as a body of coherent, interrelated statements from which can be deduced reliable hypotheses, the verdict on comparative policy analysis is neither positive nor promising. On the other hand, if one is more modest and accepts for theory a limited list of complementary questions, producing interesting and cumulative research, then the field is in rather good shape.

The cornerstones of the field of comparative policy analysis, laid in the 1960s, consist of a few pieces noteworthy for their substantively iconoclastic findings. Logic and direct observation long suggested that competitive party systems would push policies more in the direction of the have-not segments of society. Similarly logic and observation suggested that leftist governments will do things differently than rightist governments. Flawed technically and even theoretically though it may have been, the very disturbing nature of the apparent challenge of early research to that logic and those observations stimulated a qualitatively and quantitatively coherent body of research. Although the focus of this essay has been primarily on the quantitative work, there can be no doubt that much of the historically and contextually rich narrative work, such as King's or Esping-Andersen's would not have been done had good scholars not perceived the need to defend political and institutional conditions from the charge of irrelevance coming from the quantitative camp. Within that camp, furthermore, certain basic findings have stood the test of time and methodological diversity.

- Policies are not made in a socioeconomic vacuum. The choices of policy-makers are stimulated, shaped, and constrained by identifiable external conditions and patterns of demand. To ignore such consequential circumstances is to misunderstand the task of policy-making.
- Within the set of mature democracies, on which most of the comparative research has been focused, the assumed consequentiality of varying institutional structures (e.g., two- versus multi-party systems, forms of representation, federal versus unitary constitution) is yet an open question.
- The long-standing logic of democratic theory, attributing policy consequentiality to partisan conditions (party in government, competitiveness, strength of opposition, programmatic stances) has, after initial challenges, stood the test of time and ever more technically elegant inquiry.

Our argument about model specification, and particularly our guidance for a re-examination of Stonecash's pieces in the middle period of the field's development, suggests that there is room for substantial improvement in matching research tactics to theory. The drive toward statistical elegance has progressed faster than the necessary theoretical specification. In particular, the core question of the interaction between external conditions, political system characteristics, and political processes needs re-examination in terms that bring statistical procedures closer to a representation (isomorphism) of the processes they are presumed to model.

The challenge for political economy is daunting. Causation in a world of more variables than cases is an elusive target, not likely to surrender willingly to the most elegant of statistical representations. The challenge for democratic theory is less imposing. If political signals are such as to provide valid policy predictors, then much that is required to defend democracy's policy relevance is in place.

## References


Alber, J. 1982. *Von Armenhaus zum Wohlfahrtstaat.* Frankfurt: Campus Verlag.

Alt, J. E. 1971. Some social and political correlates of county borough expenditures. *British Journal of Political Science*, 1: 49–62.

Barnes, S. H.; Kaase, M.; *et al.* 1979. *Political Action.* Beverly Hills, Calif.: Sage.

Blais, A.; Blake, D.; and Dion, S. 1993. Do parties make a difference? parties and the size of government in liberal democracies. *American Journal of Political Science*, 37: 40–62.

Carmines, E. G. 1974. The mediating influence of state legislatures on the link between interparty competition and welfare policies. *American Political Science Review*, 68: 1118–24.

Castles, F. G. 1982. The impact of parties on public expenditure. Pp. 21–96 in *The Impact of Parties*, ed. F. G. Castles. Beverly Hills, Calif.: Sage.

Cutright, P. 1965. Political structure, economic development, and national social security programs. *American Journal of Sociology*, 70: 537–50.

Dawson, R. E.; James, A.; and Robinson, J. A. 1963. Interparty competition, economic variables and welfare politics in the American states. *Journal of Politics*, 25: 265–89.

Dye, T. R. 1966. *Politics, Economics, and the Public.* Chicago: Rand-McNally.

Erikson, R. S.; Wright, G. C.; and McIver, J. 1993. *Statehouse Democracy: Public Opinion and Policy in the American States.* New York: Cambridge University Press.

Esping-Andersen, G. 1985. Power and distributional regimes. *Politics and Society*, 14: 223–56.

—— 1990. *The Three Worlds of Welfare Capitalism.* Oxford: Polity Press.

Evans, P. M.; Rueschemeyer, D.; and Skocpol, T., eds. 1985. *Bringing the State Back In.* New York: Cambridge University Press.

Flora, P., ed. 1986. *Growth to Limits.* Berlin: Walter de Gruyter.

—— and Heidenheimer, A., eds. 1981. *The Develoment of Welfare States in Europe and North America.* New Brunswick: Transaction Books.

Hicks, A., and Swank, D. H. 1992. Politics, institutions, and welfare spending in industrialized democracies. *American Political Science Review*, 86: 658–74.

Hofferbert, R. I. 1966. The relation between public policy and some structural and environmental variables in the American states. *American Political Science Review*, 60: 73–82.

—— 1972. State and community policy studies: a review of comparative input-output analyses. Pp. 3–72 in *Political Science Annual: Three*, ed. J. A. Robinson. Indianapolis, Ind.: Bobbs-Merrill.

Inglehart, R. 1990. *Culture Shift in Advanced Industrial Society*. Princeton, N.J.: Princeton University Press.

Key, V. O., Jr. 1961. *Public Opinion and American Democracy*. New York: Knopf.

King, A. 1973. Ideas, institutions, and the policies of governments: a comparative analysis. *British Journal of Political Science*, 3: 293–313 and 409–23.

Klingemann, H.-D.; Hofferbert, R. I.; and Budge, I. 1994. *Parties, Policy, and Democracy*. Boulder, Colo.: Westview.

Korpi, W. 1983. *The Democratic Class Struggle*. London: Routledge and Kegan Paul.

—— 1989. Power, politics and state autonomy in the growth of social citizenship: social rights during sickness in eighteen OECD countries since 1930. *American Sociological Review*, 54: 309–28.

Kuhnle, S. 1981. The growth of social insurance programs in Scandinavia: outside influences and internal forces. In Flora and Heidenheimer 1981: 125–50.

Piven, F. F., and Cloward, R. A. 1971. *Regulating the Poor*. New York: Vintage.

Powell, G. B. 1990. Holding governments accountable: how constitutional arrangements and party systems affect clarity of responsibility for policy in contemporary democracies. Paper delivered to the annual meeting of the Amercian Political Science Association, San Francisco.

Schumpeter, J. A. 1942. *Capitalism, Socialism, and Democracy*. London: Allen and Unwin.

Sharkansky, I. 1971. Agency requests, gubernatorial support and budget success in state legislatures. Pp. 323–42 in *State and Urban Politics*, ed. R. I. Hofferbert and I. Sharkansky. Boston: Little, Brown.

—— and Hofferbert, R. I. 1969. Dimensions of state politics, economics, and public policy. *American Political Science Review*, 63: 867–79.

Shi, Y. 1995. Problems, Parties, and the Public. Unpublished Ph.D. dissertation, State University of New York at Binghamton.

Stonecash, J. 1979. Assessing the roles of politics and wealth for public policy. *Political Methodology*, 6: 461–83.

—— and Hayes, Susan W. 1980. The sources of public policy: welfare policy in the American states. *Policy Studies Journal*, 8: 681–98.

Wilensky, H. 1975. *The Welfare State and Equality*. Berkeley: University of California Press.

Chapter **26**

# Public Policy and Administration: Ideas, Interests and Institutions

Giandomenico Majone

## I Introduction

Some thirty years ago Ed Lindblom criticized the "clumsy realism" of pluralists like Arthur Bentley who dismissed the significance of ideas in politics and argued "as though minds were among the most trivial resources available to the group" (Lindblom 1965: 16). Bentley and his followers, Lindblom noted, not only bypassed questions turning on the efficiency of policy-making in a democracy, but took the irrationality of the process for granted and then accepted irrationality as the necessary cost of a system that has the essential virtue of dispersing power. Nor was Bentley's extreme reduction of human interchange to "force" and "pressure" any longer an isolated position by the mid-1960s: "The role of reason in politics is now generally obscured by the popularity in our time of the concept of political science as the study of power" (Lindblom 1965: 16).

In the mid-1990s the intellectual atmosphere is quite different. Not that political scientists have reverted to the idealistic notion that ideas by themselves can be powerful enough to determine the course of events, or that they now accept, like some early policy analysts, the rationalist fallacy of viewing policy-making as a purely intellectual exercise. For empirical political scientists the "null hypothesis" continues to be that policy outcomes are primarily determined by interests and power. What is new, rather, is that the same scholars are now willing seriously to test this hypothesis against alternative explanations, and to reject it in the face of solid contrary evidence.

Evidence in favor of an independent role of ideas, and their institutional embodiments in policy-making, has multiplied in recent years. By way of illustration only, one can mention the studies by Odell (1982), Haas (1990), Goldstein (1993) and the volume edited by Goldstein and Keohane (1994) in the field of foreign policy; Stein (1984) and Hall (1986; 1989) on macroeconomic policy-making; Derthick (1979) on policy-making for social security; Aaron (1978) and Murray (1984) on social policy; Derthick and Quirk (1985) and Temin (1987) on deregulation; Wilson (1980) and Mashaw and Harfst (1990) on regulatory policy-making. Ideas play a key role in Rose's (1993) monograph on policy learning, and in Kingdon's study of agenda-setting and policy entrepreneurship (Kingdon 1984; see also Cohen, March and Olsen 1972).

This list is merely indicative and could easily be extended, especially if one were to add journal articles and literature in languages other than English. It is, however, sufficient to make the point that the literature on ideational factors in policy-making of the 1980s and 1990s has no equivalent, either in quantity or quality, in previous decades. What are the reasons for the new emphasis on ideas? Changing academic fashion is a partial explanation at best. I submit that the deeper reason is to be found in the transformations of the process and substance of policy-making brought about by the far-reaching ideological, political, and economic changes which started in the late 1970s. In the following pages I attempt to identify those new features of public policy that are more directly relevant to our topic.

## II The changing nature of policy-making

The increased role of ideas and institutions in policy-making may be explained in terms of three relatively novel features of contemporary policy-making. These new features are: the rediscovery of efficiency as a primary policy goal; a new awareness of the strategic importance of policy credibility; and, partly as a consequence of the two previous factors, an increased willingness to delegate important policy-making powers to technocratic bodies enjoying considerable political independence.

Concerning the third factor it should be noted that although the United States has a century-old tradition of delegating significant regulatory powers to independent commissions and boards, the same is not true of Europe (Majone 1994) and of other parts of the world that follow the principle of parliamentary sovereignty. Moreover, the role of independent expert bodies is increasing not only at the national level but, what is perhaps even more significant and certainly newer, also at the international level. Striking examples of this development are the European Commission, the future European Central Bank and the World Trade Organization.

Before showing how these novel features of policy-making can explain the growing importance of ideational and institutional factors, I shall briefly examine each feature separately.

## A The rediscovery of efficiency

For several decades following the end of World War II, redistributive policies occupied the central position in the political arenas of all industrialized countries, while efficient policies were relegated to the ancillary role of providing the means to pay for the government largesse. Recall that policies or institutions are said to be efficient if their aim is to improve (with respect to the status quo) the position of all, or almost all, individuals or groups in society; the aim of redistributive policies or institutions, in contrast, is to improve the position of one group in society at the expense of another.

Several developments have contributed to the progressive erosion of the centrality of redistributive concerns. Among such developments, the fiscal and ideological crises of the welfare state have received the greatest scholarly attention, but for our purposes the parallel decline of the pluralist model is perhaps more instructive.

According to this model, public policy is the equilibrium reached in the struggle among competing group interests at a given moment. Policies change as a result of changes in the configuration of interests and power. Ideas are irrelevant: "The only reality of ideas is their reflection of the groups, only that and nothing more" (Bentley 1908/1967: 169). A group might appeal to the public interest in support of its claims, but that is nothing more than a publicity ploy to increase the attractiveness of its demands. Institutions are also essentially irrelevant. Pluralists recognized that sometimes public agencies play an independent role, but only as interest groups among other interest groups.

Pluralists were of course aware that groups wield unequal power and that access to the policy process is unbalanced. Nevertheless the general conclusion was that all active and legitimate groups in society would be able to make themselves heard at some stage in the process. This benign view of interest group competition was shattered by Mancur Olson's demonstration that commonality of interests is not a sufficient condition for the formation of active and legitimate groups. Because of the pervasive phenomenon of free-riding, the special interests of the few tend to be more readily organized into groups than the diffuse interests of the many (Olson 1965). By the same token, given a choice between efficient or redistributive policies, special-interest groups will opt for redistribution since this increases their chances of obtaining a larger share of the social output.

The first response to these criticisms of group politics was, in the words of Martin Shapiro, "an almost frantic pursuit of more and more perfect pluralism" (Shapiro 1988: 49). In America, where administrative law and public policy had been strongly influenced by the pluralist model, agencies were asked to provide public funding to the poorer groups in order to equalize access to the regulatory process. Rules of standing were expanded to include even persons who really had no special interest in an administrative decision other than that of a good citizen.

But at the time the attempts to perfect group politics were made, disillusion with pluralism was already setting in. On the one hand, given the widely varying resources of different groups, it was not clear that access to agency decision-making could ever be effectively equalized. In fact, experience had shown that the public-participation requirement could be used by powerful economic interests in order to delay regulatory decisions. On the other hand, it was becoming clear that the emphasis on redistributive issues due to the accumulation of special-interest groups could slow down economic growth and make political life more divisive by reducing the significance of common interests (Olson 1982).

Thus, a keener realization of the economic and political costs of group politics gave new plausibility to the idea that there is a public interest or a right public policy quite apart from the sum of group interests. If the processes of group politics yield public policies that are not efficient, such as subsidies to farmers or to coal producers, then those policies are substantively wrong even if the groups all struggled vigorously (Shapiro 1988). One solution for many of those dissatisfied with group struggle was to turn to efficiency as a key criterion of public policy, and to rational, even "synoptic," decision-making as the best model for policy-making. "Good" policy was no longer to be the product of group struggle but of rational policy analysis. Administrators should combine ethical discourse and policy analysis to make decisions that are substantively correct as well as democratically legitimated. Courts should require executive agencies to show that they had maximized net social benefits, subject to statutory, budgetary, and informational constraints (Sunstein 1990; Rose-Ackerman 1992).

As a result of the new emphasis on efficiency and rational policy-making, analyses produced by academic experts and by a growing number of policy think-tanks started to have practical consequences. For example, new policy instruments such as pollution taxes or emission trading, long rejected by politicians, bureaucrats and environmentalists alike, received serious attention beyond academic circles, and in some case they were actually used. In Europe, North America, Australia and New Zealand, deregulation, privatization, regulatory reform and welfare reform were preceded and prepared by intense intellectual debates.

The significance of the rediscovery of efficiency for the politics of ideas will be discussed in a later section. Now we examine another feature of the new pattern of contemporary policy-making.

## B The issue of policy credibility

Like efficiency, policy credibility became an increasingly important topic of public debate in the 1970s. It first emerged in the context of a long-running debate about rules vs. discretion in monetary policy. The question is whether governments should tailor policies to current economic conditions (discretionary policy) or conduct policy according to pre-announced rules, such as a constant rate of monetary growth.

Critics of government discretion like Milton Friedman had argued that governments and central banks lack the knowledge and information necessary for successful discretionary policy. In an important article published in 1977, Kydland and Prescott gave a new twist to the debate. The central problem of public policy, these scholars argued, is its credibility: fixed rules are preferable because they increase policy credibility while discretion leads to "time inconsistency." Time inconsistency occurs when a policy which appears optimal at time $t_0$ no longer seems optimal at a later time $t_n$. Without a binding commitment holding them to the original plan, governments will use their discretion to switch to what now appears to be a better policy. The problem is that if people anticipate such a policy change, they will behave in ways which prevent policy-makers achieving their original objectives.

It should be clear that the implications extend well beyond monetary policy. For example, assume that at time $t_0$ parliament enacts strict anti-pollution legislation. At the time, this appears to be the optimal response both to the severity of pollution problems and to the wishes of the voters. After passage of the law, however, there is a sharp economic down-turn, so that unemployment replaces environmental quality as the main concern of a majority of voters and politicians. If an election is imminent the government will be tempted to ask parliament that the law be made more permissive; or, more simply, the government may decide to reduce the level of implementation by cutting the budget of the pollution inspectorate. But industrial polluters, anticipating such policy change, will assume that they do not have to take the relevant regulations too seriously, and the original policy objectives will not be achieved. The policy lacks credibility because it is seen to be time-inconsistent: the incentives of the policy-makers at time $t_n$ differ from their incentives at time $t_0$.

As in the case of efficiency, the emergence of policy credibility as a prominent topic of research and public debate requires some explanation. Notice, first, an interesting theoretical connection between efficiency and credibility. Efficient policies tend to be more stable, and hence more credible, than inefficient ones. This is because, by definition, an efficient policy improves the position of all or almost all individuals and groups in society. If instead an inefficient outcome is reached, then a policy entrepreneur could propose an alternative that everyone would prefer. Thus, *ceteris paribus*, efficient policies tend to be stable while inefficient policies are always vulnerable to being overturned. This is true, in particular, of redistributive policies. To the extent that the policies with which an incumbent has to deal are redistributive, a challenger in the next election can always find some coalition of voters that can be made better off with some alternative pattern of redistribution. In the language of game theory, games of pure redistribution do not have cores (Ordeshook 1992: 276).

It is doubtful, however, that policy-makers were induced to take credibility seriously by theoretical considerations. A more likely reason is a new awareness of the advantages of policy credibility in a world where national borders are increasingly porous. Growing economic and political interdependence among nations has the effect of weakening the impact of policy actions on the home country and strengthening their impact on other countries. Thus, domestic policy is increasingly projected beyond national boundaries, but it can achieve its intended objectives there only if it is credible.

Some years ago, Theodore Lowi forcefully reminded pluralists that legitimate use of coercion is *the* intrinsic governmental feature. In his words, "governmentalization of a function—that is, passing a public policy—is sought because the legitimacy of its sanctions makes its social controls more surely effective" (Lowi 1979: 37). Indeed, even a nearly worthless currency can be made a legal tender by legislative fiat—but only inside the national borders. Similarly, a policy lacking credibility can be enforced by coercive means, but only domestically and only at high transaction costs. In sum, because of the growing interdependence of nations it is increasingly costly, or even impossible, to use coercive power as a substitute for policy credibility, and policy-makers realize this.

Even domestically, the growing complexity of public policy continues to erode the effectiveness of the traditional command-and-control techniques of government bureaucracy. Until fairly recently, most of the tasks undertaken by national governments were simple enough to be organized along classical bureaucratic lines. Once a program was enacted, the details of its operations could be formulated and appropriate commands issued by highly centralized command centers. By contrast, the single most important characteristic of the newer forms of economic and social regulation is that their success depends on affecting the attitudes, consumption habits and production patterns of millions of individuals and hundreds of thousands of firms and local units of government. The tasks are difficult not only because they often deal with technologically complex matters but even more because they aim ultimately at modifying expectations (Schultze 1977: 12). In this new context, credibility becomes an essential condition of policy effectiveness. In turn, credibility may be achieved by delegating powers to suitably designed institutions.

## C Delegation

The willingness of political sovereigns—legislators or political executives—to delegate important policy-making powers to independent administrative bodies is, I submit, a third distinctive feature of contemporary policy-making. As already noted, the delegation of such powers to independent commissions has a long tradition in the United States, but the same is not true for the majority of industrialized countries. Thus, traditional public administration in Europe is now challenged and to some extent transformed by the growth of independent agencies such as the new breed of regulatory offices in Britain or the French *autorités administratives indépendantes* (Majone 1994; 1995).

What is quite new, at any rate, is the extent to which powers have been, or are being, transferred to supranational bodies. Consider, for example, the powers granted to the future European Central Bank (ECB) by the 1992 Treaty on European Union (Maastricht Treaty). The ECB can make regulations that are binding in their entirety and become European and member states' law, without the involvement of the other European institutions or of the national parliaments. The Bank has a single objective—monetary stability—and the freedom to pursue this objective in complete independence. Moreover, since the governors of the central banks of the member states are members of the EC they too, according to Article 107 of the Treaty, must be insulated from domestic political influences in the performance of their task.

In practice, this implies that the central banks can no longer be players in the old game of pumping up the economy just before an election. Thus, once monetary union has been achieved in Europe, issues of macroeconomic management that have been the lifeblood of domestic politics, determined the rise and fall of governments, and affected the fate of national economies, will be decided by politically independent experts (Nicoll 1993: 28).

Why did the same politicians who always preferred to have their hands on the monetary lever, suddenly opt to delegate such far-reaching powers to an independent technocratic institution? A similar question arises in relation to the massive transfer of regulatory powers to the European Commission. Particularly in the area of social regulation (environment, consumer protection, health and safety at work, equal rights for male and female workers) the delegation of regulatory powers to the Commission has gone well beyond the functional needs of an integrated European market. For example, although the terms "environment" or "environmental policy" do not even appear in the founding treaties, today European environmental regulation includes more than 200 pieces of legislation, and in many member states the body of environmental law of Community origin outweighs that of purely domestic origin.

The most convincing explanation of the willingness to delegate policy-making powers to supranational bodies is the problem of credibility. Purely inter-governmental agreements on complex regulatory matters usually lack credibility because, in the absence of a monitoring agency, it may be quite difficult for the parties concerned to know whether or not the agreement is properly kept. But when it is difficult to observe whether national governments are making an honest effort to enforce a co-operative agreement, the agreement is not credible. The solution is to transfer regulatory powers to an independent supranational authority such as the European Commission.

As we saw above, the issue of credibility is becoming increasingly important also at the national level. Because a legislature cannot bind a subsequent legislature and a majority coalition cannot bind another, public policies are always vulnerable to reneging and hence lack credibility (Shepsle 1991). Again, delegation to an independent body is a way of achieving credible policy commitments. Thus, independent agencies are justified not only by the need of expertise in highly complex or technical matters but also because, being one step removed from election returns, they can provide greater policy continuity than government departments.

We are now ready to examine how the new features of policy-making discussed in the preceding pages can explain the increased role of ideas and institutions in the policy process.

## III Ideas and the politics of efficiency

According to Garrett and Weingast (1994: 203) "the role of ideas embodies a notion of common interest or cooperation." The statement is rather cryptic and needs some elaboration. What is meant, presumably, is that ideas matter most when collective decisions are about efficiency issues—how to increase aggregate welfare—rather than about redistributing resources from one group of society to another. Conversely, arguments are powerless when politics is conceived of as a zero-sum game. When one group's gains are another group's losses only interests and bargaining power count.

To distinguish the politics of efficiency from the politics of redistribution is to assert that politics can also be a cooperative, positive-sum game in which the members of a community engage for mutual advantage. In such a context, analysis and deliberation are important for identifying collectively advantageous solutions. Of course, arguments are also used to support or oppose redistributive policies. However, if one examines such arguments closely, one can see that they are mostly about efficiency issues. For example, they are used to show that a particular method of redistributing income, say, by lump-sum transfers, is more efficient than one that modifies relative prices, or to suggest methods of alleviating the distributional consequences of efficiency-enhancing measures.

In fact, efficient and redistributive policies are connected by what may be called a "duality relation" in the sense of the theory of linear programming (Dantzig 1963). This means that any efficient policy must also satisfy redistributive constraints in order to be politically feasible. Conversely, a sound redistributive policy must satisfy some efficiency constraints in order to avoid dead-weight losses and forms of rent-seeking which could compromise economic growth.

In spite of this duality, the logic of the two policy processes is quite different. In a democracy, decisions that improve the conditions of one group in society at the expense of another, can only be taken by majority vote since the losers cannot be expected to vote against their interest. In fact, the advocates of majority rule envisage conflictual choices in which no mutually beneficial opportunities are available. They also assume that the alternatives facing a community are single-dimensional and mutually exclusive, so that compromise proposals are not possible (Buchanan and Tullock 1962: 253).

By contrast, issues of efficiency could be settled, in principle, by unanimity rule since everybody can gain from a solution that increases aggregate welfare. Unanimous agreement, freely reached, guarantees that the solution is Pareto-efficient. The method of collective choice changes radically in this case:

> [T]he political *process* implicit in the unanimity rule argument is one of discussion, compromise and amendment, continuing until a formulation of the issue is reached benefiting all. The key assumptions underlying this view of politics are both that the game is co-operative and positive sum, that is, that a formulation of the issue benefiting all exists, *and* that the process can be completed in a reasonable amount of time, so that the transaction costs of decision-making are not prohibitive (Mueller 1989: 192; emphasis in the original).

Thus, corresponding to the two classes of public issues considered here, we have two different institutions of collective choice—majority and unanimity rule—and two different styles of policy-making. The contrast reveals the crucial importance of public deliberation for the politics of efficiency. Analysis and persuasion are needed to discover opportunities of collective gains and to elicit support in favor of the most efficient way of exploiting such opportunities.

Naturally, the unanimity rule represents an idealized model of collective decision-making. Except perhaps for very small communities, this method of collective choice entails transaction costs that are too high for everyday decision-making. Hence, advocates of unanimity such as Buchanan and Tullock and, before them, Knut Wicksell, have suggested near unanimity or some high fractional rule (say, 75 percent of the vote) as a way of preserving some of the advantages of unanimity without incurring excessive decision costs. But there are other possibilities, depending on the nature of the decision.

In a recent contribution to the debate, Buchanan (1989) develops a tripartite classification of political action: (1) enforcement of laws previously enacted; (2) collective provision of public goods; and (3) constitutional choices. Maintaining his position that the ultimate model of politics be contractarian, Buchanan argues that majoritarian determination has no place at levels (1) and (3), and is by no means a unique solution to level (2) decisions. Thus, for certain choices which may be predicted to embody potentially important consequences in expected costs and benefits, qualified majorities may be required for positive collective action. For other ranges of state activities, however, "authority may well be delegated to single agents or agencies" (Buchanan 1989: 178).

In fact, independent agencies and other non-majoritarian institutions are important actors in the politics of efficiency. Because of their insulation from the electoral cycle, their expertise and their commitment to a problem-solving, rather than a bargaining, style of decision-making such institutions sometimes succeed in resolving problems that are too complex or controversial to be dealt with by majoritarian politics. For example, American courts and independent regulatory commissions played a key role in bringing about the deregulation of basically competitive industries such as telecommunications and the transport industries. In their struggle against vested interests as well as congressional and bureaucratic inertia, these institutions relied heavily on policy analyses based on broad considerations of efficiency. In particular, the regulatory commissions "served as vehicles for converting the disinterested views of experts into public policy, even if the expert views had originated largely as criticisms of their own conduct" (Derthick and Quirk 1985: 91).

The politics of efficiency is the process by which diffuse, ill-organized, broadly encompassing interests sometimes succeed in overcoming particularistic and well-organized interests. The general conclusion of the literature cited in the introduction is that this process cannot be understood without acknowledging the role of ideas as independent variables. The significance of non-majoritarian institutions in this context is due to their ability to focus public attention on a particular issue, to diffuse policy ideas, and to translate such ideas into concrete decisions.

## IV Post-decision arguments and policy development

The significance of ideas is not limited to their role in problem-solving, however important that role may be in clarifying objectives, defining the range of possibilities for action, or in helping to select a particular outcome in the absence of a unique solution (Garrett and Weingast 1994). Because policy is made of language, arguments are used at every stage of the process. Every politician understands that ideas and arguments are needed not only to clarify his position with respect to an issue, but to bring other people around to this position. Even when a policy is best explained by the actions of groups seeking selfish goals, those who seek to justify the policy must appeal to the intellectual merits of the case (Kingdon 1984; Majone 1989).

Perhaps these are only rationalizations, but even rationalizations are important because they become an integral part of the public debate and thus can affect subsequent policy developments. Students of policy-making know only too well that ideas and analyses are often used to justify decisions already taken. When the arguments are based on considerations different from those that led to the decision, they are usually dismissed as attempts at "rationalization." However, this criticism, even if it may be justified in particular cases, misses the point that post-decision arguments can have rationally defensible uses in the overall process of policy development. A few examples will show the variety of such uses.

As my first example I take a well known episode in the history of the diffusion of economic ideas. President Franklin D. Roosevelt's policy of increased government spending to reduce unemployment and get out of the depression has been called Keynesian. But Roosevelt did not have to learn about government spending from Keynes. The idea that the influence of the British economist lay behind the policies of the New Deal began to take root fairly early, but it is only a legend (Winch 1969). The theories of Keynes only provided a sophisticated rationalization for what Roosevelt was doing anyway. The answers that these theories provided to questions about the causes of long-term unemployment and the reasons for the effectiveness of public spending were not prerequisites for Roosevelt's expansionist fiscal policy. But as these answers came to dominate the thinking of economists and politicians, they helped to make expansionist fiscal policy the core idea of liberal economic policy for several decades. In the words of a former chairman of the President's Council of Economic Advisers, "[w]ithout Keynes, and especially without the interpretation of Keynes by his followers, expansionist fiscal policy might have remained an occasional emergency measure and not become a way of life" (Stein 1984: 39).

Similarly, the Sherman and Clayton Acts establishing for the first time an American antitrust policy were not influenced by the economic theory of monopoly, then in its infancy. Rather, the present sophistication of antitrust economics in the United States owes a good deal to the early development of antitrust law, and to the consequent growth of a market in economists, whether as expert witnesses or as policy analysts (Hannah 1990: 375). But this is not to say that economics did not influence the development of antitrust policy. On the contrary, the significant changes in antitrust enforcement of the past quarter of a century have been informed by and even driven by contemporary economic analysis (Williamson 1987: 301).

To see how widespread is the use of post-decision arguments in all spheres of public life, consider the situation of a judge who decides a case on the basis of his subjective notion of fairness, a hunch that a particular decision would be right, while realizing at the same time that considerations of this kind do not count as justifications for a binding determination. Thus, the judge frames his opinion in the objective categories of legal argument, and any subsequent developments in the case (for example, an appeal) will be based on the published opinion, not on the actual process followed by the judge in coming to the conclusion. In fact, most legal systems allow the opinion stating the reasons for a judicial decision to follow rather than precede the decision. Or, again, different judges may agree on a decision but disagree about the best way to justify it; in the American system they are given the opportunity to present their positions in separate arguments.

Such procedural rules must appear absurd to somebody who assumes that a judicial opinion is an accurate description of the decision process followed by the judge in coming to a conclusion. If, however, the opinion is viewed as a report of justificatory procedures employed by the judge, then the appeal to legal and logical considerations which possibly played no role in the actual decision process becomes quite understandable (Wasserstrom 1961). In fact, the judge's opinion is not the premise of a syllogism that concludes in the decision; it is, rather, a means of exercising rational control over conclusions that may be suggested by extra-legal considerations, and of facilitating communication among participants in the legal process.

The fact that post-decision arguments are used in very different contexts—they play an important role even in the natural sciences (Majone 1989: 30–1)—is an indication that such arguments may serve important social functions beyond providing mere "rationalizations" for politically or bureaucratically determined positions. In fact, our examples suggest three such functions. First, post-decision arguments serve to *rationalize policy* in the sense of providing a conceptual foundation for a set of otherwise discrete and disjointed decisions. Policy-makers often act in accordance with pressures from external events or the force of personal convictions. In such cases arguments are needed *after* the decision is made, in order to explain it, to show that it fits into the framework of existing policy, to increase assent, to discover new implications, and to anticipate or answer criticism. Moreover, since policies exist for some time, new arguments are constantly needed to give the policy components greater internal coherence and a better fit to an ever-changing environment. The relevance of economic analysis to an antitrust policy initially conceived primarily as a *political* response to excessive market power, can be explained in such terms.

Second, post-decision arguments serve to *institutionalize ideas.* Stein's observation about the importance of Keynesian ideas in making expansionist fiscal policy "a way of life" captures the essence of the process. In a similar vein, Garrett and Weingast (1994) have shown how the idea of "mutual recognition," already present in the Treaty of Rome creating the European Economic Community, became institutionalized through the jurisprudence of the European Court of Justice and several documents of the European Commission. In this form, the idea had a powerful influence on the development and implementation of the internal market ("Europe 1992") program. It is important to note that the relationship between policy and institutionalized ideas is a dialectic one. Rather than disclosing new possibilities, such ideas only codify initial practice; at the same time, however, they serve to rationalize, evaluate and transform that same practice. Hence, our understanding of the way a policy develops cannot be separated from the institutionalized ideas and theories by which the policy is guided and, at the same time, evaluated (Majone 1989: 146–9; Krasner 1994).

The third, and perhaps most important, function of post-decision arguments is to *transform a single play into a sequential game* by making communication among the players possible. Only the judge's written opinion, not his decision as such, allows interested parties to maker further moves such as appealing the decision. It is important to keep in mind that in this as in other legal proceedings such as constitutional judicial review, the issue is what reasons *can* be given, even if those reasons are entirely *post hoc.* This shows that the purpose of the giving reasons requirement is not to improve the quality of a single decision but to facilitate the development of the entire process.

Similarly, the requirement that administrators give reasons for their decisions (as demanded, for example, by the U.S. Administrative Procedures Act and by Article 190 of the Treaty of Rome) generates a record consisting not only of the reasons actually given, but also of the statutes or treaty articles that those reasons elaborate. Thus, the giving reason requirement opens the door to a dialogue about rival statutory interpretations by court and agency. Moreover, public participation and policy deliberation are greatly facilitated if administrators have to give reasons for their decisions.

The importance of transforming a single play into an iterated game is well known to game theorists. In a Prisoners' Dilemma situation, for example, repetition allows more complicated strategies than simply "co-operate" or "defect." When the game is repeated, patterns of co-operation emerge that would be highly unlikely or even irrational in a single play. Similarly, post-decision arguments, and in particular the giving reasons requirement of western legal systems, can increase the efficiency of the policy process by facilitating communication and co-operation among the policy actors.

## V Conclusion: ideas, institutions, and the changing nature of policy-making

It may be helpful to conclude this chapter by returning to the relation between the transformation of contemporary policy-making and the new emphasis on ideas and institutions in policy studies. That transformation can perhaps be summarized by saying that we are moving away from the old state-centered, top-down view—exemplified by reliance on command-and-control instruments—and also from the bottom-up perspective so popular in the 1960s and 1970s, toward a more contractual view of policy-making. I suggest that such a view provides a convenient conceptual framework for grasping the connections between the search for efficiency, the increasing significance of policy credibility, and the rediscovery of the importance of institutions.

Briefly, the assumption is that all policy actors—among whom, it will be recalled, one may have to include foreign actors not subject to the coercive power of the state—face a situation of "incomplete contracting." Here I follow the terminology of the new institutional economics (Williamson 1985; Milgrom and Roberts 1992) in using the term "contract" to designate not only a legally enforceable promise or threat, but also an informal, even tacit, agreement among parties engaged in a joint undertaking.

Now, an incomplete contract is one where it is expected that contingencies will arise that have not been accounted for, *ex ante*, because they were not even imagined at contracting time. This is, of course, the normal situation in policy implementation (Majone and Wildavsky 1978). Incomplete contracting leads to problems of imperfect commitment. There is a strong temptation to renege on the original terms of the contract because what should be done in case of an unforeseen contingency is left unstated or ambiguous and thus open to interpretation. The problem is that the possibility of renegotiating deprives the original agreement of its credibility and prevents it from guiding behavior as intended. The problem of time inconsistency analyzed by Kydland and Prescott (1977) is the policy equivalent of imperfect commitment in incomplete contracting. In both cases the root problem is the fact that the incentives of policy actors or contractual partners in the implementation phase may no longer be the same as their incentives in the planning stage.

One response to contractual incompleteness is an arrangement, known as "relational contracting" (Williamson 1985), where the parties do not agree on detailed plans of action, but on goals and objectives, on the criteria to be used in deciding what to do when unforeseen contingencies arise, and on dispute resolution mechanisms to resolve disagreements. In adapting the relationship to unforeseen circumstances often one party will have much more authority in saying what adaptation will take place. In the absence of coercion, the other parties will be willing to delegate such discretionary authority only if they believe that it will be used fairly and effectively. An important source of this belief is reputation. The party to whom authority is delegated should be the one with the most to lose from a loss of reputation. This is likely to be the one with the longer time horizon, the more visibility, and the greater frequency of transactions (Milgrom and Roberts 1992: 140). In the context of public policy-making, this is more likely to be an expert regulator than a politician or a bureaucratic generalist. Thus, the willingness of political sovereigns to delegate important policy-making powers to expert bodies may be explained as a strategy for achieving credible commitments in situations of incomplete contracting.

However, a reputation for fairness and effectiveness cannot be established by legislative or executive fiat. It has to be based on a record of accomplishments and on the general perception that the solutions advanced by the experts are not only conceptually sound but also aimed at increasing the welfare of all parties rather than that of a particular group. It follows that ideas, arguments and persuasion have an important role to play in enhancing the reputation of those to whom policy-making powers have been delegated, and hence the credibility of their political principals.

Equally important are institutions, since an effective system of reputation cannot depend only on individual behavior, but must be supported by procedural rules capable of ensuring fairness and transparency, as well as by the administrative culture and *esprit de corps* of the entire organization.

I hope that this brief sketch of the contracting approach is sufficient to suggest its usefulness as a model of contemporary policy-making at the national and supranational levels, and also as a conceptual framework for analyzing the growing significance of ideas and institutions in the policy process.

## References

Given the limited scope of this chapter, it has been necessary to proceed very selectively. The decision to concentrate on the most recent literature and the latest methodological developments has led to the sacrifice of some older, but still valuable, studies as well as important strands of related research. Many of the older studies were concerned with the policy relevance of social scientific knowledge. Useful introductions to this topic include the National Science Foundation, *Knowledge into Action: Improving the Nation's Uses of the Social*

*Sciences* (1968), and L. E. Lynn, ed., *Knowledge and Policy: The Uncertain Connection* (1978). A penetrating critical assessment of the practical relevance of social scientific knowledge is provided by C. E. Lindblom and D. K. Cohen, *Usable Knowledge* (1979).

No student of the role of ideas in the policy process can ignore the current debate on discursive vs. instrumental rationality. In addition to my *Evidence, Argument and Persuasion in the Policy Process* (1989), the following are useful references: J. S. Nelson, A. Megill and D. N. McCloskey, eds., *The Rhetoric of the Human Sciences: Language and Argument in Scholarship and Public Affairs* (1987); and J. S. Dryzek, *Discursive Democracy* (1990).

Finally, on the political and policy relevance of philosophical concepts and conceptual change, the interested reader should consult R. E. Goodin, *Political Theory and Public Policy* (1982), and T. Ball, J. Farr, and R. L. Hanson, eds., *Political Innovation and Conceptual Change* (1989).

Aaron, H. J. 1978. *Politics and the Professors.* Washington, D.C.: Brookings Institution.

Ball, T.; Farr, J.; and Hanson, R. L., eds. 1989. *Political Innovation and Conceptual Change.* Cambridge: Cambridge University Press.

Bentley, A. F. 1967. *The Process of Government.* Cambridge, Mass.: Belknap Press; originally published 1908.

Buchanan, J. M. 1989. Contractarian presuppositions and democratic governance. Pp.174–82 in *Politics and Process,* ed. G. Brennan and L. E. Lomasky. Cambridge: Cambridge University Press.

—— and Tullock G. 1962. *The Calculus of Consent.* Ann Arbor: University of Michigan Press.

Cohen, M.; March, J.; and Olsen, J. 1972. A garbage can model of organizational choice. *Administrative Science Quarterly,* 17: 1–25.

Dantzig, G. B. 1963. *Linear Programming and Extensions.* Princeton, N.J.: Princeton University Press.

Derthick, M. 1979. *Policy-making for Social Security.* Washington, D.C.: Brookings Institution.

—— and Quirk, P. J. 1985. *The Politics of Deregulation.* Washington, D.C.: Brookings Institution.

Dryzek, J. S. 1990. *Discursive Democracy.* Cambridge: Cambridge University Press.

Garrett, G., and Weingast, B. R. 1994. Ideas, interests, and institutions: constructing the European Community's internal market. In Goldstein and Keohane 1994: 173–206.

Goldstein, J. 1993. *Ideas, Interests, and American Trade Policy.* Ithaca, N.Y.: Cornell University Press.

—— and Keohane R. O., eds. 1994. *Ideas and Foreign Policy.* Ithaca, N.Y.: Cornell University Press.

Goodin, R. E. 1982. *Political Theory and Public Policy.* Chicago: University of Chicago Press.

Haas, E. B. 1990. *When Knowledge Is Power.* Berkeley: University of California Press.

Hall, P. A. 1986. *Governing the Economy.* New York: Oxford University Press.

—— ed. 1989. *The Power of Economic Ideas.* Princeton, N.J.: Princeton University Press.

Hannah, L. 1990. Economic ideas and government policy on industrial organization in Britain since 1945. Pp. 354–75 in *The State and Economic Knowledge,* ed. M. O. Furner and B. Supple. Cambridge: Cambridge University Press.

Kingdon, J. W. 1984. *Agendas, Alternatives and Public Policy.* Boston: Little, Brown.

Krasner, S. D. 1994. Westphalia and all that. In Goldstein and Keohane 1994: 235–64.

Kydland, F., and Prescott E. 1977. Rules rather than discretion: the inconsistency of optimal plans. *Journal of Political Economy,* 85: 137–60.

LINDBLOM, C. E. 1965. *The Intelligence of Democracy.* New York: Free Press.
—— and COHEN, D. K. 1979. *Usable Knowledge.* New Haven, Conn.: Yale University Press.
LOWI, T. J. 1979. *The End of Liberalism.* 2nd edn. New York: Norton.
LYNN, L. E., ed. 1978. *Knowledge and Policy: The Uncertain Connection.* Washington, D.C.: National Academy of Sciences.
MAJONE, G. 1989. *Evidence, Argument and Persuasion in the Policy Process.* New Haven, Conn.: Yale University Press.
—— 1994. The rise of the regulatory state in Europe. *West European Politics,* 17: 77–101.
—— 1995. Mutual trust, credible commitments and the evolution of rules for a Single European Market. EUI Working Paper RSC N°95/1. Florence: European University Institute.
—— and WILDAVSKY, A. 1978. Implementation as evolution. Vol. ii, pp. 103–17 in *Policy Studies Review Annual,* ed. H. E. Freedman. Beverly Hills and London: Sage Publications.
MASHAW, J. L., and HARFST, D. L. 1990. *The Struggle for Auto Safety.* Cambridge, Mass.: Harvard University Press.
MILGROM, P., and ROBERTS, J. 1992. *Economics, Organization and Management.* Englewood Cliffs, N.J.: Prentice-Hall.
MUELLER, D. C. 1989. *Public Choice II.* Cambridge: Cambridge University Press.
MURRAY, C. 1984. *Losing Ground.* New York: Basic Books.
NATIONAL SCIENCE FOUNDATION. 1968. *Knowledge into Action: Improving the Nation's Uses of the Social Sciences.* Washington, D.C.: Government Printing Office.
NELSON, J. S.; MEGILL, A.; and MCCLOSKEY, D. N., eds. 1987. *The Rhetoric of the Human Sciences: Language and Argument in Scholarship and Public Affairs.* Madison: University of Wisconsin Press.
NICOLL, W. 1993. Maastricht revisited: a critical analysis of the treaty on european union. Vol. ii, pp. 19–34 in *The State of the European Community,* ed. A. W. Cafruny and G. G. Rosenthal. Boulder, Colo.: Lynn Rienner.
ODELL, J. S. 1982. *U. S. International Monetary Policy.* Princeton, N.J.: Princeton University Press.
OLSON, M. 1965. *The Logic of Collective Action.* Cambridge, Mass.: Harvard University Press.
—— 1982. *The Rise and Decline of Nations.* New Haven, Conn.: Yale University Press.
ORDESHOOK, P. C. 1992. *A Political Theory Primer.* London: Routledge.
ROSE, R. 1993. *Lesson-Drawing in Public Policy.* Chatham, N.J.: Chatham House Publishers.
ROSE-ACKERMAN, S. 1992. *Rethinking the Progressive Agenda.* New York: Free Press.
SCHULTZE, C. L. 1977. *The Public Use of Private Interest.* Washington, D.C.: Brookings Institution.
SHAPIRO, M. 1988. *Who Guards the Guardians?* Athens: University of Georgia Press.
SHEPSLE, K. A. 1991. Discretion, institutions and the problem of government commitment. Pp. 245–63 in *Social Theory for a Changing Society,* ed. P. Bourdieu and J. S. Coleman. Boulder, Colo.: Westview Press.
STEIN, H. 1984. *Presidential Economics.* New York: Simon and Schuster.
SUNSTEIN, C. 1990. *After the Rights Revolution.* Cambridge, Mass.: Harvard University Press.
TEMIN, P. 1987. *The Fall of the Bell System.* Cambridge: Cambridge University Press.
WASSERSTROM, R. A. 1961. *The Judicial Decision.* Stanford, Calif.: Stanford University Press.
WILLIAMSON, O. E. 1985. *The Economic Institutions of Capitalism.* New York: Free Press.
—— 1987. *Antitrust Economics.* Oxford: Blackwell.
WINCH, D. 1969. *Economics and Policy: A Historical Study.* New York: Walker.
WILSON, J. W., ed. 1980. *The Politics of Regulation.* New York: Basic Books.

Chapter 27

# Public Policy and Administration, Old and New

B. Guy Peters
Vincent Wright

## I Fundamental shifts in thinking about public administration

Of all the areas of political science, no other has undergone a transformation comparable to that experienced by public administration over the past twenty years. This reflects in large part the changing nature of the practice of governments, especially in the developed world. Almost all the essential truths that guided practising public administrators and students of administration have now been challenged and often replaced. It is unclear whether any new doctrine has been agreed upon, but it is clear that the old values and practices are now profoundly contested.

Changes in public administration also reflect changes in the intellectual approaches used to study this field. Most importantly, the changes reflect the closer linkages of many contemporary scholars of public administration with other areas of the discipline. Studies of the public bureaucracy can no longer be dismissed as "manhole counting" but are in the mainstream, and often even at the fore, of some developments in empirical political and organizational theory. The practice of traditional public administration has come under increasing attack from neo-liberal economists, interest group theorists and rational choice scholars who have provided the intellectual ammunition for receptive politicians determined to reduce the size and scope of the public sector. This is scarcely surprising, since the theoretical changes have tended to emphasize the significant extent to which public administration is political and is part of the overall process of determining "who gets what." Approaches to public administration are also embedded in wider conceptions of the state, the relationship between state and market, and of citizenship. Changes in the ideological climate are, therefore, likely to impact upon public administration.

Some of the changes occurring in public administration can be summarized in a few short phrases. The most commonly heard such phrase is the "new managerialism" (Pollitt 1993; Hood 1991), meaning that management ideas, derived largely from the private sector, have replaced the concepts of traditional public administration. While these changes have been justified in the name of the "three Es"—economy, efficiency and effectiveness—they have also had profound effects on the role of administration in making public policy and in the status of public servants. In particular, even more so than hitherto the job of the public servant has become running an organization efficiently, rather than participating in policy decisions. As well as being manifest within government, this approach has its academic advocates who argue that public and private management are essentially alike (but see Allison 1986 and Self 1995).

Associated with the new managerialism is a "new patrimonialism," in which political leaders attempt to gain greater control over appointments to public offices, and greater loyalty from civil servants. This trend has been clearest in the United States, with a number of positions in the Senior Executive Service being opened to political appointment, and has been argued to exist in other Anglo-Saxon countries. This development is less of a novelty in European Continental countries which have always allowed greater intermingling of political and administrative élites and roles, but even there questions about the degree of desirable and acceptable politicization have arisen (Mayntz and Derlien 1989).

Finally, a "new fragmentation" of government and administration is occurring in most political systems. Some of this fragmentation is occurring within central government, as ministries are subdivided into relatively autonomous organizations. Moreover, most central governments are decentralizing, transferring more powers to subnational state officials, quangos or elected officials and governments, not the least of which are powers to implement the programs of central government. These changes are cumulatively producing a much less coherent apparatus for governance in most industrialized democracies, although again the changes are usually justified in terms of efficiency. The ongoing fragmentation of governments has its theoretical advocates and forebears (Niskanen 1971) just as have the other major changes in governing, and some of the major changes in practice such as "corporatization" in New Zealand, have been strongly influenced by theorists of administration (Boston 1991).

## II The six great truths about public administration now amended

The transformation of public administration in many developed countries, and the study of public administration, with all its various manifestations, have led to the questioning of at least six fundamental "truths" that had guided both scholarship and practice (see Walsh and Stewart 1992). In their place there is, if not chaos, then certainly doubt about what constitutes both acceptable theory and good practice. Further, this transformation has made clearer than in recent years the intimate connection that exists between theory and practice in this area of the discipline. Indeed, theory appears to be more dependent upon practice than vice versa, and seemingly theories arise to help justify what is already true in practice.

### A The assumption of self-sufficiency

The idea that public administration has to be self-sufficient has been challenged by the interlinked policies of sub-contracting, privatization and competition (Wright 1994; Vickers and Yarrow 1988; Vernon 1988; Suleiman and Waterbury 1990; Gayle and Goodrich 1990; Bailey and Pack 1995). Contracting out involves the transfer to private agents, often by the process of tender, of the implementation of services previously undertaken by public officials. The policy may be compulsory or optional. Secretarial work, car services, street cleaning and garbage collection are amongst the services most commonly contracted out. More radical countries, such as the United Kingdom, have contracted out prison services, security (even of defense establishments), government computer and forecasting services. Even the contracting out of the drafting of government legislation is being contemplated. Privatization is perhaps the most spectacular and visible policy designed at "state retreat." It has become a "policy fashion" which has spread from the Chile of Pinochet to Communist China. It covers a range of policies, including the abolition or severe curtailment of public services or financial resources, on the assumption that private provision will compensate public–private partnerships in financing public infrastructure, and the transferring to the private sector of public policy responsibilities. Industrial privatization has involved the sale of subsidiaries, the recapitalization of public enterprises through private finance, and the sale of minority or majority stakes, or even the outright sale of public enterprises to the private sector. Radical governments have privatized not only competitive enterprises but also "strategic firms," national flag carriers and public sector utilities such as gas, electricity, water and telecommunications.

Linked to privatization and contracting out is the policy of competition, through deregulation (the partial or total dismantlement of state monopolies), and "marketization" or the introduction of quasi-markets into public sectors such as health, by enabling public purchasers, endowed with a degree of budgetary discretion, to seek the provider which offers the best value. Competition has also taken the form of empowering citizens, transformed into customers, by supplying them with vouchers in order to shop around for the most desirable public good available. Competition is intended to increase consumer choice, drive down prices and improve quality.

Finally, another important way in which the assumption of self-sufficiency has been challenged is through the recognition of the importance of networks of organizations within and around the public sector (Hanf and Scharpf 1978; Rhodes and Marsh 1992; Marin and Mayntz 1992). As Hjern and Porter (1981) have pointed out, the "single lonely organization" model of administration is simply no longer sustainable, either theoretically or empirically. It is now clear that to be successful any organization must coordinate its activities with those of other public organizations, frequently at multiple levels of government, as well as with organizations in the private sector. And to complicate matters, administrative regulation may now be transnational or international in character. The implementation literature has been essential in pointing out the mutual dependence of organizations in the public sector, as well as the dependence of public organizations on the private sector: the so-called "private management of public government" from the original Pressman and Wildavsky book (1994) onward through a variety of (self-proclaimed) second and third generation implementation studies (Marin and Mayntz 1992; Goggin 1990*a*; 1990*b*) the fundamental point of interdependence has remained central. This basic insight has since been fortified with a range of methodological and theoretical weapons for coping with the complexities of inter-organizational politics (Knoke and Laumann 1987), and the dynamics of populations of organizations in the public sector.

## B The assumption of direct control

A second assumption that has guided a good deal of thinking about the public sector is that of direct control, or hierarchy. Although it is sometimes associated with Weber, this principle is much older (it was one of the organizing principles of the Napoleonic administrative model of France and several other European countries), and is premised upon the need and even willingness of people in public organizations to follow orders given by their superiors within the organization. If the familiar organograms of public organizations have any validity, it is based upon that willingness of employees to exchange compliance for some monetary reward. Further, if the rule of law is to be upheld and there is to be a system of accountability within government the hierarchy becomes the crucial link between ministers and the decisions taken in their name by their numerous subordinates in the field.

Older ideas about hierarchical management are now being reassessed in the light of ideas about "empowerment," involving the granting of increased organizational power to both employees and clients of public organizations. First, the lower echelon employees of government are to be empowered, and given more control over their own jobs. While hierarchy implied direct supervision of employees and managerial controls over their decisions, empowerment is designed to give them greater latitude to make decisions and then be held accountable for them (Kernaghan 1992). Further, the ideas of total quality management, again from the private sector, have been brought into government to try to involve employees at all levels in improving the performance of their organizations (Swiss 1993).

These ideas are hardly new, and many organization theorists have long advocated greater democracy within public and private organizations. What is different, however, is the commitment of governments to this style of management—for example, in PS 2000 in Canada (Tellier 1990) or in some of the components of the National Performance Review (Peters and Savoie 1994), otherwise known as the Gore Report, in the United States. The empowerment movement also points to some of the contradictions in the current spate of reform efforts in government, and in the conceptualization of public management. On the one hand, managers are supposed to be free to manage, while on the other the lower echelons are supposed to have an increased level of organizational power.

In fairness, it is not only ideologues who have placed an emphasis on the role of lower echelon workers in public organizations. Beginning as early as the 1930s (Almond and Lasswell 1935) analysts of public organizations pointed to the importance of the lowest echelons of public organizations in determining who gets what from government. This insight later became known as "street level bureaucracy" (Lipsky 1980; Adler and Asquith 1981). The point of these empirical studies is that lower echelons of public organizations have always had a great deal of power over the public. The implication is, therefore, that rather than denying this under the fiction of traditional hierarchy and accountability, it may be better to recognize the fact, and then find ways to cope with the problems it raises (Day and Klein 1989).

As well as providing increased autonomy to lower echelon employees in public organizations, the empowerment movement also seeks to grant increased powers to the clients of those organizations. This participatory ethos and the desire for "customer driven" services has assumed a number of forms. The Citizen's Charters in the United Kingdom or their equivalents in France, for example, are largely statements of the types of services that citizens should expect from their government. Some of the reforms sponsored by the Gore Commission in the USA also force government organizations to identify their customers and attempt to serve them rather than the organization's own definitions of what those customers should have.

## C The assumption of uniformity

A common assumption of traditional public administration and governance was that all citizens should receive, in so far as possible, equal benefits from the state and bear equal burdens to support the state. This was, in part, a justification for the creation of large centralized bureaucracies in modern states, especially the welfare states of Europe. By training, rules and supervision these organizations could produce equality throughout the land. Direct control within bureaucracies discussed above was one of the means used to produce the uniformity required by this conception of good government.

Again, a number of intellectual and political forces have combined to call into question the need for uniformity and to provide alternative models about how to govern. Most importantly decentralization and deconcentration of administration, and of government more generally, have become popular antidotes for the alleged failings of centralization. Just as firms in the private sector began to break up into product lines following the consolidations of the 1970s and 1980s, so, too, have public organizations begun to differentiate themselves and become more focused on single policies and issues.

Several strands of literature in organization theory (in't Veld 1993) began to argue that centralized control and uniformity were impossible, so public organizations should not waste efforts in pursuing this will o' the wisp. Further, students of personnel management re-emphasized the old point that hierarchical control tends to alienate employees. Some theorists even began to question the extent to which equality and uniformity of services were absolute rights of citizens, or even desirable outcomes for administrative processes.

Finally, and perhaps most importantly, the implementation literature has pointed out the extent to which outcomes are dependent upon implementation by bureaucracies, and hence the crucial role these organizations play in making and enforcing allocative decisions (Pressman and Wildavsky 1974). This has often led to a passive acceptance of the idea that implementation concerns should guide policy "from the bottom up," rather than vice versa, but still points to the central position of administration in governance (Linder and Peters 1989). This literature further points out that all the policy decisions in the world amount to little if the tools are not available to put them into effect.

The result of these changes in thinking about government has been pressure for greater decentralization and deconcentration in administration. Some of the decentralization has been functional, with programs such as "Next Steps" in the United Kingdom and analogues in other countries breaking down large ministries into numerous smaller, generally monofunctional, organizations (Greer 1994). While these organizations may still strive to provide their clients equally across the territory of the state, clients of different agencies may be treated quite differently. Further, efficiency criteria may override equity criteria if the two come into obvious conflict.

The more obvious growth of inequality arises in territorial decentralization and deconcentration (de Montricher 1994). Central governments, with the notable exception of that of the United Kingdom, have become increasingly willing to permit subnational governments to make allocative decisions about who will get what from the public sector, and to permit lower level offices in their own organizations to make those decisions. This is related in part to the concept of empowerment discussed above, and the desire to enhance the role of lower echelon public servants. It is also a recognition of the principle that one size fits all may not be adequate in increasingly differentiated societies, and granting power to street level bureaucrats may therefore also increase the satisfaction of clients.

## D The assumption of accountability upward

Traditional bureaucratic models, whether rooted in colonial experience, military command structures or Weberian rationalization, underlined the principle of upward accountability to the political sovereign, which came to signify, in democratic parliamentary regimes, to ministers. Under the doctrine of ministerial responsibility, bureaucrats were garbed in anonymity, accountable to their political masters who assumed, at least in principle, responsibility for policies carried out in their departments (Day and Klein 1989). The proliferation of semi-autonomous agencies, the introduction of markets, the devolution of implementation to quangos and third sector bodies have combined to blur the traditional lines of responsibility. Cynics have even suggested that one of the basic, yet unspoken, objectives of new public management is precisely to remove ministers from the firing-line. And the early experience of the United Kingdom suggests that chief executives may take the blame for policies of their agencies: the propensity for ministers to evade responsibility has thus been aggravated. Defenders of the reforms argue that no change to the principle of political accountability has been affected. Rather, it has now merely been combined with a principle of accountability downward to the customers of public services: such innovations as the naming of public officials, under the introduction of publicized quality and performance indicators are all designed to enhance this latter principle.

## E The assumption of standardized establishment procedures

One of the central features of traditional public administration was that the civil service was a distinctive career structure and was managed according to principles that emphasized its distinctiveness from employment in the private sector. Individuals were employed as a part of a corps of public servants and received their job by demonstrating their merit to hold the position. They were rewarded for their work according to their rank in the system rather than any particular merits or demerits. Once past a probationary period they were given tenure and had a permanent post, regardless of changes in the political complexion of the government, and could be dismissed only for malfeasance.

While individual countries varied in the extent to which they conformed to one of the props of this idealized Weberian system, the principles did guide personnel policies in government. Again, much of this traditional system has come under fire and is being replaced by personnel management practices drawn from the private sector. One of the most important changes of this type has been the introduction of pay for performance in the public sector (Ingraham 1993). The performance of individual public servants is now being assessed, and differential pay, whether in the form of bonuses or increases in base pay, is being granted to better performers. This reward scheme has been introduced in almost all industrialized democracies, even those such as Sweden with long histories of pay solidarity (Sjölund 1994).

Merit pay raises a number of thorny questions for the public services. First, how is performance measured in the public service, and how can we attribute the success or failure of programs to individuals? Does measuring the measurable tend to emphasize the managerial role of public servants and de-emphasize the policy advice role? Second, is there not a conflict between individualized performance pay and group-oriented management techniques such as Total Quality Management? Also, is individual merit pay compatible with the idea of a civil service, or does it imply moving almost entirely toward a set of contract employees who simply happen to work for government?

Many of these changes call into question the "operational bases" of traditional public administration—the distinctiveness of the public service career, with its emphasis on vocation, ethos and seclusion. Of course, practice suggests that this may have been an idealized view in some countries. Nonetheless, its distinctiveness was a feature of the public administrations of many countries where the reformers have been especially active. Internal organization reforms inspired by the private sector have been combined with attempts to transform the recruitment patterns (by bringing in people from the private sector, often on short-term contracts), the objectives and the culture of public officials. These are now seen not as impartial providers of universal services to citizens, keen on respecting due process, but as managers, entrepreneurs, "doers," sensitive to the efficient and specific requirements of their clients. Whether a vast cultural change has taken place within the public administration is the subject of some skepticism.

## F The assumption of an apolitical service

The final assumption that has been brought into question by the interaction of academic research and real world developments is the idea that the civil service is, can be, or perhaps even should be, politically neutral. Of course, this principle was violated in many European countries where many top officials have openly professed their political allegiance, entered politics, or worked openly for political parties. But there appears to be a trend towards the politicization of administration in many other developed countries. The political realities of the late 20th century have been that politicians everywhere perceive themselves to have lost some of their capacity to direct government, and blame that more on the aggressiveness (or deviousness) of their civil service than their own lack of managerial and policy capacity (Peters 1991). On the other hand, civil servants are increasingly being told that they should be aggressive and entrepreneurial in pursuing the goals of their organizations. Further, they often do have policy ideas of their own that they feel would improve the services given to their clients, and therefore that they should pursue those policies. These two sets of perspectives about the role of the civil service appear destined to clash.

As we begin to discuss the changes that have occurred around this assumption, it is important first to unpack what we mean by "political" in this context. On the one hand, political often implies partisan, as when political leaders attempt to gain enhanced powers over the appointment of civil servants for the purpose of ensuring greater control over policy. Attempts to "politicize" the civil service have been noted in a number of countries with apolitical traditions (Meyer 1985). This increased partisanship need not be entirely negative, and some forms of accountability might be improved by closer linkage of political and administrative roles, but the changes will still require some rethinking of basic issues about the civil service role in governance (Day and Klein 1989).

More importantly, however, political can mean that civil servants have views about policy and about institutions that drive them towards openly advocating and promoting those views. Ministers often complain about the resistance they encounter from the "departmental view" in many well-established government organizations. This has been reported anecdotally for a number of years, but there is now more systematic evidence about the perceived roles of civil servants as policy actors (Aberbach, Putnam and Rockman 1981; Krauss and Muramatsu 1988). In almost no cases are civil servants actively attempting to thwart democratic control: rather, they are attempting to press particular views about good policy on their ministers, and often to save ministers from costly and embarrassing policy failures.

In addition to the significant empirical work on the roles of senior civil servants (Campbell 1988), there are two other important strands of theorizing about the political role of civil servants. First, the development of several versions of the "new institutionalism" in political science (March and Olsen 1989; Shepsle 1989; Thelen and Steinmo 1992) has provided another means of looking at the role of the public service as political actors. One of the premises of this approach (or at least of some variants of it) is that institutions embody values of their own that they attempt to inculcate into their members and to use as a mechanism for shaping policies. The literature on organizational culture had already made some of the same points, but the more manifestly political role of organizational values and interests has been emphasized in the institutionalist literature.

While most of the new institutionalist literature stresses the role of organizational values and ideas in shaping policy, the rational choice approach stresses the importance of organizational interests. In this view organizations develop responses to issues in order to enhance their own collective interests. They use the resources at their command, usually expertise and information, in an attempt to preserve or maximize their own budgets and to enhance their flexibility (Kato 1994; cf. Dunleavy 1985). In this view, bureaucratic organizations are self-interested actors that make coalitions with politicians or oppose them not for partisan reasons but primarily for reasons of collective aggrandizement. Such a view, of course, reifies the organizations but still points to the importance of organizational interests as motivators for decisions made within a public bureaucracy.

## III Conclusion: something lost, something gained?

The movement from the old public administration to the new public management has varied in timing, pace and intensity across developed countries. Constitutional, legal and cultural impediments have prevented its spread to several continental European countries. Where it has taken place it has created some real benefits for the employees of the public sector. Drawing on insights from empirical analysis as well as organizational theory, the managerial approach recognizes some of the realities of governing and rather than ignoring or dismissing them, accepts and even revels in them. Further, this emergent view of the public sector recognizes the central role that public administration plays in governing and is, for the most part, comfortable with that role.

One factor about the new public management that is often overlooked is the extent to which its prescriptions are very clearly contradictory. These contradictions are certainly apparent in practice, but they are also evident even for the principles that are meant to guide action. On the one hand public administrators are supposed to be autonomous and entrepreneurial, and pay attention to signals coming to them from their clients. On the other hand they are also supposed to be more responsive to the signals coming to them from their political masters (Pollitt 1995). Similarly authority is presumably to be decentralized to individual organizations, and within organizations to lower echelon employees, but at the same time tighter central financial control is supposed to save public money.

It is therefore not too much of an overstatement to say that the new public management is not a theory in any meaningful sense of the term. It does not provide a coherent and integrated set of propositions about running the public sector, but rather appears more to generate "principles" that are compatible with the political thinking of the day. The fundamental social loss arising from the change away from the old public administration, with all of its apparent rigidity and seemingly outmoded assumptions, is that at least that old model contained a consistent set of ideas that had proven their utility over time. Their antiquity is not the only defense of the ideas of the older public administration. They were able to create (in most countries) a public service with an ethos of service, and with a clear idea of the limits of its role.

Further, this more consistent approach to the public sector provided the means for linking bureaucracy and democracy in a clearer way than do the contemporary characterizations of the field. Certainly civil servants did possess powers over policy that were not totally compatible with the conventional model, but the strength was that there were always legitimate means for political leaders to recapture those powers if there were cause to do so. Certainly lower echelon workers did create informal organization and engage in behavior that bypassed hierarchical control, but again there were mechanisms for reasserting authority if necessary. If nothing else, the traditional model did provide a clearer normative standard, and one that was oriented to the public interest, than is available from much of the new public management.

## References


Aberbach, J. D.; Putnam, R. D.; and Rockman, B. A. 1981. *Politicians and Bureaucrats in Western Democracies.* Cambridge, Mass.: Harvard University Press.

Adler, M., and Asquith, S. 1981. *Discretion and Power.* London: Heinemann.

Allison, G. T. 1986. Public and private management: are they fundamentally alike in all unimportant respects? Pp. 187–208 in *Classics of Public Administration*, ed. J. M. Shafritz and A. C. Hyde. Homewood, Ill.: Dorsey.

—— 1971. *Essence of Decision.* Boston: Little, Brown.

Almond, G. A., and Lasswell, H. D. 1935. Aggressive behaviors by clients toward public relief administrators. *American Political Science Review*, 28: 643–55.

Bailey, E. E., and Pack, J. R., eds. 1995. *The Political Economy of Privatization and Deregulation.* Aldershot: Elgar.

Boston, J. 1991. The theoretical underpinnings of state restructuring in New Zealand. Pp. 1–26 in *Reshaping the State*, ed. J. Boston *et al.* Auckland: Oxford University Press.

Campbell, C. 1988. Review article: the political roles of senior government officials in advanced democracies. *British Journal of Political Science*, 18: 243–72.

Day, P., and Klein, R. 1989. *Accountabilities.* London: Tavistock.

de Montricher, N. 1994. *La deconcentration.* Paris: CNRS.

Dunleavy, P. 1985. Bureaucrats, budgets and the growth of the state. *British Journal of Political Science,* 15: 299–328.

Gayle, D. J., and Goodrich, J. N., eds. 1990. *Privatization and Deregulation in Global Perspective.* London: Pinter.

Goggin, M. L. 1990*a. Implementation Theory and Practice.* New York: HarperCollins.

—— 1990*b. Policy Design and the Study of Implementation.* Knoxville: University of Tennessee Press.

Greer, P. 1994. *Transforming Central Government: The Next Steps Initiative.* Buckingham: Open University Press.

Hanf, K., and Scharpf, F. W. 1978. *Inter-Organizational Policy-making.* Beverly Hills, Calif.: Sage.

Hjern, B., and Porter, D. O. 1981 Implementation structures: a new unit of administrative analysis. *Organization Studies,* 2: 211–29.

Hood, C. 1991. A public management for all seasons? *Public Administration,* 69: 3–19.

Ingraham, P. W. 1993. Of pigs and pokes and policy diffusion: another look at pay for performance. *Public Administration Review,* 53: 348–56.

in't Veld, R. 1993. *Autopoesis and Configuration Theory.* Dordrecht: Kluwer.

Kato, J. 1994. *The Problem of Bureaucratic Rationality: Tax Politics in Japan.* Princeton, N.J.: Princeton University Press.

Kernaghan, K. 1992. Empowerment and public administration: revolutionary advance or passing fancy? *Canadian Public Administration,* 35: 194–214.

Knoke, D., and Laumann, E. O. 1987. *The Organizational State.* Madison: University of Wisconsin Press.

Krauss, E., and Muramatsu, M. 1988. Elite attitudes and Japanese democracy. *American Political Science Review,* 82: 241–68.

Linder, S. H., and Peters, B. G. 1989. Implementation as a guide for policy: a question of "when" rather than "whether." *International Review of Administrative Science,* 55: 279–94

Lipsky, M. 1980. *Street Level Bureaucracy.* New York: Russell Sage Foundation.

March, J. G., and Olsen, J. P. 1989. *Rediscovering Institutions.* New York: Free Press.

Marin, B., and Mayntz, R. 1992. *Policy Networks.* Boulder, Colo.: Westview.

Mayntz, R., and Derlien, H.-U. 1989. Party patronage and politicization of the West German administrative elite 1970–87—toward hybridization. *Governance,* 2: 384–404.

Meyer, F. 1985. *La politicisation d'administration publique.* Brussels: Institut International d'administration publique.

Niskanen, W. 1971. *Democracy and Representative Government.* Chicago: Aldine-Atherton.

Peters, B. G. 1991. Public bureaucracy and public policy. Pp. 283–316 in *History and Context in Comparative Public Policy,* ed. D. E. Ashford. Pittsburgh: University of Pittsburgh Press.

—— and Savoie, D. J. 1994. Reinventing Osborne and Gaebler. *Canadian Public Administration,* 37: 302–22.

Pollitt, C. 1993. *Managerialism and the Public Service.* 2nd edn. Oxford: Blackwell.

—— 1995. Justifying by works or by faith: evaluating the new public management. Paper presented to European Group for Public Administration, Rotterdam.

Pressman, J., and Wildavsky, A. 1974. *Implementation.* Berkeley: University of California Press.

Rhodes, R. A. W., and Marsh, D. 1992. New directions in the study of policy networks. *European Journal of Political Research,* 21: 181–205.

SELF, P. 1995. *Government by the Market.* Boulder, Colo.: Westview.

SHEPSLE, K. 1989. Studying institutions: some lessons from the rational choice approach. *Journal of Theoretical Politics,* 1: 131–48.

SJÖLUND, M. 1994. Sweden. Pp. 144–78 in *Rewards at the Top,* ed. C. Hood and B. G. Peters. London: Sage.

SULEIMAN E. N., and WATERBURY, J., eds. 1990. *The Political Economy of Public Sector Reform and Privatization.* Boulder, Colo.: Westview.

SWISS, J. 1993. Adapting Total Quality Management (TQM) to government. *Public Administration Review,* 52: 356–62.

TELLIER, P. 1990. Public Service 2000: the renewal of the public service. *Canadian Public Administration,* 33: 123–32.

THELEN, K., and STEINMO, S. 1992. Historical institutionalism in comparative politics. Pp. 1–32 in *Structuring Politics,* ed. S. Steinmo, K. Thelen and F. Longstreth. Cambridge: Cambridge University Press.

VERNON, R., ed. 1988. *The Promise of Privatization.* New York: Council on Foreign Relations.

VICKERS, J., and YARROW, G. 1988. *Privatization: An Economic Analysis.* Cambridge, Mass: MIT Press.

WALSH, K., and STEWART, J. 1992. Change in the management of public services. *Public Administration,* 70: 499–518.

WRIGHT, V. 1994. *Privatization in Western Europe.* London: Pinter.

Part **VIII**

# Political Economy

Chapter 28

# Political Economy: An Overview

James E. Alt
Alberto Alesina

TRADITIONALLY, economic behavior meant people making utility-maximizing exchanges, and political behavior meant people voting and joining interest groups. Of course, there were institutions. Exchanges took place in markets, which are institutions. Voting and lobbying suggest the existence of legislatures and majoritarian procedures. But the institutions were all *exogenous* and moreover, the economic and political institutions were seen as *separate*, not as part of the same overall structure surrounding human interaction. No more. This chapter surveys a field that has grown out of rejecting both the exogeneity of institutions and the separation of economics and politics, which have come to be seen as not just dimly linked, but intextricably interconnected.

In this chapter "political economy" refers to research which attempts to answer simultaneously two central questions: how do institutions evolve in response to individual incentives, strategies, and choices, and how do institutions affect the performance of political and economic systems? It uses an economic approach, constrained maximizing and strategic behavior by self-interested agents, to explain the origins and maintenance of political processes and institutions and the formulation and implementation of public policies. At the same time, by focusing on how political and economic institutions constrain, direct, and reflect individual behavior, it stresses the political context in which market phenomena take place and attempts to explain collective outcomes like production, resource allocation, and public policy in a unified fashion. In contrast to either economics or political science in isolation, this positive political economy emphasizes both "economic" behavior in the political process and "political" behavior in the marketplace. Accordingly, we organize this review around institutions and policy.[1]

The *Handbook of Political Science* had no chapter entitled "political economy," though it did include ones on collective choice by Michael Taylor and on formal theory by Gerald Kramer and Joseph Hertzberg (1975). The latter listed

> the major topics of what might be called modern political economy: the theory of preference and individual choice behavior underlying much of formal theory; the notion of a multiperson behavioral equilibrium, and some of the mathematical tools needed to investigate it; some aspects of social choice theory, and related normative questions; and finally, some of the work on the theoretical analyses of political institutions (Kramer and Hertzberg 1975: 354).

The rapid growth of the field of political economy in the last two decades is reflected in the fact that one of our topics, the theoretical study of institution formation, was then little more than an afterthought, while the other, the positive analysis of public policy, did not make the list of "major topics" at all.

We devote the first half of this review to the origins, evolution, and maintenance of institutions, emphasizing Congress-style legislatures, government formation in parliamentary systems, and bureaucracies. This research program built on concepts from a generation of earlier "classic" works available by the time the *Handbook* appeared. They include Arrow's (1951/1963) discussion of cycles in majority voting and Black's (1958) median voter theorem, key elements of subsequent work on the *stabilizing properties of voting institutions.* Downs's (1957) discussion of party competition and Riker's (1962) theory of coalition formation laid the groundwork for contemporary models of *the role of distributive politics in government formation.* Coase's (1937; 1960) costly transactions, Schelling's (1960) co-ordination games, and Buchanan and Tullock's (1962) relationship between supermajorities and externalities are at the heart of recent analyses of how economic and political institutions enable actors to *capture the gains from exchange,* just as Niskanen's (1971) bureaucratic information advantages foreshadow the analysis of *agency costs* in inter- and intra-institutional relationships. The attempt to provide an integrated rational approach to the origins and policy consequences of political institutions brought new answers to older questions about institutional stability and fragility, the sources and effects of information asymmetries, and the over-

[1] Two excellent reviews, Calvert (1995) and Shepsle and Weingast (1994), helped greatly in preparing the first half of this paper. The second half draws on Alesina (1994) and Alesina and Perotti (1995*b*).

all importance of institutions for the allocation of scarce and valued resources.

The second half reviews the political economy of public policy, mostly economic and fiscal. We focus on the "new political economy" that grew at the intersection of macroeconomics, game theory, and social choice theory by making policy choice endogenous, rather than treating policy as either exogenous or chosen by the mythical "social planner." On the contrary, policy is the result of some interaction between citizen-voters and policy-makers within institutions having certain characteristics. This literature has literally exploded in a variety of directions. We selected areas of research that were active today, but in which enough had been done to enable us both to survey and evaluate progress. Our review is by no means exhaustive, and we apologize to those we slight, but to mention everyone and every work that made a contribution would leave us able to provide no more than a list of citations and keywords.

## I The political economy of institutions

Institutions help individuals deal with certain fundamental problems of exchange, collective choice, and collective action. If nothing were ever chosen by vote, there would be no problem of cyclical instability. If there were no social (i.e., Prisoners') Dilemmas, we would have less need to deal with problems of communication, co-operation, or co-ordination. If information were freely available, specialization and delegation would not produce agency costs. If there were no non-simultaneous exchange, *ex post* opportunism would not be a concern. However, all these problems exist, and institutions ubiquitously deal with the tradeoffs they create, providing opportunities for beneficial transactions that would not take place in the absence of the institutions.

Political economists largely agree that in the presence of these problems, institutions increase predictability, reduce uncertainty, or induce stability in human interactions. This view of institutions echoes one of the rare occasions when the *Handbook* did address rational choice approaches, though even in Dennis Palumbo's chapter on organization theory, the theory seemed too restrictive to be useful: "Many—perhaps most—decisions in organizations cannot be understood in terms of rational choice . . . They involve a great deal of uncertainty" (Palumbo 1975: 361).

By contrast, today's political economist would ask how, if the uncertainty were so costly, rational individuals might redesign the organization or help it evolve in a direction which would reduce uncertainty. Exactly that sort of question, how institutional changes affect the possibility of profitable transactions, is central to many recent analyses of economic and political institutions alike.

There is a sharp division between those who model institutions as (1) *equilibria* in some underlying social game and (2) rules, procedures, and choice *mechanisms* taken as pre-existing. Within the second, we connect two approaches to institutions as formal procedures to an ongoing debate about the relationship between transactions costs, institutions, and efficiency. In fact, since all of these share a transactions-costs approach to institutions, in the first half of this chapter we use this framework to examine recent research on legislatures (both the American Congress and parliaments) and bureaucracy.

## A Approaches to institutions

Some, like Schotter (1981), define institutions as equilibria in an underlying social game. This approach emphasizes the *self-enforcing* characteristic and *co-ordinating* function of institutions. If self-enforcing, they avoid the instabilities in collective choice and collective action which they are supposed to ameliorate. Models of such institutions focus on the conditions regarding common knowledge, shared information, and enforcement strategies which must be satisfied (Calvert 1995). If institutions as equilibria solve *co-operation* problems by being self-enforcing, they solve *co-ordination* problems by fostering "focal points." Kreps (1990) shows how co-ordination of beliefs offers an equilibrium strategy for dealing with unforeseen contingencies in repeated interactions. Ideally, the formal analysis of self-enforcing regimes or constitutions would treat both aspects.

Others take institutions as pre-existing, and define them to include diverse things like informal norms, complex formal organizations, and a variety of rules, procedures, and choice mechanisms that channel political and economic activity.[2] A key insight motivating this *procedural* analysis of institutions is that if individuals could make costless enforceable exchanges whenever they wished, there would often be no equilibria, and some beneficial transactions would not take place. Two particularly influential contributions are the structure-induced equilibrium (see Shepsle and Weingast 1987) and the Romer-Rosenthal "setter model" (Rosenthal 1990). In the

[2] A commonly quoted, general definition is that of North (1990: 3) who calls institutions "the rules of the game in a society or, more formally, the humanly devised constraints that shape human interaction."

former approach, vetoes, gate-keeping, or agenda power arise in jurisdictions, which are specified domains of responsibility or control, like divisions in firms or specialized committees in legislatures, perhaps backed by further rules requiring self-restraint by other actors. These institutional features are sources of friction which prevent certain comparisons from being made socially, or otherwise limit the number of enforceable policy outcomes, imposing structure on what would otherwise be disorderly choice processes. The "setter model" has a little more structure: a proposer who places a proposal in play, a defined sequence of moves which result in the proposal being subject to a "take it or leave it" choice, and a "status quo" which is the reversion point if the choice is to "leave it." This little sketch of political action turns out, like supply and demand curves, to help analysts make straightforward predictions in some institutionally complex situations. Many of the works reviewed below on Congress and bureaucracy, for example, are applications of this model, which is discussed extensively by Weingast (above: chap. 5).

### 1 Transactions costs

In all these approaches, the design of economic and political institutions affects how far transactions costs allow or prevent achieving gains from exchange. Transactions costs include the resources consumed in discerning product quality, observing performance, protecting property rights (whether with fences or lawyers), obtaining enforcement of an agreement, or even lobbying.[3] They include the costs of organization and bargaining. This sort of transactions cost is incurred within institutions whenever some individuals bargain with or act to influence the holder of a jurisdiction (Milgrom and Roberts 1990) or whenever any group of actors attempts to purchase some property right valuable to them. This second type of cost is as ubiquitous as institutions themselves, but it is not simply a resource cost. Like the incentive difficulties resulting from principal–agent problems within a firm, the possibility of costly delay or failure to reach an agreement (whether it stems from enforcement depending on laws subject to revision by majorities or specialization requiring delegation to an agent) often cannot even be overcome by the straightforward application of more resources:

> From the literature on incentive-compatible mechanisms under asymmetric information we know that there is no way to bring about an outcome to this problem corresponding to the perfect revelation of preferences without incurring some efficiency shortfall ... [T]his shortfall ... does not

[3] These costs affect firms' choices of profit-maximizing methods and levels of production (Eggertson 1990).

represent the expenditure of resources that could have been put to some next-best use, because the . . . incentive problem precludes ever doing any better (Calvert 1995).

## 2 Efficient and inefficient institutions

The argument that differing transactions costs largely frame the choice between different organizational or institutional forms has been a main theme in the study of institutions over the last two decades. It has found many areas of application: the nature, structure, and allocation of property rights, organizational differences and the efficiency of firms, the practice and consequences of regulation, the nature of democratic institutions, and the "Great Debate" itself between central planning of economies and more decentralized capitalistic forms.

From starting-points in the works of Coase (1960) and Alchian and Demsetz (1972), Williamson (1993) and others argue that, given a particular structure of transactions costs in some economic activity, there is in principle some best institutional arrangement to minimize the incidence of these costs. Williamson claims that underlying relationships between types of adaptation, governance structures, and forms of contract law mean that, where it is cost-effective, far-sighted parties will deliberately create bilateral dependencies supported by contractual safeguards of the appropriate sort.[4] For instance, parties making highly specific investments (for example, a power plant located at the mouth of a coal mine, where neither is likely to have good substitute partners) should surround themselves with distinctively contingent contracts. Joskow (1988) reviews some empirical work consistent with this argument. Some believe that market-like forces drive emerging institutions to the form necessary to achieve this "best" realistic-institutional level of economic activity (Posner 1977). Others propose evolutionary mechanisms by which such efficient institutions emerge (Rubin 1977; Nelson and Winter 1982), analogous to evolutionary mechanisms for cultural transmission where learning is costly (Boyd and Richerson 1994).

However, limited systematic empirical evidence, some experiments, and a lot of historical analysis suggest that the actual evolution of institutions looks nothing like the aggregate purposive implementation or even decentralized natural selection of efficient forms (North 1990). Once institutions have been created (for whatever reason), they generate constituencies of support in government, business, and even the public and become difficult

[4] Williamson argues that where efficiency requires co-ordinated rather than autonomous adaptation by economic agents, hierarchic and other non-market institutions offer advantages over markets (which offer high-powered incentives for individuals to learn and change), and are supported by an approach to contract law as framework rather than detailed rules.

to dislodge or improve.[5] The property-rights literature is replete with examples of institutions which reduce efficiency and whose main effect is to redistribute, as varied as rent control in Hong Kong (Cheung 1975) and salmon-fishing rights in Washington (Higgs 1982). Moreover, even an initially efficient institution may be impossible to alter, even if changing conditions make it dysfunctional, as David's (1985) account of the QWERTY typewriter keyboard standard suggests.

Political changes also have economic effects. Rosenthal (1989) shows how an "unintended" consequence of the French Revolution was to allow a lot of "efficient" water projects which had previously been politically impossible. Levi (1988) shows how the incentives of French governments in creating economic institutions worked against efficiency for long periods. However, any so-called "inefficiency" of these institutions arises from the "costliness" of the collective action that would be necessary to change them, which is itself partly determined by other political institutions. It is precisely this costliness of change that insures that institutions can last long enough to have effects on economic and political outcomes.

This division over whether institutions are efficient has been transcended by those in investigating the construction of *durable* institutions, which are designed to enforce *some* distributional outcome (and thus "inefficiency" relative to some normative standard) over time by making credible the commitment of the "enacting coalition" to this distribution. This literature extends a familiar theme in a new direction. Contracts have to be enforced, but conveying to individuals the power to enforce also conveys the power to extract. Under what circumstances will rational individuals so empowered refrain from extracting? How can the commitment to refrain be made credible? Ostrom (1990) describes a variety of transparent, simple, self-enforcing, durable arrangements for the exploitation of common-pool resources. Weingast's chapter (above: chap. 5) discusses other cases where reputation and decentralization are key elements of durable institutions.

## B The political economy of legislatures

Three "camps" of political economy theories about legislative institutions, all based on the U.S. Congress, all involving some aspect of costly exchange

[5] Milgrom and Roberts (1990) show that creating an institution creates a point of value, a jurisdiction or agenda power. Thus, the creation of any institution brings along incentives to dissipate resources in bargaining and attempting to influence decisions. A similar "industry standard" in the regulation literature is that once a regulation has been in place for some time, an industry will evolve in such a way that its leading firms become political defenders of that regulation.

or information, respectively explain institutional adaptation to secure costly but beneficial exchanges, to alleviate social dilemmas and to disperse the gains from specialization among legislators, who are assumed to have different purposes which reflect their concerns about re-election, personal views of good public policy, institutional ambitions, and also the influence of others (Shepsle and Weingast 1994). Weingast and Marshall (1988) describe how such *heterogeneous tastes* make the exchange of support necessary if legislators are to achieve their objectives. An immense, highly successful theoretical and empirical literature described how the American Congress developed to facilitate exchanges, to deal with imperfect information, to co-ordinate members' activities, and to contain agency problems (Shepsle and Weingast 1994).[6]

There are possible gains from exchange when the value a legislator places on the benefits from his or her own projects exceeds the burdens he or she must bear in supporting the projects of other legislators as part of a deal. But partners need to be identified, and deals need to be enforced, to have extended lives, and to have future benefit flows secured against unanticipated events. Replacing repeated "spot-market-like" one-at-a-time exchanges of support with an institutionalized form of deal-making economizes on these transaction costs, in Fiorina's (1987) words:

> Thus legislatures establish committees with specialized jurisdictions, permit their members to self-select positions on committees with greatest interest . . . , and adopt "norms" that lubricate the gears of the system . . . [which] allows members to exercise disproportionate influence on the formulation and implementation of policies important for their electoral interest and provides a variety of opportunities to claim political credit . . . .

To work, such a committee system must foster self-selection by legislators (honor individual preferences in the committee assignment process), retain extraordinary influence in their respective policy areas (combining gate-keeping and proposal power, relatively restrictive amendment rules, control over conference committee proceedings, and oversight authority), and adapt to the rise of new issues and changing legislative interests.

### 1 Co-operation and parties

However, decentralized jurisdictions can also create *market imperfections* like social dilemmas. For example, decoupling taxes and spending produces deficits (Kiewiet and McCubbins 1991). Such problems in turn stimulate institutional developments which smooth out the imperfections and expand the opportunities for co-operation. Thus, individual legislators

[6] Many of the earlier papers are collected in McCubbins and Sullivan (1987).

facing a collective dilemma (by acting alone they cannot capture the gains from legislation for their district) will devise mechanisms to capture potential gains. This is how to explain the development and importance to the larger institution of leadership and party organization (Cox and McCubbins 1993).

Cox and McCubbins (1993) ask: Why and with what effects are parties the basis of legislative organization? How do they interact with the committee structure and affect public policy-making? While legislators' tastes are heterogeneous, on many aspects of policy they broadly agree. The majority party, if united in political purpose, can seize legislative authority and (re)design legislative institutions and practices (like the committee system) to pursue its political agenda. Congressional parties are a sort of legislative cartel.

To gain from party labels (a collective reputation from which individual legislators can benefit), party members must create collective mechanisms for resolving disputes. Committee composition via assignments, the production and scheduling of committee products, and the control of the floor all come under party control. Parties will permit less self-selection and scrutinize appointments more carefully to those committees whose policies entail greater externalities for the re-election prospects of many partisans. Cox and McCubbins (1993) present extensive empirical evidence to support their view of the representativeness of committees.

### 2 Informational rationale for committees

Political economy also provides an alternative, *informational rationale* for legislative organization. All members stand to benefit from the provision of information by experts, and committees provide that expertise. Thus, as Shepsle and Weingast (1994) put it, committees are not only transactions-cost-economizing agents of distribution but are also specialized factors of production. Committees serve as specialized bodies of experts that collect and reveal information bearing on the relation between public policies and social outcomes.

The point, as set forth in Krehbiel's (1991) book and a series of papers summarized in Gilligan (1993), is that if legislators do not know the precise relationship between the policy instruments they select and the outcomes these instruments subsequently produce, they would like to update and improve their beliefs by collecting new information. Then institutional arrangements reflect the need to acquire and disseminate information in addition to (or instead of) the need to solve distributional issues. Majorities will provide incentives and convey resources to specialized committees who invest in becoming experts. Committees may be powerful in a legislature not (only) because they monopolize agenda power, but (also) because they monopolize information and expertise. When such new information is disseminated, it can, in turn, affect the policy choices of the whole legislature.

A host of empirically-testable implications follow. First, to limit the potential agency losses (to the legislature as a whole) from specialization, committees should in general not be composed chiefly of preference outliers (legislators far from the center of opinion in the legislature[7]) but rather should reflect a more centrist, median position. Similarly, conference committees (those resolving disputes between chambers) should reflect majoritarian sentiments. Gilligan and Krehbiel also suggest that to provide committees with incentives to become expert, majorities will often want to impose restrictive procedures on themselves to prevent tampering with committee outputs by the floor. Such self-imposed constraints should be more likely, and more likely to be effective, the more representative committees are of the parent body. Many of these implications are directly in conflict with those of the other "camps," and so the next few years' empirical work will be decisive.

This literature is dominated by the study of the U.S. Congress: it is therefore excellent but narrow. There is a parallel literature on relations between Congress, President, and bureaucracy (Moe 1984; 1990) which we discuss below, and we describe a policy model which explicitly takes into account the wishes of the President (Alesina and Rosenthal 1995). There is a nascent political economy literature on how transactions costs affect the organization of parliamentary legislatures, but with what economic consequences (Palmer 1995; Moe and Caldwell 1994; Spolaore 1993). After a quiet period, there has also been a recent resurgence of interest in the relationship between party policies, political institutions, and the process of forming governments.

## C The political economy of government formation

Early coalition models, of which Riker (1962) is the exemplar, treated coalition formation as a game played between parties for a prize denominated in Cabinet seats. Nearly all the early game-theoretic approaches assumed parties were *office*-seeking rather than policy-motivated. They also took a co-operative approach by assuming that any deals which were struck were exogenously enforced, avoiding the need to confront and model transac-

[7] The vast empirical project undertaken by Poole and Rosenthal (1991) to dimensionalize roll call votes in the U.S. Congress greatly facilitated the analysis of legislator and committee positions.

tions costs. Moreover, the lack of actors' subsequent opportunism (desire to renege on an agreement) gave these models a myopic quality. In addition, few contained any descriptions of ways in which institutions impinged on the process of government formation, in spite of rich descriptions which were readily available (Laver and Schofield 1990). Finally, even those coalition theories based on an assumption of *policy*-seeking politicians assumed that a winning coalition, if it existed, could form at any point in the policy space.

Just recently, however, there has been an proliferation of models and approaches. Some of these advance formalizations in the older, spatial-model, co-operative-game tradition. Others, however, introduce sophisticated agents and costly transactions in a variety of forms, including specific sequences in play imposed by institutions, restriction to self-enforcing (rather than exogenously enforced) bargains, indivisibilities in policy outcomes, and politicians' desire to make credible commitments in order to attract electoral support. These models attack precisely the problems of collective choice and collective action which are at the heart of the newer political economy approach.

### 1 Co-operative-game, spatial models

Schofield (1992; 1993) applies theoretical conditions for the existence of spatial equilibrium points in a weighted majority-rule setting to government formation in legislatures with political parties of various sizes. He establishes necessary and sufficient conditions for the existence of structurally stable majority cores, majority cores that are robust to small perturbations in parties' ideal points.[8] He shows that in a multidimensional policy space in which party ideal points are located, for any minimum winning coalition of parties there is a coalition's *compromise set*, bounded by the lines connecting party ideal points, the set of policy points from which departure to any other point harms at least one of the coalition parties. A policy point is a core point if and only if it lies in the compromise set of *every* minimum-winning coalition. If the intersection of these compromise sets is empty, then there is no core.[9]

A "core party" is a party whose ideal policy is a core point. When such a party exists, the government which forms will contain it, maybe as a single-party minority government or as a coalition with other parties. Either way, policy will reflect the core party's ideal point. When a core party does not

[8] Majority cores in this weighted voting problem are analogous to the median voter's ideal point in a unidimensional voting model.

[9] Moreover, if, for a particular core point, small changes in party ideals still leave it in the compromise set of every minimum-winning coalition, then that core point is structurally stable.

exist, as is often the case, Schofield's theory predicts an outcome in the cycle set, the region of the space within which cycles amongst coalitions and policies are likely to be located. Either the core or the cycle set must exist. Hopefully, the latter is reasonably compact, so that testable predictions of outcomes can be made.

## 2 Sequence and strategy

Austen-Smith and Banks (1988) provide a formal model of government formation in which parties (based on size) take turns attempting to form coalitions (an institutionally imposed sequence of moves), actors are not myopic, and their expectations take into account a reversion point if each attempt fails, and parties suffer a subsequent electoral penalty based on how far from their promises they stray in order to form a coalition. The model leads to predictions of coalitions between large and small parties. Baron (1991) uses a similar setup, with an explicit vote of confidence mechanism determining when a government forms.[10] An important feature of these non-co-operative approaches is that any "co-operative" ventures such as forming coalitions or making policy compromises which do arise do so because it is in every co-operators' interest to stick to the deal they have negotiated.

## 3 Portfolio allocations and jurisdictions

The models we have reviewed so far concentrate on the equilibrium collective policy that emerges given the balance and distribution of party strengths in the parliament. It is also possible to think of governments as collections of ministers with individual jurisdictions, rather than as a body collectively responsible for a single outcome. In the case where ministers have jurisdictions, the outcome is a bundle of individual party-preferred policies, rather than an average of spatial locations, depending on which party receives which portfolio. Jurisdiction makes a party's promises credible to its electors, for it can carry them out (only) if it receives the relevant ministry. At the same time portfolio allocation monitors party behavior, for the party receiving a portfolio has no excuse for not carrying out its preferred policy, and thus no incentives to misrepresent its desires. This approach to ministerial specialization potentially opens up models of government formation to the considerations of the costs and benefits of specialization and delegation, discussed in other sections.

Laver and Shepsle (1996) analyze a portfolio-based model of govern-

[10] Baron and Ferejohn's (1989) general model of bargaining in legislatures is used in this model. Baron (1993) treats the case where the positions of parties wishing to form coalitions are the dependent variable.

ment formation, based on a constitutionally established sequence of proposals from a historically determined status quo. They see bargaining taking place among rationally foresighted policy-motivated parties, each of which has an explicit veto over every cabinet in which it could participate, in a lattice of feasible governments (rather than a continuous space) in which every actual or potential government is a discrete entity with a particular forecast policy output. The only effective way for legislators to control a government is to credibly threaten to or actually replace it with another government. Their predictions give important roles to a "strong" party, one which participates in (and can thus veto) every cabinet preferred by a majority over its ideal, the cabinet in which the strong party takes all portfolios. Even though comparing predictions of policy outcomes with predictions of portfolio allocations is difficult, there now appear to exist significant possibilities for empirical evaluations of these contending models.

## D The political economy of bureaucracy

Niskanen (1971) models the bureaucracy as a monopoly seller of services, with private information about its costs,[11] diminishing returns to scale, the maximization of agency budgets as its source of utility, and a quiescent legislature as the buyer of its output. Each of these assumptions can be relaxed, and Niskanen's conclusion that bureaucracy grows in a democracy changes accordingly. Miller and Moe (1983) activated the legislature to highlight the information revelation that could take place in a bargaining situation. This drew scholarly attention to the question of whether the legislature or bureaucracy was "dominant." This was a somewhat black-and-white way to ask how the costs and benefits were actually shared in a situation in which some "slack" or "drift" between goals and outcomes was inevitable. However, choices over institutional design affect the extent and distribution of slack, and design choices themselves depend on assumptions about what agents maximize. Since largely the same functions were performed by state-owned enterprises in some countries and by private firms regulated by commissions in others, such choices clearly exist, and follow from the same principles we have dicussed above.

[11] Asymmetric information makes principal-agent models of legislator–bureaucrat relations inherently consistent with transactions-cost approaches. Weingast (1984) applies these models to the legislator–bureaucrat relationship.

## 1 The commitment problem

As in the "informational rationale for committees," issue complexity and the inherent consequent desire to take advantage of specialization leads representative assemblies to rely on the opinions and actions of experts. Rather than bear the costs of becoming informed, legislators who want to develop effective policies but lack the needed expertise often delegate the jobs of fact-finding and policy development to others like bureaucrats, presidents, government ministers, party leaders, as well as to legislative committees. Delegation to an expert can be an effective substitute for the acquisition of expertise. However, it can involve "agency costs" since the experts can use their expertise to take actions whose consequences are both unknown to legislators and detrimental to legislative interests.[12] In short, experts can use their expertise to seize control of the policy-making process from legislators. Thus, legislators face the problem of how to realize the potential benefits of delegation without abdicating their control over policy (Lupia and McCubbins 1994).

Moreover, the exchange between legislators and constituents is typically not simultaneous. The flow of benefits to legislators is usually more immediate than the flow of benefits to constituents. Constituents run the risk that this or subsequent legislative coalitions might undermine the benefits of legislation, quite legally, and without any prospects of compensation (Moe 1990). This is also a problem for legislators because forward-looking constituents will assess the durability of future legislative benefits and costs and reflect that assessment in the degree of electoral support that they are willing to offer.[13] In short, the expected net benefit flow from the legislation will be capitalized into present value of the legislation and therefore into the net political support offered to the enacting coalition.

It is often assumed that although legislatures cannot be bound, the enacting coalition can influence the costs that subsequent coalitions must incur to modify a deal at the legislative level as well as the administrative level. The real threat to the durability of the enacting coalition's deal is that future legislators will undermine the value of the legislation by altering the way it is administered or enforced. In particular, the enacting coalition can try to add to the durability of their deal with constituents by reducing the scope of delegated authority and by delegating that authority to an agent, like the court, that is relatively independent of the incumbent legislature (see Landes and Posner 1975).

[12] Jensen (1983: 331) defines agency costs "as the sum of the costs of structuring, bonding and monitoring contracts between agents. Agency costs also include costs stemming from the fact that it doesn't pay to enforce all contracts perfectly."

[13] The constituents in all these models are interest groups, rather than individual electors who might be assumed to be more myopic.

However, an agency problem arises again here. The enacting coalition and its constituents must delegate to implement their arrangement, but their agents do not necessarily share the objectives of the enacting coalition and its constituents, and so the enacting coalition cannot be sure that its administrative agent will administer the legislation in the manner intended at enactment. It is difficult to monitor these agents and create a system of *ex post* rewards and sanctions that will insure they act to protect the interests represented at enactment. Even imperfectly faithful implementation can only be secured with significant agency costs that include the costs of selecting administrators and monitoring their compliance (but legislative oversight is time-consuming), using *ex post* corrective devices (rewards, sanctions and subsequent legislative direction), and the cost of any residual non-compliance that produces a difference between the policy enacted and what is implemented (since the agent knows more about the merits of alternative administrative decisions than either the enacting legislature or its constituents).

## 2 Institutional choice

Precisely in this spirit, McCubbins, Noll, and Weingast (1987; 1989) argue that the creation of a new agency (or new mission for an agency) is a principal–agent problem between the enacting coalition, which seeks to have a durable arrangement implemented, and the bureaucratic entity that will implement the arrangement. While imperfect monitoring, the need to give agents some discretion, and the possibility that agents' preferences differ from those of the principals mean that there will always be agency losses, they stress how these costs can be limited (not eliminated) by features of the appointment process and by a variety of procedural controls. Analyses of some systems characterized as legislator-dominant (the U.S. in some treatments; see Weingast and Moran 1984) and others as bureaucrat-dominant (Japan in some treatments; see Cowhey and McCubbins 1995) are appearing.

Horn (1995) theorizes that legislators generally decide the type of organization that will be used (for example, state-owned enterprises for sales-financed production rather than bureaus for tax-financed production; courts rather than regulatory commissions) by minimizing the sum of the transactions costs they face in any given situation.[14] The costs are those of reaching a decision (as in "legislatures," above), agency, commitment (durability), and uncertainty (inherent in all arrangements). The key dimensions along which institutional choices are made include the extent

[14] For example, using sales-financing would economize on monitoring costs when a legislator faced a budget-maximizing agency.

to which decisions are delegated (especially the degree of legislative vagueness), the governance structure of the administrative agent (especially the way senior personnel are selected, the degree of statutory independence from the legislature and the jurisdiction of the administrative agent), the rules that specify the procedures that must be followed in administrative decision-making (including the rights of constituents to participate in the administrative process), the nature and degree of legislative monitoring and the ability to use *ex post* rewards and sanctions, and the "rules" governing the allocation and use of capital and labor; in particular, the extent to which agencies are financed by sales revenues rather than taxes and the administrators' employment conditions. Horn reviews case study evidence which is consistent with many of his conjectures, but exciting opportunities for systematic empirical work clearly exist here.

## II The political economy of public policy

We turn next to the effects of institutions on public policy, in a political economy framework. For the most part, we treat the institutions as exogenous, even though the first half of this chapter was about how institutions can be endogenous. However, assuming institutional exogeneity is justified if institutions are either self-enforcing or, at least, difficult to change because of significant costs of collective choice or collective action, at least in the short term. We focus on applications using the same set of assumptions (essentially, constrained optimizing) to analyze behavior as in the first half of the chapter. We review two research programs, one on politico-economic cycles in economic outcomes and the other on debts and deficits. In both cases, political parties and elections are important institutional elements. Also in both cases, sophisticated politico-economic models have passed successively more demanding empirical tests, which point to the importance of including particular institutional variables to explain outcomes.

### A Political business cycles

The literature on political business cycles developed in two phases. First, in the mid-1970s, Nordhaus (1975) and Hibbs (1977) identified two different types of cycles. Nordhaus emphasized an "opportunistic" cycle according to which incumbent politicians of any party stimulate the economy before each election, in order to win and continue in office. Hibbs identified a "partisan" cycle in which the Left fights unemployment even at the cost of increasing inflation, while the Right fights inflation even at the cost of higher unemployment. Both these opportunistic and partisan models were based upon a traditional, pre-rational expectations approach to macroeconomics.

The second phase took off in the mid-1980s as a branch of the game-theoretic approach to the positive theory of macroeconomic policy. "Rational" versions of both opportunistic and partisan models appeared. This second generation of models departs from its predecessors in two important ways. First, the assumption of rational expectations makes real economic activity less directly and predictably influenced by economic policy in general, and monetary policy in particular. Second, rationality implies that the voters cannot be systematically fooled in equilibrium.

**1 Traditional "opportunistic" models**

This approach is (or used to be) the most popular type of "political business cycle" model. The most famous contribution is due to Nordhaus (1975), whose model hinges on three crucial assumptions: (i) the economy is characterized by an exploitable Phillips curve with backward-looking expectations; (ii) politicians only care about winning elections; and (iii) voters are retrospective and naïve; they reward the incumbent if the economy is "doing well" (i.e., growth is high) in the period immediately before the election.

Under these assumptions, Nordhaus derives the following implications: (i) every government follows the same policy; (ii) towards the end of his term of office, the incumbent stimulates the economy, to take advantage of the "short run" more favorable Phillips curve; (iii) the rate of inflation increases around the election time as a result of the pre-electoral economic expansion; after the election, inflation is reduced with contractionary policies. Thus, one should observe high growth and low unemployment before every election and a recession after every election.

**2 "Rational" opportunistic models**

One line of research maintains the assumption of opportunistic politicians in "political business cycle" models in which voters are rational but imperfectly informed about some characteristic of incumbents, in particular, about their "competence" in handling the economy. The competence of policy-makers is defined as their ability to reduce waste in the budget process (Rogoff 1990; Rogoff and Sibert 1988), to promote growth without inflation (Persson and Tabellini 1990) or to insulate the economy from random shocks (Cukierman and Meltzer 1986). If policy-makers are more informed than the citizens about their own competence but try to appear as competent as possible, they behave in ways leading to a Nordhaus-style political business cycle. However, voters' rationality and awareness of politicians' incentives limit politicians' "opportunistic" behavior. In fact, if politicians appear too openly opportunistic, they are punished by the electorate. Thus, the political cycles in these rational models are more short-lived, smaller in magnitude, and less regular than in Nordhaus's model, and more likely to be observed in policy instruments than in policy outcomes.

## 3 The traditional partisan model

A "strong" version of the partisan model (Hibbs 1977; 1987) holds that different parties can take advantage of an exploitable Phillips curve to bring about their desired outcomes. Thus, the Left chooses a point on the Phillips curve with less unemployment and more inflation; the Right chooses a point with the opposite combination. These different preferences are due to the distributional consequences of macroeconomic outcomes.[15] These differences in macroeconomic outcomes are thus permanent, lasting for the entire term of office of the different parties.

## 4 The "rational partisan theory"

Alesina (1987) proposed a model which later became known as the "rational partisan theory." He retains Hibbs's assumption that different parties have different preferences over inflation and unemployment, but he embodies this assumption in a rational expectation model with nominal wage stickiness. This model generates a political cycle if nominal wage contracts are signed at discrete intervals and electoral outcomes are uncertain. Given the sluggishness in wage adjustments, changes in the inflation rate associated with changes in governments create temporary deviations of real economic activity from its natural level. In fact, uncertainty about electoral outcomes creates post-election unexpected inflation shocks.

The model has the following implications: (i) at the beginning of a Right (Left) government, output growth is below (above) its natural level, and unemployment is above (below) its natural level; (ii) after expectations, prices, and wages adjust, output and unemployment return to their natural level, and after this adjustment period (which lasts for no more than a couple of years) the level of economic activity is independent of the party in office; (iii) the rate of inflation should remain higher throughout the term

[15] Hibbs (1987) for instance, argues that in the United States in periods of high unemployment and low inflation the relative share of income of the lower middle class decreases.

of a Left government. The inflation rate remains higher when the Left is in office even after the level of economic activity returns to its natural level partly because of a credibility problem. In sum, this rational partisan model differs from the traditional one because it emphasizes how the differences in growth and unemployment associated with changes of government are only temporary.

## 5 The empirical evidence

A vast literature has explored the empirical implications of these models, with data from the United States and from other industrial economies. It finds that:

1) Nordhaus's model is generally and unambiguously rejected regarding its implications about unemployment and growth;
2) consistent with the "rational" version of the opportunistic model, one observes occasional short-run manipulations of policy instruments (monetary and fiscal) around elections, but these effects are small and do *not* imply regular cycles in unemployment as predicted by the traditional model;
3) partisan effects on growth and unemployment are widespread but are not persistent; they disappear about two years after the election of a new government;
4) left-wing governments are associated with permanently higher inflation than right-wing ones.[16]

*The last two observations are consistent with the implications of the "rational partisan theory."* The results of this collective effort support two very general points. First, the new theories of political cycles based on the paradigm of rational choice and rational expectations, are more successful empirically than their predecessors. Second, the partisan model out-performs the opportunistic model as an explanation of politically induced macroeconomic fluctuations of GNP growth, unemployment, and inflation.

Moreover, political business cycle models in general, and partisan models in particular, are more applicable and in fact perform better in two-party or two-bloc systems. The evidence from the U.S. is consistent with that of other two-party or two-bloc party systems. The important differences are *not* between the U.S. and parliamentary democracies in general, but between two-party or two-bloc systems and countries typically governed by "middle-of-the-road" coalition governments.

[16] These results follow Hibbs (1987), Alesina (1988), Chapell and Keech (1988) and Keech and Pak (1989) on the U.S.; and Alt (1985), Paldam (1991), Alesina and Roubini (1992) and Alesina, Cohen, and Roubini (1993) on OECD countries.

## 6 A general model of elections and the economy in the U.S.

A large, convincing, empirical literature shows that the state of the economy in an election year has strong effects on election results.[17] This observation raises two puzzles. First, why, if a strong growth performance in election years increases the probability of re-electing the incumbent, do incumbents not achieve a higher than average growth rate in election years (as argued in the previous subsection)? Then there is the effect of the economy on presidential and congressional elections in the U.S., which is much stronger on the former than on the latter. Furthermore, the President's party always loses support in mid-term congressional elections, even when growth is high in the first two years. The second puzzle is why this is so, and moreover, why even Democratic presidents suffer a mid-term loss despite their high growth performance in the first two years.

In order to answer these puzzles, Alesina and Rosenthal (1995) provide a model in which the macroeconomic cycle and electoral results are jointly determined. This model has several key building-blocks. First, the two parties are politically motivated, and the economy is described as in the rational partisan model above. Second, administrative competence influences the growth performance. Voters also have ideological preferences *and*, in addition, prefer more to less competent administrations. Therefore, it is rational for the voters to judge the incumbent's competence by observing the pre-electoral growth of the economy and by (rationally) trying to disentangle the effects of competence from those of "luck," that is, favorable exogenous shocks. Third, there is no asymmetry of information between policy-makers and voters. This assumption rules out opportunistic behavior by policy-makers. Fourth, policy outcomes depend upon which party holds the presidency, and upon the partisan composition of Congress. With a Republican president, for instance, the larger the share of the Democratic party in Congress, the more policy is pulled toward the ideal policy of the Democratic party, and vice-versa. Fifth, middle of the road voters use "institutional balancing" to bring about moderate policies. Since policy must reflect a compromise between President and Congress, moderate voters (those with preferences in between those of the two parties) balance by turning to the party opposite that of the President in congressional elections. The following example (written before the 1994 elections!) illustrates the model.

Suppose that a Democratic president, who was a slight favorite according to the polls, is elected. This tilts the electorate in favor of the Republican party in the next House elections in order to moderate the favored

[17] On the U.S., see Fiorina (1981), and Fair (1988). On other industrial economies see Lewis-Beck (1988).

Democratic candidate to the White House. The Democratic administration follows expansionary policies, which lead to an upsurge of growth in the first two years of the term, and to an increase in inflation. Therefore, in mid-term, the electorate further balances the Democratic administration by turning to the Republican party: the voters want to insure themselves against excessive inflation. This voting behavior implies a non-obvious correlation, that strong economic growth in the first half of the Democratic administration is accompanied by a poor showing of this party in mid-term.

Empirical work by Alesina and Rosenthal (1995) shows that this whole model performs remarkably well through the period 1915–89. This model is consistent with many regularities of the American political economy and provides a coherent framework which encompasses *all* the following observations, which earlier literature[18] typically studied one at a time:

1) The vote share of the incumbent president party's presidential candidate increases with the rate of GNP growth in the election year.
2) Congressional elections are less sensitive to economic conditions; the economy influences these elections only through its effect on presidential votes.
3) There is a mid-term cycle where the party holding the White House loses votes in mid-term.
4) GNP growth exhibits a partisan cycle with short-run post-electoral deviations from average growth.
5) the rate of GNP growth in election years is not systematically different from average.

## B The political economy of budget deficits

The subject of budget deficits has been a center of attention of the "public choice" literature for decades. The recent impetus to this area of research comes from the emergence of very large peacetime public debt in several, but not all, OECD economies. In fact, Grilli, Masciandaro, and Tabellini (1991) show that OECD countries can be divided into two clearly identifiable groups. In one group, including Belgium, Ireland, Italy, Greece, the Netherlands, and Spain, the ratio of total public debt to GNP is very high and/or increasing rapidly. In another group, which includes Australia, France, Germany, the United Kingdom, Japan, and Denmark, debt to GNP

[18] In addition to works previously cited, see Erikson (1988; 1990) and Alesina and Sachs (1988).

ratios are much lower and stable. By international standards, the United States belongs to the second group of relatively low debt countries. However, even in the United States, the rapid increase in the debt/GNP ratio in the 1980s reversed a downward trend which started at the end of the Second World War.

The difference in the debt to GNP ratios in otherwise relatively similar OECD economies is staggering. Belgium, Ireland, and Italy owe more than 100 percent of a year's GDP; Italy is currently pushing toward 120 percent. The debt/GNP ratio of several countries in the low debt group is about *one fourth* of that value. These large differences across OECD countries are relatively recent. They started to appear in the mid-1970s and have increased since then. We examine how far political economy approaches have been able to answer two questions, namely: why do we observe large peacetime deficits and debt, and why did they appear "now," namely in the last twenty years? Why do we observe large public debts in some countries but not in others?

## 1 The tax smoothing theory

A point of departure is the normative theory of public debt, known as "tax smoothing" theory (Barro 1979). The principle of tax smoothing is simple, but quite powerful. Budget deficits and surpluses should be used to minimize the distortionary effects of taxation, given a certain path of government spending. The implication is that tax rates should be kept approximately constant and budget deficits (surpluses) should be used as a buffer to compensate for temporary fluctuations of tax revenues (due to fluctuations in income) and spending. Thus, one should observe deficits during recessions and during periods of exceptionally high (and temporary) spending, for instance, during wars.

To be sure, the tax smoothing theory goes a long-way toward explaining the long run trend of public debt in the United States and United Kingdom. However, it cannot answer the two questions raised above. In particular, the tax smoothing theory has very little to say on the second question. It is very hard (impossible?) to explain the very large cross-country variance in debt purely based upon a tax smoothing argument.

Under perfect information, tax smoothing can minimize long-run costs of adjusting fiscal policy. Politico-economic approaches, seeking to ground positive models which encompass strategic behavior, depart from the tax smoothing theory by pointing out the possible consequences of imperfect information, social dilemmas and institutional processes, beyond the purely economic factors emphasized by the tax-smoothing model. We describe four different approaches which combine asymmetric information and collective action problems in different ways: (i) fiscal illusion; (ii) intergenerational redistribution; (iii) political conflict and instability; and (iv) institutional processes.

### 2 Fiscal illusion

The "public choice" school of Buchanan, Tullock and their associates (Buchanan and Wagner 1977; Buchanan, Rowley and Tollison 1986) made the concept of "fiscal illusion" central in their models of budget deficits. In a nutshell, the idea is that voters overestimate the benefits of current expenditures and underestimate the burden of future taxation. They do not internalize the government's budget constraint. Opportunistic politicians take advantage of this illusion by raising spending more than taxes. However, "fiscal illusion" theory does not answer either of the two questions above. The theory implies that any democracy at any point in time should be in deficit. This contradicts the evidence. Moreover, fiscal illusion theory is unconvincing in that it relies heavily on a particular form of voter irrationality, a specific bias in voter perceptions. Recent politico-economic literature on budget deficits, as well as modern political economy more generally, relies on standard notions of "rational choice."

### 3 Intergenerational redistribution

In models where the "Ricardian equivalence" does not hold, public debt generates intergenerational redistributions if the generation that is alive today, and votes, leaves the burden of the debt to future generations. Cukierman and Meltzer (1989) argue that public debt can be a way for those in the current generation who are "poor" to leave negative bequests to their offspring, since private negative bequests are not permissible. The idea that public debt redistributes in favor of the current generation, the only generation that has political influence, is quite powerful, at least at first sight. However, again, this overlapping generation model cannot explain "why now?" (we always had parents and children) nor "why in certain countries and not in others?" (parents and children are everywhere).

### 4 Political conflict and instability

These models emphasize socio-political conflicts, partisan politics, and government fragility as determinants of public debt. They separate in treating debt sometimes as a *commitment* and sometimes as the result of *government fragmentation.* In the first case, commitment, debt is used by current partisan governments to constrain and influence the choices of future governments with different preferences. Public debt is increased in polarized situations by partisan governments that expect to be followed by opponents from the opposite end of the political spectrum. A high public debt limits the choices of future governments, since a larger fraction of tax revenues must be committed to servicing the debt.[19] Empirically, this line of research implies that frequent government changes back and forth between polarized parties should be associated with large public debts. While this approach can potentially answer both questions, the evidence, though seeming generally to go in the "right direction," is often anecdotal and unsystematic.

A different way of looking at political conflict emphasizes fragmentation *within* the government, rather than polarization between one government and the next. Alesina and Drazen (1991) formalize a "war of attrition" model of delayed fiscal adjustment which occurs only when one group stops using their veto power. Delays in adjustment and debt accumulation are "rational" in this social dilemma model, since the passage of time is necessary to establish which group(s) is (are) the weakest, though costs can be avoided if all groups can agree at the outset on an equitable, co-operative rule to share the burden of adjustment.

This model implies that coalition governments are more likely than single party governments to run into fiscal imbalances (Spolaore 1993). Moreover, coalition governments (and, generally, fragmented and short-lived governments) do not *create* deficits, but rather postpone adjustment through legislative deadlock. This approach can also answer both our basic questions, and apparently does so, at least to some extent. Empirical work on OECD economies by Roubini and Sachs (1989*a*; 1989*b*) and Grilli, Masciandaro, and Tabellini (1991) shows that measures of government fragmentation are strongly associated with measure of public debt. This is a reasonable answer to the second question. As for the first question on timing, the point is that cross-country differences appeared after the major shocks of the early 1970s. Single party governments seem to have been more capable to adjusting than coalition governments.

In the U.S., the equivalent of coalition government is the common situation of divided government. Although Alesina and Rosenthal (1995) regard divided government as a balancing device, others believe that it can cause co-ordination problems with inefficient policy delays. A lively debate has sprung up within and outside academia over whether divided government in the 1980s is responsible for the rapid growth of public debt. While the evidence at the federal level is weak and rather questionable, Alt and Lowry (1994) and Poterba (1994) find much more convincing evidence at the state level, showing that states with divided government are slower in

[19] Different models of this nature include Persson and Svensson (1989), Tabellini and Alesina (1990), Aghion and Bolton (1990), Milesi-Ferretti and Spolaore (1994).

adjusting to fiscal shocks than states with unified governments, resulting in growing deficits in recessions where party control is divided. The similarity between these results and those for OECD countries (after allowing for the obvious institutional differences) is quite striking.

### 5 Institutional process

Congress scholars have analyzed in great detail how the legislature's rules influence policy outcomes in general, and fiscal policy in particular. This literature usually focuses on the level of spending, rather than deficits. In fact, its models are generally static rather than dynamic. In an influential contribution of this sort, Weingast, Shepsle, and Johnsen (1981) argue that representatives with geographically based constituencies overestimate the benefits of public projects in their districts relative to the costs of financing them which are distributed nationwide. The consequence is an oversupply of public projects.[20]

No doubt, legislative procedures have important implications for outcomes. In order to apply the insights of this typically "American" literature to issues of public debt in industrial economies, one has to tackle three issues. First, formal models need to be made explicitly dynamic to study not only the level of spending, but also the budget balance. Second, the American models focus on geographically based public "pork barrel" projects, but in fact purely redistributive programs are a larger and larger fraction of the public budgets of OECD countries. Third, one has to take into account the effects of different electoral systems and other legislative institutions in different countries. Research in this area is in an embryonic stage but in our view is quite promising. It is reinforced by other studies which show that specific debt-limiting features of constitutions affect the speed with which state governments respond to fiscal shocks (Alt and Lowry 1994; Poterba 1994). Most important, this is again a point of direct contact between the study of institutions and the study of policy.

## III Concluding remarks

Fundamental to the rational choice-based approach to the evolution and policy consequences of institutions around which we wove our selective review was a broad notion of costly transactions. The costs involve not just

[20] A large literature, which we cannot review here, has investigated how the organization of Congress and its legislative rules influence this basic insight. Kiewiet and McCubbins (1991) is a good starting place.

the problems of monitoring and enforcing contracts but the various aspects of organizing and bargaining which inevitably arise in the process of forming agreements through collective action and choice. Twenty years ago, the foundations of this approach existed, but not very much systematic, positive, empirical work had been done. Now a great deal has. Out of it has come an increased understanding of the interrelationships between delegation, jurisdiction, and agenda power on the one hand, and of the nature of politico-economic cycles on the other. Other areas in which this approach seems to be on the verge of a takeoff include trade policy, energized by a recent debate over the consequences of institutions and factor mobility for collective action and coalition formation, and whether cross-country differences in investment and economic growth are influenced by democratic institutions, political instability, income inequality and/or social polarization. Since evidence in these other areas is only just beginning to accumulate, rather than summarizing only some parts of the possible literature we recommend two recent reviews (Alt and Gilligan 1994; Alesina and Perotti 1995*a*), and we look forward to future opportunities for further retrospection!

## References


Aghion, P., and Bolton, P. 1990. Government debt and the risk of default: a political economic model of the strategic role of debt. Pp. 315–45 in *Public Debt Management*, ed. R. Dornbusch and M. Draghi. Cambridge: Cambridge University Press.

Alchian, A., and Demsetz, H. 1972. Production, information costs, and economic organization. *American Economic Review*, 62: 777–95.

Alesina, A. 1987. Macroeconomic policy in a two-party system as a repeated game. *Quarterly Journal of Economics*. 101: 651–78.

—— 1988. Macroeconomics and politics. Pp. 13–52 in *NBER Macroeconomic Annual*. Cambridge, Mass.: MIT Press.

—— 1994. Elections, party structure and the economy. Mimeo., Departments of Economics and Government, Harvard University.

—— and Drazen, A. 1991. Why are stabilizations delayed? *American Economic Review*, 82: 1170–88.

—— Cohen, G.; and Roubini, N. 1993. Electoral business cycles in industrial democracies. *European Journal of Political Economy*, 9: 1–23.

—— and Perotti, R. 1995*a*. The political economy of growth: a critical survey of the recent literature. *IMF Staff Papers*, March, 1–37.

—— —— 1995*b*. Budget deficits and budget institutions. Mimeo., Departments of Economics and Government, Harvard University.

—— and Rosenthal, H. 1995. *Partisan Politics, Divided Government and the Economy*. New York: Cambridge University Press.

—— and Roubini, N. 1992. Political cycles in OECD economies. *Review of Economic Studies*, 59: 663–88.
—— and Sachs, J. 1988. Political parties and the business cycle in the United States, 1948–1984. *Journal of Money, Credit, and Banking*, 20: 63–82.
Alt, J. 1985. Political parties, world demand, and unemployment: domestic and international sources of economic activity. *American Political Science Review*, 79: 1016–40.
—— and Gilligan, M. 1994. The political economy of trading states. *Journal of Political Philosophy*, 2: 165–92.
—— and Lowry, R. 1994. Divided government, fiscal institutions, and budget deficits: evidence for the states. *American Political Science Review*, 88: 811–28.
—— and Shepsle, K., eds. 1990. *Perspectives on Positive Political Economy*. New York: Cambridge University Press.
Arrow, K. 1951. *Social Choice and Individual Values*. 2nd edn. New Haven, Conn.: Yale University Press, 1963.
Austen-Smith, D., and Banks, J. 1988. Elections, coalitions and legislative outcomes. *American Political Science Review*, 82: 407–22.
Baron, D. 1991. A spatial bargaining theory of government formation in parliamentary systems. *American Political Science Review*, 85: 137–65.
—— 1993. Government formation and endogenous parties. *American Political Science Review*, 87: 34–47.
—— and Ferejohn, J. 1989. Bargaining in legislatures. *American Political Science Review*, 83: 1181–206.
Barro, R. 1979. On the determination of public debt. *Journal of Political Economy*, 87: 940–7.
Black, D. 1958. *Theory of Committees and Elections*. Cambridge: Cambridge University Press.
Boyd, R., and Richerson, P. 1994. The evolution of norms. *Journal of Institutional and Theoretical Economics*, 150: 72–87.
Buchanan, J.; Rowley, C.; and Tollison, R. 1986. *Deficits*. Oxford: Blackwell.
—— and Tullock, G. 1962. *The Calculus of Consent*. Ann Arbor: University of Michigan Press.
—— and Wagner, R. 1977. *Democracy in Deficit*. New York: Academic Press.
Calvert, R. 1995. The rational choice theory of social institutions. Pp. 216–67 in *Modern Political Economy*, ed. J. Banks and E. Hanushek. New York: Cambridge University Press.
Chappell, H., and Keech, W. 1988. The unemployment consequences of partisan monetary policy. *Southern Economic Journal*, 55: 107–22.
Cheung, S. 1975. Roofs or stars: the stated intents and actual effects of a rents ordinance. *Economic Inquiry*, 13: 1–21.
Coase, R. 1937. The nature of the firm. *Economica*, 4: 386–405.
—— 1960. The problem of social cost. *Journal of Law and Economics*, 3: 1–44.
Cox, G., and McCubbins, M. 1993. *Legislative Leviathan*. Berkeley: University of California Press.
Cowhey, P., and McCubbins, M. 1995. *Structure and Policy in Japan and the United States*. New York: Cambridge University Press.
Cukierman, A., and Meltzer, A. 1986. A positive theory of discretionary policy, the cost of a democratic government and the benefits of a constitution. *Economic Inquiry*, 24: 367–88.
—— —— 1989. A political theory of government debt and deficits in a neo-Ricardian framework. *American Economic Review*, 79: 713–33.

David, P. 1985. Clio and the econometrics of QWERTY. *American Economic Review*, 75: 332–7.

Downs, A. 1957. *An Economic Theory of Democracy*. New York: Harper.

Eggertsson, T. 1990. *Economic Behavior and Institutions*. Cambridge: Cambridge University Press.

Erikson, R. 1988. The puzzle of midterm loss. *Journal of Politics*, 50: 1012–29.

—— 1990. Economic conditions and the congressional vote: a review of the macrolevel evidence. *American Journal of Political Science*, 34: 373–99.

Fair, R. 1988. The effect of economic events on votes for president: 1984 update. *Political Behavior*, 10: 168–79.

Fiorina, M. 1981. *Retrospective Voting in American National Elections*. New Haven, Conn.: Yale University Press.

—— 1987. Alternative rationales for restrictive procedures. *Journal of Law, Economics, and Organization*, 3: 337–45.

Gilligan, T. 1993. Information and the allocation of legislative authority. *Journal of Institutional and Theoretical Economics*, 149: 321–41.

Grilli, V.; Masciandaro, D.; and Tabellini, G. 1991. Political and monetary institutions and public finance policies in the industrial democracies. *Economic Policy*, 6: 341–92.

Hibbs, D. 1977. Political parties and macroeconomic policy. *American Political Science Review*, 7: 1467–87.

—— 1987. *The American Political Economy*. Cambridge, Mass.: Harvard University Press.

Higgs, R. 1982. Legally induced technical regress in the Washington Salmon fishery. *Research in Economic History*, 7: 55–85.

Horn, M. 1995. *The Political Economy of Public Administration*. New York: Cambridge University Press.

Huntington, S. 1968. *Political Order in Changing Societies*. New Haven, Conn.: Yale University Press.

Jensen, M. 1983. Organization theory and methodology. *Accounting Review*, 8: 319–39.

Joskow, P. 1988. Asset specificity and the structure of vertical relationships: empirical evidence. *Journal of Law, Economics, and Organization*, 4: 95–117.

Keech, W., and Pak, K. 1989. Electoral cycles and budgetary growth in Veterans' Benefit programs. *American Journal of Political Science*, 33: 901–12.

Kiewiet, R., and McCubbins, M. 1991. *The Logic of Delegation*. Chicago: University of Chicago Press.

Kramer, G., and Hertzberg, J. 1975. Formal theory. Vol. vii, pp. 351–403 in *Handbook of Political Science*, ed. F. Greenstein and N. Polsby. New York: Addison-Wesley.

Krehbiel, K. 1991. *Information and Legislative Organization*. Ann Arbor: University of Michigan Press.

Kreps, D. 1990. Corporate culture and economic theory. In Alt and Shepsle 1990: 90–143.

Landes, D., and Posner, R. 1975. The independent judiciary in an interest group perspective. *Journal of Law and Economics*, 18: 875–901.

Laver, M., and Schofield, N. 1990. *Multiparty Governments*. Oxford: Oxford University Press.

—— and Shepsle, K. 1996. *Making and Breaking Governments*. New York: Cambridge University Press.

Levi, M. 1988. *Of Rule and Revenue*. Berkeley: University of California Press.

Lewis-Beck, M. 1988. *Economics and Elections*. Ann Arbor: University of Michigan Press.

Lupia, A., and McCubbins, M. 1994. Designing bureaucratic accountability. *Law and Contemporary Problems*, 57: 91–126.

McCubbins, M., and Sullivan, T., eds. 1987. *Congress*. New York: Cambridge University Press.

—— Noll, R.; and Weingast, B. 1987. Administrative procedures as instruments of political control. *Journal of Law, Economics and Organization*, 3: 243–77.

—— —— —— 1989. Structure and process, politics and policy. *Virginia Law Review*, 75: 431–83.

Milesi-Ferretti, G. M., and Spolaore, E. 1994. How cynical can an incumbent be? Strategic policy in a model of government spending. *Journal of Public Economics*, 55: 121–40.

Milgrom, P., and Roberts, J. 1990. Bargaining costs, influence costs, and the organization of economic activity. In Alt and Shepsle 1990: 57–89.

Miller, G., and Moe, T. 1983. Bureaucrats, legislators, and the size of government. *American Political Science Review*, 77: 297–322.

Moe, T. 1984. The new economics of organization. *American Journal of Political Science*, 28: 739–77.

—— 1990. The politics of structural choice: toward a theory of public bureaucracy. Pp. 116–53 in *Organization Theory*, ed. O. Williamson. New York: Oxford University Press.

—— and Caldwell, M. 1994. The institutional foundations of democratic government: a comparison of presidential and parliamentary systems. *Journal of Institutional and Theoretical Economics*, 150: 171–95.

Nelson, R., and Winter, S. 1982. *An Evolutionary Theory of Economic Change*. Cambridge, Mass.: Harvard University Press.

Niskanen, W. 1971. *Bureaucracy and Representative Government*. Chicago: Aldine-Atherton.

Nordhaus, W. 1975. The political business cycle. *Review of Economic Studies*, 42: 169–90.

North, D. 1981. *Structure and Change in Economic History*. New York: Norton.

—— 1990. *Institutions, Institutional Change and Economic Performance*. New York: Cambridge University Press.

Ostrom, E. 1990. *Governing the Commons*. New York: Cambridge University Press.

Paldam, M. 1991. Politics matter after all: testing Alesina's theory of RE partisan cycles. Pp. 369–98 in *Business Cycles*, ed. N. Thygesen, K. Velupillai and H. Zombelli. London: Macmillan.

Palmer, M. 1995. Toward an economics of comparative political organization: examining ministerial responsibility. *Journal of Law, Economics and Organization*, 11: 164–88.

Palumbo, D. 1975. Organization theory and political science. Vol. ii, pp. 319–69 in *Handbook of Political Science*, ed. F. Greenstein and N. Polsby. New York: Addison-Wesley.

Persson, T., and Svensson, L. 1989. Checks and balances on the government budget. *Quarterly Journal of Economics*, 104: 325–46.

—— and Tabellini, G. 1990. *Macroeconomic Policy, Credibility and Politics*. Chur, Switzerland: Harwood.

Poole, K., and Rosenthal, H. 1991. On dimensionalizing roll call votes in the US Congress. *American Political Science Review*, 85: 955–60.

Posner, R. 1977. *The Economics of Law* 2nd edn. Boston: Little, Brown.

Poterba, J. 1994. State responses to fiscal crises: the effects of budgetary institutions and politics. *Journal of Political Economy*, 102: 799–821.

Powell, G. B., and Whitten, G. 1993. A cross-national analysis of economic voting. *American Journal of Political Science*, 37: 391–414.

Riker, W. 1962. *The Theory of Political Coalitions*. New Haven, Conn.: Yale University Press.

Rogoff, K. 1990. Equilibrium political budget cycles. *American Economic Review*, 80: 21–36.

—— and Sibert, A. 1988. Equilibrium political business cycles. *Review of Economic Studies*, 55: 1–16.

Rosenthal, H. 1990. The Setter Model. Pp. 199–230 in *Advances in the Spatial Theory of Voting*, ed. J. Enelow and N. Hinich. Cambridge: Cambridge University Press.

Rosenthal, J.-L. 1989. *The Fruits of Revolution*. New York: Cambridge University Press.

Roubini, N., and Sachs, J. 1989*a*. Political and economic determinants of budget deficits in the industrial democracies. *European Economic Review*, 33: 903–33.

—— —— 1989*b*. Government spending and budget deficits in the industrialized countries. *Economic Policy*, 8: 99–132.

Rubin, P. 1977. Why is the common law efficient? *Journal of Legal Studies*, 6: 51–63.

Schelling, T. 1960. *The Strategy of Conflict*. Oxford: Oxford University Press.

Schofield, N. 1992. A theory of coalition government in a spatial model of voting. Working Paper, Center in Political Economy, Washington University.

—— 1993. Political competition and multiparty coalition governments. *European Journal of Political Research*, 23: 1–33.

Schotter, A. 1981. *The Economic Theory of Social Institutions*. Cambridge: Cambridge University Press.

Shepsle, K., and Weingast, B. 1987. The institutional foundations of committee power. *American Political Science Review*, 81: 85–104.

—— —— 1994. Positive theories of legislative institutions. *Legislative Studies Quarterly*, 19: 149–79.

Spolaore, E. 1993. *Macroeconomic Policy, Institutions and Efficiency*. Cambridge, Mass.: Ph. D. dissertation, Department of Economics, Harvard University.

Tabellini, G., and Alesina, A. 1990. Voting on the budget deficit. *American Economic Review*, 80: 17–32.

Taylor, M. 1975. The theory of collective choice. Vol. iii, pp. 413–82 in *Handbook of Political Science*, ed. F. Greenstein and N. Polsby. Reading, Mass.: Addison-Wesley.

Weingast, B. 1984. The congressional-bureaucratic system: a principal-agent perspective. *Public Choice*, 44: 147–91.

—— and Marshall, W. 1988. The industrial organization of Congress. *Journal of Political Economy*, 96: 132–63.

—— and Moran, M. 1984. Bureaucratic discretion or Congressional control: regulatory policy-making by the FTC. *Journal of Political Economy*, 91: 765–800.

—— Shepsle, K.; and Johnsen, C. 1981. The political economy of costs and benefits. *Journal of Political Economy*, 89: 642–64.

Williamson, O. 1993. The emerging science of organization. *Journal of Institutional and Theoretical Economics*, 149: 36–63.

Chapter 29

# Political Economy: Sociological Perspectives

Claus Offe

THE relationship between the academic disciplines of political economy and sociology is not settled. Mutual challenges abound, including misunderstandings and many complaints and polemics about economic "imperialism" in the social sciences. Identifying intellectually legitimate patterns of demarcation and division of labor are issues at the center of an ongoing debate (Olson 1969; Barry 1970; Oberschall and Leifer 1986), sometimes framed in terms of *homo economicus* versus *homo sociologicus* as the two sets of guiding assumptions of the two disciplines. The outlines of a synthesis are not visible on the horizon and certainly cannot be attempted here. Nor can the fading away of one of the two competing paradigms be predicted with any confidence. Sociologists have absorbed a great deal from political economists in terms of their approaches and conceptual tools, sometimes to the point of explicitly changing camps (as in much of the school of "analytical Marxism").[1] Others within the internally highly diversified discipline of sociology have tried to trump efficiency-centered economic approaches with new proto-paradigms, particularly in the subdisciplines most contested between the two: the sociology of organizations, economic sociology, political sociology, the study of inequalities and the sociological theory of institutions (March and Olsen 1989; Selznick 1992; Offe 1996).

The term "political economy" covers a great variety of social scientific approaches. The intellectual ambitions of political economists (together with those active in the related subfields of social choice, public choice, rational choice and institutional or constitutional economics) is often

[1] Compare: Roemer 1986; 1988; Wright 1985; Lash and Urry 1984.

quite grandiose: they want to understand more, or more important things, than ordinary economists (with whom they share a concern with costs, efficiency, allocation, distribution and growth), political scientists (with whom they share a concern with interest dynamics and in the regulation of conflict), or sociologists (with whom they share an interest in rational versus other kinds of action). "Political economy" is also an old term, predating all of the names of the social sciences just mentioned. What political philosophers, economists and social theorists did from the late 18th century through the 19th was routinely termed "political economy."[2] Furthermore, the term bridges not only disciplines and centuries (as well as, in the booming field of International Political Economy, countries and continents), but also the gap between positive/explanatory/predictive versus normative/critical approaches in the social sciences. Political economists have been interested in the problem of order and disorder, in the causes and mechanisms that may make a putatively solid order crumble due to internal deficiencies. Finally, political economists are probably united by the Enlightenment belief that the problem of social order is rooted in rational action—with the dual implication that rational actors can both subvert and build social order. But political economists are manifestly divided as to the value premises that must be incorporated into the institutional set-up of a "well-functioning" or "well-ordered" society, particularly with regard to the dilemmas of liberty versus equality and equity versus efficiency. Indeed, a certain political polarization can be observed, as consistent and confident practitioners of political economy seem to be amassed in the ideological camps of libertarians and Marxists, with only a few despairing social democrats in between.

It seems also fair to observe that, compared to the classical tradition in political economy that begins with the writers of the Scottish Enlightenment and ends with Karl Marx, the discipline has become more disciplined and more narrowly focused in the course of its development throughout the 20th century, more positive and less speculative. That may be considered a gain or an impoverishment. Some basic intellectual operations, however, have remained at the core of the discipline.

[2] One might object that "political economy" has been an unfortunate name for an academic discipline from the beginning, as it fails to differentiate between the corpus of knowledge and the range of phenomena that this knowledge is about. It is quite common today to speak, e.g., about the "political economy of Austria," by which mode of speaking is usually meant not the accumulated works and doctrines of the great political economists of that country, but some peculiarly Austrian institutional arrangements, such as a large public sector or the mode of co-operation between political parties and economic associations. Thus, while the discipline of biology deals with life, the science of political economy deals, strangely enough, with the "political economy" of certain periods, regions, institutions, or economic formations.

## I Components of the paradigm

This operational core is at the same time the field of contestation and mutual challenge that political economists share with sociologists. What are these core ideas?

1) *Endowments.* There is a complex assortment of independent variables that eventually will determine action and outcomes. These independent structural variables are sometimes referred to, summarily, as "endowments": the universe of resources and constraints (material, military, legal, institutional, demographic, technological, temporal, geographic, etc.) given at some point of time that are used to define the "places," the "feasible sets" with their respective "payoffs" in which the actors under consideration are located. Actors thus located within some structure of endowments can be anything ranging from the members of a legislative committee to Third World countries as a whole, from the employers of a branch of industry to some ethnic group.

2) *Interaction and interdependence in the pursuit of interests.* The actors thus endowed strive for some tangible advantage, such as wealth, income, power, tenure of office, military resources, etc. They pursue interests, which by definition they must defend against other agents pursuing competing interests. They are exposed to an environment of opportunities, incentives and threats, resulting from the action of other actors. They are involved in a "game." The universe of collective human agency consists of three families of games (Hardin 1995: chap. 2). These are co-ordination (such as all members of a collectivity speaking the same language or driving on the same side of the road), co-operation (such as going on strike or buying groceries) and pure conflict (such as war). Co-ordination is based on conventions and is costless: everyone wins, nobody loses, because no opportunity cost is involved in foregoing an alternative convention; any convention is, by definition, as good as any alternative convention. Co-operation is profitable and at the same time costly to (at least some of) those who profit: membership dues, prices, or taxes must be paid, but agents get something in return, which supposedly is more valuable than the sacrifice they have made; unless participants expected (and actually realized) some such net gain co-operation would cease. In pure conflict, there are also losses and gains, but the two are (expected to be) dissociated: wins or losses will go to either "them or us," but whether and how much each wins or loses may depend in turn on how effectively "we" or "they" co-operate.

3) *Rationality*. Within the many courses of action that are available, actors will rationally select those which maximize their interests, given their resource constraints and their knowledge and expectations about the action of others and the consequences of their own action. Actors are assumed to be both able and indeed naturally inclined to adopt this outcome-oriented and calculating approach in deciding on their course of action, while social norms, traditions, identities, loyalties to communities, as well as other such variables typically employed by sociologists play a marginal and negligible role within the paradigmatic premises of politico-economic approaches. In other words, the determinants of action are essentially original endowments plus the awareness of interactivity and interdependency plus the agent's capacity for rational utility maximization.

4) *Externalities and feedback*. The aggregate outcomes of action thus determined affect third parties (including "the future") outside the universe of rational actors under consideration. These externalities can be either positive (e.g., economizing of transaction costs, efficiency increases) or negative (e.g., exploitation, inflation). Furthermore, such repercussions can be of a "merely" distributional sort (change of the original endowment with material resources) or, beyond that, they can also be of an institutional sort (consolidation or change of institutions and rights, transition to a different regime, etc.). These feedback loops of institutional innovation are, mostly although not always (cf. North 1990), seen as stages in an evolutionary process contributing to increased efficiency of the original arrangement of "places," rights and endowments.[3]

The agenda of a full politico-economic analysis can thus be described as following a cyclical pattern. Institutions and endowments condition behavior up to a point where actors, through their rational pursuit of interests, come to alter institutions—be it through accumulated externalities, explicit renegotiation, the presence of some self-paralyzing mechanism, depletion of physical or "moral" (Hirsch 1976) resources or whatever. This cyclical model of self-subversion—institutions generating actors who generate outcomes that cannot be accommodated within existing institutions, hence leading to the generation of new and conceivably "better" institutional arrangements—is what political economists have contributed to the study of social change.

What is "political" about political economy? This question can be answered in all three dimensions that political scientists identify in political

[3] As in Marx's scheme of the throwing off of the capitalist institutional fetters of the forces of production.

life. First, the original endowment and the system of places and locations, rights and resources that are provided for by this system constitute a polity or "regime," in relation to which actors are passive "regime takers" (Krasner 1983). Within this regime, they engage in conflicts of interest ("politics") and try rationally to exploit, individually or collectively, opportunities available to them. Finally, as policy-makers or "regime makers," they form new institutional arrangements within the regime (through co-operation etc.), as well as affecting and changing, intentionally or through the blind accumulation of externalities, the future structure of the regime itself.[4]

## II Endowment-necessitated perversities

Virtually all kinds of human agency can be processed through this conceptual mill, with interesting and sometimes even robust hypotheses being generated. For instance, industrial workers being institutionally placed within generous contributory social security arrangements will engage in co-operative industrial relations but generate the externality of high unemployment, as employers are led to emphasize labor-saving technical change. Manufacturers placed in markets face uncertainties that lead them to submit to governance structures and forms of contracting that exhibit the positive externalities pointed to by transaction cost economics (Williamson 1975; 1985). Similarly, innumerable politico-economic tales have been told and tested about budget-maximizing bureaucrats (Niskanen 1971), vote-maximizing politicians, utility-maximizing voters (Downs 1957), or jurisdiction-maximizing federal states (Scharpf 1994)—all with their favorable or, more often, not-so-favorable impacts upon employment, budget deficits, inflation, growth, trust and civic culture.

For among the favorite activities of political economists has been the modeling of unpleasant surprises. One of political economy's greatest successes in the 20th century is Arrow's (1951/1963) impossibility theorem and the proof of the essential arbitrariness of aggregation rules in social choice, which sits rather uncomfortably with much of the liberal-democratic folklore concerning the "will of the people" (Miller 1993). Another great success is the modeling of the Prisoner's Dilemma as well as

[4] On endowments and their consequences, see e.g.: Sen (1981; 1983); Drèze and Sen (1989); and Dasgupta (1993). On dynamics, see the sociologically-inspired work on the political economy of positional competition (Hirsch 1976; Elster 1976; cf. Sen 1983) and inflation (Hirsch and Goldthorpe 1978) and the "fiscal crisis of the state" (O'Connor 1973; Gough 1979; Offe 1984).

Olson's (1965) related "logic of collective action," effectively demonstrating how rational actors will systematically under-utilize important resources available to them. There is an abundant supply of tragic malaise and sobering messages conveyed by titles such as "The Tragedy of the Commons" (Hardin 1968), "Why the Government Budget is Too Small in a Democracy" (Downs 1960), "Rational Fools" (Sen 1977) or "Rational Man in Irrational Society?" (Barry and Hardin 1982).

Contrary to much of classical sociology in general, and Weber and Durkheim in particular, there is also a materialist presupposition in political economy approaches in general and in particular their relatively recent and successful offshoot, known today as the "rational choice paradigm." The modeling of unpleasant surprises is often combined with the debunking of idealistic and voluntaristic theories of social action. Behavioral trajectories are determined by opportunities, incentives, interests and calculation, not by what people voluntaristically may feel or imagine they are guided by. Consistent with their "realist" perspective, political economists attach little if any explanatory value to reasons that actors themselves give for choosing a course of action. What counts in social life are not ideas, identities and norms but rather the rational pursuit of interests and the mechanisms triggered by it, with the term "interest" always implying that it derives from structural locations and that there are other players with opposed interests against whom "I" must prevail. Like Marx's identitiless "character masks," the agents that modern political economists envisage are dislodged from social and cultural contexts. Were they to follow norms uncalculatingly, chances are that they would not just fail to achieve Pareto-optimality but that they would end up with the "sucker" payoff.

If norms play a role in political economy, that role is subsidiary, derivative or "superstructural." Norms may be defined and adhered to because actors have agreed to adopt and comply with a norm in the expectation that this will further their interests; or the norm is strategically imposed by one party or class so as to further its interest against another; or the norm and its validity is the outcome of some evolutionary logic that selects norms according to their efficiency-enhancing potential (Axelrod 1984; Ullmann-Margalit 1977; Taylor 1987; Coleman 1990); or the compliance with norms is a case of arational behavior, soon to be eliminated by competitive pressures (such as racist hiring practices of employers, Becker 1971) in which people indulge in the luxury of irrationally constraining their feasibility set beyond the economic and legal constraints they "really" face.

To date, the area of thematic overlap and contestation between political economists and sociologists has remained rather narrow. Many major established fields of sociology—such as the sociology of religion, crime, culture, the family, the professions, knowledge, ethnicity, race and gender relations, education and urban communities—have not yet received much sustained attention from political economists.[5] These seem, at least for the time being, to be largely outside the field of the paradigm conflict.

## III Economists, sociologists and rationality

The proponents of a "new economic sociology" (DiMaggio and Powell 1983; Granovetter 1985; 1991; 1993; Friedland and Robertson 1990) have challenged several of the assumptions of the political economy perspective on economic and other social action, most importantly, the assumption of rational utility maximization. Picking up arguments that were developed within standard micro-economics since World War II, economic sociologists of this emerging school have argued that in order to act rationally, agents would have to have knowledge about alternative courses of action and outcomes associated with each of these alternatives. Such information, however, is hard to come by; its acquisition is costly (Stigler 1961) and often inefficient due to that costliness relative to the ("second order") uncertainty of its marginal utility; and it is asymmetrically distributed, with those who have it often failing, for strategic reasons, to disclose it (cf. Akerlof 1970). This scarcity of information is partly due to the fact that the utility I derive from any of my available courses of action is intentionally co-determined by the course of action chosen by relevant others with opposing interests. Outcomes are strategy-sensitive, and "I" do not know how "you" are going to respond to my moves, which in turn mandates caution and thereby causes the suboptimal realization of benefits that "we" could derive from mutual exchange. In short, if an economic agent does not know what the utility consequences of his or her action will be, utility maximization is no instructive guide to action.

What is called for in such situations is some substitute orientation. Apart from quite arationally adopted degrees of risk-proneness and other psychological traits, that substitute orientation is provided by social norms or what Keynes (1937/1973: 114–17) referred to as "mimesis"—"advice, fashion and habit," which already in 1937 he saw as devices to "save our

[5] But there is evidence this is changing. See, e.g., Gambetta (1988; 1994) on crime; Hardin (1985) and Hechter (1987) on group solidarities organized around racial, religious and ethnic affinities; and, from very different perspectives, Becker (1981), Carling (1991), Waring (1988), Dréze and Sen (1989: chap. 4) and Dasgupta (1993: chaps. 11–13) on gender and the family.

faces as rational, economic men." Thus, under conditions of uncertainty, the assumption of rationality of choice becomes soft beyond recognition. All options are consistent with rationality, but none is dictated by it (Elster 1986: 7).

The bone of contention between political economists and sociologists here is whether the typical situation an actor finds himself or herself in is one of risk or of uncertainty (Knight 1921; Machina 1990). Situations of risk allow probabilities to be attached to alternative outcomes. In a situation of uncertainty, probabilities of outcomes cannot be calculated, and reliance on norms—as well as reliance on the expectation that other strategic actors will also rely on and be guided by norms (Scharpf 1990)—becomes inescapable. The need to cope with uncertainty and indeed to reduce it leads actors to rely on traditions, habits, cognitive frames, stereotypes, signs, conventions, routines, orders, organizational rules, rituals, power relations, community values, styles, symbols, "non-functional" bases of association, ascriptive criteria, standards of honor and the like. Where uncertainty prevails maximizing becomes pointless, and behavior is governed instead by satisficing (Simon 1945) in accordance with social criteria of appropriateness of outcomes.

More radically, one might argue that these substitutes for rational calculation are resorted to not just under conditions of extreme complexity: rather, they are the very pre-conditions of calculability and the unfolding of economic rationality. Only after most sources of contingency have been channeled by rules is the remaining contingency susceptible to rational calculation of alternatives. Scharpf (1990) refers to "interaction orientations" that emerge from some non-negotiable "logic of appropriateness" (March and Olsen 1989) and lead to the "social construction of predictability." In this perspective, shared normative standards concerning the appropriateness of courses of action must be present before economic rational action can begin at all. Polanyi's (1944) famous "satanic mill" argument amounts to the demonstration that only a constrained market—where "not everything" is for sale—can be an efficient market. This is the substance of the notion of "embeddedness" (Granovetter 1985) or "social capital" (Putnam 1993). It is only through sharing in a set of norms which constitute and at the same time limit the universe of "economic activities" that actors know what they can trust in and count upon; and only then can calculation and utility maximization commence. To paraphrase Durkheim, "social capital" fixes the non-economic parameters of economic activity. In order to perform in markets, actors must know and respect the difference between those objects that are suitable for market transactions (or practices that are admissible in such transaction) and those that are not. *E contrario*, one might argue that early modern as well as contemporary emerging post-communist market economies suffer in their efficiency precisely from the fact that the market is unlimited and everything (salvation goods, marriage permits, violence, protection, offices, administrative favors, judicial decisions, export permits, etc.) is up for sale.

Another contested conceptual complex is that of preferences, tastes, benefits and interests. All political economists assume that these are uniquely determined, exogenously "given," or even "objective," i.e., derivatives of the endowment of actors with material and legal resources. The sociological response to this assumption of preference rigidity is that tastes are endogenous and flexible,[6] interpretive, reflexive (Goodin 1986) and multiple (March 1986), rather than fixed and uniquely determined. People have not only preferences but also second-order preferences concerning the preferences which they (counterfactually) would like to have and pursue. Persons can be ambiguous and indeed literally ambivalent—in disagreement with themselves—about whether a "benefit" is actually to be valued as beneficial. Similarly, the concept of costs is subject to interpretation and validation through social norms. Sometimes the expenditure of money, labor and other efforts, as well as the entering into "costly" commitments is experienced by actors and regarded by others with whom they are interacting as an act that yields a process benefit (Hirschman 1982) or serves to express and symbolically assert a person's identity (Pizzorno 1985).

Given the limitations of knowledge and the instability of preferences, political economists in general and rational choice theorists in particular seem to be hard-pressed to prescribe a rational course of action (Elster 1986, introduction). But a further complication emerges when it comes to the positive question of how we determine whether an act performed has been rational. Can an observer determine in non-arbitrary ways whether an act has been rational? No doubt actors can act rationally, if that is taken to mean that they are endowed with resources, stable preferences and beliefs (however erroneous) and that they then employ resources in ways they believe will result in the greatest satisfaction of their preferences. The question is how we recognize it when they do act rationally. Strangely enough, the answer seems to depend not on the actor or act, but on the perspective adopted by the observer. Focusing upon revealed preferences and beliefs, there is hardly any action that would not pass as rational, as beliefs and tastes can always be imputed from overt behavior, if often at the risk of tautology (Sen 1973/1986). However, if the observer bothers to ask the

[6] In the French and German saying that "the appetite comes with the eating."

actor and relies upon stated preferences alone (plus perhaps a reported mental process of consideration and calculation), then the rationality requirement becomes much more stringent and discriminating; but at the same time there is an increased risk that the statement of tastes and beliefs is distorted by deception, self-deception, or rationalization, with rational acts thus being miscoded as irrational or vice versa. As long as there is no uniquely "right" method to determine beliefs and tastes, the degree of rationality attributed to an observed act is virtually an artefact of the choice of observation method. If so, the answer to the question of whether the rationality of a given act is observable is "no."

Can even actors themselves know whether what they have done was rational? That also seems difficult. Here the problem is whether, after having acted rationally, actors hold fast to both their preferences and beliefs—rather than undergoing a "shift of involvements" (as in Hirschman 1982) or recognizing that their beliefs have been erroneous. If the first occurs, the experience is one of regret, if the second, of unpleasant surprise. But how much regret and unpleasant surprise is compatible with the notion of an act still "having been" rational? The very concept of rationality suddenly seems to be in need of time-indexation: the course of action I recall as my own may turn out to be $t_0$-rational, but, in the light of what I know and prefer now, $t_1$-irrational. Is rationality perhaps just an intentional quality of actors, as opposed to a retrospective quality of acts? Or is the retrospective evaluation of past action as "rational" just a function of my failure to adopt new preferences and beliefs between $t_0$ and $t_1$? That would suggest the bizarre conclusion that the more stubbornly I resist learning and stick to my tastes and beliefs, the better I pass the test of the rationality of my past action. But such taste and belief conservatism may well be conditioned, in turn, by past action itself, if only for the sake of avoiding cognitive dissonance. Who, after all, wants to look upon himself as a fool having bought a car that he now dislikes? Is it rational to commit oneself to path-dependency? Again, the very concept of rationality fails to discriminate in non-arbitrary ways.

## IV Can institutions be rational?

Institutions regulate the access to, as well as the standards of, the legitimate utilization of scarce and valued resources. In other words, they define what interests are and how interests must be pursued, given the fact that the condition of scarcity is by definition one that gives rise to contestation and conflict. Institutions regulate the interplay of interests. They define the rules of the game and the gains to be achieved in the game, but not the moves that players make in the game within some institutionally defined space of opportunities, choices and incentives. Within this space, modes of instrumental and strategic action, among others, unfold that are neither proscribed nor prescribed (but just licensed and perhaps suggested) by the institution's formal rules and liberties. Psychological traits, such as "given" preferences and degrees of tolerance for risk and ambiguity, also play a role in the selection of courses of action. Institutions define the mode in which scarcity-induced conflict must be adjudicated and resolved—the realm of freedom and action, "rational" or otherwise.

Can the quality of "rationality" also be attributed to institutions, as opposed to actors? They can be efficient, which will lead rational actors not to forgo the efficiency gains of these institutions by dismantling them. This is at least the case if there are no obstacles to co-operation in support of these institutions. The collective action problem must be resolved so that no significant subset of beneficiaries can destroy the institution by free-riding or by short-term maximization at the cost of greater long-term losses.

It is not entirely clear what political economists mean, or mean to imply, by the proposition that one institution is marginally more rational than another (Schotter 1981; North 1990). First, institutions (such as electoral laws, or banking systems or industrial relations arrangements) will typically have quite a number of effects—known and unknown, positive and negative, long-term and short-term, desired by some and undesired by others. It appears exceedingly difficult to account for all these reference criteria by a single measure of greater or lesser "efficiency." For instance, an industrial relations system may be ideally suited to generate industrial peace and labor flexibility; but it may also render industries uncompetitive because of pressures of wage costs. Similarly, a pure majoritarian electoral system is known to generate stable governments but discriminates against structural minorities. As the trade-off between the two cannot be assessed in other than arbitrary or biased ways, how are we to determine the net efficiency of that arrangement?

Second, even if that difficulty were overcome, what would an unambiguous assessment of an institution's efficiency-enhancing potential mean in practical terms? Rarely do institutions disappear because they are judged relatively inefficient, nor do others get established due to their supposed efficiency alone (Offe 1996). Rather, institutions are path-dependent, culturally embedded and reliant upon multiple justifications, efficiency being at best just one of them. They are more likely to change because their value premises have changed or because they are considered incompatible with other values and institutions than for efficiency reasons. (Selznick 1992)

Political economists and sociologists also diverge in the approaches they typically rely upon in resolving the paradigmatic conflicts that exist between their approaches. Political economists often adopt an engineering approach to institutions. They try to assess and eventually improve the efficiency not just of action but also of the institutional framework (of property rights, contracts, industrial relations, voting systems, governance structures, public administration, constitutions etc.) within which action takes place. As institutions can enhance or inhibit efficiency (most notably because of their impact upon transaction costs), they must be (re)designed so as to maximize efficiency and, thence, welfare. This approach is sometimes perceived as "imperialist" by sociologists who, lacking a specifically sociological theory of production, investment and consumption behavior, respond to it in defensive ways.

A majority of traditions in sociological theory tends to divide the social world into two. One is the sphere of the rational strategic pursuit of interest mediated by formalized codes (i.e., money and votes). The other is a sphere of an overarching social order, solidarity and intrinsically valuable forms of association and cohesion—a sphere which does not spontaneously result from market or other forms of strategic interaction, and which must in turn be protected from the onslaught of efficiency-oriented considerations.

Habermas' (1981) analytical antithesis of "systemic" versus "life world" aspects of social order is just the most refined and sophisticated version of this sociological dualism which can be traced back to Comte's call for "order and progress" at the beginning of the discipline's history (see also Etzioni 1988; 1993; Giddens 1984). Moreover, sociologists also like to point out the hidden contributions to efficiency of institutional arrangements, both as prerequisites for market transaction and as stimulants of productivity, that are not created and designed for the sake of their efficiency and therefore should not be allowed to be abolished in the name of putative efficiency gains (Streeck 1992).

From a sociological perspective, there are spheres of social life that are not constituted as "economic" (i.e., as spheres in which utility and efficiency considerations should be considered appropriate). The asymmetry between the two disciplines can be stated in the following way. Political economists do have an economic theory of institutions and tend to disregard this demarcation line separating spheres. Sociologists have perhaps only the rudiments of a sociological theory of what is going on in markets and firms (Gintis and Bowles 1986), while the most ambitious argument that sociologists do have to offer effectively demonstrates that "non-economic" spheres of society are not only constituted in different ways than the economy, but that the economy itself, for the sake of its efficiency and institutional viability, depends upon its being limited and circumscribed by non-economic spheres of action and motivation.

Promise-keeping and property-respecting are social norms of great and demonstrable economic efficiency. Where their observance is waning, transaction costs will skyrocket. But that contribution to efficiency is essentially a "latent" function of these norms. Were it to become a "manifest" function, with economizing on transaction costs becoming its only legitimating and motivating force, compliance would collapse: temptations to maximize individual utilities by violating these norms would be irresistible, as no effective external enforcement of these norms seems conceivable. By still propagating rational utility maximization as the only normatively acceptable and/or operationally real motivation in human agency, the politico-economic theory of action, once it becomes widely accepted among economic agents, would by implication risk inflicting great damage upon society. A sociologist might ask, "Just how efficient is the intellectual practice of political economy?"

## Acknowledgments

The author wishes to acknowledge the use he has made of an unpublished research paper by Jens Beckert (1995), Princeton University, in preparing this chapter.

## References

Akerlof, G. 1970. The market for "lemons": quality uncertainty and the market mechanism. *Quarterly Journal of Economics*, 84: 488–500.

Arrow, K. J. 1963. *Social Choice and Individual Values.* 2nd edn. New Haven, Conn.: Yale University Press; originally published 1951.

Axelrod, R. 1984. *The Evolution of Cooperation.* New York: Basic.

Barry, B. 1970. Sociologists, Economists and Democracy. London: Collier-Macmillan.

—— and Hardin, R., eds. 1982. *Rational Man and Irrational Society?* Beverly Hills, Calif.: Sage.

Becker, G. S. 1971. *The Economics of Discrimination.* 2nd edn. Chicago: University of Chicago Press; originally published 1957.

—— 1981. *A Treatise on the Family.* Cambridge, Mass: Harvard University Press.

Beckert, J. 1995. Uncertainty and economic action. Mimeo., Department of Sociology, Princeton University.

Carling, A. 1991. *Soical Division.* London: Verso.

Coleman, J. S. 1990. *Foundations of Social Theory*. Cambridge, Mass.: Harvard University Press.

Dasgupta, P. 1993. *An Inquiry into Well-Being and Destitution*. Oxford: Clarendon Press.

DiMaggio, P., and Powell, W. 1983. The iron cage revisited: institutional isomorphism and collective rationality in organizational fields. *American Sociological Review*, 48: 147–60.

Downs, A. 1957. *An Economic Theory of Democracy*. New York: Harper and Row.

—— 1960. Why the government budget is too small in a democracy. *World Politics*, 12: 541–63.

Drèze, J., and Sen, A. 1989. *Hunger and Public Action*. Oxford: Clarendon Press.

Elster, J. 1976. Boudon, education and the theory of games. *Social Science Information*, 15: 733–40.

—— 1979. *Ulysses and the Sirens*. Cambridge: Cambridge University Press.

—— 1983. *Sour Grapes: Studies in the Subversion of Rationality*. Cambridge: Cambridge University Press.

—— ed. 1986. *Rational Choice*. Oxford: Blackwell.

Etzioni, A. 1988. *The Moral Dimension*. New York: Free Press.

—— 1993. *The Spirit of Community: Rights, Responsibilities, and the Communitarian Agenda*. New York: Crown.

Friedland, R., and Robertson, A. F., eds. 1990. *Beyond the Marketplace: Rethinking Economy and Society*. New York: Aldine de Gruyter.

Gambetta, D. 1988. Mafia: the price of distrust. Pp. 158–75 in *Trust: Making and Breaking Cooperative Relations*, ed. D. Gambetta. Oxford: Blackwell.

Giddens, A. 1984. *The Constitution of Society*. Berkeley: University of California Press.

Gintis, H., and Bowles, S. 1986. *Democracy and Capitalism*. New York: Basic.

Goodin, R. E. 1986. Laundering preferences. Pp. 75–101 in *Foundation of Social Choice Theory*, ed. J. Elster and A. Hylland. Cambridge: Cambridge University Press.

Gough, I. 1979. *The Political Economy of the Welfare State*. London: Macmillan.

Granovetter, M. 1985. Economic action and social structure: the problem of embeddedness. *American Journal of Sociology*, 91: 481–510.

—— 1991. Economic institutions as social constructions: a framework for analysis. *Acta Sociologica*, 34: 3–11.

—— 1993. The nature of economic relationship. Pp. 3–41 in *Explorations in Economic Sociology*, ed. R. Swedberg. New York: Russell Sage Foundation.

Habermas, J. 1981. *Theorie des kommunikativen Handelns*. Frankfurt: Suhrkamp.

Hardin, G. 1968. The tragedy of the commons. *Science*, 162: 1243–8.

Hardin, R. 1995. *One for All*. Princeton, N.J.: Princeton University Press.

Hechter, M. 1987. *Principles of Group Solidarity*. Berkeley: University of California Press.

Hirsch, F. 1976. *Social Limits to Growth*. Cambridge, Mass.: Harvard University Press.

—— and Goldthorpe, J. H., eds. 1978. *The Political Economy of Inflation*. Oxford: Martin Robertson.

Hirschman, A. 1977. *The Passions and Interests*. Princeton N.J.: Princeton University Press.

—— 1982. *Shifting Involvements: Private Interest and Public Action*. Princeton N.J.: Princeton University Press.

—— 1986. *Rival Views of Market Society*. New York: Viking.

Keynes, J. M. 1973. The general theory of employment. Vol. xiv, pp. 109–23 in *The Collected Writings of J. M. Keynes*. London: Macmillan; originally published 1937.

Knight, F. H. 1921. *Risk, Uncertainty and Profit*. Boston: Houghton, Mifflin.

Krasner, S. D. 1983. Structural causes and regime consequences: regimes as intervening variables. *International Organization*, 36: 1–21.

Lash, S., and Urry, J. 1984. The new Marxism of collective action: a critical analysis. *Sociology*, 18: 33–50.
Machina, M. J. 1990. Choice under uncertainty: problems solved and unsolved. Pp. 90–132 in *The Limits of Rationality*, ed. K. S. Cook and M. Levi. Chicago: University of Chicago Press; originally published 1987.
Mansbridge, J. J., ed. 1990. *Beyond Self-Interest*. Chicago: University of Chicago Press.
March, J. G. 1986. Bounded rationality, ambiguity and the engineering of choice. In Elster 1986: 142–70.
—— and Olsen, J. 1989. *Rediscovering Institutions*. New York: Free Press.
Miller, D. 1993. Deliberative democracy and social choice. Pp. 74–92 in *Prospects for Democracy*, ed. D. Held. Oxford: Polity.
Niskanen, W. 1971. *Bureaucracy and Representative Government*. Chicago: Aldine-Atherton.
North, D. 1990. *Institutions, Institutional Change and Economic Performance*. New York: Cambridge University Press.
O'Connor, J. 1973. *The Fiscal Crisis of the State*. New York: St. Martin's.
Oberschall, A., and Leifer, E. M. 1986. Efficiency and social institutions: uses and misuses of economic reasoning in sociology. *American Journal of Sociology*, 91: 233–53.
Offe, C. 1984. *Contradictions of the Welfare State*. Cambridge, Mass.: MIT Press.
—— 1991. Introduction: The puzzling scope of rationality. *Archives Européennes de Sociologie*, 32: 81–3.
—— 1996. Designing institutions in east European transitions. Pp. 199–226 in *The Theory of Institutional Design*, ed. R. E. Goodin. Cambridge: Cambridge Univeristy Press.
Olson, M. 1965. *The Logic of Collective Action*. Cambridge, Mass.: Harvard University Press.
—— 1969. The relationship between economics and the other social sciences: the province of a "social report." Pp. 127–62 in *Politics and the Social Sciences*, ed. S. M. Lipset. New York: Oxford University Press.
Ostrom, E. 1990. *Governing the Commons*. New York: Cambridge University Press.
Pizzorno, A. 1985. On the rationality of democratic choice. *Telos*, 63: 41–69.
Polanyi, K. 1944. *The Great Transformation*. Boston: Beacon.
Powell, W. W., and DiMaggio, P. J., eds. 1991. *The New Institutionalism in Organizational Analysis*. Chicago: University of Chicago Press.
Putnam, R. D. 1993. *Making Democracy Work*. Princeton, N.J.: Princeton University Press.
Riker, W., and Ordeshook, P. C. 1973. *An Introduction to Positive Political Theory*. Englewood Cliffs, N.J.: Prentice-Hall.
Roemer, J. E., ed. 1986. *Analytical Marxism*. Cambridge: Cambridge University Press.
—— 1988. *Free to Lose*. Cambridge, Mass.: Harvard University Press.
Scharpf, F. W. 1990. Games (real) actors may play: the problem of mutual predictability. *Rationality and Society*, 4: 471–94.
—— 1994. *Optionen des Föderalismus in Deutschland und Europa*. Frankfurt: Campus.
Schelling, T. 1960. *The Strategy of Conflict*. Oxford: Oxford University Press.
—— 1978. *Micromotives and Macrobehavior*. New York: Norton.
Schotter, A. 1981. *The Economic Theory of Social Institutions*. Cambridge: Cambridge University Press.
Selznick, P. 1992. *The Moral Commonwealth: Social Theory and the Promise of Community*. Berkeley: University of California Press.
Sen, A. 1973. Behaviour and the concept of preference. *Economica*, 40: 241–59. Reprinted in Elster 1986: 60–81.
—— 1977. Rational fools: a critique of the behavioral foundations of economic theory. *Philosophy and Public Affairs*, 6: 317–44.

Sen, A. 1981. *Poverty and Famines*. Oxford: Clarendon Press.
—— 1983. Poor, relatively speaking. *Oxford Economic Papers*, 35: 153–69.
Simon, H. A. 1945. *Administrative Behavior*. New York: Macmillan.
Stigler, G. J. 1961. The economics of information. *Journal of Political Economy*, 60: 213–25.
Streeck, W. 1992. *Social Institutions and Economic Performance*. London: Sage.
Taylor, M. 1987. *The Possibility of Cooperation*. Cambridge: Cambridge University Press.
Ullmann-Margalit, E. 1977. *The Emergence of Norms*. Oxford: Oxford University Press.
Waring, M. J. 1988. *Counting for Nothing*. Wellington, N.Z.: Allen and Unwin.
Williamson, O. 1975. *Markets and Hierarchies*. New York: Free Press.
—— 1985. *The Economic Institutions of Capitalism*. New York: Free Press.
Wright, E. O. 1985. *Classes*. London: Verso.

Chapter 30

# Political Economy: Downsian Perspectives

Bernard Grofman

ANTHONY Downs's *An Economic Theory of Democracy* (1957) is one of the most influential and frequently cited works in social science of the post-World War II period (Almond 1993; Goodin and Klingemann above: 32). Most students of public choice would agree that the three most important elements of *An Economic Theory of Democracy* are the argument as to why turnout can be expected to be irrational when viewed from an instrumental perspective, the discussion of forces in (two) party competition leading to convergence to the views of the median voter, and the argument about why political ignorance can be expected to be rational.[1] Especially among those critiquing rational choice models of electoral choice, it is common to take Downs as standing for the propositions that rational voters should not vote, that in two-party plurality-based competition parties should converge toward the views of the median voter, and that rational voters should be political ignoramuses. I label this view of Downs's supposed three central propositions as the "Classic comic book" portrait of Downs (Grofman 1993: 1–16) because it does little justice to the subtleties of Downs's own portrait of political competition.

The first proposition is based on the idea that, since voting has a cost, in expected utility terms it is instrumentally rational to vote only if the consequences to you of seeing your preferred candidate elected, discounted by the probability that your vote will be decisive in electing that candidate, exceed the net expected costs. But since the likelihood of any single vote being decisive is so close to zero, then as long as the act of voting has some costs it is very unlikely that these costs will be exceeded by the expected

[1] For other important elements in Downs see discussion in Grofman (1993: 1–16) and the various essays in that collection.

benefits of voting, even if the costs themselves are quite low. Hence it would appear that no one (or almost no one) should vote.

The second proposition is based on the logic of competition over a single issue-dimension, when we posit that voters support the candidate that is closest to their own position. If one party is located further away from the median voter than the other, it can improve its vote share by taking a more centrist position. This logic leads us to expect that, in two-party competition, the parties will converge to the position of the median voter if they wish to avoid near-certain defeat.

The third proposition is based on the notion that, when it is costly to gain information, it makes sense to pay such costs only if we believe that the new information might lead us to behave differently. But new information is unlikely to be sufficiently influential in changing our cost–benefit calculations as to induce instrumentally rational turnout, given how unlikely any single vote is to change the outcome. Thus, voters would appear to have no good reason to learn about their electoral choices, since abstention will almost certainly remain the preferred option.

Of course in point of fact we know that, even in the U.S.—one of the world's lowest turnout democracies—a substantial proportion of the eligible electorate does vote, at least for the president. Similarly, we know that many recent Republican and Democratic nominees for offices of all types (including the office of president) are very far indeed from tweedledum-tweedledee in their policy platforms. Furthermore, we know that many voters do claim to have sufficient knowledge about the political choices they face to make what they consider as reasonable choices.

An increasingly frequent response to these facts is to say that rational choice is empirically nothing but a joke. As Green and Shapiro (1994) succinctly put it in their trenchant critique of three decades of empirical research on rational choice models, with particular attention to work derived from ideas in Downs's work: "The emperor has no clothes." Such a view unfairly minimizes what we have learned from rational choice modeling and exaggerates what we can expect from any single work, no matter how seminal.

## I Turnout

Both defenders and opponents of rational choice approaches to participation generally accept the claim that, because of the predictable inefficacy of any single voter's vote (i.e., the likelihood of casting a decisive vote is virtually indistinguishable from zero), and since participation invariably has some cost attached to it, purely instrumental voters should not bother to vote. However, to judge rational choice models of politics primarily or exclusively by their ability to answer questions such as "Why do people vote?" is completely misguided, even though rational choice theorists themselves often fall into this trap (Fiorina (1990) for example rhetorically asking whether "turnout is the paradox that ate rational choice theory").

It only makes sense to regard turnout as a paradox if we insist that voters must have a single-valued utility function in which the decisiveness of their vote plays a critical role. To those skeptical about the usefulness of rational choice theory (e.g., Petracca 1991), the need to invoke non-instrumental factors to "rescue" rational choice from the absurdity of its prediction, is reason enough to reject a rational choice approach to turnout as fundamentally misguided (cf. Barry 1970). But a few rational choice modelers, myself included, have come to regard the supposed paradox of turnout as a matter of having been seduced by Downs into the pursuit of a false issue. In particular, I now see the question of when it is instrumentally rational to vote as a question that has both been studied to death (see e.g., review in Uhlaner 1993; cf. Grofman 1983; Glazer and Grofman 1992) and not that useful a question to begin with (Grofman 1993: chap. 6; cf. Owen and Grofman 1984).

We do not expect micro-economists to tell us why the French prefer wine and southern Germans prefer beer. Rather, we expect economists to answer questions in "comparative statics," for example, how should the consumption of wine change when the relative price of wine and beer changes? In like manner, the proper test of rational choice models of turnout is their ability to help us answer questions such as, "How would we expect turnout among a given set of voters to vary across different types of elections?" and, "How would we expect turnout to change as specific institutional factors related to participatory incentives or participatory barriers undergo change?" Such questions admit that voters have multi-valued utility functions, but propose that a rational choice approach can help us identify factors that will affect voter choices "at the margin." Similarly, rather than simply trying to explain the puzzle of why voters should know anything, it is far more useful to use rational choice ideas to try to specify the conditions under which voters will be more likely or less likely to be well informed. When we think of rational choice models in these terms we move away from mindless rational choice bashing, on the one hand, and empty formalisms, on the other, toward testable theory.

Rather than taking Downs to be stating propositions that are written in stone, we should see him as providing reasons why, *absent countervailing forces*, certain phenomena are likely. However, while Downs can be seen as bequeathing us three important conundrums—"Why do voters vote?" and "Why don't parties in two-party competition converge?" and "Why should people know anything about politics?"—his even more important contribution is to identify variables that will be important in comparative statics terms in explaining variation.

For example, if we wish to account for variation in turnout across elections, then we may reformulate Downs's turnout calculus,

$$T = (P \times I \times B) + C - D$$

in macro rather than micro terms. Here, $T$ is turnout, $P$ is a measure of the closeness of the election, I is a measure of the importance of the election, $B$ is a measure of the difference between the candidates, $C$ is a measure of the (net) costs of participation,[2] and $D$ reflects non-instrumental factors that may motivate participation (or make it a habit). Rather than taking this equation as the choice calculus of a single voter, we interpret it in partial differential terms, i.e., in terms of

$$\frac{\partial T}{\partial P}, \frac{\partial T}{\partial I}, \frac{\partial T}{\partial B}, \frac{\partial T}{\partial C} \text{ and } \frac{\partial T}{\partial D}$$

Now Downs can be interpreted as telling us to expect that, *ceteris paribus*: for a given office, turnout will increase as elections for that office are closer, and as the differences between the candidates are seen as greater; across offices, turnout will be higher when voters see the consequences of their choices as greater (e.g., when the office is a more important one); while across countries or time periods, turnout will be higher when the barriers to participation are lower (e.g., Sunday voting, automatic registration systems); and across types of electoral systems, turnout should be higher in proportional representation systems where voters who are not in the majority can nonetheless elect candidates of choice.[3] Taking a comparative statics perspective allows us to accept that there may be many reasons why voters choose to go to the polls (including habit and social pressure) but that we nonetheless can generate testable rational choice theory by seeking to understand the conditions under which turnout can be expected to go up or go down rather than trying to predict what "baseline" turnout will be.

When we look at the kind of (mostly commonsense) propositions about

[2] $C$ may be negative if, for example, voting is legally compulsory or if voters risk denial of requests from office-holders who discount petitions from non-voters.

[3] More technically, *ceteris paribus*, turnout should increase as a function of the *threshold of exclusion* (Rae 1971; Grofman 1975).

variations in turnout generated by a comparative statics perspective we find that they are by and large strongly supported. For example the prediction that "among those who go to the polls, *ceteris paribus*, more voters should cast a ballot for top of the ticket offices than for less important ones" is strongly confirmed in U.S. elections of all types. Indeed, the result is regarded as so obvious that it is not even thought of as evidence for a rational choice approach to turnout.

Similarly, that elections which involve important offices are, *ceteris paribus*, more likely to draw voters to the polls than are elections that involve only less important offices is such a commonsensically obvious implication of a rational choice approach to turnout that its accuracy is one for which rational choice theory is rarely given credit. Yet few patterns in politics are more striking than the decline in turnout from presidential election years to mid-term elections, or the reduction in turnout in states that have shifted their gubernatorial elections to an off-presidential-year (or even more so, an odd-year) cycle (Boyd 1986). Election analysts also take for granted that local elections will, *ceteris paribus*, draw fewer voters than national ones and that elections with many offices will, *ceteris paribus*, attract more voters than those with but a single candidate. Also, it is well known that turnout will generally be low in special elections that are off the regular elections cycle.

The one exception to the predictive success of comparative statics predictions about turnout inspired by Downs is that the evidence for the link between turnout and political competitiveness is not very strong (Foster 1984). Here, however, there are a number of technical problems with the way that this proposition has been tested (see discussion in Glazer and Grofman 1992), especially failure to make use of longitudinal forms of analysis. For example, except in a few areas, the Democratic primary decided the election at the state and local level in the U.S. South for most of this century. Thus we should expect that, as the South became more competitive, turnout in the general election should rise relative to that in the Democratic primary and that Republican primaries would come into existence. Recent work of my own on relative turnout in southern primaries and general elections 1922–90 supports both expectations. Other recent work of my own on the relationship between turnout and electoral success tracked longitudinally over the course of a legislative incumbent's career also finds strong support for a turnout-closeness link.

A number of authors, including Erikson (1981) and Wolfinger (1993), have pointed out that, at least in the United States, if we are to understand turnout we must understand the decision to register, since such a high proportion of all registered voters vote in high-profile elections and since relatively few non-registered voters become registered in the period immediately before a particular election. Some of the authors who emphasize the importance of the decision to register suggest that the intermediating role of registration vitiates the Downsian approach to turnout where that approach is taken to be synonymous with an emphasis on election-specific factors (such as anticipated closeness). This view of the limitations of a rational choice approach to turnout (see, e.g., Wolfinger 1993), improperly makes Downs (1957) the sole arbiter of what is or is not *the* rational choice model to turnout. It is better to view rational choice as a style of approach (one focusing on costs and benefits) that should not be exclusively identified with any particular model of some given phenomenon. Indeed, I regard work such as that of Rosenstone and Wolfinger (1978) on the effect of registration barriers on turnout as clearly falling in the rational choice tradition. Similarly, when we turn to cross-national comparisons, the evidence is very strong that factors related to expected costs of participation (e.g., compulsory voting laws, two-day balloting, Sunday balloting) are important influences on turnout (Powell 1986; Jackman and Miller 1995; Jackman 1993).

To understand turnout in the United States, where registration barriers are high and ballots are incredibly long, we need to go beyond Downs in distinguishing three types of turnout decisions: registration among those eligible, turnout at the polls among registrants, and vote for office among those who come to the polls. The importance of keeping separate these three aspects of voter turnout has been largely neglected in the literature on political participation (see however, Engstrom and Caridas 1991, and the discussion of the four factors that affect "effective minority voting equality" in Brace, Grofman, Handley and Niemi 1988).

Once we recognize that there are three different types of turnout then our analysis of the expected linkages between turnout and competition becomes quite different from what is found in Downs. The distinction among types of turnout has a number of testable implications for the link between competition and turnout. For example, it suggests that turnout should be related to long-run rather than short-run levels of competition, since registration allows for voting in many different elections, the exact degree of competitiveness of most or all of which will at the time of registration be unknown. If registration costs are a substantial component of the total costs of voting, the decision among registrants to turn out at the polls should be only loosely coupled to the degree of competitiveness in the particular contests that might be being voted on (contrary to the usual interpretation of what a Downsian approach to turnout predicts). However, areas of the country with low *long-run* competitiveness should also, *ceteris paribus*, be characterized by low registration and low turnout.

Relatedly, the best predictors of turnout among eligibles should simply be the proportion that was previously registered or previous levels of turnout and/or the nature of the offices at issue (viz. the proposition that elections with more important offices, will *ceteris paribus*, have higher turnout).

## II Party competition

Just as turnout is *not* the paradox that ate rational choice theory, failure of parties in two-party competition to converge in no way invalidates Downsian ideas, once we recognize that we need to build on the models in *An Economic Theory of Democracy* rather than regarding them as the last word. The convergence result in Downs is contingent on a set of eight subsidiary assumptions (about there being only a single election, about myopic behavior on the part of voters, about unidimensionality, etc.) whose modification can easily eliminate the convergence result (Grofman 1993: chap. 12). Much of the more recent formal literature on two-party competition has dealt with the conditions under which convergence will *not* be expected to occur.

It is one of the truly peculiar features of *An Economic Theory of Democracy* that voters are assumed to be motivated exclusively by policy considerations, while parties/candidates are assumed to be motivated solely by desire to win elections. Why should we not allow for multiple concerns on the part of both voters and candidates if doing so is necessary to make sense of what we observe in the world? Recent models of party competition (e.g., Wittman 1973; 1977; also see Enelow and Hinich 1984; Alesina and Rosenthal 1993; and various essays in Enelow and Hinich 1990) allow politicians and party activists to care about what policies are implemented once office has been won, and not simply to be motivated by the desire to win office, *per se*.

Downs explicitly views electoral choice as a single-shot event in which voters pick the best of what is available with no concern for influencing future elections or future policies. Also, Downs does not offer any model for sequential election processes such as those in the United States involving both primaries and general elections, or for elections taking place simultaneously over multiple constituencies. More recent models such as those of Aranson and Ordeshook (1972), Coleman (1972), or Owen and Grofman (1995) generate partial divergence because of the two-stage nature of the choice process by allowing voters to eschew candidates whose positions might win a primary but lose the general election or by permitting candidates to make strategic choices of location to optimize outcomes in a two-stage election game.

For two-party competition in the U.S., the role of party activists—when combined with primaries and with the importance of durable "party images"—virtually guarantees that there will be a self-selection and weeding-out process in which candidates gravitate to and are chosen by the party whose policy positions most resemble their own. In recognizing these complicating factors, *contra* the "Classic comic book" version of Downs, we would expect that candidates in two-party competition will in general be much closer to the median voter in their own party than to the overall median voter, but will be shifted somewhat toward the views of potential swing voters. This is exactly what Shapiro *et al.* (1990) find.

Another way in which the standard Downsian approach is limited is in taking voters as consumers who must simply accept the range of options given to them by the political market and choose the best from among them, while letting politicians do little more than signal which positions from this range they advocate. Recent work avoids these limitations by considering: (1) the role of heresthetic politicians in introducing new issue dimensions and reframing old issues in new ways (Riker 1982); (2) the role of persuasion in affecting the relative weight that voters give to different issues and in changing voters' assessments of the link between proposed courses of actions and their likely policy consequences (Hammond and Humes 1993); (3) the role of political campaigns in not just characterizing the politician's own views or character traits but in (mis)characterizing those of his or her opponent (Skaperdas and Grofman 1995); and (4) the expectations that voters have of how likely candidates are to actually deliver on what they promise (Grofman 1985).

## III Information acquisition

Of the three major puzzles bequeathed to us by Downs, in my view Downs's greatest contribution is with respect to resolving the last of these: "Why should people know anything about politics?" Downs offers ideas such as the by-product theory of information acquisition, the role of simplifying cues such as ideological labeling, and the informative role of endorsements (especially party nominations). These ideas have been further elaborated by a number of subsequent authors.[4] Unfortunately,

[4] See especially Page (1978), Popkin (1991; 1993), Norrander and Grofman (1988), and Lupia (1992; 1993; 1994); also see McKelvey and Ordeshook (1985; 1986), Grofman and Withers (1993) and citations therein.

because the vast bulk of work in the Downsian tradition has been on turnout and party competition, Downs has not been given as much credit for these seminal ideas as he deserves.

## IV Conclusion

Taking potshots at rational choice models (*à la* Green and Shapiro 1994) is useful for deflating pomposity and exposing bad research, but it presents a very misleading picture of rational choice's contributions to political science. Some rational choice modelers have displayed an arrogance about elegant modeling being its own justification, an unwillingness to make the effort to explain their results in ordinary English along with reasons why anybody ought to care, and an apparent belief that no one who is not a modeler can be making a contribution to political science. Such hubris on the part of some rational choice theorists, along with exaggerated claims as to the empirical successes of rational choice models, have been key reasons for the recent backlash to rational choice modeling within the discipline.[5] Yet there is plenty that rational choice can be proud of in terms of aiding us in understanding the dynamics of electoral politics, campaigning, and voter choice, especially when we judge its empirical performance and analytic insights *relative to* the work (both theoretical and empirical) done by political scientists not working within that tradition.

## Acknowledgments

I am grateful to Dorothy Green and Chau Tran for assistance.

## References

Alesina, A. and Rosenthal, H. 1993. *Partisan Politics.* New York: Cambridge University Press.

Almond, G. A. 1993. The early impact of Downs's *An Economic Theory of Democracy* on American political science. Pp. 201–7 in Grofman 1993.

[5] Moreover, much of the modeling work on party competition, including even some of the most recent work, has suffered from an unwillingness to confront messy empirical reality by formulating testable theory. Important exceptions include Enelow and Hinich (1984), Fiorina (1992), and Alesina and Rosenthal (1993). Perhaps the single most important source of data and insights into voter choice in the multi-party context is the European Party Manifestos Project (see esp. Budge, Robertson and Hearl 1987).

Aranson, P., and Ordeshook, P. C. 1972. Spatial strategy for sequential elections. Pp. 298–331 in *Probability Models of Collective Decision Making*, ed. R. G. Niemi and H. F. Weisberg. Columbus, Ohio: Merrill.

Barry, B. 1970. *Sociologists, Economists, and Democracy*. London: Collier-Macmillan.

Boyd, R. W. 1986. Election calendars and voter turnout. *American Politics Quarterly*, 14: 89–104.

Brace, K.; Grofman, B.; Handley, L.; and Niemi, R. M. 1988. Minority voting equality: the 65 percent rule in theory and practice. *Law and Policy*, 10: 43–62.

Budge, I.; Robertson, D.; and Hearl, D., eds. 1987. *Ideology, Strategy and Party Change*. Cambridge : Cambridge University Press.

Coleman, J. 1972. The positions of political parties in elections. Pp. 332–57 in *Probability Models of Collective Decision Making*. ed. R. G. Niemi and H. Weisberg, Columbus, Ohio: Merrill.

Downs, A. 1957. *An Economic Theory of Democracy*. New York: Harper and Row.

Enelow, J. M., and Hinich, M. J. 1984. *The Spatial Theory of Voting*. New York: Cambridge University Press.

—— —— eds. 1990. *Advances in the Spatial Theory of Voting*. New York: Cambridge University Press.

Engstrom, R. L., and Caridas, V. M. 1991. Voting for judges: race and rolloff in judicial elections. Pp. 171–91 in *Political Participation and American Democracy*, ed. W. Crotty. New York: Greenwood Press.

Erikson, R. S. 1981. Why do people vote? Because they are registered. *American Politics Quarterly*, 9: 259–76.

Fiorina, M. 1990. Elections and the economy in the 1980s. Department of Government, Harvard University, Occasional paper # 90–5.

—— 1992. *Divided Government*. New York: Macmillan.

Foster, C. B. 1984. The performance of rational voter models in recent presidential elections. *American Political Science Review*, 78: 678–90.

Glazer, A., and Grofman, B. 1992. A positive relationship between turnout and plurality does not refute the rational voter model. *Quality and Quantity*, 26: 85–93.

Green, D., and Shapiro, I. 1994. *Pathologies of Rational Choice Theories*. New Haven, Conn.: Yale University Press.

Grofman, B. 1975. A review of macro-election systems. *German Political Yearbook* (*Sozialwissenschaftliches Jahrbuch fur Politik*), 4: 303–52.

—— 1983. Models of voter turnout: a brief idiosyncratic review. *Public Choice*, 41: 55–61.

—— 1985. The neglected role of the status quo in models of issue voting. *Journal of Politics*, 47: 231–7.

—— ed. 1993. *Information, Participation and Choice: An Economic Theory of Democracy in Perspective*. Ann Arbor: University of Michigan Press.

—— and Withers, J. 1993. Information-pooling models of electoral politics. Pp. 55–64 in Grofman 1993.

Hammond, T. H., and Humes, B. D. 1993. "What this campaign is all about is . . ." A rational choice alternative to the Downsian spatial models of elections. Pp. 141–59 in Grofman 1993.

Jackman, R. W. 1993. Rationality and political participation. *American Journal of Political Science*, 37: 279–90.

—— and Miller, R. A. 1995. Voter turnout in the industrial democracies during the 1980s. *Comparative Political Studies*, 27: 467–92.

Lupia, A. 1992. Busy voters, agenda control, and the power of information. *American Political Science Review*, 86: 390–403.

—— 1993. Credibility and the responsiveness of direct legislation. Pp. 379–404 in *Political Economy, Competition, and Representation*, ed. W. A. Barnett, M. J. Hinich and N. Schofield. New York: Cambridge University Press.

—— 1994. Short cuts versus encyclopedias: information and voting behavior in California insurance reform elections. *American Political Science Review*, 88: 63–76.

McKelvey, R. D., and Ordeshook, P. C. 1985. Sequential elections with limited information. *American Journal of Political Science*, 29: 480–512.

—— —— 1986. Information, Electoral Equilibria and the Democratic Ideal. *Journal of Politics*, 4: 909–37,

Norrander, B., and Grofman, B. 1988. A rational choice model of citizen participation in high and low commitment electoral activities. *Public Choice*, 59: 187–92.

Owen, G., and Grofman, B. 1984. To vote or not to vote: the paradox of nonvoting. *Public Choice*, 42: 311–25.

—— —— 1995. A two-stage model of two-party competition. Prepared for delivery at the Annual Meeting of the Public Choice Society, Long Beach, Calif., 24–26 March.

Page, B. 1978. *Choice and Echoes in Presidential Elections.* Chicago: University of Chicago Press.

Petracca, M. 1991. The rational actor approach to politics: science, self-interest, and normative democratic theory. Pp. 171–203 in *The Economic Approach to Politics*, ed. K. R. Monroe. New York: HarperCollins.

Popkin, S. 1991. *The Reasoning Voter.* Chicago: University of Chicago Press.

—— 1993. Information shortcuts and the reasoning voter. In Grofman 1993.

Powell, G. B. 1986. American voter turnout in comparative perspective. *American Political Science Review*, 80: 17–43.

Rae, D. W. 1971. *The Political Consequences of Electoral Laws.* 2nd edn. New Haven, Conn.: Yale University Press; originally published 1967.

Riker, W. 1982. *Liberalism versus Populism.* San Francisco: Freeman.

—— and Ordeshook, P. C. 1973. *An Introduction to Positive Political Theory.* Englewood Cliffs, N. J.: Prentice-Hall.

Rosenstone, S. J., and Wolfinger, R. E. 1978. The effect of registration laws on voter turnout. *American Political Science Review*, 72: 22–45.

Shapiro, C. R.; Brady, D. W.; Brody, R. A.; and Ferejohn, J. A. 1990. Linking constituency opinion and Senate voting scores: a hybrid explanation. *Legislative Studies Quarterly*, 15: 599–623.

Skaperdas, S., and Grofman, B. 1995. Modeling negative campaigning. *American Political Science Review*, 89: 62–73.

Uhlaner, C. 1993. What the Downsian voter weighs: a reassessment of the costs and benefits of action. Pp. 67–79 in Grofman 1993.

Wittman, D. A. 1973. Parties as utility maximizers. *American Political Science Review*, 68: 490–8.

—— 1977. Candidates with policy preferences: a dynamic model. *Journal of Economic Theory*, 14: 180–9.

Wolfinger, R. E. 1993. The rational citizen faces election day; or, what rational choice theorists don't tell you about American elections. In *Elections at Home and Abroad: Essays in Honor of Warren E. Miller*, ed. T. Mann and K. Jennings. Ann Arbor: University of Michigan Press.

Chapter 31

# Political Economy, Old and New

A. B. Atkinson

## I Introduction

### A "Political economy" and "economics"

Is "political economy" the same as "economics?" Not today, as is evidenced by the existence of this part of the *New Handbook*. But in the past the two terms were often treated as having the same meaning. Alfred Marshall, who is credited by Cannan (1929) with bringing about general acceptance of the term "economics," referred to it interchangeably with "political economy" on the first page of his *Principles of Economics* (1890). As Groenewegen (1985) has argued, writers at that time treated the terms as being essentially synonymous. Jevons urged the dropping of the "old troublesome double-worded name of our science" (1910: xiv) only on "grounds of convenience and scientific nicety" (Groenewegen 1987: 905). Alfred and Mary Marshall made the change in *The Economics of Industry* (1879) because the word "political" had come to have different overtones. The subject-matter had not been re-defined, and Marshall made no reference to any disjuncture between the title of his chair (Professor of Political Economy) and the title of his Inaugural Lecture ("The Present Position of Economics").

In the course of the 20th-century, the expression "political economy" came to have a slightly old-fashioned air (one of my former departments changed its title in the 1980s from Political Economy to Economics). This decline in use is described by the *Cambridge Encyclopedia*, which says that "political economy" is

> the name given to economics in the late 18th-century and early 19th-century. The term has not been much used in the present century . . . reflecting the fact that the scope of economics is today much wider, dealing with many more issues than national economic affairs and the role of government (Crystal 1990: 958).

It is interesting that this definition sees political economy as a subset of economics, and, to establish that this view is not merely held on the banks of the Cam, we may note that the *New Shorter Oxford English Dictionary* defines "political economy" as "the branch of economics that deals with the economic problems of government" (Brown 1993: 782).

In this popular conception, "political" economy appears to denote the part of economics which refers to the body politic. In contrast, in the introduction to their chapter in the *New Handbook* (above: chap. 27), James Alt and Alberto Alesina clearly see political economy as going *beyond* economics. They describe recent political economy as advancing outside mainstream economics by treating institutions as endogenous, rather than as given, by applying economic analysis to political behavior, and by bringing together economics and political science. Claus Offe (above: chap. 28) discusses the intellectual ambitions of political economy as going beyond those of both economics and political science (and sociology). Equally, Drèze and Sen in *The Political Economy of Hunger* explicitly reject the idea that political economy is a subset of economics, saying that the term is "a reminder of the breadth of the earlier tradition of the subject. Many of the analyses of the kind that are now seen as interdisciplinary would have appeared to Smith or Mill or Marx as belonging solidly to the discipline of political economy" (Drèze and Sen 1995: 14–15).

While it is important to be reminded of the broader perceptions of the classical economists, it would be wrong to present current political economy simply in historical terms. The identification in the 20th century of the term "political economy" with a distinctly different content from that of mainstream economics has been a reaction to contemporary concerns—concerns which have evolved with changing historical circumstances and intellectual trends. This is well illustrated by the political economy of the New Left of the 1960s, when "the 'revival of political economy' was the chief contribution of the neo-Marxist New Left to the radical protest movement of the 1960s" (Arndt 1984: 268). The aim of this movement was to change both the approach and the content of economics. A student attending a course on "political economy" in the 1970s would not have expected to be taught the invisible hand, welfare economics and comparative advantage, but about power, monopoly capitalism, the distribution of income, and multinationals. As put by Lindbeck (1977: 17), "the New Left criticizes economists for having neglected problems of the *interaction between economic and political factors*."

## B Public choice and public finance

Little or no reference is to be found in Alt and Alesina's chapter to the political economy of the New Left, nor is there much reference to the development, from a different area of the radical political spectrum, of the public choice approach to public finance. This is perhaps more understandable in that political economy as defined by Alt and Alesina, and by Bernard Grofman (above: chap. 29), may be seen as following in a public choice tradition: one notes among their references Buchanan and Tullock (1962), Buchanan and Wagner (1977), Downs (1957) and Niskanen (1971).

Public choice has had a major impact on public economics. This is brought out by the following passage from Baumol's *Welfare Economics and the Theory of the State* where he describes the purpose of his analysis:

> we are very little concerned with what a government does in fact do . . . and in no case have we considered the ethical question of what a state *should* do. Rather the bulk of the discussion concerns itself with an analysis of the circumstances under which government activity . . . may prove beneficial to those governed (1965: 180).

I quote this for two reasons. First it helps lay to rest the caricature of welfare economics as concerned with "a (benevolent) dictator applying the 'optimal' propositions offered by economists" (Frey 1976: 32). The purpose of the welfare economic approach is to illuminate the structure of arguments, helping us understand the relationship between instruments, constraints and objectives.

The second point which emerges from Baumol's quotation is that he did not seek to explain the behavior of government. Here the public choice approach has undoubtedly enriched modern public finance. As described by Frey (1983: 2), "Government is taken to be an *endogenous* part of the politico-economic system: it does not act autonomously but is influenced by many different forces. In this process economic and political institutions—in particular, parties, government administration and private interest groups—play an important role." Again, this may be seen as a return to earlier traditions. The latter are particularly associated with Wicksell and the Italian school (see Buchanan 1960), but in the English language literature on public finance one finds Bastable (1903: 10, 44), for instance, stating in his text that "public finance belongs to the domain of political science" and giving such examples as the role of interest groups in determining public spending.

Public choice is certainly different from the welfare-economics based public finance of the postwar period, but in my view the relation is not profitably seen in confrontational terms. The public choice and optimum taxation perspectives can be brought together, as argued in Atkinson (1995). A good illustration is provided by the analysis of targeting in the design of income transfers. Suppose that the government is concerned with the extent of poverty, and that it is considering various forms of transfer which take the form of a guaranteed amount reduced (tapered) by some percentage of the recipient's income until the transfer is extinguished. Where the taper rate is 100 percent, then we have a minimum income guarantee (MIG); where the taper rate is zero, we have a universal benefit (UB). If the total budget for the transfer is fixed, then (under certain conditions) the efficient allocation concentrates the net benefit on those with the lowest incomes via an MIG. This is moderated to the extent that people adjust their behavior. If people work less, or save less, as a result of the tax rate implicit in the MIG, then a rate of taper of less than 100 percent may be better.

So far I have simply presented a standard application of optimum tax theory. Public choice considerations may however modify the conclusions. From this perspective, concern about incentives may arise less from the quantitative effect on labor supply as from a notion of "desert." It may not be regarded as "fair" that a person is unable to improve his or her position by working more. If this is the case, then "desert," rather than an equity-efficiency trade-off, may determine the maximum rate of withdrawal and hence the acceptable level of targeting. More radically, public choice may lead us to question the notion of a fixed budget. The ability of the government to finance transfer programs may depend on the form of the transfer. There are those who suggest that highly targeted benefits to a minority of the population lack political support. The International Labour Office (ILO) report *Into the Twenty-First Century* commented that the argument that "more generous provision could be made for the poor on an income-tested basis seems at first sight to have a compelling logic" (ILO 1984: 23); but it went on to say that "people are more willing to contribute to a fund from which they derive benefit than to a fund going exclusively to the poor. The poor gain more from universal than from income-tested benefits" (ILO 1984: 23).

## C What is new?

What then is new about today's political economy? Alt and Alesina (above: chap. 28) emphasize the much greater interaction between economists and political scientists. There have long been outstanding scholars whose work lies on the borderlines, like Arrow's *Social Choice and Individual Values* (1951/1963). Downs's influential *An Economic Theory of Democracy* (1957), the subject of Grofman's chapter 30, cannot readily be classified as economics or political science. But recent years have seen a more widespread interest in cross-disciplinary debate, a good example being provided by Alesina and Carliner's 1991 collection, where the authors of the papers are political scientists and the commentators are economists. The novelty of this fraternization should not be exaggerated: the volume edited by Hibbs and Fassbender (1981) contains papers from conferences in 1978–9 bringing together economists and political scientists. And there were doubtless earlier examples. But economists are, I hope, more willing than they have sometimes been in the past to learn from other disciplines.

Alt and Alesina (above: chap. 28) are also right in referring to a rapid growth of interest. One now finds much more being written in the mainstream on the economics of politics and the politics of economics. The *American Economic Review* for 1993, for example, included articles on:

- Economic policy, economic performance and elections (Harrington 1993)
- Weapons accumulation by two adversarial countries (John, Pecchenino and Schreft 1993)
- Discipline and credibility in monetary policy formation (Garfinkel and Oh 1993)
- Tax competition and tax co-ordination (Kanbur and Keen 1993)
- Privatization and the politics of transition (Laban and Wolf 1993)

This set of articles serves to illustrate some of the key characteristics of the recent literature. These include modeling of government behavior and the pervasive influence of game theory. They include the role played by expectations, particularly with regard to future government policy, and the associated issue of credibility. In this regard, the extension of public-choice considerations into macro-economics has led to developments which can feed back into public finance.

## II New political economy in the *New Handbook*

### A Institutions

At the start of their chapter (above: chap. 28), Alt and Alesina provide a clear statement of their perception of the content of present-day political economy. The first of the themes emphasized is that of the endogeneity of institutions. As they rightly say, an important goal of modern political science and economics is to explain the emergence and evolution of institutions.

This is a research program with major implications. Taken to the limit, no institutional structure can be introduced as part of the explanation. Different political behavior in different countries cannot be attributed to institutional factors, since these factors themselves are the product of a more deep-seated set of determinants. One cannot, for example, talk of two U.S. states providing a "natural experiment" because they have different voter registration laws. The different registration laws may themselves be a reflection of more basic differences. Differences in public debt cannot be attributed to differences in state budgeting rules, since states which do not intend to borrow are more likely to pass laws limiting debt.

As an outsider, I am led to ask what can truly be regarded as exogenous variables in this analysis? What are the primitive elements in this model-building enterprise? Consideration of one possible candidate—voter preferences—suggests that the answer is not straightforward. The classical democracy model may be seen as parallel to classical consumer behavior. Preferences are a primitive concept and observed behavior follows from the properties of these preferences. Under certain assumptions the outcome is that preferred by the median voter. We may have a particular set of political institutions, but if they do not deliver what the median voter prefers, then they will be modified until these preferences are expressed. There may be political parties, but only those whose platform approximates the choice of the median voter will get elected. In concrete terms, we would not expect institutional factors to add anything to the explanation of government behavior over and above those variables which influence the preferred choice by the median voter.

As argued below, the notion of voter preferences needs to be treated with some caution. For the moment, I wish simply to emphasize the ambitiousness of the research program. Here I find sympathetic the account given by

Grofman (above: chap. 30) of the Downsian agenda. Like him, I agree that it is over-optimistic to suppose that we can explain why the French drink more wine and less beer than the Bavarians—or that it is not very illuminating to say that the difference is due to a difference in their taste-buds. Rather, as he suggests, we can hope to make statements about what happens if the relative price of wine rises. Or in the case of state borrowing, we may learn from the differential reactions to unexpected fiscal shocks (Poterba 1994).

This seems to me a more tractable agenda for two inter-related reasons. The first is that there is likely to be considerable friction in the system, and we may take a long time to converge to an equilibrium. In the long run, only voter preferences may matter, but the foreseeable future is dominated by historical experience. There is sand in the works. As Alt and Alesina (above: chap. 28) point out, once institutions have been created, they generate constituencies of support. The second reason is that there may be multiple equilibria, in which case the ultimate outcome depends on history. As a result of their starting-points, some countries may be in a high-debt equilibrium, others in a low debt equilibrium, for no apparent reason. Nonetheless, we may still be able to draw conclusions about the comparative statics.

## B Rational choice

A second major strand in the new political economy, as revealed in Alt and Alesina's discussion, is the adoption of "an economic approach, constrained maximizing and strategic behavior by self-interested agents" (above: chap. 28: 645). This, plus rational expectations, forms the basis of the rational choice approach which has apparently generated strong feelings in political science. Such reactions are perhaps not surprising, since the word "rational" carries with it a lot of baggage, some of which may be excess baggage.

It may appear strange to hear an economist questioning the rational choice approach, but I share many of the doubts expressed by Offe (above: chap. 29). The applicability of the approach clearly depends on the subject-matter. A model which is appropriate for the choice between wine and beer may not apply equally to the choice between Bush and Clinton. A decision made once every four years, under circumstances which are likely to be different each time, is not the same as one made every evening in similar circumstances. For many people, voting is an act of personal and social significance, which a visit to the liquor store is not.

Just as the decisions relate to different spheres, so too people may make choices according to different sets of preferences. As described by Sen (1977), it was in this way that Edgeworth (1881) justified adoption of the assumption of self-interest: he felt that the assumption was appropriate to the particular types of activity with which he was concerned. In the context of my earlier example of a tapered transfer system, a person deciding whether to undertake a part-time job in the face of a particular rate of taper can reasonably be modeled as consulting his or her self-interest, and attaching no weight to the implications for the government budget. However, when it comes to voting between political parties offering different programs, then the claims of others may enter the decision. The recognition of these claims may be seen as an extension of self-interest, as where a person's welfare depends on the welfare of others ("sympathy"), but it may involve people acting against their own personal interests ("commitment").

The idea of commitment "drives a wedge between personal choice and personal welfare" (Sen 1977); it means that people may rationally choose an act which yields a lower anticipated level of personal welfare. This is particularly important in the present context, since it is precisely in the case of voting that a person seems most likely to be influenced by commitment. To model voting as though it were a choice between French wine and German beer may miss the point if it ignores situations like that—rare in everyday consumer decisions—a person does not buy French wine on principle on account of their nuclear policy.

The parallel with consumer theory is also misleading in that, while it may be reasonable to assume a degree of stability of consumer tastes, this is less likely to be true of political preferences. As it is put by Buchanan,

> the assumption of given tastes in the decision-making represented by the market is essential for the development of a body of economic theory. But the extension of this assumption to apply to individual values in the voting process disregards one of the most important functions of voting itself. The definition of democracy as "government by discussion" implies that individual values can and do change in the process of decision-making (1960: 85).

Since one of the influences on such shifts in values is the writing of economists and political scientists, it scarcely behoves them to assume an immutable set of preferences.

## C Empirical evidence

In their review of political economy, Alt and Alesina are very upbeat about what can be learned from empirical evidence. We are told of "an immense, highly successful theoretical and empirical literature" about the U.S. Congress. We are told that the new theories of political cycles based on rational choice and rational expectations "are more successful than their predecessors." Later they say that "the next few years" empirical work will be decisive" (above: chap. 28).

I must confess to a degree of skepticism. Even though I very much favor looking at the empirical evidence, I feel that one should not hope too much. In economics, econometric evidence has rarely been decisive in settling disputes about economic relationships. There are very few natural experiments and the interpretation of the findings is open to dispute. To give one example, Alesina and Rosenthal (1995: 209) draw attention to the fact that their statistical tests are *joint tests* of the hypothesis of interest—rational retrospective voting—and of the other assumptions made in specifying the model. The fact that this problem is common to many econometric studies is no real consolation.

Let us take the—very interesting—rational partisan model, where Left and Right have different preferences over inflation and unemployment (hence "partisan"), where voters are rational, but there is uncertainty surrounding the electoral outcome and macro-economic contracts have to be made in advance (and cannot be contingent on the election result). With these assumptions, there will be transitory real electoral effects to an extent which depends on the degree of "surprise" associated with the election. Alesina and Roubini (1992) test the theory by introducing into autoregressive models of growth, unemployment and inflation, a dummy variable which takes the value "plus 1" for the six (or four or eight) quarters following a change of government to the right and "minus 1" following a change to the left. This variable proves to be significant in the pooled time-series cross-country dataset, for countries with two-party or two-bloc systems, with a coefficient of 0.62, with a 95 percent confidence interval of some 0.3 to 0.9, implying that 18 months after the election of a right-wing party the annual growth rate is lower by about 1.3 percent. If the regressions are performed country by country, then "all the regressions on growth, inflation and unemployment show evidence favorable to the [hypothesis], although not all the coefficients on the political variables are significant" (Alesina and Roubini 1992: 679).

There are a number of questions which a skeptic might ask. These may concern the approach to testing different theories (see the discussion by

Mirrlees of Alesina 1989). It is not obvious that electoral *surprises* are captured by *changes* in government: in the U.K. the re-election of Major in 1992 was probably more "surprising" than the election of Thatcher in 1979. The model relies on a particular macro-economic theory. Examination of individual country experience may lead readers to be dubious. In the case of the U.K., the results are "greatly strengthened" if the sample is restricted to the period post-1971, but that means that only two changes of government are covered, both of them with their special features (Wilson in 1974 after the Miners' strike; Thatcher in 1979) which mean that they are unlikely to provide much guide as to what is likely to happen if the U.K. government ever changes to the left. Finally, we may note that other researchers have been more guarded. According to Sheffrin, "the two tests of rational partisan theory . . . are not kind to theory" (1989: 256), although Alesina (1990) has responded. Paldam (1991: 323) is also cautious, saying of his tests that, "We conclude that there is an 'animal' in the data looking a lot like the Alesina RE partisan cycles, but it is, as yet, a fairly debatable creature, which should be further investigated."

## III Final remark

If my remarks have on occasion been critical, this should not be taken as implying any lack of appreciation on my part of the importance of this field of study. Quite the reverse. I believe that it should be taken very seriously. Not least of the reasons for taking it seriously is the fact that the study of political economy is itself part of the political and economic process.

## References


Alesina, A. 1989. Politics and business cycles in industrial democracies. *Economic Policy*, 4: 55–87.

—— 1990. Evaluating rational partisan business cycle theory: a response. *Economics and Politics*, 3: 63–72.

—— and Carliner, C., eds. 1991. *Politics and Economics in the Eighties*. Chicago: University of Chicago Press.

—— and Rosenthal, H. 1995. *Partisan Politics, Divided Government and the Economy*. Cambridge: Cambridge University Press.

—— and Roubini, N. 1992. Political cycles in OECD economies. *Review of Economic Studies*, 59: 663–88.

Arndt, H. W. 1984. Political economy. *Economic Record*, 60: 266–73.

Arrow, K. J. 1963. *Social Choice and Individual Values.* 2nd edn. New Haven, Conn.: Yale University Press; originally published 1951.

Atkinson, A. B. 1995. *Public Economics in Action.* Oxford: Oxford University Press.

Bastable, C. F. 1903. *Public Finance.* 3rd edn. London: Macmillan; originally published 1892.

Baumol, W. J. 1965. *Welfare Economics and the Theory of the State.* 2nd edn. London: G. Bell and Sons; originally published 1952.

Brown, L., ed. 1993. *New Shorter Oxford English Dictionary.* Oxford: Oxford University Press.

Buchanan, J. M. 1960. *Fiscal Theory and Political Economy.* Chapel Hill: University of North Carolina Press.

—— and Tullock, G. 1962. *The Calculus of Consent.* Ann Arbor: University of Michigan Press.

—— and Wagner, R. 1977. *Democracy in Deficit.* New York: Academic Press.

Cannan, E. 1929. *A Review of Economic Theory.* London: P. S. King and Son.

Crystal, D., ed. 1990. *Cambridge Encyclopedia.* Cambridge: Cambridge University Press.

Downs, A. 1957. *An Economic Theory of Democracy.* New York: Harper and Row.

Drèze, J., and Sen, A. 1995. Introduction. Pp. 13–45 in *The Political Economy of Hunger*, ed. J. Drèze, A. Sen and A. Hussain. Oxford: Oxford University Press.

Edgeworth, F. Y. 1881. *Mathematical Psychics: An Essay on the Application of Mathematics to the Moral Sciences.* London: C. K. Paul.

Frey, B. S. 1976. Taxation in fiscal exchange—comment. *Journal of Public Economics*, 6: 31–5.

—— 1983. *Democratic Economic Policy.* Oxford: Martin Robertson.

Garfinkel, M. R., and Oh, S. 1993. Strategic discipline in monetary policy with private information: optimal targeting horizons. *American Economic Review*, 83: 99–117.

Groenewegen, P. 1985. Professor Arndt on political economy: a comment. *Economic Record*, 61: 744–51.

—— 1987. "Political economy" and "economics." Pp. 904–7 in *The New Palgrave Dictionary of Economics*, ed. J. Eatwell, M. Milgate and P. Newman. London: Macmillan.

Harrington, J. E. 1993. Economic policy, economic performance, and elections. *American Economic Review*, 83: 27–42.

Hibbs, D. A., and Fassbender, H. eds. 1981. *Contemporary Political Economy.* Amsterdam: North-Holland.

ILO (International Labour Office). 1984. *Into the Twenty-First Century: the Development of Social Security.* Geneva: ILO.

Jevons, W. S. 1910. *The Theory of Political Economy.* 4th edn. London: Macmillan; originally published 1879.

John, A. H.; Pecchenino, R. A.; and Schreft, S. L. 1993. The macroeconomics of Dr. Strangelove. *American Economic Review*, 83: 43–62.

Kanbur, R., and Keen, M. 1993. Jeux sans frontières: tax competition and the tax coordination when countries differ in size. *American Economic Review*, 83: 877–92.

Laban, R., and Wolf, H. C. 1993. Large-scale privatization in transition economies. *American Economic Review*, 83: 1199–1210.

Lindbeck, A. 1977. *The Political Economy of the New Left.* 2nd edn. New York: Harper and Row; originally published 1971.

Marshall, A. 1890. *Principles of Economics.* London: Macmillan.

—— and Marshall, M. P. 1879. *The Economics of Industry.* London: Macmillan.

Mirrlees, J. A. 1989. Discussion of Alesina 1989. *Economic Policy*, 4: 89–90.

Niskanen, W. 1971. *Bureaucracy and Representative Government.* Chicago: Aldine-Atherton.

Paldam, M. 1991. Politics matter after all: testing Alesina's theory of RE Partisan cycles on data for seventeen countries. Pp. 369–98 in *Business Cycles: Theories, Evidence and Analysis,* ed. N. Thygesen, K. Velupillai and S. Zambelli. London: Macmillan.

Poterba, J. M. 1994. State responses to fiscal crises: the effects of budgetary institutions and politics. *Journal of Political Economy,* 102: 799–821.

Sen, A. K. 1977. Rational fools: a critique of the behavioral foundations of economic theory. *Philosophy and Public Affairs,* 6: 317–44.

Sheffrin, S. 1989. Evaluating rational partisan business cycle theory. *Economics and Politics,* 1: 239–59.

Part IX

# Political Methodology

Chapter 32

# Political Methodology: An Overview

John E. Jackson

Chris Achen, in 1983, referred to work in political methodology as derived largely from forays through other disciplines' attics (Achen 1983). Over the past twenty-five years this has meant primarily applications and extensions of whatever econometricians were using. The main theme of this chapter is that the importation of econometric methods has substantially advanced the practice of empirical work in political science and provided very important substantive insights. This progress, and it definitely is progress, has come with an opportunity cost. The economists' sermon of no free lunch applies in the application of econometrics as well. The cost is in the form of assumptions that are widely accepted in economics and deeply imbedded in their statistical models. The first section of this chapter reviews some of these assumptions and their implications for the empirical analysis of political behavior and institutions. The next section offers some empirical work that both questions these assumptions and illustrates some of their consequences. The chapter then concludes with a challenge to political methodologists in the form of possible new directions, one outcome of which might be to have empirical researchers from other disciplines appear at our garage sales.

## 1 Econometrics and political science

The application and refinement of econometric techniques has been the dominant theme of political methodology for the past twenty-five years. The expansion of these techniques is well documented by King (1991) and the impressive array of applications to substantive as well as methodological questions is thoroughly covered by Bartels and Brady (1993). It is easy to see the reasons for the rapid and extensive spread of these techniques. As developed by economists for studying economic behavior and testing economic theories, they are much better suited to the non-experimental nature of most political science research than the quasi-experimental techniques, such as cross-tabulation, analysis of variance, and tests of association, commonly used by earlier political methodologists. Econometric techniques, as exemplified by but not limited to the least-squares estimator, focus attention on the model of the behavior being investigated and not just on the association among a small set of variables of interest. This attention to model specification has now gone well beyond a concern for which variables belong in the analysis to include such questions as functional form, endogeneity, and measurement. All of these questions are central to the analysis of political phenomena.

The linear estimation model is very powerful in two ways that go beyond its statistical properties. First, the model itself is quite robust in the face of a wide array of difficulties. It often takes extreme deviations from the basic assumptions to produce highly suspicious results. In some instances, even with violations of its basic assumptions the generalized version of the linear model provides estimates with a smaller mean squared error than more theoretically correct, but less robust, estimators (Beck and Katz 1995).

The second and far more important form of robustness, call it adaptability, is the ability of an army of methodologists to apply and extend the basic methodological strategy to widely varied situations. Most work of political methodologists in the past twenty-five years has been a successful effort to extend the basic model to an ever broader set of problems and to overcome an increasingly wide set of violations of the basic model. As Bartels and Brady (1993) demonstrate in thirty pages of text and nearly eight-and-a-half pages of citations this has been a successful endeavor in political science. After reading the Bartels and Brady chapter, it is fair to say that work with this basic model has led to significant findings and new ways of thinking in virtually all areas of the discipline. Where once empirical methods may have been largely the province of Americanists, or even more accurately Americanists doing political behavior, methodologists are found in most subfields, and the econometric approach, in its many variations, provides the unifying language.[1]

[1] It is grossly unfair to refer to this whole body of work as econometrics and to attribute all of this work to econometricians. For example, Karl Jöreskog, a psychologist, showed the underlying model that unified factor analysis, as developed in psychology, and structural equation estimation, as developed in economics (Jöreskog 1973; Jöreskog and Sörbom 1979). Yet, I have not seen a more encompassing and less disciplinary term.

Three different applications—structural equation models, time-series analysis, and non-linear estimation—will illustrate this adaptability to sophisticated conceptual models and to a wide range of problems arising from the non-experimental nature of social science research. The rest of this section provides a quick summary of these three adaptations.

## A Structural equation estimation

Structural equation estimation treats entire sets of variables as jointly determined and stochastic. Applications derived from statistics developed for experimental situations assume that all explanatory variables are set outside the experiment and can be considered fixed and non-stochastic. This is a dubious assumption for the non-experimental setting where many of the variables of interest interact and cannot be treated as pre-determined or exogenous even though they appear as explanatory variables in some parts of the model. Econometricians discussed this problem extensively and provided a variety of approaches, such as instrumental variables, two and three stage least-squares and *k*-class estimators (Judge *et al.* 1985 offer good coverage of these methods). These methods have many applications in political science, where the examination of the asymmetric, or simultaneous, relationships among endogenous variables is central to testing important theoretical propositions. They have been used to examine party identification (Franklin and Jackson 1983), vote decisions (Jackson 1975*a*; Page and Jones 1979; and Markus and Converse 1979), the connection between economic conditions and votes (Markus 1988), and the influence of Senate leaders (Jackson 1973).

A major use of structural equation models is to incorporate systematic hypotheses about measurement error and missing variables into a wide variety of models. A major reason why the explanatory variables in an otherwise straightforward regression model are stochastic is the presence of random measurement errors. The consequence of these errors is to bias all coefficient estimates, with the direction and amount of bias dependent upon the covariances among the true and the included variables (Achen 1985). Jöreskog (1973) and Jöreskog and Sörbom (1979) showed that the general form of the structural equation model incorporated both random measurement error and simultaneity as special cases. This general model, properly referred to as the multiple indicator multiple cause (or MIMC), model also permitted estimation of sophisticated models with latent, or missing variables. One of the most significant extensions of the work with missing variables was Franklin's development of a procedure for simulating the analysis of panel data from successive cross-sections (Franklin 1990). Applications with both measurement error and latent variables are seen in most subfields, though the most extensive use has been in the analysis of survey data (see Achen 1975; Erikson 1979; Jackson 1983).

## B Time-series analysis

The second area of major innovation is the analysis of time-series data. Many of the important conceptual arguments and substantive interests concern how behavior, institutions, and outcomes vary over time and in response to specific interventions and shocks. Concerns as varied as determinants of presidential approval (Beck 1992), arms races (Williams and McGinnis 1988; McGinnis and Williams 1989), the political business cycle (Beck 1987), partisanship (MacKuen, Erikson and Stimson 1989; Smith and Box-Steffensmeier 1995), and government expenditures in Britain (Freeman, Williams and Lin 1989) are addressed with time-series data.

Rigorous examination of time-series data began with the use of simple autoregressive integrated moving average (ARIMA) models to study the over time fluctuations of individual variables (see Hibbs (1974) for the introduction of these techniques to political science). Much of this work suffered from an absence of a priori theorizing and a lack of attention to covariation among variables leading to very weak causal inferences. This work, however, has now been extended to include very sophisticated multivariate analyses of systems of variables which has substantially strengthened the resulting inferences. Strong attention is now paid to the existence of multi-period lags and to the patterns of covariations among the lags of sets of variables. In methods such as Vector Autoregression (VAR) strong inferences about exogeneity and the reaction of the system to external shocks are derived from these covariations (see Freeman, Williams and Lin 1989; Williams and McGinnis 1988; McGinnis and Williams 1989).

Some of the most sophisticated time-series analyses, referred to as cointegration and error correction models, begin with the assumption of a system of variables with a long-term equilibrium. Data are then analyzed to estimate the characteristics of this equilibrium, how the system responds to shocks that move it out of equilibrium, and the time required for the effects of the shock to dissipate (see Freeman (1994: 97–258) for an extensive set of papers discussing error correction models).

An important adaptation of the time-series model is Beck's (1989) introduction of the Kalman filter to the analysis of political data and its application by Kellstedt, McAvoy and Stimson (1995). The Kalman filter introduces latent endogenous variables and makes explicit the relationship between these variables and the observed variables, which are treated as erroneous observed variables. There is a very strong conceptual similarity between this model and the MIMC model used to analyze cross-sectional data and this demonstrates that we are now seeing the integration of various methodological innovations into models that handle several estimation problems within the same structure.

## C Non-linear models

The last extension is the development of models that are non-linear in the parameters. The early econometric work was heavily dependent on the linear model, which referred to the functional form of the parameters and to the systems of equations solved to obtain the estimates. (This model could handle a wide array of non-linear relations among the variables: see Hanushek and Jackson 1977: 96–101.) The development of cheap and powerful computers has led to the ability to estimate models with very complex functions of the parameters. This extension is important because many key theoretical propositions are about variations in the relationship among variables, not just which variables are important. Models with varying coefficients are discussed in Beck (1983), Rivers (1988), and Jackson (1992) and have a wide range of applications, such as event counts in International Relations (King 1989*b*), systematic response bias in survey data (Jackson 1995), the relationship between constituent attitudes and legislative votes (Jackson and King 1989), and most of the structural equation and time-series work just discussed.

Much of the non-linear estimation is done applying the least squares criteria to the model,

$$Y_i = \mathrm{f}(X_i, B) + u_i$$

and selecting the estimates that minimize the sum of squared errors,

$$\sum e_i^2 = \sum [Y_i - \mathrm{f}(X_i, b)]^2.$$

The minimization is done using the set of non-linear first order conditions established by the specified model (Judge *et al.* 1985). More formal methods incorporate the error term into the formal specification, possibly also in a non-linear manner, and use maximum likelihood methods to obtain the estimates. This approach requires the precise specification for the distribution of the stochastic terms (King 1989*a*).

These are only a small portion of the extensions and applications developed over recent years. They do, however, give a clear indication of the range and significance of the models and problems that have been examined through adaptations of the basic estimation model. It is this adaptability and generalizability of the econometric model and its appropriateness for non-experimentally generated data that have made the model so dominant. King (1989*a*) with only a small amount of exaggeration, can legitimately talk about unifying political methodology under the banner of the more general form of the estimation model imported from statistics through economics.

As in virtually all intellectual endeavors, this progress in the development and expansion of methodological techniques and in their application to an increasingly broad set of issues is predicated on certain assumptions. We are frequently reminded, sometimes after the fact, of some of these assumptions, such as having iid disturbances, some known distribution for the observed or unobserved variables, a priori belief in the exogeneity of certain variables, the appropriateness of some linear or non-linear form, etc. There is one assumption that is basic, but less frequently questioned, to all the current estimation models. This is the idea that a predictable stable equilibrium underlies the process being studied and generating the observed data. Stochastic terms are treated as deviations about this assumed equilibrium. These deviations may be biased in certain cases or acquire particularly noxious characteristics in others, but fundamentally the techniques are premised on the existence of this equilibrium so that appropriate techniques can then be developed to deal with the pathological deviations. Then with enough observations one can obtain reliable estimates of the parameters in this system. The remainder of this chapter explores the "What if" question posed by models of processes that may not fit this equilibrium assumption.

## II An alternative approach

The equilibrium centered model and the comparative statics analyses that are derived from it have enabled social scientists to describe and understand a wide range of important phenomena. But there are also processes and institutions that are poorly described in this manner. The motivation for this chapter is an increasing interest in these processes and institutions. Before delving into the technical aspects of the chapter, let us examine some of the situations that deviate from the static equilibrium framework.

The discussion of alternatives will focus on the class of behaviors that are frequently referred to as "path dependent." A path dependent process may well reach a stable equilibrium at some point, but the characteristics of that equilibrium are partially a function of the sequence of actions or intermediate outcomes obtained as part of the equilibrating process. The initial and exogenous factors alone are insufficient for predicting the equilibrium. One way to put this is that, "Where you end up depends upon how you got there."

Probably the simplest version of a path-dependent model is the urn problem described by Arthur and his statistical colleagues (Arthur *et al.* 1987). In this problem, one starts with an urn with one red and one white ball (could statisticians start any other way?). A ball is withdrawn, and it and a second ball of the same color are placed back in this urn. This process is repeated an infinite number of times. The process finally reaches a stable point where the proportion of red and white balls remains constant (probability limits are useful). The fundamental point of the model is that there are multiple equilibria and that the one ultimately reached is strongly related to the sequence of balls selected in the early trials. For example, drawing a red ball on the first two trials makes an equilibrium with a majority of white balls much less likely than if white balls were drawn in these trials. The urn problem becomes a nice metaphor for many path dependent type political and social processes.

## A Public preferences, political institutions, and path dependence

The two most significant explicit demonstrations of path-dependent processes in political science examine the connections between mass public issue preferences and the performance of U.S. political institutions. The first, and most ambitious, is Carmines and Stimson's study of the connection between the actions of the Congress on civil rights legislation and the evolution of public opinion (Carmines and Stimson 1989). They use time-series data to show that changes in the positions taken by Republicans and Democrats in the House and the Senate on Civil Rights legislation during the late 1950s and early 1960s led changes in public opinion on that issue. They conclude that these party shifts led to a substantial evolution in the structure and alignment of the parties.

> The model posits that the mass issue evolution begins with a "critical moment"—more visible than the creepingly slow change implied by pure secular realignment but much less pronounced than that presumed by critical election realignments. Equally significant, the initial increase in mass

> issue polarization does not complete the process but only begins it by setting in motion a change that grows over time.
>
> Dynamic evolutions thus represent the political equivalent of biology's punctuated equilibrium. The critical moment corresponds to the punctuation point—a change of some magnitude but not a cataclysmic adaptation. The slower, continuous change following the critical moment is the drive toward a new equilibrium—the semipermanent redefinition of the link between issues and mass parties (157).

They do not pursue what the new equilibrium might be or how it relates to the initial conditions and to the punctuation point, or shock, that disturbed the existing equilibrium. A fair surmise is that the link between issues and mass parties referred to is a complex one, as it must include a connection from mass opinion to the behavior of the parties and of Congress. The exact nature of these mutual relationships will determine the path from one equilibrium to the next. The analytical expressions for such a link, and the effort required to test propositions about the expected equilibrium, will be very daunting for a system as complex as one involving mass preferences, parties, and a two chamber legislature.

The second study to propose a path-dependent type process is the study of endogenous preferences by Gerber and Jackson (1993). They also examine the link between élite behavior and mass preferences, but in the context of a model of electoral competition. They argue from a series of cross-sectional analyses of survey data on civil rights and Vietnam opinions that shifts in party positions provide information that induces voters to alter their preferences. They speculate that placing endogenous preferences in a full model of electoral competition will create an electoral system that resembles a path-dependent dynamic process. Gerber and Jackson conclude that,

> Endogenous preferences introduce a dynamic element into the traditional static electoral model.... Specifically outcomes become less predictable. An equilibrium may exist under a variety of initial conditions, but any result will be partially dependent upon the sequence and timing of party platforms and upon how rapidly the parties change positions (652–3).

Jackson (1994) then pursues this speculation more formally. He shows both analytically and through simulation that a simple one-dimensional model of two-party competition with endogenous preferences has an equilibrium, defined as a stable set of preferences and party positions. The characteristics of this equilibrium, however, depend upon the sequence of party positions following any shock that disrupts the previous equilibrium—a path-dependent process. We return to this model in a later section.

These two studies raise interesting and provocative speculations about the course of U.S. political history over the past forty years—a very path-dependent process. The Carmines and Stimson and Gerber and Jackson studies describe the startling shifts in party positions and leadership behavior on the civil rights issue during the 1950s and 1960s (see also Dawson 1992 for a discussion of this history). The dramatic differences between Goldwater's and Johnson's positions on the Civil Rights Act of 1964, and with their party's previous positions, produced a radical shift in the voting patterns in that election. These strategies and shifts led eventually, but directly, to a realignment of the electorate and of the regional allocation of Congressional seats. Consider whether we would have a Newt Gingrich or have had a Democratic Senate Leader from Maine without this historical event.

## B Path dependency and models of political economy

This concept of evolution and path dependence is becoming central to the discussions and study of many other political and economic processes and institutions, with very important implications and consequences. North (1990), for example, devotes a considerable amount of attention in his book on institutions and economic performance to a discussion of the importance of path dependence. He contends, but does not illustrate empirically, that the design of economic and political institutions is central to the development process and that transactions costs, interacting with the particular institutions in place, create a strongly path-dependent process. Riker and Weimer (1993), while citing North, take the argument even further in their discussion of the importance of property rights in the former Socialist countries. They are particularly concerned about institutional design in these societies and argue that the creation of economic and political institutions begins a path-dependent process, without resorting to discussions of transaction costs.

Arguments for the presence and significance of path-dependent processes can rest on more than institutional questions and extend outside political science. Krugman's (1991) description of regional economic development includes the importance of agglomeration, or economy of scale effects, which makes the model explicitly path dependent. (This work has significance beyond the regional economic field, as it applies to differences in the growth rates of countries, and thus to international political economy as well.) His model of growth incorporates the main features of Arthur *et al.*'s urn problem and explains the geographic concentration of various industries, such as automobiles in southeast Michigan, semiconductors in northern California, etc. (See Jackson and Thomas (1995) for empirical support for this model of regional economic growth.)

My contention is that given the central interests of political scientists in the performance and evolution of economic and political institutions, models that incorporate path dependency in some form will and should become an important part of our formal theories. The question for political methodologists is how well do the empirical methods, largely imported from econometrics, perform in identifying path-dependent processes, in estimating the form and parameters in these processes, and in testing important hypotheses. My intuition is that because these econometric methods are built on the assumption of a static and exogenously determined equilibrium, they may not perform as well as desired or as needed. The remainder of the chapter tries to shed some light on this question through various analytical and simulation exercises.

## III A formal statement of the econometric model

This section presents a formal version of the general econometric model from which it is easy to see how vital the equilibrium assumptions are to the estimation method. To develop this model consider the following definitions,

$i$ = lag interval, $i = 0, \ldots, n$;

$Y_{t-i}$ = an $M$x1 vector of observations of endogenous variables lagged $i$ periods;

$X_{t-i}$ = a $K$x1 vector of observations of exogenous variables lagged $i$ periods;

$U_t$ = an $M$x1 vector of disturbances;

$A_i$ = an $MxM$ matrix of coefficients relating $Y_t$ to $Y_{t-i}$;

$B_i$ = an $MxK$ matrix of coefficients relating $Y_t$ to $X_{t-i}$; and

$C$ = an $MxM$ matrix of coefficients of relations among $Y_t$.

The general form of the model relating these variables, and the model that now constitutes the core of political methodology is,[2]

[2] For simplicity of exposition, except where noted, the discussion will use the general linear model and assume continuous and properly measured variables. Extensions to situations with limited variables and measurement errors enrich the model presented here, but maintain the basic

$$CY_t = \sum_{i=1}^{n} A_t Y_{t-i} + \sum_{i=0}^{n} B_i X_{t-i} + U_t. \quad (1)$$

We will frequently use the reduced form version of eqn. 1, which is,

$$Y_t = C^{-1} \sum_{i=1}^{n} A_i Y_{t-i} + C^{-1} \sum_{i=0}^{n} B_i X_{t-i} + C^{-1} U_t$$

$$= \sum_{i=1}^{n} Q_i Y_{t-i} + \sum_{i=0}^{n} P_i X_{t-i} + V_t \quad (2)$$

Various assumptions about each of the coefficient matrices, $A$, $B$, and $C$ or $Q$ and $P$ give the more specialized models. Let us consider several of these in turn.

## A Structural equation model: no lagged endogenous variables

With $A_i$ equal zero for all $i \geq 1$ and $B_i \neq 0$ for some $i$, eqn. 1 is the common structural equation model with joint determination among some of the endogenous variables, depending upon the non-zero values in $C$. (For the sake of this discussion, assume sufficient prior information so that $C$ and $B_i$ are identified.) With the conventional assumption that the values of $X_{t-i}$ are given exogenously, taking expectations gives,

$$E(Y_t) = C^{-1} B_0 E(X_t) + C^{-1} \sum_{i=1}^{n} B_i E(X_{t-i}) + C^{-1} E(U_t). \quad (3)$$

Standard practice either justifies the assumption $E(U_t) = 0$ or respecifies the systematic part of the model to account for deviations from this assumption. (Systematic measurement error can be handled by various respecifications (Brady 1985 and 1993; Jackson 1979 and 1995).) With this assumption or respecification one has the same expectation for $Y_t$ for given values of $X_t$ and $X_{t-i}$. This model rules out the possibility that some historical values for $Y$ or $U$ may persist and affect expectations about $Y_t$. Such circumstances may exist, but they are then assumed to be distributed independently of $X_{t-i}$ so that the basic prediction about $E(Y_t)$ holds. In other words, the assumption is that the underlying process is independent of history or the path of $Y_t$.

The model in eqn. 1 includes many possibilities for incorporating

assumption that is being discussed. Expansion to these other cases would needlessly complicate and obscure the basic point.

propositions about lagged error structures. Any, or all, of the $u_t$'s can be represented by a complete set of autoregressive and moving average terms,

$$\Phi(L)u_t = \Theta(L)\epsilon_t, \tag{4}$$

where $\epsilon_t$ is assumed to be a stationary process with $E(\epsilon_t) = 0$. Constraints, or assumptions, are usually made about the coefficients in the lag operators $\Phi(L)$ and $\Theta(L)$ so that the resulting process is stationary and well behaved. These restrictions are that all the roots of $\Phi(z)$ and $\Theta(z)$ are greater than one in absolute value. For a process with a single autoregressive or moving average term, this condition means that $\phi_1$ and $\theta_1$ are less than one. The addition of these extensions to the error-generating process does not change the basic condition that given values of the exogenous variables produce the same expected value for the current endogenous variables.

## B Lagged endogenous variables: error correction models

The models generally become more interesting for our purposes when lagged values of the endogenous variables are included. One of the most interesting and increasingly used models, often referred to as an error-correction model, occurs when the equation includes both $Y$ and $X$ from just the previous period, i.e. $A_1$ and $B_1$ are not zero but all higher level $A$ and $B$ terms do equal zero (see the symposium on error-correction models in Freeman (1994: 97–258)). Using the reduced form expression and assuming that $(I - Q_1)$ is nonsingular, which holds if the elements in each row of $Q_1$ sum to less than 1, these conditions give,

$$\begin{aligned} Y_t &= Q_1 Y_{t-1} + P_0 X_t + P_1 X_{t-1} + V_t \\ (Y_t - Y_{t-1}) &= -Y_{t-1} + Q_1 Y_{t-1} + P_0(X_t - X_{t-1}) + (P_0 + P_1)X_{t-1} + V_t \\ \Delta Y_t &= P_0 \Delta X_t - (I - Q_1)Y_{t-1} + (P_0 + P_1)X_{t-1} + V_t \\ &= P_0 \Delta X_t - (I - Q_1)(Y_{t-1} - Y^*_{t-1}) + V_t, \end{aligned} \tag{5}$$

where $Y^*_{t-1} = (I - Q_1)^{-1}(P_0 + P_1)X_{t-1}$ is the expected equilibrium value of $Y$ given $X_{t-1}$ obtained by setting $Y_t = Y_{t-1} = Y^*$ and $X_t = X_{t-1} = X^*$.

The direct interpretation of eqn. 5 is that the change in $Y$ from one period to the next is related to any change in the exogenous variables and to the deviation of $Y$ in the previous period from its expected equilibrium value in that period. In any period in which $Y_{t-1}$ exceeds its equilibrium value, e.g. a positive error, the expected change in $Y$ is negative, bringing $Y_t$ towards equilibrium, and the converse for a negative error, hence the name error-correction model. The error-correction model is a commonly used model in both economics and now political science. With the sufficiency conditions noted previously for the nonsingularity of $(I - Q_1)$, it is another model where the expected values for $Y$ depend, at least in the long run, solely on the values of the exogenous variables and are independent of previous realizations of $Y$.

## C Vector autoregression

One of the most vigorously debated versions of eqn. 1 is vector autoregression (VAR) (for a flavor of the debate and the critique of conventional models, such as eqn. 3, see Freeman, Williams, and Lin (1989)). Users of VAR models argue that our understanding of social systems is not sufficiently developed to make and sustain the rigid distinction between exogenous and endogenous variables or the necessary identifying restrictions on $C$ and $B$ required by structural equation methods. The VAR approach treats every variable in the system as endogenous and estimates the contemporaneous values of each variable as a function of the lagged values of all variables. Thus, there are no variables to be denoted by $X$, and only a reduced form version is estimated as there is no effort to specify and estimate $C$. With these assumptions and suppression of the constant term, meaning that each variable is a deviation about its mean, the model is,

$$Y_t = \sum_{i=1}^{n} Q_i Y_{t-i} + V_t, \tag{6}$$

where the elements of $V_t$ are iid with mean zero and covariance matrix $\Omega$.

In order for this system to be stationary, the elements of the $Q_i$ matrices, denoted by $q_{ijm}$, must meet certain conditions. To see this, rewrite eqn. 6 as,

$$\begin{aligned} Y_{mt} &= \sum_{i=1}^{n} q_{im1} Y_{1t-i} + \sum_{i=1}^{n} q_{im2} Y_{2t-i} + \ldots + \sum_{i=1}^{n} q_{imm} Y_{mt-i} + \ldots + \\ &\quad \sum_{i=1}^{n} q_{imM} Y_{Mt-i} + V_t \\ &= \Phi_1(L) Y_1 + \Phi_2(L) Y_2 + \ldots + \Phi_m(L) Y_m + \ldots + \Phi_M(L) Y_M + V_{mt}. \end{aligned} \tag{7}$$

Since $Y_m$ can be expressed as the sum of a set of $M$ autoregressive series, each of these series must be stationary for $Y_m$ to be stationary. This condition will be met if the absolute value of the $n$ roots of the polynomial expression $\Phi_j(z) = 0$ are greater than one for all $j$. A necessary, but not sufficient, condition for this result is that $\sum_{i=1}^{n} q_{imj} < 1$ for each $j$. If the model is formulated in terms of the levels of $Y$, then the steady state, absent continuing shocks from the $V$ terms, is for the deviations of all variables about their expected mean values to become zero.

## IV Path-dependent example: do these methods work?

It is possible, and useful, to illustrate the results obtained using the standard econometric techniques by applying them to simulated data sets where the underlying process generating the data is known. In the applications developed here, one data set will represent a system with an equilibrium determined solely by the predetermined variables and any exogenous shocks. Other data sets will be collected from a system with an equilibrium, but where this equilibrium is a result of the path taken as the system moves from one equilibrium to another in response to exogenous shocks. To further illustrate the hazards and consequences of analyzing path-dependent systems with conventional tools, these data will be generated with alternative models and with different sequences of shocks to show the different equilibria, and the different econometric results. This section first describes the model, in both its static and dynamic equilibrium forms. We then discuss the simulated data sets generated with each version of the model, and compare the results of a statistical analysis of each data set.

### A A dynamic model of two-party competition

The data used to illustrate what standard econometric techniques reveal about different classes of models are generated by a simulation model of two-party electoral competition over a single dimension. This model is selected because of its known equilibrium properties and its centrality in the formal political theory literature. The significant variation is that in one version the distribution of individuals' preferences shifts as the parties change their positions, making this distribution endogenous to the political process. In the traditional model, our baseline, preferences are fixed and can be used to predict the equilibrium outcome. In the general model there is an equilibrium outcome, but it cannot be predicted from the initial or exogenous conditions because the location of the equilibrium depends upon the sequence of positions taken by the competing parties as they seek this new equilibrium. (This example builds on and illustrates the path-dependent model linking mass preferences and political institutions discussed previously. Jackson (1994) presents a lengthier discussion of the electoral model summarized here, with both its analytical and simulation results.)

The essence of the model is very simple. In the traditional, and simplest, model of two-party competition the utility, or loss, each voter associates with a party's position is represented as a quadratic function of the distance between the person's preferred policy and that advocated by the party. The voter is then assumed to vote for the closest party, defined as the one associated with the least loss, or greatest utility. Citizens develop a partisanship based on a similar utility calculation, though partisanship is presumed to have a longer lag and to be more likely to include past issue-based utility calculations. (For a theoretical argument about the evolution of partisanship, see Achen (1992). Jackson (1975*b*) and Franklin and Jackson (1983) provide empirical results supporting this formal model.) In this version of the model, it is assumed that each voter's preferred policy remains fixed during the period in which the parties are choosing, or searching for, their optimal platforms. Under these simple conditions, the median of the distribution of voters' preferences becomes the equilibrium. (If the preference distribution is symmetric, or if there is probabilistic voting, then the mean becomes the equilibrium outcome.) If parties only want to maximize their likelihood of election, then each party eventually adopts the preference median, or mean, as its strategy. If parties place an intrinsic value on their platform, as well as on getting elected, an equilibrium still exists, but with the parties a fixed distance from the median, depending upon how strongly they value their platform relative to their proportion of the votes. (See Wittman (1983) for the development and results of a model in which parties value their platforms as well as the probability of winning elections.)

A main feature of this model is that if there is a sudden shift in, or shock to, the distribution of individual preferences, the system would reach a new equilibrium. More significantly, the main features of this equilibrium—the means of the distributions of issue preferences, partisanship, and vote evaluations and the positions of the two parties—are predictable from the size and direction of the exogenously given shock. The path, or sequence of positions, the parties take in reaching this new equilibrium will be irrelevant. Similarly, if there were a series of shocks following periods of equilibrium, we would predict the final state simply by knowing the initial conditions and the value of each shock.

Incorporating the evolutionary model of issue-preference evolution proposed by Carmines and Stimson (1989) and by Gerber and Jackson

(1993) substantially alters this version of the electoral model. In the evolutionary version, the actions and choice of the various political actors and parties exert an influence on citizens' preference. This change can be accommodated with an additional equation in the voting model. This addition is a Bayesian expression, where individuals update their issue preferences based on changes in the parties' positions. The argument here is that individuals have some uncertainty about their policy preferences and party positions provide cues to partisans about how much they might like various policies. Changes in party positions provide new information which leads these partisans to update their preferences.

If the policy preferences and partisanship of individual $i$ at time $t$ are denoted by $X_{it}$ and $P_{it}$ respectively and the two parties and their positions by $\theta_t$ and $\phi_t$, the model of preference updating is,

$$X_{it} = Z_t + B_{1_t} X_{it-1} + (1 - B_{1_t})[.5B_2(\theta_t - \phi_t)P_{it}],$$

were $B_{1_t} = e^{-B_1(\Delta\theta_t^2 + \Delta\phi_t^2)}$. (Stochastic terms are omitted from all subsequent equations in order to concentrate the discussion and comparisons on the systematic parts of the different models and analyses.) The variable $Z_t$ represents any exogenously generated shift in mean preferences that would upset an existing equilibrium. This expression is linear in the individual variables, so that dropping the $i$ subscripts gives the following expression for the mean of the preference distribution at time $t$, $\bar{X}_t$;

$$\bar{X}_t = Z_t + B_{1_t} \bar{X}_{t-1} + (1 - B_{1_t})[.5B_2(\theta_t - \phi_t)\bar{P}_t]. \tag{8}$$

The model used in the rest of the chapter concentrates on expressions for the aggregate electoral properties.

In this formulation, $[B_2(\theta_t - \phi_t)P_{it}]$ is the informational cue, and the amount of change in party positions from $t-1$ to $t$, $(\Delta\theta_t^2 + \Delta\phi_t^2)$, indicates the relative variance, or certainty, of the new information. No change in party positions indicates no new information and $B_{1_t} = 1$ and preferences remain unchanged. As the amount of change in party positions increases, the weight given to the informational cue increases and the weight given to previous preferences decreases, assuming $B_1 > 0$. If $B_1 = 0$ then $B_{1_t} = 1$ regardless of any change in party positions, which makes the previous model with exogenously determined preferences a special case of the current model.

The rest of the model contains expressions for mean partisanship, $\bar{P}$, mean vote evaluations, $\bar{V}$, and the positions of the two competing parties, $\theta$ and $\phi$. Individual partisanship and vote evaluations follow the traditional quadratic utility model, which also gives simple aggregate expressions,

$$\bar{P}_t = (1 - B_3)\bar{P}_{t-1} + 2aB_3D_t\bar{X}_t - aB_3(\theta^2_t - \phi^2_t), \text{ and} \quad (9)$$

$$V_t = (1 - B_4)\bar{P}_t + 2aB_4D_t\bar{X}_t - aB_4(\theta^2_t - \phi^2_t). \quad (10)$$

The equilibrium values for $\bar{P}$ and $\bar{V}$ are identical and are,

$$\bar{P}^* = \bar{V}^* = 2a(\theta^* - \phi^*)\bar{X}^* - \alpha(\theta^{*2} - \phi^{*2}). \quad (11)$$

Parties value their probability of winning election, the log-odds of which is linearly related to $\bar{V}$, and their platform. The desired platforms are fixed and denoted by $G_\theta = G$ and $G_\phi = -G$. The parties' utility functions are quite arbitrarily selected and contain a single parameter $\gamma$ that measures the importance of the desired platform relative to the probability of winning.[3] For a given $\bar{X}$ and $\bar{V}$ the utility-maximizing platforms are,

$$\theta^* = \frac{\alpha(B_3 + B_4 - B_3B_4)(1-\gamma)\bar{X} + \gamma G(1 + e^{\bar{V}})}{\alpha(B_3 + B_4 - B_3B_4)(1-\gamma) + \gamma(1 + e^{\bar{V}})} \quad (12)$$

$$\phi^* = \frac{\alpha(B_3 + B_4 - B_3B_4)(1-\gamma)\bar{X} - \gamma G(1 + e^{-\bar{V}})}{\alpha(B_3 + B_4 - B_3B_4)(1-\gamma) + \gamma(1 + e^{-\bar{V}})} \quad (13)$$

Parties are assumed to adjust their platform towards the utility-maximizing position, if they are not already there. The adjustment parameter is denoted by $\delta$, so that

$$\theta_{t+1} = (1 - \delta)\theta_t + \delta\theta^*_t \text{ and} \quad (14)$$

$$\phi_{t+1} = (1 - \delta)\phi t + \delta\phi^*_t. \quad (15)$$

The model with endogenous preferences, $B_1 > 0$, is important because the distribution of preferences is likely to shift with changes in party positions. The new equilibrium after a shift in preferences, i.e. a $Z_t \neq 0$, cannot be predicted by knowing only $\bar{X}_{t-1}$ and $Z_t$, as would be the case with fixed preferences, $B_1 = 0$. This difference in outcomes between the two versions of the model is shown in Figures 32.1 and 32.2. Figure 32.1 shows the paths of party positions and mean vote preference that follow a one-period shift of $Z = .75$ in voter preferences when $B_1 = 0$. Figure 32.2 shows the paths for the same exogenous shift when $B_1 = 1$. Other than $B_1$ all the other terms in the model are the same in both figures. Different values for $\delta$ in eqns. 14 and 15 produce very different equilibrium values for all the variables.

[3] The precise utility functions are:

$$U^\theta = -\gamma(G - \theta)^2 - (1-\gamma)\log(1 + \exp^{-\bar{V}}) \text{ and}$$

$$U^\phi = -\gamma(G + \phi)^2 - (1-\gamma)\log(1 + \exp^{\bar{V}}),$$

where the probability of $\theta$ winning election is $P^\theta = (1 + \exp^{-\bar{V}})^{-1}$ and the probability of $\phi$ winning is $P^\phi = 1 - P^\theta = (1 + \exp^{\bar{V}})^{-1}$.

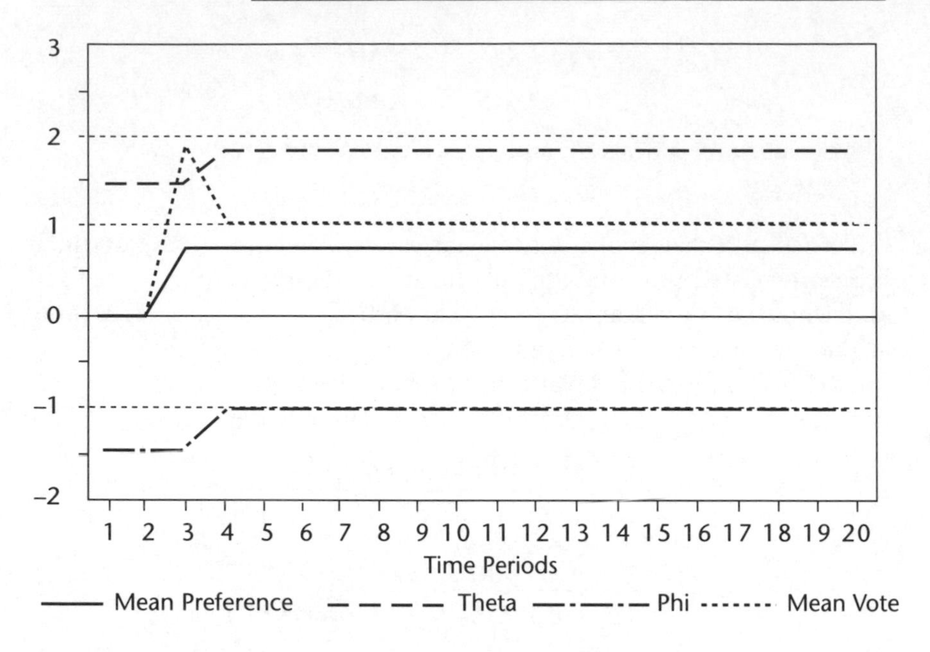


**Fig. 32.1** *Two-party electoral competition: exogenous preferences*

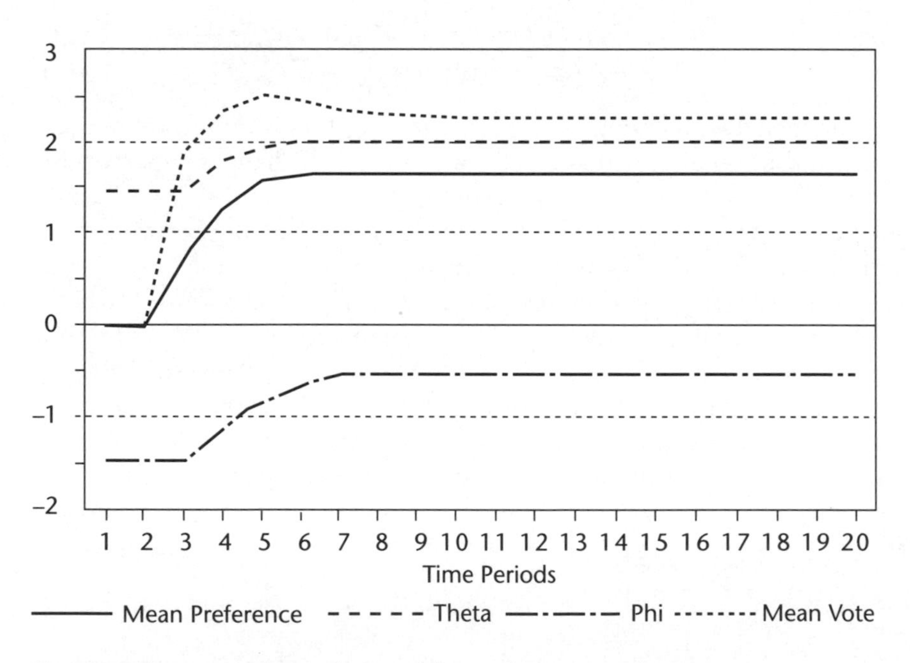


**Fig. 32.2** *Two-party electoral competition: endogenous preferences*

## B Simulations of electoral competition

The model of electoral competition just outlined was developed as a simulation model that can produce a series of elections and generate time-series data describing the outcomes of these elections, $\theta_t$, $\phi_t$, $\bar{X}_t$, $\bar{P}_t$, and $\bar{V}_t$ respectively. The basic parameters are chosen arbitrarily, with $\alpha = .5$, $B_1 = 0$ or 1, $B_2 = 1.25$, $B_3 = .5$, $B_4 = .5$, $\gamma = 1/3$, and $G = 2$ for all replications. These elections can be conducted following different scenarious, defined as different values for $\delta$ and different sequences of shocks, $Z_t$, and the data from each analyzed with various statistical methods. Comparisons of the statistical results permit evaluations of how well these methods help us understand the different processes. The prediction is that the methods will do very well at reproducing the system with exogenous preferences and much less well for the path-dependent processes.

The basic structure of the simulations is the same in all replications. The model is started at equilibrium, with the mean preference equal to zero and the two parties located at ±1.4545, respectively. Every twenty-eighth iteration a random number is drawn from a standard uniform distribution with mean zero and assigned to $\epsilon_t$. With each shock, mean preferences are assumed to follow an ARMA(1,1) process where $\bar{X}_t = .9\bar{X}_{t-1} + Z_t$ and $Z_t = \epsilon_t - .8\epsilon_{t-1}$. (The parameter values for this equation are chosen arbitrarily, but are done so as to conform closely to eqn. 8 and to generate a stationary series for $\bar{X}_t$ with a variance close to that of the stochastic term $\epsilon_t$.) This simulates the effect of a random shock to preferences that upsets the existing equilibrium. The interval of twenty-eight periods between shocks is chosen solely to insure that the system reaches equilibrium, as pictured in Figures 32.1 and 32.2. After the twenty-eighth interval, and just prior to the next shock, the values of the variables in the model are recorded as measures of the electoral outcomes. First-order conditions were computed at each iteration and printed along with the other output. These were always less than $.5 \times 10^{-6}$, indicating that an equilibrium had been reached. (Other analyses were done to assure that these were stable equilibria.) This sequence of an exogenous shock followed by periods for equilibration is repeated 100 times, giving an electoral time series of 100 observations, excluding the $t_0$ period. These observations become the data for the different estimation methods.

Three versions of the model, or scenarios, are estimated. In the first, preferences are exogenous, $B_1 = 0$. Here the predetermined variables should predict the equilibrium results perfectly. In the two other simulations preferences are endogenous, $B_1 = 1$, so the equilibrium result is expected to vary with the sequence of party positions as the parties adjust to the preference shift. The difference in these two simulations is the rate at which the parties adjust their positions in seeking a new equilibrium after the shock, $\delta = .65$ and $.80$. The larger value for $\delta$ produces larger changes in party positions from one interation to the next which should result in larger changes in preferences, and less predictable equilibria.

Each simulation consists of twenty replications of the 100 elections for a given scenario. The replications vary in the sequence of random shocks. The same sequence of random number seeds was used for each scenario, so that all three simulations have the same sequence of shocks. The replications accomplish two tasks. The first is the conventional purpose of providing a set of experiments so we can compare a distribution of results, thereby making the conclusions less sensitive to some unique feature of one replication. These replications are different from the use of replications in the conventional Monte Carlo experiment. There are no unobserved stochastic components in our model, so any variation in results across replications is the result of the inability of the statistical method to recapture the true model. The second, and less obvious, purpose of the replications is to create different histories. Since the values of the shocks, or histories, are known and included in all the analyses, the statistical results should be the same for each replication.

The most important comparisons of the statistical results are not between the static simulation and the two simulations with endogenous preferences but between the two simulations with endogenous preferences and between the corresponding replications in each scenario. Differences in the results from analyzing the two scenarios indicate the sensitivity of the methods to the path dependency of the different equilibria created by the parties adopting different strategies. Differences across replications of the same scenario demonstrate any sensitivity of the statistical results to variations in the exogenously determined histories. This is an equally disturbing result if the variations are very large, since the statistical analysis incorporates the information from each history.

## C Reduced form estimation

The first estimations are reduced form models relating the equilibrium outcomes to the predetermined and exogenous variables, $\bar{X}_{t-1}$ and $Z_t$ respectively. In the scenario with the static equilibrium model these estimations should give an accurate picture of the true model as there are no stochastic terms. For the other two scenarios, it is less obvious what to expect, though intuition suggests that results may differ, both from the sta-

**Table 32.1** Estimated coefficients in the reduced form electoral model

| | $B_1 = 0, \delta = .8$ | | | $B_1 = 1, \delta = 0.65$ | | | $B_1 = 1, \delta = 0.80$ | | |
|---|---|---|---|---|---|---|---|---|---|
| *Dept. Var.* | *Const* | $\bar{X}_{t-1}$ | $Z_t$ | *Const* | $\bar{X}_{t-1}$ | $Z_t$ | *Const* | $\bar{X}_{t-1}$ | $Z_t$ |
| $\bar{X}_t$ | | | | | | | | | |
| Mean | 0.000 | 0.900 | 1.000 | −0.006 | 0.925 | 1.195 | −0.033 | 0.850 | 0.636 |
| St. Dev. | 0.000 | 0.000 | 0.000 | 0.068 | 0.048 | 0.053 | 0.254 | 0.043 | 0.213 |
| Max. | 0.000 | 0.900 | 1.000 | 0.099 | 1.015 | 1.309 | 0.331 | 0.946 | 1.133 |
| Min. | 0.000 | 0.900 | 1.000 | −0.182 | 0.836 | 1.084 | −0.414 | 0.754 | 0.322 |
| $\bar{P}_t$ and $\bar{V}_t$ | | | | | | | | | |
| Mean | 0.000 | 1.215 | 1.350 | −0.016 | 1.264 | 1.633 | −0.069 | 1.033 | 0.821 |
| St. Dev. | 0.000 | 0.001 | 0.001 | 0.113 | 0.093 | 0.066 | 0.513 | 0.138 | 0.261 |
| Max. | 0.000 | 1.218 | 1.352 | 0.132 | 1.389 | 1.760 | 0.753 | 1.321 | 1.380 |
| Min. | 0.000 | 1.214 | 1.349 | −0.392 | 0.945 | 1.492 | −0.943 | 0.759 | 0.396 |
| $\theta_t$ | | | | | | | | | |
| Mean | 1.448 | 0.478 | 0.532 | 1.435 | 0.454 | 0.580 | 1.454 | 0.312 | 0.250 |
| St. Dev. | 0.001 | 0.007 | 0.007 | 0.048 | 0.067 | 0.101 | 0.212 | 0.152 | 0.154 |
| Max. | 1.449 | 0.490 | 0.546 | 1.513 | 0.548 | 0.684 | 1.781 | 0.540 | 0.550 |
| Min. | 1.445 | 0.462 | 0.517 | 1.312 | 0.326 | 0.390 | 1.144 | 0.093 | 0.089 |
| $\phi_t$ | | | | | | | | | |
| Mean | −1.448 | 0.478 | 0.531 | −1.451 | 0.443 | 0.562 | −1.507 | 0.282 | 0.226 |
| St. Dev. | 0.001 | 0.006 | 0.007 | 0.065 | 0.101 | 0.115 | 0.216 | 0.162 | 0.122 |
| Max. | −1.445 | 0.493 | 0.546 | −1.387 | 0.527 | 0.694 | −1.205 | 0.475 | 0.465 |
| Min. | −1.449 | 0.467 | 0.516 | −1.655 | 0.148 | 0.260 | −1.836 | 0.066 | 0.073 |

tic model and between the two scenarios because the equilibrium outcomes are not strictly related to the predetermined conditions.

The first results are the estimated equations with exogenous preferences, shown on the left side of Table 32.1. (The equilibrium values for $\bar{P}$ for $\bar{V}$ are identical, so one equation is shown.) No surprises here. The equations replicate the true model and there is virtually no variation in the estimated coefficients for the twenty replications. The fits are not shown, but the $R^2$ are 1.000 for the preference, partisanship and vote equations and above 0.998 for the party position equations. (The less than perfect fit is a consequence of the non-linear relationship between party positions and mean preference and vote evaluations, as shown in eqns. 12 and 13. More on this in the next section.) These results mean that knowing the starting conditions and the sequence of shocks allows one to reproduce the system. When the underlying model perfectly fits the assumptions of the statistical model—that the equilibrium is perfectly predictable from the predetermined conditions—the statistical methods perform very well.

The estimated equations for the path-dependent scenarios are given in the remaining parts of Table 32.1. As expected, the statistical results do not reproduce the true model and they exhibit considerable variation across replications. These differences and variations are particularly large for the scenario with very adaptive parties, $\delta = .8$. Two concerns evolve from these results. One is that variations in party strategies, which create different histories of an endogenous sort, produce different equilibria, as attested to by the fact that the correlation between the equilibrium mean preference for these two scenarios is only 0.58, even though their starting-points and sequences of exogenous shocks are identical. These different equilibria, in turn, give very different statistical results, even when averaged over twenty replications. The second concern is that even within a given scenario, where the true models are exactly identical and all the exogenous information is included in the statistical analysis, different sequences of shocks produce different results. The standard deviations and the ranges of the estimated co-efficients across the twenty replications with $\delta = .80$ indicate the sensitivity particularly well. The scenario with $\delta = .65$ exhibits substantially less sensitivity, but this scenario is designed to have less party movement, and therefore to exhibit less path dependency.

## D Structural equation estimation

The second econometric comparison estimates the structural equations of the model using these data. Again, we know the true model, so specification uncertainty is not an issue. Structural estimation is a much more demanding, but also more significant, exercise as these estimations should provide evidence about the underlying structure of the process and about what factors are endogenous and which are exogenous.

Five equations are needed to represent the structure at equilibrium, one for each party's position, eqns. 12 and 13, and one each for the means of the issue preference, partisanship, and vote evaluation distributions, eqns. 8 to 11. We obtain two separate estimates for $\alpha$ in eqn. 11, one for each of the included variables. We denote these as $\alpha_1$ and $\alpha_2$, and theoretically they should be equal.

Estimation of these equations presents several statistical challenges, and demonstrates the econometric tools discussed earlier. This is a nonrecursive structural equation system with jointly determined endogenous variables. If there were stochastic terms, as one confronts with real data, this would necessitate the use of an instrumental variables or comparable estimator to purge the explanatory variables of their confounding effects.

There are no stochastic terms here, but we will use an instrumental variables procedure nonetheless. All contemporaneous variables are treated as endogenous, with the lagged values of each variable and the shock, $Z_t$, used as predetermined variables to create instruments. Instruments will be created directly for the nonlinear variables, such as $(\theta_t - \phi_t)\bar{X}_t$. With these transformations, the equations for partisanship, vote evaluations and preferences can be estimated directly as they are linear in the parameters.

Eqns. 12 and 13 are nonlinear in the parameters, so that conventional linear estimation is not possible. In this case, we will use nonlinear least squares, which picks the values for $\gamma$ and $G$ that minimize the sum of squared differences between the observed and predicted values of the party positions. (The values for $B_3$ and $B_4$ in these expressions are determined by parameters in the individual level behavioral equations. We assume these are known from estimations done with individual-level data, such as one gets from election surveys, so they and $\alpha$ are set to their true values for these estimations.) The instruments for $\bar{X}$ and $\bar{V}$ are used in these estimations.

Table 32.2 shows the distribution of estimated structural coefficients for

**Table 32.2** Estimated coefficients in the structural equation models

| *Var.* | $B_1 = 0.000$ | | | $\delta = 0.65$ | | | $\delta = 0.80$ | | |
|---|---|---|---|---|---|---|---|---|---|
| $\bar{P}, \bar{V}$ | | | | | | | | | |
| | $\alpha_1$ | $\alpha_2$ | | $\alpha_1$ | $\alpha_2$ | | $\alpha_1$ | $\alpha_2$ | |
| Mean | .500 | .500 | | .500 | .500 | | .500 | .500 | |
| $\sigma$ | .000 | .000 | | .000 | .000 | | .000 | .000 | |
| Max. | .500 | .500 | | .500 | .500 | | .500 | .500 | |
| Min. | .500 | .500 | | .500 | .500 | | .500 | .500 | |
| $\theta$ | $\gamma$ | G | | $\gamma$ | G | | $\gamma$ | G | |
| Mean | .333 | 2.000 | | .331 | 2.002 | | .303 | 2.030 | |
| $\sigma$ | .000 | .000 | | .003 | .010 | | .032 | .026 | |
| Max. | .333 | 2.000 | | .334 | 2.042 | | .334 | 2.080 | |
| Min. | .333 | 2.000 | | .324 | 1.996 | | .222 | 1.996 | |
| $\phi$ | | | | | | | | | |
| Mean | .333 | −2.000 | | .331 | −2.001 | | .298 | −2.025 | |
| $\sigma$ | .000 | .000 | | .006 | .003 | | .035 | .031 | |
| Max. | .333 | −2.000 | | .335 | −1.995 | | .336 | −1.986 | |
| Min. | .333 | −2.000 | | .305 | −2.008 | | .232 | −2.109 | |
| $\bar{X}$ | $B_1$ | $B_2$ | $B_3$ | $B_1$ | $B_2$ | $B_3$ | $B_1$ | $B_2$ | $B_3$ |
| Mean | .900 | .000 | 1.000 | −.004 | .267 | .019 | .722 | .054 | .532 |
| $\sigma$ | .002 | .001 | .002 | .220 | .068 | .326 | .168 | .070 | .254 |
| Max. | .901 | .002 | 1.001 | .832 | .347 | 1.303 | .974 | .160 | 1.256 |
| Min. | .892 | .000 | .992 | −.299 | .002 | −.363 | .332 | −.044 | .230 |

each scenario. The constant terms have been omitted to conserve space. The estimated equations for partisanship and vote evaluations are the least interesting, as they were perfectly estimated in all three scenarios, and there was no variation in estimated coefficients across replications.

Estimation of the equations for party positions recover the true model for the scenario with exogenous preferences, with no variation across replications. This is what is expected. What is notable is that this estimation uses a nonlinear least-squares method, illustrating the adaptability noted earlier. This method does reasonably well with the scenarios with endogenous preferences, though there is a clear downward bias to the estimates of the party utility for their own platform, $\gamma$, and an upward bias to the estimate of the desired platform, $G$. Only for the scenario with $\delta = .80$ do these biases and variations across replications become particularly noticeable.

The most startling contrasts are the estimated equations for $\bar{X}_t$. As expected, with exogenous preferences, the equation is estimated perfectly. For the two scenarios with endogenous preferences, however, there are completely contradictory results. The mean of the replications for $\delta = .65$ implies that $\bar{X}$ is strictly an endogenous variable that is not affected by either its lagged value or the exogenous shock. The replications with $\delta = .80$ imply just the opposite, with the mean coefficient on the endogenous term being close to zero and the mean coefficients on the lagged and shock terms being large and positive. The coefficients on $\bar{X}_{t-1}$ and $Z_t$ have large variation over the replications for each scenario, indicating that even for the same value of $\delta$ there are results with very different implications. The estimated equations for the first replication for each value of $\delta$ and for the eighth replication for $\delta = .65$ exhibit these two different, but troubling contrasts.

$$\bar{X}_t = -.21 + .97\,\bar{X}_{t-1} + 1.26Z_t - .04(\theta_t - \phi_t)\bar{P}_t \quad \delta = .80, \text{Rep} = 1$$

$$\bar{X}_t = .00 + .00\,\bar{X}_{t-1} + .01Z_t + .26(\theta_t - \phi_t)\bar{P}_t \quad \delta = .65, \text{Rep} = 1$$

$$\bar{X}_t = -.18 + .83\,\bar{X}_{t-1} + 1.30Z_t + .00(\theta_t - \phi_t)\bar{P}_t \quad \delta = .65, \text{Rep} = 8$$

These results suggest that processes that differ only in how they move from one equilibrium to the next or in the sequence of exogenous shocks may generate data that give quite different pictures of the underlying process. This outcome creates considerable uncertainty about the reliability of traditional methods for estimating path-dependent processes where only the equilibrium outcomes are observed.

## E Vector autoregression models

The final comparison is to apply VAR methods to simulated data sets generated with models with endogenous preferences. The three replications shown above have been selected for this analysis, as there is no concise way to conduct and summarize the results of a VAR estimation on all twenty replications. The structural estimations for $\bar{X}_t$ in these three replications imply contrary exogenous influences so it will be interesting to see what patterns are discovered by the VAR method. The contemporaneous values of each variable were regressed against $Z_t$ and $\bar{X}_{t-1}$, and lagged values of itself and the other variables. A series of F-tests were conducted on sets of these variables to determine if there were any possibly significant relationships.

The results, shown in Table 32.3, are interesting and puzzling. There is no consistent pattern for the three trials. For the first replication with $\delta = .65$, all the variables are related to $\bar{X}_{t-1}$ and their own lagged values, in addition $\bar{X}_t$ is related to $\phi_{t-1}$ and $\bar{P}_t$ to the lagged position of both parties. For the replication with $\delta = .80$, $\bar{X}_t$ and $\theta_t$ are only related to $Z_t$ and $\bar{X}_{t-1}$ and $\bar{P}_t$ is only related to $Z_t$, $\bar{X}_{t-1}$ and the lagged value of itself. This might suggest that $Z_t$ and $\bar{X}_{t-1}$ determine the system's outcomes. $\phi_t$, however, is statistically related to both the first and second lags of itself and $\theta$ as well as to $Z_t$ and $\bar{X}_{t-1}$, which implies a totally different pattern of relations.

**Table 32.3** Estimated VAR model

| | *Statistically Significant Variables*[a] | | |
|---|---|---|---|
| *Equation* | $\delta = .80$, *Rep* = 1 | $\delta = .65$, *Rep* = 1 | $\delta = .65$, *Rep* = 8 |
| $\bar{X}_t$ | — | $\phi_{t-1}$ | $\theta_{t-1}, \phi_{t-1}$ |
| $\bar{P}_t$ | $\bar{P}_{t-1}$ | $\bar{P}_{t-1}, \theta_{t-1}, \phi_{t-1}$ | $\bar{P}_{t-1}$ |
| $\theta_t$ | — | $\theta_{t-1}$ | $\theta_{t-1}, \phi_{t-1}$ |
| $\phi_t$ | $\theta_{t-1}, \phi_{t-1}, \theta_{t-2}, \phi_{t-2}$ | $\phi_{t-1}$ | $\theta_{t-1}, \phi_{t-1}$ |

[a] All equations contain $\bar{X}_{t-1}$ and $Z_t$.

The eighth replication with $\delta = .65$ is the most peculiar. The structural equation estimation for $\bar{X}_t$ for this trial implies that it was only related to the predetermined variables $Z_t$ and $\bar{X}_{t-1}$. The VAR results indicate a very strong role for the lagged party positions, in addition to these first two variables. These are the results most consistent with the true model, where preferences are related to party positions, as well as vice versa. This would look like a success for VAR, relative to the structural equation estimations, except that this success is not repeated in the other trials. This casts additional doubt on the ability of any of these methods to detect and estimate path dependencies.

## V There is a problem, but is there a solution?

The different analyses of the simulated data clearly reveal problems if one attempts to use conventional methods to estimate models and test propositions about processes that contain path-dependent outcomes. All the simulations contained and reached an equilibrium, but this is not sufficient to yield consistent statistical results or results that even approximate the true structure. Even with no unobserved stochastic terms, data created by processes with identical structures and parameters, but with different sequences of shocks produced startlingly different statistical results. Also, data from simulations with identical shocks but with different adjustments to these shocks produced quite different statistical results. Even the mean parameter estimates over twenty replications of the same model did not approximate the true model. It is very unlikely that a resort to probability limits and asymptotic properties can overcome these problems. If path-dependent processes are an important part of the world of social and economic institutions we must begin to explore ways to measure and estimate this class of models.

The basic nature of the statistical problem is that outcomes reflect the values of various variables during earlier periods, and these effects do not attenuate over time. To see this, consider the model for mean preferences, $\bar{X}$, in the electoral model we have been using,

$$\bar{X}_t = B_{1_t}\bar{X}_{t-1} + (1 - B_{1_t})D_t\bar{P}_t,$$

where $D_t = (\theta_t - \phi_t)$. If the system is in equilibrium at $t = 0$ with $\bar{X} = \bar{X}^*$ and a shock $Z_0$ occurs at this point, our model for $\bar{X}$ over successive periods to time $\tau$ is,

$$\bar{X}_0 = .9\bar{X}^* + Z_0$$

$$\bar{X}_1 = B_{1_1}\bar{X}_0 + (1 - B_{1_1})D_1\bar{P}_1$$

$$\bar{X}_2 = B_{1_2}\bar{X}_1 + (1 - B_{1_2})D_2\bar{P}_2$$

$$= B_{1_2}B_{1_1}\bar{X}_0 + B_{1_2}(1 - B_{1_1})D_1\bar{P}_1 + (1 - B_{1_2})D_2\bar{P}_2$$

$$\bar{X}_3 = B_{1_3}\bar{X}_2 + (1 - B_{1_3})D_3\bar{P}_3$$
$$= B_{1_3}B_{1_2}B_{1_1}\bar{X}_0 + B_{1_3}B_{1_2}(1 - B_{1_1})D_1P_1 + B_{1_3}(1 - B_{1_2})D_2\bar{P}_2 + (1 - B_{1_3})D_3\bar{P}_3$$

.
.
.

$$\bar{X}_\tau = B_{1_\tau}\bar{X}_{\tau-1} + (1 - B_{1_\tau})D_\tau\bar{P}_\tau$$
$$= \left(\prod_{i=1}^{\tau} B_{1_i}\right)\bar{X}_0 + \sum_{i=1}^{\tau-1}\left[\left(\prod_{j=i+1}^{\tau} B_{1_j}\right)(1 - B_{1_i})D_i\bar{P}_i\right] + (1 - B_{1_\tau})D_\tau\bar{P}_\tau.$$

If we assume time $\tau$ is when the system attains its new equilibrium, then the value of $\bar{X}$ at any later time $T$, $T > \tau$ is,

$$\bar{X}_T = \left(\prod_{i=1}^{\tau} B_{1_i}\right)(.9\bar{X}^* + Z_0) + \sum_{i=1}^{\tau-1}\left[\left(\prod_{j=i+1}^{\tau} B_{1_j}\right)(1 - B_{1_i})D_i\bar{P}_i\right] + (1 - B_{1_\tau})D_\tau\bar{P}_\tau. \quad (16)$$

The estimation problem is now apparent. If all we observe are the initial conditions, the shock, and the resulting equilibrium the coefficients on the initial conditions and the shock and the summation term all contain unobserved information. Estimation of this equation suffers from a misspecification and omitted variables problem, which usually leads to biased and inconsistent estimates. This is what we have observed with the analysis of the simulated data sets. It should be pointed out that this fundamental problem exists if we are examining cross-sectional data as well as time-series data. Each of the 100 simulated elections in the previous section could be cross-sectional data taken from separate electoral units, where one knows the starting-point, shock, and outcome for each election. Eqn. 16 shows that estimation of the cross-section data also suffers from the same misspecification.

The hard task is suggesting possible remedies. Freeman's (1989) advice, and caution, is particularly useful and relevant. He forcefully pointed to the estimation problems created by aggregating distinct time intervals into single observations, as we have been doing so far. According to Freeman, if we know the exact time interval for each iteration of the dynamic process *and* if we can measure the relevant variables at each iteration, then it should be possible to estimate the system with existing techniques applied to these micro time data. Analysis of the simulated data for each iteration of the simulations confirm this point.

The three replications shown previously were rerun, recording the outcomes at each iteration until equilibrium was attained. These data are used to estimate the following version of eqn. 8 for $\bar{X}_t$,

$$\bar{X}_t = B_0 + e^{-B_1(\Delta\theta^2+\Delta\phi^2)}\bar{X}_{t-1} + .5B_2[1 - e^{B_1(\Delta\theta^2+\Delta\phi^2)}]D_t\bar{P}_t + B_3D_0\bar{X}^* + B_4Z_0,$$

where $D_0$ is a dummy variable indicating the period of a shock, $\bar{X}^*$ is the prior equilibrium value for $\bar{X}$, and $Z_0$ is the magnitude of the shock. As this expression is nonlinear in the parameter $B_1$ an iterative instrumental variable method was used in the estimation. Different values of $B_1$ were used to construct the first two right-hand side variables and then the lagged values of all endogenous variables and the values for $D_0\bar{X}^*$ and $Z_0$ were used to create instruments for these endogenous variables. (The $R^2$ in this first stage are about .98 for the term with $\bar{X}_{t-1}$ and between .4 and .5 for the term with $D_t\bar{P}_t$.) The coefficients $B_0$, $B_2$, $B_3$, and $B_4$ were estimated from the second stage regression of $\bar{X}_t$ on the instruments and the two exogenous variables. A least-squared error criteria was used to select the estimated coefficients. The results confirmed Freeman's contention. The analysis of each replication reproduced the true coefficients perfectly: $B_0 = .000$; $B_1 = 1.000$; $B_2 = 1.25$; $B_3 = .900$, and $B_4 = 1.000$.

Equations 14 and 15 for party positions were also estimated from these data, using the same non-linear least-squares method used for the structural equations, only now with the lagged values for party positions included. The results using the micro time data very closely replicated the true structure, much better than did the estimates using the time-aggregated data summarized in Table 32.2. Table 32.4 shows the estimated coefficients for the three replications using the micro time data and those from using the time-aggregated data.

These results are a small consolation. Freeman correctly points out that knowing the proper time interval *and* getting data measured at the proper points are quite difficult tasks. Without meeting these demanding criteria, one is left with the problems discussed here if there is reason to believe that the process is a path-dependent one.

**Table 32.4** Structural equations for party positions

| | $\delta = .80$, *Rep* = 1 | | $\delta = .65$, *Rep* = 1 | | $\delta = .65$, *Rep* = 8 | |
|---|---|---|---|---|---|---|
| *Parameter* | *Aggreg.* | *Micro* | *Aggreg.* | *Micro* | *Aggreg.* | *Micro* |
| $\gamma_\theta$ | .326 | .333 | .333 | .333 | .326 | .334 |
| $\gamma_\phi$ | .297 | .331 | .329 | .333 | .305 | .331 |
| $G_\theta$ | 2.039 | 2.001 | 1.999 | 2.000 | 2.042 | 1.998 |
| $G_\phi$ | 2.014 | 2.001 | 2.004 | 2.001 | 2.008 | 2.001 |
| $\delta_\theta$ | | .814 | | .649 | | .640 |
| $\delta_\phi$ | | .812 | | .648 | | .609 |

A second approach is to use highly disaggregated cross-sectional data to estimate some of the important parameters and then merge this information with time-series data. For example, Gerber and Jackson (1993) estimated the coefficients in the endogenous preference equation with cross-sectional survey data on individual voters. If one then has an appropriate aggregation rule, one can begin to construct an approximately correct system from the individual equations. (See Achen and Shively (1995) on the difficulty of developing aggregation rules.)

None of these approaches is very pretty or satisfying. They offer only a little solace. In the meantime, work should proceed to investigate alternative measures, models, and estimators. The evidence of the past quarter-century is that if the problem poses enough interesting analytical issues political methodologists can be very creative. This task offers a piece of the methodology agenda for the next several years, as I do not think the resolution of this problem is readily at hand with the current methodologies.

The good news is that it appears the problems have not yet been resolved in other disciplines either, and may not have even been discussed. The centrality of path-dependent systems to political science and the increasing interest in such models in other fields provides important opportunities for political methodologists. Resolution of the questions raised here and development of methods to identify and estimate models of path-dependent processes and to test alternative propositions within such models may have methodologists from other fields looking to political science as a source of intellectual capital and provide customers for our garage sales.

## Acknowledgments

Paper prepared for the Annual Meeting of the Political Methodology Section of the American Political Science Association, Madison, Wis., 23 July 1994 and the meeting of the International Political Science Association, Berlin, Germany, 24 August 1994. This research was supported by a gift to the University of Michigan from Mr. A. Alfred Taubman to the Program in American Institutions and by a grant from the National Science Foundation, SBR-9511469, and was partially written at the Santa Fe Institute. I want to thank Chris Achen for his many helpful and provocative comments and Elisabeth Gerber for her work on an earlier part of this project and her continuing advice and assistance. The ideas and arguments in this paper are the responsibility of the author and not those of the above organizations and individuals.

## References

Achen, C. H. 1975. Mass political attitudes and the survey response. *American Political Science Review*, 69: 1218–31.

—— 1983. Toward theories of data: the state of political methodology. Pp. 69–94 in *Political Science: The State of the Discipline*, ed. A. Finifter. Washington, D.C.: American Political Science Association.

—— 1985. Proxy variables and incorrect signs on regression coefficients. *Political Methodology*, 11: 299–316.

—— 1992. Social psychology, demographic variables, and linear regression: breaking the iron triangle in voting research. *Political Behavior*, 14/3: 195–211.

—— and Shively, W. P. 1995. *Cross Level Inference*. Chicago: University of Chicago Press.

Arthur, W. B.; Ermoliev, Y. M.; and Kaniovski, Y. M. 1987. Path-dependent processes and the emergence of macro-structure. *European Journal of Operational Research*, 30: 294–303.

Bartels, L. M., and Brady, H. E. 1993. The state of quantitative political methodology. Pp. 121–58 in *Political Science: The State of the Discipline II*, ed. A. Finifter. Washington, D.C.: American Political Science Association.

Beck, N. 1983. Time-varying parameter regression models. *American Journal of Political Science*, 27: 557–600.

—— 1987. Elections and the Fed: is there a political business cycle? *American Journal of Political Science*, 31: 194–216.

—— 1989. Estimating dynamic models using Kalman filtering. *Political Analysis*, 1: 121–56.

—— 1992. Comparing dynamic specifications: the case of presidential approval. *Political Analysis*, 3: 51–87.

—— and Katz, J. N. 1995. What to do (and not to do) with time-series-cross-section data. *American Political Science Review*, 89: 634–47.

Brady, H. M. 1985. The perils of survey research: inter-personally incomparable responses. *Political Methodology*, 11: 269–91.

—— 1993. Stability, reliability, and guessing. Paper presented at the Annual Meeting of the Midwest Political Science Association, Chicago, April 1993.

Carmines, E. C., and Stimson, J. A. 1989. *Issue Evolution: Race and the Transformation of American Politics*. Princeton, N.J.: Princeton University Press.

Dawson, M. 1992. *Behind the Mule*. Princeton, N.J.: Princeton University Press.

Erikson, R. S. 1979. The SRC panel data and mass political attitudes: change and uncertainty. *British Journal of Political Science*, 9: 89–114.

Franklin, C. H. 1990. Estimation across datasets: two-stage auxiliary instrumental variables estimation (2SAIV). *Political Analysis*, 1: 1–24.

—— and Jackson, J. E. 1983. The dynamics of party identification. *American Political Science Review*, 77: 957–73.

Freeman, J. R. 1989. Systematic sampling, temporal aggregation, and the study of political relationships. *Political Analysis*, 1: 61–98.

—— ed. 1994. *Political Analysis*, 4: 97–258.

—— Williams, J.T.; and Lin, T.-M. 1989. Vector autoregression and the study of politics. *American Journal of Political Science*, 33: 842–77.

Gerber, E. R., and Jackson, J. E. 1993. Endogenous preferences and the study of institutions. *American Political Science Review*, 87: 639–56

Hanushek, E. A., and Jackson, J. E. 1977. *Statistical Methods for Social Scientists*. New York: Academic Press.

Hibbs, D. A., Jr. 1974. Problems of statistical estimation and causal inference in time-series regression models. *Sociological Methodology*, 1973–74: 252–308.

Jackson, J. E. 1973. *Constituencies and Leaders in Congress.* Cambridge, Mass.: Harvard University Press.

—— 1975*a.* Issues, party choices, and presidential votes. *American Journal of Political Science*, 19: 161–85.

—— 1975*b.* Issues and party alignment. Vol i, pp. 101–23 in *Electoral Studies Yearbook*, ed.L. Maisel and P. Sack. Beverly Hills, Calif.: Sage.

—— 1979. Statistical estimation of possible response bias in close-ended issue questions. *Political Methodology*, 6: 393–423.

—— 1983. The systematic beliefs of the mass public: estimating policy preferences with survey data. *Journal of Politics*, 45: 840–65.

—— 1992. Estimation of models with variable coefficients. *Political Analysis*, 3: 27–49.

—— 1994. Electoral competition with endogenous voter preferences. Paper presented to the Annual Meeting of the Western Economic Association, June 1993 (revised).

—— 1995. Attitudes, no opinions, and guesses. *Political Analysis*, 5: 39–60.

—— and King, D. C. 1989. Public goods, private interests, and representation. *American Political Science Review*, 83: 1143–64.

—— and Thomas, A. R. 1995. Bank structure and new business creation: lessons from an earlier time. *Regional Science and Urban Economics*, 25: 323–54.

Jöreskog, K. G. 1973. A general method for estimating a linear structural equation system. Pp. 85–112 in *Structural Equation Models in the Social Sciences*, ed. A. S. Goldberger and O. D. Duncan. New York: Seminar Press.

—— and Sörbom, D. 1979. *Advances in Factor Analysis and Structural Equation Models.* Cambridge, Mass.: Abt.

Judge, G. G.; Griffiths, W. E. ; Hill, R. C.; Lütkepohl, H.; and Lee, T. C. 1985. *The Theory and Practice of Econometrics.* 2nd edn. New York: Wiley.

Kellstedt, P.; McAvoy, G. E.; and Stimson, J. A. 1995. Dynamic analysis with latent constructs. *Political Analysis*, 5: 113–50.

King, G. 1989*a. Unifying Political Methodology.* New York: Cambridge University Press.

—— 1989*b.* Event count models for international relations: generalizations and applications. *International Studies*, 33: 123–47.

—— 1991. On political methodology. *Political Analysis*, 2: 1–29.

Krugman, P. 1991. *Geography and Trade.* Cambridge, Mass.: MIT Press.

MacKuen, M. B.; Erikson, R. S.; and Stimson, J. A. 1989. Macropartisanship. *American Political Science Review*, 83: 1125–42.

Markus, G. B. 1988. The impact of personal and national economic conditions on the presidential vote: a pooled cross-sectional analysis. *American Journal of Political Science*, 32: 137–54.

—— and Converse, P. E. 1979. A dynamic simultaneous equation model of electoral choice. *American Political Science Review*, 73: 1055–70.

McGinnis, M. D., and Williams, J. T. 1989. Change and stability in superpower rivalry. *American Political Science Review*, 83: 1101–23.

North, D. C. 1990. *Institutions, Institutional Change and Economic Performance.* Cambridge: Cambridge University Press.

Page, B. I., and Jones, C. C. 1979. Reciprocal effects of policy preferences, party loyalties and the vote. *American Political Science Review*, 73: 1071–89.

Riker, W. H., and Weimer, D. L. 1993. The economic and political liberalization of socialism: the fundamental problem of property rights. *Social Philosophy and Policy*, 10: 79–102.

Rivers, D. 1988. Heterogeneity in models of electoral choice. *American Journal of Political Science*, 32: 737–57.

Smith, R. M., and Box-Steffensmeier, J. M, 1995. The macrofoundations of aggregate partisanship: a fractional integration analysis of heterogeneity and permanence. *American Political Science Review*, forthcoming.

Williams, J. T., and McGinnis, M. D. 1988. Sophisticated reaction in the U.S.–Soviet arms race: evidence of rational expectations. *American Journal of Political Science*, 32: 968–95.

Wittman, D. 1983. Candidate motivation: a synthesis of alternative theories. *American Political Science Review*, 77: 142–57.

Chapter 33

# Political Methodology: Qualitative Methods

Charles C. Ragin
Dirk Berg-Schlosser
Gisèle de Meur

La méthode comparative est la seule qui convienne à la sociologie.
(Durkheim 1894/1988: 217)

## I Introduction

While Durkheim's early claim clearly overstates the case, it remains true today that systematic comparative analysis constitutes one of the primary means for establishing social scientific generalizations in macropolitical inquiry. Comparisons in political science are, however, confronted with a special dilemma. At the level of whole political systems or "nation-states" there are relatively few cases to observe (for example, the present 180 or so U.N. member states) while, at the same time, each case exhibits a bewildering complexity that may confound even the best-informed country expert. This "many variables, few cases" dilemma (see also Lijphart 1971; 1975) lies at the heart of qualitative methods in macropolitical research, especially research that seeks to establish generalizations across cases.

We use the term "qualitative" in this essay to refer to an interest in presence or absence of specific characteristics or specific configurations of characteristics pursued by means of systematic comparison of multiple cases. As the term is used here, therefore, it should not be confounded with the variety of qualitative methods used at the microlevel (e.g., the method of participant observation in sociology) or with qualitative interpretation (e.g., hermeneutic methods). The realm of qualitative methods in macropolitical inquiry is situated between the extremes of analyzing a single case with the help of one central explanatory variable (e.g., world systems analysis; Wallerstein 1980) and an attempt to cover all existing political systems on a global scale with as many variables as possible (e.g., "World Handbooks" inspired by Karl Deutsch such as Taylor and Jodice 1983). It also lies between conventional case-study research (countless variables, but only one case) and conventional quantitative research (many cases, relatively few variables).[1]

The appeal of qualitative methods is two-fold. First, qualitative methods can be used to great advantage in research situations where theories are underdeveloped and concepts are vague. Thus, qualitative investigations are often at the forefront of theoretical advancement (Feagin, Orum, and Sjoberg 1991). For example, a common practice in qualitative inquiry is to "double-fit" a theoretical concept and an empirical category, thereby sharpening and clarifying the content of both the concept and the category (e.g., the concept "anti-neocolonial revolution" and the set of cases that qualify as instances). Second, researchers often ask questions that simply cannot be addressed with conventional, quantitative methods. For example, most questions about historically or culturally significant phenomena concern empirical categories that are bounded in time and space and thus contain a finite, usually small, number of known instances. The demands and assumptions of quantitative methods are very difficult to meet in small-N situations. Research questions may also conflict with other requirements of quantitative methods, which variously include: a well-defined population of observations, the possibility of sampling, an interest in pre-conditions and outcomes that clearly vary across cases, confidence that the cases included in a study are homogeneous with respect to hypothesized relationships, and so on. Rather than alter their research questions to fit the demands of conventional quantitative methods, many researchers choose qualitative approaches.

While there is a considerable overlap between qualitative methods, as defined here, and the central concerns of the field of comparative politics, the recent literature has been more concerned with substantive issues than with methodological questions.[2] However, there is clear evidence that methodological concerns, especially the problem of "many variables, few cases," have an impact on problem choice in comparative social science. Surveys of relevant literatures by both Sigelman and Gadbois (1983) and

[1] See Ragin 1994a.

[2] See, e.g.: Wiarda 1985; Mayer 1989; Rustow and Erickson 1991; Keman 1993; Lane and Ersson 1994; Dogan and Kazancigil 1994; exceptions include Przeworski 1987; Ragin 1987; Teune 1990; Collier 1993.

Bollen *et al.* (1993) reveal a U-shaped frequency distribution of number of studies plotted against the number of cases investigated in each study (see also Ragin 1989). Most studies address one or two cases, on the one hand, or more than twenty cases, on the other.

Departures from this U-shaped relation certainly exist.[3] Still, such studies are the exception. This lacuna does not indicate, however, that this type of research is by any means inferior to the more common single-case studies or large-N quantitative studies. On the contrary, studies with intermediate-sized Ns may very well prove to be more relevant both for building a systematic knowledge of cases and for advancing theory.

In this chapter, we address some of the basic features and problems of this type of research, which we call "qualitative-comparative research"—the cross-case analysis of configurations of similarities and differences. We first deal with both the cases and the variables sides of the above-mentioned dilemma and then sketch some recently developed techniques such as "most similar" and "most different systems" designs (MSDO and MDSO) and "Qualitative Comparative Analysis" (QCA). A subsequent section addresses basic epistemological issues involved in investigations of this kind.

## II Framing the area of investigation

Each empirical field of study can be described by the cases ("units") analyzed, the characteristics of cases ("variables") being considered, and the number of times each unit is observed ("observations").[4] We focus especially on the first two aspects, units and variables. For many analyses it is indispensable to examine variation over time, and such longitudinal study can be conducted using both quantitative (e.g., Petersen 1993) and qualitative (e.g., Heise 1989; Griffin 1993) methods. However, studies of this type pose their own special problems, which would require lengthy elaboration, and we set them aside for now.

### A The selection of units

In stark contrast to large-scale statistical analysis, with seemingly unlimited numbers of cases or random samples based on these, qualitative-comparative

[3] e.g., broad-based studies such as discussed in Skocpol (1984) and recent studies using some of the techniques discussed below; see, e.g., Ragin (1991); Griffin and Ragin (1994).

[4] For a similar description see, e.g., King *et al.* 1994: 51 ff.

studies always deal with a limited universe (with twenty or so cases a relatively large number). "Increase the number of cases as much as possible" by extending the analysis both geographically and historically (as recommended by Lijphart (1971: 686)) thus is usually neither feasible nor necessarily desirable. Instead, the selection of cases must be based on compelling theoretical and substantive criteria because the inclusion or exclusion of any single case may significantly alter the investigator's conclusions. At the same time, the findings for each case are considered important in themselves. No case is ignored as a negligible "outlier" in a frequency distribution.

At the outset of any investigation, therefore, an area of homogeneity must be defined which establishes boundaries for the selection of cases. Cases must "parallel each other sufficiently" and be comparable along certain specified dimensions. The specification of relevant cases at the start of an investigation amounts to an explicit or implicit hypothesis that the cases selected are in fact alike enough to permit comparisons. In the course of the research, the boundaries of the investigation may shift as more is learned about the similarities and differences among cases.

The primary consideration in delimiting cases for a qualitative-comparative study is the dependent variable in the investigation. For example, the breakdown or survival of democratic regimes in inter-war Europe presupposes the prior existence of some form of democracy in the selected cases. In addition, some limitations in time and space can also constitute the homogeneity and thus the comparability of the cases examined, or other cultural or historical factors relevant to the problem may be considered. For example, certain kinds of colonial or other forms of external domination or religious-cultural influences may be useful criteria for selecting a specific group of cases. Conversely, it does not make sense for most purposes of current theory, except at the most abstract level, to compare, for example, ancient Greek city-state "democracy" with acephalous forms of consensual decision-making in parts of traditional Africa (as among the Maasai in Kenya, for example) or with the present system of government in, say, the United States or the Federal Republic of Germany.

A second consideration concerns the extent of diversity within the selected universe. In this regard, a maximum of heterogeneity for a minimum number of cases should be achieved. Taking the above example again, both survivors and breakdowns of democracy can be considered, and among the latter perhaps some more specific variants such as fascist versus more generally authoritarian outcomes can be examined. The full extent of variation, however, may become apparent only when the selected cases are analyzed more closely. In this sense, the initial "hypotheses" concerning the selection of cases may be revised in the course of research and lead to the elimination or addition of cases.[5]

## B The selection of explanatory variables

In a similar manner, the selection of variables must be guided by theoretical criteria. Here, however, we are confronted with the opposite embarrassment, namely, a potential abundance of factors to be considered. Given the state of social science theory for most empirical questions, a large number of variables often cannot be excluded *a priori*. This predicament is exacerbated if we seek to go beyond "universalizing" explanations and explore diversity (i.e., "variation-finding"; see Tilly 1984) or address patterns of "conjunctural causation."

The conventional way to select a limited number of variables is to test any relevant hypothesis for the problem concerned in a strictly "Popperian" falsificatory manner. Thus, for example, the well-known Lipset thesis that "the more well-to-do a nation the greater the chances that it will sustain democracy" (Lipset 1963: 31) can be tested in this way. In the contemporary world, this hypothesis is verified for about 70 percent of the cases that are relatively well-established and consolidated democracies. However, the hypothesis does not account for poor countries with relatively stable democracies like Botswana, India, and Papua New Guinea or for the failure of democracy in relatively well-developed ones like Weimar Germany. (A specific test of this hypothesis for the inter-war period in Europe produced a score of 10 out of 18 "correct" results; see Berg-Schlosser and De Meur 1994.)

Such a hypothesis must be modified and specified more closely if it is to satisfy any epistemological criteria more demanding than vague "probability." The next step might be to test "conjunctural hypotheses," in which the selection of variables is guided by explanations that are combinatorial in nature (see, e.g., Amenta and Poulsen 1994). In this way, certain constellations of factors, for example concerning conditions favorable to democracy in a number of poor countries or unfavorable in some richer countries, can be identified and tested.

Widening the horizon still further, investigators may adopt a "perspectives" approach, that is, supplying a mixed bag of variables derived from

[5] Sometimes it is difficult to come to any agreement regarding relevant "negative cases." For example, it might prove difficult to specify the set of countries lacking social revolutions that are relevant to the analysis of positive instances of social revolution. In such studies, analysis may focus only on positive instances, and the researcher may attempt to ascertain the multiple paths to a common outcome. In other research, however, there is less question about relevant negative cases.

the main theoretical perspectives in the empirical literature. This approach is probably the most common way of dealing with complex problems in empirical social research. The investigator takes a thorough look at the "state of the art" in any given area and then develops a specific research design which takes the wider range of these factors into account. At the same time, he or she develops a way to adjudicate between competing explanations and to allow for "interaction effects" among certain factors. To stay within our previous example of empirical democratic theory, the works of Dahl (1971; 1989) and the overview by Lipset (1994) discuss a wide range of factors that are conducive to more stable forms of democracy in the modern world. These reviews do not, however, provide any specific "weights" of these factors or consider their interaction effects.

But even such broad overviews may not address all possibly relevant factors or all relevant interactions. Often, therefore, researchers must adopt a comprehensive approach, relying on all extant theories, hypotheses, and explanations. Even though the "all" in this formulation can never be fully satisfied, it points to the potential for complexity on the variables side of the "many variables, few cases" dilemma. Such a comprehensive approach can be structured with broad "systems" models of the Parsonsian or Eastonian kind which potentially comprise all relevant aspects and interactions to be considered. The different sub-systems or categories of such a model can be "filled" using a theoretically and historically informed listing of variables (as in the "perspectives" approach). Still, the comprehensive approach offers no assurance that all relevant factors and interactions have been taken into account.

## III Macro-qualitative techniques

On the basis of a carefully selected set of cases and a usually still wide range of variables, qualitative comparative methods can be applied. At the core of most comparative analyses, especially the less formal sort, is the simple idea that similar outcomes follow from similar causes. Most users of this simple idea, in turn, have relied on arguments presented by John Stuart Mill in *A System of Logic* (1843), especially his "method of agreement" and his "indirect method of difference" (Cohen and Nagel 1934). While these methods have inspired generations of researchers, they are quite rudimentary and cannot address certain basic features of social life. Specifically: (1) outcomes often follow from combinations of causes, not from single causes (see also Lieberson 1991), and (2) it is often the case that a given outcome may follow from several different combinations of causes. This degree of causal complexity, which is characteristic of human social life, especially at the macropolitical level, simply cannot be addressed with most conventional techniques of data analysis. Qualitative comparative methods, however, are fundamentally holistic and configurational in nature and thus are well suited to addressing the many variables, few cases dilemma.

In the remainder of this section, we discuss three ways of coping with the problem of "many variables, few cases." The techniques we discuss complement each other. We first examine "most similar" and "most different systems" designs. These designs use the entire range of variables without looking at all of their many different combinations and interactions. We then present a brief overview of "Qualitative Comparative Analysis" (QCA) which is based on a somewhat limited range of variables (a maximum of twelve in the presently available software, QCA 3.0; see Drass and Ragin 1991), but which considers all logically possible combinations and interactions among variables and arrives at a reduced representation of information on cases. The third and final topic is the problem of reducing complexity by eliminating some variables and reconstructing others in a more comprehensive manner.

## A "Most different" and "most similar" systems designs

"Most different" and "most similar" designs have been in use for some time, but most applications to date have been haphazard and unsystematic. The approach was formalized in 1970 by Przeworski and Teune:

> The most similar systems design is based on a belief that a number of theoretically significant differences will be found among similar systems and that these differences can be used in explanation. The alternative design, which seeks maximal heterogeneity in the sample of systems, is based on a belief that in spite of intersystemic differentiation, the populations will differ with regard to only a limited number of variables or relationships (1970: 39).

In this way, a systematic matching and contrasting of cases can be attempted which allows identification of key distinguishing or common variables while controling for the others.

It is important to note, however, that such procedures, like all qualitative comparative methods, should never be applied in a purely mechanical way. They direct the attention of the researcher to specific variables which then must be interpreted in the light of relevant theoretical propositions and, depending on the problem analyzed, specific historical knowledge.

Until very recently, "most different" and "most similar" designs were not fully operationalized (see also Przeworski 1987). An attempt to apply them systematically was made in the previously mentioned project on democracy in inter-war Europe (De Meur and Berg-Schlosser 1994*a*). In this research the central dependent variable is the survival or breakdown of democratic systems. The cases included in this investigation offer three decisive comparisons: (1) Among the survivors it is instructive to pinpoint the commonalities shared by the *most different* cases. (2) In a parallel fashion, among cases of breakdown it is instructive to identify the commonalities shared by the *most different* cases. (3) Finally, comparing across instances of breakdown and instances of survival, it is instructive to identify the differences distinguishing the *most similar* pair of cases with contrasting outcomes (a case of survival paired with a similar case of breakdown). The first two comparisons are called "most different with same outcome" (MDSO); the third is called "most similar with different outcome" (MSDO).

Before applying these designs, each case was characterized using a mixture of a "perspectives" and a "comprehensive" approach on the basis of an Eastonian system model (for details see Berg-Schlosser and De Meur 1996). Altogether, sixty-three causally relevant variables in seven broad categories for eighteen cases were identified, each case now being defined by its particular configuration of variables. At this point, several problems must be addressed which follow from the necessity of measuring the proximity or remoteness of pairs of cases in the heterogeneous, multidimensional space defined by the independent variables. These distance measures provide the basis for determining the "most different" and "most similar" pairs of cases—the key comparisons in MDSO and MSDO designs.

The two main issues are: (1) choosing from among a variety of different ways of measuring the distance between pairs of cases in a multidimensional space (e.g., Euclidian or "city block" measures), and (2) assigning relative weights to the variables that define this space. The second problem is clearly a theoretical concern but also involves practical issues. For example, an investigator might wish to group variables into separate domains (e.g., seven domains based on Easton's model) and then weight the domains equally in the computation of distance scores. Thus, if some domains contain more variables than others, then variables in these domains would receive smaller weights. Without entering into the details here, it should be noted that several levels of similarity or dissimilarity have been maintained—both within and across domains—so that the "most (dis-) similar" pairs of cases for a few domains are taken into account as well as the somewhat less (dis-) similar ones across a larger number of domains. In this way the complexity of the data is retained in the complexity of the proximity measure.[6]

These procedures make it possible to identify "most different cases with the same outcome" (MDSO) and the "most similar cases with different outcomes" (MSDO). The causes of different outcomes (e.g., breakdown versus survival) may now be assumed to lie in the commonalities that remain among MDSO cases and the differences that distinguish MSDO cases. In the project on the fate of democracy in inter-war Europe, for example, the cases of Finland and Estonia emerged as the most similar systems with different outcomes. This pair turned out to be different on sixteen of the sixty-three considered variables. When Germany was added as another quite similar breakdown case and Sweden as another quite similar case of democratic survival, some additional idiosyncracies of specific cases could be eliminated and only six distinguishing variables remained. Among these some more distinct causal factors, such as, in this example, specific differences in political culture, may now be assumed to lie. Subsequently, these broad background conditions can be considered the "givens" against which the specific impact and extent of a crisis as a common external "stimulus" and the reactions of the major social and political actors can be assessed. In this way, longer-term "structure-oriented" approaches and specific "actor-oriented" analyses or, in Jon Elster's (1989) terms, the "opportunity set" in any given situation and the specific choices made, can be integrated into a more encompassing "quasi-experimental" perspective.

In contrast to this method, which focuses on extreme configurations of cases and the entire range of selected variables, a different method, described in the following section, considers the entire range of cases and a delimited set of variables relevant to a given outcome. Furthermore, as we show, it focuses explicitly on complex patterns of multiple conjunctural causation.

## B Qualitative Comparative Analysis (QCA)

Qualitative Comparative Analysis is a new analytic technique that uses Boolean algebra to implement principles of comparison used by scholars engaged in the qualitative study of social phenomena (Ragin 1987; 1994*b*). In qualitative comparative analysis each case is conceived as a specific configuration of pre-conditions and an outcome. The simplest type of analysis

[6] For this purpose some specific software has been developed by De Meur and Quenter (1994).

involves dichotomous variables, but multichotomies can be used as well, and interval-scale variables can be transformed into dichotomies (Ragin 1994*c*) or into multichotomies. In qualitative comparative analysis, pre-conditions are conceived as variables that define the different configurations that are possible within the limits of the analysis. For example, the specification of seven dichotomous pre-conditions provides for 128 (i.e., $2^7$) logically possible, qualitatively distinct configurations.

Once pre-conditions have been selected, cases conforming to each combination of values on the pre-conditions are examined to see if they agree on the outcome. If there are any combinations with cases that disagree on the outcome variable, the investigator may take this as a sign that the specification of pre-conditions is incorrect or incomplete. The close examination of cases that have the same values on the pre-conditions, yet display contrasting outcomes, is used as a basis for selecting alternate or additional conditions. The investigator moves back and forth between specification of pre-conditions (using theory, substantive knowledge, and substantive interests as guides) and examination of cases, as long as contradictory combinations remain, to build a combinatorial model with a minimum number of cases having the same combinations of values but contrasting outcomes.[7]

Once a satisfactory set of pre-conditions has been identified, data on cases can be represented as a truth table and then the truth table can be logically minimized. A truth table lists the different combinations of values on the pre-conditions and the value of the outcome variable for the cases conforming to each combination. An analysis with three dichotomous conditions yields a truth table with eight rows; four conditions produce a truth table with sixteen rows; five conditions, thirty-two rows; and so on. The goal of the logical minimization is to represent in a logically shorthand manner the information in the truth table regarding the different combinations of conditions that produce a specific outcome.

The first step taken by the algorithm is to compare rows with each other and simplify them through a bottom-up process of paired comparison. These paired comparisons follow a simple rule which mimics a series of *ex post facto* experimental designs: combine rows that differ on only one pre-condition, but produce the same outcome. The process of paired comparisons culminates in the production of *prime implicants*. Often there are more prime implicants than are needed to cover all the original combinations for an outcome. In QCA, a *prime implicant chart* is used to map the correspondence between the prime implicants (just derived) and the

[7] Alternatively, the investigator may use probabilities to construct the truth table; see Ragin and Bradshaw 1991; Ragin, Mayer, and Drass 1984.

original combinations for the outcome of interest drawn from the truth table. Use of the prime implicant chart is the final phase of logical minimization and culminates in a minimal logical formula for the outcome of interest.

When the number of pre-conditions is greater than four, it is difficult to perform these operations by hand, and computer algorithms are necessary, especially for the simplification of large prime implicant charts. In the 1950s several algorithms were developed by electrical engineers (Quine 1952; McCluskey 1956; see McDermott 1985; Mendelson 1970; Roth 1975). They have been adapted for social science data by Drass and Ragin (1991).

One of the riches of QCA lies in the fact that it can be used in different ways for different purposes. (1) It can be used to examine a universe of cases with regard to a given output in an exhaustive manner. This description will be the most concise, including all possible alternate minimal expressions which allow for the most parsimonious interpretations. (2) It can be used to test hypotheses in the literature by showing whether they are consistent with the combinations described. If many contradictions occur, the hypothesis fails. (3) It can be used to deduce the shortest possible formulas, describing the simplest subset of actual, potential, or counterfactual cases not contradicting the respective outcome. In this way, more general expressions can be deduced by using "simplifying assumptions." QCA thus serves some of the most important aims of any science, namely systematic description and the falsification and construction of theories.

QCA shares some of the problems and limitations concerning the loss of information and a certain amount of arbitrariness in setting cutting-points for dichotomized variables. In contrast with common statistical procedures, however, it retains the full complexity of all cases and all variables considered. For example, in the study of the fate of democracy in inter-war Europe, on the basis of ten central reconstructed variables and making a number of explicit "simplifying assumptions," the following four alternate formulas for breakdown of democracy were derived. (Upper-case letters indicate the presence of a condition; lower-case letters indicate the absence of a condition; multiplication indicates intersecting conditions; addition indicates alternate combinations of pre-conditions.)

$$\begin{aligned} \text{Breakdown} = {} & 1)\ M + h \cdot U \cdot w \\ & 2)\ M + e \cdot h \cdot U \\ & 3)\ M + C \cdot U \cdot w \\ & 4)\ M + e \cdot C \cdot U \end{aligned}$$

This can be shortened to:

$$\text{Breakdown} = M + U \bullet \begin{bmatrix} h \bullet w \\ e \bullet h \\ C \bullet w \\ e \bullet C \end{bmatrix}$$

The much reduced factor "political role of the military" ($M$) thus emerged and accounted for nine out of the ten breakdown cases. Only Estonia, again, is a case apart and could not be accounted for in this manner. Instead, it is described by four alternate expressions, all including a *high level* of social unrest ($U$) involving armed militias and, in different combinations, factors like *low levels* of economic development ($e$) and world market integration ($w$), social homogeneity ($h$) and *strong* commercial interest representation ($C$).

As this example shows, non-unique formulas may emerge, but this must be considered a strength and not a weakness of the technique. It deals explicitly with "outliers" and covers them fully in their original complexity. It can arrive at several "conjunctural" constellations which explain the same outcome. When several different formulas cover a single case, the technique forces the researcher (as the MSDO technique) to take a closer look at the results and to *interpret* them in light of the intensive historical knowledge of the case. In macropolitical inquiry, this approach is preferred to a purely mechanical procedure which, in many statistical analyses, entirely obscures the fate of particular cases. Here, in fact, begins the real *qualitative* work, depending very much on the training and quality of researchers, their in-depth knowledge of cases, but also their sensitivity and understanding.

## C **Reduction and reconstruction of variables**

To reduce the enormous complexity of the original "area of investigation," especially the "many variables" side of the "many variables, few cases" dilemma, a number of procedures have been proposed (see also De Meur and Berg-Schlosser 1994*b*). This issue is especially important in qualitative comparative work because of the allowance for conjunctural and multiple conjunctural causation.

Several strategies focus on the relation between the pre-conditions and the outcome variable. One technique is to look for "constants" across the observed cases with regard to the outcome variable. If one variable turns out to be consistently linked to a particular outcome, it becomes a strong candidate for inclusion in any explanation as a necessary, but perhaps not a sufficient condition. A related approach focuses on correlations between the independent variables and the dependent variable. Like the search for constants just described, the examination of correlations may provide strong hints about broad patterns. More sophisticated techniques include discriminant analysis and logistic regression (applied to dichotomous outcome variables). A relatively large number of variables can be simultaneously treated in this way. These techniques typically are used to assess the net, additive contribution of each independent variable to some outcome. Variables that are uncorrelated with other variables but strongly correlated with the outcome are favored by these methods. Thus, these techniques are biased toward causal factors that act independently, not conjuncturally.

These statistical approaches are useful for identifying broad patterns, but they are of very limited utility in situations of causal complexity. Most complex causal combinations are hidden from analyses based on correlations. For example, if a variable must be present in some conjunctures for an outcome but absent in others, the correlation between this cause and the outcome may be 0. QCA can be used in situations of causal complexity. In contrast to the linear techniques just described, QCA focuses on configurations of variables. In QCA, an independent variable can be eliminated from an analysis if it does not uniquely distinguish any case that manifests an outcome (e.g., breakdown of democracy) from at least one case lacking the outcome (viewing cases as configurations of pre-conditions). However, QCA's current maximum of twelve independent variables limits its utility as a technique for eliminating independent variables.

Other techniques for reducing the number of variables focus on patterns among the variables. In the light of relevant theoretical and empirical knowledge, investigators may select a more limited number of major variables or reconstruct them in different ways. Such reductions may proceed statistically. For example, researchers often use factor analysis to combine correlated variables into single indexes. However, when the number of variables exceeds the number of cases, as is usually the case in macropolitical inquiry, it is not always clear how to interpret the results of such analyses.

Qualitative techniques for constructing more encompassing causal variables are more theoretically driven (see Ragin and Hein 1993 and the exchange between Nichols 1986 and Skocpol 1986). For example, in the inter-war study the existence of a rural proletariat and large-scale landlords has been combined into the more encompassing concept of "feudalism." Instead of receiving separate scores on the presence or absence of a rural proletariat and the presence or absence of large-scale landlords, cases would receive a single presence or absence coding indicating whether or not both factors are present. Similarly, a Boolean "addition" of factors, constituting alternative constellations in combination with another variable, may be possible. For example, the presence of ethnic, religious *or* regional social cleavages in the absence of any overarching (*verzuiling*) structures was combined to the variable of "social heterogeneity" in the study of democratic breakdowns and survivals. Use of ideal typic constructs (a common practice in macropolitical inquiry) can also be conceived of as reductions of complexity emphasizing some characteristics and de-emphasizing others (Weber 1949). In this way, the overall number of variables may be reduced considerably while still retaining much of the original information.

## IV Epistemological issues

The discussion of methods and techniques in macropolitical inquiry is still steeped in profound controversy (e.g., Lieberson 1994; Savolainen 1994; King *et al.* 1994). The positions scholars take in these debates are strongly shaped by their different paradigms (Kuhn 1962). It is no wonder that this debate is fraught with misperceptions and misunderstandings. Without pretending to have the last word on these matters, some basic issues involving qualitative methods in macropolitical inquiry should be clarified.

### A Variable versus case orientation

The specific location of qualitative comparative methods (few cases, many variables) has already been sketched (Section II). It is precisely from this location that many of the limitations of these methods and attempts to overcome them originate. To use evidence about micro-events involving many cases and few variables to "refute" these methods (e.g., Lieberson's 1991 use of car accidents) entirely misses the point and simply trivializes core issues. Similarly, to argue that all scientific evidence must be variable-based is equally pointless. All scientific observations are both variable-*and* case-based. Here, however, the accent lies on the small-N situation where each case is given direct consideration in the explanation of phenomena. For this reason, both the selection of cases and the selection of variables must be guided by theory.

## B Universal versus conjunctural explanations

A second issue concerns the universality or specificity of explanations. Most large-scale statistical inquiries attempt to isolate a small number of variables capable of "explaining" variation in some outcome. Of course, only part of the variation can be explained, and researchers must rely on probabilistic arguments. (The requisite assumptions for such arguments are rarely, if ever, fulfilled in macropolitical inquiry.) By contrast, a nuanced case-oriented explanation might do a much better job, even though the resulting explanation would fall short of universality. As a simple illustration, imagine a two-way scatterplot with three clusters of cases, the first in the lower left quadrant, the second in the middle of the top half of the plot, and the third straddling the upper-right and lower-right quadrants. A least-squares fit, as in linear regression analyses, would give a poor "universalizing" explanation of the observed data. By contrast, a close inspection of the distribution of the cases and their identities might lead to a much better fit and differentiate the explanation according to different kinds of cases. Of course, this problem is recognized by those who attempt to construct more "robust" statistics (Hampel *et al.* 1986). Our point is that it is a much more common concern of those engaged in macropolitical research.

## C Causality

Causality is central to any scientific inquiry, but the concept presents complex and often intractable issues, depending on the degree of empirical specificity the investigator seeks (see also King *et al.* 1994: 75 ff.). As shown in the examples discussed in this chapter, qualitative comparative methods identify broad conditions for the occurrence of particular outcomes. The patterns identified using these methods do not involve automatic mechanistic processes that generate the outcomes in question. Rather, these methods are heuristic devices—tools that organize and systematize the dialogue of ideas and evidence (Ragin 1987: 164–71). To speak of "deterministic causes" (e.g., as Lieberson 1991 does), therefore, grossly distorts the issue of causation. Furthermore, the methods we sketch often provide several alternatives which direct the attention of the researcher in possible directions. In the end, researchers must make final decisions and interpretations in light of their theoretical knowledge and their empirical understanding of the cases in question.

Only rarely in the social sciences will explanations involving conditions of occurrence cite distinct "invariant" relationships. (King *et al.* 1994 speak of the "systematic and non-systematic components" of any inference.) The "fundamental problem of causal inference" (Holland 1986) remains. In the words of King *et al.*: ". . . no matter how perfect the research design, no matter how much data we collect, no matter how diligent the research assistants, and no matter how much experimental control we have, we will never know a causal inference for certain" (1994: 79).

The broad "structural" conditions identified with qualitative comparative methods set the stage for examining specific "actor"-oriented factors, delineating their respective scope and weight. In this manner, a *sufficient* explanation may be achieved. The closer we come to the explanation of any particular event, the more idiosyncratic factors concerning psychological aspects of the personalities involved and the particular circumstances come into play leading, finally, to an "individualizing" historical explanation (Tilly 1984). But this is far removed from what macro-qualitative methods can or should achieve.

## D **Induction versus deduction**

A great deal of epistemological debate has centered on this issue. Naïve inductionism certainly has been refuted by now (cf., e.g., Cohen and Nagel 1934 and, of course, Popper 1968). It has been convincingly demonstrated that any theoretically meaningful operation requires deductive steps and assumptions, however minimal. Similarly, scientific results are always preliminary; only falsification and no permanent verification of any theoretical proposition is possible. But to mobilize only the "falsificatory killer instincts" (see von Beyme 1992: 27 ff.) of any discipline and not to acknowledge what has been and can be achieved by certain methods, in our view, would be going too far. O'Hear comments:

> Popper always tends to speak in terms of *explanations* of *universal* theories. But once again, we have to insist that proposing and testing universal theories is only part of the aim of science. There may be no true universal theories, owing to conditions differing markedly through time and space; this is a possibility we cannot overlook. But even if this were so, science could still fulfill many of its aims in giving us knowledge and true predictions about conditions in and around our spatio-temporal niche (1989: 43, original emphasis).

In this regard, macro-qualitative procedures are closer to what has been called "analytic induction" (see also Blalock 1984: 86 ff.). The systematic investigation of and constant reflection on a limited number of cases with regard to a particular problem may lead to "medium-range" theories in Merton's sense, bound both in time and space. These "islands of theory" (Wiarda 1985) may be expanded, following strict criteria of comparability, to other (but still limited) cases and events. Given that we are dealing with a malleable "plastic" matter, situated between the extremes of deterministic mechanical "clocks" on the one hand and completely diffuse intractable "clouds" on the other (Almond and Genco 1977), this may be the best we can hope for. Assessing the limits of such inferences returns us to our initial problem—namely, the theoretically informed selection of cases and variables. In our view, this process has to involve long-term feedback between our hypotheses and our understanding of complex patterns of causality, with only certain approximations possible over the longer run.

## V Conclusion

The preceding discussion surveys some of the literature and highlights key epistemological issues in qualitative comparative methods. We address the selection of cases and variables, specific techniques like "most different" or "most similar" systems design and Qualitative Comparative Analysis, and problems of explanation and theory-building using qualitative methods. We did not consider more longitudinal forms of analysis which also may be based on formal qualitative methods.

In closing, it is important to emphasize that the qualitative comparative approaches we advocate here must be *supplemented* with other forms of analysis. Studies at the micro-level as well as macro-quantitative methods all have their proper and legitimate place, if their specific problems and limitations are well understood and respected. As David Collier has observed:

> With good communication, country specialists and experts in qualitative small-N comparison can push the comparative quantifiers toward more carefully contextualized analysis. Likewise, the comparative quantifiers can push the country specialists and experts in qualitative comparison toward more systematic measuring and hypothesis testing (1993: 116; see also Teune 1990: 48).

## References

AAREBROT, F. A., and BAKKA, P. H. 1992. Die vergleichende methode in der politikwissenschaft. Pp. 51–69 in *Vergleichende Politikwissenschaft*, ed. D. Berg-Schlosser and F. Müller-Rommel. Opladen: Leske and Budrich.

ALMOND, G. A., and GENCO, S. J. 1977. Clocks, clouds and the study of politics. *World Politics*, 29: 489–522.

AMENTA, E., and POULSEN, J. D. 1994. Where to begin: a survey of five approaches to selecting independent variables for qualitative comparative analysis. *Sociological Methods and Research*, 23: 22–53.

BARTON, A. H. 1955. The concept of property space in social research. Pp. 45–50 in *The Language of Social Research*, ed. P. F. Lazarsfeld and M. Rosenberg. Glencoe, Ill.: Free Press.

BERG-SCHLOSSER, D., and DE MEUR, G. 1994. Conditions of democracy in inter-war europe: a boolean test of major hypotheses. *Comparative Politics*, 26: 253–79.

—— —— 1996. Conditions of authoritarianism, fascism and democracy in inter-war europe: systematic matching and contrasting of cases for 'small N' analysis. *Comparative Political Studies*, forthcoming.

BEYME, K. V. 1992. *Die politischen Theorien der Gegenwart. Eine Einführung.* 7th edn. Opladen: Westdeutscher Verlag.

BLALOCK, H. M., JR. 1984. *Basic Dilemmas in the Social Sciences.* Beverly Hills, Calif.: Sage.

BOLLEN, K. A.; ENTWISLE, B.; and ALDERSON, A. S. 1993. Macrocomparative research methods. *Annual Review of Sociology*, 19: 321–51.

COHEN, M., and NAGEL, E. 1934. *An Introduction to Logic and Scientific Method.* New York: Harcourt, Brace.

COLLIER, D. 1993. The comparative method. Pp. 105–19 in *Political Science: the State of the Discipline II*, ed. A. W. Finifter. Washington, D.C.: American Political Science Association.

DAHL, R. A. 1971. *Polyarchy: Participation and Opposition.* New Haven, Conn.: Yale University Press.

—— 1989. *Democracy and its Critics.* New Haven, Conn.: Yale University Press.

DE MEUR, G., and BERG-SCHLOSSER, D. 1994*a*. Comparing political systems: establishing similarities and dissimilarities. *European Journal for Political Research*, 26: 193–219.

—— —— 1994*b*. *Reduction of Complexity for a Small-N Analysis—A Systematic Example from Inter-War Europe.* Paper presented at the World Congress of the International Sociological Association. RC 18. Bielefeld.

—— and QUENTER, S. 1994. *MSDO/MDSO Designs.* Brussels, Marburg (software available from the authors).

DOGAN, M., and PELASSY, D. 1984. *How to Compare Nations.* Chatham, N.J.: Chatham House.

—— and KAZANCIGIL, A., eds. 1994. *Comparing Nations: Concepts, Strategies, Substance.* Oxford: Blackwell.

DRASS, K. A., and RAGIN, C. C. 1991. *Qualitative Comparative Analysis 3.0.* Evanston, Ill.: Center for Urban Affairs and Policy Research, Northwestern University.

DURKHEIM, E. 1988. *Les Règles de la Méthode Sociologique.* Paris: Flammarion; originally published 1894.

ELSTER, J. 1989. *Nuts and Bolts for the Social Sciences.* Cambridge: Cambridge University Press.

FEAGIN, J. R.; ORUM, A. M.; and SJOBERG, G., eds. 1991. *A Case for the Case Study.* Chapel Hill: University of North Carolina Press.

GRIFFIN, L. 1993. Narrative event-structure analysis and causal interpretation in historical sociology. *American Journal of Sociology*, 98: 1094–133.
—— and RAGIN, C. C., eds. 1994. *Sociological Methods and Research*, 23.
HAMPEL, F. R. *et al.* 1986. *Robust Statistics: The Approach Based on Influence Statistics.* New York: Wiley.
HEISE, D. 1989. Modelling event structures. *Journal of Mathematical Sociology*, 14: 139–69.
HOLLAND, P. 1986. Statistics and causal inference. *Journal of the American Statistical Association*, 81: 945–60.
KEMAN, H., ed. 1993. *Comparative Politics: New Directions in Theory and Method.* Amsterdam: VU University Press.
KING, G.; KEOHANE, R. O.; and VERBA, S. 1994. *Designing Social Inquiry: Scientific Inference in Qualitative Research.* Princeton, N.J.: Princeton University Press.
KUHN, T. S. 1962. *The Structure of Scientific Revolutions.* Chicago, Ill.: University of Chicago Press.
LANE, J. E., and ERSSON, S. O. 1994. *Comparative Politics.* Oxford: Polity Press.
LIEBERSON, S. 1991. Small N's and big conclusions: an examination of the reasoning based on a small number of cases. *Social Forces*, 70: 307–20.
—— 1994. More on the uneasy case for using Mill-type methods in small-N comparative studies. *Social Forces*, 72: 1225–37.
LIJPHART, A. 1971. Comparative politics and the comparative method. *American Political Science Review*, 65: 682–93.
—— 1975. The comparable cases strategy in comparative research. *Comparative Political Studies*, 8: 157–75.
LIPSET, S. M. 1963. *Political Man.* New York: Doubleday.
—— 1994. The social requisites of democracy revisited. *American Sociological Review*, 59: 1–22.
MAYER, L. C. 1989. *Redefining Comparative Politics.* Newbury Park, Calif.: Sage.
MCCLUSKEY, E. J. 1956. Minimization of Boolean functions. *Bell System Technical Journal*, 35: 1417–44.
MCDERMOTT, R. 1985. *Computer-Aided Logic Design.* Indianapolis, Ind.: Howard A. Sams.
MENDELSON, E. 1970. *Boolean Algebra and Switching Circuits.* New York: McGraw-Hill.
MILL, J. S. 1974/75. *A System of Logic.* Vols. vii and viii in *Collected Works of J. S. Mill.* London: Routledge and Kegan Paul; originally published 1843.
NICHOLS, E. 1986. Skocpol and revolution: comparative analysis versus historical conjuncture. *Comparative Social Research*, 9: 163–86.
O'HEAR, A. 1989. *Introduction to the Philosophy of Science.* Oxford: Clarendon Press.
PETERSEN, T. 1993. Recent advances in longitudinal methodology. *Annual Review of Sociology*, 19: 425–54.
POPPER, K. R. 1968. *The Logic of Scientific Discovery.* New York: Harper and Row.
PRZEWORSKI, A. 1987. Methods of cross-national research, 1970–1983. Pp. 31–49 in *Comparative Policy Research: Learning from Experience*, ed. M. Dierkes, H.N. Weiler and A.B. Antal. Aldershot: Gower.
—— and TEUNE, H. 1970. *The Logic of Comparative Social Inquiry.* New York: Wiley.
QUINE, W. V. 1952. The problem of simplifying truth functions. *American Mathematical Monthly*, 59: 521–31.
RAGIN, C. C. 1987. *The Comparative Method: Moving Beyond Qualitative and Quantitative Strategies.* Berkeley: University of California Press.
—— 1989. New directions in comparative research. Pp. 57–76 in *Cross-national Research in Sociology*, ed. M. L. Kohn. Newbury Park, Calif.: Sage.
—— ed. 1991. *Issues and Alternatives in Comparative Social Research.* Leiden: E.J. Brill.

Ragin, C. C. 1994*a*. *Constructing Social Research: The Unity and Diversity of Method.* Newbury Park, Calif.: Pine Forge Press.

—— 1994*b*. Introduction to qualitative comparative analysis. Pp. 299–319 in *The Comparative Political Economy of the Welfare States*, ed. T. Janoski and A. Hicks. New York: Cambridge University Press.

—— 1994*c*. A qualitative comparative analysis of pensions systems. Pp. 320–45 in *The Comparative Political Economy of the Welfare States*, ed. T. Janoski and A. Hicks. New York: Cambridge University Press.

—— and Bradshaw, Y. 1991. Statistical analysis of employment discrimination: a review and critique. *Research in Social Stratification and Mobility*, 10: 199–228.

—— and Hein, J. 1993. The comparative study of ethnicity: methodological and conceptual issues. Pp. 254–72 in *Race and Ethnicity in Research Methods*, ed. J. H. Stanfield and R. M. Dennis. Newbury Park, Calif.: Sage.

—— Mayer, S. E., and Drass, K. A. 1984. Assessing discrimination: a Boolean approach. *American Sociological Review*, 49: 221–34.

Roth, C. 1975. *Fundamentals of Logic Design.* St. Paul, Minn.: West.

Rustow, D. A., and Erickson, K. P., eds. 1991. *Comparative Political Dynamics.* New York: HarperCollins.

Savolainen, J. 1994. The rationality of drawing big conclusions based on small samples: in defense of Mill's methods. *Social Forces*, 72: 1217–24.

Sigelman, L., and Gadbois, G. H. 1983. Contemporary comparative politics: an inventory and assessment. *Comparative Political Studies*, 16: 275–305.

Skocpol, T., ed. 1984. *Vision and Method in Historical Sociology.* Cambridge: Cambridge University Press.

—— 1986. Analyzing configurations in history: a rejoinder to Nichols. *Comparative Social Research*, 9: 187–94.

Taylor, C. L., and Jodice, D. A. 1983. *World Handbook of Political and Social Indicators.* 2 vols. New Haven, Conn.: Yale University Press.

Teune, H. 1990. Comparing countries: lessons learned. Pp. 38–62 in *Comparative Methodology: Theory and Practice in International Social Research*, ed. E. Oyen. London: Sage.

Tilly, C. 1984. *Big Structures, Large Processes, Huge Comparisons.* New York: Russell Sage Foundation.

Wallerstein, I. 1980. *The Modern World System.* New York: Academic Press.

Weber, M. 1949. *The Methodology of the Social Sciences*, trans. and ed. E. A. Shils and H. A. Finch. Glencoe, Ill.: Free Press.

Wiarda, H. H., ed. 1985. *New Directions in Comparative Politics.* Boulder, Colo.: Westview Press.

Chapter 34

# Political Methodology: Research Design and Experimental Methods

Kathleen M. McGraw

THE first important extensive discussion of experimentation for a political science audience was provided by Richard Brody and Charles Brownstein in their chapter for the 1975 *Handbook of Political Science.* (McConahy wrote a related piece for the 1972 *Handbook of Political Psychology.*) In the final sentence of their chapter, Brody and Brownstein (1975: 254) concluded that "the more general application of the logic of experimentation is long overdue in political research." Recent evidence suggests that political scientists heeded this call during the ensuing twenty years. Experimentation is alive and well, increasingly prominent and well-regarded in the discipline (Kinder and Palfrey 1991; 1992*a*; 1992*b*; McGraw and Hoekstra 1994). Two decades ago, experimentation was largely limited to research in the substantive areas of political psychology and policy evaluation, with practitioners often forced to be defensive about their methodological choice. But the acceptance of experimentation is increasingly evident in the "mainstream" political science research community, suggesting its heightened legitimacy (see McGraw and Hoekstra 1994, for documentation of publication trends). By way of additional evidence: in the 1993 *State of the Discipline* volume (Finifter 1993), the chapters on formal rational choice theory (Lalman, Oppenheimer and Swistak 1993), political methodology (Bartels and Brady 1993), public opinion (Sniderman 1993), and political communication (Graber 1993) all contain positive and even enthusiastic discussions of the contribution experimentation has made to those substantive areas and the discipline more generally.

I have two goals in this chapter. In the first half, I describe the basic principles that characterize experimentation, point to recent controversies and concerns about these defining attributes, and provide suggestions as to how experimentation in political science might be improved. This introductory discussion is deliberately aimed at quite a general level, as the conceptual issues involving experimentation are not unique to particular disciplinary fields. In the second half of the chapter, I turn to the key question guiding this effort: what important substantive contributions have resulted from political science experiments in the past two decades?

## I Experimentation defined

### A Control and random assignment

Experiments can be characterized both structurally and functionally (Cook and Shadish 1994). Structurally, experiments are marked by a deliberate intervention in the natural, ongoing state of affairs. In one of the earliest and most influential works on experimental design, Fisher (1935) described experimentation as "experience carefully planned in advance." Although experiments can be remarkably diverse, what they have in common that distinguishes them from other types of scientific inquiry is that the researcher creates the conditions necessary for observation, rather than passively observing naturally occurring situations. Thus, experimentalists "share an interventionist spirit" (Kinder and Palfrey 1992*b*: 6) because they dare to intrude on nature. This spirit carries with it certain hazards:

> The scientist cannot lead us into nature's secret retreats unless he will risk having her slam the door in his face; experiment knocks on the door. The cardinal principle of experimentation is that we must accept the outcome whether or not it is to our liking . . . After all the planning and preparation, a time comes when the voice of the experimenter is stilled while nature speaks (Kaplan 1964: 145, 155).

Functionally, experiments test propositions about cause and effect relationships—whether the "intervention" or "treatment" is causally responsible for some observed outcome "in the restricted sense that the change would not have occurred without the intervention" (Cook and Shadish 1994: 546). Without doubt, the ability to test causal hypotheses is the foremost advantage of experimentation, in comparison to other methodologies, and indeed virtually all experimentation is motivated by the desire to test some causal proposition(s).

The structural and functional attributes of experiments are linked by the principles of control and random assignment. Control manifests itself in two key ways. First, experimentalists control the operationalization of the causal (independent) and outcome (dependent) variables under investigation; and second, they control the myriad of extraneous factors—both known and unknown, plausible and implausible—that may be linked to the phenomenon of interest. This latter type of control is achieved through random assignment of subjects to conditions. Random assignment is so important that it is generally viewed as the critical attribute for defining a particular study as an "experiment" (Carlsmith, Ellsworth and Gross 1976; Cook and Campbell 1979). Random assignment is the "great equalizer," in that a single, simple procedure eliminates (within statistical limits) factors that are extraneous to the specified cause-and-effect hypothesis as alternate causal explanations of any condition-based differences. In other words—and this is the strength of the randomized experiment—the expected mean difference between randomly created groups is zero, given the null (no-treatment-effect) hypothesis. Any subsequently observed differences are validly attributable to the manipulated independent variable. Consideration of the central role of random assignment highlights an additional feature of experimentation. It is inherently a comparative methodology, requiring at least two groups of participants: those receiving the treatment(s) of interest, and those who do not, known as the "control" group.

Using random assignment as the defining criterion of experiments is not always held in the political science literature.[1] Many studies that do *not* include random assignment would be considered experiments by many political scientists, and indeed are labeled as such in journal article titles and abstracts. These studies are usually of two types: quasi-experiments (Cook and Campbell 1979) where the researcher has no control over which "treatments" the participants received, and random assignment is therefore not possible; and "laboratory" studies, frequently incorporating materials and procedures common to those used in experiments, but where all participants are treated in exactly the same way (i.e., there are no manipulated conditions to which participants are randomly assigned). There are certainly valuable lessons to be learned from quasi-experiments and non-randomized laboratory studies, but consumers (readers) need to be aware that the causal inference advantages resulting from random assignment do not accrue to these kinds of designs.[2]

[1] Kinder and Palfrey (1992), for example, include both "true" experiments, as defined by random assignment, as well as other types of designs in their collection.

[2] On a related point, all too often published political science studies involving multiple conditions do not explicitly mention whether random assignment was or was not used to assign subjects to conditions (McGraw and Hoekstra 1994). Researchers conducting "experiment-like" studies would do their readers a service by clarifying this procedural point.

## B Internal and external validity

Donald Campbell is the social scientist most responsible for working through the special issues of validity which arise when testing causal hypotheses in experimental research, particularly in complex social settings (Campbell 1957; Campbell and Stanley 1966; Cook and Campbell 1979). Indeed, his resulting conceptual distinctions regarding different types of validity have entered the lexicon of most of the social sciences. Random assignment lends well-designed experiments a high degree of what Campbell coined *internal validity,* or the "approximate validity with which we infer a relationship between two variables is causal" (Cook and Campbell 1979: 37); an experiment "has" internal validity if the treatment yields a significant effect and there is no reason to believe that effect is due to an artifact or anything other than the manipulated independent variable. *External validity* refers to "the approximate validity with which we can infer that the presumed causal relationship can be generalized to and across alternate measures of the cause and effect as well as across different types of persons, settings and times" (Cook and Campbell 1979: 37). Although these concepts are familiar to most social scientists, they have also engendered controversy and debate among methodologists and philosophers of science (see Cook and Shadish (1994) for a brief review). This debate moved Campbell (1986) to devise new labels for the two kinds of validity (specifically, *local molar causal validity* and *proximal validity*), terms that have noticeably failed to catch on among social scientists. I relate these developments not in order to urge political scientists to adopt a new methodological language, but rather to point out the existence of a lively and provocative literature on the various types and meanings of validity that are evoked by experimentation.

## C Beyond internal validity: identification of mediators and moderators

Although internal validity is the crown jewel of experimentation, it is not the solution to all research problems, or even the pathway to complete causal understanding. Although well-designed randomized experiments can provide information about *whether* a causal relationship exists, they do not necessarily (and in practice rarely) provide information about the underlying processes accounting for the connection between the treatment and outcome. Understanding of underlying processes requires consideration of *mediators,* or "the generative mechanisms" that explain *how* or *why* manipulated treatments have an impact on some outcomes (Baron and Kenny 1986). A sterling example of experimentation in political science exhibiting sensitivity to uncovering mediating processes is Iyengar and Kinder's (1987) work demonstrating the impact of television news on public opinion. In addition to establishing a causal link between the news and public opinion, Iyengar and Kinder identified a psychological process mechanism accounting for that link—namely, "priming," defined as changes in the standards or criteria used to make political judgments.

Well-executed randomized experiments also rarely provide a full explanation or total prediction of some effect, and in fact most do not even provide satisfactory levels of explanation or prediction (i.e., explained variance).[3] There are a number of related concerns here. Although experimentation necessarily requires isolating and manipulating a single or limited number of causal variables, many political scientists would argue that this approach is misguided because other relevant variables—that is, those linked to the effect in the world outside the experimentalist's control—are deliberately omitted from consideration, resulting in underspecified or misspecified models. But experimentalists are rarely motivated by "large $R^2$"-type concerns (that is, prediction), focusing instead on the dichotomous question of "does the manipulated variable have a significant effect or not?" (that is, explanation). As Brody and Brownstein (1975: 222) noted, "Experimentation is unrivaled in its capacity to aid in distinguishing features that are merely present from those that are essential to the phenomena of interest." Paradoxically, then, it is through the experimental isolation of potentially critical variables that the causal determinants of complex social and political phenomena are eventually understood.

Experiments can be made more fully explanatory than they typically are, however. This can be achieved by focusing on mediating processes, as noted above, or by moving away from simple "main effect" models to those that ascertain the "conditions under which" a given causal relationship holds (Greenwald *et al.* 1986; McGuire 1983).[4] This approach identifies

[3] Methodologists and philosophers of science have discussed the differences between explanation and prediction. Space does not permit a discussion of this literature here; see Kaplan (1964) for an introduction.

[4] Underlying this recommendation is an assumption about the state of the "real world." Simple linear models—regardless of estimation technique (that is, linear regression or analysis of variance)—assume a "real world" characterized by main effects. In contrast, contemporary philosophers (particularly Mackie (1974) ) view the real world as much more complex, with cause and effect relationships highly contingent upon a range of variables embedded in the social context. From this perspective, the challenge social scientists face is to devise empirical tests that are appropriately sensitive to this conditional complexity. I acknowledge here my intellectual debt to Tom Cook (whose influence is evident throughout this essay), who repeatedly emphasized the "pretzel-like" nature of social phenomena in my graduate training in methodology and philosophy of science at Northwestern University.

*moderators*, or constructs that explain *when* manipulated treatments have an impact (Baron and Kenny 1986); evidence for moderation occurs when a statistical interaction between the focal manipulated variable and some other factor (e.g., participant or situation characteristic) is obtained.

Delineating "the conditions under which" causal relationships exist typically does not occur in a single experiment, but rather is the result of a cumulative, even collective undertaking, requiring multiple experiments that systematically probe the boundary conditions under which certain relationships occur. A good, admittedly self-centered, example from political science concerns the evolution of the "on-line" model of candidate evaluation (Lodge, McGraw and Stroh 1989). In our initial paper, we demonstrated that when voters have the goal of forming an impression of a candidate (the experimental manipulation), they construct a "running tally," with each new piece of information used to update the judgment. Consequently, when asked to express an opinion (e.g., in the voting booth or in a survey context), they simply retrieve this tally and do not need to retrieve the specific bits of information from memory upon which the tally is based. The original exposition of the model was quite general and completely silent about contingencies. Subsequent experiments have moved toward specifying the conditions under which on-line processing of candidate information occurs. For example, McGraw, Lodge and Stroh (1990) demonstrated that political sophisticates are much more likely to process candidate information on-line than are non-sophisticates (a participant characteristic contingency), whereas Rahn, Aldrich, and Borgida (1994) showed that the structure of the information environment interacts with sophistication in influencing information processing strategies (a situation characteristic contingency). The discovery of these contingencies has limited the generalizability of the on-line model of candidate evaluation to be sure, but at the same time has strengthened the explanatory power of the model by clarifying the circumstances under which the key principles hold.

## D Pursuing external validity: priorities, replication, and synthesis

In practice, external validity—the ability to generalize—is "the Achilles heal" of political science experimentation (Iyengar 1991). All experiments are contextually specific, characterized by samples, procedures, settings, and time that are unique to a particular research undertaking. But experimentalists also aspire to generalizable results, particularly in political science, where the relevance of research findings are of greatest interest when they bear on workings of actual political processes (Kinder and Palfrey 1992*b*). It should be noted that—implicitly or explicitly—experimentalists tend to give a higher priority to internal validity than to external validity concerns. As Campbell and Stanley (1966) so influentially argued, concern about the generalizability of a cause-and-effect relationship can only be meaningful if we are confident about the validity of the causal connection. This preference for internal validity is not without its critics (for example, Cronbach 1982), and the more general lesson to be learned from this controversy is that it is important for researchers to be explicit about their goals when planning any experiment. "Pure" abstract theory testing typically places a high priority on internal validity concerns with relatively little attention to external validity, whereas more applied research requires closer attention to external validity concerns (Cook and Campbell 1979).

Priorities aside, it is still the case that most political science experiments could strive for higher degrees of external validity, as available resources permit. The majority of published political science experiments rely on the always-convenient "college sophomore" (McGraw and Hoekstra 1994), which, as Sears (1986) cogently argued, is a rather unusual member of the real population of interest (the "public"). Moreover, experimental settings and materials are often scandalously artificial in comparison to their manifestation in the real world (for example, information about candidates and campaigns). These generalizability concerns are real, and experimentalists need to be cognizant of them. But no matter how diligent and resourceful a researcher is in remedying these common problems (see Iyengar and Kinder (1987) and Iyengar (1991) for laudable examples), it is still the case that the ability to generalize beyond the specific research context is limited. Generalizability ultimately requires a long-term perspective involving "carefully chosen and selective replication" (Kinder and Palfrey 1992*b*: 28; Cook and Campbell 1979). Because each new experiment involves a different population, setting and time, as well as the likelihood of new investigators and variations in method and measurement, generalizability is attained when results converge despite the heterogeneity of these conceptually irrelevant research dimensions. As Cook and Campbell argue, "a strong case can be made that external validity is enhanced more by many heterogeneous small experiments than one or more large experiments" (1979: 80).

Replication is a necessary starting-point for generalizability, but probably insufficient for confident causal generalization. Once a series of experiments on a similar topic have been completed, it is then possible to integrate the conclusions reached in those studies in systematic fashion through the statistical technique known as meta-analysis (Cook 1992;

Rosenthal 1988). Meta-analysis permits a determination of whether causal relationships are robust across a series of studies as well as identification of the conditions under which the effect is strengthened or attenuated. Political scientists as a rule do not have a history of valuing systematic research reviews;[5] and meta-analysis in particular is, as far as I know, quite rare (see Krosnick and Berent (1993) for an exception). As the empirical science of politics continues to develop, and as experiments in particular become more prominent, such synthetic research reviews will hopefully become more common.

## II Lessons learned: contributions to political science from experimentation

If experimentation has indeed gained increased legitimacy and visibility since Brody and Brownstein's 1975 *Handbook* chapter, as I have argued here and elsewhere (McGraw and Hoekstra 1994; see also Kinder and Palfrey 1992*b*), then it should be possible to point to noteworthy studies or programs of research based on experimentation that have provided important substantive lessons to the discipline—principles about the political world that were not, and perhaps could not have been, known prior to the experimental investigation. My goal in the second part of this chapter is to identify specific experimental studies that have made such a contribution. Readers should be forewarned that this kind of exercise is inherently subjective, idiosyncratic, and in this case probably biased by my training as an experimental social psychologist and research expertise in the area of political psychology. I have also slanted this review toward book-length contributions. The organizational scheme owes much to that used by Kinder and Palfrey (1992*b*), but the boundaries are also fuzzy. Other commentators would undoubtedly include studies that I have omitted and omit studies that I have included. Despite these caveats, I am confident that the discussion that follows provides a representative sample of the variety, richness and importance of political experiments conducted in the past two decades.

[5] It is rare, for example, for political science journals to publish articles that are limited to reviews of existing research. Psychologists, on the other hand, have a journal—*Psychological Bulletin*—largely devoted to just those kinds of articles.

## A Public opinion

Kinder writes that although the "literature devoted to American public opinion alone is gigantic, . . . laced with claims and counterclaims, [this] should not distract us from appreciating that our understanding of public opinion is, in fact, deeper and more sophisticated than was true when Key was writing and this is due, in some measure, to experimentation" (1992: 44). There has been an explosion of experimental research that can be subsumed under the public opinion umbrella. Let us begin with what is probably on everybody's "Top 10" list of public opinion experiments, and deservedly so, namely the Sullivan, Piereson and Marcus (1978) "question wording" experiment. Designation as a "question wording" experiment actually does a disservice to the study, as it resolved an important substantive controversy regarding Converse's (1964) "ideological innocence" thesis. Although Converse claimed that Americans were innocent, even ignorant, of ideological concepts and lacking true opinions on most policy questions, critics argued, with compeling "naturalistic" data, that public opinion in America exhibited greater cohesion beginning in 1964—a finding attributed to the hotly contested ideological election campaigns of 1964 (Nie, Verba and Petrocik 1979). But Sullivan, Piereson and Marcus noted that 1964 also saw a change in the National Election Study questions used to probe the ideological competence of the electorate, and so they undertook an experiment wherein half of the respondents were asked the pre-1964 questions and the other half the format adopted in 1964. Their results verified that the enhanced attitudinal constraint observed in 1964 was not due to any real change in the electorate but rather to the changes in the survey questions, a most elegant demonstration of a cause-and-effect relationship that would have eluded detection in the absence of experimentation.

A more recent contribution to the public opinion literature based largely on experimental studies is Sniderman, Brody and Tetlock's *Reasoning and Choice: Explorations in Political Psychology* (1991). This book illustrates in a variety of domains how ordinary people, even those with limited knowledge and interest, reason in a sensible and reliable manner about politics through the use of judgmental heuristics. The theoretical and substantive insights contained in the volume are important, but perhaps the more lasting contribution of this work lies in its successful development and employment of a new research technique, integrating randomly assigned experimental variations into the traditional public opinion interview through computer-assisted telephone interviewing (CATI). The CATI technique provides the opportunity to combine the methodological advances of the true experiment (that is, random assignment to multiple conditions of theoretical importance) with the strength of the omnibus public opinion survey (representative samples). As such, it is has potential to blur the traditional distinctions made between experimental and survey research.

Sustaining tolerance is a critical challenge to a pluralistic democratic society, and empirical investigations of intolerance have a long and distinguished history in political science. The most recent contribution to this literature is *With Malice toward Some: How People Make Civil Liberties Judgments* (Marcus *et al.* 1995). In the context of a richly nuanced model of tolerance drawn from the existing literature, Marcus *et al.* demonstrate experimentally how information about social groups (such as threat), information about democratic principles, and the citizen's "state of mind" (that is, attention to thoughts versus feelings) combine to influence tolerance judgments. The research enterprise has a number of the desirable properties of experimentation discussed earlier: replication of the experimental design and results across three independent samples, identification of individual characteristics that moderate the influence of the manipulated information factors, and consideration of the psychological processes that mediate the links between information and tolerance opinions. As such, the book represents an exemplary case of experimentation on a deeply important political topic.

The final example of experimental research informing our understanding of public opinion is probably the best-known and influential program of experimentation in political science, namely Iyengar and Kinder's (1987) *News that Matters* and Iyengar's *Is Anyone Responsible* (1991). The Iyengar–Kinder television news research deserves to be singled out, as it has successfully maximized the strengths and advantages of experimentation (identification and testing of politically important causal relationships between the media and public opinion, as well as identification of psychological mediating processes) while at the same time minimizing the common threats to external validity (through the use of realistic materials, settings and non-student samples, as well as replication over multiple political issues). We have learned from the Iyengar–Kinder work that "news matters" in three consequential ways: first, it sets the public agenda, determining what issues the public deems to be important; second, it "primes" voters to rely on different criteria when evaluating political candidates; and third, it shapes our views about political responsibility through the manner in which social and political problems are portrayed.

## B Decision-making and information processing

Choice is critical to politics, and research on individual political decision and judgment utilizes a variety of methodologies, including experimentation. Recent advances incorporating experimentation have been influenced by two related theoretical perspectives in psychology. The first draws on the "bounded rationality" perspective associated with Herbert Simon (1978; 1985), and the larger behavioral decision theory literature which provides descriptive models of judgment and choice that depart from the normative prescriptions of rational choice theory in many substantial respects (Abelson and Levi 1985; Dawes 1988; Kahneman, Slovic and Tversky 1982). An informative set of mini-experiments illustrating the applicability of principles from behavioral decision theory to political choice is provided by Quattrone and Tversky (1988).

Whereas the behavioral decision theoretic perspective tends to focus on decision rules and heuristics, the information-processing approach is broader in scope, concerned with specification of the cognitive processes by which people acquire, store, retrieve, transform and use information. *Political Judgment: Structure and Process* (Lodge and McGraw 1995) is a collection of ten articles, all informed by experimental data, that illustrate the many contributions the information processing approach has made to understanding political choice. Our reading of the burgeoning literature on political information processing and choice is that the past decade has been marked by both remarkable achievements as well as empirical contradictions and holes, the latter a natural consequence of a young, emerging field (see McGraw and Lodge (1996) for a discussion).

## C Collective action

Collective action problems, or social dilemmas, occur when personal and group interests conflict: there is a disincentive for individual potential beneficiaries of some public good to contribute resources, but if all individuals respond in the self-interested way the collective suffers. In other words, these dilemmas are defined by two simple properties: (1) each individual receives a higher payoff for social defection or "free-riding" but (2) all individuals are better off if all co-operate (Dawes 1980). Understanding the actual prevalence of the free-rider problem and the motivations underlying co-operation and defection, as well identifying methods for eliciting and sustaining co-operation are problems of tremendous theoretical and practical importance in the social sciences. Social dilemma experiments have a long history in psychology (see Dawes (1980) for a review) and more recently have made an appearance in the political literature, often the result of interdisciplinary collaborations (for example, between political scientists and economists (Ferejohn *et al.*, 1982) or between political scientists and psychologists (Dawes *et al.*, 1986)). The Ferejohn *et al.* (1982) experiment compares the performance of different institutional rules that vary the structure of individual incentives, whereas the Dawes *et al.* (1986) compares incentive modifications that serve to minimize psychological motivations for defection, such as "fear" and "greed." One of the more robust findings in this literature concerns the salutary consequences of communication within the group on co-operation rate, with recent attention paid to probing the meaning and limits of this positive communication effect (see Ostrom, Walker, and Gardner (1992) and citations therein).

In a separate line of inquiry, Frohlich and Oppenheimer (1992) utilize experimentation to explore a key question in social ethical theory: how do groups reach decisions regarding questions concerning distributive justice? In particular, their experimental designs allow them to evaluate the claims made by two leading philosophers, Rawls's (1971) principle that fairness results when the utility of the worst-off member of a society is maximized and Harsanyi's (1955) claim that fairness results when the group's utility is maximized. Frohlich and Oppenheimer's experiments provide descriptive evidence for these normative claims, their evidence suggesting that what people prefer is somewhere in the middle of the Rawls and Harsanyi positions, reflecting a desire for some kind of social "safety net" for the least-able members of a group.

## D Formal theory and public choice

Palfrey (1992: 389) notes that, given the fact that the earliest formal theoretical work in political science was concerned with agendas and other procedures on committee decisions, it is not surprising that most of the early experimental work designed to test the predictions of formal theories are concerned with committees. Charles Plott, a leader in this tradition, provided a useful discussion and defense of experimental methods as applied to the study of committee behavior (Plott 1979); and Plott's research (for example, Plott and Levine 1978) "represents a systematic attempt to identify which classes of models predict well in which classes of institutions" (Palfrey 1992: 390). Among the more influential studies in this area for political scientists is Fiorina and Plott's (1978) investigation of committee decision-making under majority rule, testing the predictions of sixteen different theoretical models. Palfrey notes that startling results from one of Fiorina and Plott's experiments, where a core (majority rule equilibrium) failed to exist, has spawned a voluminous experimental literature that is central to the "new institutionalism" in public choice theory (Palfrey 1992: 392).

A second important line of inquiry falls under the category of "laboratory elections," evaluating the implications of formal models, in particular the spatial model (Downs 1957; Enelow and Hinich 1984), developed to understand the theoretical underpinnings of competitive democratic elections. For example, McKelvey and Ordeshook (1985) considered two-candidate elections under incomplete information. Their experimental data provided some support for a model predicting an equilibrium under these conditions comparable to what would occur under full information conditions. Thus, McKelvey and Ordeshook argue, in this article and elsewhere, that a poorly informed electorate poses little threat to democratic functioning.

## E **Public policy**

When Brody and Brownstein wrote their earlier chapter on experimentation, they noted that "the almost complete failure of nonexperimental techniques to provide adequate information about the comparative impact and relative effectiveness of alternative policies and programs has encouraged experimentation" in the area of policy analysis (Brody and Brownstein 1975: 254). Campbell's call for an "experimenting society" (1969) advocated randomized or quasi-experimental tests of the effectiveness of social programs before adopting them on a large scale, and a number of such social experiments in the areas of health, education, social welfare, criminal justice, and medicine were conducted in the 1960s and 1970s (see Boruch, McSweeney and Soderstrom (1978) for a listing). Perhaps the best known, and modestly successful, of these was the New Jersey Negative Income Tax (or income guarantee) experiment (Kershaw and Fair 1976; Pechman and Timpane 1975). The value of social experimentation at this level, particularly in comparison to other program evaluation methodologies, has been and remains controversial (Hausman and Wise 1985), because random assignment to policy conditions entails considerable ethical, political, and legal problems. Interestingly, after a period of limited use in the 1980s, the pendulum has apparently swung back in favor of randomized field experiments for policy analysis. A recent review concludes "the last decade has witnessed a powerful shift in scientific opinion toward randomized field experiments and away from quasi-experiments or non-experiments" and that as a consequence "field experiments are much more commonplace than they were twenty years ago," particularly in the fields of labor economics and community health promotion (Cook and Shadish 1994: 557, 575). Consequently, we might expect to see increased dissemination of the results of these policy evaluation studies through scholarly outlets.

Experiments have contributed to public policy and social change in another way, although these contributions are rarely, if ever, acknowledged in discussions of experimentation. Specifically, the results of experimental research are often included in the *amicus curiae* (or "friend of the court") briefs submitted to the courts by social science organizations, in particular those prepared by the American Psychological Association (APA).[6] Many of these briefs concern issues of concern to political scientists, particularly those interested in the law and socio-legal change, as well as those more generally interested in the utilization of social science knowledge. Consequently, political scientists should be aware of their existence, and in particular the fact that the arguments put forth in many *amicus* briefs are supported by experimental findings. Recent briefs submitted by APA to the United States Supreme Court that include experimental findings include those dealing with the death penalty (*Lockhart* v. *McCree,* 1986; see Bersoff 1987), jury size (*Ballew* v. *Georgia,* 1978; see Tanke and Tanke 1979), testimony of child witnesses (*Maryland* v. *Craig,* 1990; see Goodman *et al.* 1991), and sex discrimination (*Price Waterhouse* v. *Hopkins,* 1989; see Fiske *et al.* 1991). The Court's responses to these briefs and the empirical evidence contained within them have been mixed, but it is also clear that the presence of well-presented scientific data has played an important role in how the Court fashioned its decisions in these cases (Grisso and Saks 1991).

## III Conclusion

As I have tried to convey in this limited overview, experimentation in political science has come of age. The landscape is diverse, and the work innovative and sophisticated. Most critically, experimentation has generated, or contributed to, bodies of knowledge of vital and central importance to the concerns of the discipline. Of course, political science experiments could

[6] It is my understanding that the American Political Science Association does not have a comparable history or policy regarding the submission of *amicus* briefs.

be done better (as is true of experimentation in any discipline) and I have tried to suggest principles that might facilitate this goal. Nevertheless, the foundation is strong. I believe that the experimental work of the past twenty years, reviewed above, will stand the test of time, and that scholars twenty years hence will continue to recognize their contributions. We can look forward with anticipation to what the next generation of experimentation in political science will bring.

## References

This chapter was not intended to provide a "how to" manual in experimentation. I concur with Kinder and Palfrey's (1992b) advice that those interested in the philosophy and logic of experimentation, as well as the pragmatic "nuts-and-bolts" issues to be faced in designing an experiment would do well to consult Campbell and Stanley's *Experimental and Quasi-Experimental Designs for Research* (1966), Cook and Campbell's *Quasi-Experimentation* (1979), and Carlsmith, Ellsworth and Aronson's *Methods of Research in Social Psychology* (1976). Although written by psychologists, these volumes provide an accessible and thorough overview of the experimental method for all social scientists. See the underappreciated Brody and Brownstein 1975 *Handbook* chapter for an explicit discussion of the logic of experimentation for political scientists. For concrete examples of how political scientists conduct experiments, I recommend Kinder and Palfrey's *Experimental Foundations of Political Science* (1992*a*), which brings together fifteen previously-published, and influential, articles utilizing experimental, quasi-experimental, and laboratory methods. Palfrey's (1991) *Laboratory Research in Political Economy* consists of ten original articles describing experimental investigations of formal theory models. Plott's (1979) chapter is also an important source for experimentation in public choice/formal theory. Finally, for a listing and review of published randomized experiments in top American political science journals from 1950–1992, see McGraw and Hoekstra (1994); an updated (post-1992) listing of published experiments is available from the author on request.

Abelson, R. P., and Levi, A. 1985. Decision making and decision theory. Pp. 231–310 in *Handbook of Social Psychology*, ed. G. Lindzey and E. Aronson. New York: Random House.

Baron, R., and Kenny, D. A. 1986. The moderator-mediator distinction in social psychological research: conceptual, strategic, and statistical considerations. *Journal of Personality and Social Psychology*, 51: 1173–82.

Bartels, L. M., and Brady, H. E. 1993. The state of quantitative methodology. In Finifter 1993: 121–59.

Bersoff, D. N. 1987. Social science and the Supreme Court: Lockhart as a case in point. *American Psychologist*, 42: 52–8.

Boruch, R. F.; McSweeney, A. J.; and Soderstrom, E. J. 1978. Randomized field experiments for program planning development and evaluation. *Evaluation Quarterly*, 2: 655–95.

Brody, R. A., and Brownstein, C. N. 1975. Experimentation and simulation. Vol. 7, pp. 211–63 in *Handbook of Political Science*, ed. F. I. Greenstein and N. W. Polsby. Reading, Mass.: Addison-Wesley.

Campbell, D. T. 1957. Factors relevant to the validity of experiments in social settings. *Psychological Bulletin*, 54: 297–312.

—— 1969. Reforms as experiments. *American Psychologist*, 24: 409–29.

—— 1986. Relabeling internal and external validity for applied social scientists. Pp. 67–77 in *Advances in Quasi-Experimental Design Analysis: New Directions for Program Evaluation*, ed. W. Trochim. San Francisco: Jossey-Bass.

—— and Stanley, J. C. 1966. *Experimental and Quasi-experimental Designs for Research*. Chicago: Rand McNally.

Carlsmith, J. M.; Ellsworth, P. C.; and Aronson, E. 1976. *Methods of Research in Social Psychology*. Reading, Mass.: Addison-Wesley.

Converse, P. E. 1964. The nature of belief systems in mass publics. Pp. 206–62 in *Ideology and Discontent*, ed. D. E. Apter. New York: Free Press.

Cook, T. D. 1992. *Meta-Analysis for Explanation: A Case Book*. New York: Russell Sage Foundation.

—— and Campbell, D. T. 1979. *Quasi-Experimentation*. Chicago: Rand McNally.

—— and Shadish, W. R. 1994. Social experiments: some developments over the past fifteen years. *Annual Review of Psychology*, 45: 545–80.

Cronbach, L. J. 1982. *Designing Evaluations of Educational and Social Programs*. San Francisco: Jossey-Bass.

Dawes, R. M. 1980. Social dilemmas. *Annual Review of Psychology*, 31: 169–93.

—— 1988. *Rational Choice in an Uncertain World*. New York: Harcourt Brace.

—— Orbell, J. M.; Simmons, R. T.; and van de Kragt, A. J. C. 1986. Organizing groups for collective action. *American Political Science Review*, 80: 1171–85.

Downs, A. 1957. *An Economic Theory of Democracy*. New York: Harper.

Enelow, J., and Hinich, M. 1984. *The Spatial Theory of Voting*. Cambridge: Cambridge University Press.

Ferejohn, J. A.; Forsythe, R.; Noll, R. G.; and Palfrey, T. R. 1982. An experimental examination of auction mechanisms for discrete public goods. *Research in Experimental Economics*, ii ed. V. Smith. Greenwich, Conn.: JAI Press.

Finifter, A. W. , ed. 1993. *Political Science: The State of the Discipline II*. Washington D.C. : American Political Science Association.

Fiorina, M., and Plott, C. 1978. Committee decisions under majority rule: an experimental study. *American Political Science Review*, 72: 575–98.

Fisher, R. 1935. *Design of Experiments*. New York: Hafner Publishing.

Fiske, S. T.; Bersoff, D. N.; Borgida, E.; Deaux, K.; and Heilman, M. E. 1991. Social science research on trial: use of sex stereotyping research in *Price Waterhouse v. Hopkins*. *American Psychologist*, 46: 1049–60.

Frohlich, N., and Oppenheimer, J. A. 1992. *Choosing Justice: An Experimental Approach to Ethical Theory*. Berkeley: University of California Press.

Goodman, G. S.; Levine, M.; Melton, G. B.; and Ogden, D. W. 1991. Child witnesses and the confrontation clause: the American Psychological Association brief in *Maryland* v. *Craig*. *Law and Human Behavior*, 15: 13–29.

Graber, D. A. 1993. Political communication: scope, progress, promise. In Finifter 1993: 305–32.

Greenwald, A. G.; Pratkanis, A.; Leippe, M.; and Baumgardner, M. 1986. Under what conditions does theory obstruct research progress? *Psychological Review*, 93: 216–29.

Grisso, T., and Saks, M. J. 1991. Psychology's influence on constitutional interpretation: a comment on how to succeed. *Law and Human Behavior*, 15: 205–211.

Harsanyi, J. C. 1955. Cardinal welfare, individualistic ethics, and interpersonal comparisons of utility. *Journal of Political Economy*, 63: 302–21.

Hausman, J. A., and Wise, D. A., eds. 1985. *Social Experimentation.* Chicago: University of Chicago Press.

Iyengar, S. 1991. *Is Anyone Responsible?* Chicago: University of Chicago Press.

—— and Kinder, D. R. 1987. *News that Matters.* Chicago: University of Chicago Press.

Kahneman, D.; Slovic, P.; and Tversky, A. eds. 1982. *Judgment under Uncertainty: Heuristics and Biases.* Cambridge: Cambridge University Press.

Kaplan, A. 1964. *The Conduct of Inquiry.* San Francisco: Chandler.

Kershaw, D., and Fair, J. 1976. *The New Jersey Income-Maintenance Experiment: Vol. 1: Operations, Surveys, and Administration.* New York: Academic Press.

Kinder, D. R. 1992. Coming to grips with the Holy Ghost. In Kinder and Palfrey 1992*a*: 43–52.

—— and Palfrey, T. R. 1991. An experimental political science? Yes, an experimental political science. *Political Methodologist,* 4: 2–8.

—— —— eds. 1992*a*. *Experimental Foundations of Political Science.* Ann Arbor: University of Michigan Press.

—— —— 1992*b*. On behalf of an experimental political science. In Kinder and Palfrey 1992*a*: 1–42.

Krosnick, J. A., and Berent, M. K. 1993. Comparisons of party identification and policy preferences: the impact of survey question format. *American Journal of Political Science,* 37: 941–64.

Lalman, D.; Oppenheimer, J.; and Swistak, P. 1993. Formal rational choice theory: a cumulative science of politics. In Finifter 1993: 77–104.

Lodge, M., and McGraw, K. M., eds. 1995. *Political Judgment: Structure and Process.* Ann Arbor: University of Michigan Press.

—— —— and Stroh, P. 1989. An impression-driven model of candidate evaluation. *American Political Science Review,* 83: 399–420.

Mackie, J. L. 1974. *The Cement of the Universe.* Oxford: Oxford University Press.

McConahay, J. B. 1973. Experimental research. Pp. 356–82 in *Handbook of Political Psychology,* ed. J. K. Knutson. San Francisco: Jossey-Bass.

McGraw, K. M., and Hoekstra, V. 1994. Experimentation in political science: historical trends and future directions. Vol. iv, pp. 3–30 in *Research in Micropolitics,* ed. M. Delli Carpini, L. Huddy and R. Y. Shapiro. Greenwood, Conn.: JAI Press.

—— and Lodge, M. 1996. Review essay: political information processing. *Political Communication,* 13: 131–42.

—— —— and Stroh, P. 1990. On-line processing in candidate evaluation: the effects of issue order, issue importance, and sophistication. *Political Behavior,* 12: 41–58.

McGuire, W. J. 1983. A contextualist theory of knowledge: its implications for innovation and reform in psychological research. Vol. xvi, pp. 1–47 in *Advances in Experimental Social Psychology,* ed. L. Berkowitz. New York: Academic Press.

McKelvey, R. D., and Ordeshook, P. C. 1985. Sequential elections with limited information. *American Journal of Political Science,* 29: 480–512.

Marcus, G. E.; Sullivan, J. L.; Theiss-Morse, E.; and Wood, S. L. 1995. *With Malice Toward Some: How People Make Civil Liberties Judgments.* New York: Cambridge University Press.

Nie, N. H.; Verba, S.; and Petrocik, J. R. 1979. *The Changing American Voter.* Cambridge, Mass.: Harvard University Press.

Ostrom, E.; Walker, J.; and Gardner, R. 1992. Covenants with and without a sword: self-governance is possible. *American Political Science Review,* 86: 404–18.

Palfrey, T. R. 1991. *Laboratory Research in Political Economy.* Ann Arbor: University of Michigan Press.

Palfrey, T. R. 1992. Agendas and decisions in government. In Kinder and Palfrey 1992*a*: 389–98.

Pechman, J. A., and Timpane, P. M., eds. 1975. *Work Incentives and Income Guarantees: The New Jersey Negative Income Tax Experiment.* Washington, D.C.: Brookings Institution.

Plott, C. 1979. The application of laboratory experimental methods to public choice. Pp. 14–52 in *The Application of Laboratory Experimental Methods to Public Choice*, ed. C. S. Russell. Baltimore, Md.: Johns Hopkins Press.

—— and Levine, M. 1978. A model of agenda influence on committee decisions. *American Economic Review*, 68: 146–60.

Quattrone, G. A., and Tversky, A. 1988. Contrasting rational and psychological analyses of political choice. *American Political Science Review*, 82: 719–36.

Rahn, W. M.; Aldrich, J. H.; and Borgida, E. 1994. Individual and contextual variations in political candidate appraisal. *American Political Science Review*, 88: 193–9.

Rawls, J. 1971. *A Theory of Justice.* Cambridge, Mass.: Harvard University Press.

Rosenthal, R. 1988. *Meta-Analytic Procedures for Social Research.* Beverly Hills, Calif.: Sage.

Sears, D. O. 1986. College sophomores in the laboratory: influence of a narrow data base on social psychology's view of human nature. *Journal of Personality and Social Psychology*, 51: 515–30.

Simon, H. A. 1978. Rationality as a process and product of thought. *American Economic Review (Papers & Proceedings)*, 68/5: 1–16.

—— 1985. Human nature in politics: the dialogue of psychology with political science. *American Political Science Review* 79: 293–304.

Sniderman, P. M. 1993. The new look in public opinion research. In Finifter 1993: 219–46.

—— Brody, R. R.; and Tetlock, P. E. 1991. *Reasoning and Choice: Explorations in Political Psychology.* New York: Cambridge University Press.

Sullivan, J. L.; Piereson, J. E.; and Marcus, G. E. 1978. Ideological constraint in the mass public: a methodological critique and some new findings. *American Journal of Political Science*, 22: 233–49.

Tanke, E. D., and Tanke, T. J. 1979. Getting off a slippery slope: social science in the judicial process. *American Psychologist*, 34: 1130–8.

Chapter 35

# Political Methodology, Old and New

Hayward R. Alker

He who by his nature and not simply by ill-luck has no city, no state, is human only like [Homer's] war-mad man . . . [who] is a non-co-operator like an isolated piece in a game of draughts . . . Nature, . . . for the purpose of making man a political animal . . . has endowed him alone among the animals with the powers of reasoned speech . . . For the real difference between man and other animals is that humans alone have perception of good and evil, right and wrong, just and unjust. And it is the sharing of a common view in these matters that makes a household or a city.
ARISTOTLE, *Politics* (quoted in Kratochwil 1989: 265)

[S]ince politics uses the rest of the sciences, and . . . it legislates as to what we are to do and what we are to abstain from, the end of this science must include those of the others, so that this end must be the good for man [or, even finer and more god-like] for a nation or for city-states . . . [P]olitical science aims at . . . the highest of all goods achievable by action.
ARISTOTLE, *Nichomachean Ethics*, 1094a and b (McKeon 1941).

## I The inauthenticity malaise in political methodology

ALTHOUGH thriving materially within political science departments, many political methodologists are now suffering from a certain, not easily diagnosed malaise. Part of this unease comes from criticisms of their scientific pretensions, especially of their unexamined epistemological and ontological presuppositions. As a result, political methodologists tend to be either defensively dismissive or even antipathetic towards the intellectual sources of such criticisms, which have been variously labeled as "interpretive," "constructivist," "post-structuralist," "post-positivist," or "postmodern." In conversations with their more receptive colleagues, political methodologists rightfully claim pride in recent, modest, technical achievements of primarily a statistical sort. But among fellow methodologists, they also mention feelings of disciplinary inferiority. I suggest therefore, and shall now explore, the diagnosis of this malaise as "disciplinary inauthenticity."

Achen's metaphoral abhorrence at being present at the garage sales of other disciplines' methodologists (Achen 1983) seems to have struck a responsive chord,[1] the replayings of which I take to be evidence for the "inauthenticity" diagnosis just given. It seems that no one from other disciplines buys our products, even when drastically reduced for quick, tax-free clearance. Achen does have a point: who would want to go to a political methodology garage sale only to find used, or remodeled, tools developed by other methodologists for other disciplines' key substantive problems?[2]

Bartels and Brady (1993), in their subsequent, more qualified, but authoritative overview of quantitative political methodology, again express a sense of inauthenticity, of inferiority. But, paradoxically, they also indirectly suggest a way of transcending it. Responding to (Achen 1983: 69), they argue that there is a new, if modest, line of indigenous products we political methodologists can now sell from our own, but still small shops:

> While political methodologists have still "done nothing remotely comparable" to the invention of factor analysis by pyschometricians or structural equation methods by econometricians . . . , they have invented, adopted, or further developed an impressive variety of useful techniques for dealing with event counts . . . , dimensional models . . . , pseudo-panels . . . , model misspecification . . . , parameter variation . . . , aggregated data . . . , selection bias . . . , non-random measurement error . . . , missing data . . . , and time series data . . . (Bartels and Brady 1993: 121)

Simultaneous equations technology was and is a brilliant way of handling the difficult problem of estimating simultaneously operative supply and demand relationships. Multiple factor analysis allowed the empirical determination and measurement of not-directly-observable, distinctive, *multiple* dimensions of human intelligence—a central issue in the securing of scientific respect and societal support for statistically regulated psycho-

[1] Thus, like Gary King (1989) and Larry Bartels and Henry Brady (1993), John Jackson (above: chap. 32) is worried—if not preoccupied—with missed opportunities, with roads not taken.

[2] Thinking qualitatively, one could easily add Ragin, Berg-Schlosser and De Meur's (above: chap. 33) imaginative use of logician's tools—Boolean algebra, as developed and applied to the qualitative sociological analyses of small samples of social or political units—to a list of borrowed techniques.

logical research on human abilities. Unfortunately, however, the main message that shoppers are likely to derive from Bartel and Brady's heterogeneous list is that political methodology's new products still lack substantively central, discipline-linked, distinctive characteristics. The advertisement, "made by political methodogists," will not help their sales.

An important part of this malaise appears to be connected with specificational uncertainties: how to represent and model the substance of the political phenomena at hand. Thus, later in their review, Bartels and Brady (1993: 140) recognize that, "Ironically, as political methodologists have become more sophisticated, fundamental problems of specification uncertainty have become increasingly pressing." They then list several of the tools mentioned above, invented by econometricians and psychometricians and statisticians—"complex simultaneous equation, factor analysis, and covariance structure models," which they describe as "increasingly commonplace in various areas of political science . . ." Then, consistent with the present interpretation, they note that, as a side effect of this complexity, a multiplication of the "difficult, often arbitrary, specification decisions" required by their use (Bartels and Brady 1993: 140).

Bartels and Brady's long list of retoolings, from event counts to time-series data, does not seem clearly and compellingly to be linked to the fundamental substantive problems of our discipline. These include the issues of power and influence assessment, justice and injustice, capability-enhancing public assistance, the substantive and institutional fulfillment of democratic aspirations and the overcoming of their corruption, internationally justifiable coercive interventions, and sustainable social, economic and political development within a globalizing world economy—issues which concern both citizens and professional political scientists alike.

Rather, the accomplishments mentioned by Bartels and Brady reflect a variety of statistical problems of special relevance to different clusters of social or natural scientists from many different, or no particular, disciplines. Thus "statistics," which once meant "the empirically based study of states" and was grounded in different theoretical discussions of community and national development (Alker 1975), seems to have lost its political roots. Let us hope that without becoming lost in the past we can help rediscover, reappreciate and also (where appropriate) redefine these roots.

## II A remedy: philosophically informed specificational innovation

### A From Poli(s)metrics to Political Methodology

As a way of trying to ground political methodology in the classics of political inquiry, I proposed in the 1975 *Handbook of Political Science* calling this field "polimetrics," a name containing (like the "psyche" in "psychometrics") an explicit reference to the "polis," the Greek city-state—a brief but potent symbol for the larger, more varied class of polities whose "metricizing" we wished to accomplish (Alker 1975). Since that grounding was not obvious, I called for a close, phenomenological look at the life of, and in, the polity, and a rethinking of our subdiscipline's descriptive foundations. A closely related concern, given the treatment of political activity as a special kind of social action, was with our subdiscipline's people-related, interpretive, inferential practices.

As also argued elsewhere (Alker 1974; 1984), I emphasized a Weberian priority of concern with getting right our "understanding" (of what we were studying, in human, interpretive terms) before causally "explaining" (what had previously been properly understood). The most distinctive feature of my 1975 *Handbook* chapter, perhaps, was its offering of some relatively novel, rich and powerful specificational possibilities relevant for such Weberian purposes. I favored an ontological commitment to context-sensitive social action perspectives—Dahl, Deutsch, Habermas, Lasswell, Parsons, and Weber included—as a broadly acceptable basis for such efforts.

In discussing these ideas, many political methodologists have cited the view, attributed to the late William Riker, that for their metricizing subdiscipline "poli-" is too undisciplined a root. When the implicit "s" of "polis" is dropped, one can indeed become confused, thinking of "poly" (or "many") things, not just politics. Although the city-state is surely an obsolete preoccupation in the modern era of nation-states—indeed, Aristotle's student Alexander made them obsolescent units of political inquiry in the imperial age he himself introduced—the issue here is deeper than the mere choice of words. Accepting the newer term "political methodology" because of its now common usage, I would like to suggest some deeper reasons for my original advocacy of, and the subsequent rejection of, "poli(s)metrics."

One only needs to don one's rigorous analytical glasses and read through my introductory quotations from the most important originator of political science in Western civilization, Aristotle, to see how the rejection worked. Aristotle, one is analytically tempted to say, has an anti-individualist bias, confuses the normative and the empirical, links the ethical and political (which realpolitik has supposedly taught us to separate).

Were I given the chance of reply, I would have noted that (having not yet been convinced by economists that "talk is cheap") Aristotle points phenomenologically, politically, toward a focus on reasoned speech about the public good. Relevant methodological skills suggested by this subdisciplinary foundational account are as likely to come from the old fashioned disciplines of logic, rhetoric and dialectics as from modern (or postmodern?!) versions of hermeneutics, linguistics, discourse analysis, and even critical philosophy. Econometrics is not on either list! Moreover, because Aristotle points towards collectively oriented action as part of the constitution and fulfillment of human nature, he takes us away from the study of political behavior conceived as the result of randomly varying dispositions, of exogenously fixed political identities, with individually different utility calculi.

Reflecting on this simulated interchange—not unlike those heard at contemporary political science gatherings—I am struck by how much the anti-poli(s)metrics position embodies the malaise of inauthenticity already referred to. Econometrics plus utilitarian economics will only satisfy the political analyst who has discarded from his or her scientific lexicon the notions of political substance (the just or unjust practices of a social entity or political community), of political speaking and thinking, of ethical collective will formation, of social action in the service of virtuous ends that Aristotle described as the essence of the political and the fulfillment of the political dimension of human nature.

Phenomenologically lacking such a substantial core in their technical work, rejecting postmodern claims as to their ontological dissipations, unable to engage their colleagues in an informed fashion on contemporary epistemological debates, political methodologists seem to have nothing left to do but subsist in the world of Aristotle/Homer's dehumanized, polis-deprived, solitary, grasping, "war-mad" men.

## B Some virtues of neo-Aristotelian foundations

Von Wright (1971: chap. 1) traces the foundations of social scientific inquiry back beyond the *verstehen-erklaren* controversy ("understanding" versus "explanation") to two older traditions, both valuable: the

Aristotelian and the Galilean. While King, Keohane and Verba (1994), in a much-discussed contemporary methodological text, also give serious attention to problems of interpretive or "descriptive" inference, they pay little or no attention to the phenomenological, epistemological, methodological and specificational literatures following from both pre- and post-Weberian versions of this traditional debate. Substantial neo-Aristotelian notions—of purposive speech about right and wrong, of purposive and practical agency in collectivities, of feedback and adaptive or maladaptive functioning, and of organized complexity more generally—are therefore not made central concerns of political methodology in their otherwise very Galilean account. While a creative synthesis of both Aristotelian and Galilean root concerns is clearly preferable in trying to spell out the methodological components of a *political* science, a corrective emphasis here on the contribution of the former tradition seems therefore highly appropriate.

Aristotle's phenomenological assurance, evident in the quotation from the *Nichomachian Ethics* above, allows his political methodology students to be cosmopolitan shoppers without inferiority complexes. Since politics is, or can be, the realm of human beings' highest forms of individual and collective self-realization, the contributions of the other disciplines to that end are to be sought where appropriate, but not on a fire-sale or garage-sale basis.

To put this point in modern terms: since the products of scientific work entered the public domain with their public announcement or publication, disciplinary legitimization and priorities of discovery do not seem to have been such problems; good legislation should be built on the findings of any and all other disciplines. As Aristotle's great-grand-students, we political scientists, like our predecessors in other methodological subdisciplines, are entitled to use *whatever is helpful for our substantive concerns and our methodological purposes* from any of dozens of possibly relevant representational and inferential literatures. We do not have his luxury of ontological assurance, however; we must be prepared, in contemporary times, to address phenomenological issues carefully, constructively and in an informed, open-minded way.

Because philosophers of (social) science and social scientists have been wrestling with problems of political causality, functionality and collective purposes, for a very long time, it is important to have a good idea of their track records.[3] Knowing the past contexts of methodological innovations

[3] Among my favorites, for teaching students in political science and international relations, are: Mill 1843; Diesing 1971; Dallmayr and McCarthy 1979; Elster 1989; Walton 1990; Hollis 1994; and Schiffrin 1994.

by neo-Aristotelians and neo-Galileans helps us understand their possibilities and limitations in newer and different contexts of application. It gives us more freedom not to repeat past practices, when they are inappropriate to present or future concerns. And we do not all have to be neo-Aristotelians, to appreciate these reflections.[4]

What I tried to do in the 1975 *Handbook of Political Science* now seems especially relevant. It was to suggest novel ways of relooking at the phenomenological and interpretive foundations of political methodology. I thought that the right response to the successes of psychometricians and econometricians in their own disciplines was to be similarly problematically sensitive and methodologically creative within our own. New, exciting opportunities for doing so were evident in fields like political psychology and sociology, computational linguistics, cognitive science and political discourse analysis.

## C Reconnecting political methodology with communicatively oriented political phenomenologies

Building on Donald Moon's (1975) phenomenologically perceptive, philosophically informed, synthetic approach to positivistic and hermeneutic (critically interpretive) logics of political inquiry, and his use of Von Wright and Habermas' writings on explanation and understanding, I focused on examples where power and influence, systems of such relationships, or justifications for collective action were a central political concern. Retrospectively, I favored new or old communicatively oriented representational or specificational strategies, like those one would look for from an Aristotelian or Habermasian "reflectivist" perspective (Kratochwil 1989). My attempt to resurrect "reason analysis" from early European market research by political sociologist Paul Lazarsfeld and others anticipated some of the more methodologically innovative work on reasoning, choice and political action (Sniderman, Brody and Tetlock 1991), although their work might have benefited from more ambitious representational specifications of the sort explored in Hudson (1991) and Slade (1994).

[4] See Alker (1974; 1984; 1993; and especially 1996) for an elaboration of neo-Aristotelian themes. These works are important sources of many of the arguments in this chapter. Except when citations are obligatory, I shall not cite them further here.

## D The promise of alternative formal representations

Similarly, the most important directional suggestion implicit in the Bartels and Brady quotation above is its expressed need for new and better, empirically researchable specifications of political relationships. A communicatively oriented set of process specifications is evident, *inter alia*, in Crecine's (1969) and Alker and Greenberg's (1977) use of governmental problem-solving-process modeling ideas. As these were suggested by the work of Herbert Simon and his colleagues in public administration and cognitive science, there were no problems of disciplinary inauthenticity associated with them! Similarly, the leap into Schank–Abelson conceptual dependency formalisms for better representing the ways in which political ideologies about right and wrong (Alker 1975; Schank and Abelson 1977) inform our political speaking and acting seemed to me, at the time, to represent a very innovative collaboration between a social/political psychologist and a computational linguist. This line of work has subsequently been taken up, and reformulated, by a number of writers using artificial intelligence tools (see the contributions to, and bibliographies in, Walton 1990; Hudson 1991; Taber 1992; Duffy and Tucker 1995). Stephen Slade (1994), a student of Abelson and Schank, has applied their formal representations or specifications of political psychological reasoning to congressional voting decisions with impressive results.

John Jackson's critical self-awareness concerning the limitations of individualist, equilibrium-assuming modeling approaches to the study of political-institutional change has been similarly pointed towards specificational innovations of a neo-Aristotelian sort. His chapter (Jackson above: chap. 32) gives considerable attention to his unconventional specifications of political phenomena in intriguing, possibly non-equilibriating, path-dependent terms; nor does he ignore the sometimes difficult statistical estimation issues raised by this approach.

Jackson's chapter is particularly impressive because of the substantive content of its all-too-brief references to the changing composition of the Democratic and Republican parties. Attending to new specificational possibilities suggested by Russian cyberneticians, American economists, system theorists, and earlier work on party system dynamics, he has taken a significant step toward the rehistoricizing of political methodologies relevant to issues of organized complexity.

The treatment of qualitative methods of macropolitical inquiry by Ragin, Berg-Schlosser and De Meur (above: chap. 33) has a similar, innovative, philosophically sophisticated flavor. Their interest in case-specific and period-specific political concerns, as well as in conjunctural causation, points away from statistically modeled universalistic, timeless generalities; it has a rather Aristotelian sense of the contextual and contingent quality of most political truths. These authors illustratively apply a computerized version of Ragin's earlier (1987) reworking of Boolean algebraic representations to the problem of redescribing, in an explanatorily suggestive way, the qualitative determinants of democratic regime breakdowns in interwar Europe.

One is struck by the similarity of Ragin's and his collaborators' reflections on causal complexity to those of Alexander George (1979) in his effort to develop new methods for historical case studies. A good starting-point in each case is John Stuart Mill's informed discussion of the "inverse deductive, or historical method" in Book VI ("On the Logic of the Moral Sciences") of his *A System of Logic* (Mill 1843)—a discussion that is skeptical about the simple extension of experimental methods to complex historical phenomena, which also helped to initiate the *verstehen-erklaren* debate.

In each of the above examples, language-processing capacities generally, political reasoning in particular, and/or socially-organized historical processes have been specified, using representational formalisms other than the probabilistic, statistical models familiar to statisticians or econometricians. Yet inauthenticity problems do not seem very pressing—the studies in question touch on issues of obvious political substance, and do so communicatively, historically, creatively and critically. These seem good standards for the future of political methodology, combining Galilean and Aristotelian themes.

## III The future of Political Methodology

Having suggested that political methodology already has a series of methodological innovators from Aristotle to Weber (and many others not mentioned in between), I have mentioned, from among the more influential recent writers, some contributions of Robert Abelson, Alexander George, Jürgen Habermas, Paul Lazarsfeld and Herbert Simon. It would not have been hard to elaborate this (partly personal) list with citations of specific writings by Robert Axelrod, Lincoln Bloomfield, Donald Campbell, Noam Chomsky, Karl Deutsch, Paul Diesing, Harold Guetzkow, George Lakoff, Harold Lasswell, Livia Polanyi, William Riker and Anatol Rapoport; or to go overseas and examine the philosophically erudite, immensely learned and provocative contributions of writers like Jon Elster,

Antonio Gramsci, Michel Foucault, Johan Galtung, Martin Hollis and Karl Popper. With lists of their progenitors like this, political methodologists need not have inferiority complexes.

Political methodologists must, as well, get over their allergy to philosophies of social science and to theoretical and practical debates inspired by the classics of the field of political inquiry. A conception of political methodology broad enough to include the contributions of these writers is the best way, in my view, to get beyond the inauthenticity malaise of the political methodology field, too narrowly defined as political statistics. That conception might be to think of methodology as applied epistemology, as applied philosophy of research. Realizing that there are a variety of philoso*phies* of social and political research points toward the sister realization that these philoso*phies* are usually linked—as done clearly in the writings of Arendt, Aristotle, Hobbes, Marx, Weber, Lasswell and Popper, for example—to *political phenomenologies* and *political theories.*

Although theory and practice can usefully be distinguished, the difference is not absolute. Political theories are usually connected to distinctive perspectives on, or advocacy of, particular programs of political activity. The connection of political theories to specific political contexts also serves to bring concepts and issues of substance into their view, linking back to more-or-less fully elaborated philosophies of political inquiry, applied as political research methodologies. Since Donald Moon (1975) did an excellent introductory job in establishing some of these links, and I have attempted to explore these questions at length elsewhere, I shall not review them further here.

Rather, I would like to suggest, in closing, a broader conception of the "science" in political science which further supports the vision of political methodology's future which I have in mind. Not unlike some of the ideas of Aristotle and Lasswell,[5] it is a pedagogical schema bridging the gaps linking interpretive and explanatory, policy-oriented and "scientific," constructivist and naturalist styles of political inquiry. It is a broad, deeply humanistic conception of the social sciences, including peace research and political science, suggested by Johan Galtung.

In a too infrequently read collection of Galtung's methodological essays, my favorite among many gems is his "Empiricism, Criticism, Constructivism: Three Aspects of Scientific Activity." There, he elaborates a trilateral conception of science as sharing these aspect (Galtung 1977–88: vol. i, pp. 41–71 at 60–3). Geometrically, his separate figures and tables synthesize into the schema shown in Figure 35.1.

[5] For a Lasswellian application see also Namenwirth and Weber 1987.

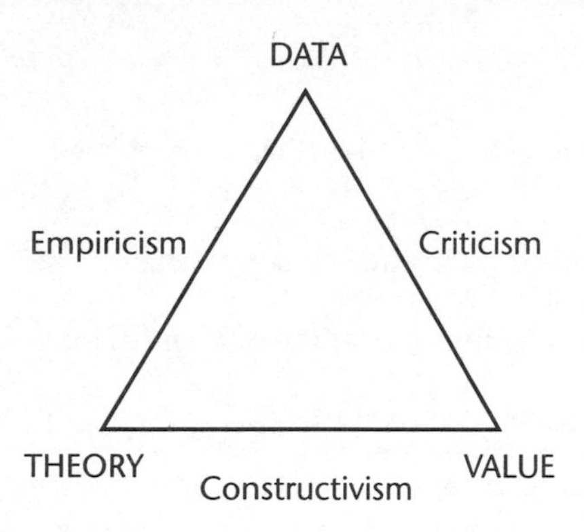


**Fig. 35.1** *Galtung's integrated conception of social science*

There are six stages to the normal sequence of practices associated with Galtung's integrated, critical-empirical-constructivist social science. After phenomenologically sensitive *criticism* of the links between empirical data and relevant values deriving from "critical awareness," classical empiricist analysis attempts to *understand* why the empirical world is as it appears and *forsee, ceteris paribus,* its probable future. Next, value and theory are combined in *goal-creation and theory-creation* tasks, creating an image of a preferred future and a theory able to account for a potential world including that image as a possibility. Fifth, *constructivist analysis* combines these theories and values to ascertain the viability and attainability of the preferred world, through the making and analysis of proposals for change. Finally, and partially, there is the "action part," which Galtung calls *reality-creation* or *invariance-breaking,* but which I might call, hopefully more modestly, *reality-ammendation* or potentiality fulfillment. Here science and politics merge, as efforts to achieve higher degrees of consonance between the observed, the foreseen and the preferred.

The volumes from which this idea in question emerges go on further to engage Western and non-Western ideas of cosmology and social science, including Christian and Buddhist ideas of truth-seeking, action and contemplation, development and peace. Broadening our conception of political science like this, and centering it in the common and varying political aspirations, achievements and failings of our fellow citizens around the globe, is a good recipe for the future of political methodology.

## References

ACHEN, C. H. 1983. Towards theories of data: the state of political methodology. Pp. 69–94 in *Political Science: The State of the Discipline*, ed. A. W. Finifter. Washington, D.C.: American Political Science Association.

ALKER, H. R., JR. 1974. Are there structural models of voluntaristic social action? *Quality and Quantity*, 8: 199–246.

—— 1975. Polimetrics: its descriptive foundations. In Greenstein and Polsby 1975: vol. vii, pp. 139–210.

—— 1984. Historical argumentation and statistical inference: towards more appropriate logics for historical research. *Historical Methods*, 17: 164–73, 270.

—— 1993. Making peaceful sense of the news: institutionalizing international conflict-management event reporting using frame-based interpretive routines. Pp. 141–59 in *International Event-Data Developments: DDIR Phase II*, ed. R. L. Merritt, R. G. Muncaster and D. A. Zinnes. Ann Arbor: University of Michigan Press.

—— 1996. *Rediscoveries and Reformulations: Humanistic Methodologies for International Studies*. Cambridge: Cambridge University Press.

—— and GREENBERG, W. J. 1977. On simulating collective security regime alternatives. Pp. 263–305 in *Thought and Action in Foreign Policy*, ed. G. M. Bonham and M. J. Shapiro. Basel: Birkhäuser Verlag.

BARTELS, L. M., and BRADY, H. E. 1993. The state of quantitative political methodology. Pp. 121–59 in *Political Science: The State of the Discipline II*, ed. A. W. Finifter. Washington, D.C.: American Political Science Association.

CRECINE, J. P. 1969. *Governmental Problem-Solving: A Computer Simulation of Municipal Budgeting*. Chicago: Rand McNally.

DUFFY, G. N., and TUCKER, S. A. 1995. Political Science: Artificial Intelligence Applications. *Social Science Computer Review*, 13/1: 1–19.

DALLMAYR, F. R., and MCCARTHY, T. A., eds. 1979. *Understanding and Social Inquiry*. Notre Dame, Ind.: University of Notre Dame Press.

DIESING, P. 1971. *Patterns of Discovery in the Social Sciences*. Chicago: Aldine-Atherton.

ELSTER, J. 1989. *Nuts and Bolts for the Social Sciences*. Cambridge: Cambridge University Press.

GALTUNG, J. 1977–88. *Essays in Methodology*. 3 vols. (Vol. i: *Methodology and Ideology*; Vol. ii: *Papers on Methodology*; Vol. iii: *Methodology and Development*). Copenhagen: Christian Ejlers.

GEORGE, A. L. 1979. Case studies and theory development: the method of structured, focused comparison. Pp. 43–63 in *Diplomacy: New Approaches in History, Theory, and Policy*, ed. P. G. Lauren. New York: Free Press.

GREENSTEIN, F. I., and POLSBY, N. W., eds. 1975. *Handbook of Political Science*. Reading, Mass.: Addison-Wesley.

HOLLIS, M. 1994. *The Philosophy of Social Science: An Introduction*. Cambridge: Cambridge University Press.

HUDSON, V. M., ed. 1991. *Artificial Intelligence and International Politics*. Boulder, Colo.: Westview Press.

KING, G. 1991. *Unifying Political Methodology: The Likelihood Theory of Statistical Inference*. Cambridge: Cambridge University Press.

—— KEOHANE, R. O.; and VERBA, S. 1994. *Designing Social Inquiry: Scientific Inference in Qualitative Research*. Princeton, N.J.: Princeton University Press.

KRATOCHWIL, F. V. 1989. *Rules, Norms, and Decisions: On the Conditions of Practical and Legal Reasoning in International Relations and Domestic Affairs*. Cambridge: Cambridge University Press.

Mill, J. S. 1843. *A System of Logic, Ratiocinative and Inductive.* London: Longmans, Green.
Moon, J. D. 1975. The logic of political inquiry: a synthesis of opposed perspectives. In Greenstein and Polsby 1975: vol. i, pp. 131–228.
McKeon, R., ed. 1941. *The Basic Works of Aristotle.* New York: Random House.
Namenwirth, J. Z., and Weber, R. P. 1987. *Dynamics of Culture.* Boston: Allen and Unwin.
Ragin, C. C. 1987. *The Comparative Method: Moving Beyond Qualitative and Quantitative Strategies.* Berkeley: University of California Press.
Schank, R., and Abelson, R. 1977. *Scripts, Plans, Goals and Understanding: An Inquiry into Human Knowledge Structures.* Hillsdale, N.J.: Lawrence Erlbaum.
Schiffrin, D. 1994. *Approaches to Discourse.* Oxford: Blackwell.
Slade, S. 1994. *Goal-Based Decision Making: An Interpersonal Model.* Hillsdale, N.J.: Lawrence Erlbaum.
Sniderman, P. M.; Brody, R. A.; and Tetlock, P. E. 1991. *Reasoning and Choice: Explorations in Political Psychology.* Cambridge: Cambridge University Press.
Taber, C. S. 1992. POLI: an expert system model of U.S. foreign policy belief systems. *American Political Science Review,* 86: 888–904.
Von Wright, G. H. 1971. *Explanation and Understanding.* London: Routledge and Kegan Paul.
Walton, D. N. 1990. *Practical Reasoning: Goal-Driven, Knowledge-Based, Action-Guiding Argumentation* Savage, Md.: Rowman and Littlefield.

# Notes on Contributors

ALBERTO ALESINA is Professor of Economics and Government at Harvard University.

HAYWARD R. ALKER is John A. McCone Professor of International Relations at the University of Southern California.

GABRIEL A. ALMOND is Emeritus Professor of Political Science at Stanford University.

JAMES E. ALT is Professor of Government at Harvard University.

DAVID E. APTER is Henry J. Heinz II Professor of Comparative Political and Social Development at the Department of Political Science, Yale University.

A. B. ATKINSON is Warden of Nuffield College, Oxford.

BRIAN BARRY is Professor of Political Science at the London School of Economics.

DIRK BERG-SCHLOSSER is Professor of Political Science, Philipps-University, Marburg, Germany.

EDWARD G. CARMINES is Professor of Political Science at Indiana University, Bloomington.

DAVID LOUIS CINGRANELLI is Associate Professor and Chair of the Department of Political Science at the State University of New York at Binghamton.

RUSSELL J. DALTON is Professor of Political Science at the University of California at Irvine.

MATTEI DOGAN is Director of Research at the Centre National de la Recherche Scientifique in Paris and Professor of Political Science at the University of California, Los Angeles.

GAVIN DREWRY is Professor of Public Administration at Royal Holloway and Bedford New College, University of London.

PATRICK DUNLEAVY is Professor of Government at the London School of Economics.

KJELL GOLDMANN is Professor of Political Science at the University of Stockholm.

ROBERT E. GOODIN is Professor of Philosophy at the Research School of Social Sciences of the Australian National University, Canberra.

BERNARD GROFMAN is Professor of Political Science at the University of California at Irvine.

RICHARD I. HOFFERBERT is Professor of Political Science at the State University of New York at Binghamton and Recurring Visiting Professor at the Wissenschaftszentrum Berlin für Sozialforschung.

ROBERT HUCKFELDT is Professor of Political Science at Indiana University, Bloomington.

JOHN E. JACKSON is Professor of Political Science and of Business Administration and Research Scientist at the Institute for Social Research at the University of Michigan.

ROBERT O. KEOHANE is Stanfield Professor of International Peace in the Department of Government, Harvard University.

HANS-DIETER KLINGEMANN is Director of Research Unit III at the Wissenschaftszentrum Berlin für Sozialforschung and Professor of Political Science at the Free University, Berlin.

PETER MAIR is Professor of Political Science at the University of Leiden.

GIANDOMENICO MAJONE is Professor of Comparative Public Policy at the European University Institute in Florence.

KATHLEEN MCGRAW is Associate Professor of Political Science at the State University of New York at Stony Brook.

GISELE DE MEUR is Professor of Mathematics for the Social Sciences at Université Libre de Bruxelles.

WARREN E. MILLER is Regents' Professor of Political Science at Arizona State University.

BARBARA J. NELSON is Vice President for Academic Programs at Radcliffe College.

CLAUS OFFE is Professor of Political Sociology and Social Policy at Humboldt University, Berlin.

FRANZ URBAN PAPPI is Professor of Politial Science at Mannheim University.

BHIKHU PAREKH is Professor of Politics at the University of Hull.

B. GUY PETERS is Maurice Falk Professor of American Government at the University of Pittsburgh.

CHARLES C. RAGIN is Professor of Sociology and Political Science at Northwestern University.

Bo Rothstein occupies the August Röhss chair of Political Science at Göteborg University.

David Sanders is Professor of Government at the University of Essex.

J. Ann Tickner is Associate Professor in the School of International Relations at the University of Southern California.

Klaus von Beyme is Professor of Political Science at the University of Heidelberg.

Barry R. Weingast is Senior Fellow at the Hoover Institution and Professor of Political Science at Stanford University.

Laurence Whitehead is Fellow in Politics at Nuffield College, Oxford.

Vincent Wright is Fellow in Politics at Nuffield College, Oxford.

Iris Marion Young is Professor in the Graduate School of Public and International Affairs at the University of Pittsburgh, with affiliated appointments in the Departments of Philosophy and Political Science.

# Subject Index

# Name Index